International Conference on Communication and Media: An International Communication Association Regional Conference (i-COME'16)

SHS Web of Conferences Volume 33 (2017)

Kuala Lumpur, Malaysia
18 - 20 September 2016

Editors:

Bahtiar Mohamad
Hassan Abu Bakar

ISBN: 978-1-5108-3660-0

Printed from e-media with permission by:

Curran Associates, Inc.
57 Morehouse Lane
Red Hook, NY 12571

Some format issues inherent in the e-media version may also appear in this print version.

Printed by Curran Associates, Inc. (2017)

For additional information, please contact EDP Sciences – Web of Conferences at the address below.

EDP Sciences – Web of Conferences
17, Avenue du Hoggar
Parc d'Activité de Courtabœuf
BP 112
F-91944 Les Ulis Cedex A
France

Phone: +33 (0) 1 69 18 75 75
Fax: +33 (0) 1 69 28 84 91

contact-edps@webofconferences.org

Additional copies of this publication are available from:

Curran Associates, Inc.
57 Morehouse Lane
Red Hook, NY 12571 USA
Phone: 845-758-0400
Fax: 845-758-2633
Email: curran@proceedings.com
Web: www.proceedings.com

TABLE OF CONTENTS

SHS Web of Conferences
Open access conference proceedings
in Human and Social Sciences
A flexible, fast and efficient
publication service,
for an open access recording
of your event
and a wide dissemination of
your proceedings
www.shs-conferences.org
edp open
Open Access Publishing
from EDP Sciences

ISSN (electronic edition): 2261-2424

PROGRAMME BOOK

i-COME'16

International Conference on Communication and Media

AN INTERNATIONAL COMMUNICATION ASSOCIATION REGIONAL CONFERENCE

EXPLORING COMMUNICATION: BEYOND NATION CULTURAL ADAPTION, IMAGES AND IDENTITY

Organized by: UUM

Sponsored by: TENAGA NASIONAL, Taylor & Francis

Supported by: Malaysia

Partner Institutions: PURDUE, Infrastructure University

September 18–20, 2016

Istana Hotel, Kuala Lumpur MALAYSIA

i-COME'16

International Conference on Communication and Media

AN INTERNATIONAL COMMUNICATION ASSOCIATION REGIONAL CONFERENCE

Organised by: UUM Universiti Utara Malaysia

EXPLORING COMMUNICATION: BEYOND NATION CULTURAL ADAPTION, IMAGES AND IDENTITY

September 18–20, 2016 · Istana Hotel, Kuala Lumpur, MALAYSIA

THE International Communication and Media Conference 2016 (i-COME'16): An International Communication Association Regional Conference provides a broad platform for the discussion of emergent and interconnected issues shaped by the meaning of communication, cultural, images and identity as understood in the South East Asian region as well as in other parts of the world. The 2016 ICA regional conference theme "Exploring Communication: Beyond Nation Cultural Adaption, Images and Identity" is aimed to advance scholarship in the field of communication that can be of benefit to communities experiencing change.

At the end of 2000, East Asia (South Korea, Malaysia, Thailand, and Indonesia) was faced a financial crisis that had affected business worldwide. Prior to the crisis up to now, these countries were the center for international investments in receiving 25% of world's foreign direct investments. From communication viewpoint, these are interesting times. Although the crisis has changed the political climate in East Asia, some interesting phenomena has also evolved. China and India eventually become prominent in South East Asia countries economic and security policy, while Australia and New Zealand (perceived as the Western countries) start to look to South East Asian region as their strategic business partners. Together with ASEAN countries (Association of South East Asian Nation), China, Japan, South Korea, India, Australia, New Zealand, and United States, jointly formed the APEC (Asia-Pacific Economic Cooperation) to facilitate businesses in the Asia-Pacific region. The interesting aspects during these phenomena, from communication viewpoint, are not only businesses and the (mis-) communications of the various participants. Prejudice and stereotyping of other cultures is already an important subject within the field of communication, and have attracted scholars to deliberate the applications of these concepts in real setting. In fact, communication scholars have already talked about applying knowledge derived from indigenous perspectives to research and practice.

The conference provides a broad platform for discourse on the crucial concerns (and their potential solutions) in relation with beyond national cultural, identity and image, and interconnected elements from various communication perspectives that influence our ways of thinking in the context of national cultural setting.

Secretariat – i-COME'16: An ICA Regional Conference
Communication Department
School Of Multimedia Technology and Communmication
Universiti Utara Malaysia
06010. UUM SINTOK. KEDAH. MALAYSIA.
www.i-comeuum.com

Organising Committee

PATRON	Professor Dato' Seri Dr. Mohamed Mustafa Ishak Vice Chancellor Universiti Utara Malaysia
ADVISOR	Dr. Norsiah Abdul Hamid Dean, School of Multimedia Technology and Communication
EXECUTIVE DIRECTOR	Associate Prof. Dr. Hassan Abu Bakar
DIRECTOR	Mohd Baharudin Othman
DEPUTY DIRECTOR	Dr. Nik Adzrieman Abdul Rahman
SECRETARY	Dr. Syarizan Dalib Nafiza Ahmad
PROCEEDINGS	Dr. Bahtiar Mohamad Associate Prof. Dr. Hassan Abu Bakar Prof. Dr. Che Su Mustaffa Dr. Joyce Lynn-Sze Cheah Dr. Jamilah Jamal Mohamad-Noor Sharipudin Shamsu Dauda Abdu Syed Hassan Raza
INTERNATIONAL RELATIONS AND COOPERATION	Associate Professor Dr. Hassan Abu Bakar Associate Professor Dr. Norhafezah Yusof
PROMOTION AND PUBLICITY	Dr. Jamilah Jamal Adibah Ismail
LOGISTIC AND HOSPITALITY	Dr. Nik Adzrieman Abdul Rahman Mohd Azrul Othman
FINANCE AND SPONSORSHIP	Mohd Baharudin Othman Dr. Mohd Khairie Ahmad Sharifah Adawiyah Syed Yahya
REGISTRATION AND PARTICIPANTS	Dr. Romlah Ramli Dr. Suhaini Muda
PROTOCOL AND SOUVENIRS	Nor Hayati Mohd Jalil Prof. Madya Dr. Norhafezah Yusof Sharina Samsudin Norhayati bt. Mohamad Hussain

Scientific Committee

Head of Scientific Committee

Scientific Committee Members

Reviewers

Abdul-Aziz Haruna, *Ahmadu Bello University, Zaria, NIGERIA*
Adamu Muhammad Hamid, *The Federal Polytechnic Bauchi, NIGERIA*
Ainurliza Mat Rahim, *Universiti Sains Islam Malaysia, MALAYSIA*
Ajoke Adebiyi, *Moshood Abiola Polytechnic, NIGERIA*
Ali Salman, *Universiti Kebangsaan Malaysia, MALAYSIA*
Amira Firdaus, *University of Malaya, MALAYSIA*
Arumugam Balasubramanian, *Amrita Vishwa Vidyapeetham, Coimbatore, INDIA*
Bahiyah Omar, *Universiti Sains Malaysia, MALAYSIA*
Bee Chen, Crystal Ooi, *Universiti Tunku Abdul Rahman, MALAYSIA*
Bolanle Ismail, *Crescent University, Abeokuta, NIGERIA*
Bradley Freeman, *American University in Dubai, UNITED ARAB EMIRATES*
Changsong Wang, *Xiamen University, MALAYSIA*
Cheng Ean (Catherine) Lee, *Sunway University, MALAYSIA*
Christiantine Della, *Infrastructure University Kuala Lumpur, MALAYSIA*
Eli Jamilah Mihardja, *Universitas Persada Indonesia YAI Jakarta, INDONESIA*
Endang Setiowati, *Universitas Indonesia, INDONESIA*
Evie Ariadne Shinta Dewi, *Universitas Padjadjaran, INDONESIA*
Ganiu Okunnu, *Crescent University, Abeokuta, NIGERIA*
Genina Mariel Arceo, *The Outstanding Students of the Philippines Alumni Community, PHILIPPINES*
Halina Sendera Mohd. Yakin, *University of Sussex, UNITED KINGDOM*
Haliza Harun, *Universiti Sains Islam Malaysia, MALAYSIA*
Holger Briel, *Xi'an Jiaotong Liverpool University, CHINA*
Irwansyah, *Universitas Indonesia, INDONESIA*
Isma Rosila Ismail, *Universiti Malaysia Terengganu, MALAYSIA*
Ismail Sualman, *Universiti Teknologi MARA, MALAYSIA*
Jamilah Jamal, *Universiti Utara Malaysia, MALAYSIA*
Joyce Lynn-Sze Cheah, *Universiti Utara Malaysia, MALAYSIA*
Kalyani Suresh, *Amrita Vishwa Vidyapeetham, INDIA*
Khaizuran Abdul Jalil, *Athromed Sdn Bhd, MALAYSIA*
Khatijah Omar, *Universiti Malaysia Terengganu, MALAYSIA*
Kiranjit Kaur Pritam Singh, *Universiti Teknologi MARA , MALAYSIA*
Lateef Adelakun, *University of Malaya, MALAYSIA*
Latifah Abd Latib, *Universiti Industri Selangor, MALAYSIA*
Lucy Supratman, *Telkom University, INDONESIA*
Marc Pacoma, *Adamson University, PHILIPPINES*
Martha Lestari, *Telkom University, INDONESIA*
Masduki, *Indonesian Islamic University, INDONESIA*
Melicent Jalova, *Mindanao State University-Iligan Institute of Technology PHILIPPINES*
Minho Ju, *Sungkyunkwan University, REPUBLIC OF KOREA*
Mohamad-Noor Sharipudin, *Universiti Utara Malaysia, MALAYSIA*

Mohd Baharudin Othman, *Universiti Utara Malaysia, MALAYSIA*
Mohd Khairie Ahmad, *Universiti Utara Malaysia, MALAYSIA*
Mohd. Nor Shahizan Ali, *Universiti Kebangsaan Malaysia, MALAYSIA*
Mudassar Hussain, *University of The Punjab, PAKISTAN*
Nik Rozaidi, *Rutgers University, UNITED STATES*
Nor Azura Adzharuddin, *Universiti Putra Malaysia, MALAYSIA*
Norealyna Misman, *University of Malaya, MALAYSIA*
Norhafezah Yusof, *Universiti Utara Malaysia*
Normazla Ahmad Mahir, *Universiti Sains Islam Malaysia, MALAYSIA*
Norzita Yunus, *Infrastructure University Kuala Lumpur*
Olawale Ismail, *Ahmadu Bello University, Zaria, NIGERIA*
Patrick Chin Hooi Soh, *Multimedia University, MALAYSIA*
Pijar Suciati, *Universitas Indonesia, INDONESIA*
Pin Lick Soo Hoo, *Universiti Utara Malaysia, MALAYSIA*
Rizwanah Souket, *UCSI University, MALAYSIA*
Roro Wulan, *Telkom University, INDONESIA*
Saidathul Nizah Mat Tazin, *Universiti Teknologi MARA, MALAYSIA*
Samuel Hatane, *Petra Christian University, INDONESIA*
Shamsu Dauda Abdu*, Federal Polytechnic, Bauchi, NIGERIA*
Siti Nor Amalina Ahmad Tajuddin, *The University of Queensland, AUSTRALIA*
Stephen Croucher, *University of Jyväskylä, FINLAND*
Sumekar Tanjung, *Islamic University of Indonesia, INDONESIA*
Syed Hassan Raza, *Bahauddin Zakariya University, PAKISTAN*
Thanaseelen Rajasakran*, Universiti Tunku Abdul Rahman, MALAYSIA*
Thinavan Periyayya, *Universiti Tunku Abdul Rahman, MALAYSIA*
Wan Norbani Wan Noordin*, Universiti Teknologi MARA, MALAYSIA*
WANG Yan, *National Chengchi University, TAIWAN*
Yeen Lai Khong, *Universtiy Tunku Abdul Rahman MALAYSIA*
Zalinah Ahmad, *Universiti Utara Malaysia, MALAYSIA*

Keynote Speakers: Abstract

SYED ARABI IDID

International Islamic University Malaysia, MALAYSIA

Differential Development of Two Theories: Agenda Setting and Diffusion of News

This paper discusses the development of theories over a period of years by selecting two theories, namely Agenda Setting and News Diffusion. McCombs, M.E. and Shaw, D.L first brought the Agenda Setting into prominence in 1972 while Miller, D. first wrote the first piece on the Diffusion of News in 1945. Since then, Agenda Setting has recorded more than 500 articles, News Diffusion has on record of no more than 70 articles to its credit with scholars giving differential attention to both of these two theories. This paper provides some of the reasons for the differential development by looking at the growth of these two theories. This paper traces some of the studies being done using the two theories, highlighting the concepts and the methodologies used.

PATRICE M. BUZZANELL,

Purdue University, UNITED STATES

"Beyond Prejudices and Stereotyping of Other Cultures": Constituting Interventions for Inclusion through Design Thinking

Taking cues from the theme for i-COME'16, this keynote presentation positions diversity and inclusion as wicked problems for which the typical remedies are inadequate. This positioning enables us to rethink diversity and inclusion efforts as naturally contradictory and messy rather than as rational and with singular solutions that match the problems. Design thinking provides the theoretical and practical underpinnings for a robust process of change. This change process requires attention to dialogue, images and identity, cultural humility, and deviation amplifying efforts.

DONGHEE SHIN

Sungkyunkwan University, SOUTH KOREA

Human Experience, Technology, and Interaction

This talk will discuss some fundamental cognitive psychological mechanisms in human technology interaction focusing on user experience along with quality/heuristics, and explain how those mechanisms can be effectively applied to the design of successful ICT products and interfaces. The exponential growth of services via the Internet of Things (IoT) is making it increasingly important to cater to the quality expectations of end users. Quality of experience (QoE) can become the guiding paradigm for managing quality provisioning and application design in the IoT. This study examines the relations of human experience and quality perception of IoT and develops a conceptual model for QoE in personal informatics. It first characterizes quality of service (QoS) using a field experiment and subjective evaluation to compare QoS to QoE. It then performs a survey to identify user behavior factors in personal informatics. It finally proposes a user experience model, conceptualizing QoE specific to personal informatics and highlighting relationships with other factors. The model establishes a foundation for future IoT service categories through a heuristic quality assessment tool from a user-centered perspective. The results provide a ground truth basis for developing future IoT services with QoE requirements, as well as for dimensioning the underlying network provisioning infrastructures, particularly with regard to mobile access technologies.

CHEN LING

Hong Kong Baptist University, HONG KONG

Globalization, Biographic Uncertainty and Cultural Identity

Globalization, Biographic Uncertainty and Cultural Identity

"The individual's biography, if she is to maintain regular interaction with others in the day-to-day world, cannot be wholly fictive. It must continually integrate events which occur in the external world, and sort them into the ongoing 'story' about the self (Giddens, 1991, p. 54)." Globalization has linked peoples not only economically but also culturally and affected individuals in a full spectrum of choices they make and must make and in choices they may have or wish they have. Moreover, the increased options available create uncertainty at many levels, from material necessities to associations and identifications at social and cultural as well as ethnic levels including the individual's life course and biography, whether they are mobile or stay put. This paper explores implications of what is known as late modernity to cultural identity, a core element of an individual's social experience and biography. The discussion will direct toward related day-to-day interactions in people's life, with a few secondary examples from East Asian societies.

SUMMARY OF TIME TABLE FOR i-COME'16 (18-21 SEPTEMBER 2016)

Organized by: UUM, ICA

Sponsored by: Tenaga Nasional Berhad, Adabi, Taylor & Francis

Supported by: Malaysia Convention & Exhibition Bureau, Malaysia Truly Asia

Partner Institutions: Purdue Liberal Arts, Chulalongkorn University, Infrastructure University Kuala Lumpur

TIME	8.00 – 9.00 am	9.00 am – 11.00 am	11.00 – 12.30 pm	12.30 – 2.00 pm	2.00 – 4.00 pm	4.00 – 4.15 pm	4.15 – 6.15 pm	8.00 – 11.00 pm
SUNDAY, 18 SEPTEMBER 2016	REGISTRATION & REFRESHMENTS *Mahkota Lobby*	PARALLEL SESSION 1A – Mahkota 3 1B – Baiduri 1C – Berlian 1D – Delima	KEYNOTE ADDRESS 1 *Mahkota 3* *Syed Arabi Idid & Patrice Buzzanell*	LUNCH *Mahkota 1*	PARALLEL SESSION 2A – Mahkota 3 2B – Baiduri 2C – Berlian 2D – Delima	REFRESHMENTS *Mahkota 1*	PARALLEL SESSION 3A – Mahkota 3 3B – Baiduri 3C – Berlian 3D – Delima	WELCOMING DINNER & OPENING CEREMONY *Safir Room*
TIME	**8.30 – 10.30 am**	**10.30 – 11.00 am**	**11.00 – 12.30 pm**	**12.30 – 2.00 pm**	**2.00 – 4.00 pm**	**4.00 – 4.15 pm**	**4.15 – 6.15 pm**	
MONDAY, 19 SEPTEMBER 2016	PARALLEL SESSION 4A – Mahkota 3 4B – Baiduri 4C – Berlian 4D – Delima	REFRESHMENTS *Mahkota 1*	KEYNOTE ADDRESS 2 *Mahkota 3* *Donghee Shin & Chen Ling*	LUNCH *Mahkota 1*	PARALLEL SESSION 5A – Mahkota 3 5B – Baiduri 5C – Berlian 5D – Delima	REFRESHMENTS *Mahkota 1*	PARALLEL SESSION 6A – Mahkota 3 6B – Baiduri 6C – Berlian 6D – Delima	
TIME	**8.30 – 10.30 am**	**10.30 – 11.00 am**	**11.00 – 1.00 pm**	**1.00 – 2.00 pm**	**2.00 – 4.00 pm**	**4.00 – 4.15 pm**	**4.30 – 5.00 pm**	
TUESDAY, 20 SEPTEMBER 2016	PARALLEL SESSION 7A – Mahkota 3 7B – Baiduri 7C – Berlian 7D – Delima	REFRESHMENTS *Mahkota 1*	PUBLICATION WORKSHOP Asian Journal of Communication *Mahkota 3*	LUNCH *Mahkota 1*	PARALLEL SESSION 8A – Mahkota 3 8B – Baiduri 8C – Berlian 8D – Delima	REFRESHMENTS *Mahkota 1*	CLOSING CEREMONY AND AWARDS *Mahkota 3*	
TIME	**8.30 am**	**9.00 – 2.00 pm**						
WEDNESDAY, 21 SEPTEMBER 2016	Waiting at Istana Hotel Lobby	Post tour conference – Putrajaya						

SUMMARY OF PARALLEL SESSION THEMES FOR i-COME'16 (18-21 SEPTEMBER 2016)

Organized by: **Sponsored by:** Taylor & Francis Taylor & Francis Group **Supported by:** **Partner Institutions:**

1A – Political Communication Politics and Communication: Understanding the Democratization and Participations	**1B – Journalism** Journalism as Dialogue Platform	**1C – Communication and Technology** Media and Technology: The Roles and Challenges	**1D – Intercultural** Intercultural Competence: Constructing the Cultural Identity
2A – Panel Session - Purdue University Developing New Knowledge about Peace building and Communication through Engaged Scholarship	**2B – New Media** Media Technology: The Demands and The Challenges	**2C – Intercultural** Accessing the National Identity : Exploring the Possibilities in Intercultural Communication	**2D – Government & Communication** Challenges in Global Economic Trend: The Communication Perspectives
3A – Panel Session - Hong Kong Baptist University Health Promotion and Disease Prevention in Cultural Context	**3B – New Media** Credibility in Social Media: Quo Vadis?	**3C – Organizational Communication** Communication in Organization: Matter of Contexts	**3D – Intercultural** Demographic Issues in Media Studies: Hypothetical or Reality?
4A – Panel Session - Chulalongkorn University ASEAN Integrated Media Practice: A Model for Transcultural Media Education	**4B – New Media** New Media: The Intergenerational Issues and Perspectives	**4C – Corporate Communication** Reflecting The Internal Communication in Workplace	**4D – Intercultural** Matters of Diversity: The Intercultural Communication Adaption
5A – Panel Session - University of Philippines ICT-Mediated Communication and Education	**5B – New Media** Media: Traditional vs New Media	**5C – Film Study** Film and Global Media: Reality and the Challenges	**5D – Business Communication** Advertising and Consumer: Communicating Issues Trough Media
6A – Communication for Social Change Communication for Social Change: Exploring a Local Contexts	**6B – Journalism** Cultural Challenges in News Reporting and Writing	**6C - Organizational Communication** Challenges in Workplace: Organizational Culture, Reality and Practice	**6D - Social Issues** Theme: Social Issues: From Traditional to Contemporary Approaches
7A – Crisis Communication Managing Crisis Communication: Are We Doing Enough?	**7B – New Media** Multi Perspectives in Social Media Research: Unbundling the Main Issues	**7C – Theory & Strategy** Theory and Strategy: Challenging the Communication Approaches	**7D – Intercultural** Leadership, Competency and Social Setting: Matters of Intercultural Communication
8A - Political Communication Political Communication and Media: Bringing the Perspectives in Communication	**8B - Health Communication** Health Communication: Promoting an Effective Message	**8C – Broadcasting** International Broadcasting: The Mythology in Communication Research	**8D– Organizational Communication** The Impact of Social Media: Issues and Trends

List of Participants

A

A.A.I Prihandari Satvikadewi, University of 17 Agustus 1945, **INDONESIA**, vika@untag-sby.ac.id
Aaron Gazley, Victoria University of Wellington, **NEW ZEALAND**
Abdul Latiff Ahmad, Universiti Kebangsaan Malaysia, **MALAYSIA,** alba@ukm.edu.my
Abdul Mua'ti @ Zamri Ahmad, Universiti Putra Malaysia, **MALAYSIA**, abmuati@upm.edu.my
Abdul-Aziz Haruna, Ahmadu Bello University, **NIGERIA,** abdulharunas@yahoo.com
Abdullah Abdulrahman Alfares, Capital Market Authority, **SAUDI ARABIA**, Afnan.Shehri@cma.org.sa
Adibah Ismail, Universiti Utara Malaysia, **MALAYSIA**, adibah@uum.edu.my
Aep Wahyudin, Islami State University of Sunan Gunung Djati, **INDONESIA,** fwahyu2000@yahoo.com
Ah Choy Er, Universiti Kebangsaan Malaysia, **MALAYSIA**, eveer@ukm.edu.my
Ainurliza Mat Rahim, Universiti Sains Islam Malaysia, **MALAYSIA**, ainurliza@usim.edu.my
Aishah Mohammad, Bond Pricing Agency Malaysia Sdn Bhd, **MALAYSIA**, aishah@bpam.com.my
Aishat Abdulrauf, Universiti Utara Malaysia, **MALAYSIA**, ajokeaisha@yahoo.com
Ali Salman, Universiti Kebangsaan Malaysia, **MALAYSIA**, asalman@ukm.edu.my
Alina Ali Durrani, Punjab University, **PAKISTAN,** alina.durrani@gmail.com
Alison Yu Wei Chong, University of Malaya, **MALAYSIA**, alisoncyw@gmail.com
Amira Sariyati Firdaus, University of Malaya, **MALAYSIA,** amira.firdaus@gmail.com
Amli Hazlin Ahmad Mokhtar, Infrastructure University Kuala Lumpur, **MALAYSIA**, amli@iukl.edu.my
Anindita Dasgupta, Taylor's University, **MALAYSIA,** Anindita.Dasgupta@taylors.edu.my
Antoon De Rycker, Taylor's University, **MALAYSIA,** antoon.derycker@taylors.edu.my
Arlianny Sayrol, National University of Singapore, **SINGAPORE,** arlianny@nus.edu.sg
Arum Priadi, Universitas Ahmad Dahlan Yogyakarta, **INDONESIA,** arumpriadi@gmail.com
Arumugam Balasubramanian, Amrita University, **INDIA,** a_balasubramanian@cb.amrita.edu
Arunima Krishna, Boston University, **UNITED STATES,** sconnaug@purdue.edu
Atsushi Hirasawa, Chuo University, **JAPAN,** atsushi@tamacc.chuo-u.ac.jp
Azizi Aziz, Universiti Utara Malaysia, **MALAYSIA,** aziziaziz@uum.edu.my
Azlee Johar, Universiti Utara Malaysia, **MALAYSIA**, azlee@ymail.com
Azman Azwan Azmawati, Universiti Sais Malaysia, **MALAYSIA**
Azman Samsudin, Universiti Sais Malaysia, **MALAYSIA,** azman@cs.usm
Azreena Hani Abd Aziz, Universiti Utara Malaysia, **MALAYSIA,** reenhaniaziz@yahoo.com

B

Baek Seob Kim, Sungkyunkwan University, **REPUBLIC OF KOREA**, bskim3296@gmail.com
Bahiyah Omar, Universiti Sains Malaysia, **MALAYSIA,** bahiyah@usm.my
Bahtiar Mohamad, Universiti Utara Malaysia, **MALAYSIA**, mbahtiar@uum.edu.my
Basuki, UPN "Veteran" Yogyakarta, **INDONESIA,** basuki_upnjogja@yahoo.com

Bee Chen Ooi, Universiti Tunku Abdul Rahman, **MALAYSIA,** ooibc@utar.edu.my
Bee Chuan Sia, Universiti Tunku Abdul Rahman, **MALAYSIA,** siabc@utar.edu.my
Bei Ju, Beijing Normal University Zhuhai Campus, **CHINA,** jubei@bnuz.edu.cn
Bradley Freeman, American University in Dubai, **UNITED ARAB EMIRATES,** bfreeman@aud.edu
Brett Robertson, California State University-Fullerton, **UNITED STATES,** rober186@mail.chapman.edu
Brittany Hanna, Lieberman Research Worldwide, **UNITED STATES,** hanna117@mail.chapman.edu
Bushra Hameedur Rahman, University of the Punjab, **PAKISTAN,** bushra.ics@pu.edu.pk

C

Catherine Lee Cheng Ean, Sunway University, **MALAYSIA,** catherinelee@sunway.edu.my
Changsong Wang, Xiamen University, **MALAYSIA,** chanson_wang@hotmail.com
Charles T. Salmon, Nanyang Technological University, **SINGAPORE,** salmon@ntu.edu.sg
Che Haniza Che Noh, Universiti Malaysia Terengganu, **MALAYSIA**
Che Mahzan Ahmad, International Islamic University Malaysia, **MALAYSIA,** che_ahmad@yahoo.com
Che Su Mustaffa, Universiit Utara Malaysia, **MALAYSIA,** chesu402@uum.edu.my
Chong Sheau Tsuey, Universiti Kebangsaan Malaysia, **MALAYSIA,** stchong79@gmail.com
Christiantine Della, Infrastructure University Kuala Lumpur, **MALAYSIA,** chris@iukl.edu.my
Cyntia Dewi Anggraini, Telkom University, **INDONESIA,** cyntiadewia@gmail.com

D

Dahiru Baba, Ahmadu Bello University, **NIGERIA,** dahirugombe@gmail.com
Dan Fleming, University of Waikato, **NEW ZEALAND,** dfleming@waikato.ac.nz
Dayang Aizza Maisha Abang Ahmad, Universiti Malaysia Sarawak, **MALAYSIA,** aadamaisha@unimas.my
Di Zhu, University of Missouri, **UNITED STATES,** dz78d@mail.missouri.edu
Dian Mariesta, Universiti Kebangsaan Malaysia, **MALAYSIA,** dian.mariesta@gmail.com
Dil Froz Jan Sayed Halem Shah, Infrastructure University Kuala Lumpur, **MALAYSIA,** dilfrozjan@iukl.edu.my
Dinu Das, Amrita Vishwa Vidyapeetham, **INDIA,** mailittodinu@gmail.com
Dolores Borras, St. Bridget College, **PHILIPPINES,** dborras@sbcbatangas.edu.ph
Dong-Hee Shin, Sungkyunkwan University, **REPUBLIC OF KOREA,** dshin@skku.edu
Dwi Sulisworo, Universitas Ahmad Dahlan Yogyakarta, **INDONESIA,** sulisworo@gmail.com

E

Eli Jamilah Mihardja, Universiti Sains Malaysia, **MALAYSIA,** jamilah.mihardja@gmail.com
Elira Turdubaeva, American University of Central Asia, **KYRGYZSTAN,** eliraturdubayeva@gmail.com
Elmie Nekmat, National University of Singapore, **SINGAPORE,** elmie@nus.edu.sg
Emma Mohamad, Universiti Kebangsaan Malaysia, **MALAYSIA,** emmamohamad@ukm.edu.my
Endang Setiowati, Universitas Indonesia, **INDONESIA,** setiowati@gmail.com
Estrella Arroyo, University of Saint Anthony, **PHILIPPINES,** stararroyo.dean@yahoo.com
Evan Swandie G. Pedrosa, University of Philippines Open University, **PHILIPPINES,** evanswandie@yahoo.com

Evie Ariadne Shinta Dewi, Universitas Padjadjaran, **INDONESIA,** evieasd@gmail.com
Ezhar Tamam, Universiti Putra Malaysia, **MALAYSIA,** ezhar@upm.edu.my

F

Faridah Ibrahim, Infrastructure University Kuala Lumpur, **MALAYSIA,** faridah@iukl.edu.my
Ferdy Fachrudin, Combine Resource Institution, **INDONESIA,** ferdhiputra@gmail.com
Firouzeh Keshavarzi, University of Malaya, **IRAN,** fyfy80@yahoo.com

G

Ganiu Okunnu, Crescent University, **NIGERIA,** okunnu.ganiu@crescent-university.edu.ng
Genina Mariel Arceo, Adamson University Alumna, **PHILIPPINES,** geninamarielarceo@gmail.com
Grace Javier Alfonso, University of Philippines Open University, **PHILIPPINES,** jsaludadez@upou.edu.ph
GuanGim Yeoh, Universiti Tunku Abdul Rahman, **MALAYSIA,** yeohgg@utar.edu.my
Gv Nair, Universiti Tunku Abdul Rahman, **MALAYSIA**

H

Hadi Purnama, Telkom University, **INDONESIA,** hadipurnama21@gmail.com
Hamedi Adnan, University of Malaya, **MALAYSIA,** hamedi@um.edu.my
Hameedullah Azimi, Universiti Kebangsaan Malaysia, **MALAYSIA,** hameedullahazimi@yahoo.com
Hamisah Hassan, Universiti Putra Malaysia, **MALAYSIA,** hamisah@upm.edu.my
Hanaa Kadum Kassed, Ministry of Science and Technology, **IRAQ**
Hang Fu, Huazhong University of Science and Technology, **CHINA,** vsheer@hkbu.edu.hk
Haslina Halim, Universiti Utara Malaysia, **MALAYSIA,** lina@uum.edu.my
Hasrina Mustafa, Universiti Sains Malaysia, **MALAYSIA,** hasrina@usm.my
Hassan Abu Bakar, Universiti Utara Malaysia, **MALAYSIA,** abhassan@uum.edu.my
Hassan Raza Syed, Bahauddin Zakariya University, Multan, **PAKISTAN,** sherazibzu@gmail.com
Hatikahaty Binti Mohamad Diah, Puteri Nursing College Sdn Bhd, **MALAYSIA,** hatikahaty@kpjuc.edu.my
Hee Jhee Jiow, Singapore Institute of Technology, **SINGAPORE,** jhee.jiow@singaporetech.edu.sg
Holger Briel, Xi'an Jiaotong-Liverpool University, **CHINA,** holger.briel@xjtlu.edu.cn
Hong Wen NG, National University of Singapore, **SINGAPORE,** nghongwen@gmail.com
HuiNee AuYong, Universiti Tunku Abdul Rahman, **MALAYSIA,** auyonghn@utar.edu.my
Husna Afifi Mohd Yusoff, Universiti Malaysia Perlis, **MALAYSIA,** husnaafifi@unimap.edu.my
Hyeonjin Kim, Sungkyunkwan University, **REPUBLIC OF KOREA,** hyeonjin.k@skku.edu

I

Idha Saraswati, Combine Resource Institution, **INDONESIA,** idha@combine.or.id
Inge Hutagalung, Universitas Mercu Buana, **INDONESIA,** inge_hutagalung@yahoo.com

Irwansyah, Universitas Indonesia, **INDONESIA,** irwansyah09@ui.ac.id
Isma Rosila Ismail, Universiti Malaysia Terengganu, **MALAYSIA,** ismarosila@umt.edu.my
Ismail Sualman, Universiti Teknologi MARA, **MALAYSIA**

J

Jamilah Jamal, Universiti Utara Malaysua, **MALAYSIA**, jamilah@uum.edu.my
Janice Aurellia, Petra Christian University, **INDONESIA,** m36412055@john.petra.ac.id
Japen Sarage, Universitas Ahmad Dahlan Yogyakarta, **INDONESIA,** japen.sarage@pbi.uad.ac.id
Jasmine R. Linabary, Purdue University, **UNITED STATES,** sconnaug@purdue.edu
Jean A. Saludadez, University of Philippines Open University, **PHILIPPINES,** jsaludadez@upou.edu.ph
Jen Sern Tham, Universiti Tunku Abdul Rahman, **MALAYSIA,** thamjs@utar.edu.my
Jennifer Gibbs, Rutgers University, **UNITED STATES,** jgibbs@rutgers.edu
Jessada Salathong, Chulalongkorn University, Bangkok, **THAILAND,** jess.salathong@gmail.com
Jessica Pauly, Purdue University, **UNITED STATES,** sconnaug@purdue.edu
Jia Wang, Hong Kong Baptist University, **HONG KONG,** vsheer@hkbu.edu.hk
Jirayudh Sinthuphan, Chulalongkorn University, Bangkok, **THAILAND,** jirayudh.S@chula.ac.th
Joane V. Serrano, University of Philippines Open University, **PHILIPPINES**, jsaludadez@upou.edu.ph
Joanna Samuel, Consultant, **UNITED KINGDOM**, josamuel@mail.com
Joanne Lim, Monash University Malaysia, **MALAYSIA**, julian.hopkins@monash.edu
Joyce Lynn-Sze Cheah, Universiti Utara Malaysia, **MALAYSIA,** joycecheah@uum.edu.my
Julia Eisenberg, Pace University, **UNITED STATES,** jeisenberg@pace.edu
Julian Hopkins, Monash University Malaysia, **MALAYSIA,** julian.hopkins@monash.edu
Julinawati Suanda, Universiti Malaysia Perlis, **MALAYSIA,** julina@unimap.edu.my
Jusang Bolong, Universiti Putra Malaysia, **MALAYSIA,** jusang@upm.edu.my

K

Kalyani Suresh, Amrita University, **INDIA,** s_kalyani@cb.amrita.edu
Kara Chan, Hong Kong Baptist University, **HONG KONG,** vsheer@hkbu.edu.hk
Kara Patricia Agbayani, University of the Philippines Baguio, **PHILIPPINES,** karapatgoagbayani@gmail.com
Karen Jennifer Golden, Monash University Malaysia, **MALAYSIA,** julian.hopkins@monash.edu
Kavitra Devi Muniandy, Universiti Kebangsaan Malaysia, **MALAYSIA,** devi.kavitra@gmail.com
Kelly Vibber, University of Dayton, **UNITED STATES,** sconnaug@purdue.edu
Kellyn LEE, Singapore Management University, **SINGAPORE,** elmie@nus.edu.sg
Kenneth Rogerson, Duke University, **UNITED STATES,** rogerson@duke.edu
Kenneth Woo, Singapore Institute of Technology, **SINGAPORE,** Kenneth.Woo@singaporetech.edu.sg
Kerk Kee, Chapman University, **UNITED STATES**, kerk.kee@gmail.com
Kim-Shyan, Victoria University of Wellington, **NEW ZEALAND**
Kiranjit Kaur Pritam Singh, Universiti Teknologi MARA, **MALAYSIA,** kiran126@salam.uitm.edu.my
Klan Yeik Koay, Multimedia University, **MALAYSIA**

Kok Eng Tan, University Tunku Abdul Rahman, **MALAYSIA,** tanke@utar.edu.my
Kok Wai Chew, Multimedia University, **MALAYSIA**

L

Lateef Adelakun, University of Malaya, **MALAYSIA,** delak058@yahoo.co.uk
Ling Ling Chew, Singapore Institute of Technology, **SINGAPORE,** LingLing.Chew_2014@SIT.SingaporeTech.edu.sg
Lubna Zaheer Zaheer-ud-din, University of the Punjab, **PAKISTAN,** lubna.zaheer91@gmail.com
Lucy Supratman, Telkom University, **INDONESIA,** doktorlucysupratman@gmail.com

M

Marc Pacoma, Adamson University, **PHILIPPINES,** marc.pacoma@adamson.edu.ph
Mario Antonius Birowo, Universitas Atma Jaya Yogyakarta, **INDONESIA,** mbirowo@yahoo.com
Martha Lestari, Telkom University, **INDONESIA,** martha.djamil@gmail.com
Marzura Ibrahim, Universiti Utara Malaysia, **MALAYSIA,** marzura@uum.edu.my
Masduki , Indonesian Islamic University, **INDONESIA,** masduki@hotmail.com
Maylanny Christin, Telkom University, **INDONESIA,** maylannychristin@gmail.com
Mazeyanti Mohd Ariffin, Universiti Teknologi Petronas, **MALAYSIA,** mazeyanti2012@gmail.com
Melicent Jalova, Mindanao State University-Iligan Institute of Technology, **PHILIPPINES,** melicent09@yahoo.com
Melinda De La Pena-Bandalaria, University of the Philippines Open University, **PHILIPPINES,** evanswandie@yahoo.com
Meng Yoe Tan, Monash University Malaysia, **MALAYSIA,** julian.hopkins@monash.edu
Mikaila Ajaga, Lagos State University, Lagos, **NIGERIA,** ajagajigi@yahoo.com
Min Ho Ju, Sungkyunkwan University, **REPUBLIC OF KOREA,** mhju@skku.edu
Minah Harun, Universiti Utara Malaysia, **MALAYSIA,** minn@uum.edu.my
Miriam Park, Monash University Malaysia, **MALAYSIA,** julian.hopkins@monash.edu
Mohamad-Noor Sharipudin, Universiti Utara Malaysia, **MALAYSIA,** mnsaleh@uum.edu.my
Mohammed Rezal Hamzah, Universiti Malaysia Perlis, **MALAYSIA,** rezal@unimap.edu.my
Mohd Adnan Hashim, Universiti Teknologi MARA , **MALAYSIA**
Mohd Baharudin Othman, Universiti Utara Malaysia, **MALAYSIA,** bahar@uum.edu.my
Mohd Faizal Kasmani, Universiti Sains Islam Malaysia, **MALAYSIA,** faizalkasmani@usim.edu.my
Mohd Khairie Ahmad, Universiti Utara Malaysia, **MALAYSIA,** khairie@uum.edu.my
Mohd Nor Shahizan Ali, Universiti Kebangsaan Malaysia, **MALAYSIA,** shahizan@ukm.edu.my
Mohd Rizal Bin Majid, Puteri Nursing College Sdn Bhd, **MALAYSIA,** rizalmajid@kpjuc.edu.my
Mohd Sobhi Ishak, Universiti Utara Malaysia, **MALAYSIA,** msobhi@uum.edu.my
Mohummad Kamrul Hassan, **BANGLADESH,** khassan75@gmail.com
Moniza Waheed, Universiti Putra Malaysia, **MALAYSIA**
Mudassar Abdullah, Universiti Utara Malaysia, **MALAYSIA,** mudassar80d@yahoo.com
Mudassar Hussain, University of the Punjab, **PAKISTAN,** mudassar.hussain29@yahoo.com

Muh Saeful Effendi, Universitas Ahmad Dahlan Yogyakarta, **INDONESIA,** Muh.effendi@pbi.uad.ac.id
Muhammad Anshari, Universiti Brunei Darussalam, **BRUNEI DARUSSALAM,** anshari.ali@ubd.edu.bn
Muhammad Ashraf Khan, Bahauddin Zakariya University, Multan, **PAKISTAN,** ashraf_1963@bzu.edu.pk
Muhammad Helmi Elias, Singapore Institute of Technology, **SINGAPORE,** Helmi_2014@SIT.SingaporeTech.edu.sg
Munirah Ilias, Infrastructure University Kuala Lumpur, **MALAYSIA,** munirah@iukl.edu.my
Mus Chairil Samani, Universiti Malaysia Saawak, **MALAYSIA,** smchairil@unimas.my
Muzayin Nazaruddin, Islamic University of Indonesia, **INDONESIA,** begawan.nazaruddin@gmail.com

N

Natalie Collie, The University of Queensland, **AUSTRALIA,** n.collie@uq.edu.au
Neva Štumberger, Purdue University, **UNITED STATES,** sconnaug@purdue.edu
Nicole Yang Lai Fong, Taylor's University, **MALAYSIA,** laifong.yang@taylors.edu.my
Nik Adzrieman Abdul Rahman, Universiti Utara Malaysia, **MALAYSIA,** adzrieman@uum.edu.my
Nik Ahmad Rozaidi, Rutgers University, **UNITED STATES,** nrozaidi@rutgers.edu
Nikmah Suryandari, UTM Bangkalan East Java, **INDONESIA,** nikmahsuryandari@gmail.com
Nina Lim, Singapore Institute of Technology, **SINGAPORE,** JiaHui.Lim_2014@SIT.SingaporeTech.edu.sg
Noor Adzrah Ramle, Universiti Sains Islam Malaysia, **MALAYSIA,** nooradzrah@usim.edu.my
Nor Azura Adzharuddin, Universiti Putra Malaysia, **MALAYSIA,** zurh@upm.edu.my
Nor Hayati Mohd Jalil, University Utara Malaysia, **MALAYSIA,** n.hayati@uum.edu.my
Noratikah Mohamad Ashari, University Malaysia Sarawak, **MALAYSIA,** manoratikah@unimas.my
Norealyna Misman, University of Malaya, **MALAYSIA,** elmisman@gmail.com
Norhafezah Yusof, Universiti Utara Malaysia, **MALAYSIA,** norhafezah@uum.edu.my
Norzita Yunus, Infrastructure University Kuala Lumpur, **MALAYSIA,** norzita@iukl.edu.my
Novell Lyndon, Universiti Kebangsaan Malaysia, **MALAYSIA,** novel@ukm.edu.my
Nurzihan Hassim, Universiti Malaya, **MALAYSIA,** nurzihan@siswa.um.edu.my

O

Oki Cahyo Nugroho, Muhammadiyah University of Ponorogo, **INDONESIA,** okicahyo@umpo.ac.id
Olalekan Akashoro, Al-Hikmah University, Ilorin, Kwara State, **NIGERIA,** lekkan2003@yahoo.com

P

Panji Dwi, UPN "Veteran" Yogyakarta, **INDONESIA,** panjidwi.pd@gmail.com
Paonrach Yodnane, Chulalongkorn University, Bangkok, **THAILAND,** uhuminky@hotmail.com
Patrick C.H. Soh, Multimedia University, **MALAYSIA,** chinhooi8@gmail.com
Peck Ling Tee, Universiti Tunku Abdul Rahman, **MALAYSIA,** teepl@utar.edu.my
Peter Simmons, Charles Sturt University, **AUSTRALIA,** psimmons@csu.edu.au
Pijar Suciati, University of Indonesia, **INDONESIA,** pijar.suciati@gmail.com
Pin Lick Soo Hoo, University Utara Malaysia, **MALAYSIA,** eileensoohoo@gmail.com

Pramela Krish N. Krishnasamy, Universiti Kebangsaan Malaysia, **MALAYSIA,** pramela@ukm.edu.my
Puckpilai Khuptawathin, Chulalongkorn University, Bangkok, **THAILAND,** kpuckpil@chula.ac.th

Q

R

R. A. Noer Doddy Irmawati, Universitas Ahmad Dahlan Yogjakarta, **INDONESIA,** witha_irmabs@yahoo.com
Ramachandran Ponnan, Taylor's University, **MALAYSIA,** ramachandran.ponnan@taylors.edu.my
Ranggabumi Nuswantoro, Unversitas Atma Jaya, **INDONESIA,** ranggabumi_nuswantoro@mail.uajy.ac.id
Rayvinder Athwal, Singapore Institute of Technology, **SINGAPORE,**
Rayvinder_2014@SIT.SingaporeTech.edu.sg
Ridwan Adetunji Raji, Universiti Utara Malaysia, **MALAYSIA,** rajiridwanadetunji@gmail.com
Rizwanah Souket, International Islamic University Malaysia, **MALAYSIA,** rizwanahsouket@gmail.com
Romlah Ramli, Universiti Utara Malaysia, **MALAYSIA,** romlah@uum.edu.my
Ronavel Mei Bañaga, University of the Philippines Baguio, **PHILIPPINES,** ronavelmeicbanaga@gmail.com
Roro Retno Wulan Roro, Telkom University, **INDONESIA,** rorowoelan28@gmail.com
Roshidi Din, Universiti Utara Malaysia, **MALAYSIA,** roshidi@uum.edu.my
Rosidayu Sabran, Universiti Sains Islam Malaysia, **MALAYSIA,** rosidayu@usim.edu.my
Rosya Izyanie Shamshudeen, Universiti Putra Malaysia, **MALAYSIA,** rosya@upm.edu.my
Rosyidah Muhamad, Universiti Malaysia Terengganu, **MALAYSIA,** rosyidah@umt.edu.my
Rozita Abdullah, **MALAYSIA,** rozita@usim.edu.my
Ruri Supadmi, Universitas Ahmad Dahlan Yogyakarta, **INDONESIA,** sulisworo@gmail.com
Ruth Wan Ching Lee, Universiti Tunku Abdul Rahman, **MALAYSIA**

S

Sabariah Mohamed Salleh, Universiti Kebangsaan Malaysia, **MALAYSIA,** sabariah@ukm.edu.my
Sadaf Zahra, Bahauddin Zakariya University, Multan, **PAKISTAN**
Saidathul Nizah Mat Tazin, Universiti Teknologi Mara, **MALAYSIA,** saidathul06@yahoo.com
Saima Nisar, Universiti Utara Malaysia, **MALAYSIA,** saimaanisar@gmail.com
Salizatul Aizah Binti Ibrahim, Universiti Tunku Abdul Rahman, **MALAYSIA,** salizatul@utar.edu.my
Samsudin Abd Rahim, Universiti Kebangsaan Malaysia, **MALAYSIA,** samsudinukm@gmail.com
Sangmee Lee, Sungkyunkwan University, **REPUBLIC OF KOREA,** sangmee7@gmail.com
Sabrina Mohd Rashid, Universiti Utara Malaysia, **MALAYSIA**, sabrina@uum.edu.my
Sara Al Bastaki, University College of Bahrain, **BAHRAIN**
Semuel Hatane, Petra Christian University, **INDONESIA,** samy@petra.ac.id
Shamsu Dauda Abdu, Universiti Utara Malaysia, **MALAYSIA,** shamsud.abdul@gmail.com
Sharina Samsudin, Universiti Utara Malaysia, **MALAYSIA,** sharina@uum.edu.my
Shawaludin Anis, University College of Bahrain, **BAHRAIN,** sanis@ucb.edu.bh
ShawHong SER, Chulalongkorn University, Bangkok, **THAILAND,** shaw.h@chula.ac.th
Shun-itsu NAKASAKO, Chuo University, **JAPAN,** nakasako@tamacc.chuo-u.ac.jp

Siti Aisyah Saat, Universiti Malaysia Terengganu, **MALAYSIA,** s.aisyah@umt.edu.my
Siti Nor Amalina Ahmad Tajuddin, Universiti Pendidikan Sultan Idris, **MALAYSIA,** sitinoramalina@fbk.upsi.edu.my
Siti Suriani Othman, Universiti Sains Islam Malaysia, **MALAYSIA,** suriani@usim.edu.my
Siti Zanariah Yusoff, Universiti Sultan Zainal Abidin, **MALAYSIA,** zanariahyusoff@unisza.edu.my
So-Jeong Kim, Sungkyunkwan University, **REPUBLIC OF KOREA,** ai.gimso@gmail.com
Solahuddin Ismail, Universiti Utara Malaysia, **MALAYSIA,** solahuddin@uum.edu.my
Sona Varghese, Amrita Vishwa Vidyapeetham, **INDIA**, angelmarysona@gmail.com
Stacey L. Connaughton, Purdue University, **UNITED STATES,** sconnaug@purdue.edu
Stephen Croucher, University of Jyväskylä, **FINLAND,** stephen.m.croucher@jyu.fi
Suffian Hadi Ayub, Sunway University, **MALAYSIA,** suffianay@sunway.edu.my
Suhaimee Saahar @ Saabar, Universiti Teknologi MARA, **MALAYSIA,** aimeesavvy@gmail.com
Suhaini Muda, Universiti Utara Malaysia, **MALAYSIA,** suhaini@uum.edu.my
Sukma Ari Ragil Putri, University of 17 Agustus 1945, **INDONESIA,** sukmaariragilputri@gmail.com
Sumekar Tanjung, Islamic University of Indonesia, **INDONESIA,** sumekar.tanjung@yahoo.com
Sunariya Utama, Universiti Utara Malaysia, **MALAYSIA,** sunariya.utama@gmail.com
Supaporn Phokaew, Chulalongkorn University, Bangkok, **THAILAND,** jirayudh.S@chula.ac.th
Suziah Sulaiman, Universiti Teknologi Petronas, **MALAYSIA,** suziah@petronas.com.my
Syamimi Ariff Lim, Universiti Brunei Darussalam, **BRUNEI DARUSSALAM,** syamimi.ariff@ubd.edu.bn
Syarizan Dalib, Universiti Utara Malaysia, **MALAYSIA,** syarizan@uum.edu.my
Syed Arabi Idid, International Islamic University Malaysia, **MALAYSIA,** sarabidid@iium.edu.my
Syed Siraj, Allama Iqbal Open University, **PAKISTAN,** professorsiraj58@yahoo.com

T

Takuya Sakurai, Tokyo Denki University, **JAPAN,** takuyasakurai@icloud.com
Thanaseelan Rajasakran, Universiti Tunku Abdul Rahman, **MALAYSIA,** thanak@utar.edu.my
Thinavan A/L Periyayya Periyayya, Universiti Tunku Abdul Rahman, **MALAYSIA**, thinavan@utar.edu.my
Tony Wilson, London School of Economics and Political Science, **MALAYSIA**, tonyjwilson@yahoo.com

U

V

Vandana Saxena, Taylor's University, **MALAYSIA,** Vandana.Saxena@taylors.edu.my
Victor Yew-Cho Yen, Nanyang Technological University, **SINGAPORE,** yyen001@e.ntu.edu.sg
Vimala Govindasamy, Amrita University, **INDIA,** veemala21@gmail.com
Vincent Tao-Hsun Chang, National Chengchi University, **TAIWAN,** Vince.T.Chang@gmail.com
Vivian C. Sheer, Hong Kong Baptist University, **HONG KONG,** vsheer@hkbu.edu.hk

W

Wan Fatimah Wan Ahmad, Universiti Teknologi Petronas, **MALAYSIA,** fatimhd@petronas.com.my
Wan Rozaini Sheik Osman, Universiti Utara Malaysia, **MALAYSIA,** rozai174@uum.edu.my
WANG Yan, National Chengchi University, **TAIWAN,** talktowangyan@gmail.com

X

Xin-Yu Dai, Hong Kong Baptist University, **HONG KONG,** vsheer@hkbu.edu.hk
Xuan Xie, Hong Kong Baptist University, **HONG KONG,** 14485354@life.hkbu.edu.hk

Y

Yeen Lai Khong, University Tunku Abdul Rahman, **MALAYSIA,** khongyl@utar.edu.my
Yeon Oh Hong, Sungkyunkwan University, **REPUBLIC OF KOREA,** boystomkig@naver.com
Yesuselvi Manickam, Sunway University, **MALAYSIA,** yesuselvim@sunway.edu.my
Yiming Chen, Jinan University, **CHINA,** jidayiming@163.com
Yit Phing Loh, Universiti Putra Malaysia, **MALAYSIA,** lyitphing@gmail.com
Yun Bae Kim, Sungkyunkwan University, **REPUBLIC OF KOREA,** kimyb@skku.edu
Yunxia Zhu, The University of Queensland, **AUSTRALIA,** y.zhu@business.uq.edu.au
Yunya Song, Hong Kong Baptist University, **HONG KONG,** vsheer@hkbu.edu.hk
Yves Peter Carlo Medina, University of the Philippines Baguio, **PHILIPPINES,** ypcmedina@gmail.com

Z

Ziinine Abdesselam, Universiti Teknologi MARA, **MALAYSIA,** robbie_mk@yahoo.com
Zulkifli Abd.Latif, Universiti Teknologi MARA, **MALAYSIA,** robbie_mk@yahoo.com

Sponsors

- Adabi Consumer Industries Sdn Bhd
- Aifar Tours & Vacations Sdn Bhd
- Asa Mesra Printer
- College of Arts and Sciences, Universiti Utara Malaysia
- Department of Communication Studies, Hong Kong Baptist University
- Dewan Bahasa & Pustaka
- Faculty of Communication Arts , Chulalongkorn University
- Hotel Istana, Kuala Lumpur
- Insfrastructure University Kuala Lumpur
- International Communication Association (ICA)
- Kementerian Komunikasi dan Multimedia Malaysia (KKMM)
- Kementerian Pelancongan dan Kebudayaan Malaysia
- Malaysia Convention and Exhibition Bureau (MyCEB)
- McGraw Hill Education
- School of Multimedia Technology and Communication, Universiti Utara Malaysia
- Sungkyunkwan University
- Taylor & Francis Group
- Tenaga Nasional Berhad
- Purdue Peace Project, The Brian Lamb School of Communication, Purdue University
- Unit Kenderaan, Universiti Utara Malaysia
- Unit Komunikasi Korporat, Universiti Utara Malaysia
- Universitas Ahmad Dahlan, Yogjakarta
- Universiti Pendidikan Sultan Idris Press
- University of Saint Anthony, Iriga City
- University of the Philippines Open University, Los Banos

edp sciences

SHS Web of Conferences

Statement of Peer review

In submitting conference proceedings to SHS Web of Conferences, I certify to the Publisher that adhere to the **Policy on Publishing Integrity** of the journal in order to safeguard good scientific practice in publishing.

1. All articles have been subjected to peer review administered by the proceedings editors.
2. Reviews have been conducted by expert referees, who have been requested to provide unbiased and constructive comments aimed, whenever possible, at improving the work.
3. Proceedings editors have taken all reasonable steps to ensure the quality of the materials they publish and their decision to accept or reject a paper for publication has been based only on the merits of the work and the relevance to the journal.

Title, date and place of the conference

INTERNATIONAL CONFERENCE ON COMMUNICATION AND MEDIA
AN INTERNATIONAL COMMUNICATION ASSOCIATION REGIONAL CONFERENCE
SEPTEMBER 18-20, 2016
ISTANA HOTEL, KUALA LUMPUR, MALAYSIA

Proceedings editor(s):

BAHTIAR MOHAMAD
HASSAN ABU BAKAR

Date and editor's signature

1/12/2017

Happiness" as Constructs of Reality for Selected Fourth Year Students of the University of the Philippines Baguio

Yves Peter Carlo de Rivera Medina[1,*]
[1] University of the Philippines Baguio, Governor-Pack Road Baguio City 2600, Philippines

Abstract. This paper examined how social media such as the Instagram, provides an avenue for individuals to construct their world as they deem fit and modify it based on subjective truths. Utilizing the method of textual analysis, this paper examined the photographs posted in individual Instagram accounts by selected students of the University of the Philippines Baguio. What appears to be real for them, in this curated form of diary, is that their constructed world is "happy." This paper then addressed these questions: What is their perspective of reality as communicated by the selected Instagram users? How is this depicted in their Instagram accounts? The researcher then has found out that reality in the net space particularly in Instagram is always changing, incoherent, and in determinant.

1 Introduction

In this day and age, social media has already been part of everyone's daily routine. The emergence of social networking sites gave people an extended platform to express, interact, and share experiences with other people all over the world. One of the rising social media platforms utilized by the younger generation is the Instagram, a photograph social application designed for smartphone that allow its users to share daily experiences through uploading photos to their individual and personalized accounts. This phenomenon sparked the interest of the author to identify and describe the constructs of reality of selected Instagram users. As Poster stressed that "the individual becomes aware of participating in something greater than him/herself, something where the individual does not have a fixed vantage point [1]". In his book entitled "Mode of Information," Poster wrote that in the presence of the net space "the self is decentered, dispersed, and multiplied in continuous instability [2]". Taking off from Poster's statement then, the net space, specifically in the use of social media such as the Instagram, provides an avenue for individuals to construct their world as they deem fit and modify it based on subjective truths. This study then, utilized textual analysis and examined the photographs posted in individual Instagram accounts by selected fourth year students of the University of the Philippines Baguio. Presenting a curated form of diary, Instagram accounts compiled and read in this research, narrates on the Instagram users' constructed notions of reality.

* Corresponding author: ypcmedina@gmail.com/nehemiahbangalan@gmail.com

1.1 The Problem

Reading the Instagram accounts of selected fourth year students of the University of the Philippines Baguio, the study examined the constructed notions of reality by reading the bevy of photographs posted and uploaded in their individual profile accounts. Addressing these two questions: 1) What is their perspective of reality as communicated by the selected Instagram users? and, 2) How is this depicted in their Instagram accounts? Upon answering these questions, the researcher presented the perceived reality of the Instagram users by attempting to explain how their reality becomes incoherent and indeterminant.

This study used Mark Poster's The Mode of Information as a theoretical underpinning in the reading of constructs of reality in the online space. Coining the term mode of information, Poster proposes three general stages of the mode of information that corresponds to the manner in which messages are communicated. These stages are, face-to-face, print age, and lastly the electronic mediated stage [3]. This paper, however, gave focus on the electronic mediated stage. Poster posits that communication in the electronic age is instantaneous and is not restricted by time and space [4]. With the existence of electronically mediated communication, communication as a field now can be described as the transfer of information across space and time. It usually comes in representation of the language. It can be a symbol, text, and/or images. Communication plays vital role in the construction of reality. It is important to note that reality is made through messages sent by an individual. In the case of Poster's electronic age, the constructs of reality are established through the representation of language an individual post in an electronic medium. These representations convey meaning and shape the constructs of reality.

2 Reality as Depicted by Photographs in the Instagram Accounts

In determining the constructs of reality, the researcher has identified twenty-two emerging themes from the photos of the respondents. This table illustrates the number of photos per theme, the frequency of the appearances of these themes to the respondents and their corresponding ranks according to frequency. From this table, the top five emerging themes are identified. The top emerging themes were: a) Selfie and Self Portraits, b) Groupfies and Friends c) Food and Drinks d) Nature e) Academics. Two salient subject photographs were also on: a) feet and b) motivational and inspirational quotes. From these constructs then, the researcher established the notions of reality of the respondents.

Table 1. Frequency, Percentage and Ranking of Themes

Themes	No. of Photos	Appearances in the Profile of 30 Respondents	Percentage Based on Appearances	Rank Based on the No. of Appearance in the 30 Respondents
Selfie and Self Portraits	282	27/30	90%	1
Groupfies and Friends	226	24/30	80%	2
Food and Drinks	252	23/30	76.76%	3
Nature	139	22/30	73.33%	4
Academics	118	21/30	70%	5
Random Objects	112	21/30		
Others	86	20/30	66.67%	6
Quotes	55	18/30	60%	7
Book	21	17/30	56.67%	8

 DOI: 10.1051/shsconf/20173300004

Family	72	16/30	53.33%	9
Artwork	41	15/30	50%	10
Feetfie	32	15/30		

2.1 Constructed Reality in Instagram

Reality is everything that an individual considers to be real. And in Instagram, the respondents have set a certain kind of reality by posting different photos of self, friends, food and drinks, nature, academics, feet and quotes.

Selfie and Self Portraits. In this theme, respondents represent their self-image in varied ways but there are common traits that can be traced in each photo. The respondents wish to create a self-image in Instagram which is free from weakness, failures, and sadness. Moreover, it is important to note that, since this is the top theme, respondents find their self- image as the utmost priority. They put primacy to their self -image. This reflects the priorities of the youth today. Furthermore, the projection of their lives as happy and positive attempts to convince the online community that they are living a perfect life in a perfect world. They choose not to expose their weakness and that they are living their lives with full of comforts. In doing so, they somehow internalize this behavior in their real lives and actually believe that they have this kind of life.

Groupfies and Friends. "Tell me who your friends are, and I will tell you who you are" still holds true today. By posting online who your circle of friends is, is like telling the people what kind of person you are. Moreover, the manner in how the respondents present their friends also depicts a positive and happy life. This is regardless of the context of the photos. Respondents frame everything in a positive angle including moments under pressure and stress. This affirms that respondents see Instagram as platform to share only happy moments. Lastly, this theme placed second to selfie and self-portraits which shows that aside from their self, friends are the second top priority of the respondents. Aside from themselves, people around them are what make them.

Food and Drinks. Food and drinks' reality in the online world should be anything that is fancy; something that looks expensive. For the respondents, typical meals prepared at home are not the food qualified for Instagram. They have established the notion of only food with good plating is real in the application. Moreover, they stressed that these good foods actually serve as their comfort and source of happiness in times of stress.

Nature. In this theme, respondents view nature as something that is directly connected to their personal lives. They do not post nature just to give its appearance dignity rather they post nature photos to actually represent anything that can symbolize what they are feeling or thinking.

Academics. Their idea of academics is the real-life application of what are learned in class. They show to people the career that they might be pursuing after college. Furthermore, they also view academics to be stressful but they understand that these are part of the struggles in reaching their dreams.

Feetfie and Quotes. Respondents actually see their feet as good representation of themselves. Their state, fashion and mindset in life can be substituted by their feet in the online world. Furthermore, they Feetfie photos show that they believe that life is a journey. On the other hand, respondents see quotes as pieces of advice they need to hear. They use inspirational messages to communicate what they are feeling or what they need. Moreover, they see quotes as something that can uplift not only them but also the people who they share their accounts with.

From the descriptions of constructs of reality, reality for them is anything that is happy and positive. Moreover, respondents post different aspects of their lives in their Instagram accounts; and all of these aspects are used to strengthen their self-construct and identity. They use self- portraits, group pictures, food and drinks, nature, academics, feet, and quotes to establish the reality they want. Furthermore, in presenting these aspects, they tend to frame them as happy and optimistic.

Looking at the age group of the respondents (18 to 21 years old), it is not surprising that their notion of reality is reflected on the photos they post. For them, life is full of happiness, hope, positivity, and adventure. However, these things are not actually real. In life, there is a whole spectrum of emotions and experiences that are not only enjoyable but also unsatisfying. And these were not seen on the feed of the respondents in their Instagram accounts.

The physical reality is composed of all things that one can experience---happiness, sadness, anger, fear and the like. On the other hand, constructs of reality in Instagram are filtered. Daily experiences were captured by the respondents using their smartphones and posted them in their Instagram accounts. The danger here is that, photos posted in the application are just highlights of a moment and/or of an experience which cannot actually narrate the whole experience of the respondents. Moreover, reality in the posted photos are already fabricated- from framing the photo, adding filters, partnering it with captions and choosing which experience only is to post in Instagram. These mentioned additional elements to the photos contribute in bridging the gap between the new shape of reality to that of the tangible reality.

Looking at it from a different light, adding filters and effects on photos, putting captions, and framing the photos are essential in establishing a notion of reality these respondents want to set. These behaviors are tools in achieving that Internet reality they wish to show. And in the case of this study, respondents do such behavior to establish a reality which is full of happiness and positivity.

The construct of reality in the application will tell a lot about the respondents. In this modern age, aside from friends, photos posted on Social Media can define what kind of an individual a person is. Moreover, since the online communities of the respondents are almost the same as their physical community, there will be a blur line when the internet reality is penetrated to the physical reality. The notion of reality constructed in their Instagram accounts could affect, in one way or another, the physical reality of the respondents.

Ultimately, looking at the two salient themes- *Feetfie* and quotes, reality for them can also be represented by their feet and words. These two are great metaphors to sum up one's life; feet as symbol of a person's being and journey and words as symbol of the individual's philosophy. Indeed, reality is too hard to define and cannot have one definition. It varies from context, culture and even age generation. Yet, it is still important to attempt to determine a certain group's construct and notions of reality in order to understand certain behaviour and know where they are coming from.

2.2 Mediation of the Image in Reality Construction

In this study, the physical reality is represented by the photos taken and then posted by the respondents in their Instagram accounts. Today, one way of proving that a certain incident really happened is by taking a picture of it and uploading it in different social networking sites, one of which is the Instagram; the interactions are done mostly through posting and exchanging of photos online. Users find this more convenient than actually narrating things through text.

Photos then become the ultimate language of reality in the online space since the users find it more convenient to narrate using photos. Using Poster's posits that a language once represented through the electronically mediated communication, losses its authenticity. It is interesting to see how the physical reality is mediated in Instagram using the selected photos of the users.

While it is true that photos posted by the respondents happened in real life, it is important to look at what kind of scenarios are only posted, how did they frame it and what alterations did they use in enhancing the photos. These behaviors are essential in establishing the subjective reality of the users. From these photos, there emerges the blurry line between the users' notion of reality and truth.

2.3 Reality as Incoherent

Incoherency of reality in Instagram is two pronged. One incoherency is observed through the quotes posted by the respondents. They post photos that are inspiring to share and inspire other people. But in reality, they may not be really happy and are currently down so they use inspiring quotes to keep them positive and be reminded that they need to have positive outlook in life.

Second, happiness is the main language of the respondents in their accounts even if the respondents know that their real time and space is not perfect. Their offline space gives them all kinds of experiences, not only happiness. However, the respondents refuse to share these negative and distressing experiences to their online community which makes their reality incoherent.

2.4 Reality as in determinant

In this study, it can be observed that the constructs of reality of the respondents are greatly influenced by their environment, particularly their friends. Having a positive and happy life is the reality that is imposed to the respondents. This is why photos posted by the respondents reflect this reality. From their selfie to the quotes that they post, everything depicts the reality as happy and positive. On the other hand, what is also interesting is the theme Feetfie. They actually believe that life is a journey and that they need to keep their feet on the ground.

With these observations, it can be assumed that the reality established in the net space is always changing. In the presentation of photographs of feet which symbolizes direction, goal and being grounded, this sets an articulate language of their reality in the online space; to be and become happy. But, in the real time and space, there are setbacks and hindrances to the path of happiness which they may encounter all the time. To counter-act this real time and space everyday problems, the respondents post quotes to motivate and encourage their own selves. In a way, they continue on in re-constructing their reality as happy in the net space.

Poster said that self-identity and image online is constantly changing.[†] But, in this case, because the Self is the main topic in the respondents' Instagram accounts, the goal and reality aimed for is happiness. Meaning, their self-identity should always be anchored to being and becoming happy.

Indeterminacy of constructed and re-constructed reality in the online space becomes apparent when individuals, in this case, the respondents keep on constructing and re-constructing their goal to become happy only in the online space. Because they can believe they are indeed happy whenever they see their posted photographs but in the offline space they are not. A problem in this arena of indeterminacy poses another challenge. The respondents are determined to be happy in the online space, what if they remain there because the offline space presents reality that is far from their own constructed reality in the online space.

Fragmentation becomes consequential to indeterminacy of the self and constructed reality as two different spaces (online and offline) collide and will not merge. The identity of the respondents may result to multiple identities, i.e., happy in the online space but frustrated in the real world, etc. This fragmentation of identity raises these questions then: Can these respondents re-align themselves and become happy so as to merge the offline and online space realities? Meaning, can they actually control their own happiness in the offline space? If reality in the net space can always be altered depending on the trend in the net space and on what the online community of the users are showing then, what will remain real for these respondents?

As Poster explained, a language once represented through an electronically mediated communication will form new shapes of it; it is the same when the physical reality is represented through photos in Instagram. Photos posted are already filtered by the respondents. They don't usually post everything that happens in their everyday lives. As what the photos collected show, only those that depict happy moments are posted in the Instagram. Moreover, they use filter functions to enhance the aesthetic appeal of the photos. Thus, the researcher poses this question: What do they believe as real?

The selecting, framing, and editing of photos create the reconstructed reality. This reconstructed reality is a reflection of the subjective reality of the users in which they wish to manipulate and shape. From this subjective reality, there is the existence of blurry lines within indeterminacy and incoherency of the users' notion of reality and truth.

It is mentioned earlier that the notion of reality of the users is anything that is happy and positive. This may be true in the online space, but not in the offline space. The offline space provides spaces for failures, problems and negativism. However, users prefer not to show this in the online space. There is no problem with this but with the users' massive subscription to the online space, there might be a consequential creation of a gap in their notions of reality and truth.

Since the online space affords them to create a reality that is more convenient to them, the indeterminacy and incoherency of their truth will arise when the individuals spend more time interacting in their online communities than in their physical ones. An individual may be incoherent with his/her physical and internet reality. When this happens, the quality of relationships of the individual with his/her physical community might be compromised.

3 Conclusions

Based on the identified constructs and notions of reality in the respondents' Instagram accounts, Instagram, therefore, is seen as the curated form of diary in the sense that only happy experiences of everyday life are posted. Moreover, unlike the classic diary, Instagram is shared to chosen group. Like any other social media websites, gives its users the power to create one's ideal self; or create multiple versions of self. This phenomenon's role in communication field, however, is until now arguable. Respondents' subscription to the notion of defining a successful life with happy and positive outlook is seen in their

DOI: 10.1051/shsconf/20173300004

Instagram accounts. Their notion of reality is a great reflection of the principles and the things they value in life.

Furthermore, apart from the conflict that Internet and Physical reality bring to the lives of the respondents, what is really interesting to discover in this study is the respondents' ultimate pursuit of happiness. Their notion of reality greatly revolves around the idea of becoming and being happy and gaining a positive life. This notion, perhaps, is greatly influenced by the environment of the respondents. Since childhood, a person is always reminded to always follow his/her dreams and to live a happy life. As a person grows, life gets more complicated and again the society reminds him/her that a secret to a successful life is to stay happy and positive. This notion that has been repeatedly imposed to the individuals may have influenced these respondents that regardless of the whole spectrum of emotions and feelings, one prefers to create a reality that is full of happiness and positivity.

From the researcher's observations on the Instagram users' behavior, question now arises, for whom and what are they doing this for? On one hand, it can be assumed that they do this to convince the Instagram community that they actually live this kind of life. On the other hand, this behavior can also be interpreted as the respondents are trying to convince themselves that they actually live this lifestyle, one who has a positive outlook in life and adventurous, by posting it online. It is as if whatever posted in Instagram did really exist in real life or will really exist in real life. In reality, people sometimes complain and get mad and the respondents surely experience these things but these things were not seen in their accounts.

References

1. Mark Poster. *The Mode of Information* (Chicago: The University of Chicago Press 2007).
2. S.J. Murray, M. Poster, Qui Parle, **14** (2003)
3. R.E. Babe, E. Comor, Canadian Journal of Cultural Studies, **16** (2006)
4. Poster, Mark. *The Mode of Information*, (The University of Chicago Press, Chicago, 1990)

Application of Motion Correction using 3D Autoregressive Model in Kinect-based Telemedicine

Baek Seob Kim[1], *Yun Bae* Kim[2], *So-Jeong* Kim[2,*] and *Dong*-Hee Shin[3]

[1]Smart Content Research team, ETRI, 11-41 Simin-daero 327 beon-gil, Dongan-gu, Anyang, Korea,
[2]Department of Systems Management Engineering, Sungkyunkwan University, 2066 Seobu-ro, Jangan-gu University, Suwon, Korea
[3]Department of Interaction Science, Sungkyunkwan University, 25-2 Sungkyunkwan-ro, Jongro-gu, Sungkyunkwan University, Seoul, Korea

Abstract. In telemedicine, where the convergence of different types of medical treatment occurs, it is very important to establish credibility regarding the mutual communication between patients and medical workers by acquiring and sharing more accurate data. For rehabilitation treatment in particular, where motion data are required, auxiliary equipment such as a Kinect sensor is being more widely used. This study proposes a methodology for improving the motion recognition rate by compensating the noise from a Kinect sensor using a 3D autoregressive model. Moreover, this study investigates the methods applied for vitalizing the area of telemedicine under this particular trend.

1 Introduction

As people are placing increasingly high expectations on their health, the need for acquiring and recognizing relevant information is growing. Along with this growth in expectations, in the area of health communication, which helps with smoothing the interaction between patients and medical workers, the acquisition of personal health-related data and a precise delivery of the significance of such data are playing a greater role [1]. Specifically, interlinked with the development of IT technologies such as high-speed Internet, network security, video conferencing, and collaboration, telemedicine is regaining attention because it enables personally customizable health treatment, diagnosis, and remedies to be administered by connecting patients and medical workers within a network. Telemedicine is playing an appropriate role in disease protection, diagnosis, curing, and recovery by attracting the proper behavioral changes from patients based on their personal health data acquired from credible communication between the patients and medical personnel [2]. For this reason, along with proper communication between patients and medical workers, the means of acquiring and delivering data can be a crucial factor.

* Corresponding author: ai.gimso@gmail.com

DOI: 10.1051/shsconf/20173300005

With the development of information and communication technologies, the manner in which personal health data are acquired and delivered through telemedicine may be transformed. In general, medical personnel recognize, determine, and diagnose a patient's status through 2D video clips, images, and text-centric data acquired from the patient. However, patients have the responsibility to gather and deliver the data related to their own health status and conditions. However, different medical departments may require different forms of data, and the ease in acquiring such data may also vary. For instance, for the psychiatric care of anxiety or mood disorders, or the administering of internal or family medicine, which deal with chronic diseases such as high blood pressure and diabetes, telemedicine can be applied without much difficulty, and does not necessarily require support through physical contact or additionally acquired data. In contrast, rehabilitation medicine such as orthopaedics faces a difficulty when treating patients without a clear understanding of their body motions. As a way to resolve this issue, a rehabilitation program, KiReS (a kinect-based telerehabilitation system), through which a medical service provider sends a remote treatment program to a patient, allowing the patient to rehabilitate themself using a Kinect sensor, is being introduced into the market [3]. Motion recognition devices can be categorized into two types, contact-based or touch-less. Representative technologies for these two groups include OptiTrack and Kinect. OptiTrack can acquire more accurate motion data than Kinect, but is very expensive and not easily available for purchase. On the other hand, Kinect, which was originally developed for gaming but whose applications have been expanded into many areas such as medical auxiliary devices such as KiReS, has become very popular [4].

In this light, this research aims at finding a new methodology for gathering motion data in a precise manner by reducing the noise from a Kinect sensor through the application of a 3D autoregressive model. To do so, using a mathematical model, we tried to compensate the motion data in real-time by improving the motion recognition rate, and compared the experimental results with the results from a highly precise OptiTrack to see how the data acquisition can be improved. In addition, we tried to study how receiving and delivering personal health data can impact a patient's awareness and attitude toward their behavioral changes, and along with this, we tried to determine how to motivate the telemedicine industry to aggressively adopt auxiliary medical devices such as a Kinect sensor.

2 Literature Review

Health communication not only facilitates mutual communication between patients and medical personnel to improve patient health based on precise data, it also stimulates behavioral changes and active participation of the patient for the prevention, diagnosis, and treatment of diseases [1,2,5]. Therefore, the importance of health communication is receiving more attention, and its overall concept includes all processes such as information gathering, data sharing through wearable devices, and face-to-face remote counselling between medical personnel and patients. Although not regarded as significantly more effective than any other methods, this research is heavily focused on telemedicine [6], which enables the ubiquitous delivery of medical information in voice, data, and video communication formats, because owing to advancements in ICT technologies the amount of attention being paid to this field is growing.

Korea has conducted a pilot project since September of 2014 to determine the safety and validity of telemedicine. The first phase (with six-month duration) showed a high satisfaction rate of 77%. The second phase of the project started from March of 2015. Telemedicine service centers have increased in number from 18 to 50, and services have been provided for crews on deep-sea fishing vessels, soldiers, prisoners, and medically underserved people in remote areas [7]. This trend implies that telemedicine could be highly vitalized if it can be proved that the benefits for patients and the opportunity costs for medical personnel are comparably better than those from face-to-face medical treatment.

As a friendly and heart-felt manner of interpersonal communication [8], telemedicine is said to improve our understanding of mutual interactions between patients and medical workers, and support decision-making based on behavioral changes in a more effective manner [2]. This implies that telemedicine can have an influence on the level of satisfaction regarding various medical practices. Furthermore, it is difficult to tell decisively whether telemedicine is clinically better than traditional face-to-face medical services, but its patient satisfaction has been shown to be consistently superior [9].

However, without the help of auxiliary devices in some medical departments, telemedicine experiences difficulties in achieving efficient communication, achieving medical goals, and inducing behavioral changes [2]. Without the support from auxiliary devices such as a Kinect sensor, rehabilitation therapists have had to treat patients based on their own experience and personal decision-making through physical treatments and comprehending the improvement in a patient's status with their naked eye. Clinical tests using auxiliary devices such as a Kinect sensor have proved that their use has led to a two-fold better performance compared to cases in which such devices were not used, and it was proved that a low-cost approach leads to high rehabilitation outcomes [10]. For this reason, the increasing number of applications using auxiliary medical devices such as a Kinect sensor is a natural trend and it is necessary to find a way to acquire, deliver, and share data from such devices in an efficient manner. In particular, a Kinect sensor needs to accurately recognize the motions of the patient and must be stable for use in rehabilitation treatment [10]. Medical treatment can be easily practiced when obtaining the natural body motions of the patients. Therefore, this study focuses on a way to improve the motion recognition rates by using a Kinect sensor as a telemedicine supplement for recognizing more natural body movements.

3 Methodology

3.1 Data Collection

A person has 20 Kinect-recognizable joints, two of which were selected as data acquisition targets in the present study. The two hands of the experiment participants were marked, and each participant's clapping motions were captured using both a Kinect and an OptiTrack to obtain the 3D coordinates of the target joints. A Kinect records at 30 fps, whereas an OptiTrack records at 120 fps. Thus, the amount of collected data differs between the two devices. We conducted an experiment to compensate the Kinect coordinates. Therefore, OptiTrack's frame intervals were adjusted to be the same as those of the Kinect. Specifically, data collection was conducted using a Kinect SDK 1.7 and OptiTrack Motive 1.8.1.

3.2 Data Calibration

To measure the offset in the alignment of the Kinect coordinates, it was necessary to adjust the OptiTrack, which was assumed to acquire accurate measurement data, based on the Kinect-captured 3D coordinates.

DOI: 10.1051/shsconf/20173300005

$$\widehat{x_o} = y_o(n) + mean(y_k(n)) - mean(y_o(n))$$
$$\widehat{y_o} = y_o(n) + mean(y_k(n)) - mean(y_o(n)) \quad (1)$$
$$\widehat{z_o} = z_o(n) + mean(z_k(n)) - mean(z_o(n))$$

Here, $x_o(n), y_o(n),$ and $z_o(n)$ are the coordinates of the n-th frame measured using OptiTrack, and $x_k(n), y_k(n),$ and $z_k(n)$ are the same coordinates as observed by the Kinect. Using formula (1), the OptiTrack coordinates were converted into Kinect coordinates.

3.3 Outlier Detection

Among the 3D coordinates observed using a Kinect, noise occurs when a person's joints are hidden behind an obstacle, or when the person is out of view of the Kinect's infrared depth vision camera; these coordinates are therefore estimated based on the coordinates of adjacent joint. However, during a small timeslot of $1/30^{th}$ of a second, some intervals may have a large amount of displacement, and after analyzing the graphs, the frames may show an obstruction of the person's natural movements. We therefore searched the outliers in all of the datasets. With the exception of those frames that contain estimated coordinates, the Inter Quartile Range (IQR) of the other frame's coordinates was calculated and the intervals outside of this range were regarded as frames with noise.

$$IQR = Q_3 - Q_1 \quad (2)$$

$$Lower\ Bound = Q_1 - (1.5 \text{ x } IQR)$$
$$Upper\ Bound = Q_3 + (1.5 \text{ x } IQR) \quad (3)$$

Here, Q_1 and Q_3 represent an accumulated percentile of 25% (the first quartile) and 75% (the third quartile), respectively. When the 3D coordinates observed by a Kinect sensor are smaller than the lower bound or bigger than the upper bound of equation (3), the frame belonging to the coordinates is treated as a noise-added frame.

3.4 3D autoregressive model program

When continuous observations, $\{Z_j^1, Z_j^2, Z_j^3\}_{j=1,\cdots,N}(Z_{-j}^a = Z_{N-j}^a, \mathrm{a} = 1 \cdots 3)$, are given, a 3D auto regressive model of order m can be represented through formulae (4) and (5).

$$\textstyle\sum_{a=1}^{3} \frac{\sigma^a}{2} \hat{Z}_j^a = \sqrt{-1} \sum_{k=1}^{m} [\sum_{a=1}^{3} \frac{a^a}{2} Z_{j-k}^a, \sum_{b=1}^{3} \frac{\sigma^b}{2} A_k^b] = \sum_{k=1}^{m} \sum_{c=1}^{3} \frac{\sigma^c}{2} (Z_{j-k} \times A_k)^c \quad (4)$$

$$\hat{Z}_j^a = \textstyle\sum_{k=1}^{m} (\mathbf{Z}_{j-k} \times A_k)^c \quad (5)$$

The 3D coordinates observed by a Kinect have the same attributes as continuous frame-based data, namely, time-series data, and are thus suitable for an autoregressive model. We developed a program that compensates the frames containing the estimated values and the additional frames defined in Chapter 3.3 using an autoregressive model that compensates the 3D data. First, the program loads an Excel file where the 3D coordinates are stored. The frame numbers of the estimated frames that need to be compensated, and the total number (N) of frames prior to the frames to be compensated, are used as the input data. Then, order m of the model is provided to the program. After clicking the 'compensation button', vectored 3D coordinates in the autoregressive model are calculated. Finally, using formula (5), the compensated 3D values are calculated. Following these procedures, the coordinates in a frame containing noise are compensated, and the program updates the dataset to include the adjusted coordinates to compensate the following noisy frames.

3.5 Experiment Environment

To conduct the experiment on minimizing noise from a Kinect sensor, the device was placed on top of an OptiTrack to record the same motions simultaneously, and the data were compared. The lenses of the two cameras were then adjusted to look in the same direction. Given that the depth sensor camera on a Kinect is set as the origin, the 3D axes provided by Kinect SDK were set as follows. The left side of the Kinect is the positive direction of the x-axis, the right side is the negative direction of the x-axis, the upper side is the positive direction of the y-axis, the lower side is the negative direction of the y-axis, the front side is the positive direction of the z-axis, the back side is the negative direction of the z-axis, and the unit of length is measured in meters. On the other hand, the origin of the OptiTrack is set after recognizing the three markers attached to the OptiTrack origin recognition tool, and thus we aligned OptiTrack's axes with those of the Kinect before carrying out the experiments. To simplify the data observation process, the wrist of the subject was placed at the OptiTrack's origin, and we then executed the skeleton information-gathering program provided by Kinect SDK. Heuristically, we found the point where the x-axis becomes zero, and adjusted the tool's position just to the right of this point.

4 Result

We arranged the experimental environment according to the experimental methods described in chapter 3. We then collected data from the two motion recognition cameras of OptiTrack and Kinect, and converted the OptiTrack coordinates into Kinect coordinates using formula (1). Lastly, we assumed these coordinates to be the experimental values and compared them with the values from a 3D autoregressive model and based on the root mean squared error (RMSE).

Table 1. RMSE values and improvement rate after applying 3D autoregressive model

	Axis	RMSE (before)	RMSE (after)	Improvement Rate
Hand Right	X	0.023539	0.016845	28.43%
	Y	0.020090	0.018858	6.13%
	Z	0.012727	0.008986	29.39%
	Average	0.018785	0.014896	21.31%

As Table 1 shows, by comparing the experimental values with the RMSE values, the x-axis, which represents the left or right, and the z-axis, which represents the front or back, showed no distinctive differences. However, the experiments recorded an average improvement of 21.31%. This result proves that the non-contact and touch-less camera used in a Kinect sensor can be applied to recognize natural motions when any errors are appropriately minimized.

5 Conclusion

In this study, we tried to determine whether an auxiliary device such as a Kinect sensor can be used to recognize motions more precisely when accompanied with a mathematical compensation model in the field of telemedicine, which allows patients and medical workers to communicate with each other remotely. This research proved that the motion recognition rate can be improved when the noise from a Kinect sensor is compensated using a 3D autoregressive model, and we examined certain possibilities for improving the role of health communication in the area of telemedicine.

We looked into the role and need for a Kinect sensor to be used as an auxiliary device in telemedicine based on cases taken from the industry as well as from previous studies. Being able to recognize body motions at a low cost, Kinect is a suitable device for rehabilitation treatment in terms of accuracy, processing speed, and expandability from previous researches [3-10]. Many patients seem to feel that face-to-face medical treatment is not an absolute solution for disease prevention, diagnosis, cures, and recovery, which is an attitude that is growing in magnitude. Owing to a limitation of medical laws in Korea, telemedicine is currently being carried out only for crews on deep-sea fishing vessels, soldiers, prisoners, and medically underserved people in remote areas who have very limited access to medical services; however, it is expected that when the benefits to patients such as cost reductions and improvements in medical quality increase, telemedicine will become vitalized. Moreover, the ratio of people who have a greater interest in obtaining comparative information related to their own health status will increase with the expansion of diverse health communication media, and for disease diagnosis and cures, patients will not rely solely on personnel one-to-one medical communication, but will participate equally in communication regarding their ailment with the help of additional data. In such a case, the experience and personal opinions of medical workers are important, but will not have an absolute impact on their credibility as deemed by the patients. Therefore, acquiring more objective data among medical workers and patients, along with the real-time sharing of data, is necessary, and there should be an environment that helps patients voluntarily participate in this process.

However, motions with a high level of difficulty were unable to be captured in our experiments because only one OptiTrack was used. Further researches using more than two OptiTracks to capture very difficult motions as a way to compare errors are highly recommended, and if conducted more accurate results might be made available, with a validation of the improvement rates expected. During our experiments, we realized that immersive and fun environments, along with an improvement in motion recognition, are very important factors for patients. Thus, future researches are better off applying virtual or augmented reality technologies to create a more immersive environment.

In summary, this study indicated that the proposed method for improving the motion recognition rate through a mathematical model using a Kinect sensor can be applicable to the field of telemedicine without additional costs; in addition, this method can be used to acquire and share objective data between patients and medical personnel in an effective manner. Moreover, if it is possible to elicit the participation of the patients and engage their interest by creating a virtual or augmented-reality environment that can be equipped using Kinect SDK-based 3D vision software that feels similar to real face-to-face medical

DOI: 10.1051/shsconf/20173300005

treatment, such changes will be a catalyst vitalizing the field of telemedicine.

References

1. J.M. Bernhardt, Am J Public Health, **94** (2004)
2. V.S. Freimuth, S.C. Quinn, Am J Public Health, **94,** 12 (2004)
3. D. Antón, A. Goni, A. Illarramendi, J.J. Torres-Unda, J. Seco, KiReS: A Kinect-based telerehabilitation system. *In e-Health Networking, Applications & Services (Healthcom), IEEE 15th International Conference on IEEE,* (2013)
4. B.S. Kim, *Motion Correction of 3D Autoregressive Model Using Kinect Skeletal Joint Coordinate.* Master dissertation, Sungkyunkwan University, Korea (2016). (to be published)
5. M.L. Clayman, J.A. Manganello, K. Viswanath, B.W. Hesse, N.K. Arora, J Health Commun, **15**, (2010)
6. D.J. Kim, Korea Institute for Health and Social Affairs Issue & Focus, **270** (2015)
7. C.S. Park, *Report* (*Korea Ministry of Health and Welfare*, 2015)
8. A. Duggan, Journal *of health communication,* **11,** (2016)
9. W.D. Leach, *If You Bill It, They Will Come: A literature review on clinical outcomes, cost-effectiveness, and reimbursement for telemedicine,* (California Telemedicine and eHealth Center, Sacramento, CA, 2009)
10. H. Mousavi Hondori, M. Khademi, Journal of Medical Engineering, (2014)

Matter of Similarity and Dissimilarity in Multi-Ethnic Society: A Model of Dyadic Cultural Norms Congruence

Hassan Abu Bakar[1,*], *Bahtiar* Mohamad[1]

[1]Department of Communication, School of Multimedia Technology & Communication, 06010 UUM Sintok, Kedah, Malaysia

Abstract. Taking this into consideration of diver cultural norms in Malaysian workplace, the propose model explores Malaysia culture and identity with a backdrop of the pushes and pulls of ethnic diversity in a Malaysia. The model seeks to understand relational norm congruence based on multiethnic in Malaysia that will be enable us to identify Malaysia cultural and identity. This is in line with recent call by various interest groups in Malaysia to focus more on model designs that capture contextual and cultural factors that influences Malaysia culture and identity.

1 Introduction

In Malaysia, ethnic divisions are coterminous with its linguistic, cultural, religious and economic differences. Malaysian society, individuals hold differing cultural orientations, norms and values according to their ethnicity (i.e., ethnic Malays, ethnic Chinese, ethnic Indian). While there is very little cultural homogeneity between the three ethnicities in Malaysia, their divergent value differences converge under the nation of Malaysia. This convergence can be considered as contributing to the unique face of Malaysian identity. Thus, we see ethnic divisions in the nature of work, occupations and social structure in Malaysia. For example, a majority of Bumiputra work in manufacturing and the public sectors, the Chinese dominate management and professional positions along with a small number of Indians, and a majority of Indians work in the plantation sector. Such unique heterogeneity (with homogeneous subdivisions) helps highlight the complexity of cultural norms in Malaysia, However, to date not many studies have address this research lacuna, therefore this study will explore the Malaysia identity and culture from various perspectives.

Taking this into consideration, the propose study will explore Malaysia national culture and identity with a backdrop of the pushes and pulls of ethnic diversity in a Malaysia. This article seeks to understand relational norm congruence based on multiethnic in Malaysia that will be enable us to identify Malaysia cultural and identity. This is in line with recent

* Corresponding author: abhassan@uum.edu.my

call by various interest groups in Malaysia to focus more on research designs that capture contextual and cultural factors that influences Malaysia culture and identity. In paper we question how cultural relational norms and values in Malaysia may provide conditions for the congruence model of multiethnic society in Malaysia? and we believe this is a fundamental question that deserves further systematic investigation. We feel that this salience question to explore characteristic of the Malaysian national cultural and identity.

2 Theory and Model Development

Malaysia is a multiethnic society in which three major ethnic groups (Malays/Bumiputra, Chinese, Indian) interact with a significant degree of harmony in their everyday living. Each of these ethnic groups has also, to a large degree, managed to preserve their divergent ethnic identities via continued practice of customs, behavior, languages spoken, norms, values, and beliefs. Malaysia's society and work force has been found to be reflective of this cultural, behavioral, and value diversity. While there is very little cultural homogeneity between the three ethnicities in Malaysia, their divergent value differences converge under the nation of Malaysia. This convergence can be considered as contributing to the unique face of "Malaysian identity". However, to date not many studies have addressed this research lacuna, therefore this study will explore the Malaysia identity and culture from various perspectives. For this reason, we argue that understanding and developing Malaysia national cultural and identity is a significant and powerful issue, and can be viewed as a salient issue for diverse multiethnic society like Malaysia. In fact, to date little is known about what underlies Malaysia national culture and identity in a truly multiethnic society like Malaysia. Thus, this offers unique opportunity for investigation as aims to explore Malaysia national culture and identity by capturing diversity in a society that is inherently diverse. Our proposition here is in line with a recent call by various interest groups in Malaysia to focus more on research designs that capture contextual and cultural factors that influences Malaysia culture and identity that will lead to national unity. We question how cultural relational norms and values in Malaysia may provide conditions for the congruence model of multiethnic society in Malaysia? And we believe this is a fundamental question that deserves further systematic investigation. We feel that this salience question to explore characteristic of the Malaysian national cultural and identity.

Two conceptual frameworks underlie most ethnicity and cultural identity research [1]. The first of these is similarity-attraction/identification, while the second is relational norm congruence. Similarity-attraction theory was largely developed devoid of cultural considerations and is based on social cognitive processes of "like attraction" [2], social identification [3, 4], and self-categorization [5], The general idea behind similarity-attraction is that people like (and are attracted to) others who are similar (vs. dissimilar) to themselves. This "birds of a feather flock together" concept posits that similarity leads to interpersonal attraction and shared group identity, which produces in-group favoritism or positive bias. These biases and favoritisms are then associated with various relational and task-related outcomes. Social scientific research has provided significant support for tenets of similarity attraction theory since the mid-1900s. A. S. Tsui [6] argued that similarity in relationships demographic attributes would be associated with more positive outcomes and affect toward multiethnic relationships, and less role ambiguity and conflict. Their empirical test was partially supported. To more fully understand relational demography and its impact on workplace interactions, later studies went beyond social cognitive explanations and focused on our second underlying theory - relational norm congruence [7, 8]. Relational norm congruence specifies that multiethnic demographic similarity or dissimilarity, when congruent with the cultural norm for that demographic category, is associated with more positive relational and task outcomes [9]. It is worth noting that while perceived similarity has been shown to perform better than relational demographic variables in many Western contexts, some research suggests that they may be less useful in non-Western settings where relational norm congruence is a key concern [10]. Research has shown that the effect of relational demography is, to a large extent, a function of the relational norms of a particular culture [11]. To further understand this dyadic (dis)agreement in a particular cultural context, we argue that it is necessary to incorporate relational demography as an important factor in our analysis. This is especially true in cultures where relational norm congruence and societal cultural norms are best characterized by relational demography. Thus, we see that understanding relational norm congruence based on multiethnic in Malaysia will be enabling us to identify Malaysia cultural and identity. This can be done through identifying similarity and dissimilarity characteristics of cultural and identity that are congruence with norms of each ethic in Malaysia.

Hierarchical differentiation is an important part of ongoing interpersonal interactions. Without question, the supervisor plays a large role in shaping a subordinate's "organizational reality," and in many ways the supervisor can be the face of the organization for the subordinate. The supervisor is often the information provider to the subordinate, the connection to the "upper ranks" of the organization, and the provider of the subordinate's job definition and realities. The supervisor-subordinate relationship quality (ranging from low quality to high quality relationship exchange) can affect a supervisor's decisions concerning resource allocation, delegation of responsibilities, and evaluative information about job behavior (i.e., performance evaluations). All of this can have a profound effect on the subordinate because of their lower hierarchical status in the dyad. Taking the above into consideration, we argue that the dynamics (in terms of demographic diversity) that help shape supervisor-subordinate relationship quality exchanges will eventually affect subordinate perceptions of the exchanges, as well as attitude about the job itself [12]. While we recognize that several studies have found simple employee demographic characteristics to correlate with outcomes such as satisfaction, commitment and performance [13, 14] these studies do not connect demographic diversity variables in a specific cultural context. Therefore, this study will focus our investigation of demographic attributes that are tied closely to the Malaysian organizational context. They include (dis)similarities in ethnicity, religion, gender, age, and tenure. In the following paragraphs, we develop hypotheses with regard to each attribute, respectively. Ethnicity and Religion Malaysian society is primarily comprised of the following ethnic groups; Malay 50.4%, Chinese 23.7%, indigenous 11%, Indian 7.1%, others 7.8% [15]. Each of these ethnic groups maintains their own strong ethnic identity and their own cultural customs, practices, language, values and beliefs [16]. Religion is closely tied to ethnicity in Malaysia. For example, the Malays are Muslims, most Indians follow Hinduism, and the majority of Chinese practice a combination of Taoism and Buddhism, mixed with values associated with Confucianism. A small minority of Indian and Chinese follow Christianity and Islam, respectively [17, 18]. Unlike most Western societies where religion is kept out of the workplace, Malaysian work settings readily accommodate those observing various religious rituals and practices [17]. In sum, ethnic difference is a salient relational demographic attribute actively maintained and highly visible in both secular and religious affairs [19]. Perhaps not surprisingly, when team building in Malaysian organizations, strong ethnic identities tend to cultivate in-group favoritism and trust [18]. As such, managers may be more likely to prefer members from the same ethnic group as we would expect from the similarity attraction, social identity and self-categorization arguments discussed earlier. Motivated by self-enhancement and uncertainty-reduction, ethnic similarity reinforces a stable social identity and facilitates in-group social interaction. According to social exchange theory [20], reciprocity - as a primary basis of exchange - is culturally conditioned. Leader-member dyads with the same ethnic identity share similar norms of reciprocity and are more likely to achieve mutual agreement with regard to the levels of exchange resources, such as organizational commitment, job satisfaction and performance ratings. In contrast, leader-member dyads comprised of those with differing ethnic identities are more likely to have different cultural expectations, and are also likely to have difficulties in achieving agreement about their relationship quality. As mentioned, religion is closely related to ethnicity in Malaysia. Unlike what one might see in many Western societies, religion in Malaysian workplaces is visibly practiced and highlights differences in cultural values. Similar to ethnicity, then, we would expect that dyads sharing the same religion would find it easier to achieve agreement and reciprocate benefits and obligations. Prior research speaks to this point, as similar dyads in ethnicity were found to reinforce members' perceptions of agreement about organizational values [14] and contribute to performance [21]. Hence, we predict that:

Proposition 1: Within the ethnic similarity dyad will be positively associated with the dyad relationships quality ratings
Proposition 2: Within the ethnic dissimilarity dyad will be negatively associated with the dyad relationships quality ratings

3 Limitation and Future Directions

Although the proposed model has advanced our knowledge about relationships in diverse workplace, a few limitations of this study need to be noted. First, when interpreting the proposed model, one must recognize that the model was limited to a single country. While we strongly feel that this multi-ethnic organization is quite typical of large Malaysian businesses, this does remain the model limitation. Second, statements of causality based on the positive or negative relationships must be treated with caution given the co-relational nature of the variables. Future research might examine the relational demography paradigm within a variety of organizations that have different demographic distributions (e.g., expatriates working in Malaysian organizations) to increase the generalizability of the present model. More cultural variables should be considered. Third, our current model limits itself to similarity and dissimilarity and relationships outcome variables. These outcome variables are an important component of organizations, and they are interrelated. Therefore, the correlation between relational demographic and outcome variables must be treated with caution. While the model has uncovered many useful findings, future studies may include other outcome variables such task delegation and absenteeism. Lastly, the present study was limited to five demographic variables. Future research should try to include as many demographic characteristics as possible, including work group context and educational differences. In summary, despite this model's limitations, the model herein unquestionably extends our understanding of relational demography and supervisor-subordinate relationships quality. We have done so by identifying specific demographic characteristics as important components of the social context in which the dyad operates in a workgroup. While the pattern of the model raises some questions about the generalizability of the relational demography paradigm to all demographic characteristics, our model strongly suggests that similarities in the dyad are critical to the dyad relationship and work attitudes. In order to continue providing knowledge useful for organizations and their managers, we urge researchers to continue their efforts to identify specific demographic characteristics within the supervisor-subordinate relationship and work group that may influence work relationships outcomes.

DOI: 10.1051/shsconf/20173300006

References

1. M. M. Steward, Garcia-Prieto, J Organ Behav, **29** (2008)
2. D. Byrne, *The attraction paradigm*. (Academic Press, New York, 1971)
3. B. E. Ashforth, F. A. Mael. ACAD MANAGE REV, **14**, (1989)
4. M. A. Hogg, D. Abrams, D. *Social identifications: A social psychology of intergroup relations and group processes*. (Routledge, London 1988)
5. M. A. Hogg, D. J. Terry. (2000). ACAD MANAGE REV, **25**, (2008)
6. A. S. Tsui, C. A. O'Reilly. ACAD MANAGE J, **32**, (1989)
7. R. C. Liden. Asia Pacific Journal of Management, **28**, (2011)
8. R. P. Vecchio. D. M. Brazil. Personnel Psychology, **60**, (2007).
9. C. Goldberg. C. M. Riordan. L. Zhang. Group & Organization Management, **33**, (2008)
10. A. S. Tsui. H. Wang. K. R. Xin. Management & Organization Review, **2** (2006)
11. D. C. Lau. L. W. Lam. S. D. Salamon. Journal of Social Psychology, **148** (2008)
12. M. A. Wesolowski. K. W. Journal of Organizational Behavior, **18** (1997)
13. D. R. Avery. P. F. McKay. D. C. Wilson. *Journal of Applied Psychology,* **92** (2007)
14. D. R. Avery. P. F. McKay. D. C. Wilson. *Personnel Psychology,* **60** (2007)
15. K. T. Bhal. N. Gulati, N. M. A. Ansari. LEADERSHIP ORG DEV J, **30** (2009)
16. Abdul Rashid, M. Z., Sambasivan, M., & Abdul Rahman, A. LEADERSHIP ORG DEV J, **25** (2004)
17. A. Abdullah. S. Singh. S. K. Gill. *Understanding the Malaysian workforce: Guidelines for managers*. (Malaysian Institute of Management, Malaysia, 2001)
18. K. Ahmad, K. International Journal of Commerce & Management, **11** (2001)
19. C. Selvarajah. D. Meyer. Leadership & Organization Development Journal, **29** (2008)
20. R. Cropanzano. M. S. Mitchell. J MANAGE, **31** (2005).
21. P. F. McKay. D. R. Avery. M. A. Morris. Personnel Psychology, **61** (2008)

Exploring the Educational Effectiveness of Culturally-enhanced Serious Game in Education

Mazeyanti Mohd Ariffin[1,*], *Wan Fatimah* Wan Ahmad[1], and *Suziah* Sulaiman[1]

[1]Computer & Information Sciences Department, 32610 Sri Iskandar, Perak, Malaysia

Abstract. The main objective of this study is to explore the efficiency of utilizing a culturally - enhanced serious game in education particularly in Information Technology course. To attain this objective, the academic achievement of 61 participants was measured by using quantitative Quasi-Experimental research design approach. Sets of pre- and post-test questions were administered to all participants, where descriptive analysis and hypothesis evaluation were performed to the collected data. The finding revealed that there were significant increase on the mean values of participants who played culturally-enhanced serious game as compared to those who played contemporary serious game (M= 3.00). Therefore, the study suggested that culturally-enhanced serious game has potential to support education particularly in Information Technology course.

1 Introduction

Our demands, inclinations and anticipations are inspired by our own upbringing culture [1]. These demands, inclinations and anticipations determine our choice of applications or products that we use either for work, school or even at home. For example, Google simple landing page (Fig 1) is very popular in western countries. In fact, Google dominates almost 90% in most western countries and almost 80% globally [2]. However, this extremely popular search engine is struggling to grasp the attention of South Korea's users; which merely 3% of South Koreans enjoy using it. The South Koreans are more incline to use native search engine, called Naver.com (Fig 2) over Google, where Naver.com dominates 77% of the South Korean market [2]. This disproportion was influenced by culture. The issues rely not only on the language used but also on the presentation of the interface, the selection of colours and its functionalities. It was found that the western users are more incline to use a simple interface while Eastern users anticipate the search engine interface to have highly contrasting and bright colours with many hyperlinks [3, 4].

Interestingly, gamer's behaviours are also influenced by culture. Their demands and anticipation is manifested on their selection of the type of game they play as well as their behaviour during the gameplay. For example, Western gamers favour to play more challenging game type as compared to Eastern gamers [5]. Additionally, Eastern gamers prefer to avoid in-game challenges whenever they have chances (e.g. through in-app

* Corresponding author: mazeyanti2012@gmail.com

 DOI: 10.1051/shsconf/20173300007

purchases) as compared to Western gamers [5]. These tendencies and preferences may be attributed to the nature of Westerners' culture that enjoys doing challenging tasks and activities.

Fig. 1. Google.

Fig. 2. Naver.com.

Furthermore, studies have revealed that culture also had influence on an individual's needs, preferences and expectations towards education [6–8]. For example, Western learners expect constructivist learning approach in contrast to Eastern learners, thus Western learners prefer to utilize social constructivist tool such as e-learning, Wikipedia and Facebook [6, 9, 10]. Additionally, the instructional expectations among learners are also vary across culture [6]. Learners from Nordic countries (e.g. Sweden, Norway and Denmark) expect constructive feedbacks as compare to learners from South Africa countries. Hence, utilizing cultural aspects in education particularly in serious game could assist to address the learner's learning needs, preferences and expectations; thus improves learner's academic performance.

The objective of this paper is to explore on the efficiency of a culturally-enhanced serious game as compared to a contemporary serious game. This paper has several parts: Part 1 is the introduction, Part 2 is the review of the literature, Part 3 is the methodology, Part 4 discusses on the findings and finally the conclusion in Part 5.

2 Literature Review

2.1 Culture

Culture can be defined as a shared pattern of mind that distinguishes the members of a group from one another [1]. These shared patterns include thoughts, manners and actions [11]. In essence, a collective or a society will have similar and distinct behaviours patterns, preferences, tendencies and expectations which differentiate them from other collective or society [6]. To describe these behavior patterns, Value Survey Model (VSM) is introduced [1]. VSM is used to distinguish one culture from another; which clarifies on the differences of a society's selections and preferences towards their own symbols, heroes, rituals and values. The VSM is extensively utilized in varies domains stretching from ICT to interpersonal communication [12,13]. This study used a serious game which was constructed based on VSM. Details of the serious game design model can be found in [4].

Current researches disclose that culture and learning are interrelated [6,14]. Evidences show that presenting learning environment that resemblance learner's culture will help to boost learners' motivation due to perceived likeliness and familiarity [9]. In spite of the connection between culture and learning, research works in this domain is still lacking particularly related to serious game [4,6,9]. With limited number of research works in this domain, its impact on educational effectiveness is scarce and sparse.

Though the use of VSM as a basis to generalize a culture receives critiques from some scholars, many well documented studies suggested that VSM could explain on a society's choices of symbols, preferences, behaviour, interfaces and game strategies preferences [13]. Additionally, VSM is also one of the most establish, widely replicate, has better level of validity as compare to other alternative models as well as repeatedly refer by scholars [12], [13,16]. For example, VSM was deployed in designing websites [16] while others incorporated it into user interfaces [16]. Hence, VSM is justifiable to be used in this research work in order to tackle the limited studies on culturally-enhanced serious games.

2.2 Serious games

Generally, computer games have story or goals, game mechanics, rules, conflict, interactivity, graphical representation and a sense of challenge or competition [17]. Meanwhile, serious games are computer games which emphasize on educational purpose instead of amusement [18]. Serious games are appropriate to be leveraged in education since they (1) provide secure and harmless environment for learners to employ their recently learnt knowledge (2) allow learners to become part of the game scenario by actively involved in the game environment and (3) motivate learners [6].

Currently, there are only a few serious games which attempt to incorporate culture aspects in its game design and game strategies. For example *Smoke?* [15], Trinbago Adventures of L Macawell (TALM), and Caribbean Conquest [14]. With limited number of existing culturally-enhanced computer games, its impact on educational effectiveness is scarce and limited. Thus, this paper aims to explore the effectiveness of a constructed culturally-enhanced serious game (called my-CEN) over a contemporary serious game (called CyberCIEGE).

3 Methodology

3.1 Research design

This study employed Quasi-Experimental research design approach. 61 participants were involved in this study and they were randomly divided into two groups; experimental 1 (X1) and experimental 2 (X2). Participants in X1 were required to play CyberCIEGE while participants in X2 were required to play my-CEN. All participants were also required to take pre- and post-test in order to assess their academic performance. The pre- and post-test questions have 3 parts: Part A is the demography, Part B is about serious game experience and finally Part C is formulated to assess the participants' knowledge in computer security principles and malicious threats. The academic performance scores are analysed by using two approaches; descriptive analysis and hypothesis evaluation. A null hypothesis ($H_0 1$) was created and stated as below:

$H_0 1$: There is no significant difference on acquisition of knowledge between X1 and X2 group.

To assess the academic performance of each participant, the pre-test scores were subtracted from the post-test scores and Independent Samples t-Test using α=0.05 was performed to test $H_0 1$.

3.2 CyberCIEGE serious game

CyberCIEGE is a serious game developed by Rivermind Inc. with collaborators from academicians and developers from United States [19]. CyberCIEGE serious game is a real-time strategy (RTS) serious game that requires player to administer computer systems of a company. Player is given an initial budget to buy setup and sustain the computer systems from malicious code attack. Player must balance the computer security level and the employee's productivity within the given budget and time frame. Computer security implementation which is extremely strict ensures 100% security of the computer systems however will make workers incapable to conduct their task, which indirectly contribute to the reduction of the company's productivity.

3.3 my-CEN serious game

My-CEN was customized and enhanced from CyberCIEGE. It was customized and extended based on author's game design model called GADEM. Generally, GADEM is a model which facilitates game designers and game developers to develop a culturally-enhanced serious game. GADEM was grounded on Intercultural Communication and Social Construction of Technology theories. Detail information on GADEM can be found in [4]. My's-CEN game environment was developed to be similar to the learner's culture which in this case is Malaysian culture. Therefore, my-CEN is considered as a culturally-enhanced computer game.

4 Results and discussion

61 first year students from Computer & Information Sciences Dept. in Universiti Teknologi PETRONAS participated in this study. 30 participants were randomly assigned to X1 while the remaining 31 participants were assigned to X2. The number of participants participate in this work was justified since according to ref [20], the sample size from 30 to 500 was sufficient to produce significant result.

67.00% of participants in X1 are male while only 20.00% of participants in X2 are male. On average, participants in X1 spent approximately 9.33 hours per week playing computer games while participants in X2 spent about 9.45 hours per week for the same activity. This result suggested that both groups had quite similar time spent and computer game exposure.

Additionally, the result also revealed that "Strategy" and "Role-Playing" game were the most popular game genre played by the participants regardless of gender. Meanwhile, "Racing" and "Dance" were the least popular game genre. These patterns were demonstrated in both groups. Since majority of the participants preferred to play "Strategy" game genre, hence it could be inferred that these participants were accustomed with game interfaces components, navigation scheme and input mappings of strategy games.

4.1 Descriptive analysis result

The descriptive analysis result of total increment scores for each group is depicted in Figure 3. About 65% of participants in X1 showed improvement in their post-test while an astounding of 90% of participants in X2 improved their scores on the same test. Hence, it clearly showed that both groups performed better in their post-test as compared to their pre-test regardless the type of serious game that they played.

Figure 4 provides more detail evidence on the performance of the participants of both groups. Majority of participants in X1 showed low improvement in between 1 to 5 marks (78.9%) as compared to only 44.4% in X2. Another 15.8% of participants in X1 improved moderately in between 6 to 10 marks while 48.1% of participants in X2 improved moderately. Finally, 7.4% of participants in X2 demonstrated sheer improvement in their post-test as compared to only 5.3% in X1. Although both groups demonstrated positive improvement in their post-test, however participants in X2 scored higher in terms of marks distributions. Almost half (49%) of participants in X2 improved their scores in between 6 to

DOI: 10.1051/shsconf/20173300007

10 marks as compared to participants in X1. Additionally, more participants in X2 obtained more than 10 marks as compared to participants in X1. Hence, it could be inferred that participants in X2 demonstrated better achievement as compared to participants in X1.

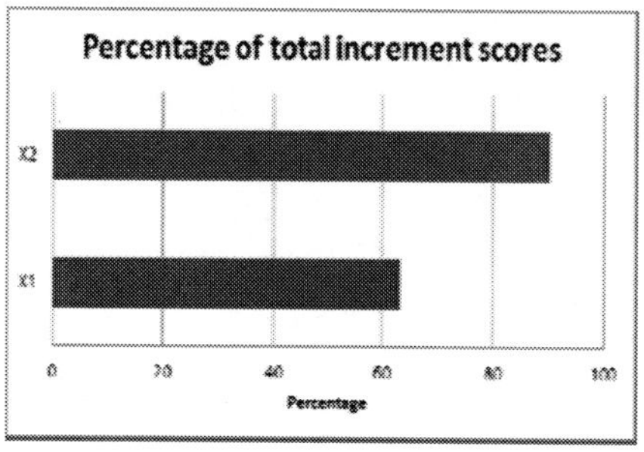

Fig. 3. Percentage of total increment scores.

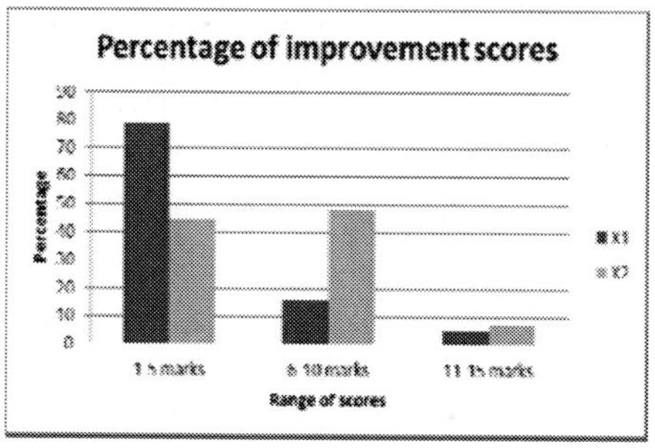

Fig. 4. Percentage of improvement scores.

4.2 Hypothesis evaluation

In order to provide more significant evidence, an Independent Samples t-test was conducted to compare the total increment scores for X1 and X2. Table 1 shows the Independent Samples t-test result of $H_0 1$.

Table 1. Independent Samples t-Test result for $H_0 1$

Group statistics				
	Group	N	Mean	Std. Deviation
Marks	X1	30	2.0667	3.66656
	X2	31	5.0000	3.26599

Independent Samples Test				
t-test for Equality of Means				
	t	df	Sig. (2-tailed)	Mean Difference
Marks	-3.296	57.730	.002	-2.93333

From Table 1, it showed that the mean value of X2 is higher (M=5.00) as compared to mean value of X1 (M=2.00). Additionally, Independent Samples t-test showed that there is significant difference in scores for X1 (M=2.00, SD=3.66) and X2 (M=5.00, SD=3.26); t (59) =3.296, p=0.002 (two-tailed). This finding suggests that the null hypothesis is rejected. Therefore, this result indicates that participants in X2 demonstrated higher academic achievement as compared to the participants in X1.

5 Conclusion

The findings from both descriptive analysis and hypothesis evaluation suggested that $H_0 1$ was substantiated. It showed that there is significant improvement in terms of academic performance between participants playing culturally-enhanced serious game and participants who played a contemporary serious game. Hence, it could be concluded that participants in X2 performed better as compared to participants from X1. This finding suggests that learning from environment that is culturally similar to the learners may influence the performance of the learners due to familiar and similar cultural context. This is consistent with recommendation made by scholars in Intercultural Communication who highlighted on the deniable connection between learning and culture. Hence, this study concludes that culturally-enhanced serious game has potential to increase student's academic achievement as compared to learning by using contemporary serious game.

References

1. G. Hofstede and G. Hofstede, *Cultures and organizations: Software of the mind.* (MacGraw Hill, 2005).
2. K. Reinecke, "Culturally Adaptive User Interfaces," Unpublished PhD thesis, University of Zurich, Germany, (2010).
3. K. Reinecke and A. Bernstein. *ACM Trans. Comput. Interact,* **18**, 2 (2011).
4. M. A. Mazeyanti, W. A. Wan Fatimah, and S. Suziah. *J. Eng. Appl. Sci.*, **10**, 23 (2015).
5. InMobi, "Mobile Gaming Cross-Market Analysis: First Edition, 2014 Q1," (2014).
6. M. A. Mazeyanti, W. A. Wan Fatimah, and S. Suziah, "Culturally Enhanced Computer Game: The Impact on Educational Effectiveness," *.Mediterranian J. Soc. Sci.*, vol. in press, 2015.
7. R. C. S. Yee, *"Perception of Online Learning in an Australian University: Malaysian Student's Perspectives,"* Queensland University of Technology, (2011).
8. D. Subramony, *"Socio-Cultural Issues in Educational Technology Integration," Build. Blocks Soc. Found*, **6**, 1, (2011).
9. P. Mohammed and P. Mohan, *Caribb. Teach. Sch.*, **1**, 1 (2011).
10. W. Littlewood. *Lang. Teach. Res.*, **5**, 1 (2001).
11. Jupit, Amelia Jati Robert, et al. "Cross-cultural awareness in game-based learning using a TPACK approach." *The 10th International Workshop on Internationalisation of Products and Systems.(11-14 July 2011)*. (2011).
12. I. Zurida, J. Azizah, and A. Azlina. *Procedia - Soc. Behav. Sci.*, vol. **65**. (2012).
13. K. Reinecke and A. Bernstein. *MIS Q.*, **37**, 2 (2013).
14. P. Mohammed and P. Mohan, *"Combining Digital Games with Culture: A Novel Approach towards Boosting Student Interest and Skill Development in Computer Science Programming,"* in *Proceedings of International Conference on Mobile, Hybrid, and On-Line Learning*, (2010).
15. R. Khaled, P. Barr, R. Biddle, R. Fischer, and J. Noble, *"Game design strategies for collectivist persuasion,"* in *Proceedings of ACM SIGGRAPH Symposium on Video Games - Sandbox*, (2009).
16. B. Duygu, "Cultural Factors in Web Design," *J. Theor. Appl. Inf. Technol.*, **3**, 7 (2009).
17. A. Derryberry, "Serious games : online games for learning." (Adobe Ltd, 2007).
18. Ariffin, Mazeyanti Mohd, Wan Fatimah Wan Ahmad, and Suziah Sulaiman. "Guideline for Designing an Effective Serious Game by Using Cultural-Based Game Design Model." *International Visual Informatics Conference.* Springer International Publishing, (2015).
19. M. Thompson and C. Irvine, "Active Learning with the CyberCIEGE Video Game," in *Workshop on Cyber Security Experimentation and Test*, (2011)
20. U. Sekaran, *Research method of business: A skill-building approach*, (Fourth Edi. John Wiley & Sons, 2006).

SHS Web of Conferences **33**, 00008 (2017) DOI: 10.1051/shsconf/20173300008
i-COME'16

Cultural Factors in the Flow of International News: A Review of the Literature

Takuya Sakurai[1,*]

[1]Tokyo Denki University, 5 Senju-asahi-cho, Adachi, Tokyo 1208551, Japan

Abstract. This paper offers a review of the literature of international news flow, and surveys how the previous studies have attempted to capture "cultural factors" influencing the flow. The factors are grouped into four types of variables: language, former colonial tie, ethnicity, and geographical proximity. This paper argues that cultural factors cannot be significant without economic interests in the era of post-Cold War, that the structure of international news is imbalanced because a few powerful countries dominate the international news market, and that international news reduces cultural varieties to the singular international realities disseminated by the media of such countries.

1 Introduction

Our images of globalization are mostly mediated through many institutions such as science, education, and the media [1]. Few people understand how much the world is globalized directly. Instead, most people rely largely on international news, one of the most effective and useful media to know how the world is getting globalized. International communication studies have attempted to examine international news since the inception of the New World Information and Communication Order (NWICO) in the late 1960s [2]. Against the backdrop of the United Nations Educational Science and Cultural Organization (UNESCO), the studies have begun to indicate that the structure of international news flow is imbalanced [3], and that news from developing countries is often biased and misrepresented [4]. A number of studies have also investigated the determinants of international news flow, especially since Scandinavian scholars such as Galtung and Ruge [5] and Ostgaard [6] proposed theoretical foundations that focus on "factors" influencing the flow of international news. The studies have proposed various factors influencing international news flow and attempted to identify what the best factors are to predict the flow of international news.

As factors influencing the flow, cultural affinity and cultural proximity are believed to be able to predict and describe how some international news is selected over others. They basically assume that if a guest country is culturally closer to a host country, the host country's media cover the guest country more often than those countries that are culturally less close. The factors demonstrate that the flow of international news help Western country locate at the center of the process to select international news, and that the culturally closer

* Corresponding author: tsakurai@cck.dendai.ac.jp

countries are to those Western countries, the more likely they are selected. They are also believed as one of the factors that reveal the dominance of Anglo-Saxon and the imbalance between Western developed countries and non-Western developing countries in the flow of international news. This paper thus intends to summarize and explain how cultural aspects of the flow of international news have been studied and to reveal some underpinning elements of the studies. More specifically, it is our task to recap how culture is conceptualized and operationalized in the flow of international news and to unfold what is assumed behind the idea of "culture" in the studies.

2 Cultural Factors: A Review of the Literature

There are roughly two types of factors that influence the flow of international news: intrinsic (event-oriented) and extrinsic (context-oriented). The former refers to newsworthiness of events such as importance, sensationalism, prominence, conflict, negativity, timeliness, and proximity [7-8]. The event-oriented approach examines news values intrinsic in the news itself. In other words, news values of events are considered to determine what news to be selected over others. The latter mentions extra-media components such as political and economic elements, population, geographical proximity, cultural proximity, and press freedom. The context-oriented approach examines how extrinsic predictors influence news coverage among countries. The following components are usually used as extra-media variables: communication access and technologies [9], cultural proximity [6], economic development [10], elite status [11], former colonial ties [12], ideological groupings [13], international news services [14], geographical proximity [15], GNP per capita [16], language [17], population and size of nation [18], and regionalism [19].

The previous studies have operationalized cultural factors as cultural proximity and/or affinity. Some of them employ cultural factors as an independent determinant influencing international news flow. Others tend to employ cultural factors as a determinant interrelated with other determinants influencing international news flow. The former attempts to see how well cultural factors predict international news while the latter evaluates what role cultural factors play in determining international news flow. Hester [20] argues that cultural affinity involves language, migration, travel, and former colonial relations between countries. Language is employed to evaluate how cultural affinity works in international news flow [17-18], while language is also used to assess cultural proximity in the research on international news flow [7-21] Cultural affinity, furthermore, is measured in terms of ethnicity [8-22], former colonial relationships [12-19], or geographical proximity [23]. Cultural proximity is also evaluated by some concepts such as ethnicity [5, 24, 25, 26] and geographical proximity [6]. While those usages and definitions of cultural factors vary across the studies, the cultural factors employed in the previous studies can be categorized into the following four groups according to how they are conceptualized and operationalized under the names of either proximity or affinity: language, colonial tie, ethnicity, and geography.

2.1 Language

Language is employed as a component of cultural factors influencing international news flow. Wu [4-27] hypothesizes, if a common language is shared between a guest country and a host country, the host country's media cover the guest country more often than those countries that do not share the language. Kim and Barnett [18] employed language as a means of capturing how cultural affinity influenced international news flow and of predicting the structure of international news flow. They propose that the world should be

 DOI: 10.1051/shsconf/20173300008

divided into the following eight geographical-linguistic groups in terms of what language they use: "European-North American, Chinese, Portuguese, Greek-Tuck, Latin American-Spanish, Middle Asian-Indian, North African, and overseas French groups" (p. 346).

While Kim and Barnett's and Wu's [4-18] studies employed language to examine international news flow in the larger framework of global news system, Chang, Shoemaker, and Brendlinger [7] used it to identify the best predictor determining what international events were covered in the U.S. media. Kariel and Rosenvall [17] employed language to compare English and French newspapers in Canada as to see how cultural affinity influences international news natures. Comparing two types of Canadian daily newspapers, English and French newspapers, Kariel and Rosenvall [17] found that both the newspapers used the same international news sources from the Canadian press. While Kim and Barnett [18] argue that language comes within the context of geographical proximity, Kariel and Rosenvall [17] mention that each newspaper tended to choose what international news to print in terms of certain cultural affinities of its readers. French newspapers tended to choose news items from France, whereas English newspapers those from the United Kingdom.

2.2. Colonial Tie

While language is a powerful tool when modern nation states unify territories and when a country colonizes other countries, former colonial tie itself is also considered to be an important factor influencing international news flow [12]. Dupree [10] argues that the best predictors of international news are former colonial ties and economic comparative figures. Robinson and Sparkes [11] discuss that both the United States and Canada sent a large number of foreign correspondents to England. Mayer indicates that the former British colonies obtain their international news from the British wire services and the former French colonies used those from the French wire services. Former colonial tie is also closely related to technological investment and presence of international news agencies. Since it is much more economical to use news copies provided by news services than to send correspondents to foreign countries, the presence and the size of news agencies stationed in a former colonized country were important in determining the amount of international news on that country [28]. In the study of international news in cyberspace, Kang and Choi [29] unveiled that the inequity in the structure of international news flow should be interpreted as technological dependency. The technological resources that could make the distribution of information possible were available to only the limited developed powerful nations. Hicks and Gordon [30] also argue that television coverage of international events was more likely controlled by technological problems such as availability of camera crew and cost of transmission of stories than actual geographical proximities.

In a comparison of the directions and volumes of news exchanges between Canada and the United States, however, Robinson and Sparkes [11] found few technological differences between the two countries. In order to explain a huge imbalance between the two countries in terms of being international news, the presence of international news agencies could be considered to be the largest factor. The global news wire services such as AP and UPI, Agence France-Press, and Reuters dominate international news market as wholesalers, selling their products to ninety agencies in the United States and Canada [11]. Mayer [13] also points out that the Big Four international news agencies—AP, UPI, AFP, and Reuter—control "the bulk of international news" from the perspective of neoimperialism in global news flow (p. 256). The dominance of the Big Four news syndicates allows the United States, France, and the Great Britain to carve out their hegemonic spheres of international influence. In a similar sense, Masmoudi [2] argues that the influence of the "Big Four" Western news agencies is the major problem of the imbalances in international news flows.

2.3. Ethnicity

Ethnicity explains how immigrants affect cultural factors influencing international news flow and coverage. Golan and Wanta [22] counted the numbers of immigrants in the United States from different foreign countries as an indicator of cultural affinity. Johnson [25] examined the quantity of U.S. newspaper coverage of Mexico to understand how Mexican immigrants affected the amount of international news on U.S. newspapers. She found that, although cultural proximity affected the selection of foreign news, it did not play a main role in deciding to cover the subject. Zaharopoulos [8] investigated the Greek coverage of the 1988 U.S. presidential campaign to examine whether the Greek media would find cultural proximity in one of the candidate, the son of Greek emigrants. The study concludes that the local gatekeepers had control in selecting some news over others on the coverage of the 1998 U.S. presidential election campaign, recognizing that cultural proximity played an important role in the way two major Greek newspapers covered the campaign.

While many studies have conducted content analysis to examine cultural factors, a few studies interview journalists to see how cultural proximity influences journalists' decisions of selecting some international news over others with regard to their personal and professional backgrounds [24-26], for example, interviewed AP Interbureau wire editors and Wisconsin newspaper wire editors. The Interbureau editors believed their clients to be "more interested in news from foreign nations which share common cultural and ethnic backgrounds with the United States" and stated that "they would give first priority to stories from developed Anglo-Saxon nations over items from non-Anglo-Saxon developing countries if news values were held constant (p. 37).

2.4. Geographical Proximity

Some studies have attempted to conceptualize cultural proximity in a geographical sense and to propose geographical proximity as one of main factors influencing international new flow. While Kim and Barnett [18] employed language as a measurement of cultural affinity, the study also shows that language cannot be considered without the context of geographical proximity. Ostgaard [6] did not set up a clear boundary between cultural and geographical proximities. He argues that cultural proximity has more influence on news flow than geographic distance because "for whatever reason the news is published, it will be intended to reach an audience" (p. 25). Wang [23] used cultural affinity as one of the factors determining international news on the foreign countries' coverage of the 1989 Chinese student pro-democracy movement. Her study shows that neither cultural affinity nor geographical proximity necessarily affected the extent and the direction of the coverage as much as the political ideology and diplomatic sensitivity did. Meyer [13] advocates the concept of "news dependency" (p. 243), which explains how developing countries are dependent upon the West for international news. Almost half of international news in the Latin American newspapers was sent from the American news agencies (AP and UPI).

3 Discussion and Summary

Galtung and Ruge [5] developed cultural factors as psychological aspects of international news—how audience perceived international realities. They were originally concerned with how audiences could find their interests in international news; therefore, they conceptualized some sort of familiarity that audiences might find in foreign countries as cultural factors. These factors, however, cannot be the best predictors of international news flow when they are examined with other factors such as political and economic factors [4, 19, 27]. Therefore, Rosengren [31] added extra-media (context-oriented) factors such as political and economic variables to the Galtung and Ruge's intra-media (event-oriented) factors [5]. Once cultural factors were widely identified as one of extra-media factors, the studies began to seek more comparative and testable variable to examine cultural factors. Methodologically, the studies have reduced cultures to the terms either proximity or affinity first, and examined the flow of international news through language, colonial tie, ethnicity, and geographical proximity. Not only does this kind of operationalization, however, reduce diversified cultures around the world to distinguishable and testable accounts, but also it incorporates cultural differences into the singularly comparative perspective. The singular perspective reduces cultural differences to some elements that support the values and norms of a few powerful countries, such as politics, economy, trade, power relationships. Thus, the following conclusions are made: No factors can be as influential as political ideology on global political coverage [23], factors of presence of international news agencies are more crucial element of news gathering [4-28], and economic interests play significant roles in international news flow [4]. Wu [27] indicates that economic interactions between countries are important to be covered because economic interests produce more coverage such as politics and culture in foreign media. Ishii [16] also argues that since cultural factors always contain economic and political implications, the flow of international news is much about how the international establishments sharp and disseminate their powers over the world.

As part of international communications, thus, international news plays significant roles in promoting and supporting Western cultural values. Relationships between countries are reduced to a comparative "economy" paradigm, and cultural realities are being integrated into economy-centered perspective. According to Wu [4], one of the reasons why economic factors are the best predictor of international news flow is that economy tends present a country rather than ideological and/or political orientations on the era of the post-Cold War. In a similar sense, Wasburn [32] describes the post-Cold War world as an international circumstance where "power depends as much on the ability of a nation's banks and corporations to capture market share and defeat foreign trade rivals as it does on the capabilities of its military power" (p. 191). This economy-centered tendency in the flow of international news is the result of justifying the Westernized market expansion. Even though the studies identify the imbalances among between Western and non-Western countries as the problems to be minimized, international news itself is a product of the global market; in other words, international news both supports the market and is part of the market. International news is a commercial product and is selected to meet the needs of customers (audience and advertisers), and cultural factors thus tend to be reduced to economic interests. Seeing international news as "a commodity" or "a cultural artifact" enables us to understand how the media disseminate those globalized realities [18]. Interestingly, this nature of international news flow still remains in Internet news services [33] and social media networks [34], while the latter notes that there emerges some divergence in the specific news events that gather attention. Accordingly, future studies need to examine if the idea of "culture" has remained or changed in the flow of international news with the Internet.

In sum, we have reviewed the literature of international news flow and found that cultural proximity and affinity are examined through language, colonial tie, ethnicity, and geographical proximity. As seen above, in the flow of international news, while powerful countries play central roles, less powerful countries have to accept the power structure of international news as marginalized group. Although the flow of international news presents itself as a "gatekeeper" model [35], selection is a construction as well. Selecting some international news over others is based on how media construct news as social artifacts in order to reflect and reinforce the international establishment's status quo. Frequency to be covered in international news just encourages many countries to join the competition where a few Western powerful countries set up the rules. Guest foreign cultures within the singular perspective have to be compared quantitatively and seen in the scope that one is "better" than another. This legitimates the "one-way flow" from a tiny rich group to the rest of the vast majority and requires the majority to see the world that way the flow does. International news thus plays a significant role in constructing certain globalized realities centralizing a few powerful countries [4-36]. As Mattelart [37] insists, an emerging international network of information and communication might bring not a fragmentation of the world, but a uniform world culture.

References

1. J.V. Ginneken, *Understanding global news: A critical introduction.* (Sage, Thousand Oaks, CA, 1998)
2. M. Masmoudi, Journal of Communication, **29** (1979)
3. R.L. Stevenson, Journalism Quarterly, **69**, 3 (1992)
4. H.D. Wu, Journal of Communication, **50** (2000)
5. J. Galtung, H. Ruge, Journal of Peace Research, **2** (1965)
6. E. Ostgaard, Journal of Peace Research, **2** (1965)
7. T. Chang, Shoemaker, P. J., & Brenlinger, N. Communication Research, **14** (1987)
8. T. Zaharopoulos, Journalism Quarterly, **67** (1990)
9. A.J. de Verneil, *A correlation analysis of international newspaper coverage and international economic, communication and demographic relationships.* (Transaction, New Brunswick, NJ 1977)
10. J.D. Dupree, International Communication Gazette. **17** (1971)
11. G.J. Robinson, V.M. Sparkes, International Communication Gazette, **22** (1976)
12. T. Nnaemeka, J. Richstad, International Communication Gazette, **26** (1980)
13. W.H. Meyer, Comparative Political Studies, **22** (1989)
14. J.F. Larson, Journal of Communication, **29** (1979)
15. H.D. Wu, International Communication Gazette, **60** (1998)
16. K. Ishii, Gazette Leiden then Deventer then Dordrecht, **57** (1996)
17. H.G. Kariel, L.A. Rosenvall, Journalism Quarterly, **59** (1983)
18. K. Kim, G.A. Barnett, Communication Research, **23** (1996)
19. W.A.E. Skurnik, International Communication Gazette, **28** (1981)
20. A. Hester, International Communication Gazette, **19**, (1973)
21. Y. Chang, K. Chang, International Communication Bulletin, **38**, (2003).
22. G. Golan, W. Wanta, International Communication Gazette, **65**, (2003)
23. S. Wang, International Communication Gazette, **49** (1992)
24. A. Hester, International Communication Gazette, **17** (1971)
25. M.A. Johnson, Journalism and Mass Communication Quarterly, **74** (1997)
26. S. Peterson, Journalism Quarterly, **58** (1979)
27. H.D. Wu, International Communication Gazette, **65** (2003)
28. T.J. Ahern, Jr. (1984). *Determinants of foreign coverage in U.S. newspapers. In R. L. Stevenson & D. L. Show (Eds.), Foreign news and the new world information order* (Iowa State University Press, Ames, 1984)
29. N. Kang, J.H. Choi, Communication Research, **26** (1999)
30. R.G. Hicks, A. Gordon, Journalism Quarterly, **51** (1974)
31. K.E. Rosengren, Journal of Peace Research, **11** (1974)
32. P.C. Wasburn, Political Communication, **14** (1997)

DOI: 10.1051/shsconf/20173300008

33. T.K. Chang, I. Himelboim, D. Dong, International Communication Gazette, **71**, (2009)
34. N.T. Ekeanyanwu, Y. Kalyango, A.S. Peters, European Scientific Journal, **8**, (2014)
35. D.M. White, Journalism Quarterly, **27** (1950)
36. H.W. Park, Journal of Communication, **48** (1998)
37. A. Mattelart, *Mapping world communication: War, progress, culture*, (University of Minnesota Press, Minneapolis, 1994)

DOI: 10.1051/shsconf/20173300009

The Function of Interpersonal Communication in Conflict Management Organization

Inge Hutagalung[1,*]

[1]Postgraduate Communication Programme, Universitas Mercu Buana, Indonesia

Abstract. The research aims to find out how the effectiveness of interpersonal communication functions in conflict management organization. The research results showed that interpersonal communication is taking place can be more effective to make the employee close each other, and creating a favorable climate in the workplace. In organization where a favorable climate is created, the conflict would be solved with deliberation. The conflict is seen as a natural life dynamics and something humane. There are two dimensions of interpersonal communication that is most instrumental in minimizing organizational conflict. The first dimension is the "openness", and the other is the "empathy".

1 Introduction

Communication functions within the organization are a means to modify behavior, affect change, convey information, and to achieve the goal. "Communication is the lifeblood of an organization" is a phrase that emphasizes that the organization is inseparable from communication. The communication is contributing ideas about how to think and act in the organization. If there is no communication, employees cannot know what is done by other employees, the leadership cannot accept the input of information from his subordinates, and supervisors cannot give instructions to the staff. Coordination of work is not possible and organizations will be disbanded due to lack of communication.

Patterns of communication within the organization can be grouped into two [3], which is a communication network that follows the organizational structure. This communication process flows through formal channels. The second is the communication patterns of informal communication networks. The communication process usually takes place outside formal channels and called grapevine. Informal communication arises from the interaction of individuals with the spread of a message that cannot be suspected, more informal communication leads to interpersonal communication.

*Corresponding author: inge_hutagalung@yahoo.com

2 Effectiveness of Interpersonal Communication

According to the relational perspective view [1], interpersonal communication is defined as that communication occurs between two people who have seen a clear relationship between the two, such as the communication of two brothers, teachers and students, lovers, and others. Because it involves the relationship of two people interacting, then often the interpersonal communication is defined as the dyadic approach.

Dimensions effectiveness of interpersonal communication in a humanistic perspective includes openness, empathy, supportive behavior, and similarity. In general, these properties will help the interaction becomes more effective. Meanwhile, in a pragmatic perspective include self-confidence, togetherness, interaction management, expressive behavior, and orientations to the other party. In this study, researcher used the size of the effectiveness of interpersonal communication with the humanistic approach. The argument was that the humanistic approach in accordance with the conditions of interpersonal communication in an organization.

3 Research Methodology

This study used a qualitative approach. Methods of data collection in qualitative research can be done through observation, participation, interviews, and ethnography. The qualitative method is a method that produces descriptive data in the form of words written or spoken by people with observed behavior. Arguments researcher chose a qualitative method is the consideration to be able to adjust to the reality on the ground in an effort to collect data as detailed as possible.

The unit of analysis in this study is a working group within banking organizations, namely the unit front office and back office. As the subjects were selected employees with work restrictions, minimum of two years, starting from the level of the leadership and staff in the unit. The subjects were divided into groups according to the observations of the working structure, composed of 6 unit leaders, 6 front office staff and 6 back office staff. The main data source of this study was obtained through Focus Group Discussion (FGD) through a structured question.

4 Research Results

Based on the research results it can be stated that the interpersonal communication contribute to conflict management organization. There are three reasons that can be concluded from the answers of the subjects observations with regard to existing roles, namely (a) the interpersonal communication that exists effectively will produce good relations and cooperation between units in the organization, the clarity of message can be more accountable, the channel to ask or discuss something it is more open; (b) interpersonal communication which take place effectively can make more staff closer to each other, so familiarity could be entwined; (c) the presence of effective interpersonal communication, all disagreements and conflicts can be solved conversationally.

On the other hand, the leader also realize that in business services such as banking, interpersonal communication is very helpful in coordinating and monitoring tasks, build a comfortable working atmosphere, away from the stiffness and boredom.

The description of the dimensions of the effectiveness of interpersonal communication in the unit of analysis is as follows:

- **Openness**: The leader always wants to interact with the staff to discuss the issues facing either by phone or face to face. Through the interactions, the staff will know the opinions, thoughts and ideas of the management.
- **Empathy**: the existence of consciousness of employee that banking is a working system, which means the work unit intertwined in the unity of the organization. Empathy attitude is manifested in the ability and willingness of the employee to put their self in the other person's position or role. If there is a problematic work units then the corresponding work units will help to find suitable solutions
- **Supportive behavior**: the subject of observation realizing that they really love the organization and have to work in one system according to the vision and mission of the organization. This makes the behavior of individuals who persist in not being an issue.
- **Positive behavior**: the subject of observation to be positive view of others, self and organizational situations. This positive behavior encourages the employee to communicate openly, relaxed, and friendly to each other.
- **Similarity**: the subjects of observation had an equivalent position in sending and receiving messages. On the other hand, the dimension of "similarity" also means the employee has the value of communication; attitude; behavior and experience equivalent.

Furthermore, the observation of the subject admits there are two dimensions of interpersonal communication that is most instrumental in minimizing organizational conflict. The first dimension is the "openness". The arguments given were (1) the openness made staff feel comfortable, do not be afraid to ask, while leaders can provide direction and monitor the progress of a given task; (2) openness can be developed through face to face, in writing or using electronic media. The second dimension is empathy. The arguments given were (1) the interpersonal communication can be effective if interpersonal communication can feel and put ourselves in the position of those who communicate; (2) can be a good listener and do not interrupt so that conflict can be avoided.

5 Discussion

The organization consists of actions and interactions that involve people. The organization was created and nurtured through continuous interactions change made by people in the organization. The behavior of people in organizations will form organization [5]. In short, communication is a basic human activity. The process of effective communication in an organization will facilitate the operations of the organization, and vice versa, the lack or absence of communication will make the operations of an organization become less successful.

The research results showed that interpersonal communication are interwoven in an effective manner will result in a good cooperation between environmental organizations, the clarity of the message can be more accountable, because the channel to ask questions or discuss something more open. This is in accordance with what is expressed by Robbins [7] that poor communication is most often referred to as the source of the conflicts which ultimately will cause stress and inhibit the performance of the work. Familiarize to communicate effectively at work; employees will be able to minimize the stress of work. The same opinion was also expressed by Munandar [2] that an important contribution toward stress management is the quality of the relationship (communication) between the employees or supervisor. Pace [5] also stressed that any difference shape can be muted or minimized by the existence of a supportive climate in order for interpersonal communication can be maintained and improved.

On the other hand, the model of the Dance and Larson [4] can be used to explain how interpersonal communication can minimize conflict within the organization. Through its communication model, Dance and Larson asserts that there are three functions of communication, namely linking function, the function mentation, and regulatory function. If the third communication function is used in an organization then it is seen that the communications function is the center of operational organizations. When employees use communication to connect with other employees (linking function) then the partnership will be realized. Cooperation established will help to achieve the common goal, and generate creative problem solving from a variety of ideas and debate that comes up. The principle of linking function is through an effective communication, togetherness as well as support for the achievement of a common goal can be realized immediately.

As for mentation of the communication function, according to the model of the Dance and Larson can be seen when people communicate to solve complex problems. The more an employee communicating with other people, the more effective problem solving will emerge. Lastly, the regulatory function is the most important part of the communication process in performing the prediction and understanding regulations within an organization to achieve a common goal.

In short, interpersonal communication makes the employees are actively creating the ' world ' in which they can participate and cooperate embodies common goals.

Further, the researcher also assume that the proper functioning of interpersonal communication in minimize conflict also due to cultural factor. According to Ting Toomey [6] in "Communication Across Cultures", cultural in collectivism (as in Indonesia) tend to be goal-oriented groups, using "we" identity, enshrine the obligation in the group rather than personal rights, tend to emphasize the internal relationships of the group, paying more attention to the consideration of the group rather than personal. While an individualism cultural identity "I" (I), more personal, goal-oriented emphasis on personal rights, tend to emphasize the relationships between individuals, paying more attention to personal considerations rather than groups, have a relationship that is voluntary.

Based on the view of collectivism by Ting Toomey, having interpersonal communication effectively in the organization, employees will be able to strengthen the mutual relationship as part of the organization.

6 Conclusion

The role of communication within an organization is an absolute. Communication is the blood that turned the organization, the glue that unites all members of the organization, and connecting all systems within the organization. Most of the communication activities carried out in both formal and informal organizations is done through interpersonal communication. Effective interpersonal communication will be very helpful for each employee to be familiar with each other, and to prevent a conflict that would lead to labor conflict.

There are two dimensions of interpersonal communication that play a dominant role in minimizing labor conflict, namely openness and empathy. Openness, making the employee feels comfortable and appreciated, while the employee can feel empathy create togetherness among employees.

Effective interpersonal communication within the organization will make the conflicts that arise in organization seen as part of the dynamics of the job rather than problems of a personal nature.

References

1. A. Joseph De Vito, *The Interpersonal Communication Book* (Harper & Row Publishers, New York, 1995).
2. Ashar Sunyoto Munandar, *Psikologi Industri dan Organisasi* (Penerbit Universitas Indonesia, Jakarta, 2001).
3. C. Daniel Fieldman, H.J. Arnold, *Managing Individual and Group Behavior in Organization* (McGraw Hill International Book Company, Tokyo, 1983).
4. Dance. F.E. X, C.E. Larson, *The Functions of Human Communications: A Theoretical Approach* (Holt, Rinehart & Winston, New York, 1976).
5. R. Wayne Pace, F. Faules Don, *Komunikasi Organisasi* (PT. Remaja Rosdakarya, Bandung, 1998).
6. Stella Ting Toomey, *Communicating Across Cultures* (The Guildford Press, New York, 1999).
7. Stephen P. Robbins, *Organizational Behavior* (Prentice Hall, New Jersey, 2001).

Students' Adoption of Facebook in Higher Education: A gender-based Comparison

Cheng Ean (Catherine) Lee[1,*], and *Alison Yu Wei* Chong[2]

[1]Department of Communication and Liberal Arts, Faculty of Arts, Sunway University, Malaysia
[2]Faculty of Education, University of Malaya, Malaysia

Abstract. This paper aims to examine the gender differences of undergraduate students' educational use of Facebook. The objectives of the study include identifying male and female students' Facebook usage and assessing their perception on the usefulness and ease of use of Facebook in higher education. Using a quantitative research method, results reveal that male students have a higher frequency of Facebook visit (93.9%). Male students are more competent (48.5%) and more satisfied (42.2%) in using Facebook in comparison to female students. Although there were marginal differences between males' and females' perception for both usefulness and ease of use of Facebook (with mean differences ranging from 0.01 to 0.33); the differences in their overall perception on the usefulness and ease of use of Facebook were not significant ($p<.05$). In conclusion, male students had a higher overall score on the perceived usefulness of Facebook in higher education, while female students had a higher overall score on the perceived ease of use of Facebook.

1 Introduction

With the increasing number of research on social media in higher education, the most common and popular social media tool used in higher education is Facebook [1]. Recent years have seen the proliferation of empirical research on Facebook use in higher education particularly in Malaysia [2, 3, 4, 5, 6, 7, 8, 9, 10,11]. Besides, prior literature revealed that the use and influence of social networking sites in students' daily lives are pervasive as they used it not only for social purposes but also for educational reasons [4]. Drawing on the literature, Facebook has the potential to be used as a platform for online academic discussions [2]; Facebook is used for informal learning [8]; Facebook complements the current teaching and learning practices [3, 7]; Facebook enhances teaching and learning practices in large classes [6]; and Facebook has a positive impact on students' academic performances [9, 10, 11].

In addition, four research [12, 13, 14, 15] focus on investigating gender differences on academic uses of Facebook among undergraduate students because [12] claimed that 'differences in male and female communication are widely known, but peer reviewed research is difficult to find for social networking communication' (p. 89). Although [9] and [10] found no significant gender differences regarding the use of Facebook for academic

* Corresponding author: catherinelee@sunway.edu.my

DOI: 10.1051/shsconf/20173300010

activities, [5] state that an academically viable sense of direction for future research on exploring Facebook especially information about gender differences is much needed.

Therefore, our study aims to examine the gender differences of students on the educational use of Facebook and to gauge their attitude towards adopting Facebook in the classroom. The two research questions guiding this study are: How male and female students differ in terms of Facebook usage? and What are the differences among male and female students' perception on the usefulness and ease of use of Facebook in higher education?

2 Method

We utilised a self-administered questionnaire with undergraduate students of a private university in the Klang Valley from 26 to 29 January 2015. The survey respondents were selected through purposive random sampling from students who use Facebook for educational purposes. The respondents were to provide their usage of Facebook, level of competency and satisfaction of using Facebook as well as perception on the effectiveness of adopting Facebook in higher education. The result of the survey was analysed using both descriptive and inferential statistics. A total of 71 questionnaires were collected with 64 questionnaires were usable (90.14%). Of the total of 64 respondents, 33 (51.6%) were males and 31 (48.4%) were females. Majority of the respondents aged 18 to 24 years old (87.5%, n=56), Chinese (71.9%, n=46) and studying in the Faculty of Arts (70.3%, n=45). Table 1 illustrates the demographic profile of the respondents participated in this study.

Table 1. Respondents' demographic profile (n=64).

Gender	**Frequency**	**Percentage (%)**
Male	33	51.6
Female	31	48.4
Age Group		
Below 18	7	10.9
18 – 24 years old	56	87.5
25 – 44 years old	1	1.6
Race		
Malay	5	7.8
Chinese	46	71.9
Indian	4	6.3
Others	9	14.0
Faculty		
Pre-U	10	15.6
Faculty of Arts	45	70.3
Sunway University Business School	3	4.7
Faculty of Science and Technology	6	9.4
Total	**64**	100.0

 DOI: 10.1051/shsconf/20173300010

3 Results and discussion

3.1 Differences in the use of Facebook between male and female students

In answering research question one, Table 2 shows the differences in terms of Facebook use between male and female students. In general, nearly all male students (93.9%, n=31) reported a higher frequency of visit to Facebook in comparison to female students (80.6%, n=25). This result contrasts with [12, 15] because in their study, female students reported spending a greater percentage of daily Internet time on Facebook than their counterpart. However, in terms of their frequency of updating profile page on Facebook, the number of female students outrides the male students. 13 (42.0%) female students claimed that they update their profile page quite often with at least once a week, whereas only ten (30.3%) male students often updated their profile page. In regard to students' self-reported level of satisfaction and competency in using Facebook, the results unveil that male students were more satisfied (42.2%, n= 14) and more competent (48.5%, n=16) in using Facebook for educational purposes compared to female students.

Table 2. Male and female students' use of Facebook.

	Male (n=33)		Female (n=31)	
	Frequency	Percentage (%)	Frequency	Percentage (%)
Frequency of visit (n=64)				
3+ times daily	27	81.8	16	51.6
1-3 times a day	4	12.1%	9	29.0
Once every few days	2	6.1%	4	12.9
Once a week	0	0	1	3.2
Once every few weeks	0	0	1	3.2
Less than once a month	0	0	0	0
Others	0	0	0	0
Frequency of profile update (n=64)				
More than once a day	1	3.0	3	9.7
Once a day	2	6.1	0	0
Once every few days	6	18.2	3	9.7
Once a week	1	3.0	7	22.6
Once a month	10	30.3	9	29.0
Once a few months/ Once a year	10	30.3	7	22.6
Others	3	9.1	2	6.5
Level of satisfaction of using Facebook (n=64)				
Highly dissatisfied	3	9.1	1	3.2
Moderately dissatisfied	4	12.1	2	6.5
Neutral	12	36.4	16	51.6
Moderately satisfied	10	30.3	10	32.3
Highly satisfied	4	12.1	2	6.5
Level of competency of using Facebook (n=64)				
Not competent	0	0	2	6.5
Somewhat competent	7	21.2	7	22.6
Uncertain	10	30.3	7	22.6
Competent	10	30.3	10	32.3
Highly competent	6	18.2	5	16.1

3.2 Differences between male and female students' perceived usefulness and ease of use of Facebook

Finally, the second research question examines the gender differences of students' perception on the usefulness and ease of use of Facebook. Table 3 shows that both genders highly believed that Facebook is easy to use with the mean not less than 3.40 for each category. While the perception on the ease of use of Facebook was fairly moderate for both males and females, with the mean ranging from 3.15 to 3.48 on the former and 2.87 to 3.35 on the latter, neither males nor females held any of the perception categories with low scores.

Although there were differences between males' and females' perception for both usefulness and ease of use of Facebook, such differences were marginal, with the mean differences ranging from 0.01 to 0.33. Despite these considerable differences, the results reveal that male students and female students did not significantly differ in their overall perception on both the usefulness and ease of use of Facebook ($p<.05$). Even though no significant differences were found, males had a slightly higher overall score on the perceived usefulness of Facebook, while females have a higher overall score on perceived ease of use of Facebook.

Besides, as compared to female students, male students reported a higher score in all six (100%) statements on the perceived usefulness of Facebook in higher education. On the other hand, females reported a higher score in four out of six (66.7%) statements on the perceived ease of use of Facebook, excluding "it is easy for me to become skilful when using Facebook" and "learning to operate Facebook is easy for me".

Table 3. Differences in male and female students' perceived usefulness and perceived ease of use of Facebook.

	Male (n=33)		Female (n=31)		*t*	*p*-value
	M	S.D	M	S.D		
PU Perceived Usefulness	3.33	0.60	3.16	0.80	1.007	0.318
PEOU Perceived Ease Of Use	3.81	0.72	3.92	0.62	-0.661	0.511
PU1: Using Facebook enables me to complete the tasks more quickly.	3.48	0.76	3.35	0.99	0.095	0.554
PU2. Using Facebook improves my academic performance.	3.15	0.62	2.87	0.89	1.478	0.145
PU3. Using Facebook increases my academic productivity.	3.36	0.78	3.03	1.11	1.387	0.171
PU4. Using Facebook enhances my learning effectiveness.	3.24	0.79	3.23	0.88	0.079	0.937
PU5. Using Facebook makes it easier to learn.	3.33	0.74	3.16	1.00	0.786	0.435
PU6. Overall, I find Facebook useful as an educational tool.	3.42	0.79	3.29	1.13	0.551	0.583

PEOU1. Learning to operate Facebook is easy for me.	4.03	0.88	4.03	0.80	-0.009	0.993
PEOU2. I find it easy to get Facebook to do what I want it to do.	3.79	0.70	3.81	0.95	-.090	0.929
PEOU3. I find Facebook flexible to interact with.	3.73	0.84	4.06	0.63	-1.810	0.075
PEOU4. My interaction / activity on Facebook is clear and understandable.	3.67	0.89	3.94	0.77	-1.287	0.203
PEOU5. It is easy for me to become skilful when using Facebook.	3.58	0.97	3.48	0.96	0.380	0.705
PEOU6. Overall, I find Facebook easy to use.	4.06	0.83	4.19	0.54	-0.755	0.453

4 Conclusion

Our study presents two main findings that explored the use of Facebook by both male and female students of a private university and their perception on the effectiveness of adopting Facebook in higher education. Firstly, majority of the students were heavy users of Facebook who paid frequent visit to Facebook for more than three times in a day with male students reported having a higher frequency of visit, more satisfied and more competent in using Facebook for educational purposes in comparison to female students. Secondly, there were marginal differences between males' and females' perception for both usefulness and ease of use of Facebook for educational purposes; yet it did not significantly differ in their overall perception on both the usefulness and ease of use of Facebook in higher education ($p<.05$).

To conclude, in view with the above, it is clear that there was no gender difference in using Facebook in higher education. Besides, there were no significant differences in their overall perception on the usefulness and ease of use of Facebook as an educational tool. Yet for effective use of Facebook in higher education, positive attitude of both male and female students are crucial.

This paper is part of the research project funded by Sunway University Internal Research Grant Scheme 2014 (INT-FOA-DCLA-0114-11).

References

1. S. Neier, L.T. Zayer, J. of Marketing Edu, **37** (2015).
2. T. Lim, Turkish Online J. of Distance Edu, **11** (2010).
3. S. Hamid, J. Waycott, S. Chang, S. Kurnia, Ascilite, (2011).
4. N.M. Almadhoun, F.W. Lai, P.D.D. Dominic, *An examination of social networking sites usage among the students in the Malaysian universities. ICMEF Proceeding*, (2012).
5. H.H. Hamsan, M.N. Kumar, Pertanika J. of Social Science & Humanities, **21** (2013).
6. R.R. V. Rasiah. Social and Behavioral Sciences, **123** (2014).

DOI: 10.1051/shsconf/20173300010

7. C.E. Lee, S.V. Sangaran Kutty, S.P. Wong, *Exploring the use of Facebook in the classroom: A Malaysia case study. Proceedings for 2nd European Conf. on Social Media ECSM*, (2015)
8. A. Hamat, M.A. Embi, H.A. Hassan. Int. Edu Studies, **5**, 3 (2012)
9. A.M. Helou, N.Z. Ab. Rahim, *The influence of social networking sites on students' academic performance in Malaysia. Proceedings for the Int. Conf. on Internet Studies,* 1-18 (2011)
10. A.M. Helou, Z.Z. Ab. Rahim, Z.Z. & Oye, N.D. Int. J. of Social Networking and Virtual Communities, **1** (2012)
11. N. Din, S. Yahya, S. Haron, Information retrieval and academic performance among Facebook users. *AicE-Bs 2012 Cairo ASIA Pacific Int. Conf. on Environment-Behaviour Studies*, **68** (2012)
12. S.H. Thompson, E. Lougheed. College Student Journal, **46** (2012)
13. G. Grosseck, R. Bran, L. Tiru, Dear teacher, what should I write on my wall? A case study on academic uses of Facebook. *WCES 2011* (2011)
14. A.K. Alhazmi, A. Abdul Rahman. *AISS*, **5** (2013)
15. A. ul Haq, S. Chand. *Bulletin of Edu and Research,* **34,** 2 (2012)

Exploring the Construction of 'National Unity' in Malaysia: Framing Analysis of Texts and Audiences

Siti Nor Amalina Ahmad Tajuddin[1,*], *Natalie Collie*[2], and *Yunxia Zhu*[3]

[1] Faculty of Languages & Communication, Universiti Pendidikan Sultan Idris, 35900 Perak, Malaysia
[2] School of Communication and Arts, The University of Queensland, St Lucia Qld 4072, Australia
[3] UQ Business School, The University of Queensland, St Lucia Qld 4072, Australia

Abstract. This paper examines the construction of 'national unity' within the culturally diverse society of Malaysia. It does so through a framing analysis of 102 recent Malaysian government advertisements. Audience responses, based on a series of focus group interviews, were also analysed. Although the power of visual advertisements comes from its capacity to blend fact and emotion, to engage audiences, and to add the narrative complexity of ethnicities, this paper also discovers a struggle over the meaning when the frames become contested. The findings suggest that multiple, often conflicting frames are involved in making sense of 'national unity' for different stakeholders, yet contested narratives of nationhood and ethnic identity is a central theme of the analysis. This paper contributes to a critical understanding of 'national unity' beyond culture, images and identity of multi-ethnic groups from two different narratives: the government through which discourse is constructed and the presence discourse of ethnicity.

1 Introduction

The discourse related to power, identity, and 'national unity' has emerged across the social sciences as an appropriate theoretical framework [1] that forms the connection between the theoretical approach and practical aspects of the investigation undertaken. This paper examines the construction of 'national unity' beyond culture, images and identity of people within a multi-ethnic contemporary Malaysia through framing analysis of texts and audiences. The discussions about 'national unity' is not something novel as scholars in various disciplines have raised concerns about the idea of class, gender, identity, ethnicity or race, and also nationality [2, 3, 1]. The use of government advertisements to promote 'national unity' is understood to only represent one particular, albeit dominant, representation of what 'national unity' might constitute and how it should be achieved. The diverse, multicultural nature of Malaysia however, suggests the need to examine other perspectives beyond the State's point of view to understand the multiplicity of perspectives

* Corresponding author: siti.ahmadtajuddin@uq.net.au , sitinoramalina@fbk.upsi.edu.my

DOI: 10.1051/shsconf/20173300011

of the Malaysian public. The paper argues that the people's voice including minority ethnic groups are important for fully understanding 'national unity' as this concept is far broader and more complex than the meaning in common usage. In doing so, this paper not only concentrates on an analysis of the language of powerful elites but also takes into account focus group discussions of people from different ethnic groups.

Although Engstrom [4] argues that 'national unity' is a highly problematic notion that presumes the existence of 'single-nation-states' instead of 'multinational states', this paper approaches 'national unity' differently. 'National unity', according to Amienyi [5] is a process that seeks to unite people of different ethnic, culture, religious, and socio-economic backgrounds for reciprocally beneficial goals. Correspondingly, it occurs when people from the same nation think, feel and care for one another and are willing to sacrifice the individual interest for the nation [6]. We, however, acknowledge that the meaning of 'national unity' is not fixed, as it is also negotiated and contested through its communication. Despite the reality of having a unique multi-ethnic Malaysian society representing different cultures and religions, ethnic diversity has always posed a challenge to 'national unity'[3]. Realizing the tensions among ethnic groups as a result of ongoing issues such as special privileges and rights for Bumiputera (People of the Soil) and the distribution of economic wealth [3], the following questions will be examined:

- How 'national unity' is constructed in Malaysian government advertisements?
- What 'frames' are created by the Malaysian audiences who view government advertisements for 'national unity' and in what ways have the 'frames' become contested?

This paper contributes to these debates by providing an in-depth analysis of precisely how the concept of 'national unity' has been constructed, tracing the dominant narratives, metaphors, and imagery used to promote a particular version of what a unified nation might entail. It is to the theoretical and methodological framework of the paper that we now turn.

2 Framing: A theoretical and methodological framework

This study adopts framing as a theoretical and methodological framework for examining the construction of 'national unity'. Framing studies that focus on 'audiences' have the tendency to restrict their analysis to the effects of specific frames on the information processing and decision-making processes of audiences (See, examples, [7,8]). Yet, very few studies examine the framing process in terms of construction (i.e. how a certain issue is constructed and how the frame-makers choose certain frames over the others to present a specific issue). Specifically, the voluminous literature on framing effects has ignored elite control – control by those with power in society and access to the means of producing and distributing mass communications – over framing or how the frames become contested and negotiated [9]. Through the theoretical framework of framing, this study focuses on how 'national unity' is constructed by frame-makers and how the frames get contested by audiences.

Methodologically, framing by definition is an interpretive process [10, 11, 12]. As such, it guides an interpretation of the construction of 'national unity' that embraces visual and textual elements, with readings and counter readings of words, images, as well as metaphors through a systematic interpretation of framing devices [10, 11, 12]. These devices provide a rigorous interpretation of how the selection of particular words, images, and visuals was constructed by the elite power and how they are received by people of different ethnic groups. For this study, we analysed 102 Malaysian government advertisements in the form of print media (stamps, postcards, magazines, and billboards), broadcast media (television), and online media (from government official websites). Examining the construction of 'national unity' in the data begins with asking questions about "how are particular words or

DOI: 10.1051/shsconf/20173300011

images given specific meanings" [13] to the nation that people are identifying with – in specific the object of identification. Table 1 shows how this paper systematically applies the framing devices as the guiding principle to develop the coding process and conduct a coherent analysis of the government advertisements.

Table 1. The coding process for interpreting government advertisements

Codes	Themes/ Phrases/ Slogan	Message of the ads	Depictions of characters	Logo / Emblem	National instrument	Historical and cultural celebration
Sub-code	*Rakyat Didahulukan, Pencapaian Diutamakan* (People First, Performance Now)	Unity / ethnic relations	Non-characters	Malaysian emblem	National flag	*Merdeka* Day
Sub-code	*Janji Ditepati* (Promises Fulfilled)	Nilai-nilai murni (moral values)	Celebrity endorser	Ministry's logo	Country's map	Malaysia Day
Sub-code	*Bahasa Asas Perpaduan* (Language is the foundation of unity)	Patriotism	Adults (man and woman) -Live and animated-	Govt's department logo	National anthem	Chinese New Year
Sub-code	*Perpaduan dalam kepelbagaian* (Unity in diversity)	National identity	Children (Live and animated)	Theme's logo	Cultural artefacts	*Hari Raya*
Sub-code	Transformation Successful, People Prosperous	-	Metaphors	1Malaysia logo	Buildings, monument, and sites	*Deepavali*

Furthermore, a total of six focus group interviews of different Malaysian ethnic groups (Malay, Chinese, and Indian) were used to gather data to discover their responses towards the construction of 'national unity' in government advertisements. The interview questions were divided into two parts. In the first part, participants were asked about their understanding of 'national unity' and general questions about government advertisements. In the second part of interviews, participants were shown twenty-three advertisements and the remaining questions were asked to them about their responses towards the advertisements. Using Atlas.ti, a comprehensive coding structure was developed from the advertisement and interview data for an in-depth analysis.

DOI: 10.1051/shsconf/20173300011

3 Results and Discussion

3.1 Media text

There are three frames emerged from the analysis of 102 government advertisements: 1) Nationalistic-based frame; 2) Ideology-based frame; and 3) Ethnic identity-based frame. The descriptions of each frame are explained below.

3.1.1 Nationalistic-based frame

The nationalistic-based frame explores the ways in which the contents of the advertisements construct 'national unity' from the Malaysian national manifestation such as the use of slogans/catchphrases, exemplars, metaphors, and visual images to communicate nationalistic ideas. In the examined government advertisements, one of the national symbols that appear repeatedly throughout the data is the Malaysian flag. The use of catchphrase of *Perpaduan dalam kepelbagaian* (Unity in diversity) reflects the idea of 'national unity' from the perspective of ethnicity as shown in Figure 1.

Fig. 1. A living interaction among people of different ethnic groups

3.1.2 Ideology-based frame

The ideology-based frame in this paper ascertains the use of certain words and images in the ways that promote the ideological power of the State. The ideology-based frame in this paper refers to a system that prioritizes idea, opinion, and beliefs of a given social group, and legitimizes certain ones as true, proper, natural and correct [14]. By identifying recurring images, words, and narratives through government advertisements, this analysis gain a greater understanding of how ideologies, in particular relating to 'national unity', are produced in the society [15]. The ideological framework that is discursively constructed through government advertisements are the recurring images of *1Malaysia* logo, *1Malaysia* narratives, and interestingly the images of children. An illustrative advertisement is provided in Figure 2.

Fig. 2. The use of *1Malaysia* logo as an ideological power of the State

3.1.3 Ethnic identity-based frame

Ethnicity can be associated with a group of people who share a culture, language, nationality, heritage, or religion [16] and thus, the meaning of ethnic identity is continuously negotiated, reviewed and rejuvenated. The analysis of the government advertisements reveals that the idea of 'unity' is always related to individuals of the different ethnic group. The images of individuals from different ethnic groups are not only in the form of 'real people', but also metaphors. Some illustrative data which are empirically analysed are shown below:

Fig. 3. Images of people of different ethnics in the form of metaphors and real people

3.2 Media audience

The findings show that the meaning of 'national unity' was explicitly interpreted as 'unity among different ethnic groups' by audiences. Some of the participants' responses are:

> *Participant G4-YY: From my own perspective, I think 'national unity' means that everybody comes together regardless of races, age, sex, religion, and others.*

> *Participant G1-SHH: 'National unity' is usually associated with a nation. In Malaysia, we always refer to Chinese, Indian, Malay and we would say that we must unite and co-operate each other.*

However, further analysis reveals that competing frames exist when audiences construct certain frames that opposed the meanings of preferred frames as the outcome may not necessarily consistent with the intent of the frame-maker (See Table 2). Here, 'elite-framed reality' refers to the preferred frames for understanding 'national unity' produced by those with power – in this case, the Malaysian government.

 DOI: 10.1051/shsconf/20173300011

Table 2. The competing/conflicting frames (Elite frames vs. audience frames)

Elite-framed reality Illustrative textual data (ads)	**Audience-framed reality** Powerful foam of counter interpretations
The recurring images of people of different ethnic groups are manifested through the advertisements in several ways: • people stand side by side and close to each other; • their harmonious relationships; • the idea of togetherness; • a living interaction between each other; and • the sense of national pride These manifestations celebrate the idea of 'national unity' and ethnic harmony from the authority-defined context.	The analysis of audience responses suggests that there is an intense struggle over the meaning of 'national unity' as the frames become contested as below: • *Participant G3-PP: The government wants people to watch these ads and get the message of the ads, but the ads don't show the reality of the society...not telling what is happening in the society.* • *Participant G2-TN: The ad is trying to identify people according to their ethnic groups such as through their skin colour and the use of dialect. But those criteria don't necessarily refer to that particular ethnic group.*

4 Conclusions

The use of framing analysis for analysing both advertisements and interview data summarize and define 'national unity' as the way people of different ethnic groups can work together in achieving a united nation and looking towards 'unity in diversity', is massive and still far from the finished plan. What can be concluded here is, much of the communication efforts of meaning-maker or (frame-maker) belong to the realm of producing 'frames' that are not realistic here-and-now, but are anticipated to become real in the future (there-and-then). Furthermore, frames attempt to convince individuals (essentially the general public) that a given frame is, in fact, the best interpretation of reality and thus, it appears to be the one and the only truth to approach a specific issue [10]. The ideal representation of the nation (people of different ethnics) rests on the construction of 'imagined community' [17] where people are ideally projected as practicing a harmonious relationship, a living interaction between each other and 'unity' through the findings.

References

1. R. Wodak, R. De Cillia, M. Reisigl, and K. Liebhart, *The discursive construction of national identities*. (2009)
2. D. Hartmann, Sociol. Q., vol. **56** (2015)
3. R. Hashim and C. Tan. *Citizsh. Teach. Learn.*, **5** (2009)
4. J. Engstrom. *Democracy and democratisation*, *In Routledge handbook of ethnic conflict*, pp. 103–111 (Routledge, London, 2011)
5. O. P. Amienyi, *Communicating national integration: Empowering development in African countries* (Ashgate Publishing Company, Farnham, 2005)
6. S. Mathew and M. A. Ansari, *An introduction to education.* (2012)
7. M. A. Cacciatore, D. A. Scheufele, and S. Iyengar. Mass Commun. Soc., **5436**, (2015)
8. D. A. Scheufele and D. Tewksbury. J. Commun., **57**, (2007)
9. S. J. Baran and D. K. Davis, *Mass Communication Theory: Foundations, Ferment, and Future*, (Nelson Education, Scarborough, 2011).

10. S. Kwan and J. Graves. *Framing fat: Competing constructions in contemporary culture*. (Rutgers University Press, London, 2013)
11. B. Van Gorp. *Strategies to take subjectivity out of framing analysis, in Doing news framing analysis: Empirical and theoretical perspectives (p. 84-109).* (Routledge., New York, 2010)
12. B. Van Gorp. *J. Commun.*, **57** (2007)
13. G. Rose, *Visual methodologies: An introduction to researching with visual materials.* (Sage, LA, 2012)
14. R. Wodak and J. E. Richardson,. *Analysing fascist discourse: European fascism in talk and text.* **5,** (Routledge,, London, 2013)
15. M. Chan. *Discourse Commun.*, **6** (2012)
16. K. J. Fitzgerald, *Recognizing race and ethnicity: Power, privilege, and inequality* (Westview Press, Boulder, 2014)
17. B. Anderson, *Imagined communities: Reflections on the origin and spread of nationalism.* (Verso Books, London, 2006)

Positive Impacts of Social Media at Work: Job Satisfaction, Job Calling, and Facebook Use among Co-Workers

Brittany Hanna[1], *Kerk F. Kee*[2,*], and *Brett W. Robertson*[3]

[1]Lieberman Research Worldwide, Huntington Beach, California 92647, USA
[2]Department of Communication Studies, Chapman University, Orange, California 92866, USA
[3]Department of Human Communication Studies, California State University, Fullerton, California 92834, USA

Abstract. The number of Facebook users grew rapidly since its conception. Within today's workplace, employees are increasingly connecting with each other on Facebook for interpersonal reasons. Due to sensational reports by media outlets of inappropriate social media use, many organizations are taking extreme measures about how their employees who utilize Facebook to connect with colleagues. Contrary to the negative assumptions, McAfee [1] states that social media within the workplace can promote positive dynamics. The present study uses McAfee's argument to examine if a positive connection exists between colleagues who use Facebook to connect with each other. An online survey with questions involving Facebook use with co-workers, job satisfaction, and perceived job calling was completed by employees (N=70) at two high-tech companies in Northern California, USA. Results revealed that job satisfaction is positively correlated with intensity of Facebook use among co-workers. Furthermore, feeling called to one's line of work was statistically significantly higher for the group of employees who spent the most amount of time interacting on Facebook with their co-workers than the group that spent the least amount of time. These results suggest that companies could begin to explore the positive benefits of social media use within the workplace.

1 Introduction

It is no surprise why many people hold negative assumptions about social media use within the workplace; most related cases in the media are sensational stories about employees being fired from the jobs due to careless Facebook posts. Love [2] cites seventeen such cases where this in instance has been displayed: a waitress was fired from her job for complaining about two customers who did not leave a tip, thirteen Virgin Airlines crew members for discussing details of their job on their Facebook accounts, and many others. After having a bad day at work, it is not unusual for one to vent about their day to their family and friends. When done offline and with trusted listeners, most people are able to

* Corresponding author: kerk.kee@gmail.com

rant without worrying about getting fired for expressing such human emotions and desire for social support [3]. However, because private social media postings are easily searchable and accessible, many people experience high levels of stress [4]. This leads to a wide assumption that it is not wise to carelessly connect with co-workers on Facebook and allow professional colleagues into one's private online space.

Even with the negative light that the media has placed on Facebook use within the workplace, some hold a more positive position. McAfee [1] boldly puts forth the argument that not only can the internal use of social media promote positive dynamics within the workplace, banning such uses can lead to missed opportunities for innovations and problem-solving. Furthermore, in a study conducted by Brandtzag et al. [4] it was found that a users' experience with Facebook expands their relationships with their friends on the website. Finding the balance between posting careless comments on Facebook, on one hand, and utilizing it for innovation communication, problem solving, and relationship building, on the other, is a delicate task for workers in the second decade of the 21st century. Given the realization that the internal use of Facebook within an organization can promote positive outcomes, this paper sets out to investigate the connection between intensity of Facebook use, job satisfaction, and job calling among today's knowledge workers.

2 Literature Review

2.1 Job Satisfaction

The concept of *job satisfaction* can be understood as employees' overall attitude towards their work and work environment [5]. Job satisfaction has been conceptualized to express how someone feels about himself or herself as an individual in the context of their work, which often stems from different psychological needs experienced by the workplace environment [6-7]. The human relations movement of organizational management laid the foundation for the belief that happy and satisfied employees work harder than those who are not [8]. Therefore, job satisfaction is often regarded as a key driver of productivity in many modern organizations [7].

Among several theories of the human relations movement, Frederick Herzberg's Motivation-Hygiene Theory examines job satisfaction [8]. Proposed in the 1950's and having been modified since, Herzberg claims that there are factors that promote job satisfaction. These factors, also known as *motivators*, include responsibility and career advancement. Conversely, *hygiene factors*, which prevent someone from being dissatisfied, include environmental conditions and interpersonal relations at work. These needs are not static, but can change over time [9]. For employees to keep a positive attitude toward their work, organizations must maintain motivators and hygiene factors at a high level for their employees through various services and strategic programs. Therefore, it is common for companies to invest in ways for employees to feel more satisfied with their jobs so that productivity could be constantly promoted.

However, job satisfaction can be inter-subjective, thus influenced by social networks in organizations. Boren and Johnson found that if an employee overhears resentment-filled messages from their co-workers, such an employee is less likely to use some of the services the company provides to their employees to increase job satisfaction [10]. Malone and Hayes noted that backstabbing-type behaviors from co-workers negatively impacted the overall workplace climate [11]. These sour outcomes can decrease organizational morale. Since job satisfaction is crucial, organizations must be louder than those employees with resentment. Making their strategic programs visible might require more effort, but could help build job satisfaction among employees. Not everyone will remain positive about his or her work, but unsatisfied employees could negatively impact the satisfaction of other employees.

2.2 Perceived Job Calling

Dik and Duffy define *perceived job calling* as a combination of finding a career that goes along with ones sense of personal meaning, is motivated by their social values, and comes to someone in a transcendent way [12]. In addition, Duffy and Sedlacek report that 40% of people in college feel a calling for a particular line of works, while another 30% are striving to find their calling during their college years [13]. Duffy and Autin argue that the possibility of living one's job calling is not determined by education and income levels [14]. Those ranging from 'no college degree' to a 'graduate degree' reported similar results of getting to live their job calling. However, there was a difference in the perception of if someone else should try to live out their job calling; the more educated, and higher the income one reports, the more one believes that someone should live their job calling [14].

During the human relations movement of organizational management, another prominent theory that emerged was Maslow's Hierarchy of Needs [8]. Proposed in the 1940s as a general theory and later applied to the organizational context, Maslow argues that everyone has the drive to fulfill five basic needs of physiological survival, safety, affiliation, esteem, and self-actualization in a progressive fashion. The notion of perceived job calling discussed in this paper is akin to the needs of self-actualization conceptualized by Maslow as the ability to excel and become what one is truly capable of. Self-actualization comes more easily when one feels that his/her work has personal meaning, is motivated by their social values, and comes to someone in a transcendent way.

This theory remains informative today, as having a form of monetary income is vital to human survival. While many talk about the desire of having a 'dream job,' for a vast majority of people, the need to earn an income comes from simply having a job, whether they personally enjoy the position or not. This situation is akin to simply fulfilling the physiological survival and safety needs in Maslow's theory. In the notion of having a 'dream job,' ideally people will hold a job that is of interest to them, one that allows them to fully express who they are and all that they are capable of becoming, as if being called into a particular career or field. Therefore, perceived job calling is similar to the notion of self-actualization in Maslow's model.

When people feel the sense of a job calling, and they are able to obtain a position within the scope of their calling, they feel self-actualized and more meaningful at work. If a manager recognizes employees' inherent talents and places them to a job that utilizes the talents, boredom and anxiety levels will lower [15]. Being able to find a job where you can utilize your talent is something many people attempt to acquire. Hirschi shows that the factors that lead to this feeling of meaningfulness are that people experience more work engagement, dedication, and absorption at their job [16]. These positive impacts on employees are what organizations often look for in the people they hire.

2.3 The Role of Social Media at the Workplace

While the use of social media to connect with co-workers generally has had a negative connotation [2], positive organizational scholarship has attempted to study the role technology can play within the workplace [17]. In recent years, social media, such as Facebook and Twitter, have become increasingly popular. As of 2015, 1.49 billion people actively use Facebook on a monthly basis [18]. The design of Facebook allows one to connect with their friends and family on a free web-based service and online platform. An overwhelming majority of people tend to use social media websites to keep in touch with old friends or people they have met recently [19]. Besides simply connecting with friends and family, it is becoming common to add co-workers as 'friends' on Facebook.

Some companies have experienced positive outcomes as the results of social media use within the workplace [1]. Positive communication using social media can a form of engagement for members of an organization [17]. "By engaging in dialogue with other members, sharing information, building public goods, and becoming engaged, individuals involve themselves substantively and symbolically in the sense-making activities of the organization and move closer together in the sense of a more common understanding and a greater sense of belonging to the organizational community" (p. 572) [17].

In this study, we investigate the relationships among internal use of Facebook with co-workers as well as employees' job satisfaction and perceived job calling. We recognize an opportunity to extend theories within the human relations movement to the digital age of social media. Since working adults spend majority of their waking hours at work, feeling a psychosocial connection with one's co-workers and the outcomes of such connections motivated the present study.

In a college-student context, it was found that online extroversion influenced positive satisfaction with interacting with their peers in a face-to-face context [20]. If people know more about whom they are around physically at work, uncertainty about their co-workers can be reduced [21]. Therefore, we argue that the greater the intensity an employee engages in on Facebook interacting with their co-workers, the greater the job satisfaction they experience. Furthermore, due to today's economic state, it is true that many have a job or career path for which they do not feel a strong sense of calling or enjoyment [12]. However, the more time one spends interacting with their co-workers on Facebook, the more the use of Facebook can lead to positive social affiliation with their co-workers as 'friends' on the site, leading to a sense of community at work. A sense of community at work can lead to the development of meaning about their work as a community of practice emerges [22]. Therefore, we proposed that the greater the amount of time one spends socializing with their co-workers on Facebook, the greater the sense of job calling one experiences. Below, we advance two hypotheses to be tested in this study:

H_1: Employees' job satisfaction is correlated with their intensity of Facebook use with co-workers.

H_2: Employees' perceived job calling varies across groups of employees who spent different amounts of time on Facebook with their co-workers.

3 Methods

3.1 Participants

For this study, we contacted two high-tech companies with main offices located in Northern California, USA. After gaining access to the companies, the gatekeepers of the organizations sent out an invitation email to employees in their divisions. Company 1 yielded 31 respondents and Company 2 had 39 respondents. There was a 25% response rate for Company 1, while the percentage is unknown for Company 2 due to additional instructions Company 2's gatekeeper sent with the survey.

3.2 Instrumentation

The questionnaire asked participants to indicate the social media platforms (e.g., Facebook, Twitter, LinkedIn, etc.) they use to connect with their co-workers, followed by a specific set of questions related to their Facebook use, job satisfaction, and job calling. We adapted the Intensity of Facebook Use Scale [23-24] to measure the psychosocial attachment through the use of Facebook with friends to the work context. The adapted scale is reliable (cronbach's alpha=0.814) and included the following questions: "Facebook is part of my everyday activity with people from work (including during breaks, after work, during the weekend, etc.)"; "I am proud to tell people from work I am on Facebook"; "Facebook has become part of the daily routine in my work life (during breaks, after work, during the weekends, etc.)"; "I feel out of touch with people from work when I haven't logged onto Facebook for a day"; "I feel I am a part of the Facebook community with people from work"; "I would be sorry if Facebook was banned in my organization (including during breaks, after work, during the weekends, etc.)"; and finally we added to the adapted scale a reversed question of "I limit my engagement with people from work on Facebook in order to maintain a work-life separation". Furthermore, we also measured the approximate amount of time individuals spend on Facebook communicating and interacting with their co-workers. The question asked was "In a typical work week, approximately how much time per day (including during breaks, after work, during the weekend, etc.) do you spend on Facebook interacting with people from work?" with the response being in different time increments (i.e., less than 10 minutes, 10-30 minutes, 31 minutes to 2 hours, etc.).

For job satisfaction, questions were adapted from Brayfield and Rothe [25] and Judge et al.'s [6] validated scale. The job satisfaction scale in our survey (cronbach's alpha= 0.755) included these items: "I feel fairly well satisfied with my present job"; "Most days I am enthusiastic about my work"; "Each day of work seems like it will never end" (reversed coded); "I find real enjoyment in my work"; and "I consider my job rather unpleasant" (reversed coded). To test one's total perceived job calling, the Brief Calling Scale created by Duffy, Bott, Allan, Torrey, and Dik was used [26]. The scale in our survey (cronbach's alpha=0.859) contains two Likert-type questions: "I have a calling to a particular kind of work" and "I have a good understanding of my calling as it appears to my career."

3.3 Procedure

We drafted a persuasive invitational email to the gatekeepers of the two organizations. This email offered the companies the option to receive the statistical results we found at the completion of this study as an incentive to participate. Once they agreed to allow their companies to participate, these gatekeepers sent a pre-crafted invitation email on behalf of the researchers to their employees and co-workers. This invitation email contained a link to an online survey, which was completed by 70 corporate workers in these two high tech companies.

4 Analysis

A correlation was run to test the relationship between job satisfaction and intensity of Facebook use for Hypothesis 1. A one-way ANOVA was used for Hypothesis 2 to test if employees' perceived job calling varies across three groups of employees who reported having spent different amount of time on Facebook interacting with their co-workers.

5 Findings

A Pearson correlation was performed to assess the relationship between job satisfaction and intensity of Facebook use to test Hypothesis 1. There was a positive correlation between the two variables, $r= 0.320$, $n= 55$, $p= 0.017$. Hypothesis 1 was supported.

A one-way between groups (independent groups) ANOVA was conducted to compare the mean scores of perceived job calling across three different groups of employees categorized based on the amount of time their reported to have spent on Facebook interacting with their co-workers on a daily basis. The three groups have the following different time increments: Group 1 (less than 10 minutes), Group 2 (10 to 30 minutes), and Group 3 (31 minutes to 2 hours). An ANOVA test showed that there was a statistically significant difference in job satisfaction across groups defined by the amount of time someone spends on Facebook with their co-workers, [F (2, 51) = 4.824, p=.012].

Post hoc comparisons using the Tukey HSD test indicated that the mean score for Group 1 (M= 3.17, SD= .75) was significantly different from Group 3 (M= 4.56, SD= .62). However, Group 2 (M= 3.73, SD= .56) did not significantly differ from either Group 1 or Group 3. Given the ANOVA test result, Hypothesis 2 was supported.

6 Discussion & Conclusion

The results from this study could help shift the outlook on how companies and businesses view their employees' use of Facebook between co-workers. Generally, social media gets a negative reputation for uses related to the workplace. Work-life balance, or keeping one's personal life and work life separate, is what society deems as acceptable. While that has worked for previous generations, this study suggests that, now, it is much easier for people in today's workplace to connect with their social circles via social media, including co-workers. The findings from Hypothesis 1 indicate job satisfaction is positively correlated with how intense an individual uses Facebook to connect with their co-workers. This finding suggests that organizations should look into potentially promoting the use of Facebook within the workplace. While it might seem unconventional for companies to embrace their employees using Facebook, it could be an important strategy to success for a company that is yet to be fully tapped.

Hypothesis 2 revealed that the more time someone spent on Facebook interacting with their co-workers, the more they felt a sense of calling for their job. This finding could also be the result that these workers already feel like having something (e.g., a sense of job calling for their work) in common, therefore, they use Facebook to interact with each other with a greater degree of intensity. Connecting outside of work on a website might have provided an opportunity for these workers to further share their common interests in their line of work. Furthermore, it might also be possible that a greater desire to connect with one's co-workers comes from having other interests in common other than just work alone.

A limitation to this study is the small sample size. For future research, a larger sample size would help yield more conclusive results. It would also be interesting to investigate companies outside of a technology-based industry. It is possible these results were discovered due to the culture of high tech companies. Investigating companies where every employee does not use a computer regularly to complete their job could yield different results.

Another area to look at for future research would be the internal social network platforms companies are creating for their employees to utilize. These are becoming an upcoming trend since use of social media is constantly growing. Companies want their employees to be social with each other, but sometimes only on company-based internal websites. These workplace social networks can give employees a chance to talk more openly about workplace-related experiences, while getting to know each other outside of the traditional work environment.

References

1. A. P. McAfee, *Enterprise 2.0: New collaborative tools for your organization's toughest challenges*. (Boston, MA: Harvard Business School Press, 2009)
2. D. Love, *17 People who were fired for using Facebook.* http://www.businessinsider.com/facebook-fired-2011-5 (2011)
3. D. A. Olson, J. Liu, K. S. Shultz, *J. of Organizational Psychology*, **12**, 133 (2012)
4. P. Brandtzæg, M. Luders, J., Skjetne, *International J. of Human-Computer Interaction*, **26**, 1006 (2010)
5. F. Xu, J. Shen, *Chinese Education & Society*, **40**, 86 (2007)
6. T. A. Judge, E. A. Locke, C. C. Durham, A. N. Kluger, *J. of Applied Psychology*, **83**, 17 (1998)
7. N. Ichikura, S. Kumagai S, *J. of Integrated Design & Process Science*, **14**, 51 (2010)
8. K. Miller, *Organizational communication: Approaches and processes*. Belmont, CA: Wadsworth Publishing Company (2012)
9. J. A. Medved, *Educational Leadership*, **39**, 555 (1982)
10. J. P. Boren, S. L. Johnson. *Southern Communication J.*, **78**, 128 (2013)
11. P. Malone, J. Hayes, *Communication Studies,* **63**, 194 (2012)
12. B. J. Dik, R.D. Duffy. *The Counseling Psychologist*, **37**, 424 (2009)
13. R. D. Duffy, W. E. Sedlacek, *Career Development Quarterly*, **59**, 27 (2010)
14. R. D. Duffy, K. L. Autin, *J. of Counseling Psychology*, **60**, 219 (2013)
15. M. F. Steger, B. J. Dik, In P. A. Linley, S. Harrington, N. Garcea (eds). *Oxford handbook of positive psychology and work*. Oxford: Oxford University Press, (2010)
16. A. Hirschi, *J. of Counseling Psychology*, **59**, 479 (2012)
17. L. Browning, G. H. Morris, K. F. Kee, In K. S. Cameron, G. M. Spreitzer (eds), *Oxford handbook of positive organizational scholarship*. Oxford: Oxford University Press, (2012)
18. Facebook, Company Info at http://newsroom.fb.com/company-info (2015)
19. J. Raacke, J. Bonds-Raacke, *Cyberpsychology & Behavior*, **11**, 169 (2008)
20. X. Liu, R. LaRose, *CyberPsychology & Behavior*, **11**, 310 (2008)
21. K. Subrahmanyam, P. Greenfield, *Future of Children*, **18**, 119 (2008)
22. E. Wenger, *Communities of practice: Learning, meaning, and identity*. New York, NY: Cambridge University Press (1999)
23. N. B. Ellison, C. Steinfield, C. Lampe. *J. of Computer-Mediated Communication*, **12**, 1143 (2007)
24. S. Valenzuela, N. Park, K. F. Kee, *J. of Computer-Mediated Communication*, **14**, (2009)
25. A. H. Brayfield, H. F. Rothe. *J. of Applied Psychology*, **35**, 307 (1951)
26. R. D. Duffy, E. M. Bott, B. A. Allan, C. L. Torrey, B. J. Dik, *J. of Counseling Psychology,* **59**, 50 (2012)

 DOI: 10.1051/shsconf/20173300013

Two Sun over One Firmament Culture: A Political-Communication Analysis on the Contestation of Jakarta and Kuala Lumpur over "Indonesian Malay Communities"

Evie Ariadne Shinta Dewi[1,*]
[1]Universitas Padjadjaran, Indonesia

Abstract. This paper tries to extent a personal experience on an interesting discussion concerning the communication issues in Medan, North Sumatra. There is the inclination of the North Sumatrans to the news broadcasted by Kuala Lumpur and Medan's tends to identify with its cultural similarities. For Jakarta is officially the center of political authority over the 'Indonesian Malay Communities', this dual cultural loyalty creates an imaginary phenomenon: 'Two Suns' in term of authoritative news resources that applies upon 'One Firmament Culture'. This means that there is a divided news orientation among the Indonesian Malay Communities' that put Jakarta and Kuala Lumpur in a tacit contestation. While until today this kind of contestation is still going on, the same features have brought me to the phenomenon of hierarchical communication structure. This vividly seen when we realize how lopsided is in nature the communication relations between the global authorities who have controlled the strategic means of communications as well as their contents with the rest of the world. For, in exception to technological matters, the Kuala Lumpur-inclined of 'the Indonesian Malay Communities' in communication practices is based more on cultural aspects than their technical and political necessities words.

1 Introduction

There something stuck the thought of an interesting fact related to dual cultural loyalty that prevails within the people of North Sumatra as by Gelora Viva, Head of Postel Infocomm Department of North Sumatra office:

> *"Di Sumatera Utara ini posisinya dekat dengan Malaysia, sedangkan Jakarta tidak. Sementara semua peraturan dibuat di Jakarta. Akibatnya kita tidak punya benteng di pesisir timur dari segi Televisi sementara TV Malaysia dapat diterima dengan baik. Informasi dari Luar bebas masuk sementara yang dari dalam justru masih terhambat. Akibatnya kita akan terganggu secara Ipoleksosbudhankam (ideologi, politik, ekonomi, sosial, budaya, pertahanan dan keamanan). Soal Pilkada (pemilu kepala daerah) di*

[*]Corresponding author: evieasd@gmail.com, evie@unpad.ac.id

DOI: 10.1051/shsconf/20173300013

Medan ini malah dibandingkan dengan pilkada di Malaysia tidak dibandingkan dengan Jakarta. Hal lain, soal pertumbuhan ekonomi, di Sumut tidak tergantung Jakarta, makanya pertumbuhan ekonomi di sini lebih tinggi jika dibandingkan di Jakarta." (North Sumatra, is close to the position of Malaysia, while Jakarta is not. While all rules are made in Jakarta. As a result we do not have a fort on the east coast of Malaysia in terms of the TV while the Malaysia's TV can be received well. Information from Outside entry freely while on the inside it is still hampered. As a result, we would be disturbed by Ipoleksosbudhankam (ideology, politics, social, culture, defence and security). People prefer compare Pilkada (regional election) with local election in Malaysia does not compare to Jakarta. Economic growth in North Sumatra does not depend Jakarta, so that's why economic growth is higher here than in Jakarta.) "

As will be mentioned later, taking a different historical path with their Javanese fellow, the North Sumatran people share their cultural traits with other people that geographically separated from the official lines of Indonesia. This, at its turn, creates a kind of dual cultural loyalty within this people. It is at this juncture, it deserves to raise a conceptual question: Is the Jakarta's influence —as a center of government and the center of national political authority— penetrative enough, on the matters of spreading cultural information, into its national peripheral areas, especially in North Sumatra?

Exploring the Sundanese and the Javanese, and compared with other Indonesia's myriad ethnic groups, which both of these people share much more similarities. And unconsciously, both of the peoples see themselves as the center of Indonesia's culture. Cultural communication between them, therefore, would be going much smoothly, simply because the shared cultural values running within the core geographical boundaries of Indonesia. When this perspective is used to see the North Sumatran and its adjacent peoples whom would call them as, "the Indonesia's Malay communities", we would fine an interesting phenomenon in terms of cultural communication and cultural loyalty. By going deeper into this subject matter, I think the scene is worth enough to be shared here.

2 Global power and the lopsided communication contestation

It is undeniably that the dominance influence of the developing countries' capital cities as a sole source of cultural information are increasingly challenged by the outside forces or actors. In other hand, the field of communication is the structures of information current that naturally run in the form of mutual exchange. In a sense that the communication processes cannot take place in a unilateral way, or in other words, transmitted information must getting responses from the communicants [1]. So, Jakarta as a capital cities and source of information have to perform the credibility's communicators to the community, except people will leave and looking for other credible source of information, like neighbor country, Malaysia. Furthermore since the 21st century, information is the central theme of the researchers. People currently live in the information age that provides a variety of information technology such as the internet and mobile phones. Manuel Castell [2], try to define and explain the information age between the years 1996-1998 through the theory he had made and presented network society as a social structure characterized by a network of communication and information-processing technology, "the network society as a social structure is ccharacterized by networked communications and information processing technologies", said Castell.

To observe North Sumatran's people's behavior, aside from Castells's information theory, we can also use the concept of Media Flow, which is built from a series of studies since 1970-1980an, which researchers claim their unbalanced, unidirectional flows of TV Programs and foreign news from the "center" to the "periphery". It is evident that one of the phenomena of intense from the North Sumatrans networking society is inequality of information dissemination. Analyzing information about inequality, we can use the Media Imperialism Concept. Where the concept was built "within a broader analysis of cultural imperialism and dependency theories". As stated by Barrett "the process whereby the ownership, structure, distribution of the content of the media in any one country are singly or together subject to substantial external pressures from the media interests of any other country or countries without proportionate Reciprocation of influence by the country so affected [3]. Referring to the concept of Media Imperialism, we can see Jakarta as the center of power and in some cases; also the culture of course is one of the actor's disseminators of information. But faced with global actors with great imagination and control of it makes the position even Jakarta serves only as the recipient of the communication flow transmitted by actors giants.

There is unquestionable that those who involve in the communication process are tend to be pluralistic, the quality, and above all, the power of the communication contents are increasingly concentrated in the hands of a handful centers of actors. This has brought about its qualitative nature, namely a communication situation that freely place every participant mutually exchange within the framework of equal status, simply does not work. In other words, through the fast development of technological information and communication, the quality, the contents or resources of communication have been massively shifting to a more world-wide conglomerated centers that they are even able to dictate the structure of world peoples' imagination.

Quoting Paul Kennedy, these MNCs stress their interests more with international rather than national that structurally put their influence beyond the borders of their national geographic areas. It is the growth of these MNCs that bringing the world to the communication technology revolution. This phenomenon is clearly depicted by Kennedy:

They existed, in embryonic form, in the cosmopolitan private bank of the late nineteenth and early twentieth centuries, whose growth was assisted by the earlier 'communication revolution' of the telegraph and by the absence of major Great Power coalition wars. But today's globalization is distinguished from those earlier examples by the sheer quantity and extent of the multinational firms in our expanded and integrated global economy. As noted above, they emerged in the postwar international economic order that reduced protectionism and encouraged a recovery of world trade, and were further stimulated in the 1970s by the United States' decision to abandon the gold standard followed by a general liberalization of exchange controls, at first only in a few countries, later in many others. This not only provided more liquidity for world trade, but increased the flow of transnational capital investments, as companies invested abroad without constraints imposed by central banks [4].

At least, there are three points that could be drawn from these paragraphs. First, the laden economic interests of the MNCs had driven them to pursue further communication technological developments. Through these technologies, the MNCs were able to develop an effective communication in controlling their capital investments. Second, although the spread of information seemingly confined strictly within the space of economic interests of those corporations, it's by product had gone beyond these frontiers. The needs to spread the capital investments world-wide, for instance, invariably require a wider and a more sophisticated communication networks at global level. Third, through such a typical process, we witness the creation, though had gone gradually, a hierarchic structure in the practice of communication. It is within the third element drawn from those paragraphs we see the autonomous expansive development of the power of a handful center of actors, who possess the means of technological communication, in dictating the quality and contents of communication —that then influence the rest of the world. In other words, it is they who have the actual right to determine what kind of information should be spread out to the world public.

The hierarchical structure of information transmission running through the uneven power in commanding the communication technologies. As the consequence, the forms, contents and the quality of information that are determined to be disseminated by them practically biased to their cultural background. By possessing and commanding the technological information, it is they who define the superiority of expressions. Namely, the expressions that can be found in various instructive novels, movies or other cultural creations that confiningly derived from a single world of culture. Since the rest of the world possess no control over the means of communication, the world-wide spread of these superior expressions —centered among a handful of actors— have gradually and inevitably become a global norms. Having superior in contents and being spread through the sophisticated means of communication, the messages they send to the world public will unconsciously tend to be accepted as the "final truth".

This lopsided disseminating information power had clearly analysed by Edward Said his monumental book *Orientalism* [5] where which all of the miseries produced by the Middle-East perennial conflicts were blamed into the hands of the Arabs, chiefly the Palestinian, responsility. And the terrorist stigma putting upon the shoulders of the Arabs had been perceived as a "truth" at the global public arena.

It is in this context we are witnessing the decline of power of the developing countries' capital cities in controlling the swirl communication contents around them, simply because they possesses no authoritative communication sources as well as and effective and sophisticated technological communication to compete with. Instead of being able to create their own communication power, these developing countries' capital cities have become the powerless recipients of information fanned by the global central actors who command both technologies and information resources. This weakening position is also the case with Jakarta, as the core of Indonesia's power and political authority —and to some extent is fighting to get the status of the center of its own national culture too.

Theoretically, as the center of centers within national boundary of Indonesia, Jakarta should take a central role in communication affairs for the whole of Indonesian people. In practice, however, this limitedly applies to the current of information concerning merely on political issues, administrative regulations that centrally transmitted from Jakarta. But, in the field of cultural issues and system of ideas, the centrality of the Jakarta's position remains problematic. For, in reality, instead of being the main actor in the field of communication, Jakarta less articulative and is often merely being able to act as a recipient before the huge wave of world-wide transmitted communication contents.

This is vividly seen when it is being faced against the global actors who control both of the world cultural imaginations and expressions and technical means of communication. Movies such as *Spider Man*, *Iron Man*, *Frozen* or intellectual phrases such as Schumacher's *small is beautiful* and Anthony Giddens' *Third Wave* take their gound much deeper into the national public mind compared with a series of the Indonesia's president or the elite of political parties addresses or the national strategic groups' point of views.

These all provide an undeniably fact that the aforementioned actor pluralism's assumption in the communication processes, in a sense the dyadic mutual exchanges in ideas and information, stands on a weak ground. What really happening on the stage are the sheer majority of recipients and the handful centers or actors who hold and control the disseminators of information within the system of communication affairs. By looking at this phenomenon, it could hypothetically be stated that —as the case with other developing countries' capital cities— the Jakarta's influence position in this world-wide communication field is less penetrative over the national public. And that all of the prevailing public discourses are borrowed from those that centrally produced by the global actors.

3 Cultural Contestation between Jakarta and Kuala Lumpur

Through the global phenomena described earlier, instead, the distinctive structures of the North Sumatra's physical geography that makes *Semenanjung Melayu* (Malay Peninsula) of Malaysia much closer compared with other areas within the span of Indonesian frontiers. This specific geographical location thus facilitates and heightened the frequencies of huge wave back and forth movements of both North Sumatran and Malaysian populations. As a result, these undisturbed geographical people movements strengthen the old and historic cultural pattern of relations between the North Sumatrans and the Malaysian people. At its turn, this refreshing and strengthening cultural relations provides the North Sumatran population access into another actors who control the source of cultural values and information alternative to Jakarta.The North Sumatran, the Acehnese, the Jambi people, and the West Sumatran are looked culturally more as "the Indonesian Malay communities" than a general feature that generally marking "Indonesians". Sharing the cultural system, these "Indonesian Malay communities" established deep and long emotional bound with the people of *Tanah Semenanjung*, whose path development is not shared with both Jakarta and Javanese-Sundanese people as a whole.

The unavoidable contestation between Jakarta and the center of the Malay Culture in the area of communication soon comes into existence as Kuala Lumpur managed to gain "supremacy" in terms of its national economic performance at regional level, namely among the ASEAN countries. In this case, Kuala Lumpur, as the center of Malay culture, has not only proven itself to be one of the leaders among the ASEAN level, but also has successfully transformed the Malaysian society in accordance to modern ways. This is remarkably shown by looking at the legacy of Mahathir Muhammad's development programs which a tremendous success. Leading this country since the early 1980s, Mahathir ultimately brought the state and the Malaysian society into an unprecedented development success story. It is at this point Jakarta potentially fail to be a single penetrative communication power upon the "the Indonesian Malay communities", although undeniably they are geographically exist within its sovereign boundaries. For, as has been stated, the success of Kuala Lumpur in economic development refreshing the collective memory among the Malay constituent culture about their past emotional bound. In other words, through the impressive Kuala Lumpur economic progression, "the Indonesian Malay communities" found their modern cultural model alternative to Jakarta. Kuala Lumpur, therefore, presents itself as a dignified model for these communities.

It is in this context we see the Jakarta's position in the communication realm becomes problematic. On the one hands, "the Indonesian Malay communities" perceive the success of Kuala Lumpur leadership as a signal that their cultural system possesses value network that supportive to modern development goals. As a consequence, Kuala Lumpur could potentially be one of the centers of cultural loyalty of these "Indonesian Malay communities". On the other hands, since having cultural differentiation, "the Indonesian Malay communities" regard Jakarta confiningly functions as the center of political and administrative authority. As has been mentioned earlier, this happens because historically their cultural relations with Jakarta and Java had gone less deeper compared with the *Tanah Semenanjung*. The effective colonial administration (1830-1942) did integrate, in the past, the geographical parts into one nation what is known today as Indonesia. The government, however, could not go beyond this line. One of its consequences is the Jakarta's failure to shift the cultural loyalty of these "Indonesian Malay communities" totally to itself.

However, strong political relations with Jakarta have been taking a different way with their cultural orientations. It could be assumed, then, that "the Indonesian Malay communities' relations with Jakarta is characterized more "political nationalism" and, thereby, less cultural in nature. In other words, while the national political integration between both of them is conspicuously seen, its cultural integration remains to be seen. It means that these cultural orientation and loyalty are still being contested between Kuala Lumpur and Jakarta.

It seems "the Indonesian Malay communities" relatively independent from Jakarta in terms of communication materials, for they have greater information access from the neighbouring countries, chiefly Malaysia. Entertainment programs, international news, consumer goods advertisings and global economic situation that are being broadcasted from the neighbouring country's mass media endlessly intrude into the families' rooms of "the Indonesian Malay communities". For the time being, this all could be used as the indicators of the Indonesian Malay communities' preference to the information produced by Malaysian Kuala Lumpur news networks. Is should be stressed here that sharing the historic common cultures between "the Indonesian Malay communities" and the Kuala Lumpur-centered *Tanah Semenanjung* communities plays its own role.

It mean for Jakarta in the context of communication, is that the Jakarta's effectiveness in presenting itself into "the Indonesian Malay communities" could not achieve its total goal —for the absent of cultural-emotional bound. The still Kuala Lumpur slanted of the Indonesian Malay communities' cultural orientation gives a vivid example about the continued "divided" cultural loyalty among these communities between Jakarta and Kuala Lumpur. It means that the information dissemination between Jakarta and Kuala Lumpur remains in the endless contestation process within "the Indonesian Malay communities". In this case, "the Indonesian Malay communities" face or possess *two suns* of cultural orientations: Jakarta and Kuala Lumpur that naturally presents a dual cultural leadership. For this has been going decades-long, it seems that the contestation would not end in the near future.

Conclusions

To win this contestation, Jakarta must build a communication strategy that fully garbed with cultural contents. Only by formatting its strategy full with cultural contents could Jakarta replacing the supremacy position of Kuala Lumpur and superimposing its own terms in the realm of effective communication. As the core of a sovereign state, culturally, Jakarta must be *the only sun* upon "the Indonesian Malay communities".

References

1. E.A. Griffin, A First Look at Communication Theory, (5th ed, McGraw-Hill, 2003)
2. M. Castell, The Network Society, The Power of Identity, and the End of the Millennium, (1996, 1997a, 2000)
3. O.B. Barrett, Media Imperialism Concept, (1977)
4. P. Kennedy, Preparing for the Twenty-First Century, (New York: Vintage Books, 1993)
5. E.W. Said, Orientalism : Western Conception Of The Orient, (London : Penguin Books, 1978)

Foreign News as Cultural Expression: Media, Perspective, and Consciousness

Takuya Sakurai[1,*]

[1]Tokyo Denki University, 5 Senju-asahi-cho, Adachi, Tokyo 1208551, Japan

Abstract. This paper attempts to challenge our dualistic understanding of foreign news and to discuss it as a cultural expression in terms of different structures of consciousness as well as different types of media. Previous foreign news studies seldom explain what foreign news is and how we experience it, and leave what assumptions and presuppositions constitute our understandings of foreign news unanswered. In order to unfold what essentially constitutes foreign news and how it comes to appear to human consciousness, hence, this paper reviews the literature of foreign news and seeks to explain foreign news as a cultural expression.

1 Introduction

Whenever new media emerge, presentations of foreign news have been in transition. Not only are the forms of foreign-news reportage being changed, but also, more important, the experiences of foreign countries reported are ever being in change. Meaning of foreign news is processed on the basis of contexts bounded multiplicity, including cultural, social, and historical constitution that influences audiences' particular identities or consciousness in processing mediated messages. Different media convey different foreign news in terms of how news stories are packaged in a form of reportage. McLuhan's [1] statement "the medium is the message" might be partly right. However, it is a little too simple to explain the medium power in foreign news. The historical and cultural contexts also give foreign news the differences. Foreign news cannot be simply a medium form of reportage of what is happening in foreign countries; rather, it is reported and consumed in relation to audiences' identities and images of what their country is. In this sense, foreign news can be seen as what Pilotta [2] calls "analogical relatedness." With this analogical relatedness, audiences can be interchangeable and transformative to the media, and they can be united as a group of audiences.

What we categorize into foreign news is relatively a recent formalized type of foreign experiences. In the perspectival sense, foreign news is formed to meet the demand of a rational/mental explanation on foreign events or countries. Although such inventions as telegraphs, television, and the Internet are making foreign news speedier and timely ever than before, reflecting the desires audiences seek in foreign news, perceptivity only allows us to see the history of foreign news in a deterministic manner. Because foreign news has ever since reported in the modern professional forms, the shift of main medium of foreign

* Corresponding author: tsakurai@cck.dendai.ac.jp

news from newspaper to television and the Internet reflects what Jean Gebser [3] calls "temporal anxiety," a modern human condition. As Gebser [3] and Lewis Mumford [4] claims, the shift is rather discontinuous and mutational. Foreign news has its own meanings to be reported in different media in different histories and cultures.

Foreign news is a culturally expressive form that allows audiences to find their identities and values in what foreign countries are reported through what channel. Different cultural horizons explore different meanings of foreign news. The horizons lead to a variety of interpretations of the same phenomenon. Differences are perspectives [5]. Differentiating themselves from others are the emerging individual identities [6]. While different horizons are struggling for discursive truth, foreign news is always already being redefined. While post-structuralism focuses on discourse in the metaphysical and linguistic-orientation, Gebser [3] deals with discourse as a "cultural expression," making it free from the articulation of psycho-cultural orientation.

Different values manifested in different media signify underlying differences in a cultural logic, though which some values are predominated over the others. The cultural logic cannot be reduced to any binary opposition or quantification; rather, it is a form with which people can make sense out of what they experience. Gebser [3] calls this logic a "structure of consciousness." People cannot experience anything without their own cultural forms. What they express in their cultural artifacts such as foreign news is a form of how they make sense of the world. All cultures contain these structures of consciousness, but one structure is predominant over the others, and the uniqueness is manifested in their cultural expressions.

Previous studies have attempted to exclusively define foreign news in either behavioristic or political-economic terms. Such an approach is more descriptive than explanatory. In order to grasp foreign news as a form of cultural expression and to explain what foreign news is in terms of our experiences of it, this study attempts to provide a new avenue to illustrate what foreign news is. Based on the idea of perspective, we unfold what constitutes foreign news and how it comes to appear into human consciousness. For the reasons, we review the literature of foreign news, and then seek to explain foreign news as a cultural expression.

2 Foreign news studies: a review of the literature

In the realm of communication research, especially in journalism and mass communication, foreign/international news has become an independent field of study. International communication scholars see foreign news from the viewpoint of a world-information order and attempt to unveil the Anglo-American dominance over the flow of foreign news and the imbalances between Western developed and non-Western developing countries [7-8]. Media effect studies pay attention to the processes of how foreign news shapes our view of the world [9]. The studies analyze how journalists disseminate international realities and how the audiences make sense of the disseminated realities. Furthermore, the so-called critical studies deal with international news as an ideological product maintaining the status quo [10]. They consider that the news media internalize and legitimize the government's foreign policies within a presentation of international news.

2.1 International News Flow Studies

International communication studies examine international news since the inception of the New World Information and Communication Order (NWICO) in the late 1960s [11]. Against the backdrop of the United Nations Educational Science and Cultural Organization (UNESCO), the studies have begun to indicate that the structure of international news flow is imbalanced, and that news from developing countries was often biased and misrepresented [8]. A great number of studies have investigated the determinants of international news flow, especially since Scandinavian scholars such as Galtung and Ruge [12] and Ostgaard [13] proposed theoretical foundations of international news flow. The studies propose various factors influencing international news flow and attempted to identify what the best factors are to predict the news flow. The structural theory of international news flow points out that international news we are exposed to is far from a direct reflection of international reality. International news is selected, edited, and deleted through a multitude process of the news media. It also addresses that international news flow is quantitatively imbalanced between Western industrialized and the Third World developing countries, while revealing the Anglo-American dominance of information flow in the world.

International news flow is a major means of information exchange. Since the studies on the structure of international news hardly provide a systematic analysis of a whole-world picture of international news flow, some studies employ the world-system school in order to figure out a global-interaction system that is structured into a center-peripheral relationship. Based on the world-system theory, these studies attempt to identify the structure and process of international news flow within the context of the global network. The world-system theory is an extension of the dependency theory. The dependency theory captures the global system at a two-tier level with the Western developed countries at the center and the rest of the world at the periphery [14]. In his framework, Wallerstein [15] sees the world as a global system in which countries are interdependently linked within the capitalist system, locating countries in the three interactive spheres or zones of cultural, economic, social, and political relations: core, semi periphery, and periphery [14].

The studies employing the world-system theory show a clearer picture of dependency relations with international news flow in a larger structure of international communication. Especially in the age of information technology, the structure of international communication in which countries are located could be a better indicator than their geographical locations. At the same time, however, the growth of global communication systems has brought the emergence of transnational media corporations [8]. Today, these media giants themselves might have the power to set up their own structures of international news flow beyond the boundary of nation-states, while the studies [7-16] employ nation-states as the units of analysis. In this sense, since one of the main premises of the world-system theory is that dependency causes underdevelopment on the nation-state level, their findings may be limited to the extant frameworks of dependency, ignoring the impact of global media corporation such as CNN and Star-TV. Furthermore, the world-system theory, as Chang [7] points out, provides little exploration on "whether there is a casual or nonlinear relationship in the news flow" (p. 557). Above all, the "dependency" relationship shows the dynamics of international news flow in a larger formation; however, it leaves the complexity of news industries and the nature of international news unanswered.

2.2 Media Effect Studies

Media effect studies attempt to analyze international news, paying attentions to news media's contributions to the constructed meanings within a society [17,18,19]. The theories focusing on the processes through which our views of the world are shaped by the mass media have strong impacts on the ways of studying foreign news. These studies see international news as a social artifact constructing a social reality rather than a mirror of reality and support the notion that international news is framed in particular ways by journalistic practices and ideological beliefs.

Public opinion is believed as one of the major factors that affect foreign policy. Since most people do not have personal knowledge of foreign events, the agenda-setting effect of foreign news is believed to be stronger than that of domestic news. Salwen and Matera [20], for example, found the correlation between foreign news coverage and public opinion to demonstrate agenda-setting effects on international news. Cohen [21] also argues that international news could possibly cause changes of foreign policies and international relations. Thus, these scholars have found that international news coverage is often consistent with the national foreign policies. With regard to the possible effects on coverage on foreign policies, the agenda-setting theory plays an important role in international news studies. Scheufele [22], however, argues that there are two main constraints preventing agenda setting from being called a theory: the lack of inferring causal relations and the ambiguous criterion of measuring of perceptions of issue importance. Especially, the latter constraint becomes a weakness of second-level agenda setting in a comparison with framing. By contrast, "framing influences how audiences think about issues, not by making aspects of the issue more salient, but by invoking interpretive schemas that influence the interpretation of incoming information" [22], p. 309.

2.3 Critical Studies

Critical scholars see foreign news as a social and ideological belief. This perspective considers that the news media support the status quo, especially in cases of international conflicts and any challenges from outside the power. News studies focusing on media's ideological roles rely largely on Antonio Gramsci's idea of hegemony, which refers to the predominance of one social class over others [23]. Hegemony tends to be explained with the three key dimensions: common sense, process, and system. Hall [19] explores this view of hegemony to "a web of meaning" and "a terrain of ideology." The idea of hegemony allows researchers to see the news media as "a site for ideological and cultural struggle" [24], p. 80.

The hegemonic ideology shapes the production of news and the news media are seen as part of the process of constructing an accepted symbolic reality. Since the news media are seen as part of the dominant value system, international news reinforces the value of national unity [25]. These studies consider the role of the news media as maintaining the status quo and characterize foreign news as tied to home country's ideology. For example, the studies examining the coverage of social movements show how the news media portrayed the movements and how they delegitimized those who we are outside the dominant groups [18-9]. Thus, the studies employing the hegemony thesis attempt to ask large and important questions about the role of foreign news in a society, and they show us some critical views of how our perceptions are framed hegemonically by the news media.

However, there is a lack of empirical evidence to support these identified roles of the new media [26]. Along with the empirical shortcomings, Altheide [27] also argues that the hegemony thesis fails seeing the contradictions within news messages and its contradictory social effects. Altheide continues to claim that although hegemony is easily used as an interpretive framework for news studies, the news media's characteristics of legitimizing and maintaining the status quo, which the hegemony thesis illustrates, is one-way and arbitrary. As a result, these studies fail examining the complicated interactions between audiences and media texts. This interaction is rather complicated because audiences' interpretations of media contents rely largely on their own identities.

3 Discussion

As seen above, foreign news studies have been conducted not only with a number of theoretical and methodological bases, but also with a variety of contexts; however, these studies seldom explain how foreign news is interpreted and consumed. All the three schools reviewed above have attempted to exclusively define foreign news in either behavioristic or political-economic terms. Such an approach is more descriptive than explanatory. The following section discusses foreign news as a form of cultural expression and unfolds what constitutes foreign news and how it comes to appear into human consciousness.

3.1 Foreign News as a Cultural Expression

The idea of cultural expression is concerned with how discourses are manifested in human consciousness. Gebser [3] articulates that the significance of discourse varies across cultures and that cultural expression provides us with an assessment of how facts and artifacts are incorporated into a discourse. Modern discourse is a form of rationalization, in which signifiers and signified are arbitrarily integrated in the technological/rational milieu. With newspapers, foreign countries are transformed into a modern form of discourse. It has a strong impact on the way people understand foreign countries. Foreign experiences are reduced to perceptivity. The perspectival foreign experience is not only manifested in the modern and rational awareness, but also is a reduced form of knowledge. This transformation of foreign knowledge means that journalism has become institutionalized and individualized [28-29]. News has also become "individualizing activities" and began to be read individually. This perspectival domination over magic/mythical ways of interaction with foreign countries indicates that foreign news becomes "my" story. You can choose some foreign news over others based on your identities and interests. Institutions have also sought to be professionalized as journalism and formalized foreign news into a certain type of discourse [29].

Along with newspaper, television further promotes the perspectival consciousness in foreign news. Kramer contends [30], "Television fosters this process of shifting the sign from being symbolic to being signaled via repetition and false presence—pseudo nature/truth" (pp. 258-259). People are dissociated from the real and extended themselves to what they see and what they image. They become less emotional from their existences. In the perspectival world, human consciousness is dominated by such dichotomous relationships as subject/object and body/mind. Foreign countries emerge in that consciousness as objects of our country. Our country and foreign countries no longer appear complementarily; rather, foreign countries are reduced to others, not "us." The more rationalized foreign news is the more accuracy and precision it requires.

3.2 Foreign News as an Integrating Interpretation

Perspective is a rational form of how we make sense of the world. Whereas there are no points of view in the magic and mythical structures of consciousness, a point of view emerged with the perspectival consciousness. Perspective, however, has to be magic and mythical because people cannot do or invent anything without perspective. While foreign news is a form of perspectival consciousness, a perspectival form always contains magic and mythical elements [3]. Magic and myth provide human beings with the fundamental power and connection to something foreign.

Magic is the most fundamental source of power to make something unimaginable imaginable and unidentifiable identifiable. Magic manifestation is the efficacy of not cause, but "making" (Mickunas 1972, p. 183). Magic's *mah(g)* has the same root as "make," "machine," "mechanism," and "man," which refers to power in Indo-European language [3,5,6] Foreign news is a way of revealing our being-in-the-world. This is the power of "making." It not only makes us available to locate in the world, but also makes foreign countries available to interpret with our identifications. Calling television "a technological expression of logocentrism," Kramer [31] claims, "broadcasting is a civilizational expression of the magic impulse in humankind" (p. 35).

The mythical structure is audio-oriented. The word "mythos" means "to speak, say, instruct" [32], (p. 181). Gebser [3] points out the connection between the English word *mouth* and the Greek *mythos*. The "mythos" root is also related to the Greek *myein*, which refers to close oneself [32]. To muse in the mythic structure does not mean thinking in the perspectival sense; rather it implies "belonging" and "thanking." Although this sound/mute might be opposite in the perspectival sense, the relation is rather complementary in the mythical structure. The mythical consciousness serves a basis of providing the hermeneutical foundations of human communication such as comprehension, understanding, interpretation, and translation of foreign countries.

The mythic consciousness also involves imagination: *imago,* Latin "image" [3]. International images elicit emotional responses on a mythic level like flags, national anthems, and totemic animals. Kramer [30] claims, "Reification of television images is facilitated by the Western bias of logocentrism (p. 257), a cultural bias that assumes that the truth is what we see [31]. Newspaper is also being a logocentric medium, co-existing with television and the Internet. In order to attract television-generation audiences, more visual-centered layouts and story constructions are required in the newspaper. USA Today is a good example to succeed in capturing this trend. Kramer, furthermore, phenomenologically investigated international images, and contends, "International images are deeply rooted in the doxic beliefs in worlds 'out there,' not metaphysically, but geographically, historically and culturally" [30], p. 252.

Foreign news provides the images of foreign countries the way they appear in audience's consciousness. All of the constituents of the images on foreign countries are part of the natural fact-world [30]. However, the images do not necessarily correspond to anything actual; rather, they are the reflections and manifestations of the structure of consciousness as to how meanings and significations are manifested. Images of foreign countries exist the way one believes them to be; therefore, one understands presuppositionally what foreign countries are in accordance with his/her images of them. In order to bracket our natural attitude of imaginary countries, thus, we need a reflective attitude toward the foreign images in the doxic awareness.

4 Summary

In sum, the present study has attempted to challenge the perspective view of foreign news and to explore it as a cultural expression. As discussed above, foreign news is not just about what is happening in foreign countries; rather, it is about how we interpret ourselves through foreign countries. Cultural expression intends to be theoretical and explanatory unlike most instrumental and prescriptive communication theories. As opposed to a stimuli-react, process-orientation view, it points out that communication is multidimensional rather than unidimensional, and meanings exchanged in communication are ever-presently creative rather than transmission. Cultural expression thus illuminates foreign news as a cultural manifestation as to how we make sense of a given discourse within our cultural milieu. Ways of understanding foreign news reveal a particular belonging to a certain cultural milieu, and cultural expression provides us with a critical means of unfolding what creates and sustains our perspectival understanding of foreign news.

References

1. M. McLuhan, *The medium is the massage*. (Random House, New York, 1967).
2. J.J. Pilotta, *Media power working over the body: An application of Gebser to popular culture. Consciousness and culture: An introduction to the thought of Jean Gebser*, (Westport, Greenwood, 1992)
3. J. Gebser, N. Barstad, A. Mickunas, *The ever-present origin* **1** (Ohio University Press, Athens, 1985).
4. L. Mumford, *Technics and civilization*. (Harcourt, Brace & World. Orlando, 1963)
5. E.M. Kramer, *Modern/postmodern: Off the beaten path of antimodernism*. (Westport, Praeger, 1997)
6. E.M. Kramer, *Dimensional accrual and dissociation: An introduction. In J. Grace & E. M. Kramer (Eds.), Communication, comparative cultures, and civilizations* **3** (Hampton, New York, 2013)
7. T. Chang, Communication Research, **25**, 5 (1998)
8. R.L. Stevenson, Journalism Quarterly, **69**, 3 (1992)
9. P.J. Shoemaker, S.D. Reese, *Mediating the message: Theories of influences on mass media content*. (Longman. New York, 1991)
10. C. Lee, J.M. Chan, Z. Pan, C.Y.K. So, *Global media spectacle: News war over Hong Kong*. (State University of New York Press, Albany, 2002)
11. M. Masmoudi, Journal of Communication, **29**, 2 (1979)
12. J. Galtung, H. Ruge, Journal of Peace Research, **2** (1965)
13. E. Ostgaard, Journal of Peace Research, **2** (1965)
14. A.Y. So, *Social change and development: Modernization, dependency and world-system theories*, (Sage, CA: Newburg park, 1990)
15. I. Wallerstein, *The modern world system*. (Academic Press, New York, 1974)
16. K. Kim, G.A. Barnett, Communication Research, **23,** 3 (1996)
17. K.M. Carragee, Journalism Monographs, **128** (1991)
18. T. Gitlin, *The whole world is watching: Mass media in the making and unmaking of the New Left*. (University of California Press, Berkeley, 1980)
19. S. Hall, *The rediscovery of 'ideology': Return of the repressed in mass media studies. Culture, society, and the media. (*Methuen, New York, 1982)
20. M.B. Salwen, F.R. Matera, Journalism Quarterly, **69**, 3 (1992)
21. B.C. Cohen, *The press and foreign policy*. (Princeton University Press, Princeton, N.J., 1963).
22. D.A. Scheufele, Mass Communication & Society, **3** (2000)
23. A. Gramsci, *Prison notebooks*. (International Publishers, New York, 1971)
24. H.A. Park, Journal of Communication, **48** (1998).
25. T. Chang, Shoemaker, P. J., & Brenlinger, N. Communication Research, **14** (1987)
26. K.M. Carragee, Western Journal of Communication, **57**, 3 (1993)
27. D.L. Altheide, *Public Opinion Quarterly, **48***, 2 (1984).
28. T. Glander, *Origins of Mass Communications Research during the American Cold War: Educational Effects and Contemporary Implications*. (Lawrence Erlbaum Associates, Mahwah, N.J., 2000)
29. M. Schudson, *Origins of the ideal of objectivity in the professions: Studies in the history of American journalism and American law*, (Gerland Publishing, New York, 1990)
30. E.M. Kramer, *Phenomenology of international images. In Philip Blosser, Eiichi Shimomisse, Lester Embree, & Hiroshi Kojima (Eds.), Japanese and Western phenomenology*. (Kluwer Academic Publisher, Dordrecht, 1993)
31. E.M. Kramer, *The origin of television as civilizational expression. In John Deely (Ed.), Semiotics 1990: Sources in semiotics*. (University Press of America, Lanham, M.D, 1993)

32. Mickunas, Algis. Civilizations as structures of consciousness. Main Currents in Modern Thought, **29**, 5 (1972)

Visual Information Priming in Internet of Things: Focusing on the interface of smart refrigerator

Hyeonjin Kim[1], *Sangmee* Lee[1], and *Donghee* Shin[1,*]

[1]Department of Interaction Science, Sungkyunkwan University, 314 International Hall, 53 Myeongnyun 3-ga, Seoul, Korea

Abstract. Internet of Things (IoT) supports an environment where users are able to interact and get information by sharing visual informational cues. The purpose of this study was to investigate what effect was made on the perceived usefulness, perceived ease of use, intention to use and self-efficacy about use of smart refrigerator when visual information related to the food and function is provided in advance in the interface. A data was collected from a between-subjects experiment (N = 26) with an independent variable (priming visual information vs. non-priming). According to the experiment results, users who were primed about the visual cues perceived higher usefulness (PU), ease of use (PEoU), intention to use (IU), and self-efficacy (SE) toward the fridge compared to the users in a non-primed group. The study proposes theoretical and practical implications for designing food-related smart home appliances' interface and visual affordances.

1 Introduction

As information and the environment where objects are connected to the wireless network are shared, the objects inside households are also being evolved [1]. Smart refrigerator which applies various on-line network services provides better in-life object internet experience by supporting functions such as the temperature adjustment, conserved food management, grocery shopping, memo board and recipe. A display screen installed in the front of the product possesses potential value enlightening the convenience and efficiency of the product user [2]. However, while the information that the user should accept increases, the heuristic lacks about use of various functions and thus there are limitations in that the user use various functions voluntarily [3,4]. According to the system interface studies, it has been reported that this problem is caused by the conventional drop-down menu selection method.

Especially, since the various icons shown in the menu screen of the smart refrigerator have limitation in explain detailed functions that each of the sub-information contains, they obstruct the natural information search of the user when the user accesses the target information and there is complexity of use. In order to solve this problem, a study about

* Corresponding author: dshin@skku.edu

new precipitating cue is needed which support easy and natural information search of the user and induce use of functions of the smart refrigerator. This study examines if self-efficacy increases when the information image that the refrigerator display proposes is exposed as the visual stimulus which reminds of the use of its functions. Also, this study has verified if the priming versus non-priming level of the corresponding information affects the reception of smart refrigerator technology.

2 Theoretical Backgrounds

2.1 Information Priming

A psychological concept of priming effects explains that whether the informational cue previously exposed to users affects their perception and recognition toward target stimulant and causes their behavioural changes [5]. According to the food advertisement studies, when consumers are exposed to the sense-related cues about food, their food consumption increases [6,7]. One of the food advertisements focuses on increasing instant satisfaction of consumers' sense by exposing them to the characteristic visual stimulus which provoked their appetite (e.g. hot vapor) [8]. When they are exposed to the stimulus in advance, complex physical behavior is vitalized and even in the cases where there is no special intention or when the consumer does not perceive appropriately the corresponding stimulus, such vitalization may occur due to the external stimulus [9]. Especially, consumers go through the priming process of various sensual stimulus (e.g. vision, hearing) when accepting food related information. In such cases, information recipients perceive the information (e.g. product benefits, ingredient) more effective and thus positive perception and attitude about products are formed [10]. Since smart refrigerator shows food and product related images based on on-line system, it can effectively use the food related image stimulus which induces use of refrigerator. Based on this rationale, we assume that when a smart refrigerator user is primed on the sensual stimulus, he or she can be induced to use the system voluntarily without a need to learn additionally about the use of sub-menu. Thus, the user who has been primed on visual stimulus will show positive reaction about use of instant system.

2.2 Technology acceptance model (TAM) in effective interface

The representative factors which affect the acceptance of new technology are perceived usefulness, perceived ease of use and intention to use technology. The perceived usefulness shall mean the perception that when the user uses the system, the individual outcome will improve, the perceived ease of use shall mean the belief that the user can use a particular system with no big efforts and the intention to use shall mean the acceptance of the user about a new information system based on the user's belief about a specific system [11].

According to the conventional studies, the TAM is appropriate to measure the positive effect of the visual image stimulus on the consumer. According to Hausman et al [12], there is a study result that when a user performs an on-line shopping task, the perception of usefulness improved according to if there were human factors (e.g. visual images). Especially, in the case of the internet based contents such as the digital archive resources, if effective interfaces such as user-friendly search tool are provided to the user on the interface, the perception of the user (perceived ease of use & perceived usefulness) is affective positively [13]. Thus, this study intends to verify what effect the priming about visual cues has on the usefulness, predicted ease of use and intention to use when using interface of the smart refrigerator.

- H1. The user's perceived usefulness (H1.1), perceived ease of use (H1.2) and intention to use (H1.3.) will be higher when the user is primed on the visual cues than when he is not primed.

2.3 Self-Efficacy

According to the social cognitive theory, self-efficacy is the belief that an individual's judgement of their ability to perform courses of action or achieve specific task. Such belief affects one's cognitive and behavioral outcome (i.e. motivation and performance) about use of a system [14]. According to Igbaria [15], it was said that when the user cannot access easily the necessary functions due to a system that he is not accustomed to, if help or interest is given about use of the system, the self-efficacy can be improved. Additionally, Hill and his colleagues [16] found that when self-efficacy is high, the technology acceptance about a new technology product rises. Thus, the self-efficacy is an important factor to affect user when accepting the new technology and thus it shall be measured along with the TAM model. In the case of the smart refrigerator, since its penetration rate is yet low, and thus there is a limitation in assessing the technology acceptance of the users who have not used a smart refrigerator, but since the self-efficacy also affects the prediction of use of the technology [17], it shall be used as a criteria for a user who has not used in person to assess the product.

- H2. The user's self-efficacy will be higher about use of smart refrigerator when the user is primed on visual cues than when he is not primed.

3 Research Method

The experiment participants were selected at random from the 26 applicants of students in their 20s through 30s who go to four year private universities in Seoul (M = 23.53, SD = 2.31). The recruitment of the participants was made against the people who have used a smart device for more than a year.

3.1 Apparatus

This study was conducted using the 13.3" screened laptop (LG 13ZD950 LX20K) in the laboratory. By excluding the characteristics appearance and design phase which represent specific brands, the study was made not affected by the manufacturer. The content used in the experiment was produced based on the LG Dios smart refrigeration, and the prototype interface was proposed (Figure 1). The experiment was composed in such a way that the experiment participants experience each given content according to the priming versus non-priming level.

3.2 Manipulation

The priming level of the visual cues about use of smart refrigerator interface was divided into two experiment groups. (1) Home screen with no visual cue (2) home screen with visual cue priming. And difference of user's perception about each group was experimentally verified. In the visual cue with priming (Figure 2), in the left side, there is image of the materials kept at the refrigerator (e.g. mushroom, pork, clam etc) and in the right side, there is a image of recommended dish which can be made with the material kept. The image of the material kept in the refrigerator has been proposed in sliding method in turn with the text(e.g. the expiry date for the mush room is 2 days now.)

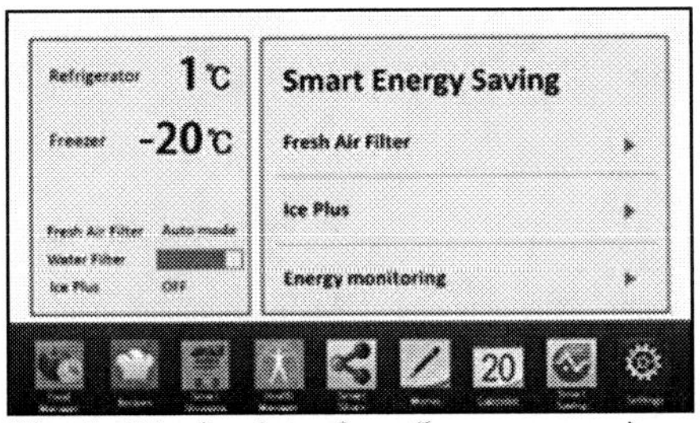

Fig. 1. Display interface (home screen) on smart fridge

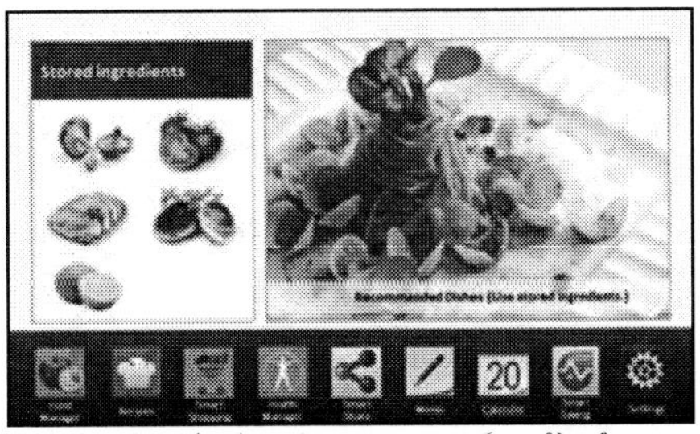

Fig. 2. A priming images on the display panel.

3.3 Procedure

The experiment participants were distributed at random at laptops of which the visual priming levels were differently set and they participated for 15 minutes. The experiment procedure was done in the order of the introduction to the experiment, the main experiment and questionnaire about visual priming cue after experiment. The subject group which was exposed to the smart refrigerator interface with visual cue priming was ordered to stare at the screen and follow the voice instruction. Each subject freely surveyed the functions (e.g. food management, recipe, shopping and energy management etc.) of the refrigerator using the icon on the bottom of the display. Other group subjects were exposed to directly to the icons on the bottom of the display without being proposed the image cue in advance. They surveyed the functions of the smart refrigerator freely just as the subjects with the visual cue priming did. After all the experiment was terminated, the experiment participants went through the questionnaire using a paper and a pencil.

3.4 Measurement items

As the perceived ease of use (Cronbach's α = .754) of the Technology acceptance model [11], a total of three items including "I think that how to use the smart refrigerator is easy." And "To learn the functions of the smart refrigerator is easy to me" were measured, as the perceived usefulness(Cronbach's α = .854), a total of three item including "I can do that task I want effectively when I use the smart refrigerator," "I think that use of smart refrigerator is valuable overall," and etc. were measured and as the intention of use (Cronbach's α = .877), a total of three items including "I have the intention of using the various functions of the smart refrigerator" and "I will continuously use the various functions of the smart refrigerator" etc. were measured on a five-point Likert scale and all the scores were summed up and made an average.

As the self-efficacy [17] (Cronbach's α = .873) a total of five items including "I will be able to use the smart refrigerator well even if there is no person explaining its functions" and "I will be able to use the functions well even if I have never used the smart refrigerator before" etc. were measured on a five-point Likert scale and the results were summed up and made average.

DOI: 10.1051/shsconf/20173300015

4 Results

This study has used one-way analysis of variance (ANOVA) and analyzed in order to assess the effect of the priming of visual cue. It was verified that the main effect of the perceived usefulness according to the priming condition was higher in the primed condition (M = 4.692, SE = .099) than the non-primed condition (M = 3.674, SE = .099). It was verified that the main effect of perceived ease of use according to the priming condition was higher in the primed condition (M = 4.256, SE = .092) than in the non-primed condition (M = 3.712, SE = .092). (F = 17.544, $p < .05$, $\eta 2 = 0.423$). It was verified that the main effect of the intention to use according to the priming condition was higher in the primed condition (M =4.654, SE = .105) than in the non-primed condition (M = 3.697, SE = .105). (F = 41.428, $p < .05$, $\eta 2 = 0.633$) It was verified that the main effect of the self-efficacy according to the priming condition was higher in the primed condition (M = 3.385, SE = .168) than in the non-primed condition (M = 4.518, SE = .168). (F = 22.717, $p < .05$, $\eta 2 = 0.487$).

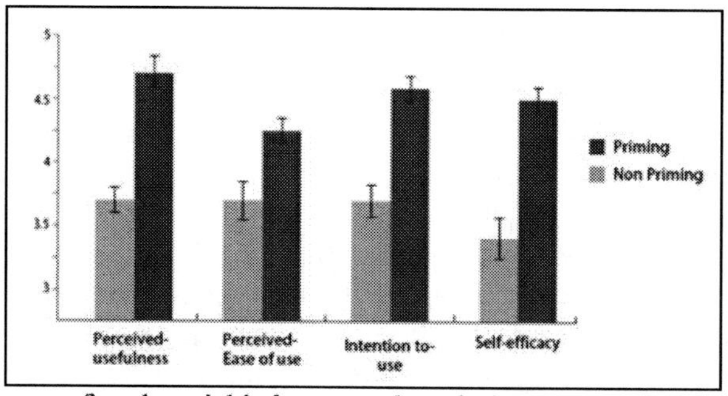

Fig. 3. Differences of each variable between the priming and non-priming level.

5 Discussion and conclusion

This research examines the perception and behavioral change of the user about the use of the smart refrigerator which is proposed later on according to the priming condition in advance of the visual information about use of the food and smart refrigerator. Based on this, thus study has proposed the smart refrigerator interface including the priming information based on the conventional smart refrigerator interface. According to the experiment result, a significant effect has been observed in terms of the perceived usefulness, perceived ease of use, intention to use and self-efficacy etc. according to the priming condition.

Based on the study result, two interpretations in two phases (TAM & self-efficacy) are possible about visual information priming about use of smart refrigerator. First, as the study by Harris and colleagues has proposed, when the visual information about food has been primed for the user, the perception of the user about the product was more positive than in the non-primed condition of the visual information about food. Also, appetizing food images could be an attractive or appealing visual human factors in the technology use process enhanced the behavioral intention of the user. Further, it was revealed that when visual information related to the use of the smart refrigerator system has been proposed together, the visual priming affected the technology acceptance such as perceived ease of use, perceived usefulness and intention to use of the user.

Second, as in the H2, when the user has been primed about visual cue, the self-efficacy was very effective. As in the study asserted by Igbaria and his colleague [16], this was because the visual cue provided help and interest when the user used the technology that the self-efficacy was high. The users who have used smart refrigerator first time could find a cue which enhances their ability to use new function from watching expiration dates and recommended dishes. This can be seen as the fact that the prior priming of the visual information about use of smart refrigerator invoked provision of motive about use of smart refrigerator with the induction of interest or effective proposal of information and as a result of it, belief about the task performance ability about use of refrigerator system has been improved. This study has some limitations in future studies which will be conducted later on. First, since the experiment participants were in their 20s and 30s only and did not include people in their 40s and 50s, there is a possibility that such participants can not represent the whole users of the smart refrigerator. However, in the case of the university students in their 20s and 30s, they have a high possibility of being potential users and they have possessed the experience of using smart machine, and thus it is considered that an ample result has been gained in measuring the smart refrigerator use experience. In future studies, based on these findings, the age and scope of the user layer will be expanded and more qualitative study will be conducted through in-depth behavioral study. Second, since the priming effect of the visual cue was measured only in the home screen currently, it will be needed to examine additional functions (e.g. food manager, smart shopping) in future studies. Through this, the priming of the visual cue can be deducted about effective functions and eventually, the users will be induced into use various functions of the smart refrigerator more. Although this study has been performed based on the smart refrigerator interface based on this content above, in the case of the smart appliances (e.g. smart oven) which have similar attributes based on the internet and which are related to food also, the method of proposing in advance the visual cue related to food and use of products can be applied and the use of smart appliances can be made automatically activated.

References

1. H. H. Lee, A. M. Fiore, J. Kim, *International Journal of Retail & Distribution Management*, **34**, 8 (2006)
2. D. Uckelmann, M.Harrison, F. Michahelles. Springer Berlin Heidelberg. (2011)
3. L. Swan, A. S. Taylor). *CHI'05* extended abstracts on Human factors in computing systems, ACM. (2005)
4. P. Y. P. Chi, J. H. Chen, S. Y. Liu, & H. H. Chu, *Interaction Platforms and Techniques*. (Springer Berlin Heidelberg, 2007)
5. M. Rothensee. (Springer Berlin Heidelberg 2008)
6. D. Logan, *Cognitive psychology*, **12**, 4 (1980)
7. D. FEDOROFF, J. Polivy, C. P. Herman, *Appetite*, **28**, 1(1997)
8. L. Terrenghi. (Springer Berlin Heidelberg, 2005)
9. R.Loewen, P. Pliner, *Appetite,* **32**, 3 (1999)
10. A. Dijksterhuis, J. A. Bargh, *Advances in experimental social psychology*, **33**, (2001)
11. J. L. Harris, J. A. Bargh, K. D. Brownell, *Health psychology*, **28**,4 (2009)
12. F. D. Davis, Perceived usefulness, perceived ease of use, and user acceptance of information technology. *MIS quarterly*, (1989)
13. A. V. Hausman, J. S. Siekpe, *Journal of Business Research*, **62**, 1 (2009).
14. J. C. Hong, M. Y. Hwang, H. F. Hsu, W. T. Wong, M. Y. Chen, *Computers & Education*, **57**, 3 (2011)
15. Bandura, *Psychological review*, **84**, 2 (1977)
16. M. Igbaria, J. Iivari, *Omega*, **23**, 6 (1995)
17. T. Hill, N. D. Smith, M. F. Mann, *Journal of applied psychology*, **72**, 2 (1987)
18. Bandura, *American psychologist*, **37**, 2 (1982)

DOI: 10.1051/shsconf/20173300016

Glocalizing' the Hijab: A Malaysian Perspective

Nurzihan Hassim[1,*]

[1]Department of Media Studies, Faculty of Arts and Social Sciences, Universiti Malaya, Malaysia

Abstract. This article examined the hijab, or the act of veiling as a transformative socialization agent for Malay-Muslims in Malaysia's multicultural state. Through the media, the hijab was purported as a progressive lifestyle with representations of veiled Muslim women who are predominantly Malay. Apart from upheavals of Islam and the globalization of veiling across Muslim-majority countries, local socio-political developments have commodified the hijab as a product of Malay sovereignty rather than a symbol of religious obligation. Malay-Muslim women are predisposed to media messages that altered conservative Malay-Muslim stereotypes through convergence of old and new media platforms. The article explores the local hijab phenomenon by examining its media movement and its influence towards attitudes and behaviours of audiences. The evolution and visibility of the hijab are analyzed in relations to multiplicity of the Malay-Muslim identity and its integration with transcultural ideas of the East and West. Findings indicate that acceptance of this dynamic cultural identity resonate through diversity of content and multitudes of accessibility.

1. Introduction

The `hijab', or the act of veiling in its true intention protects a Muslim woman's modesty, both physically and mentally. Translations from the Quran describe cloaking the body to avoid harassment from the opposite sex and displaying religious devotion [1], where the An-Nur verse read, "...O Prophet! Tell thy wives and daughters, and the believing women, that they should cast their outer garments over their persons (when abroad): this is most convenient, that they should be known (as such) and not molested (33: 59)" [2]. In recent years, Muslim-majority countries have revived female modesty as a uniform for cultured, modern Muslim women to illustrate Islam as a progressive religion in comparison to its Western counterparts; in addition to using the hijab as a form of solidarity against Islamophobia. *Al-Jazeera*, a news organization based in Qatar responded to the headscarf ban controversies in France through the hijab worn by its presenters and prevalence of similar issues in its news content [3]. Consequently, in Europe and parts of Australasia where Muslims are the minority, "others" have grown accustomed and appreciative of hijab as an ethno-religious signifier that created a window for them to learn more about Islam rather than condemn it [4]. The emergence of the hijab as a cultural artefact has created a unique global movement for female modesty and lent a voice to Muslim women whom

* Corresponding author: nurzihan@siswa.um.edu.my

were mostly cloaked in patriarchal stereotypes and "backwardness" from the ancient practice of veiling.

On the other hand, anxieties in cultivating more bohemian and inclusive Muslim society saw a series of capitalist ventures that reversed the media flows from the East to West. Evidence from a Thomson Reuters 2014-2015 report suggest that Muslim fashion is growing rapidly with USD 266bil spent on clothing and footwear, while this figure is expected to multiply by 2019 [5]. International retailers such as *H&M, DKNY* and *Dolce & Gabbana* have acknowledged its Islamic consumers by introducing clothing lines that are designed for the Middle-Eastern market in efforts to elevate the status of the hijab and have it eliminated from prejudicial perceptions [5,6]. Meanwhile, in Asia, Indonesia is renowned for neo-traditional Islamic fashion that are endorsed by celebrities and clerics in mass media, of which is also being exported back to the Middle East [7]. Subsequently, Dian Pelangi, a leading designer of modest fashion was named one among the top 500 people in fashion from by UK-based magazine, *Business of Fashion* through her self-named hijab label [8]. It can be seen that the reach of modest fashion have extended to even non-Muslim Asian regions as illustrated by Japanese company *Uniqlo* in their collaboration with Muslim designer Hana Tajima in a worldwide launch of scarfs, inners and kaftans that are modeled by Yuna, an international, Malaysian-born singer that is considered as a role model for hijab fashion enthusiasts.

Similarly, Malaysia is experiencing a climate of `hijab elite' through the widespread growth of hijab brands. As an example, *Fareeda* recorded revenue of MYR30million (USD8mil) in 2013 alone [9] by marketing luxurious hijab designs that are catered to high-profile and affluent wearers. In the same way, several other hijab-specific business empires such as *Naelofar Hijab* and *dUckscarves* established by notable media personalities; Neelofa Mohd Noor of and Vivy Yusof respectively, have produced exclusive hijab that are catered to the millennial middle-class Malays in the aim to remove unfavourable perceptions upon veiling that were normally donned by elderly, unfashionable Malay-Muslim "makciks" (aunties). For *Naelofar,* the family business gained profits through 700 distributors nationwide and has now reached buyers in both Asian and European countries and soon, the United States, through the availability of social networking [8]. Meanwhile, the scarcity of hijab with upscale, international styles inspired Vivy Yusof of *dUckscarves* to design modern, Malaysian-flavoured scarves for empowered, professional women that is also applicable for style-savvy non-Muslims. Another essential factor that contributes to the steadfast hijab movement is the move by media figures such as Neelofa and Vivy to start veiling after achieving their celebrity statuses. Practicing modesty after reaching a certain stage of adulthood is considered common in Malaysia as veiling remains an option to Malay-Muslim adults despite Islam being a birth religion to all Malays [10].

In addition to this recursive pattern, advent of networking sites like Facebook, Twitter and Instagram corresponds to the use of traditional media in Malaysia. Hijab-centric television programmes such as *"Hijabku Gayaku"* and ground-breaking magazines like *Hijabista* [11] displayed veiling as a widely-accepted regime for urban Malay-Muslims, although its more contemporary depiction is relevant to a limited pool of Malay audiences. In brief, the unyielding competition of hijab marketing in Malaysia may have excluded the working-class and reconstructed the Malay-Muslim identity in the public sphere as a determinant of social status. This article reviews the diminishing role of hijab as the observance of faith, whereby audiences' desires of social inclusion is slowly replacing the importance of modesty.

2. Redefining 'Malayness'

More often than not, urban hijab representations in Malaysian media today are encouraging young Malay women to be liberated and experimental with their veiling choices in the emphasis to appear modern and unique. While Islamic modesty conceals sexual elements of the female body, artistic interpretations of the hijab are going beyond boundaries that are not necessarily sharia-compliant. As seen in depictions of hijab in magazine spreads, the focus was on emulating international fashion magazines in its feminist stance and resistance towards traditionalist views of Malay-women [12]. The traditional role of Malay women as head of the household and purveyor of moral values to young children is absent in the reproductions of the hijab as a symbol of feminist manifestations that seek to explore her sexuality in the hijab.

Despite bearing strong colonial influences, Malaysia has negotiated multicultural identities since its independence but the mainstream media often compartmentalized the core-cultures of Malays that define the bumiputera (sons of the soil) under legislations of Barisan Nasional that are often Malay-centric [13]. Malay-Muslim women often relinquished their religious upbringing from the villages in favour of job opportunities and more modern lifestyles during the New Economic Policy that favoured progress of the Malays in the 1970s, but have since transitioned to the former. In its place, the revival of Islam in the 1980s motivated Malay undergraduates to participate in religious social movements with *Angkatan Belia Islam Malaysia* (ABIM) and dress in modest clothing such as the *jubah* (loose dress) and *tudung* (veil) [14]. As Islam further exposed audiences to religious content on broadcast and print media, the integration of Islam in administration and politics through Islam Hadhari that was introduced by former Prime Minister, Dato Seri Abdullah Ahmad Badawi had possibly magnified the need for Malays to be represented as Muslims. In a study by Martin [15], the proliferation of Muslim-living keenness in Malaysia observed that youths now no-longer acknowledge themselves as Malays per se, but as new Muslims instead.

As marketing of halal products in Malaysia became a norm, emergence of the hijab in the media was agency to the group identity of Malay `hijabis'. Authors have contended that despite flailing numbers, the excess of womens' magazines are catalyst to consumerism and commodification [16–18]. These publications present the construct of ideal womanhood through glamorous clothing and makeup, and subsequently wearing the hijab is considered as a privilege where its significance is measured from its connection to popular culture and materialism [19]. Loo established that capitalist media retain the essence of Malay-Muslims through shared language, cultural heritages, and social history [20], therefore the negotiation of Malayness in the new economy of veiling lies upon the desire for young Malay-Muslims to remain ethnically dominant within a changing community of imported fads and trends; particularly Islamisation and Arabicization of the nation state which now in its coercion reassures "others" that hijab is no more backward. The exclusivity of the hijab is perceived by highly educated women as proponents of egalitarian Islamic ideologies as the objectifications of women's bodies in media determine audiences' expectations of their self and govern their self-confidence [21]. As described by Khoo in *Reclaiming Adat* [22], Malay women were established as "bumigeois" where they struggle to maintain their traditional ways whilst integrating their self-image with current forms of Islam that are pervasive in local media experiences.

3. `Glocalizing' the Hijab

After examining the conceptual term of `glocalisation' discussed by Roland Robertson [23], it can be said that the integration of hijab in the Malay-Muslim society is a phenomenon that blended oppositions. Considering global outlook on the hijab and its immersivity with local conditions, the hijab can be linked to the concept of hybridity that matched best practices from different cultures or patterns that form new subcultures that are anticipated to inspire improved life qualities. The acculturation of Malay-Muslims towards Islam are from conditions regulated by its state policies, while glocalisation mediated 'cultural pluralism' to accommodate social and cultural differences that arise in the multi-ethnic landscape. An illustration of glocalization is Rupert Murdoch's acquisition of Star TV, a merger that was pervasive with Anglo-American content but sustained Asian presenters to display a sense of ethnic loyalty to its audiences and transcend boundaries of culture in its delivery of information [24]. In the same way, the hijab converged with Malay traditions to form a more demure, Pan-Arabic identity to project Islamic solidarity whilst also being equally progressive and tolerant of others, outspoken and adaptable to any challenges as their Western peers despite the limitations of veiling.

Even so, the hijab is not often accepted as positive reinforcement. In contrast to religious teachings that subdue materialistic wants and needs; merchandising of the hijab saturated into a form of brand identity and sought attention in the public sphere. The idiosyncrasy of subjugating oneself to God and yet displaying the very ornaments that the religion aims to conceal from the opposite sex, has streamlined the hijab as a malevolent representation of Malay-Muslim altogether. Marina Mahathir, a womens' rights activist criticized *glocalization* of Arabic identity as a form of colonialization in itself thus undermining Malay women, particularly those in the hijab from freely participating in social activities as there are morals to observe behind their veiling [25]. Even in social media, a status post by the Facebook page of *MizanSisters,* an online community for Muslim women contained an `open letter' criticizing *Hijabista* magazine from an anonymous writer named `A Worried Muslimah'. The magazine was accused by netizens as a mockery of religion by fashionizing of the hijab in garish visuals that are only beautiful aesthetically but fail to prove its functionality to the lives of young Muslim women [26]. Further to this, the status mobility offered by veiling to urban audiences have, in some cases backfired ; specifically in how proprietors of the *Fareeda* company had taken pride in disconcerting behaviours of its patrons in the unforeseen mob that ensued during the recent launch of its premium hijab design called *Sweet Love.* Although the video of the incident posted on its Facebook page meant to illustrate the credibility and impact of its brand identity, affected *Fareeda* hijab fans were likened to lowly herds of mindless `cattle' in responses by opinion leaders and audiences [27]. It can be seen that from these few examples that media audiences are active participants in the discourse of female modesty in Malaysian media, specifically on the Internet that allows instantaneous feedback and channels of rapid information delivery.

Commodification of the hijab as a symbol of wealth and stature is now a moral dilemma as to how Malay-Muslim women incorporate wealth and material desires in their submission to Allah. Although veiling is presently common among Malay-Muslim women, it takes self-consciousness to proclaim the hijab as a symbol of freedom for each individual and it also takes into account the level of participation by the hijabi audience in media. Reinforcement of the hijab identity through media recommendations and exchange of messages with other members of the community through the media determines whether the voice heard in the public sphere is meant to benefit the modesty movement as a spiritual initiative or merely for personal gain to boosting one's ego and emotional well-being. Kulenovic asserts that this is a form of `projective identity' [28], where social agencies from cultural materials redefine an individual's place in the society and transform the overall social structure altogether.

4. Conclusion

The resurgence of Islam in many parts of the world ordained female modesty as a symbol of retaliation against Islamophobia and criticizes Muslim stereotypes of female oppression [10,14]. Similarly, modern hijab representations in Malaysian media are preoccupied with the reinvention of the Malay-Muslim community through their cultural resistance to colonial ties. By projecting a hybriditized identity of the hijab, the identity projects Malay supremacy and underterring gender empowerment. While Ang contended that the import of external practices and forms supplemented by media developments are re-organized around "the commercial principle of production of culture for profit" or commodification [29], the growing influence of hijab in a progressive Malaysian environment with increasing media capacities have re-traditionalized the role Malays as modern and more intellectual Muslims [30]. With a more globalized outlook on the hijab as signifier of female Muslims, the accessibility of the Internet played a key role in resonating the modern hijab as a form of self-expression in the minds of urban Malay-Muslim women. The hijab phenomenon in Malaysia can be seen a relapse of identity influenced by the country's Muslim-centric progress. The term "Malay-Muslims" is a form of Malay supremacy in its symbolism, and the shift in power from the last thirty years have also amplified the hijab as a culmination of Malayness in its negotiation with a multicultural society. Through their persistence as a hijabi, veiled Malay women seek to be treated equally and most capably whilst remaining relevant in the global media capitalization on the hijab.

References

1. M. Woldesemait, The Rhetoric of the Modern Hijab, (2012)
2. T. F. Ruby, Listening to the voices of hijab, *Womens. Stud. Int. Forum*, **29**, 54-66, (2006)
3. S. Cherribi, *From Baghdad to Paris: Al-Jazeera and the Veil*, The Harvard Int. J. Pres/Politics, **11**, 121-138, (2006)
4. E. Tarlo, *Hijab in London: Metamorphosis, Resonance and Effects*, J. of Material Culture, **12**, 131-156, (2007)
5. F. Naib, "Sweden's 'hijabista': Selling Muslim fashion," *ALJAZEERA*, (2015)
6. N. Agrawal, "Dolce & Gabbana Launched A Line Of Hijabs And Abayas, But Something's Off," *Huffpost Style*, (2016)
7. E. F. Amrullah, *Indonesian muslim fashion Styles & Designs*, ISIM Review, **22**, (2008)
8. Kate Mayberry, "BBC - Capital - How Muslim headscarves became a fashion empire," *BBC*, 2015. [Online]. Available: http://www.bbc.com/capital/story/20151214-how-muslim-headscarves-became-a-huge-fashion-empire. [Accessed: 12-Apr-2016]
9. S.L. Boo, "Tudung Industry in Malaysia: Cashing in conservative Islam," *The Malay Mail Online*, 2015. [Online]. Available: http://www.themalaymailonline.com/malaysia/article/tudung-industry-in-malaysia-cashing-in-on-conservative-islam. [Accessed: 01-Apr-2016].
10. A. Ong, State versus Islam: *Malay Families, Women's Bodies, and the Body Politic in Malaysia*, Am. Ethnologist, **17**, 258-276, (1990)
11. S. Harian, "HIJABISTA, majalah pertama dunia fesyen berhijab di Malaysia - Nasional - Sinar Harian," 2012. [Online]. Available: http://www.sinarharian.com.my/nasional/hijabista-majalah-pertama-dunia-fesyen-berhijab-di-malaysia-1.41541.
12. N. Hassim and N. L. Khalid, *"Stailo & Sopan : Modesty and Malay-Muslim Women*, Int. Conference on Trends in Soc. Sci. Bus. and Humanities (TSSH-2015*)*, (2015)
13. M. Stivens, 'Family values' and Islamic revival: Gender, rights and state moral

projects in Malaysia," In *Women's.* Stud. Int. Forum, **29**, 354-367, (2006)
14. R. Khalid and M. O'Connor, *The Hijab: Representation among the Muslim Women in Malaysia*," in The Second Asian Conference on Arts and Humanities, (2011)
15. D. Martin, Redefining Malayness: *Expectations of Young- Adult Malaysian Muslims," in Thinking Through Malaysia* : Culture & Identity in the 21st Century, Petaling Jaya: Strategic Inf. and Res. Dev. *Centre*, 31–49, (2012)
16. U. Khattab, "*Saya Happy: Re-reading the Promotion of Female Identity in Local and International Women's Magazines–a semiotic study*," e-Journal Soc. & Behav. Res. in Bus., **3**, 14-28, (2012)
17. F. Boni, *Framing media masculinities men's lifestyle magazines and the biopolitics of the male body*, European. J. Comm., **17**, 465–478, (2002)
18. S. Kassam, *Marketing an imagined Muslim woman: Muslim Girl magazine and the politics of race, gender and representation*, Soc. Identities, **17**, 543-564, (2011)
19. B. Gökarıksel and A. J. Secor, *New transnational geographies of Islamism, capitalism and subjectivity: the veiling-fashion industry in Turkey*," *Area*, **41**, 6-18, (2009)
20. H. C. Loo, G-SEC WORKING PAPER No. 18 Corrupting the culture of denial?: A trans-ethnic cinematic representation and transformation of Malay/sian nationalism in independent film-making, (2007)
21. S. Wok and S. Mohd, *The impact of TV and magazine on fashion and dressing of urban women of different ages*, Jurnal Pengajian Media Malaysia, **10**, 157-170, (2008)
22. G. C. Khoo, *Reclaiming adat: Contemporary Malaysian film and literature.* Vancouver: UBC Press, (2006)
23. R. Robertson, Glocalization: Time-space and homogeneity-heterogeneity, *Glob.* Mod., **25**, 25-45, (1995)
24. F. Terry, *Understanding Global Media.* London: Palgrave MacMillan, (2007)
25. S.L. Boo, "Marina Mahathir: Malaysia undergoing 'Arab colonialism' | Malaysia | Malay Mail Online," 2015. [Online]. Available: http://www.themalaymailonline.com/malaysia/article/marina-mahathir-malaysia-undergoing-arab-colonialism. [Accessed: 04-Oct-2015].
26. N. Hassim, *A Comparative Analysis on Hijab Wearing in Malaysian Muslimah Magazines*, J. South East Asia Rese. Center Comm. and Humanities, **6**, 79-96, (2014)
27. N. Bakeri, "Pelanggan 'merusuh' rebut giliran beli tudung, promosi Fareeda dipertikai," *Astro Awani*, 2016. [Online]. Available: http://www.astroawani.com/gaya-hidup/pelanggan-merusuh-rebut-giliran-beli-tudung-promosi-fareeda-dipertikai-100847. [Accessed: 08-Apr-2016].
28. T. Kulenović, *A veil (hijab) as a public symbol of a Muslim woman modern identity*, Collegium antropologicum, **30**, 713-718, (2006)
29. I. Ang, *Living room wars: Rethinking media audiences for a postmodern world.* (Psychology Press, 1996).
30. A. Ong and M. G. Peletz, *Bewitching women, pious men: Gender and body politics in Southeast Asia.* Universiti of California Press, (1995)

 DOI: 10.1051/shsconf/20173300017

Hero and Villain on a Foreign Land: A Textual Analysis on U.S. Newspaper Coverage of China's Uighur Unrest

Di Zhu[1,*]

[1]School of Journalism, University of Missouri, 65211, Columbia, Missouri, United States

Abstract. This study aims at understanding U.S. media coverage of China's relationship with its mostly Muslim minority group Uighur during the period of several deadly attacks conducted by members from Uighur ethnic group. The study looks into 52 news stories and 3 opinion pieces in the *New York Times* and the *Washington Post* in a six-month period. Drawn from the theories of news framing and news narratives, the analysis found that the news discourse about China's Uighur unrest presented a clear-cut vision of a repressive government and a prisoned intellectual Ilham Tohti. This image conjures two major mythical elements: the repressive villain and the persecuted hero. By attributing the ultimate blame unanimously to government control and policy, the articles employed a resentment frame to portray China's ethnic relationships and downplayed the severity of actual attacks.

1 Introduction

Humans seek to comprehend and make sense of individual events, including tragedies in our world. News, as one major source of public knowledge, provides us with narratives and helps construct a social reality that is coherent and meaningful [1, 2]. International news, in particular, opens a window for people to understand things that happened on the other side of the world. They might be the only source for people to learn about the political institutions and social conflicts that are different from their own. At the same time, journalism, as an "authored text" [1] is never free from values. Journalists embed individual facts into bigger frames and use storytelling techniques to pass on assumptions, attitudes and values.

Two different lines of theoretical development and empirical studies, namely, news framing and news narrative studies, focus on these underlying elements of news and try to explain how journalists use consistent themes, frames, myth and archetypes to diagnose causes and provide moral reasoning [3, 4, 5, 6]. Although framings and narratives research come from different intellectual lineage, they both offered lenses to examine the power of

* Corresponding author: dz78d@mail.missouri.edu

journalism to give meanings, pass on values and provide moral judgments. In this sense, they could be used together to understand news portrayal of international events and offer an insightful account of ethnicity, conflicts and politics, because international reporting often requires journalists to provide more contexts and explanations. This study tries to combine theoretical contributions of news framing and news narrative research with an analysis of the coverage of deadly attacks conducted by ethnic minority groups in China.

In the year of 2014, several deadly attacks conducted by people from Uighur ethnic minority group in China have drawn international attention. Among them, the Kunming Train Station mass stabbing in March left more than 30 innocent passengers dead and more than 140 injured. In May 40 civilians were killed ruthlessly in a market attack in the capital city of Xinjiang Province. Following attacks happened in rural areas of Xinjiang province later, and the two most deadly ones happened in the county of Shache and Luntai, where groups of attackers started with assaulting government offices and police stations, and both Uighur attackers and civilians were killed in these clashes. Uighur is one of China's 55 ethnical minority groups, and most of its people live in the country's northwestern province Xinjiang and practice the religion of Sunni Islam.

Chinese government and media called these attacks terrorist attacks incited by separatism forces while the U.S. news coverage of these incidents and China's relationship with its ethnical minorities paints a different picture. Scholars have argued that news discourse often does not and cannot capture an objective reality. Instead, it highlights certain aspects of an issue and downplays others by employing frames and telling tales [3, 7, 8]. U.S. newspaper coverage of these attacks and why they happened can reflect and confirm not only the nature of these events, but also the values and beliefs that are consistent with a shared reality about China in the American public.

This article studied the *New York Times* and the *Washington Post*'s coverage of Uighur groups against the background of the unrest and terrorist attacks. The article will first trace out how terrorist and terrorism were often portrayed and how issues about China were discussed in U.S. media. The concept of news myth and news framing were adopted to analyze important events under each category. Then the fifty-five news stories and opinion pieces on the Uighur attacks will be examined in a textual analysis to explore how the story of terror in China were framed and retold in U.S. mainstream media.

2 Literature Review

2.1 Media coverage of Muslim group and terrorism

Previous studies on U.S. media's coverage on Muslim groups found that news media tend to quickly associate terrorism with Muslim and Islam and feed a narrative emphasizing fear and violence after the terrorist attacks of 9/11 [9, 10]. This negative association in media magnified the reality of "Islamic extremism" and reinforced an Orientalism ideology that widened the divide between "Self (Western world)" and "Others (So called Orient that are different from Western world)" [11, 12].

One way to look at this misrepresentation of terrorism is the framing theory. News media constantly employ "frames" to tell stories and making sense of complicated events. Scholars argue that news frame works as an interpretive structure and help journalists identify problems, assign values and provide solutions [13]. Intentionally or unintentionally, news writers emphasize certain aspects of the issue while excluding other factors that they consider irrelevant or less important. The coverage of international tragic events like terrorist attacks often adopted a frame that is consistent with U.S. foreign policy or administrative stand. Certain policy catchphrase invented by government administration, such as the "War on Terror" label, can be internalized by journalists in their daily discourse, reified as an uncontested thing and naturalized as a taken-for-granted worldview in the news coverage [14].

As media become an integral part of American politics, journalists adopt the language and perspective of the political establishment, particularly in foreign affairs coverage Sometimes the news coverage can in turn influence politics. Entman examined U.S. media coverage of the shooting down of two passenger planes in 1980s, one shot down by a Soviet fight plane, another shot done by a U.S. Navy ship, and found that U.S. media outlets employed different frames in telling the stories [4]. By eliciting moral evaluations in the former story and framing the later one as a technical accident, these two kinds of frames reinforced and strengthened public belief of "Soviet's moral guilt" and legitimized U.S. military policies in the Gulf [4].

The way news stories are structured and produced can also be understood through the perspective of myth. Myth exists in journalists' storytelling traditions and underlies and sustains certain ideologies and social values [8]. Bird and Dardenne [3] believed the journalistic use of myth, the illusion to heroes, villains, scapegoats and archetypal principles and figures, has a function of explaining and comforting. They argued that, "one function of myth is to explain which cannot be easily explained [...] as well as more intangible things, such as notions of morality, appropriateness and fairness. This [...] was a function of people's intolerance for randomness, inexplicability and ambiguity. [3]" U.S. media have constantly resorted to myths to make sense of traumatic events in its own history. Lule [8] studied the major myths used in the New York Times editorials after the September 11th terrorist attack. He pointed out that the allusion to shattered innocence, the victims, the heroes and the anxiety about future constructed a strategy to proclaim and sustain social order. Berkowitz [15] explored the mythical narratives in news stories and opinion pieces after the Virginia Tech shooting and found that the reference to the Holocaust and the allusion to an ironic hero who survived the Holocaust but died in the shooting attack have served a function in the public healing process.

Most of existing studies on Muslim and terrorist attacks focused on the narrative of incidents happened on the soil of U.S., where there are major tensions between American people and Muslim individuals (or groups). U.S. news discourse on Muslim in other countries might adopt a different strategy. The present research looked into several disastrous attacks happened in China and explored how U.S. media employ frames and myth in making sense of a terrible event that happened outside United States.

2.2 Media coverage of China

An analysis of U.S. media coverage of attacks happened in China cannot overlook the important tradition of how China was often portrayed in U.S. news discourse. U.S. media coverage of issues in China have been found to be related to the international relations between the two countries and U.S. foreign policy at different times [16, 17, 18]. Retzinger and Scheff [19] pointed out that a country's ideology both produce and is produced by the narrative that it tells to itself and the others about conflicts. Therefore, a look into the news rhetoric about China cannot be fully understood without acknowledging the two countries political, economical and ideological differences and competition. U.S. International Relation theorists argues that China's growing capabilities has become a potential threat for

U.S. economy and in the competition for global powers [20, 21]. Pan [22] proposed that the "China threat" literatures are "not value-free, objective descriptions of an independent, preexisting Chinese reality out there, but are better understood as a kind of normative, meaning-giving practice that often legitimates power politics in U.S.-China relations and helps transform the 'China threat' into social reality."

Turner [23] conducted case studies on the representations of China in different historical periods and showed that the image of a dangerous China helps reaffirm U.S. identity from time to time. Their studies on IR literatures suggest a myth of "China threat" that was imagined by U.S. politicians and scholars and legitimized U.S. ideology as universal and indispensable. Compared to international relation study literature and political speeches, news media play an even more important role in constructing reality for American public and reinforcing the ideology and moral values of U.S. political system in the world. Therefore, the rhetoric of an internationally influential elite media such as the *New York Times* and the *Washington Post* can be seen as a relatively subtle, powerful and natural way to influence public beliefs and values. Therefore, the ways that this article addresses and implies about China's policy and ideology will be another focus of this study.

Besides these studies on International Relations literature, scholars also studied the U.S. news discourse on several important events in China and argued that media framing and invocation of these singular incidents should be seen as a representative part of the overall media framing about China [24, 25]. Lee et al. [26] examined *New York Times* and *Washington Post*'s editorial use of and reference to the 1989 Tiananmen protests and its crackdown as a "news icon" over a period of 20 years and found that this event was initially used as a powerful symbol of totalitarian dictatorship in China, then becomes a symbol of China's human rights violation in 1990s, and at last turn into a less-powerful ritualistic memory of China's repression in news editorials. Another analysis of media frames employed on reporting the Hong Kong's reunification with China in 1997 found that the coverage of the event can be summarized into ideological packages that projected Hong Kong's loss of freedom under Chinese rule and its potential role as a "Trojan horse" to spread political and economic changes in China [27].

The previous researches on media coverage of both terrorism and China have informed the coding and analysis of this study. The present study thus asked two general research questions: 1). How were the Uighur attacks and violence in China framed in U.S. mainstream media? 2). To view all the news stories and opinion pieces on this issue as integral parts of a constructed reality, how was blame and responsibility attributed and the hero identified in the form of myth? A textual analysis will be used to answer these research questions.

3 Methodology

This study examined 52 news stories and 3 editorial pieces on the *New York Times* and the *Washington Post* about the relationship between Chinese government and Uighur minority during the period of the four most deadly Uighur attacks in 2014 (from the first attack in March 1st to one week after the fourth attack on September 21st). Both the *New York Times* and the *Washington Post* have long been regarded as the most influential national newspapers that influences and represents U.S. foreign policies and set the political agenda for public issues [28, 29, 30]. Therefore, their coverage of China and its ethnic conflicts can be seen as an indication of U.S. foreign policy toward China and a reflection of the public narrative about China in the mass media.

The search term "Xinjiang" and "Uighur" are used together to find as many related articles as possible and yielded a result of 99 stories in total. The researcher then went through every article and eliminated all the articles that is not related to the attacks or violence. Words like "unrest," "tension," "clashes" and "violence" were all considered as an allusion to the attacks; articles with these words were thus included in the final group of articles.

The texts were then analyzed with a coding sheet designed both deductively and inductively (See Appendix A). Questions such as "Did the article use a thematic or episodic frame to describe the attack(s)," and "How was the blame of the attack(s) attributed?" were drawn from the framing theory literature [7, 31]. Questions such as "Who was identified as the villain in this article?" "Is there a hero in the story, how was he/she described?" were drawn from the news myth literature [3, 8, 15]. Quotations in all the articles were closely examined, too, as journalists also constantly filter their sources and make decisions on whom to quote. Quotations, as well as background information provided in a story, help build the theme and support a point a view in news stories [31]. Due to the specific subject of this study, the coding sheet also inductively include items about the image of Uighur people, the image of Xinjiang region, the label of the attacks (whether it used the label of terrorist attacks or not) and the phrase used to describe the relationship between Chinese government and Uighur minority group to fully capture the constructed reality about the violence, the hero and an "repressive" China in U.S. media.

Because one person coded all the selected articles based on the code sheet, the study examined individual words and phrases to ensure that the data were analyzed rigorously. Representative wordings in the article are also presented in the analysis to ensure transparency. For example, to code for the question concerning villains in the article, the researcher looked for specific phrases that directs blame or depicts an oppressive power. Sentences and words that are ambiguous are not included in the analysis.

4 Analysis

The articles in New York Times and the Washington Post covered a variety of aspects in the unrest. The themes of the articles ranged from the attacks themselves, Chinese government's response toward the attacks, feature stories about the region's Xinjiang and Uighur minority groups, the arrest and trial of suspects, and interestingly, the arrest and trial of a Uighur scholar Ilham Tohti. Out of 55 articles, only 15 focused on the violent acts themselves. And even in the article that talks about individual attack, the journalist tend to adopt a thematic frame instead of an episodic frame to discuss the causes and background information. Violent attacks in China and Uighur minority groups are both distant and unfamiliar concepts to most of American readers, therefore it is necessary for the journalists to provide some context to make sense of the incidents and help readers understand potential factors of the crime. In this sense, most of the articles offered some explanation on the attacks and its consequences. A detailed description of the myths and frames used in these explanations will be discussed as followed.

4.1 The Repressive Villain

In a majority of the news stories and opinion pieces, the blame of violence and unrest is partially or fully attributed to the policies of Chinese government or the Communist party. A picture of an oppressive central government is the repeating theme in all the stories. The unrest was described as "an inevitable result of Beijing's continued repression" ("Pushing the Uighurs too far," *NYT*, March 19). Some articles even put it more openly, saying, "The escalation of violence is the direct result of China's repression" ("The Price of China's Repression," *NYT*, September 26). China's policies were often described as "Beijing's iron rule" or "Beijing's heavy-handed tactics" in a lot of articles, while Uighur people were depicted as victims "chafed under Chinese rule" ("Ethnic Tension in China's Xinjiang

DOI: 10.1051/shsconf/20173300017

Province," WP, March 3) and "backed into a corner by Beijing's relentless antagonizing tactics" ("Pushing the Uighurs too far," *NYT*, March 19). Even in articles that tried to show two sides of the arguments, as Chinese government blame the attacks on Islam extremists while Uighur exile groups blame them to Chinese government's crackdown, the Chinese official's accounts were always put into less important position and called into question later with an indication that there's something more severe and scary than the recent deadly attacks on civilians:

Chinese authorities described the mass slaughter as a "premeditated, violent terrorist attack" perpetuated by separatists from Xinjiang. [...] Clashes over the past year have claimed more than 100 lives, many of them Uighurs killed by security forces during what officials describe as terrorist attacks. Uighur exile groups blame for the bloodshed on paramilitary police officers they say have been given the green light to use deadly force against unarmed protest. ("China Remodels an Ancient Silk Road City, and an Ethnic Rift Widens," NYT, March 6)

It is interesting to note that the government account of the attacks were always put in quotation marks while the Uighur groups and Western experts' point of view were often presented without a quotation mark. In this way, the reporters kept their distance from whatever Chinese government and media says about the attacks and legitimize the narrative from Uighur exile groups.

Related to this issue is the emphasis on China's propaganda efforts, which suggest a deceitful agenda behind all the investigation and responses to the attacks. In all of the articles, the blame to the separatist (or terrorist) groups was always immediately followed by "says government-run Xinhua News Agency" or "according to a government-run website." In this way, the intricate relations between Chinese government and any media outlet were always identified openly. However, the background of other sources, such as the Radio Free Asia (a U.S. government funded radio station), was not explained.

Furthermore, lengthy descriptions about Chinese government's propaganda schemes become a recurring context in a lot of the articles. Some stories went in-depth about the specific propagandas the government has executed: "the state media showcased a new 'harmonious village' where the two ethnicities would live side by side" ("Uighur Scholar's Life Sentence is Seen as China's Effort to Stifle Debate on Minorities," *NYT*, September 25). Chinese government's efforts to educate Uighur women and provide them with training classes were described as "Beijing has spent heavily wooing women in Xinjiang" and "cheerful propaganda" ("Veils Emerge as Sign of Uighur Protest in China," *NYT*, August 8).

One of the *New York Times* article devoted all its content on how "party-run history machine" made up historical stories from the real cause of country's disastrous famine to the blaming of United States for starting the Korean War. It summarized, "China's Communist Party has devoted enormous resources to composing historical narratives that seek to legitimize its rule and obfuscate its failure" ("In China, Myth of Social Cohesion," *NYT*, August 19).

By identifying most of the governmental policies and programs as propaganda and making the propaganda as the most salient background of the attacks and unrest, these news coverage and editorials discredit explanations made by Chinese government and suggest that the repression and propaganda is something more important and more horrible than attackers killing innocent people on the street. And the attacks were just an extreme form of "Uighurs resistance to rule by Han majority."

Although some of the articles did mention an "Islamist militant group" blamed by the Chinese government for the attacks, but the author always went back to blame Chinese government's suppression as the ultimate explanation for all the violence. For example, in one of the *Washington Post* article that focused on the jihadist threat posed by Uighur

DOI: 10.1051/shsconf/20173300017

extremists, the author spent the first half of the articles talking about how U.S. and China might have a common interest in cooperation against militant from Islamic State, and then spent the second half of the article condemning China's domestic policy:

But China's focus on the external threat inspiration for some of the violence in Xinjiang is missing the bigger picture, experts say. The long-running and brutal suppression of Uighur rights, culture and nationalist sentiment has bred deep resentment here while an intensified campaign to educate people by turning them away from Islam — preventing women from wearing veils and students from attending mosques for example — seems to have caused even more anger ("China Faces Jihadist Threat at Home and Abroad," WP, September 8).

In sum, most of the articles employed a frame of repression and fear in the coverage over the course of six month in between these four attacks. Interestingly, the fear is not directed towards the attackers (if not terrorists), in fact, the attacks were often used merely as a background to the portrayal of China's ruthless domestic policies and silencing of dissidents. The narrative of most of the articles identified a bigger source of fear, a villain government, who carried out crackdowns, surveillance and propaganda, a villain who pursued economic and political policies that were unfair to its minority. And this villain is the real cause for all the resentment that leads to violence.

4.2 The Imprisoned Hero

The hero in this narrative is, naturally, the oppressed. But he is not any kind of Uighur under government surveillance and control. The hero, Ilham Tohti, is the intelligent and politically active scholar who was found guilty of "inciting ethnical separatist" and prisoned by the Chinese government. Most of the articles identified Ilham Tohti as a "Uighur economist" or "the most prominent Uighur scholar," which give him a sense of intellectual credibility of speaking out for Uighurs and calling for political changes. He taught economics at the Central University of Nationalities and had a website that tried to encourage communication between Uighur and Han people; the latter is the majority ethnical group in China. He was arrested in January for inciting students to violence. Tohti is a recurring character in most of the articles, not only in the fourteen articles that focused on his trial and conviction. His name is also frequently mentioned in other articles that talk about the attacks and governments' strategies to tamp down the violence as an archetypal figure of the victim under government repression.

The news stories described his careers and works as "building bridges between Han Chinese and Uighurs" and "a model of moderation." By portraying him as a bridge builder and moderate dissident who does not want to do any harm to the unity of China, the repressive villains' allegation and strategy seems even more irrational and outrageous: "The sentence [of Ilham Tohti] is a sign of confusion and desperation behind the governments' policies toward Uighurs" ("Fruitless Repression," NYT, September 29). The articles that focused on his trial and sentence also consistently gave Tohti benefit of doubt by devoting a majority of their content quoting his layer, wife and his daughter who studied in United States. Foreign experts and groups were also frequently quoted to condemn China's unfair treatment and support Tohti's innocence, while the actual charges against him were often accompanied by the word "reportedly."

In addition, the situation of Tohti's imprisonment was described in detail with similar structures across publications. Articles in both the *New York Times* and the *Washington Post* talked about him being shackled, his hunger strike, and how prison officials have refused him food in the following days. To this point, the news discourse further changed the direction of attention from attackers' constant mass slaughtering during these six months to the dreadful accusation of a single intellectual by the repressive government.

4.3 The Attacks and the people as a backdrop

As mentioned earlier, brutal attacks were not the focus in most of the articles and were used just as a background to discuss other issues. Sometimes they were mentioned as a lead to the criticism of China's "anti-terrorism steps;" other times they were brought up to explain Ilham Tohti's arrest. In one of the articles that focused on Ilham Tohti's imprisonment, the whole article only mentioned one attack once (the Kunming train station deadly stabbing) at the end of one sentence:

[Ilham Tohti] went on a hunger strike in January for 10 days because his food was not halal — prepared according to Muslim religious traditions — and was denied food by prison authorities for a similar period after Uighur separatists carried out a deadly knife attack in the city of Kunming in March ("Uighur Scholar Faces Chinese Trial that Critics Say Highlights Repression," WP, September 17).

In fact, during this period of six months, and out of 55 articles, only 15 news stories were talking about specific attacks. The rest of the articles gave the attacks a secondary role and a majority of them rejected the label of terrorist attacks. The incidents were called "big, bloody race riot," "a deadly wave of ethnic violence," "mass slaughter," "train station rampage." The description of the attacks and the restive situation still invokes fear, but by dissociating them from terrorism, the responsibility of the violence were wholly attributed to bad government policies.

The attacks were also constantly associated with the label "protests" and framed as a response to political repression instead of malicious sabotage. This kind of frame can be best summarized in one article, the reporter claimed, "their protests are a reaction" ("Clashes, anti-terrorism steps worsen China's Uighur Ties," *WP*, May 11). Another article went back to an earlier attack in 2009 and elaborated the course of events in detail: "after a heavy-handed police response, the protests spiraled into ethnic bloodletting, claiming nearly 200 lives, most of them Han"("Chinese Officials Seek to Shift Attention From Rampage," *NYT*, March 5). The attacks and violence, according to the news narratives, becomes some kind of last resort and a side effect of peaceful protests for a just political cause.

Besides the backdrop of constant violence, these articles' depiction of the relationship between Chinese government and Uighur minority groups focused on a frame of resentment and fragmentation. The word "resentment" and "hatred" were used frequently when talking about Uighur people's attitudes toward both central government and the majority Han people. The articles also suggested a fundamental divide instead of cohesion between Uighur minority groups and the rest of the country: "Xinjiang, rich in resources and strategically important, lies on the borders of Central Asia and is home to the mainly Muslim Uighurs, who speak a Turkic language and are culturally distinct from China's ethnic Han majority" ("Market Attack Kills 31 in Far Western China," WP, May 23). Similar depiction of Xinjiang and Uighur people appears again and again in articles to make the cultural, linguistic and religious divergence between Uighur and Han the most salient characteristic of the region and the people.

The background information about Xinjiang as an "ethnically divided region that has been convulsed by mounting violence" and about the Uighur-Han relations as "an ethnic rift" seems to suggest that the region and the people have irreconcilable differences with the rest of the country. This narrative downplays any similarities, friendships and cooperation and domestic unity among different ethnical groups or discredits them as government propaganda. In this way, the call for independence and "autonomy" were legitimized and the strategies to identify and restrict separatist forces were portrayed as a sign of intrusion and outside repression.

5 Discussion

This study present the stories told in the *New York Times* and *Washington Post* presented about China's Uighur unrest. These stories offered a clear-cut vision of a repressive government and a prisoned intellectual Ilham Tohti. The depiction conjures two major myth elements: the repressive villain and the persecuted hero. By attributing the ultimate blame unanimously to government control, unfair policies and even the problematic political system, these articles framed the attacks as unpreventable tragedies caused by the government's repression. The attackers' role as villain was blurred and downplayed; China's crackdown on what they believed as terrorist movement was so severe and inhumane that it surpassed the priority of covering the actual events. Even in the actual coverage of singular events, to understand the cruel tragedies happened in another country, U.S. media resorted to myth to make sense of the event, and by portraying China as an authoritarian society under moral scrutiny of U.S. democratic standards, the villain myth also helped U.S. to distance itself from this kind of tragedy. This finding is consistent with what Lee et al. [26] found in their study about how U.S. mainstream media has associated the Tiananmen as a symbolic reference to an Authoritarian government and its human rights abuses. The coverage of the Uighur unrest reinforces and reuses this already constructed image of a repressive China and its human rights problems.

In addition, the articles also employ a frame of fear and resentment in its coverage of the relationship between Uighur groups and the Han-dominated government. The narrative indicates a hostile situation among all the groups in Xinjiang and project a future of dreadful backslash and more severe clashes. At last, different from the common negative portrayal of Muslim people in U.S. media, the image of the Muslim Uighur are depicted in a sympathetic light as they, as an entire ethnic group, undergone the heavy-handed repression of Chinese government. An intellectual from Uighur group, Ilham Tohti, were given great attention in the coverage and his argument about the unfair economic and political conditions of Uighurs were reiterated in a lot of articles. This positive and sympathetic depiction of Uighur people's life and their marginalization as a minority group provided counter narrative for Muslim groups in U.S. media.

The conventional frames adopted in U.S. media's coverage of terrorism often provided a one-sided context that make one simple explanation seem natural and render contradictory interpretations hard to fit into the preexisting viewpoint [32, 33]. The finding of this study suggests that there could be alternative frames in covering violent attacks. By empathizing with Muslim minorities who faced economic and political predicament and individuals who were imprisoned by government due to national security reasons, the news media can provide a better platform for multiple communities to express the meaning and interpretation of similar events.

The authors of these articles from time to time employed different frames in their arguments. For example, one article in the New York Times pointed out in the end that, "management of diversity and pluralism is a pressing world issue, from Scotland, Ukarine to Ferguson, Mo" ("Fruitless Repression," NYT, September 29). To see the problem from an international perspective and to compare it with other similarly complicated issues, even problems facing United States itself, is to veer away from targeting a single villain and ignoring other aspects of the factors that caused violence. However, most of the times, even in this editorial piece, the author came back to imposing an U.S. dominant ideological to evaluate foreign issues.

This study has several limitations. First, it only analyzed U.S. mainstream newspapers' coverage of Uighur attacks happened in China. Future studies could use a comparative method to see if there are differences between the coverage of attacks conducted by minority groups in China and attacks happened in U.S. or other Western countries. This study of the *New York Times* and the *Washington Post* coverage of foreign issues also raise further questions on how it corresponds with U.S. foreign policy or how it can influence public opinion. A more detailed study that combined an examination of Sino-U.S. relationships and with the media coverage of attacks (the earliest deadly attack started from 2009) over the years can be conducted to investigate this relationship.

References

1. B. McNair, *The sociology of journalism*. New York, NY: Arnold. (1998)
2. M. Schudson, *The power of news*. Cambridge, Mass: Harvard University Press. (1996)
3. S.E. Bird, R.W. Dardenne Bird, Myth, chronicle, and story: Exploring the narrative qualities of news. In J. W. Carey (Ed.), *Media, myth and narratives*. Beverly Hills,CA: Sage. (1988)
4. R.M. Entman, Framing U.S. coverage of international news: Contrasts in narratives of the KAL and Iran air incidents. J. Commun. **41**, 4, 6-27 (1991)
5. W. A. Gamson, A. Modigliani, Media Discourse and Public Opinion on Nuclear Power: A Constructionist Approach. AJS **95**, 1, 1 (1989)
6. S. Hall, The narrative construction of reality: An interview with Stuart Hall. S. Rev. **17**: 3-17 (1984)
7. R. M. Entman, Framing: toward clarification of a fractured paradigm. J. Commun. **43**, 4: 51-58 (1993)
8. J. Lule, Myth and Terror on the Editorial Page: The New York Times Responds to September 11, 2001. JRMC **79**, 2, 275–293 (2002)
9. K. A. Powell, Framing Islam: an analysis of U.S. media coverage of Terrorism since 9/11, Commun. Stud. **62**, 1, 90-112 (2011)
10. P. Morey, Y. Amina, *Framing Muslims: Stereotyping and Representation after 9/11*. Cambridge, Mass: Harvard University Press. (2011)
11. E. W. Said, *Covering Islam*. New York, NY: Pantheon Books. (1981)
12. H. K. Karim, *Islamic peril: Media and global violence*. New York, NY: Black Rose Books. (2000)
13. R. M. Entman, J. Matthers, L. Pellicano, Nature, sources and effects of news framing, in Karin Wahl-Jorgensen & Thomas Hanitzsch (ed.) *Handbook of Journalism Studies*. New York, NY: Routledge. (2009)
14. S. C. Lewis, S. D. Reese, What is the War on Terror? Framing through the Eyes of Journalists. Journalism Mass Comm. **86**, 1, 85–102 (2009).
15. D. Berkowitz, The Ironic Hero of Virginia Tech: Healing Trauma through Mythical Narrative and Collective Memory. Journalism. **11**, 6, 643–59 (2010)
16. T. K. Chang, The impact of presidential statements on press editorials regarding U.S.-China Policy, 1950-1984. Commun. Res. **16**, 4, 486-509 (1989)
17. Y. Yu, D. Riffe, Chiang and Mao in U.S. news magazines. Journalism Quart. **66**, 913-914 (1988)
18. Z. Peng, Representation of China: An across time analysis of coverage in the *New York Times* and *Los Angeles Times*. Asian J. Commun. **14**, 1, 53–67 (2004)
19. S. Retzinger, T. Scheff, Emotion, alienation, and narratives: Resolving intractable conflict. Mediation Quart. **18**, 1, 71–85. (2000)
20. H. Harding, The concept of "Greater China": Themes, variations and reservations, in David Shambaugh (Ed.) *Greater China: The next superpower?* 31-50 Oxford: Oxford University Press. (1995)
21. D. Roy, Rising China and U.S. interests: Inevitable vs. Contingent hazards," *Orbis 47*. **1**, 130-151 (2003)

DOI: 10.1051/shsconf/20173300017

22. C. Pan, The "China Threat" in American Self-Imagination: The Discursive Construction of Other as Power Politics, Alternatives: Global, Local, Political **29**, 3, 305-331 (2004)
23. O. Turner, "Threatening" China and US security: the international politics of identity. Rev. Int. Stud. **39**, 4, 903–924 (2013)
24. G. C. Stone, Z. Xiao, Anointing a New Enemy: The Rise of Anti-China Coverage after the USSR's Demise. Int. Commun. Gazette **69**, 1, 91–108 (2007)
25. C.-C. Lee, Established Pluralism: US Elite Media Discourse about China Policy. Journalism Stud. **3**, 3, 343–57 (2002)
26. C.-C. Lee, H. Li, F. L. Lee, Symbolic Use of Decisive Events: Tiananmen as a News Icon in the editorials of the elite U.S. press. Int. J. Press/Politics, **16**, 3, 335–356 (2011)
27. C.-C. Lee, Z. Pan, J. M. Chan, C. Y. K. So, Through the eyes of U.S. media: banging the democracy drum in Hong Kong. J. Commun. **51**, 2, 345–365 (2001)
28. S. Hess, *International News and Foreign Correspondence*. Washington: Brookings. (1996)
29. D. Graber, *Mass Media and American Politics*. Washington: Congressional Quarterly Press. (1997)
30. H. J. Gans, *Deciding What's News*. New York: Pantheon. (1979)
31. Z. Pan, G. M. Kosicki, Framing analysis: an approach to news discourse. Polit. Commun. **10**, 55-75 (1993)
32. K. S. Johnson-Cartee, *News narratives and news framing: constructing political reality*. Lanham: Rowman & Littlefield Publishers. (2006)
33. S. Iyengar, *Is Anyone Responsible? How Television Frames Political Issues?* Chicago, IL: University of Chicago Press. (1991)

DOI: 10.1051/shsconf/20173300018

Carl Jung's Archetypes in Malayalam Film: A Case Study on The Film 'Urumi'

Sona Varghese [1,*] and *Arumugam* Balasubramanian[2]

[1,2]Department of Communication, Coimbatore, Amrita Vishwa Vidyapeetham University, Tamil Nadu, India

Abstract. Movies are the visual- auditory symbolic narrative that explains the living reality of people. Films employ Carl Jung's concept of archetypes-prototypical characters. This research provides an insight about the theory of archetype based on the assumption that archetypes resides deep inside human mind. The researcher proposes that the Malayalam film industry in India showcases the western concept of archetype using the movie Urumi as a case study. For this study, different clips are taken from the movie to represent six essential archetypes, namely, Hero, Anima, Animus, Mentor (sage), Shadow, and Trickster. The psychoanalysis of this movie carried out in this paper provides evidence for the existence of all six essential archetypes identified by Carl Jung. Thus this article will be a fruitful resource for future research in the field of psychoanalysis of movies.

1 Introduction

The mass media has gone through various development and refinement in past centuries. The contemporary information age is dominated by mass communication media such as movies, television and internet [1]. The stories and characters presented in this mass media are important for sharing human experience. People use movie plot analogies and metaphors in their life due to its effect on the mass audience [6]. The phenomena of using the examples of movies and other mass media in ordinary life, points to the existing internal psychological forces that allows people to connect with these commonly shared metaphors and analogies.

The shared cultural phenomenon which unites people in a thread is the archetype. The concept of archetypes is taken from collective unconsciousness, a theory developed by the Psychologist Carl Gustav Jung. In collective unconsciousness, Jung suggest that, all human beings share a deep level of unconscious mind universally which is hidden beneath the personal psyche [16]. The collective unconsciousness contains modes of behavior and some contents which are identical in every human being that include a common psychic content of a universal nature present in individuals across the globe. Thus, human beings shows a tendency to perceive common meanings embodied in a particular symbol as same, even at the unconscious level [2].

* Corresponding author: a_balasubramanian@cb.amrita.edu

Jung developed the concept of archetypes in 1964. Archetypes were defined as the components of collective unconsciousness which is an inborn tendency acquired in the unconscious mind. They exist in psych and prepare individuals to deal with life experiences psychologically that are common for everyone [22]. An extreme analogy of describing archetypes will be as the structure of psyche, which is similar to organs inside physical body [13].

Archetypes are represented as ancient motifs and preconceptions about the behavioral patterns that are symbolically manifested as archetypes in all cultural forms, arts and in dreams. Jung formulated and identified archetypes from his patients and from his own emotions by verbalizing the visualized images for certain concepts [16]. Drawing Mandala archetype symbol from Hinduism helped Jung to represent 'The self'. Mandala is also considered as a telematic design of cosmos and consciousness which represents high spirituality among humans [11]. The other archetypes include, Anima, Animus, Sage and Mother, and they have been identified as the major elements in the common language which involves the stories people communicate with others [6], [18]. Archetypal characters assert certain characteristic motives and other qualities to promote actions that everyone can recognize.

In modern day media, these archetypal figures take form through visual narratives. Movies are the visual- auditory symbolic narrative that explains the living reality of people [4]. Based on Jung's theory, movies have become a resource for studying archetypes in popular culture and contemporary society. Cinema offers the content and also the means for psychotherapeutic sessions which enables the viewer to identify their psyche with the projection in screen. The cinema also claims to be a platform in the contemporary society where people can set apart from their daily life and connect with their unconsciousness similar to a session of hypnosis or dreaming [9].

A number of studies have proposed and proved the presence of archetypes in the stories of various cultures [14]. The archetypal analysis on movies has been conducted in the western film industries to have an insight on the psychoanalysis of movies. The studies include archetypal analysis of Jim Casey in the Grapes of wrath. In the analysis the author has compared Jim Casey with Christ. The Christ is considered as a leader, forerunner and a sacrificial figure who sacrifices his life for cleansing the sins of others and was in a constant battle with the Satan [7]. The other study is on the famous adventurous movie series Indiana Jones. The heroic myth by Joseph Campbell is influenced by the archetypal situation of Jung. This was later shortened by Christopher Vogler and is present in Indiana Jones than in any other movies. In the movie, Indiana Jones; Raiders of Lost Ark', the first stage called ordinary world is established through showing the life of Jones as a college professor [5].

The movies across the globe have some universal characteristics especially in the context of psychology and therefore can be applied in the Malayalam film industry in India. The study focuses on conducting psychoanalysis of Malayalam movie Urumi based on Carl Jung's Archetypes and monomyth.

2 Urumi

Urumi is the Malayalam movie filmed, directed and co-produced by Santhosh Sivan. The movie is the recent release in the genre, Indian epic historical drama of Malayalam film Industry. The film is an adoption from the real life of a patriotic warrior Kelu Nayanar in 16th century whose father was killed by the explorer Vasco da Gama and his troops. The movie is set on two different centuries, the contemporary 21st century and 16th century. The main characters and their mirror reflections along with similar situations connect one era to the other. The story mainly focuses and narrates the journey of Kelu Nayanar (Prithviraj) and his journey to avenge the death of his father and to redeem the freedom of Indians fromVasco, who have taken the authorization of Indian Ocean and his motherland. He is accompanied by his friend Vavali (Prabhu Deva) and a warrior princess Ayesha of Arakkal (Genelia D'souza). The story set in 16th century is connected to the contemporary world through the predecessor of Kelu Nayannar and his mirror reflection Krishna Das in the 21st century.

3 Objectives

The researcher through this study aims to identify and analyze portrayal of Carl Jung's six archetypes in the Malayalam film Urumi.

4 Analysis and Interpretation

Table 1. Archetypes character identified in the movie

Archetypes	Character identified in the movie	Time duration of clips selected
Hero	Chirakkal Kelu	00:14:20-00:22:44, 00:28:00-00:30:20, 00:34:25-00:35:30 00:41:00-00:45:00, 00:58:00-01:02:00, 01:22:50-01:24:00 01:47:00-1:50:30, 01:54:11-01:54:29, 01:57:00-01:59:25 02:15:22-02:15:42, 02:26:20-02:29:05, 02:33:15-02:41:00 02:41:28-02:42:48, 02:44:30-02:44:42, 02:45:07-02:45:07
Mentor	Chirakkal Kothuval	00:15:00-00:15:52, 00:17:00-00:17:30, 00:22:00-00:22:40 00:24:30-00:25:00, 00:34:30-00:35:00, 00:41:00-00:45:00
	Bhoomi	00:08:15-00:11:15, 01:05:30-01:10:00
	Vavvali	00:35:00-00:35:20, 00:42:30-00:42:40, 01:03:30-01:04:50 01:26:00-01:26:40, 02:30:00-02:31:00.
	Arackkal Ayesha	01:44:30-01:45:45
Trickster	Chennichery Kurup	00:37:45-00:38:20, 00:45:00-00:45:30, 01:13:20-01:13:40 01:14:54-01:18:00, 01:31:30-01:32:20, 01:40:30-01:43:26 01:52:30-01:52:44, 01:53:40-01:54:10, 01:54:30-01:55:48 02:04:36-02:06:20, 02:10:35-02:11:25, 02:12:45-02:15:20 02:22:20-02:24:43, 02:42:50-02:43:45, 02:44:15-02:44:30 02:44:43-02:45:07
Shadow	Vasco da Gama	00:17:40-00:20:45
	Estavio da Gama	00:54:25-00:56:40, 01:31:10-01:31:30, 02:21:45-02:23:30 02:29:10-02:29:50, 02:39:40-02:40:20
Anima	Chennichery	00:36:00-00:36:20, 00:37:05-00:38:22,

	Kurup	01:11:10-01:12:30 02:13:00-02:15:00, 02:22:20-02:24:20
Animus	Arackkal Ayesha	00:50:30-00:53:25, 00:57:30-00:58:23, 00:58:55-01:00:10 01:20:55-01:22:00, 01:24:00-01:24:15, 01:26:45-01:27:50 01:33:15-01:35:21, 01:43:34-01:45:45, 01:49:35-01:50:10 01:53:21-01:53:36, 01:57:08-01:59:15, 02:18:40-02:20:20

The table 1 shows the collection of clips of each character, under the six archetypes based on which we conducted the analysis. The hero, mentor, trickster, shadow, anima and animus are identified and the clips are taken based on the definitions given by Jung. The analysis is based on each clip mentioned in this table with supporting statements of definitions.

4.1 Hero

The word hero has connotations with adversity, challenges, honour, strength and victory [3]. It is the duty or responsibility of the hero to make the wrong things right and make his world into a utopian paradise. Kelu Nayannar (Krishna Das in mundane world) is the hero archetype identified in the movie Urumi. The Hero archetype has an alluring character since childhood. He has to contend with forces that would deny him his own independence and self-direction [12]. In the first part at magical world, Kelu Nayannar's childhood is shown when Vasco Da Gama arrives with warships to attack Kerala. In this part the situations which change the entire life of Kelu and make him an orphan is portrayed keeping his father Chirakkal Kothuval as the central figure and an ideal for his son Kelu. The cruelty imposed by Gama and the murder of his father shakes his roots and force him to give over his independence and direction of life. Thus he develops contend for Gama and his troops which took his independent life and father. Kelu grows up to avenge for the death of his father and to redeem his and his motherland's freedom from whites.

The hero is faced with some countless struggles for the remainder of his life which involve the epic life and death proportions [12]. Kelu after getting rescued by his father is left alone in the sea banks where he sees the scattered dead bodies of his loved ones. He stands in between the proportions of life and death on the seashore with the gold gifted by a dying mother. Later the young orphan Kelu makes his life with his friend Vavali and turns the gold into a magical weapon, a golden Urumi (whip sword). He receives education from different places and learns about war tactics, weapons and literature. This is shown in the scene where Kelu has a talk with Chirakkal King. These situations acted like a constant guide light for the hero.

The warrior modality in the hero is so strong that slaying others is considered a necessity for protecting himself and the things he value. The outcome of each fight is not just the victory over evil but it also shows how personal convictions can shape the world for a better tomorrow if people take charge of their life [19]. Kelu is a warrior from the moment he is shown in the 16th century. His warrior capability is established when he saves the Chirakkal princess Bala from the attackers. Later his warrior capabilities can be seen when he kills and attacks the Portuguese troops arrived to convict Arackkal Baliyahasanan at Arackkal fort. He also rescues Ayesha from Chirakkal guards and then rescues other women of Arackkal from the jail. Throughout the movie we get various other instances when Kelu is shown slaying the enemies with his Golden Urumi. It includes the battles with Portuguese army men who comes to rescue Estavio, and also with Estavio himself and his troops. In the climax Kelu dies in such a battle but only to resurrect inside the hearts of people and also inside Krishna Das, Kelu's mirror image and predecessor in the mundane world. It is from the wisdom gained from the mentors and Kelu's story that opens the insight of Krishna Das and forces him to sacrifice everything for the people living in his ancestral property. Thus the hero becomes the master of both the magical and mundane world.

4.2 Mentor

The immature hero needs counseling to face the adventures and tests in the journey for quest. This is where the mentor archetype comes into scenario. The mentor archetype exemplifies the hidden strength which the hero needs to carry out through the journey. The mentor may not always live in the main stream society but in a secret abode like a hermit [16]. In the movie there are two main mentors and two side kick mentors. The main mentor is Chirakkal Kothuval(Hero's father in magical world)/ Thankachan (The tribal head in mundane world) played by actor Arya. The other important mentor is Bhoomi, the NGO head in mundane world and a nameless mythical magician in magical world. The sidekick mentors are Vavali and Arackkal Ayesha.

Thankachan is the mirror reflection of Chirakkal Kothuval in the 16th century. The mentor (sage) through his wisdom and moral insight is able to induce the self reflection in hero. Who am I? Where am I? What is the purpose of my journey? -These are the questions posed by the mentor to make the hero stop and think before action [15]. In the first part of the movie where hero confronts Thankachan, the tribal head informs Krishna das about his royal lineage and about the ancestor Kelu Nayannar who fought for the motherland and father in the soil which Krishna das intend to sell. Thankachan narrates the story of Kelu to Krishna Das to make him understand the importance and value of the soil which he wants to sell for money.

Krishna das is the mirror reflection of Kelu in the mundane world. The real mentor image of Thankachan comes in the magical world as Chirakkal Kothuval. The mentor archetype can also be compared to that of a father archetype because they act as a role model for the hero. The father-mentor archetype is often pictured as a strong and powerful leader. Sometimes he disciplines his subjects with a benevolent hand and at other instants he will enforce power with iron fists [17]. Chirakkal Kothuval, the knight at Chirakkal kingdom is the protector of people under his realm. He is also seen as the teacher and guide for his son and other soldiers under the Chirakkal Kingdom in the first part of movie. He enforces his power against Vasco da Gama and his Portuguese army men who tries to attack his land. Even after his death, Kelu is always guided by the knowledge and wisdom passed to him through his father. When hero returns to the mundane world, Thanachan provides the magical weapon, the golden Urumi to Krishna Das who undergoes purgation through the story of Kelu. This aspect of mentor providing a magical weapon to hero according to Jung's mentor archetype is the final feather to the main mentor character in the movie.

The mentor Bhoomi is the mirror reflection of the mystical women in the magical world who comes as a prophet into the journey of Kelu and Vavali. In first part, she comes as a social worker who is currently taking care of the NGO started by late Chandramani Thampuratti, the mother of Krishna Das. Bhoomi tries to tell Krishna das about the social issue related to the sale of his ancestral property but fails in the beginning. She presents him the pen used by his mother as emotional magical weapon which might help the hero to stay for a moment and think. This connects to the characteristics of mentor character who leads the hero to stop and think before action. The main role of Bhoomi lies in the mystical oracle character in the 16th century.

 DOI: 10.1051/shsconf/20173300018

Kelu after capturing Estavio Da Gama is in confusion whether or not to kill Estavio and also about going back to Chirakkal. The mystical character of Bhoomi arrives as a mentor at that time. Sleeping on the problem is as important as engaging in the warrior mode, when combined both qualities together, it will ensure the hero's success in the battle [17]. The mystical character comes to provide a sleeping time for hero as the hero and his loyal retainer sleeps while listening to the oracle both in symbolic and practical aspect. The mentor posses' great magical abilities often referred to as a magician can communicate ideas and suggestions through dreams [17]. The mystical character announces the prophecy that Kelu will first become the knight of Chirakkal and later King. The magical mentor also asks them to be brave and go through the way lighten by the light inside the heart through the song. After the song sequence, the hero and his loyal retainer get into the confusion whether it was a dream or reality. The sidekick mentor Vavali guides Kelu when he loses his path by providing wisdom to him. Ayesha on the other hand can be seen as a mentor when she suggests Kelu to create troops from villages for an army to fight against Gama. Because fighting alone against Portuguese would have been foolish. The mentor can only show the way to hero and cannot force him to take the way. Fortunately in all stories hero listens to the mentor and acts according to the suggestions given by the mentor [4].

4.3 Trickster

The medieval description of devil is considered as an ape of God 'Simia dei' and his characterisation in folklore is that of a simpleton who is a cheater and trickster. The trickster is known for his fondness towards sly jokes, malicious pranks, dual nature and his exposure to all kinds of torture which makes him a negative figure [15]. In this movie, Minister of Chirakkal, Chennichery Kurup is recognized as the trickster. Chennichery is the ape of God in the movie who is a cheater and a trickster. In the first introductive narration given for Chennichery narrates how he is the only minister who escaped from the Ministers held as captives under Gama. In the description of trickster given by Jung, he states that the trickster must have gone through tortures which makes him wicked and leads to spread destruction around him. He also shows the characteristics of a homosexual which indicates the duality of nature. Chennichery kurup is seen to have a fond of sly jokes and malicious pranks throughout the movie.

The trickster tempts to attack hero with his sly jokes and pranks on hero [15]. In the scene where Prince Bhanu Vikraman brings Kelu and Vavali as the ones who tried to kill his sister, Chennichery is first seen as saying sly jokes at Kelu. It is even carried out when Kelu and Vavali are served food at Chennichery's residence after declared innocents by Chirakkal princess Bala. In the following scene where Arrackal women are brought in front of the Chirakkal King, Chennicheri follows similar sly jokes. Chennichery is also seen playing malicious pranks and shows a hidden support towards the Portuguese and Gama. Tricksters may be sentiments or allies working for the shadow or they may be independent agents with their own skewed agendas [15]. Chennichery, in the first scene of him shows a hidden support and slant towards Gama. He educates the Chirakkal King about the profit Portuguese can provide in return of Pepper and Black gold of Kerala. Trickster Kurup even tries to kill the hero who is sent to capture Gama arriving in Arackkal by Chirakkal King. The minister is also seen as taking care of the son of Vasco Da Gama, Estavio Da Gama. He even tricks the Chirakkal prince Bhanu to support Gama in return of showing the final victory, the throne of Chirakkal. The trickster Kurup is sole shadow who makes a son to kill his father. Thus Bhanu kills the Chirakkal King with the support of Kurup.

Trickster is seen as the personification of chaos and mischief for its own sake [3]. Chennichery destroys the Chirakkal kingdom for the ownership of Lakshadweep Islands offered by Gama. This gets revealed only towards the Climax of the movie when Kelu confronts Chennichery after the death of Chirakkal King. In this scene we also see the shaman in Kurup. Trickster from medieval period is considered to practice Shamanism [15]. Kurup is shown applying black medicine in beetle leaves which is fed to King when the King gets out of the influence of Kurup. But the intention behind the black medicine is understood when in the climax Kelu confronts Kurup. Thus all these facts prove Chennichery to be a trickster and it becomes evident when at last Kurup watches Prince Bhanu getting killed by Estavio and makes Kurup the governor of Lakshadweep.

4.4 Shadow

The shadow archetype is the opposite of light and provides the tension, anxiety and fear in the story. The shadow is often seen opposing the hero and typically the main antagonist [21]. The movie Urumi has two shadow figures, Vasco da Gama and his son Estavio Da Gama. The shadow operates in the prelogical thoughts governed by strong desires and animalistic needs [10]. In the movie, the first part when Gama returns to Kerala with war ships provides evidence for Gama's prelogical thought process to have a war with Samoodri and Kolathiris, the main Kingdoms in Kozhikode, Kerala. His strong desire towards the black gold, pepper became an obsession for Gama and his troops. For attaining the desire Gama left his humanity. His animalistic characteristic can be seen when the Brahmin mediator approaches Gama with the child Kelu in the ship. Gama creates a scar on the face of Kelu for looking straight into Gama's eyes. He is also seen cutting the tongue and chopping down the ear of Brahmin mediator. Later Gama stitches the ear of Gama's dog to the mediator. Here we can see the face of shadow figure who teaches discipline to the people who come as an obstacle in his line of attack.

Estavio Da Gama is also seen as the shadow figure, like his father. It is visible when Estavio arrives at Arackkal fort to crucify Baliyahasan. Estavio brutally injures the rebel captured for fighting against Portuguese with Baliyahasanan and also hurts Baliyahasan by trying to pull of his ears and hair. Estavio even takes revenge to Vavali by shooting him down to trap him at Chirakkal as a cover for rescue held by Kelu to save Arackkal women captives. Towards the end it is Estavio who leads to the death of Kelu Nayannar by shooting the hero from back. The shadow figures are often seen as breaking the rules and norms set by society for their personal needs.

4.5 Anima

Anima is an archetype based on gender. In the unconscious of the male, the archetype anima is the feminine inner personality where as animus is expressed as the masculine inner personality in the unconscious of female. The anima is the most important factor in the psychology of man for emotions and its affects [16]. In this movie the character of Jagathi Sreekumar as the Minister of Chirakkal Chennichery Kurup is identified as the archetype anima. The anima intensifies and exaggerates the soft but sharp emotions to generate fantasies and entanglements [16]. In the movie Chennichery enacts the role as a homosexual. The character dressed up partially like a woman shows the characteristics of a woman. The mannerism adopted by Chennichery clearly indicates towards his inner feminine overpowering the masculine. The character intensifies the jealousy and insecurities of Prince Bhanu Vikraman and exaggerates the emotions of Chirakkal King and Prince to fulfill his own fantasies. For that he creates entanglements around the King and Prince.

According to Jung, when anima is strongly constellated, it softens the character of a man and makes him touchy, irritable, jealous and moody with vain. He is then in a discontent state and spreads discontent all around him [16]. Chennicherry displays the irritation and jealousy when Kelu Nayanar and Vavali get importance in the Chirakkal kingdom over him. The aftermath of this is visible when he tries to talk to Chirakkal king about them. The mood variations and his vain over Kelu and Vavali can also be seen in the way Chennichery nicknamed them based on their caste as Nayanar (Nair) to Kelu and Mohammaden (Muslim) to Vavali. The anima characteristic of Chennichery overrules him and associates him to greedy and jealous women. Kelu's confrontation scene with Chenichery after Chirakkal King's death provides a proof for this. The moody jealous and irritated Kurup later is seen as a negative figure that spreads discontent all around him by doing treachery against his own motherland for his personal needs.

4.6 Animus

In the female unconscious, she is compensated by a masculine element and therefore has a masculine imprint which is called as animus. Animus is the counterpart of anima in the psyche of female. This influences woman's interaction and attitude towards men. The woman will attain masculine characteristics like stubbornness, lust for power, chivalry and willfulness [15]. The movie Urumi, has an animus archetype in the character of Princess of Arackkal, Ayesha. She exhibits strong masculine qualities like that of a warrior who killed the people who tried to attack her and her family. In the narrative part of Ayesha's introduction, the author narrates that the stubbornness displayed by the princess is the reason for her abduction by Portuguese army men. Arackkal Ayesha is also one among the rivals fighting for freedom even after strong oppositions from her own family. It shows her willfulness and commitment towards her motherland.

The overwhelming demeanor and always right attitude can test the limits of the patience of a man to the part where he drives unstable because of this incredibly strong Animus [8]. The overwhelming demeanor of Ayesha can be seen the way she first reacts to Kelu. She blames him for the cruelty happened to her baby sister by the Prince of Chirakkal in the Chirakkal Kingdom. Kelu saves them and takes the female captives of Arackkal to a safe zone; even then she refuses to show the sign of any gratitude or apology for her behavior. In the scenes followed Kelu gives up in front of her attitude and allows her to do according to her wish. The Ayesha later when the movie attains syzygy by accepting her both anima and animus when she falls in love with Kelu and therefore avoids any discontent.

5 Conclusion

Carl Jung's archetypes have been subject for study by many researchers, mostly due to its universality. The researchers from west proved that archetypes, globally accepted symbols and its interpretations are widely used in mass media. The movie Urumi is one of the latest releases in the Indian historical epic genre of Malayalam film industry in India. The psychoanalysis of this movie carried out in this paper provides evidence for the existence of all six essential archetypes identified by Carl Jung. Thus this article will be a fruitful resource for future research in the field of psychoanalysis of movies.

References

1. L.Bogart, Med. Stu. J, **5**, 13-33 (1991)
2. S.Bradshaw, L. Storm, Int. J. of Jun. Stu (2013)
3. J. Campbell, *The hero with a thousand faces* (Princeton, N.J.: Princeton University Press. (2013)
4. H.M. Chang, L.Ivonin,M. Diaz, A J. of Spe. Res..**11**, 99-113 (2013)

5. R.Crespolini , Indian Jones Master Thesis (2002)
6. C.Downing, Mirrors of the self: Archetypal images that shape your life. Los Angeles: Jeremy P. Tarcher. (1991)
7. Y.Fan, TPLS. **4** (2014)
8. F.Fordham, *An introduction to Jung's psychology*. Harmondsworth: Penguin books. (1966)
9. C.Hauke,I. Alister, *Jung and film: Post Jungian takes on the moving Image*. East Sussex, UK: Brunner- Routledge. (2001)
10. R.Hopcke, *A Guided Tour of the Collected Works of C.G. Jung*. Boston: Shambhala. (1999)
11. J.A.Huh, TA: A J. of Spe. R (2010)
12. J.Iaccino, *Jungian reflections within the cinema a psychological analysis of sci-fi and fantasy archetypes*. Westport, Conn.: Praeger. (1998)
13. J.Jacobi, *The psychology of C.G Jung*. London, Uk: Yale University. (1973)
14. L.Jadot. Rev. Psy. Edu., **10**, 367–374 (1975)
15. C.G Jung,Jung and archetypes,*The collected works of C.G Jung*. Princeton, N.J.: Princeton University Press. (1968)
16. C.G.Jung,*The archetypes and collective Unconscious*. Princeton, NJ:Princeton University Press. (1981)
17. C.G. Jung, *Memories, dreams, reflections*. New York, NY:Vintage. (1989)
18. M. Mark, C. S.Pearson, *The hero and the outlaw: Building extraordinary brands through the power of archetypes*. New York: McGraw Hill. (2001)
19. C.S. Pearson, *The hero within: Six archetypes we live by* San Francisco: Harper & Row. (1989)
20. C.S.Pearson,H.K .Marr, *Pearson-Marr Archetype Indicator*. Gainesville,FL: Center for Applications of Psychological Type. (2001)
21. .C. Vogler, M.Montez, *The writer's journey: Mythic structure for writers*. Studio City, CA: Michael Wiese Productions. (2007)
22. S.Walters, J. Soc. Evo. Sys **17,** 287-306 (1994)

DOI: 10.1051/shsconf/20173300019

Traditional and New Media Convergence on Philippine Noontime Show, Eat Bulaga

Genina Mariel M. Arceo[1,*]

[1]Communciation, Adamson University, Manila, Philippines

Abstract. The Philippines is still moving towards the direction of merging Traditional and New Media. Some TV formats had been affected by the rise of social media although a big part of the Filipino populace is still into watching Television as their source of information. Almost all of the shows in the Philippines are available online and in mainstream television to cater viewers who wish to watch their favorite shows anytime and anywhere. Social Media Trends had helped and invaded the mainstream media on creating content for the viewers. Even social media sensations have paved its way to mainstream media by being televised on the news and segments as guests to a show. Giving these shows high ratings with an online social media star is a sign of things to come. This study is a condensed version of a research that examines how a 36-year-old noontime show being viewed by older generation has penetrated social media and has attracted younger viewers. It also aimed to know what really happen when traditional and new media integrates. Initial findings show that media convergence is absolutely more than a technology shift as most people know it.

1 Introduction

In July 2015, after 36 years in Philippine noontime slot, Eat Bulaga has finally found a way to penetrate social media and attract younger viewers through introduction of the AlDub segment. Eat Bulaga has captured the attention of 40 million netizens and garnered more than double the number of its household viewers in 2015. The market potential for noontime shows yet its television show viewers are relatively high considering that not everyone is watching TV during noontime [1].

Eat Bulaga is the longest running Philippine noontime show which started in 1979 and led mostly by 1960's comedians. It has been around for 36 years and its viewers have aged with the show. The show being viewed by the older generation got a boost in television viewers from Generation Y and the millennials, the 18 years old to late 20s generation after the noontime show brought in Maine Mendoza in one of its segment, Juan for All, all for Juan and when accidentally, the AlDub phenomenon has been created.

It is inconceivable for the millennials to take interest in Eat Bulaga as they are hooked more on their smart phones, tablets which give them instant gratification comparing to television. Noontime shows require them to sit before a television set at the time when they

* Corresponding author: geninamarielarceo@gmail.com

 DOI: 10.1051/shsconf/20173300019

are in school or at work. In fact, Eat Bulaga's social media presence pales in comparison to those of Anne Curtis and Vice Ganda, the stars in its rival noontime show, It's Showtime! [2]. But by getting Maine Mendoza, it was a game-changing move.

1.1 The Problem

Television has been traditionally considered to be a popular source of information and important facilitator of social interaction. Morrison and Krugman states that the television content is the conversation trigger while television is the medium that physically connects two or more people into a confined space. However, the unstoppable rise of new media allows television viewers to enjoy communal viewing experience even without being physically present. This shows that the media landscape is changing at a fast pace and moving in the direction where traditional and new media have no choice but to coexist. Given this case, this might seem to be a problem for decades long TV shows. Viewers who have aged with the show might not be as active as the young viewers are. This study hopes to determine and examine the media convergence of the oldest noontime show in the Philippines. The study asks the following questions:

a. How Philippine Noontime Shows being viewed by the older generation, have gotten a boost in television viewers from Generation Y and the millennial?
b. What happens when traditional and new media converge?
c. What are the effects of media convergence and the combination of diverse media forms?

2 Traditional Versus New Media

New media is changing the overall media landscape. The new media refers to content that is easily accessible via many different forms of digital media but it cannot fly as high as traditional media. It only creates things from the traditional media more powerful. New Media has no clear business model. One can be a superstar on social media but they won't earn their keep. That's what mainstream media or television has. The online digital universe especially in terms of entertainment has not yet on the level of business like that of the television.

3 Eat Bulaga: the longest philippine noontime show

Eat Bulaga being on the noontime slot for 36 years must have experienced all the rapidly changing media landscape on its run for years. Being on air must have been an indication that they have effectively adapted the significant change and sudden shift of media. It is evidently seen on their approach to their recent segments which entails that the show adapted the norms of digital age and cater new viewers for the show. The viral Dubsmash application was also integrated to one of Eat Bulaga's segments. But this time, the show casted one of its viral dubsmasher, Maine Mendoza for their Juan for All, all for Juan, a segment where the hosts visits villages in Metro Manila to hand out food and prizes.

4 AlDub phenomenon

Maine Mendoza is a superstar recruited by mainstream TV [3]. Mendoza became a Facebook sensation after she posted a compilation of her Kris Aquino Dubsmash videos. Kris Aquino is a famous television host and movie actress in the Philippines. Aquino has been tagged into doing horror movies in the past years that's why her famous movie lines have been used mostly by netizens in the viral Dubsmash application. According to Google Play, Dubsmash is a mobile application used to create short selfie videos while doing lip sync to famous sounds.

The "Queen of Dubsmash" tagged by netizens suddenly became part of Eat Bulaga's Juan for All, All for Juan segment. By bringing Mendoza in, Eat Bulaga has ensnared her by then a million Facebook Friends, about a million Twitter Followers and almost 100,000 YouTube subscribers to her site [4]. The Juan for All, all for Juan segment introduced the character of Mendoza as 'Yaya Dub" (Nanny Dub) still doing dubsmash. On July 16, 2015 episode, Mendoza was obviously tickled pink and could not concentrate after noticing that Alden Richards, a Filipino Actor was watching her do live impersonation on the show's live split screen. The hosts teased them and both of them played along and "flirted" with each other live on TV – this marked the beginning of the first KalyeSerye or street series in the Philippines, starring the accidental love team turned phenomenon called AlDub, a portmanteau of the tandem's character names [5]. Much to everyone's surprise, the new tandem took the internet by storm and millions of netizens continued their fascination over it. The online buzz eventually got the attention of print and radio, which fanned the AlDub romance even more.

4.1 Kalye Serye

Philippines has been known to many television series or what they usually called as "Teleserye" which inculcates good Filipino values and carries entertainment factor for its viewers. But #AlDub is not your usual teleserye. It is a kalyeserye, happening right in the streets intimately familiar to our people which brought the fun out from the noontime studio into the villages and people's homes [2]. Aside from the onscreen chemistry of the two, Kalye Serye follows a no strict script and filled by Filipino values.

5 Traditional and new media convergence

Cheskin research explained in a 2002 report that the old idea of convergence was that all devices would converge into one single product that did everything for you. Now, these devices carry multiple functions suited to your needs. But convergence is more than this technological shift. Jenkins argued that it is a cultural shift where consumers are encouraged to seek new information, interact with one another and create new meanings of old concepts at a dispersed media.

In the realm of progressive digitalization, it has been perceived that viewers start switching to internet-based media thus, affecting the even fragmented television market. New technologies allow the viewer to change from a passive receiver of television broadcasts to an active participant [6]. The phenomenon of convergence forces broadcasters to modify the way content is constructed, both in terms of technology and concept [7].

6 Convergence culture

Convergence occurs within the consumers' minds, within the same devices, within the same company, and within the same fandom [7]. The consumers have more control over media consumption, production and distribution. Henry Jenkins sees that the process of convergence supports the development of a participatory culture for there is a change in the way the media is consumed. Without you knowing, it is not just the flow of content being consumed overtime but also the life, relationships, aspirations, and memories flows across

DOI: 10.1051/shsconf/20173300019

media channels and consumed in this process. There has been a change in the way media is produced as consumers are already considered media producers themselves. Vincent Miller described in his book "Understanding Digital Culture" the producers as those that create content and cultural objects, and the consumers as the audience or purchasers of contents or objects. The blurring of the producer and consumer roles is called "prosumer" as coined by Alvin Toffler. Convergence then is not an endpoint but a process and there will be no single box that controls the flow of media into our homes [7]. The way Jenkins sees it was that the power of participation will really come from writing over the content, modding it, modifying it, expanding and adding a different perspective, then re-circulating it and feeding it back to the mainstream media. The conversation created by users creates buzz and accelerates the circulation of the media content [7]. The consumption then, as Jenkins described it, has become a collective process or what French Cyber Theorist Pierre Levy has coined as the Collective Intelligence. When all active participants or users' knowledge have been combined and published online, a certain community is created and the result of it is the collective intelligence which can be a source of alternative strength over media control and consumption.

Henry Jenkins referred to this whole transition as Convergence Culture. Jenkins described it as a cultural process where media creation and content also converges and removing the boundary on the roles of the consumer, producer, organization and the advertisers. Ready or not, we are already living in a Convergence Culture [7].

7 Eat Bulaga versus it's showtime

Based on the data provided by AllFamous Digital Philippines, Eat Bulaga obviously outshone its rival noontime show program, It's Showtime after taking over 96.9% (1, 996, 716 mentions) of the total number of online mentions from July 13 to August 13, 2015. In contrast, a mere 3.1% (63, 451 mentions) of these online posts came from its closest competitor. Most viewers hooked on the segment of Eat Bulaga called KalyeSerye, considers Saturday as "AlDub Day "given the fact that some of the viewers doesn't have any school or work by then. The Tweet volume for the #AlDub hashtag continuously increases as the days passed on with still relatively million tweets for AlDuB every Saturday. The #AlDubTamangPanahon hashtag earned a whooping of 41 million tweets surpassing the previous Aldub Twitter record of 26 million Tweets when Alden Richards visited Maine Mendoza at the mansion on September 25 as well as the 35.6 million register of 2014 Fifa World Cup semi-final match between Germany and Brazil last July 8, 2014.

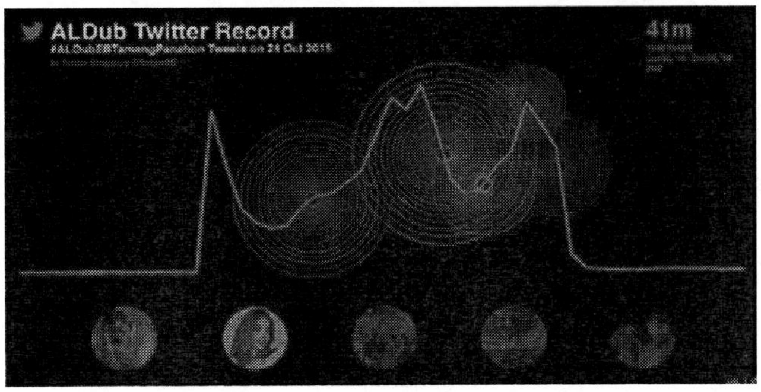

Fig. 1. ALDub Twitter Record

Fig. 1. Twitter shared that there were three peaks in the #ALDub EBTamang Panahon conversation being the highest at more than 48,000 Tweets per minute at 2 :29pm on October 24, 2015 when Alden Richards sang "God Gave Me You" to Maine Mendoza and the couple hugged on stage in front of 55,000 fans at the charity concert. The most shared Tweet for #ALDubEBTamangPanahon came from Mendoza which had over 74,000 Retweets and was viewed nearly 2.4 million times on Twitter.

In terms of online popularity, Eat Bulaga has now usurped It's Showtime. It Showtime had always been the netizens favorite. The churn out hashtags that reach the top trending list at noontime and the different fad produced by the hosts created funny dialogues all over the social media stream. But it seems that the tables turn quickly. The popularity of It's Showtime on the Internet fizzled during the week the AlDub love team was born [5].

Eat Bulaga is not just successful in social media but also in the television landscape with high ratings compared to their rival program. Prior to the #AlDub, the popular belief is that the CDE television market - the same audience claimed by Eat Bulaga – isn't as immersed in social media as other market segments [2]. But the AlDub Phenomenon set it straight and prove that the CDE are in fact online as well into sharing and watching videos of AlDub.

8 Theoretical framework and methodology

Henry Jenkins describes convergence culture as the by-product of the collision between traditional and new media. Jenkins identified this as the relationship between three concepts, the media convergence, the participatory culture and the collective intelligence. This study conducts a content analysis of the effects of media convergence on the noontime show, Eat Bulaga using the theory of convergence culture. This follows the narration of how participatory culture and collective intelligence emerge in one of the show's segment and created a buzz across media platforms and content creation.

9 Discussion / Findings

Media Convergence is the rise of user-created content. When the Dubsmash application was launched, several users have gaged themselves estatically and even mastered the art of lip syncing. Maine Mendoza is one of the many users of the application. Her hobby of making facial expressions suddenly turned into an opportunity after she has caught the attention of Eat Bulaga producers. Mendoza has been drafted to be part of the longest noontime show in the Philippines at the Juan for All, All for Juan Segment as Yaya Dub (Nanny Dub). Her character only speaks through lip syncing of famous television and movie lines, and even making exaggerated expressions. Having a somehow familiar face at that moment, the show has slightly captured the attention of the viewers.

The next day albeit accidentally, AlDub has been created and become a massive hit. This content has achieved virality and reached a higher level of engagement in accordance with the desires of social media users for the two characters to meet. Each episode, the users are waiting for this moment to happen for the two characters haven't met even before. It's only in the noontime show and live split screen where they first met. Days passed and still their desires haven't been granted, the users started to express their imagination by editing old photos of Alden and Yaya Dub looking as if they've come across with each other. It is in this way they are being media producers themselves. It is also a crowdsourcing strategy on what viewers would want to see on the next episodes. The results, a whooping top tv ratings and top trending issue worldwide when both of the characters finally meet. Henry Jenkins termed this as Participatory Culture. It is the state where the media producer and the consumer both become participants, interacting with one another to create new concepts and eventually make it happen. Now, both of them occupy the same role. But the media producer can only gather concepts into a television content that will be eventually offered to the viewers. The power over the received content is gradually transferred from the media producer to the viewers, who learn how to control the received content using technology [7].

The circulation of media content depends heavily on consumers' active participation [7]. The flow of this content eventually reaches several media channels. The same concept has been even adapted on several TV commercials and magazine cover shoots. Several companies adapted the concept of the characters doing dub smash acts and split screen method strengthened the content more after the advertising materials have been released. It has even more contributed to the overall flabbergasting feeling of waiting on the next progression of the content. Likewise, attracting new viewers and spurred curiosity on it. It is as if they become participants as well on the viral content and used it for marketing purposes. It supported the claim of Henry Jenkins where the media was forced to modify their content in order to garner viewers' attention.

Much information has sprouted with all the reactions of the users. This became a content worthy for the producers of the noontime show to control what and what not the users can look forward to on the next episodes. Many were curious about the real voice of Mendoza and with the budding romance between the two characters. This other information added an incentive on the consumer's' part as they begun to talk to themselves, put their hands on it and hope that the producers will deliver it the way they want it. Another term was identified by Henry Jenkins when the temporary knowledge of the user about a content was combined to other temporary knowledge of other users, it will be resulting to a community which he called as Collective Intelligence. This community is relevant and increasingly valued by the media industry because they are those who heightens and strengthens the content as the days have passed. The Collective Intelligence can be related to social communities who have actually built their own fandom.

10 Conclusion

Gone are the days that the audience is sitting back and enjoying shows in the television. The emergence of media convergence has made the audiences visible and has allowed participation for various channels. The nonstop comments, photo sharing, tagging, online discussion and recommendation system are associated with greater user control over media. There is a blurring line between the producer and the consumer but they have not necessarily shifted roles. Users have only been given a chance to become co-creators of the show.

Maine Mendoza has already online followers before she joined Eat Bulaga. It has gotten the attention of the Digital-savvy Millennials while at the same time retaining their loyal viewers which comprises the Generation X (34-54 years old) spreading the "AlDub" romance online and across the world. Richards and Mendoza's writing of sweet nothings on white paper to communicate to each other mimics the propensity of millennials to communicate via social media or mobile phones [4]. Sharon Ross declared in her book, beyond the box that the Millennial generation is particularly "migratory" as they are multitasking across the internet to wander and hunt down some content without them knowing it. It will be realized that the forms of participatory culture may be summarized into expression, affiliation, collaboration and circulation. Inside the participatory culture, the active users have more power than others due to their skills and desire. Growing fans of AlDub started to express their desires by producing a creative content which affiliates to other users who share the same aspirations. Thus, forming a social community where combined efforts are made to develop creative shared content and eventually escalates the media circulation. This participatory culture has also been used as a marketing strategy to join the trend and eventually led to promoting brands. The good Filipino values portrayed in the show flows as well along with media transmission. The convergence of traditional and new media cut across the social media classes and bridged the generation barrier. Since this is an accidental love team, people want to be part of it as witnesses of their journey. The relatable experiences of Mendoza's Characters and Richards provided the entertainment enjoyed by people across all social classes.

What happened in Eat Bulaga showed that traditional and new media results are successful in both sides when they converge. The success of the show's rejuvenation lies from the acquired new viewers of the show and the continued existence interaction with old media. The high ratings showed us that more people still turned to television as their main source of information. But like new media content, the temporary acquired and merged content being shown in the television may fall out of fad at a particular time. Likewise, the new viewers may realize that they become new customers mostly due to a different value proposition [8]. On the other hand, from the perspective of the now fans of AlDub, they will remain loyal and passionate for it even outside the noontime show.

References

1. J. Go, Is the AlDub phenomenon losing its appeal? (2016)
2. T. Cruz, Long Live the #AlDub Nation (2015)
3. B. Abunda, Maine Mendoza is a superstar recruited by mainstream media. Retrieved from http://entertainment.inquirer.net/179041/boy-abunda-maine-mendoza-a-superstar-recruited-by-mainstream-tv
4. A. D. Nocum, 'AlDub magic' – marriage of old and new media (2015)
5. P. Quimbo, AlDub : Blurring the line between traditional media and digital media (2015)
6. Kurczewska, Izabela. The Influenece of Media Convergence on Talent Show Programs. 2013
7. Jenkins, Henry. Convergence Culture: Where Old and New Media Collide. New York and London. New York University Press. (2006)
8. I. Gan, The Social Media Phenomenon that is AlDub (2015)

DOI: 10.1051/shsconf/20173300020

South China, East Vietnam or West Philippine? Comparative Framing Analysis of Regional News Coverage of Southeast Asian Sea Disputes

Bradley C. Freeman [1,*]

[1]Associate Professor of Communication & Information Studies, Mohammed bin Rashid School for Communication, American University in Dubai, United Arab Emirates

Abstract. For years, the world paid scant attention to the sporadic skirmishes and sovereignty squabbles in what is known by many as the South China Sea. However, in the past few years, there have been several noteworthy happenings that have drawn the attention of media outlets, including a Chinese oil rig placement near Vietnam, a Philippine-international court case, a U.S.–ASEAN summit, and Chinese reclamation projects. Many of the countries involved are members of the regional grouping known as ASEAN (Association of Southeast Asian Nations). This study examined Asean's English-language press, looking at the amount, patterns, and tone of coverage given to regional maritime disputes. Results indicate that use of the term 'South China Sea' prevails in all countries except Vietnam. The Asean newspapers have devoted a consistent and substantial amount of space to covering the topic. Indonesia's *Jakarta Post* tended to use neutral or positive language when mentioning Asean and the US, while China saw more negative associations in the coverage. Further research is suggested, taking into account mass communication theories and perspectives.

1 Introduction

To any casual observer of world events and international relations, it would have been difficult in the past few years to avoid news items mentioning disputes related to what is often referred to as the South China Sea region. While the issues involved have existed for many decades, and researchers have raised concerns about potential conflict previously [1, 2, 3] international news media have tended to cover the mainly territorial disputes through the lens of each incidences' newsworthiness. While news items certainly exist going back many years, the recent increase in the amount and kinds of coverage clearly signals a change in the situation.

Rival countries have argued for years over strategic reefs and atolls in the disputed waters of the South China Sea, a marginal water body that is part of the Pacific Ocean,

* Corresponding author: bfreeman@aud.edu.my

covering an area from the Malacca Straits to the Strait of Taiwan. By now, and due in part to the increased news coverage, members of the international community are more familiar with the issues at stake in this affair. Notably, media audiences have come to know about China's "9-dash line," several intermittent dotted lines on maps that demarcate the area where China claims sovereignty. Several countries have overlapping claims to the same area, namely Vietnam, the Philippines, Brunei, Malaysia, Taiwan, and Indonesia. The United States has also become involved in the matter based on the issue of safety of maritime trade routes, its historical presence, as well its alliances with other parties and countries. Other countries have also had a say based on international relations, including Japan, Cambodia, India, and more notably, the regional grouping of countries known as the Association of Southeast Asian Nations (henceforth, Asean).

Countries use many different means to address disputes in an attempt to resolve them. Diplomatic channels, legal cases, and economic means are typically brought to bear on the issue, with military action likely only preferred as a last resort. Recently, China has become more assertive regarding its claims, ramping up island-reclamation and increasing naval patrols. These activities have been met with various responses; increased media attention has followed. For example, a front-page banner headline on February 27, 2016, in *The Manila Times* declared "China taking de facto Sea control" [4]. In 2013, the Philippines requested that an international tribunal in The Hague hear its arguments against China's activities. At a "U.S. – Asean Summit" the name "South China Sea" was never used in a joint statement - the "Sunnylands Declaration" calling for "respect of each nation's sovereignty and for international law" [5]. In July 2016, the court ruled for the Philippines; a few months later, Philippine President Duterte raised the issue with a Chinese delegation; at the same time, President Obama called the ruling 'binding' [6].

In the information age, countries are aware that how an event plays out in the media can have an effect on awareness, attitudes, and policies. While it may not always be the case, media coverage can create certain frames or perceptions and thus play a role in how governments react and respond. Indeed, Kaplan suggested as much when he wrote that the Philippines, for example, might be "better off employing restraint and an appeal to world opinion" in their disagreement with China [7]. Thus, for the current case, there are (in a way) two "courts" involved: the international tribunal (a matter of law and international relations), and the so-called court of public opinion (a matter of the media and mass communication researchers). The former is an institution in the Netherlands; the latter is an imagined community made up of domestic and international media audiences. This study examines the way that the dispute has been covered and portrayed in Asean's newspapers and discusses what it may mean for the countries concerned, as well as the global community.

2 Literature Review

The literature on this topic is vast and comes from several disciplines. Researchers in international relations, military studies, diplomacy, history, international law, geography, Asian studies, and others, have all written about South China Sea disputes. Several prominent books have been published in this area [7, 8].

There is not so much in the way of mass communication studies that have taken on this subject. As this study deals with news coverage and country imaging, several mass communication theories might be applied, and would include agenda-setting (second level), priming, and media framing [9, 10, 11, 12]. These theories hardly need an introduction. Over the past several years, they have been used numerous times in many studies [13]. Framing, in particular, implies that news articles can influence the way readers perceive an issue based on the way that the information is presented [14]. In a way, it is the kind of spin that a media outlet (or individual journalist) may give (knowingly or unknowingly) to a piece of information – highlighting some aspects, and downplaying other attributes. There are many different influences in the creation of media content and on the end product of news [15]. In cases where the government has an influence (either directly or indirectly), we expect that foreign policy would play a role in the types and kinds of frames that are ultimately written into newspaper articles. In this regard, Reese correctly advised that "framing suggests more intentionality on the part of the framer and relates more explicitly to political strategy" [16].

Studies referencing these theories have taken different approaches and utilized varied methodologies in an attempt to deconstruct and define the phenomena surrounding the subject matter of media coverage and national imaging as formed or supported via mass media. Investigations of the media coverage of individual nations and business companies are more common than on collective regional country groupings [17, 18]. Some public relations studies have also tested whether hiring such counsel alters coverage in any significant way [19, 20]. Wang wrote that "one of the reliable indicators of a nation's reputation is the opinion of the given nation as expressed by foreign publics" [21]. Brewer, Graf, and Wilnat acknowledged that people depend on the media to gather information about international affairs, and, therefore, the media play a significant role in shaping "mass perceptions" of other nations [22]. These images come in many different forms, from television and press reports, travel features in newspapers, talk radio discussions, and foreign films, to brief, seemingly innocuous comments about a country in a prime time television drama series -- ala Desperate Housewives and the Philippines [23]. Each item represents an attribute of a state, all of which collectively inform audiences what to think and how to think about foreign lands; borrowing from Lippmann it allows for the basic pictures of a country to be shaped in Americans' heads [24].

In examining the literature to date, a clearer picture is forming as to which items will best determine or predict news coverage and the frame of one country in another, given the specific circumstances (i.e., which country and what type of media outlet) and also how this might be affected by or even what effect this may have on policy decisions or other activities [25]. For the present study the frame categories identified by Valkenburg and Semetko [26] are utilized; they are, conflict, human interest, economic impact, morality, and responsibility. Based on previous literature, and given the specifics of the South China Sea disputes, the following research questions are posited:

RQ1: What geographic term is preferred in the Asean print media when it comes to reporting on the marginal sea area that many of its member countries surround? What's in a name? As those working for Google Maps will likely tell you "a lot." There are many areas in the world over which map makers must tread carefully [27]. Use of the name "South China Sea" is one such area of contention [28]. Although a recent Malaysian newspaper suggested that the body of water be renamed as the Southeast Asia Sea, most of the world refers to it by one of three titles: South China Sea, East Sea, and West Philippine Sea. A cursory search revealed that "South China Sea" is clearly the dominant term found in the English-language global press; will that be true for Asean's newspapers?

RQ2: What are the coverage patterns like for Asean newspapers regarding South China Sea disputes? Put simply, how important is this issue to these six countries? The expectation is that newspaper coverage increases as newsworthiness increases. Thus, how many articles have been written in the six countries selected print media outlets during the past five years? This research question is interested in the numbers involved, with the understanding that amount of coverage provides one vantage point towards understanding the situation.

RQ3: What are the dominant thematic frames uncovered in *The Jakarta Post's* coverage when it comes to the sea disputes? Several possible frames might be present in the stories. Although arguable, prevailing thought has been that Indonesia had reached a level of bilateral understanding with China concerning South China Sea issues. However, with the fishing boat incidences of 2016, there seemed to be a slight shift in the situation.

RQ4: How does *The Jakarta Post's* coverage depicts the three principal actors of China, Asean, and the United States? An article in *The Yomiuri Shimbun* (Japan) stated that China viewed the United States "as an "extra-regional state" in the South China Sea" and refused to accept U.S. intervention in the region [29]. Yet, the same article reported on the US-ASEAN summit and indicated a strengthening of ties between the U.S. and Asean over this matter. Does *The Jakarta Post's* coverage tends to agree with this position? There is an expectation that it does.

3 Method

This study is based on a content analysis of press coverage of newspapers in six of the ten ASEAN member countries: The Philippines – *The Manila Bulletin*, Vietnam – *Vietnam News Agency Bulletin*, Malaysia – *New Straits Times*, Indonesia – *The Jakarta Post*, Singapore – *The Straits Times*, and Thailand – *Bangkok Post*. The newspapers are known and valued in their respective media markets. The purpose of the analysis is largely descriptive in nature, to identify what the reports say, and how this may reflect countries' positions.

The study involved two distinct phases. In the first phase, all six newspapers were examined (2011-2016) to address research questions 1 and 2. In the second phase, relevant articles were reviewed in Indonesia's *Jakarta Post* (2015-2016) to address research questions 3 and 4. It is anticipated that the corpus of articles will undergo coding – and this represents the next step in this research project. A manual content analysis approach was used to examine coverage based on several key variables. Searches using the Factiva database were carried out using different keywords "South China Sea," "East Sea," and "West Philippine Sea." To investigate coverage of 'Asean' and 'United States' involvement, relevant terms were added and searched within the overall corpus of articles during the second phase which again selected articles only from Indonesia's *Jakarta Post*. In this phase, 135 articles were initially collected, however after the filtering process, 52 articles were determined to be most relevant and thus underwent the reading and coding process.

The unit of analysis is the article; the recording unit is at the sentence assertion level. Assertions within each article were identified and coded on several levels. For the current conference paper, and due to space and time limitations, only two variables are assessed and are reported here: [1] *Dominant Frame* – conflict, human interest, economic impact, morality, and responsibility (as used in Valkenburg & Semetko, 2000). And, [2] Valence (e.g., tone of coverage towards three principal actors): positive, neutral, negative -- dictionary of words as suggested by previous research [30, 31, 32].

4 Results and findings

The first phase of the search was meant to uncover which geographical term was preferred in Asean newspapers. Similar to global realities, the term "South China Sea" was the term receiving the most use in Asean English-language newspapers, Figure 1. The exceptions are found in the newspapers of Vietnam and the Philippines, where we see greater use of the

 DOI: 10.1051/shsconf/20173300020

terms "East Sea" and "West Philippine Sea" respectively. Yet, in the case of the Philippines, the term "South China Sea" is still seen more often than "West Philippine. Sea." In Vietnam, "East Sea" dominates, and the term "West Philippine Sea" is never used at all. Thus in answering RQ1, the results indicate that "South China Sea" is the preferred term outside of Vietnam and to a lesser extent the Philippines.

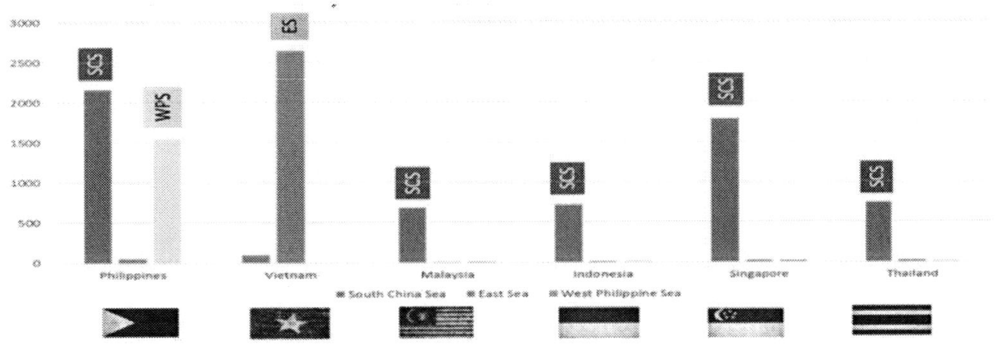

Fig. 1 Preferred geographic term for the main body of water in S.E. Asia, 2011-2016

Figure 2. What we can see are some spikes in coverage, for example, Vietnam's coverage went up in 2014 – and this would be expected due to the issue of a contested oil rig in the waters off of Vietnam's coast (i.e., Hải Dương – 981; [33]). Also noticeable is the large number of articles in the Singapore paper, *The Straits Times*. One might suggest that only using one newspaper per country might not show the coverage effectively, however, when we added second news sources to the mix, we saw that the coverage was similar to other papers. Thus, focusing on one newspaper from each country was sufficient for the purposes of the study.

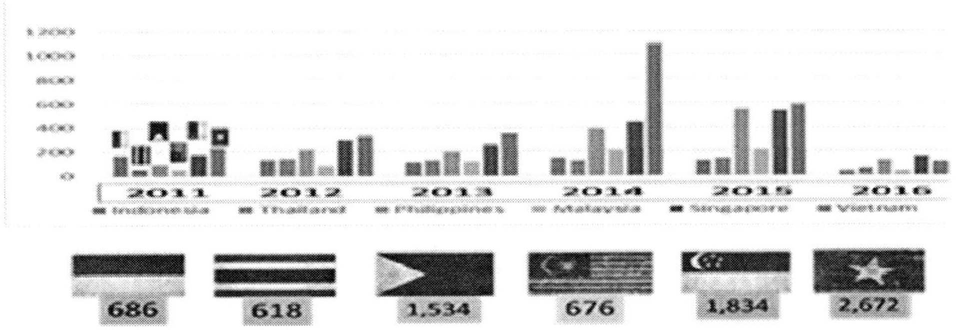

Fig. 2 South China Sea articles, select Asean newspaper, 2011-2016

The results of the coding on the frames found within the *Jakarta Post's* articles show a bifurcated result, Figure 3. The frames of 'Conflict' and 'Responsibility' were the two that featured most prominently in the pages of the Post. The conflict frame tended to reflect conflict and disagreement among principal actors. The 'Responsibility' frame is defined as "a way of attributing responsibility for [a] cause or solution to either the government or to an individual or group" [26]. Coding was done on all 52 articles by both coders with a final agreement figure of 86%. In these articles, there was a slight yet noticeable tendency to attribute responsibility to China for increasing tensions in the region. It should be noted, that during the time period under investigation, Indonesia caught a Chinese fishing boat in its territorial waters off of the Natuna islands.

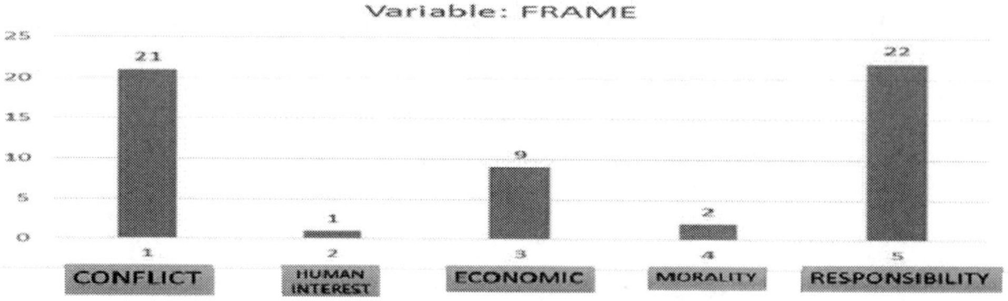

Fig. 3 Dominant Frame, Jakarta, post April 2015 to April 2016, N=52

In the study's collected articles, the tendency was for negative words to appear in and around the sentences that mentioned the country of 'China,' Figure 4. This happened more often than was the case for neutral or positive word pairings. The reverse was true for 'Asean' and 'the United States', where neutral and positive word pairings were noticed. It is clear, though that China and Asean received far greater mentions in the coverage.

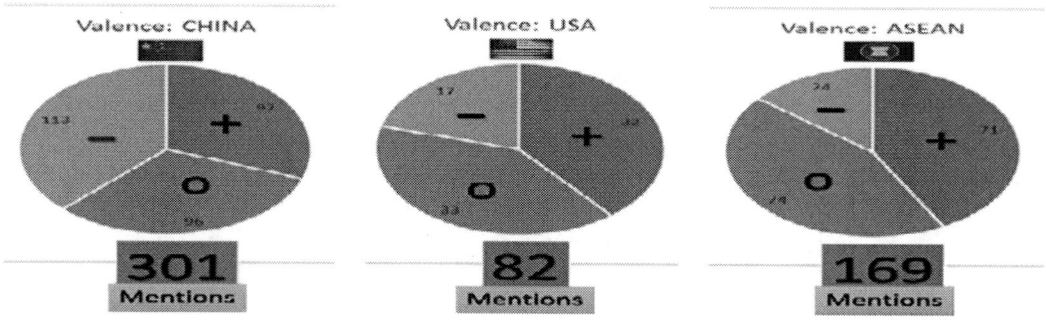

Fig. 4 Variable : Valence. Jakarta post, April 2015 to April 2016, N=52

5 Discussion and conclusion

There is no doubt that the issues surrounding the South China Sea are gaining more and more attention from scholars and journalists the world over. The study found that the term 'South China Sea' is the dominant title used in Southeast Asia's regional English-language newspapers – with the exception of Vietnam's press, where 'East Sea' is preferred. One might expect that the Philippine papers would follow in the same footsteps by preferring to use the term 'West Philippine Sea.' Use of certain labels and terms may have a 'branding' effect and can lead to a 'spiral of framing' in an intermedia agenda-setting way – where the use of one term over another may come to be preferred.

The Asean newspapers clearly afford topics surrounding this body of water a lot of coverage. Interestingly, Singapore's newspaper had the most amount of coverage (next to Vietnam), though it does not have a territorial dispute with China on the matter. There may be a few reasons for this. Singapore may see itself, arguably true, in a unique position to comment on this matter. Singapore settled a court case with Malaysia over the territory of Pedra Branca (International Court of Justice in 2008). The case between the Philippines and China is unlikely to settle issues as succinctly.

Following this preliminary study, more research will be necessary to dig deeper into the specifics of this coverage – to uncover the precise nature of all of the relevant coverage. The articles have been collected from the Asean papers, and the task of coding is expected to proceed. The degree to which the topic receives attention does tend to match closely specific cases that occur with regards to this body of water. That is to say, when the

DOI: 10.1051/shsconf/20173300020

Philippines brought the international court case, when a summit has been held, or when the Vietnamese protested the Chinese oil rig placement – we see a corresponding increase in coverage of the matter. In recent years, there seems to be a discernable increase in activity in the South China Sea region.

As is typical for the news media, 'Conflict' (the conflict and disagreement that exists in or around an issue) was shown to be a major theme in the coverage, but that was also matched by the frame of 'Responsibility.' It will be interesting to examine and see if this holds true for other years, and for other papers in the region. The current study represented a preliminary effort to investigate the South China Sea issue from the perspective of the media (newspaper coverage). Again, further investigation into the coverage is warranted – particularly when the link between media coverage, soft power, and public relations is known to impact on media audiences and potentially, public policy.

References

1. M.J. Valencia, South China Sea: Present and potential coastal area resource use conflicts. *Ocean Management, 5*, 1-38. (1979).
2. M.S. Samuels, *Contest for the South China Sea.* New York: Methuen. (1982).
3. C.K. Lo, *China's Policy towards territorial disputes: the case of the South China Sea Islands.* New York: Routledge. (2003).
4. China to secure. China to secure 'de facto' control of S. China Sea: US admiral. *Agence France Presse.* (2016, February 26). Retrieved using the Factiva database.
5. R. Murphy, J. Marx, Obama's ASEAN summit ends with business, security pact. *The Desert Sun.* (2016, February 16). Retrieved from: http://www.desertsun.com/story/news/2016/02/16/sunnylands-obama-Asean-statement/80457706/
6. H.Y. Tan, Tribunal ruling on South China Sea is binding, stresses Obama. *The Straits Times.* (2016, September 9). Retrieved from: http://www.straitstimes.com/asia/tribunal-ruling-on-s-china-sea-is-binding-stresses-obama
7. R.D. Kaplan, *Asia's Cauldron: The South China Sea and the End of a Stable Pacific.* New York: Random House. (2015).
8. B. Hayton, *The South China Sea: The struggle for power in Asia.* Yale University Press. (2014).
9. E. Goffman, *Frame Analysis: An Essay on the Organization of Experience.* New York: Harper & Row. (1974).
10. T. Gitlin, *The whole world is watching.* Berkley, CA: University of California Press. (1980).
11. R.M. Entman, Framing: Toward clarification of a fractured paradigm. J. Commun *43*(4), 51-58. (1993).
12. J.W. Tankard, The empirical approach to the study of media framing. In D. Reese, O. Gandy, and A. Grant (Eds.), *Framing public life: Perspectives on media and our understanding of the social world*, 95-106. New York: Routledge. (2001).
13. D.H. Weaver, Thoughts on agenda setting, framing, and priming. J.Commun, *57*(1), 142-147. (2007).
14. D.A. Scheufele, Framing as a theory of media effects. J. Commun *49*(1), 103-122. (1999).
15. P.J. Shoemaker, S.D. Reese, *Mediating the message*, 3rd ed. Thousand Oaks, CA: Sage. (2013).
16. S.D. Reese, The framing project: A bridging model for media research revisited. J. Commun *57*(1), 148-154. (2007).

17. S. Yousaf, Representations of Pakistan: A framing analysis of coverage in the U.S. and Chinese news media surrounding Operation Zarb-e-Azb.J Commun, ***9***, 3042–3064. (2015).
18. C.E. Carroll, M. McCombs, Agenda-setting effects of business news on the public's images and opinions about major corporations. *Corporate Reputation Review,* ***6***(1), 36-46. (2003).
19. J. Kim, S. Yang, Effects of government public relations on international news coverage. *Public Relations Review,* ***34***(1), 51-53. (2008).
20. S. Lee, An analysis of other countries' international public relations in the US. *Public Relations Review,* ***32***(2), 97-103. (2006).
21. J. Wang, Managing national reputation and international relations in the global era: Public diplomacy revisited. *Public Relations Review,* ***32***(2), 91-96 (2006).
22. P. Brewer, J. Graf, L. Wilnat, 'Priming or Framing: Media Influence on Attitudes toward Foreign Countries', J.Commun ***65***(6), 493–508 (2003)
23. E. Regullano, Filipinos Depicted in American Culture. *e-Research:* AJUR, ***3***(1), 6. (2014).
24. W. Lippmann, *Public opinion.* Transaction Publishers. (1946).
25. C. Naveh, The role of the media in foreign policy decision-making: A theoretical framework. *Conflict and Communication online,* ***1***(2), 1-14. (2002).
26. P. Valkenburg, H. Semetko, Framing European politics: a content analysis of press and television news. J.Commun, ***50***(2), 93–109. (2000).
27. M. Sparkes, Revealed: how Google moves international borders. *The Telegraph.* (2014, June 24). Retrieved from: http://www.telegraph.co.uk/technology/google/10922595/ Revealed-how-Google-moves-international-borders.html
28. Google maps. Google Maps drops Chinese name for disputed area in the South China Sea. *South China Morning Post.* (2015, July 14). Retrieved from: http://www.scmp.com/news/ china/diplomacy-defence/article/1838609/google-maps-drops-chinese-name-disputed-area-south
29. U.S., ASEAN warn Beijing. U.S., ASEAN warn Beijing over outrageous actions in S. China Sea. *The Yomiuri Shimbun.* (2016, February 18). Retrieved from: http://the-japan-news.com/news/article/0002756678
30. H.D. Lasswell, J.Z. Namenwirth, J. Z. *The Lasswell value dictionary.* New Haven: Yale University Press. (1969).
31. C.E. Osgood, The nature and measurement of meaning. *Psychological bulletin*, ***49***(3), 197. (1952).
32. C.E. Osgood, G.J. Suci, P.H. Tannenbaum, *The measurement of meaning.* Champaign, Illinois: University of Illinois Press. (1957).
33. J. Perlah, China, Vietnam Oil Rig Spat: China Says It 'Will Never Send Military'. *International Business Times.* (2014, June 13). Retrieved from: http://www.ibtimes.com /china-vietnam-oil-rig-spat-china-says-it-will-never-send-military-1601252

 DOI: 10.1051/shsconf/20173300021

Exploring Approaches to Drivers and Barriers of Corporate Social Responsibility Implementation in Academic Literature

Vimala Govindasamy[1,*], *Kalyani* Suresh[2]

[1,2]Department of Communication, Amrita School of Engineering, Coimbatore, Amrita Vishwa Vidyapeetham [Amrita University], India

Abstract. CSR implementation is a relatively untapped area of study. Barriers are components that block and test the execution of CSR in an organization while drivers are positive CSR performance indicators. There is a dearth of academic literature of exploration of barriers and drivers of CSR in developing economies. This paper focuses on synthesizing existing scholarly research literature on drivers and barriers of Corporate Social Responsibility from the perspectives of developed and developing economies. With most companies around the world joining the CSR bandwagon in a bid to positively influence their relationships with employees and stakeholders, it is imperative that CSR research addresses the barriers and drivers that companies are likely to face during the implementation of their CSR strategies.

1 Introduction

Notions of CSR have undergone vast changes in the conception - while early definitions looked at CSR as a tool for maximizing profits for the company's shareholders, recent definitions position CSR as a tool for societal and organizational development. CSR is a self-regulating integrated business model that is also known as corporate conscience, corporate citizenship, social performance, or a sustainable responsible business. In 1990s, CSR began to be defined by terms such as corporate social performance (CSP), stakeholder theory, and business ethics theory [1, 2, 3]. Barriers are components that block and test the execution of CSR in an organization while drivers are positive CSR performance indicators.

2 Theoretical Approach

Notwithstanding the theoretical advice, in order to analysis the drivers and barriers, the concept of CSR should be understood. There are many definitions of corporate social responsibility (CSR). Drawing on an impressive history associated with the evolving definition and conceptualization of corporate social responsibility, [4] established for its

* Corresponding author: veemala21@gmail.com

evolution and asserted that evidence of business community's concern for society has existed for centuries. The meaning of CSR may now also refer to activities which appear to advance a social agenda beyond that which is required by law (Siegel and Vitaliano, 2007)[5]. The definition may also include "responsible business practices related to sustainable economic growth and prosperity, social cohesion and equity, and environmental integrity and protection" [6].

Carroll's CSR are divided into four components; economic, legal, ethical and discretionary where an organization and they are required to fulfill all the four areas for CSR obligation. Philanthropic responsibilities is to be a good corporate citizen, ethical responsibilities, obligation to do what is right, fair and avoid any harm. Whereas for legal responsibilities is to obey the law, as they say the law is society's codification of right and wrong and lastly economic responsibilities is to be profitable and the foundation up which all others rest.

There are several definitions of institutional theory. An institutional theory of organization gives an affluent and intricate view of the organization. [7] has characterized institutional theory as components that assume an organization, either in a standard procedure or hierarchical structures. It could be sub-divided into two components either internal or external or qualities that may affect the decision making process either from the internal and external to the organization.

3 Literature Review

This paper focuses on synthesizing academic research literature on drivers and barriers of corporate social responsibility. With most companies around the world joining the CSR bandwagon in a bid to positively influence their relationships with employees and stakeholders; it is imperative that CSR research addresses the barriers and drivers that companies are likely to face during the implementation of their CSR strategies.

According to the study by Mirvis and Googins [8], generally all organizations are not in the same stages of corporate citizenship, but this can be explained by factors that can be divided into external pressures, or drivers and internal motivators, or drivers. The socio-economic and socio-political factors are the external forces that influence the organization. Traditions and values are the internal motivators, or drivers and community and politics are external forces.

When it comes to the drivers for development there are instances the drivers come from the middle and up to the senior management, but in normal circumstances is from bottom to the top and vice versa. Essential internal drivers are company's image as well as a reputation for fostering CSR as expected by the international and the national society. The investment in a community is basically influenced by governmental and the community at large.

Motivations for pursing CSR practices by firms based on items selected from CSR studies [9, 10] were found to be primarily ethical and value orientated followed by enhancing profits and reputations and satisfying stakeholder demands. Graafland and Van [9] states that strategic motives and moral motives are the two important elements which motivate organizations to practice CSR.

There is a dearth of academic literature of exploration of barriers and drivers of CSR in developing economies. With reference to the corporate cultures of the West and South East Asia, according to Raman [11] majority of CSR oriented studies are Europe and the US-centric, while [12] and other authors such as [13, 14, 15] discuss the shortage of theoretical and empirical literature available on the subject in the emerging economies.

Arevalo and Aravind [16] offer insights into models, motivations and barriers of CSR within Indian firms. They referred [13, 17, 18] in the delineation of four models of social

DOI: 10.1051/shsconf/20173300021

responsibility operating within Indian firms: the ethical (Gandhian) model; (focus on community welfare and ethical awareness) the statist (Nehruvian) (driven by legal requirement) model, the liberal [19] (focus on shareholder objectives) model and the stakeholder [20] model (stakeholder centric-focus). They categorized these motivators as caring versus profit driven or moral versus strategized. Their study also revealed the most significant barriers in the implementation of CSR study by Indian firms were insufficient resources (training and financial), lack of management support and training.

According to Mitra [21] despite the above scholars attesting to the four operational models of CSR the latest Indian context strongly focuses on a neoliberal profit centered model. Mitra [21] deals with a CSR discourse from the point of view of culture-centered approach (CCA) and outlines five motivations of CSR namely nation-building façade, underlying neoliberal logics, CSR as voluntary, CSR as synergistic, and a clear urban bias. (p. 131).

However, he integrates main stream CSR discourse through the culture - (e.g. faith, Swadeshi, Swaraj, Ahimsa, Ghandian ethics) - structure (state policy, organizational strategic, global/local flows) - agency (Subaltern, reframing of institutional responsibility etc.) triad of CCA. This study opens up the vast gap in the literature in terms of social reality and corporate rhetoric and highlights an alternative view of interpreting culture, business and society relations. Few studies examine barriers of CSR implementation with national approach.

Valmohammadi [22] investigated the concepts in an Iranian context pertaining to the understanding of CSR domain alongside with the analysis of CSR drivers and barriers. Iran is not prepared or responsive to CSR at the moment. However, they could see CSR emerges as an opportunity over the years. Respondents in the study revealed that CSR as the "forgotten element of businesses in Iran". For instance, community welfare and organization conscious, branding and image, profitability is constantly referred as CSR drivers. Therefore, environmental preservation, quality improvement and customer retention are the prominent key drivers.

In this study, Valmohammadi, [22] concentrates on the ISO 26000," code of conduct and contracts his investigation to seven intensive issues (supported by 105 feedback acquired from Iranian organizations): authoritative administration, human rights, work practices, the environment, reasonable working practices, end-users issues, and public participation and advancement.

Meanwhile, [23] recorded CSR perceptions among Bangladeshi managers, and CSR endeavours and barriers were investigated through face-face interviews. Perceived barriers fell under three categories, i.e. psychological, moral and systemic, with least important being systemic.

Further, [24] present results from 23 semi-structured interviews with senior corporate managers from Bangladeshi organizations on social reporting in Bangladesh and concentrate in particular on the non- disclosure on three issues: child labour, equal opportunities and poverty alleviation. The findings suggest that the main barriers to CSR non-disclosure are low resources, the profit angle, lack of knowledge about legal requirements, bad performance and the fear of bad publicity.

Nevertheless, [25] try to bridge the gap in existing Southeast Asian academic literature on barriers to CSR in the textile industry. They identify 12 common barriers as lack of stakeholder awareness, lack of training, lack of information, financial constraints, lack of customer awareness, lack of concern for reputation, lack of knowledge, lack of regulation and standards, diversity, company culture, lack of social audit, and lack of top management commitment. They propose and apply a model framework using fuzzy multi-criteria decision making (MCDM) to reveal the relatively most important barrier in the Indian textile sector as financial constraints and the least as diversity.

One way to deal with recognizing CSR barriers is to associate them to a firm's size. A study by Laudal, (2011)[26], investigating the Norwegian clothing sector, examined the drivers and barriers of CSR and then differentiated the changes of these elements within the SMEs and MNEs. His study found that in terms of public policy, the main focus for SMEs should be ensuring their CSR performance capabilities. Firms in the process of transition between an SME and an MNE should focus on strengthening market incentives for CSR and to disseminate best practices. As for large MNEs, cementing partnerships between governments and business representatives for furthering long-term interests, gains importance.

With the help of literature reviews and semi-organized meetings, [27] investigated the possibilities and barriers of CSR in Irish origin. He likewise contrasts the circumstances of both possibilities and barriers of CSR between big organizations and SMEs.

According to Mont and Leire [28] investigated the CSR drivers in supply chains from the perspective of Swedish corporate. Internal drivers ranged from risk reduction, protecting brand name, peaceful working environment for preserving the national image internationally, while the external drivers typically were fulfilling stakeholder (NGO, media) expectations, external evaluation and rating, reporting on sustainability issues.

Early literature on CSR in Malaysia [29, 30], suggested that CSR in Malaysia was still in the early stages. Active communication about CSR did not match awareness about it. It is believed that CSR practices were introduced to Malaysia through multinational organizations [31]. Post-independence, Malaysia was dependent on foreign investment, and joined the CSR fad. This was a major driver for adoption of CSR in Malaysia [32].

A 2013 report by UNICEF Malaysia (UNICEF Malaysia, 2013)[33] on the Malaysian government's implementation of Vision 2020 outlines four CSR related strategic challenges: (i) establishing a fully moral and ethical society, (ii) establishing a fully caring society and a caring culture, (iii) ensuring an economically just society, and (iv) establishing a prosperous society, with an economy that is fully competitive dynamic, robust and resilient. CSR policies and practices in Malaysia have shown tremendous participation from various sectors which include the Malaysian government, Bursa Malaysia and Companies Commission of Malaysia to create awareness and encourage companies to promote CSR in their business community and environment to uplift the status of CSR.

Based on [34] in their study identify internal and external enablers of CSR in sustainable supply chain practices in the Malaysian context as people's issues; strategic issues and functional issues internal drivers include management commitment and employee involvement, competitive advantage, reputation and risk management, organization size, purchasing and supply. External drivers include government, customers, suppliers, investors and NGOs. Internal barriers are lack of management commitment, resource costs, and lack of training. External barriers are listed as government, customers, media organizations and technology.

Table 1. Summary of CSR drivers and barriers identified by academics

Author	CSR drivers and barriers	Country in context
Grafland and van de Ven, (2006) and Hemingway and Maclagan, (2004)	Strategic and moral motivators - ethics, values, profit orientation, reputation, satisfying stakeholders.	Netherlands
Mirvis and Googins, (2006)	Companies at different stages of CSR implementation have different external (government and society) and internal (company image) motivating forces	Finland

	affecting them.	
Belal and Owen (2007)	Main barriers - non-disclosure, low resources, the profit angle, lack of knowledge about legal requirements, bad performance and the fear of bad publicity.	Bangladesh
Sweeney, (2007)	Possibilities and barriers of CSR in Irish origin. Contrast the circumstances of both possibilities and barriers of CSR between big organizations and SMEs.	United Kingdom
Amran and Susela, 2008	Driver - post-independent Malaysia dependent on foreign investment, and joined the CSR fad.	Malaysia
Mont and Leire, (2009)	Drivers in supply chains - internal drivers - risk reduction, protecting brand name, peaceful working environment for preserving the national image international; external drivers - fulfilling stakeholder (NGO, media) expectations, external evaluation and rating, reporting on sustainability issues.	Sweden
Duarte and Rahman, (2010)	Perceived barriers fell under three categories, i.e. psychological, moral and systemic with least important being systemic.	Bangladesh
Arevalo and Aravind, (2011) referring to Balasubramaniam et.al. (2005), Mehta et.al. (2006) and Kumar et. al. (2011)	Motivations and barriers of CSR within Indian firms – Motivators - Moral versus Strategic/caring versus Profit driven. Barriers - insufficient resources (training and financial), lack of management support and training.	India
Valmohammadi, (2011)	CSR as the "forgotten element of businesses in Iran". Prominent key drivers are environmental preservation, quality improvement and customer retention.	Iran
Laudal (2011)	Norwegian clothing sectors, examined the CSR drivers and barriers and differentiated the changes of these elements within the SMEs and MNEs.	Norway
Mitra, (2012)	Culture-centered approach (CCA). Outline five motivations - nation-building façade, underlying neoliberal logics, CSR as voluntary, CSR as synergistic, and a clear urban bias. Strong focus on the neoliberal profit centered model.	India
Shen, Govindan, and Shankar (2015)	Barriers in the textile industry - lack of stakeholder awareness, lack of training, lack of information, financial constraints, lack of customer awareness, lack of concern for reputation, lack of knowledge, lack of regulation and standards, diversity, company culture, lack of social audit, and lack of top management commitment.	Southeast Asian
Tay, Rahman, Aziz, and Sidek, (2015)	Identify internal and external enablers in sustainable supply chain as - people's issues; strategic issues and functional	Malaysia

DOI: 10.1051/shsconf/20173300021

	issues. Internal drivers - management commitment and employee involvement, competitive advantage, reputation and risk management, organization size, purchasing and supply. External drivers - government, customers, suppliers, investors and NGOs. Internal barriers - lack of management commitment, resource costs, and lack of training. External barriers - government, customers, media organizations and technology.	

4 Conclusion

A synthesis of literature available on drivers and barriers in CSR implementation in the West reveals a vast gap in the literature in terms of social reality and corporate rhetoric, scholarly research in Southeast Asia the need for highlighting an alternative view of interpreting drivers and barriers to implementation of CSR taking into account the business, culture, and society triad in which the corporation is situated. This study also comprehends that the drivers and barriers of CSR vary according the size of corporations and the strategic versus moral motives adopted by them in the pursuance of CSR.

References

1. Porter, M. Kramer, *Strategy and Society: The link between Competitive Advantage and Corporate Social Responsibility*. Harv. Bus Rev. Dec; **84(12):** 78:92, 163 (2006)
2. C.B. Bhattacharya, D. Korschun & S. Sen, *"Strengthening Stakeholder – Company RelationshipsThrough Mutually Beneficial Corporate Social Responsibility Initiatives"* S. J Bus Ethics **85:**257. doi: 10.1007/S10551-008-9730-3, (2009)
3. D. Jamali, *A stakeholder approach to corporate social responsibility: A fresh perspective into theory and practice*. J Bus Ethics **82:213**. doi: 10.1007/s10551-007-9572-4 (2008)
4. A. Carroll, A. *Corporate Social Responsibility: Evolution of a definitional construct*. Bus Soc **38**(3): 268-295. (1999)
5. D.S. Siegel, D.F. Vitaliano. *An empirical analysis of the strategic use of corporate social responsibility*. J. of Econ. and Mgmt. Strgy., **16**(3), 773-792 (2007)
6. N.A. Torugsa, W. O'Donohue, R. Hecker. *Proactive CSR: An Empirical Analysis of the role of its economic, social and environmental dimensions on the association betweencapabilities and performance*. J Bus Ethics **115**:383, doi: 10.1007/s10551-012-1405-4 (2013)
7. L.G. Zucker, *Institutional Theories of Organisation*. Annu. Rev. So. Vol **13:**443-464, doi: 10.1146/annurev.so.13.080187.002303 (1987)
8. P. Mirvis, B. Googins, *Stages of Corporate Citizenship: A Development Framework*. Cal. Mgmt. Rev. **48:2**, pp.03-126 (2006)
9. J. Graafland, B. van de Ven, *Strategic and moral motivation for corporate social Responsibility*. J. of Corp.Ctznship., **22**:111-123, (2006)
10. C.A. Hemingway, P.W. Maclagan, *Managers' personal values as drivers of corporatesocial responsibility*. J. Bus Ethics Vol. **50(1)**:33-44 (2004)
11. S.R. Raman, *Corporate social responsibility in India: a view from the top.*Gobal Bus Rev. Vol. **7(2)**:313-24 (2006)

12. B. Arora, R. Puranik, *''A review of corporate social responsibility in India'', Society for International Development.* Vol. **47(3):** 93-100 (2004)
13. N.K..Balasubramanian, D. Kimber, F. Siemensma, *Emerging opportunities or traditions reinforced?''*, J. of Corp. Ctzship., Vol. **17**:79-92 (2005)
14. M. Mitra, *It's only business! India's corporate social responsiveness in a globalized world. New Delhi, India*: Oxford Uni. Press (2007)
15. M. Narwal, T. Sharma, *"Perceptions of corporate social responsibility in India: An Empirical study"*. J. of Knowledge Globalization Vol. **1**:61-79 (2008)
16. J.A. Arevalo, D. Aravind, D. *"Corporate social responsibility practices in India: approach, drivers, and barriers"*, C.G. Vol. **11(4):** 399-414 (2011)
17. Mehta, V., John, P., Kumar, A., Maitra, I., Puranik, R., Shrivastava, S., et al. *Delivering value: An exploration of community development vehicles adopted by corporate in India. New Delhi: Partners in Change* (2006)
18. R. Kumar, D. Murphy, V. Balsari, *"Altered images: the 2001 state of corporate social responsibility in India"*. Business-Social Partnership: Beyond Philanthropy Conference IAM Calcutta, India, Dec. (2011)
19. M. Friedman, *The social responsibility of business is to increase profits.* The NY Times Mag., September 13:32-33 (2007)
20. R.E. Freeman, *Strategic Management: A Stakeholder Approach*, Pitman Series in Bus And Public Policy, Boston, MA.
21. M. Mitra, *"My Country's Future": A Culture-centered Interrogation of CSR in India.*J. Bus. Ethics (2012) **106**:131-147. doi:10.1007/ss10551-011-0985-8. Recd. 29 July 2009/Accepted:21 July 2011/Pub. online: 4 Aug. 2011@Springer Science+ Bus. Media B.V. (2011 and 2012)
22. C.Valmohammadi, *'Investigating corporate social responsibility practices in Iranian organisations: an ISO 26000 perspective'.* Bus. Strgy. Series, Vol. **12(5):** 257-263 (2011)
23. F.P. Duarte, S. Rahman, *Perceptions of corporate social responsibility by Bangladeshi Managers: An exploratory study.* Int. Rev. Bus. Res. Pap. 6(5), 119–136 (2010)
24. A.R. Belal, D.L. Owen, *"The views of corporate managers on the current state of, and future prospects for, social reporting in Bangladesh: An engagement-based study"*,Accting, Auditing & Acctbility J. Vol. **20(3):** 472-494 (2007)
25. L. Shen, K. Govindan, M. Shankar, *Evaluation of barriers of corporate social responsibility using an analytical hierarchy process under a fuzzy environment – A Textile Case,* Sustainability 7 (3):3493-3514; doi: 10.3390/su7033493 (2015)
26. T. Laudal, *Drivers and barriers of CSR and the size and internationalization of firms.* Soc. Responsib. J. **7**: 234-256 (2011)
27. L. Sweeney, *Corporate Social Responsibility in Ireland: barriers and opportunities experienced by SMEs when undertaking CSR.* Corp. Gov.: The Int. J. of Bus. In Soc. 7 (4): 516–523 (2007)
28. O. Mont, C. Leire, *"Socially responsible purchasing in the supply chain: drivers and barriers in Sweden".* Soc. Responsib. J. **5** (3): 388-407 (2008)
29. AB. Amran, S.S. Devi, *Corporate social reporting in Malaysia: An institutional perspective.* WREMSD Vol. 3(1): 20-36 (2007)
30. M. Nejati, A. Amran. *CSR and SMEs: Exploratory study on motivations from a Malaysian perspective.* Bus. Strategy Series **10** (5): 259–265 (2009)
31. A. Amran, M.M. Zain, M. Sulaiman, T. Sarker, S.K. Ooi, *Empowering society for better corporate social responsibility (CSR): The case of Malaysia.* (2013)
32. A. Amran, S.Susela Devi, *The impact of government and foreign affiliate influence on corporate social reporting: The case of Malaysia.* Mgrial Auditing J. **23**(4):386–404 (2008)

 DOI: 10.1051/shsconf/20173300021

33. UNICEF Malaysia, *Corporate Social Responsibility Policies in Malaysia: Enhancing the child focus. Unite for children* retrieved from http://www.unicef.org/malaysia/Unicef_CSR_Msia_110713_lowres.pdf on 30.08.2015 (2013)
34. M.Y. Tay, A.A. Rahman, Y.A. Aziz, S. Sidek, *A Review on Drivers and Barriers towards Sustainable Supply Chain Practices.* Int. J. of Soc. Sci. and Humanity, Vol. **5** (10)**:** 892-897 (2015)

Relationships between the Advertising Appeal and Behavioral Intention: The Mediating role of the Attitude towards Advertising Appeal

Syed Hassan Raza[1,*], *Hassan* Abu Bakar[2], and *Bahtiar* Mohamad[3]

[1]School of Multimedia Technology & Communication, 06010 UUM Sintok, Kedah, Malaysia/ Lecturer Bahauddin Zakariya University, Multan, Pakistan
[2,3]Department of Communication, School of Multimedia Technology & Communication, 06010 UUM Sintok, Kedah, Malaysia

Abstract. The link between the advertising appeal, attitude and behavioral intention is dynamic in nature, however, little research has been pursued towards explaining the mediating role of attitude towards the advertising in link of the advertising appeal with the behavioral intention. Researchers have identified the importance of the attitude in the connection of the advertising appeal within the framework of the (TPB) which remains unaddressed in the explicating the mediation of the attitude in relationship of advertising appeal and behavioral intention. This paper aims to provide insight of the attitude towards the advertising as the mediating factor in explaining the relationship between the advertising appeal and behavioral intention. In addition, this paper is only a theoretical exploration, it is expected that this work lead towards an explanation of the mediating role of the attitude to address the relation of advertising appeal and intention which may be studied further to determine the empirical finding about the other dynamic of the (TPB).

1 Introduction

Despite, the fact that the dynamic relation of attitude and Advertising appeal is widely explained in the body of the literature in different perspectives, but remains less clear with the mediating aspect of the attitude in relationship of advertising appeals and behavioral attention [1-2]. Hence, the connection of the attitude with the behavioral intention and its effect in the sphere of advertising is consistency with the Theory of Planned Behavior (TPB) [3-4]. Thus, it remains minimal attention paid in the literature to address the issue how the dynamics of attitude mediate the relationship between Advertising appeals and to predict behavioral intention.

Moreover, in the previous studies mediation of the attitude to the Advertising appeal towards the behavioral intention is specifically based on the attitude towards advertising brand or product [5]. Therefore, it remains lack of the systematic inquiry in the context of the empirical explaining the mediation of the attitude to predict the behavioral intention and

* Corresponding author: sherazibzu@gmail.com

Advertising appeal [5]. Previous literature indicates that attitude does not solely predict the behavioral intention in consistent with the theory of planned behavior, where as there are many antecedents of attitude which are explored in the past studies towards behavioral intention but remains detached [6]. The link between the advertising appeal, attitude and behavioral intention is dynamic in nature [7]. Consequently, mediation of attitude in the Advertising appeal and behavioral intention lacks empirical evidence in the body of knowledge as less attention is paid towards this specific dimension which is the contribution of this study.

Thus, mediating role of the attitude in predicting behavior intention can be determined altogether with the message contained in the advertising and its appeal [8]. Therefore, it provides another argument to investigate the mediating direction of attitude towards predicting behavioral intention as it is affected by the Advertising appeal but remains minimal in the past literature. Sallam and Wahid, [9] noted that advertising effects on behavioral intention is routed through the attitude as it serves as the mediating factor between the behavioral intention and Advertising appeal.

2 Review of Literature

2.1 Theory of planned behavior

In explaining the relationship of attitude and behavioral intention key concepts, the Theory of Planned Behavior (TPB) proposed by the Ajzen and his contemporaries discussed that the relationship of attitude and behavioral intention serves as a wide range of explanations and predictions towards the behavioral intentions on the account of the attitude [3,4,10,12]. TPB hypothesizes that, on the description of the accuracy of the attitude contribute as a potential antecedent to predict the behavioral intention. Therefore, behavioral intention is considered as the depending factor on the motivation and likeness emphasize by the measure of the attitude and premise behavioral intention as a rationale process in which individual's course and make use of the presented information to reach at a intention consideration [10-12].

What's more the previous studies indicate, there is a variety of the inconsistency in the results of relationship between the attitude and behavioral intention in terms of contradictory in variance which suggests to conduct more research and contribution in this area [13–16]. Armitage and Conner [14] found after the Meta-analysis of 185 previous studies carried out on the TPB and found variance ranging from 27% t0 39%. Furthermore, he suggests that additional investigations are essential to give improved understanding and explanation of the other potential factors which may involve in the relationship of TPB key constructs. Therefore, this current article addresses the aforesaid issue as a contribution to test the important forerunner factors like Advertising appeal involved in the development of the attitude-behavioral intention link.

2.2 Advertising appeal and attitude

Advertising appeal is the defined as the management and arrangement of the advertisements to the potential purchasers [17]. Advertising appeal is also considered as the key objective of the advertising as it tempted to give appeal to the targeted audience. The Advertising appeal is given by several modes of the advertising ranging from the television advertising to the billboards. Therefore, this is also the aim of the any advertiser to enhance the number of the Advertising appeal as it can probably increase the frequency to attain the basic objectives of the advertising [18]. Furthermore, repetitive Advertising appeal help to seek the objective to motivate the consumers to pursue towards the product or the services desired by the advertiser as it directly affects the attitude of the individual by manipulating ideas [19].

Whereas, attitude is considered as an expression of the goodwill or the disapproval for the thing, idea or the person. Attitude is also described as the position of the individual based on its internal evaluation in respect of the tendency or the orientation towards the person, character or the object and depends on the experiences and motivation [9]. Furthermore, attitude differs on the bases of the motivation asked towards the specific stimulus in the form of an idea, person or product so that the attitude formulation is highly dependent on the position of the asked phenomena. Therefore, the attitude of an individual learns and gets affect from the Advertising appeal as it motivates the individual towards the desired position of the attitude by the manipulating the thinking of an individual [20].

Past research has also suggested that the advertising impacts the attitude of the audience due to the persuasive nature of the advertising. Furthermore, Advertising appeal leads the attitude which indeed affects the behavioral intention [21]. However, Inadequate inquiries have been done in the past to address the relationship of the attitude as a driver of behavioral intention in result of Advertising appeal [13-22]. Advertising appeal can be the potential influential factor in the development of the behavioral intention as it has a direct effect on the attitude, but it remained unclear in the past literature as systematic inquiry is needed to understand this mediated relationship.

Therefore, on the basis of the above mentioned literature, it may be postulated that attitude and Advertising appeal have direct dynamic link to define the relationship of advertising and behavioral intention as it remains an uncertain issue in the research. Hence, if the relationship of attitude and the Advertising appeal positive it positively affects the behavioral intention, we would expect that at the same time if it is negative it can affect vice versa. Thus we propose the following preposition.

Preposition 1: There is a direct positive relationship between Advertising appeal and attitude.

2.3 Attitude and behavioral intention

Behavioral intention is defined as the individual's perceived chance of the likeness or subjective prospect to adopt or engage in the certain behavior. Moreover, it is supposed as the instant forerunner of the behavior as it designates the individual's willingness to execute the certain behavior [23]. Furthermore, behavioral intention is also considered as the particular predictor of an individual's behavior, but on the other hand formulation of the behavioral intention depends on some predictors as mention in the (TPB), the attitude towards the behavior is one of the significant predictor of the intention [24].

Despite the fact that the direct relationship of attitude and behavioral intention is highly examined and in many previous studies [25–27] and found it significant. While on the other hand, there is enough evident literature which indicates the role of certain antecedents to formulate the attitude towards the intention [28-29]. Thus, these facts improve the initiative to examine the mediating role of attitude in to explain the relationship of behavioral intention with the factors involved in the attitude formation like the Advertising appeal. Therefore, to explain the said phenomena we propose that:

Preposition 2: There is a positive relationship between attitude and behavioral intention.

2.4 Advertising appeal and behavioral intention

There are enough traces in the literature providing evidence to understand and explain the relationship between advertising and behavioral intention[26,30,31] however, at the same time it remains less explanatory in term of mediation of the attitude in the dynamic link between both concepts [10,32,34]. In previous studies it is found that Advertising appeal results to have positive behavioral intention as it positively influence the attitude or vice versa which is evident of the fact that there is a link between advertising and behavioral intention with the mediation or facilitation of the attitude which need more explanation by considering it for further empirical research [35].
It is indicated in a study that the elements of the advertising like information to the people about the advertised thing or idea work in the positive direction to have an influence on the behavioral intention of the people. He also suggested that the perspectives of the entertainment and credibility of the advertising results significant effect on the behavioral intention, but he also found some other factors which may influence the relationship of Advertising appeal and intent. Therefore, it may be assumed on the basis of many studies accessible in the previous literature which indicated involvement of certain factors like attitude as the predictor or mediator between the link of the Advertising appeal and behavioral intention but remains detached [36]. Based on this, we propose that:

Preposition 3: There is a positive relationship between the Advertising appeal and behavioral intention.

2.5 Attitude as Mediation

The attitude of the individual is the essential determinant of the behavioral intention as it serve as the considerable predictor of it as suggested in (TPB), but at the same time it is also found in studies of many researchers that it is a significant fact that attitude also have role to explain the relation of the behavioral intention and Advertising appeal [20,36,37]. There is enough evidence about the effects of antecedents on attitude towards the intention within the framework of (TPB) which strengthen the idea of attitude's mediation inquiry in connection of the Advertising appeal and intent [11,16,25,38]. However, no empirical compilation and enough studies in the past have been done to explain the mediating role of the attitude in the connection of the Advertising appeal to the behavioral intention which is proposed in this current study. Based on this, we propose that:

Preposition 4: The relationship between the Advertising appeal and behavioral intention is mediated by attitude.

On the basis of the above arguments, we propose the following preposition.

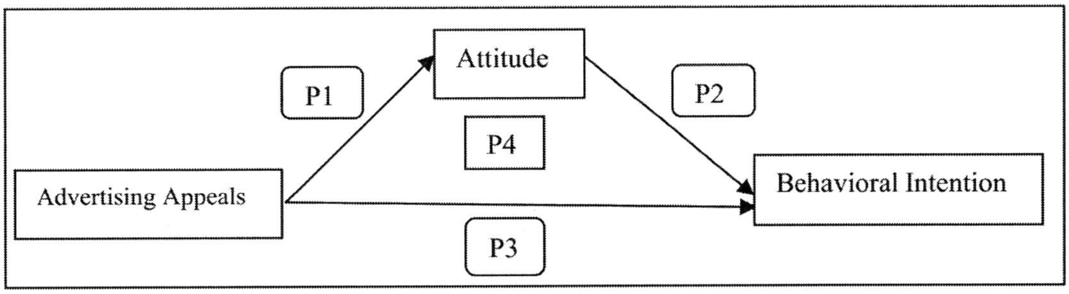

Fig. 1 Proposed Model of Mediation of Attitude in relationship of Advertising and Behavioral Intention.

References

1. U. C. Eze and C. H. Lee, *Int.* J. Bus. Manag., **7**, 94–108, (2012)
2. E. Kaynak, A. Kara, and C. S. F. Chow, *J. Asia-Pacific Bus.*, **14**, 88–106, (2013)
3. I. Ajzen, *Psychol. Health*, 26, 1113–1127, (2011)
4. I. Ajzen, *Orgnizational Behav. Hum. Decis. Process.*, **50**, 179–211, (1991)
5. L. Zarantonello and B. H. Schmitt, *J. Brand Manag.*, **17**, 532–540, 2010.
6. M. G. Kenney and N. M. Khanfar, *Serv. Mark. Q.*, **30**, (2015)
7. J. D. Martín-Santana and A. Beerli-Palacio, *J. Euromarketing*, **17**, 159–181, (2008)
8. U. R. Orth, H. F. Koenig, and Z. Firbasova, *Eur. J. Mark.*, **41**,327–348, (2007)
9. M. A. A. Sallam and N. A. Wahid, *Int. Bus. Res.*, **5**, 55–66, (2012)
10. I. J. Donald, S. R. Cooper, and S. M. Conchie, *J. Environ. Psychol.*, **40**, 39–48, (2014)
11. M. F. Chen and P. J. Tung, *Int. J. Hosp. Manag.*, **36**, 221–230, (2014)
12. R. J. Vallerand, P. Deshaies, J.-P. Cuerrier, L. G. Pelletier, and Et Al, *Journal of Personality and Social Psychology*, **62**, 98–109, (1992)
13. L. R. Glasman and D. Albarracín, *Psychol. Bull.*, **132,** 778–822, (2006)
14. C. J. Armitage and M. Conner, *Br. J. Soc. Psychol.*, **40**, (2001)
15. T. L. Webb and P. Sheeran, *Psychol. Bull.*, **132**, 249–268, (2006)
16. G.-J. de Bruijn, K. Verkooijen, N. K. de Vries, and B. van den Putte, *Psychol. Sport Exerc.*, **13**, 771–778, (2012)
17. S. Schmidt and M. Eisend,*J. Advert.*,**14**, (2015)
18. K. Y. Wang, E. Shih, and L. a. Peracchio,*Int. J. Advert.*,**32**, (2013)
19. W. D. Crano and R. Prislin, *Attitudes and attitude change.* **18**, (2008)
20. C. Dianoux, Z. Linhart, and L. Vnouckova, *J. Compet.*,**6**, 87–103, (2014)
21. A. Morris and L. A. Kahlor, *Howard J. Commun.*, **25**, 415–430, (2014)
22. H. N. T. E.Ma.Tsung-kuang , J.Y Dong, *Int. J. Organ. Innov.*, **2**,190–198, (2009)
23. B. Kidwell, *Adv. Consum. Res. Consum. Res.*, **34**, (2007)
24. S. E. Zemore and I. Ajzen, *J. Subst. Abuse Treat.*, **46**, 174–182, (2014)
25. B. Schreurs, E. Derous, E. a J. van Hooft, K. Proost, and K. de Witte, *J. Organ. Behav.*, **30**, pp. 761–783, (2009)
26. W.-H. Hsiao and T.-S. Chang, *Behav. Inf. Technol.*, **33**,730–742, (2013)
27. C. R. Nigg, S. Lippke, and J. E. Maddock, *Psychol. Sport Exerc.*, 10, 219–225, (2009)
28. O. M. & N. L. R. Peter, *Int. J. Soc. Sci. Enterpreneursh.*, **3**, (2014)
29. D. Sari and B. Kusuma, *Asian Mark. J.*, **6**, 50–63, (2014)
30. A. Muk and C. Chung, *J. Bus. Res.*, **68**, (2014)
31. Y. J. Kim, D. Njite, and M. Hancer, *Int. J. Hosp. Manag.*, **34**, 255–262, (2013)
32. J. I. Rojas-Méndez, G. Davies, and C. Madran, *J. Bus. Res.*, **62**, 947–954, (2009)
33. Y. Wang and S. Sun, *Int. Bus. Rev.*, **19**, 333–344, (2010)

34. M. de Mooij and G. Hofstede, *Int. J. Advert.*, **29**, 85–110, (2010)
35. S. D. Myers, M. B. Royne, and G. D. Deitz, *J. Public Policy Mark.*, 30, 110–118, (2011)
36. Ling, K. C., Piew, T. H., & Chai, L. T. *Canadian social science*, 6(4), 114-126.(2010)
37. A. Tehrani and A. G. Tehrani, *J. Bus. Manag. Rev.*, **3**, 54–67, (2014)
38. H. H. Cheng and S. W. Huang, *Int. J. Inf. Manage.*, **33,** (2013)

Amateur Radio Communication Technology Contingency Communication in Emergency Situation

Azlee Johar[1,*], *Wan Rozaini* Sheik Osman[2], and *Azizi* Ab. Aziz[3]

[1,2,3] School of Computing, Universiti Utara Malaysia, 06010 UUM Sintok, Kedah, Malaysia

Abstract. Disasters cannot be avoided, but efforts can be made to reduce death, injuries and property damages through better communications and better planning to deal with it before it happen. Thus we need to adopt available technology, like the Amateur Radio Communication Technology (ACT), to face this problem. These include providing awareness, training and support for the communities who adopt the technology for Emergency Communication (EC). Through having more qualified Amateur Radio Communities (ARC) in the adoption of the ARCT, we would be able to improve the communication before, during and after the disasters struck. There are many challenges and issues in the adoption of the ARCT that would part of this study. This article also describes several case studies of communication failure during Hurricane Katrina (2005) in the USA, as well as disasters in several other countries and Malaysia. The tragedies led to failure of existing communication infrastructure, and were reprieved by the help of the voluntary ARC to support the relief agencies in the evacuation and assistance the disaster victims. Therefore, this article will highlight the significant issues that influence the adoption of ARCT in EC.

1 Introduction

Emergency Communications (EC) may include early warning, command on the evacuation, curfews and other self-protection measures and other matters which influence relief, response and recovery activities [2, 3, 18]. Many disasters led to the breakdown of existing infrastructure, and resulted in the loss of many live and damages to properties [1]. Indeed, a wireless communication technology like Amateur Radio communication Technology (ARCT) as a contingency communication is suitable to facilitate the dissemination and sharing of information among stakeholders and role players. ARCT is important role in restoration in emergencies due to easy to operate and easily accessible simultaneously everyone within the same time in form of radio waves. This fact has been proven in several emergency incidents that have occurred around the world and its role cannot be denied. However, [23, 26, 29] argued that the adoption of ARCT is still having gaps and not widespread use, particularly among developing country.

* Corresponding author: azlee@gmail.com

As researchers in the field of wireless communications technology it is interesting to see the slogan from the International Amateur Radio Community (IARC), *"When all else fails, ham radio works"*. The slogan belongs to the American Radio Relay League [5]. These statements indicate that ARCT is important and can be a very useful as it works every time when existing communication fails to operate in emergency situation.

The Emergency Situation (ES) could be pre and post of most disasters [15]. That means when every existing communication facilities have been destroyed due to any huge disaster, the contingency communication in ES should be performed. In this situation the role of Amateur Radio Community (ARC) become important in adoption and use of ARCT to support relief agencies. The importance of ARCT as contingency communication has proved during the disasters struck in most country and in Malaysia.

2 Methodologies

This article is a qualitative research by reference related books and through studies related journals. The outline and angle of this article is only focus to the adoption the ARCT. The methodology involved casual interview approach to certify usability former to get perfect evidence. Casual interviews with several individual among Amateur Radio Community (ARC) and individuals involved in disaster management and emergency communications are conducted to obtain an issues related to the adoption of ARCT. A casual interview is to realize their perception of the issues at hand also conducted. Some of significance articles and reference books which are related to ARCT were obtained from them.

3 ARCT contingency communications

The most apparent and unique feature regarding the adoption of ARCT is the parallel volunteer work by the same ARC during Hurricane Katrina in New Orleans (2005) in USA. ARC provides much information to the relief agencies in rescue missions and evacuation missions. Hurricane Katrina strikes the New Orleans city in USA and cause the failure of the existing communications. An issue is about the failure to prepare an early warning before a disaster strike that caused major loss in lives and properties. This tragedy is a great disaster for a strong government as USA [4, 20, 21].

Communications became an increasingly critical function in the emergency management during that time. The sharing and dissemination information of timely and accurate to the general public, elected and community officials and the media is a most important role in the effective management of disaster response, however, the relief and recovery action was totally lost in New Orleans city after Hurricane Katrina struck. Communications failures by government responders in Hurricane Katrina (2005) were noted in the report prepared by the United States House of Representatives. According to [6, 8, 10] the president of USA during that time, George W. Bush lamented about the communication failure when Hurricane Katrina struck;

> *"What happen with communications structure during that time, why coordination and information sharing and dissemination among local, state and federal governments too miserable and why situational awareness was so foggy, for so long and why unsubstantiated rumours and uncritically repeated press reports at times fuelled by top officials were able to delay, disrupt, and diminish the response"*.

The failure of existing communications mechanism among local, state and federal governments, relief agencies and community in sharing and dissemination information in ES, due to the violence storm, caused huge community concerns and further delayed relief during Hurricane Katrina in New Orleans that year. Fortunately there was an ARC who volunteered to build a tower and setup the antenna to communicate via ARCT among them and relief agencies to report on conditions that enable further action by the government to assist the victims of the disaster [20]. Thus, the adoption and use of ARCT as the

DOI: 10.1051/shsconf/20173300023

contingency communications is effectively used for the EC in the USA after Hurricane Katrina.

Many Malaysians remember the landslide disaster that occurred at the Highlands Tower apartment in Bukit Antarabangsa, Kuala Lumpur in 1999. The incident not only trapped many people as the main road was buried by the loose soil but also claimed the loss of many lives of the Highlands Tower apartment residents. The emergency situation has led the country into panic and confusion caused by the irregular communication [22, 25].

The whole country was in shocked, safety and medical officers were deployed to support and relief for those who were trapped. The main problem that arises is how to coordinate emergency communications among relief agencies at the time due to traffic communications congestion. However, according to [21] disaster management and emergency communication will not execute smoothly without effective communication mechanism. In such instances, common communication system like mobile phones, fix line and wireless communication will not function properly due to traffic congestion.

The ARC is experienced in adopting, using and handling traffic congestion wireless communications devices when required to support. Therefore, the government relief agencies have decided to request volunteers among ARC to support and to coordinate the EC among agencies [9, 22].

They were a group of volunteers from ARC, which is known as Malaysian Amateur Radio Emergency Service Society (MARES). They were able and successful to provide contingency communication via ARCT during Highlands Tower tragedy [22]. Thus, when in emergency situation, systems and mechanism communication via ARCT which are adopted and used among ARC are able to play the role.

According to [5] many factors contribute to the common communication lines disrupted, including faulty tower or substation communication station power outages. However, the disorder does not occur among ARC in adoption and use of ARCT, because to them, it's not too difficult to setup and to operate but can be difficult to ordinary people. This is because ARCT is unique hobbies among ARC. The passion and close rapport between the ARC members are hard to define and compare [26, 27].

According to [22] in 1999 MARES were called to support and to coordinate communication between Jabatan Pertahanan Awam Malaysia (JPAM), Polis (PDRM), Department of Civil Aviation (DCA) and the Royal Malaysian Air Force (RMAF) in the forest in Mantin, Negeri Sembilan, due to the Cessna Aircraft crash. ARCT base station and repeater communication tower was setup near the aircraft crash location to support relief agencies. So, [9. 19] argued that the advantages of communicating through ARCT were fully utilized during, search and rescue activities. During the incident, others wireless communications such cell phones do not work due to out of coverage area except ARCT.

In addition, each relief agencies use trunk radio communication which is in different frequency and therefore, communication is limited [9, 17]. Thus, MARES was the network who connected and coordinated the communication order among relief agencies.

Most significantly, the importance of ARCT can be seen at the time of the tsunami in Aceh, Indonesia in year 2004 [30]. Even far away, radio frequency from Aceh to request aid has been received by the MARES in Kuala Lumpur. MARES have answered the emergency call from ARC in Aceh that was supposed to be received by the Indonesian communication and relief team in Medan. As a solution, MARES resend the emergency message to the Indonesian authorities. The issues may be due to terrain of mountains, hills and the weather that has deflected the radio signals to Kuala Lumpur from Aceh, even though the distance to Medan is nearer [22].

Consequently, the contingency communication provided during the ES and the contributions of MARES in Kuala Lumpur, made MARES a mediator to help in sharing and dissemination of information between the ARC in Aceh and the Indonesian authorities in Medan. MARES were also involved in the rescue operation for tsunami victims in Aceh by providing ARCT equipment to the affected residents.

However in normal situations, MARES or other ARC are forbidden to intrude authorities' frequency or any *'Air band frequency'*. ARC in Malaysia only can use those frequencies with the order and permission of the Malaysia Multimedia and Communication Commission (MCMC) during emergency and in a controlled manner and procedure [24]. This is subjected to the Communications Policy and Acts in order to avoid disruption of communication between the authorities [28, 29].

4 The Significance Contingency Communication

The adoption and use of ARCT among ARC as contingency communication are very significant. In the world history, the starting point of ARCT as contingency communication in ES was seen from January 31 to February 1, 1953. The combination of high tides and severe weather caused huge flooding in Britain, Holland and Belgium that caused more than 2000 loss of life, crops destroyed and huge facilities damage. In the early time using ARCT is a unique hobby among ARC in rural area in Britain and Holland. Most members are farmers and fisherman. Nevertheless their hobby has changed to a more important role in the aftermath of the massive tide that struck their countries. This is the beginning point of ARCT work as contingency communication in the world [7, 11, 12, 18].

This leads to the establishment of a group of ARC as contingency communication in ES in their country. Until now, two ARCT stations are active in the UK and Holland as benchmark and to maintain public awareness about the role of ARC in ES [13, 14, 16, 23].

ARC is the key player in the emergency response recorded in Britain and the response leads to official gratitude of ARCT as a source of information sharing and dissemination in emergency situation during any disasters.

According to [14, 25, 27], in developing countries, ARCT can be the contingency communication among the ARC, stakeholder and public to share and dissemination of information, mostly to use in a rural area and generally in emergency situation.

During the ES in disaster areas, information sharing and dissemination through ARCT is an ideal solution to support and evacuate the victim. Usually, they cannot travel to get the necessary information or use the services, because of lack of transportation, disaster factor, time, or bad weather [3, 27, 29].

It is quite evident that, ARCT services have many benefits such as mobility, ubiquity, ease of use, on-time information delivery, improving emergency management, quick access to role player and stakeholder services and who do not have access to cellular communication and telecommunication services provider. Similarly there are many significant researchers who consider the adoption and use of ARCT in disaster areas as the effective solution to interact with government services, particularly for support and relief team during emergency situations. The most important is the cooperation and contribution of developed countries to underdeveloped countries in providing a framework for effective communication in ES during disaster. Thus ARCT has become a main source of EC in ES during most disaster.

It is obviously that ARCT role as a contingency communication in ES is significance. [6, 8, 21] stated an effectively EC through contingency communication is to support and to ensure public safety, protect property, facilitate the response, prompt cooperation, public confidence and help families to reunite during any emergency situation. The extent to which community to respond to an EC via contingency communication is influenced by several factor, including individual characteristics and perceptions, whether the message comes from a trustworthiness source, compatibility technology and how the information sharing and dissemination among role player and stakeholders.

DOI: 10.1051/shsconf/20173300023

5 Conclusion

We have many communication devices to adopt as contingency communication in ES, including in-person events, mass media and broadcast media, internet and social media but the importance thing is to provide trustworthiness information and accurate to generate public safety. Each contingency communication device has advantages and disadvantages depending on the communication impartial and the intended to public. Whatever communication tools to adopt and use is to ensure EC are clear, specific and acceptable information, are synchronise with other information shared and disseminated and to the public during disaster.

Actually the usage of ACRT among ARC are very widespread as only a hobby throughout the world, but their capacity and capability is still less adopted as a contingency communication in ES in most country. Even the military of some country strictly control and gives special emphasis to the communication methods and tactical for special security forces in using ARCT and the public are prohibited from using ARCT.

Lastly, the researchers will propose a study to identify the factors for adoption of ARCT among the ARC in Malaysia in order to highlight their role in ES. Other researchers should also carry out related study on ARCT for the benefits to the stakeholders, the public and the ARC community to cope with ES in the future.

References

1. A. Majid, M.Sonia, and A. Kazi. Post - disaster communications: A cognitive agent approach. ICN 2008. *Seventh International Conference on IEEE*, (2008)
2. A.B. Sutiono, H. Suwa and T. Ohta. "Designing an emergency medical information system for the early stages of disasters in developing countries: the human interface advantage, simplicity and efficiency. J MED SYST 34.4, 667-675(2010)
3. A.Qiantori, A.B. Sutiono, H. Hariyanto, H. Suwa and T. Ohta. An emergency medical communications system by low altitude platform at the early stages of a natural disaster in Indonesia. J MED SYST 36.1, 41-52(2012)
4. ADPC. *Emergencies Communication during Disasters*. Asian Disaster Preparedness Centre. UN World Conference on Disaster Reduction. South Publication, (2013)
5. ARRL. Amateur Radio Emergency Service USA and Canada (ARES). Emergency communication via amateur radio. "when all else fails ham radio works" .J.FEMA(2001)
6. C. Coile and C Russell. The role of amateur radio in providing emergency electronic communication for disaster management. *Disaster Prevention and Management: An International Journal* 6.3: 176-185(1997)
7. D. Kraker and M.J. Adriaan. Flood events in the southwestern Holland and coastal Belgium, 1400–1953. HYDROLOG SCI J. 51.5: 913-929(2006)
8. E. C. Codes. Communicating in an Emergency. Environment of Care® News, 13(7). USA Journal of Federal Emergency Management Agency, (2010)
9. F.A . Ashriq. *Radio Amatur Komunikasi Luar Jangka*. Kuala Lumpur : Malaysian Amateur Radio Emergency Services (MARES). Utusan Malaysia Publication, (2011)
10. Phillips, D. Brenda D, M. N. David and W. Gary. *Introduction to emergency management*. CRC Press, (2011)
11. H. Haddow, B. Jane, and P. C. Damon. *Introduction to emergency management*. Butterworth-Heinemann, (2013)
12. I. Poole, and M. Dennison. *Amateur Radio: The First 100 Years: A Photographic History*. Potters Bar: Radio Society of Great Britain, (1999)

13. IARU. *Amateur Radio - 60th Anniversary of North Sea Floods*. The International Amateur Radio Union Publication ,(2013)
14. J.Blackman, K. Colin, and S. Lara . *Telecommunications regulation handbook, World Bank Publications*, (2011)
15. K. Kovacs, R. Gyongyi, and M. Spens. "Humanitarian logistics in disaster relief operations. Int. J. Phys. Distrib. Logist. Manag. 37.2, 99-114 (2007)
16. L. Benjamin. Working it out together: Radio policy from Hoover to the Radio Act of 1927. J. Broadcast Electron. Media 42.2: 221-236, (1998)
17. L. Lundgren, E. Regina, and A. H. McMakin. *Risk communication: A handbook for communicating environmental, safety, and health risks*. John Wiley & Sons, (2013)
18. M. Acharya. *Amateur Radio, A potential tool in emergency operations*. I Magazine 3.1, (2005)
19. M. Basiron, and N. Fazilah. Pengurusan bencana gempa bumi menurut perspektif Al-Quran dan sains moden. Diss. University of Malaya, (2014)
20. M. Burby & J. Raymond. Hurricane Katrina and the paradoxes of government disaster policy: Bringing about wise governmental decisions for hazardous areas." The Annals of the American Academy of Political and Social Science 604.1:171-191(2006)
21. M. Leonard, T. Michael, S. Graham, and D. Bonacum. The human factor: The critical importance of effective teamwork and communication in providing safe care. Quality and Safety. i85-i90(2004)
22. M. N. Abdul Aziz. *Pengenalan kepada Radio Amatur dan Komunikasi Ketika Bencana*. Seminar Malaysian Amateur Radio League. Kuala Lumpur: MARL Publication, (2014).
23. M. Scheffe, J. Bascompte, W.A. Brock, V. Brovkin, S.R. Carpenter, Dakos and G. Sugihara. Amateur Radi0 - Early-warning signals for critical transitions, 53-59(2009)
24. M. Abu Hassan. *Memanfaatkan teknologi maklumat dan komunikasi (ICT) untuk semua*. Universiti Putra Malaysia Press, (2008)
25. MARL. Gabungan Persatuan Radio Amatur Malaysia. Vol. 4. Sekudai Johor : Malaysian Amateur Radio League, (2014)
26. P.E. Jacobsmeyer. Amateur Radio Often overlooked, but vitally important. Extremely important communications, especially during and after emergencies. IWCE's. J. W.Commun. IEEE Communications Magazine, 49(6), 26-30, (2011)
27. R. Edwards and E. Lynn. *The Utilization of Amateur Radio in Disaster Communications*. DIANE Publishing, (1994)
28. R.L. Mathew, J.E. Dolly. Information technology and public health management of disasters a model for South Asian countries. Prehospital and disaster medicine 20.01, 54-60(2005)
29. T.Schmitt , Rao, R. Ramesh and J. Eisenberg. Improving disaster management: the role of IT in mitigation, preparedness, response and recovery, National Academies Press, (2007)
30. Y. Irwandi.*Mengenang Mangsa Bencana Gempa Bumi dan Tsunami di Aceh 26 Desember 2004. Indonesia,* Provinsi NAD, (2009)

A Reception of Muslim Images in Magazines: British Residents View the Identities of Muslim Women

Ainurliza Mat Rahim [1,*]

[1]Communication Programme, Faculty of Leadership and Management, Universiti Sains Islam

Abstract. In the context of multi-ethnic Britain, the major concern lies in the diversity and complexity of Muslims living in the West, which somehow is misrepresented in the western media as a frozen, static population, fixed in time and space. This misrepresentation dominates mainstream media through the hegemony of western superiority. The operation and role of Muslim media organisations are still under-researched yet potentially constitute an integral part of accommodating the minority population within the wider society. This paper discusses on the reception of images published in two British Muslim magazines by taking views from Muslims and non-Muslims into account. The results show that both groups recognize the identities of British Muslims via visual representations in the Muslim media and that the representations challenge the mainstream images of Muslims.

1 Introduction

Multiculturalism in Britain, according to Aziz [1] was the result of post-war immigration. Historically after the Second World War, people from the post-colony countries were encouraged to migrate to England to fulfil work demands in the pottery and textile industries. Immigrants from the formerly colonized countries such as India and Pakistan arrived in the UK rightly after the countries gained independence in 1947. The number of immigrants dramatically increased between 1965 and 1972 when the flux of East African immigrants, particularly from Uganda and Kenya entered the UK unskilled-labour market. Gradually, the community of immigrants grew larger when families from the original countries joined in British settlement.

It has been suggested that the politics of racism in Britain has now moved from biological descendants to cultural inheritance. Yasmin Ali [2] points out that at the beginning of the 1980s, 'black community' was a normalised concept of anti-racism that preclude other non-black communities. At that time, early Muslim communities who came from the Indian subcontinent were known as the 'Asian community' [2]. The campaign of War on Terror, nonetheless, has ignited new racism of Islamophobia, which Meer & Modood [3] argue as a form of 'cultural racism'. This new racism modifies the British

* Corresponding author: ainurliza@usim.edu.my

discourse on racialized minorities. It has been transformed from the racism of "colour" in the 1950s and 1960s; to "race" in the 1960s, 1970s, and 1980s; to "ethnicity" in the 1990s; and to "religion" in the new millennium [4,5]. This claim will now be examined in relation to media representations of Islam and Muslims in the UK.

2 Literature review

In examining the media representation of Muslims in the British context, Edward Said's critique [6] on Orientalism provides an important background to understand this. Said argued that *Orientalism is a style of thought based upon an ontological and epistemological distinction made between "the Orient" and (most of the time) "the Occident"* (p.2). It often involves seeing the Oriental/Arab/Eastern/Muslim culture as exotic, backward, inferior, uncivilized, and at times dangerous. From this point of view, the West has defined themselves as having such the opposite qualities to the Orients as modern, intelligent, civilized, and superior. Most often than not, the West projects inaccurate and stereotyped cultural representation of the East.

Many researchers have argued that there is homogeneity in the media representation of Muslims. For example, in a research that highlights the degree of Muslim visibility in British print media before and after the 9/11 attack, Poole [7] observes that issues concerning British Muslims exhibited continuation of certain dominant discourses, which link Muslims with terrorism, extremism and cultural difference over 15 years (1994-2008). Likewise, Elgamri [8] indicates that the coverage of world important events in four decades – the Iranian Revolution in 1979, the Salman Rushdie Affair (1989), the massacre of foreign tourists in Egypt (1997), and the 9/11 attack (2001) – cumulatively universalise Islam as a monolithic entity and synonymous with religious hysteria. Recent study by Baker et al. [9] also proves that the British press associate the word 'Muslim' in their reporting with conflict by putting Islam as problematic and central to aggression. The prolonged homogenization of bad Muslims by the press therefore has a tendency to shape a reality of this population.

In a more recent study, Allen [10] concludes that all visible manifestations of Islamic attire including hijab, niqab, abaya, or jilbab has been perceived as 'wrong, problematic and threatening.' Henceforth, every devoted Muslim woman has now become a construction of 'otherness.' "The veil has become an image of otherness, of a refusal to integrate and an example of the 'failings' of multiculturalism" [11]. Indeed, Muslim women have been politicised in a gendered representation to reinforce the centrality of white culture in the discourse of multiculturalism.

For the purpose of this research, the major concern lies in the diversity and complexity of Muslims living in the West, which somehow is misrepresented in the western media as a frozen, static population, fixed in time and space. In this regards, Said [12] urged for more subjective and critical fashion in evaluating the Muslim populations. So far, however, there is no reliable evidence of the diversity of Muslim population in the mainstream western media and no informed data on how people perceive their heterogeneity. To investigate the heterogeneity of Muslim population, the paper will focus on the reception of Muslim women images in British Muslim magazines among Muslim and non-Muslim informants.

In examining the media representation of Muslims in the British context, Edward Said's critique [6] on Orientalism provides an important background to understand this. Said argued that *Orientalism is a style of thought based upon an ontological and epistemological distinction made between "the Orient" and (most of the time) "the Occident"* (p.2). It often involves seeing the Oriental/Arab/Eastern/Muslim culture as exotic, backward, inferior, uncivilized, and at times dangerous. From this point of view, the West has defined themselves as having such the opposite qualities to the Orients as modern, intelligent,

DOI: 10.1051/shsconf/20173300024

civilized, and superior. Most often than not, the West projects inaccurate and stereotyped cultural representation of the East.

Many researchers have argued that there is homogeneity in the media representation of Muslims. For example, in a research that highlights the degree of Muslim visibility in British print media before and after the 9/11 attack, Poole [7] observes that issues concerning British Muslims exhibited continuation of certain dominant discourses, which link Muslims with terrorism, extremism and cultural difference over 15 years (1994-2008). Likewise, Elgamri [8] indicates that the coverage of world important events in four decades – the Iranian Revolution in 1979, the Salman Rushdie Affair (1989), the massacre of foreign tourists in Egypt (1997), and the 9/11 attack (2001) – cumulatively universalise Islam as a monolithic entity and synonymous with religious hysteria. Recent study by Baker et al. [9] also proves that the British press associate the word 'Muslim' in their reporting with conflict by putting Islam as problematic and central to aggression. The prolonged homogenization of bad Muslims by the press therefore has a tendency to shape a reality of this population.

In a more recent study, Allen [10] concludes that all visible manifestations of Islamic attire including hijab, niqab, abaya, or jilbab has been perceived as 'wrong, problematic and threatening.' Henceforth, every devoted Muslim woman has now become a construction of 'otherness.' "The veil has become an image of otherness, of a refusal to integrate and an example of the 'failings' of multiculturalism" [11]. Indeed, Muslim women have been politicised in a gendered representation to reinforce the centrality of white culture in the discourse of multiculturalism.

For the purpose of this research, the major concern lies in the diversity and complexity of Muslims living in the West, which somehow is misrepresented in the western media as a frozen, static population, fixed in time and space. In this regards, Said [12] urged for more subjective and critical fashion in evaluating the Muslim populations. So far, however, there is no reliable evidence of the diversity of Muslim population in the mainstream western media and no informed data on how people perceive their heterogeneity. To investigate the heterogeneity of Muslim population, the paper will focus on the reception of Muslim women images in British Muslim magazines among Muslim and non-Muslim informants.

3 Focus groups

In this study, the focus groups were composed of full-time students at Staffordshire and Keele Universities† as well as part-time students at Suyuti Institute. To examine the reception of images, data were gathered from focus group discussions, which is distinguished between Muslim and non-Muslim participants, who are comprised of British and international citizens. They compose 36 respondents. Throughout the discussion, pseudo names are used to keep the confidentiality of the participants.

A total of 6 images were selected from *Emel* and *Sisters* magazines to stimulate discussions on Muslim women. The selection of pictures was made on the basis of multiple appearances of Muslims where participants could see the women in full. Four images are pictured singly (Picture 1A, Picture 1C, Picture 1D, and Picture 1E), one as part of a pair (Picture 1B), and one in a group (Picture 1F) respectively. The pictures encompass depictions of women generally in private and public domains from both magazines. From the selection, it was expected to gain an overview of Muslim women from the participants' perspective. The participants were asked how they interpreted these images. How they defined the women? Did they find them positive or negative, and why? Did the images have an impact on a multicultural society?

4 Findings and Discussions

In the context of Muslims living in British society, Meer [13] observes that Muslim consciousness is a dynamic phenomenon that can be interpreted from an intersectional approach of theology and sociology. In other words, Muslim identity comes into existence within an interplay of religious and ethno-cultural elements. This intersection may provide an explanation about the Muslim presence in Britain.

The manifestation of Muslimness emerged in all focus group discussions concerning the images in Muslim media. Four sub-themes were observed in the group interviews, namely ethnic diversity, religious identity, dimensional visibility, and incorporation into western society.

Diverse ethnicity: All participants from both Muslim and non-Muslim groups identify the Muslim community as colour-blind and inclusive of all races. The results were consistently obtained from the participants when they were asked what they thought of all Muslim women images in general. For Muslim groups, Muslim women are not identifiable by their race or skin colours.

> *Fizzy: You can show the woman can be anyone from any colour, any culture, any society, any background.*

Another non-Muslim participant perceived racial diversity as a feature of a multicultural society such as Britain.

> *Deevan: First, I think every picture has been pretty much seen in different race. Let say white male, white female each of them seen before, it's all part of normal life I've seen. That could be a model of someone else. So I think they are all pretty much in the standard of culture here and fit in to the culture of England.*

Religious identity: The next important theme that has been frequently discussed is religious identity. By realizing the dress code of Muslim women should be modest, all agree that regardless of their skin colours or ethnicities, the hijab is the first identification for a Muslim woman. To one Muslim participant, hijab acts as an identity marker that constitutes her personality as an obedient Muslim.

> *Shereen: To me the picture does not have facial feature (Picture 4A). It's kind of that woman is defined by the hijab. She doesn't seem double personality because she's not smiling, without expression and the fact that it's all faded away, have face a kind of conscious Muslim, she's wearing hijab.....kind of getting a personality from that.*

The view that hijab is a symbol of identity for Muslim women was affirmed by non-Muslim respondents.

> *Ember: (Picture) 1A, if I am saying correctly, it does seem a police officer. However, A representing the Muslim culture. They put a hijab on, they have a headscarf on to show their religion.*

To non-Muslim participants, hijab is only a manifest symbol of religion. They held a view that the variegated images of women shows that the religion and the hijab do not stop the women from participating in society.

> *Isaac: I think what they're trying to do by these photographs is it shows that it doesn't really make any difference whether not to wear scarves or what your religion is, you can just...I mean, you would just functional as society does.*

Dimensional visibility: Besides the hijab, the majority of participants also commented on the diverse appearances of Muslim women in the presented images. They observed that Muslim women are represented in a continuum of the veil-less, headscarf and full-faced veil. Muslim participants consider this continuum as a positive representation of Muslim women to show the freedom they have when they appear in public.

> *Mareesa: Well, pic 1C I could see myself in there. It's a young Muslim girl and I can see likes fashion at the same time she likes to follow her religion so by putting an identity on her that she's Muslim practising. Pic 1D as the niqabi I mean masyaAllah, a lot of Muslim sisters are wearing the niqab and working all time and I really respect that they are struggling on that. So I'm thinking that's something else that represents our society today. Pic1F is the woman without hijab but that doesn't mean she is not Muslim and also you do not know how her relationship with her god but she still a Muslim and she is also one of the part of our society.*

The same view was shared by non-Muslim participants alike. For the image of unveiled woman, one participant perceived her as sociable and adopting a western appearance.

> *Valerie: In general, I think it shows Muslim women in a variety of settings, variety roles, it shows diversity within the religion that in D a woman could be completely veiled or B like very social, western.*

Incorporation into western society: Muslim incorporation has been rigorously discussed in two aspects – their contributions towards British society (labour market participation); and their physical attributes. The participants on the whole demonstrated that all images presented in the focus group sessions are representing Muslim women as part of British society. Through a portrayal of Muslim women in different roles and contexts, Muslim groups regarded this as a contribution towards society, thus being part of the society they work for. As one participant said:

> *Shah: Going back to our pictures what they represent, they do different sort of activities and different roles whether the police woman, driving instructor, or politician or even a mother at home.*

The same view was identified within non-Muslim groups. They recognize Muslim women's existence in multicultural Britain as having a normal life like others, pursuing careers and without over emphasizing their religious identity.

> *Jason: I think if they would share images more like A and F, that would be different. Because I think both, when you say about multicultural society, they show people getting on and living a lot, doing a job without regards necessary like overwhelmingly to their religion. I think one of the things what multicultural things should be, involved in and to respect with each other, that's not what to be a specification of the white, Christian majority.*

Besides having a job, participants also observe that the way the women dress demonstrates a union of Islamic and western styles, through a combination of headscarves with pants and shirts. The dress meets a normative standard:

> *Asma: Just to say that in a comparison, if you can hold the pictures except for 1D, it seems that all of them are sort of conforming to the western standard. Even 1B, I guess she's a mother but technically she's still working, sort of the western kind of civilization like all women are independent in everything coz they work. 1A she's working. She's a Muslim woman but obviously she's wearing a uniform. 1C she's wearing western clothing. 1E she's an academic and she's still wearing English clothing. 1F they're all wearing English clothes.*

For some non-Muslim participants, Islam is seen as non-western and strange to British culture, so the incorporation into western society which is indicated through a sartorial range is positive.

> *Sita: 1C is really, really cool representational cultural harmonisation. You can see someone who is embracing the religious identity and also the modern western fashion and I think that's really cool because it's kind of like saying you don't have to be of any extreme and take one of the middle spots of the bar.*

5 Conclusion

One major finding is manifest in the focus group discussions. It is apparent that the reception of Muslim women imagery receives a quite similar response from Muslim and non-Muslim participants in the discussion of the manifestations of Muslimness. They all recognize the British Muslim consciousness as having a blended identity of being Muslim and British. All images presented during the focus groups cognitively resonate with the participants' knowledge.

The fact that Muslim women have an heterogeneous identity is congruent with Meer's [13] thesis on the Muslim consciousness. The Muslimness is constituted within dynamic categories of religiosity, culture, ethnicity, sexuality, personality, economic status and so forth. Although religious identification was seen as a problem by egalitarians and a threat to multiculturalism [14], intersected categories of Muslim identification have been recognized by all participants and this undermined a claim that Muslims are not compatible with British society.

All participants generally agree that Muslims and non-Muslims are similar and lead normal lives, thus blurring the boundary between the Muslim community and the wider society. The boundary between 'us' and 'them', 'we' and 'others' seems to be lessened in the discussions. In other words, the boundary of Muslimness is becoming less obvious, hence showing a tendency to integrate with the majority.

The majority of respondents perceived that the representation of Muslim women in the two magazines extenuates the dichotomy of superiority and inferiority, the gap between the West and the East. To them, the Muslim consciousness of maintaining Islamic identity while at the same time conforming to western standards is negotiated in the magazines by the construction of hyphenated identity. Muslim media construct what Edward Said [12] suggested:

'Rather than the manufactured clash of civilizations, we need to concentrate on the slow working together of cultures that overlap, borrow from each other, and live together in far more interesting ways than any abridged or inauthentic mode of understanding can allow.'

References

1. M.A. Aziz, *Debating Multiculturalism* ***1***, 59-74 (2012)
2. C. Allen, *Islamophobia* (Ashgate Publishing Limited, Surrey, 2010)
3. N. Meer, T. Modood, *Arches Quarterly: Islamophobia and Anti-Muslim Hatred: Causes & Remedies*, **4**, 116-126 (2010)
4. A. Kundnani, Eur J. of Cult Std, **2**, 155 (2012)
5. T. Abbas, American Journal of Islamic Social Sciences, **21**, 26 (2004)
6. E. Said, *Orientalism* (Pantheon, New York, 1978)
7. E. Poole, Global Media Journal, **4**, 49 (2011)

8. E. Elgamri, *Islam in the British Broadsheets: The Impact of Orientalism on Representations of Islam in the British Press* (Ithaca Press, Reading, 2010)
9. P. Baker, C, Gabrielatos, T. McEnery, Applied Linguistics, **34**, 255 (2013)
10. C. Allen, J. of Muslims in Eur, 3, 137 (2014)
11. M. Williamson, G. Khiabany, Race & Class, **52**, 85 (2010)
12. E. Said, *Orientalism* (Penguin Books, London, 2003)
13. N. Meer, Ethnicities, **12,** 178 (2012)
14. T. Modood, F. Ahmad, Theory, Culture & Society, **24**, 187 (2007)

SHS Web of Conferences **33**, 00025 (2017) DOI: 10.1051/shsconf/20173300025
i-COME'16

The Practice of Traditional Rituals in Naga Aradhana (Snake worship): A Case study on Aadimoolam Vetticode Sree Nagarajaswami Temple in Kerala, India

Dinu Das[1], and *Arumugam* Balasubramanian[2,*]

[1,2] Department of Communication, Coimbatore, Amrita Vishwa Vidyapeetham University, Tamil Nadu, India

Abstract. Snakes are worshiped in different parts of the world based on the culture and traditions. Snake worship is a part of Hindu religion and plays an important role in cultural as well as in ritual aspect in India. This study explores the ritual practices in Naga aradhana in the Indian state Kerala, by taking Aadimoolam Vetticode Sree Nagarajaswami temple as a case study. This paper aims to understand the importance of Naga aaradhana and the practice of rituals related to Naga Dosha (curse of snakes), like Kalamezhuthu, Pulluvanpaatu, Sarpamthulal , Sarpabali, the annual festival Ayilyam Mahotsavam and the importance of Sarpakaavu (sacred groove) in the temple. This research is conducted using the methodology of personal interviews and video analysis. This study concludes that the Sarpakavu has a unique meaning and believers from different caste and religion have respect and admiration for this divinity.

1 Introduction

Snakes are animals that fascinate many people while frightening others. Good or bad, most people have strong feelings about snakes, but few people remain neutral [11]. The rituals devoted to the propitiation and supplication of the *sarpa*, as the common snake is called in Sanskrit, as well as the snake's supernatural counterpart the *Naga* are present in the Indian sub-continent for more than two millennia [9].Worship of the Snake Gods (Naga) has taken a prominent place in various forms of literature, traditional paintings, art, sculpture, folklore, architecture and so on. The ancient civilizations throughout the world treated Naga's as a part of society. The snake therefore can be seen in old traditional cultures where snakes are the entities of strength and renewal. There are different kinds of festivals associated with Naga/snake throughout India. In northern India king of snakes, Nagaraja is worshiped whereas in southern India; the snakes are worshiped as whole.

Snakes are considered as demigod in the southern India, especially in Kerala. People in Hindu religion worship Snakes in temples and in their natural habitats. The devotees offer snake god with milk, incense, and prayers. In Hindu rituals and spiritual tradition, the

* Corresponding author: a_balasubramanian@cb.amrita.edu

snakes represent divinity, eternity, materiality, life and death, and time as well as timelessness. It symbolizes the three processes of creation, namely creation, preservation and destruction [8]. Throughout Kerala there are temples devoted to snake worship. The major temples are Mannarsala, Vettikode, Pambumekkatu mana, Pulikal Shankaramadam, Pambady Nagaraja Temple, Kumaranalloor Nagaraja Temple and Madanmarukavu Nagaraja temple in Alanadu. The presence of these temples and the number of devotees visiting these temples point to the deep rooted spiritual beliefs and religious values related to snake worship. This religious tradition in Kerala is directly connected to the ancient mythology and literature of Hinduism in India.

The ancient literature and the epics like Mahabharata and Ramayana give ample documents on the Nagas. The characterization of Naga in art is common during the historical period. It is hard to track down when and how Naga aradhana began in Kerala. Mythology through Keralolpathi, a history of Kerala points to the period when Brahmins settled in Kerala after the lord Parasurama. On the request of Brahmins, Parasurama allotted a different residential area for the Nagas to eliminate troubles they might cause to human beings [2], [3]. In Hindu mythology Nagas play a prominent role, through various legends. Shesha also called as Aadi shesha or Ananda is the snake on whom lord Vishnu does his Yoga nidra (sleep), otherwise called as Ananda shayana. The snake Vasuki is considered to be the king of Nagas, who allowed demons and gods to use her body as a rope to bring out the elixir for saving people. The Lord Shiva, the god of gods wears a snake around his neck. It is believed that Shiva has given an important place for Nagas.

Several researchers have conducted studies on Snake and snake worship in India and also in other countries. A researcher pointed out that the Snakes, Goddesses, and Women's Ritual Responses in Contemporary South India have a focus on the rituals in an Amman temple of Tamil Nadu [1]. Ethnographic study conducted on snake (naga) worship and traditions associated with naga doṣam (snake blemish) are also an important aspect in the study. It studies the astrological flaw related to delay in marriage, skin diseases, infertility in women and children born with low intelligence and autism.

In India, there are families who worship the Naga as their family god and have monument around their home and sometimes even inside their home. The families devoted to Naga worship thus follow rituals to please the snake God. A researcher pointed out how snake worship and protection of the environment correlates to one another and also about different varieties of snakes present in Kerala [4].

2 Aadimoolam vetticode sree nagarajaswami temple

Aadimoolam Vetticode Sree Nagarajaswami Temple is located near Kayamkulam in Alappuzha district of Kerala. Hindu mythology says that, Sree Nagaraja Swami Temple was consecrated by Lord Parasurama. The temple was built as a gratitude to Anantha, the snake god, who asked to suck out the unwanted elements in Kerala to turn it into a fertile land for the residents. The temple is located on a six-acre land with a Sarppakavu (sacred groove) which has Nagaraja and Nagayakshi as principal deities. The temple has Sri Parameswaran Nampoothiris as the Karanavar (chief), who has devoted his life to Naga Aaradhana. The temple, devoted to snake worship, practices certain rituals and traditions.
The major reason for practicing these rituals is Naga Dosha or curse of snake. Naga dosha is caused when a person knowingly or unknowingly hurts a snake or destroys the place where they reside. The personal interview with the philosopher Madhav Sreekumar and Astrologer Joseph Nedumkandam related to a general concept that Naga dosha means anger of God. For getting rid of Naga dosha people practice various pooja or offerings in the Temple. The major rituals are Sarpabali, Sarpamthulal along with Kalamezhuthu and Pulluvanpaatu. Sarpabali is conducted using a sketch or Kalam drawn for Naga Raja (snake

god) on the day of Aayilyam (a zodiac sign according to Malayalam calendar) and the priest performs various offerings along with religious chants. Sarpamthulal is a long procedure and happens near the sacred groove inside the temple compound. In this ritual a sketch is drawn on the ground based on the need of pooja. Pulluvanpaatu, the hymns by a tribe devoted to Nagas, sing songs to please and invite Naga god into the sketch. It is said that the Pulluvanpaatu reaches the heaven where the Naga Gods reside. Young virgin girls receive the soul of Naga god during the procedure and talks to people and try to redeem them from curse of snake. Later these people clear the sketch and conclude the pooja.
The temple also offers various other rituals to cure minor Naga dosha based on astrological calculations, it includes Ashta Naga pooja, Dhara, Aadiya Enna, Uruli Kamezhuthu and Rahur Dasha shanti. Ashta Naga pooja, is to please the eight major snake gods. Dhara is an offering for both Nagaraja (the king of snake God/ anantha) and Nagayakshi (the queen of snake god) using an abhisheka jala (sacred water). Aadiya enna is an offering where devotees give bath to the snake deities using gingely oil, mainly to cure skin disease. Uruli Kamezhuthu is carried out by married couple who are childless. In this they keep a round vessel called Uruli in reverse position in front of Nagaraja. The rahur dasha shanti is for clearing the Naga dosha in the astrological calendar.

The following are the objectives of this study:

- To understand the significance of Naga (snake worship) culture and rituals.
- Study of Sarppakkavu or Sacred groves for Snakes with relation to the Naga aradhana and Snake song, Pulluvanpattu.

3 Methodology

To study the practice of traditional rituals in Naga aradhana, the qualitative method has been considered where personal interviews along with visual documentation are the tools taken for collecting data. For the personal interview, priests, astrologers, anthropologists and believers belonging to different caste and religion were selected. The process of data collection was spread over three months. During the in-depth interview, the researchers encouraged the participants to describe their belief, experience and what they found exciting about the traditional rituals in Naga aradhana. All the in-depth interviews were conducted one-on-one and visually documented. Visuals were viewed and the important segments to the interview questions were transcribed.

4 Analysis & interpretation

Every religion has rituals and it varies according to the traditional and cultural beliefs. The Naga worship in Aadimoolam Vetticode Sree Nagarajaswami temple is a proof for the existence of primitive culture of snake worship in Kerala. The legend of Parasurama and the oral culture transferred through the Karanavar (chief) of the temple, points to the fact that this temple is the first temple built for snake worship by Lord Parasurama. There is also a legend stating that Lord Parasurama himself created the idol of snake god, Anantha out of a rock using his axe. This temple is maintained by the traditional family called Meppallil Illam. The Chief of temple, Parameswaran Nampoothiri is the head of that family.

According to the astrologer Joseph Nedumkandam, Naga is considered to be the manifestation of nature and psychological quality of intelligence in Hinduism. Therefore Naga Aaradhana or snake worship is worshiping the nature and also the intelligence. The head priest of Nagarajaswami Temple said that, in India people worship Nagas because of the fear that they are capable to bless and curse.

4.1 Naga dosha

The curse of Naga is called as Naga dosham. Naga dosham in general according to astrologer Joseph is the anger of snake god. The head priest defined the reason for Naga dosham as harming Nagas or any sacred object related to snake worship. Another aspect of sarpadosha according to astrologer Joseph is associated to the harms against the earth, like digging or cutting of trees and plants. This is associated to another myth related to Adisesha, personification of force which keeps earth stable along the path way of sun. The Naga dosham causes health problems like white patches on skin, belated marriage and pregnancy, eye disease, handicapped or disabled birth, unnecessary dispute and quarrel among family members and other incurable diseases. The temple offers various rituals (poojas) as a solution to get rid of Naga dosham.

4.2 Sarpabali

Sarpabali is the ritual of worshiping Lord Nagaraja or Anantha on Ashlesha Nakshatra (an Indian Zodiac star) otherwise called as Ayyilyamstar. Sarpabali is performed to appease Ashta Nagas and to fend off evil effects of Sarpa dosha (curse of snake). As part of Sarpabali, a colorful sketch of Padma (Lotus) is drawn on the ground of the temple. In the middle portion there is Ashtadala (eight petals) and its outward triangle represents nature and Man. The upward and downward triangles of ashtadala indicate the close association between nature (prakrithi) and man (Purusha). The extreme outer part of the sketch has Sarpakettu. This again is in a lotus shape. The padma(lotus) for the sarpabali can be seen only from that point. The priest offers prayers and sacrifices (bali) to please naga through various mantras and chants near this sketch for the people having Naga dosha. The people who witness this holy performance will redeem themselves from the anger of Nagas.

4.3 Ashta Naga pooja

The ashta naga pooja is a special ritual performed for ashta Nagas (eight snakes). The main eight snakes are Anantha, Gulika, Vasuki, Shankapala, Thakshaka, Mahapadma, Padma and Karakodaka. The pooja offers Noorum Paalum through Thelichukuda ritual to each snake deity separately. The Noorum paalum includes ingredients like, tender coconut, fresh milk of cow, rose water, turmeric and the seeds of tender coconut (pookula). The juice of these ingriedients together is called as Noorum Paalum.

4.4 Rahur dasha Shanti

The poojas or rituals related to Sarpa dosha changes according to the astrological aspects of a human and the severity of the anger of snake god. The other poojas (rituals) related to Naga dosha includes Rahur dasha Shanti. The ritual concentrates on offerings to get rid of all diseases and ailments due to the presence of Nagas in Rahu dasha (considered to be the worst time according to the Hindu astrological calendar).

4.5 Dhara

The ritual Dhara, (pouring sacred water on the devotees) is another ritual for Sarpa dosha. The ritual's main element is Abhisjeka Jala (the sacred water). It is a special offering to the main deities, Nagaraja and Nagayakshi. The Abhisheka Jala or dhara is made out of tender coconut water, milk and plain water. This water is poured on people with Naga dosha.

4.6 Uruli Kamizhthu

The ritual is conducted for people without children due to the anger of snake gods. The couples with faith purity in mind will undertake this ritual. The ritual starts with Pradikshana(circling the main temple of snake God Anantha) and Namajapa (chants) with Panchavadyamelam (music created using five temple musical instruments). The couples carry a Uruli (wok) covered with red coloured silk cloth in front of the idol of Nagayakshi (diety). They later come to the idol of Nagaraja for the act of Uruli Kamzhithu (turning the wok upside down). The nivedya (banana and milk) is offered to both the deities for their blessing. After the birth of child, the couple need to come back and perform Uruli Nivarthu (turning the wok upwards) with the child.

4.7 Aadiya enna

Aadiya enna (bath to deities using gingely oil) is another ritual for people suffering from skin disease. The deity Nagaraja is given a bath using gingelly oil on Brahma Muhurtha (auspicious time) to get rid of skin disease caused by Sarpakopam (anger of snake god).

4.8 Sarpamthulal

Sarpamthulal is prerformed in this temple during the day of Ayilyam star in Malayalam calander. The ancient legend of Nagas in Kerala according to the astrologer Padmanabhan Sharma narrates that Nagas who were the masters of forests left their home and moved to a narrow area (sacred grooves) for the peaceful life of humans. Thus once in a while based on the astrological calendar, the snake devotees perform Sarpamthulal to reveal the gratitude towards snake gods and also to seek their blessings. The Sarpamthulal includes rituals like Kalamazhuthu and Pulluvan paatu. In the temple they perform Sarpamthulal for the Ashta Nagas. The sarpapattu is a spiritual dance performance done by the devotees who get incarnated with the soul of eight snake gods during the holy ceremony. The Sarpamthulal is performed nearby sacred groove (Sarpakaavu) inside the temple compound.

The ceremony starts with Kalamezhuthu. Kalams are the combination of two dimensional and three dimensional designs drawn on the floor using fine powder extracted from nature [13]. The colours extracted from fruits are used to draw sketches related to snakes. Through these sketches, snakes are invited into the pooja (sacred ritual) through various chants and prayers. The sketches of snake vary from the need for Pooja. In the temple we can see the sketch related to Ashta Naga. For the ashta naga sketch, they used five natural colours. They are white (rice flour), red (dry saffron's powder), yellow(turmeric powder), green(the powder of leaves) and black (the powder from burned rice barks). The colours are added to the sketch drawn out of white powder (rice flour). The drawing is done using hands and the art of shading, especially the background which gives a velvet touch. Combinations and permutations are used for foregrounding or to create illusions. The later addition is the influence of oil lamps around the Kalam where rituals are known as *thiri uzhichil*, a form of worship [13]. The Ashta naga sketches include the sketches of Nagaraja (snake god), Nagayakshi, Anjanamani Nagam, Mani Nagam, Para Nagam and Kari Nagam. After finishing the sketch the Pulluvans start their hymns and invite the tridevs in Hinduism, that is Lord Brahma,Lord Vishnu and Lord Shiva.

Later the pulluvans start singing hymns for deities including snake gods and other deities to invite the snake gods. The songs illustrate the myth behind the Naga accompanied with unique instrument called *Pulluva Veena* .A virgin girl waits with a Pookula in her hand to receive the soul of Snake god in her body. After a while the virgin receives the soul and starts the spiritual dance called as Sarpamthulal. The energy passed on to the girl through the soul makes the virgin to appear supernatural and powerful. The virgin later blesses the devotees and redeem the people from the curse of snakes. Last stage is called *Kalam Mayickal.* In this stage the virgin erases the sketch drawn on the floor and the pooja gets completed.

4.9 Pulluvan Pattu

The Pulluvan Patti is the praising hymns sang by Pulluvans. Pulluvans are people dedicated to sing hymns and songs for snake gods in Kerala. Vinayakan Nampoothiri said that Pulluvans Paatu reaches the eternal world of snakes. These people are the predecessors in snake worshipers in Kerala. Pulluvans travel from house to house and sing hymns for naga during the Malayalam month of Kanni and Makaram to cure Naga Dosha [14]. They sing hymns for the Lord Anantha and all other Nagas along with the assistance of musical instruments including Veena and Kudam (Pot). The one stringed violin known as the Pulluvan veena, the Pulluvan kudam, an earthen pot with a string fastened to it, Pulluvan mizhavu and cymbals. Out of these, the Pulluvan veena plays a major part during the rituals. The music produced with the combination of this hymns and musical instruments of the Pulluva community praise 'holy serpents' such as Ananthan, Manimudgaran, Adiseshan and Mahapadman. The pulluvan Paatu is a major element and is one of the major worship practices of Naga called Sarpamthulal.

4.10 Ayiliya Mahotsavam

Another important aspect of this temple is the annual Ayiliya Mahotsavam festival. The festival is performed on the Ayiliyam day of Malayalam month Kanni. The celebrations start seven days prior to the Ayiliyam day. There are various rituals and traditions associated with the festival, which will be carried out during these days. This is performed to increase the power of deity and to shower blessings on the devotees. The Nadabhrama with instruments like Kombu and Kuzhal (wind instruments) and Nadaswara instruments like Maddallam, Chenda, Edakka and the Thakil will be played to please the deities. This is followed by Sarpabali and Ezhunnalathu (procession). The procession starts from temple to Meppalil Illam then back to temple after certain poojas in the Illam. The significance of procession is that it distributes and spreads blessing across the way to thousands of devotees witnessing it. Later a sarpabali will be conducted in the temple for the devotees with sarpa dosha or sarpa kopam. The temple remains closed till Brahma Muhoortha and after that Shudhi Kriyas (Purification rituals) the festival is concluded with the Dhaara (pouring sacred water on deities).

4.11 Sacred grooves/Sarpakaavu

Kavu in Malayalam vernacular language means a garden or a group of trees. The sacred groves which are dedicated to the snake deities are called Sarpa kavu. Sacred grooves are maintained nearby temples and also in the families of Naga devotees. In the sacred grooves, *Chithrakoodu kallu* (a specially designed stone) is the ideal of Naga God. The Sarpa Pooja or offerings for Naga god will be conducted on the day of Aayilyam Star according to malyalam calendar. Naga devotees carry out the traditional Ayiliyam diet and conduct the pooja to please Naga gods [14]. Sarpa kaavu according to the priest Rajan namboothiri maintains the ecological balance. The predecessors cut down the forest for agriculture and therefore attacke the ecosystem as a whole. So they preserved a part of land allotted for Naga gods to maintain the environmental balance and also for the Snake gods. The sacred

 DOI: 10.1051/shsconf/20173300025

grooves are blessed with a wide variety of trees and medicinal plants. It is also said to be a major source for underground water

5 Conclusion

In India people worship. Naga is considered to be the manifestation of nature and psychological quality intelligence in Hinduism. Offerings in this temple are given only once in a year in the Malayalam month of Thulam. The Sarpakavu has a unique meaning and believers from different caste and religion have respect and admiration for this divinity.

References

1. A.Allocco,. RSAR, **7** (2013)
2. S.Achyuthavaryar, *Kerala Samskaram*, Thiruvananthapuram Kerala Bhasha Institute, (2008)
3. K.K.N. Kurup, *Aryan and Dravidian Elements in Malabar Folklore*. Trivandram: Kerala Historical Society, (1977)
4. M.Suresh, *Sarpparadhanayum Prakriti Samrakshanavum*, (2000).
5. K.Murugan, V.Ramachandran, K.Swarupanandan, M.Remesh, Ind. J. Trad. Know., **7**, 455-462,(2008)
6. R.Bala, *Snake Worship in Kerala In R. M. Sarkar, Through the Vistas of Life and Lore* Folkloric Reflections on Traditional India, Calcutta: Punthi Pustak. (1946).
7. J.P.H. Vogel, *Indian Snake Lord or the Nagas in Hindu Legend and Art*. Delhi: AES Books. (1998)
8. J. V, The Symbolism of Snakes and Serpents in Hinduism retrieved from. http://hinduwebsite.com/buzz/symbolism-of-snakes-in-hinduism.asp
9. Laurie Cozad. (2004). DG, **2**.
10. Jones.G, OJR RSRO, 89-119, (2010)
11. J.W. Gibbons, M. E. Dorcas, *Snakes of the southeast*. Athens, GA: University of Georgia Press.(2005)
12. J.W. StanleyOMICS **2**(2):42-58. (2008).
13. Jayaraj. P, Kalamezhuthu: A ritual performance, (2010) retrieved from http://www.narthaki.com/info/articles/art280.html
14. M.V. Vishnunamboothiri, *Pulluvapattum Nagaradhanayum*, (National book stall, Kottayam. (1977)
15. E. Thurston - *Castes and Tribes of South India*, Cosmo Publications, Delhi. (1970)

Sexual Objectification of Pinoy Males in Billboard Advertisements

Marc Agon Pacoma[1,*]

[1]Department of Communication – Adamson University, Manila, Philippines

Abstract. Male models in skimpy undergarments showing their Adonis-like physique, presented in sexualized poses with disturbing bulges or crotches is one of the common landscapes in the Philippine advertising today. The author found the conflicting idea of a nation rich in morals, values and principles but endure high profanity or obscenity through sexualized male bodies as seen in billboards of fashion brands and gratification of visual desire among its spectators. Advertising promotes this phenomenon stimulating the idea that sex truly sells, male models are objectified as sexual objects capturing consumers' attention resulting to persuasion and eventual consumption of the product. Ads showing Pinoy males' chiselled bodies in silhouettes communicating sexual innuendos constitute a contemporary marketing technique; a ploy for companies to earn profit yet an avenue for public scrutiny and moral outcry.

1 Introduction

Advertising is a marketplace of ideas normally transferred through various media platforms, with the goal of communicating distinct characteristics of products and services among consumers [1]. Furthermore, it helps companies and brands build their identities thus, being recognized and recalled among its target audiences [2]. Presently, numerous types of advertising are evident [3] employing various techniques in order to earn profits or revenues, as advertisers continuously face difficulties in terms of sending across the messages about their products [4]. With this, most of them resorted to numerous marketing techniques and tactics. One of the most pervasive advertising practices is displaying the images of gender as part of products which typically draw in consumers' attentions, resulting to sales and earnings among brands and companies that exploit this marketing style.

This paper focuses on Bench Body's strategy of promoting its undergarment products through billboard advertisements. Specifically, it describes how Bench as a company utilizes images of male models to attract attention of heterosexual consumers, resulting to persuasion and possible product consumption. The paper argues that men just like women experiences commodification or objectification, portraying them as sexualized individuals or objects of sexual desires and fulfilment. It further delves into the idea that men same with its feminine counterpart experiences negative portrayals or representations in various

* Corresponding author: marckiepacoma@yahoo.com.

DOI: 10.1051/shsconf/20173300026

media platforms. Also, the paper describes how male models are presented in these billboard ads and how Bench pleasures out from this marketing ploy. Using the Cultural Studies of Stuart Hall, the researcher seeks to answer the subliminal messages or interpretations presented in the billboard advertisements.

Bench Body is the focus of the paper due to its current position as the market leader in the Philippine undergarments and the pioneer in the undergarment billboard advertising. Since its inception, it produces roughly 100 billboards in various themes and concepts, featuring local and international-renowned models and celebrities. This move make them a top choice among Filipino consumers, posting roughly P2 billion in sales, way ahead of its closest apparel competitor, Penshoppe [5]. The figures imply that consumers are into enormous patronage and high consumption of Bench and Bench Body's merchandises, which makes it ideal to investigate. But what makes it more worthy of scrutiny is its constant usage of male models who are sexually-posed and sparsely-dressed with provocative captions or headlines in its outdoor advertising. With this, the study was conjured in an attempt to examine the sexual objectification of Filipino males in billboard advertisements of Bench Body. More so, it seeks to describe how these billboard advertisements are sexualized in nature and present the elements that make them bearers of male commodification.

2 Sexual Objectification: a contemporary advertising technique

The quest in finding the perfect way to reach the right consumers seemed to be evasive. New advertising strategies came to light in order to cater the various potential customers [6]. Portrayals of gender thrived in media and considered to be the most prevalent one. This new way of selling is characterized by images of men and women appearing side by side with the product; performing less professional and subordinate social roles; showing highly sexualized poses or depicting demeaning stances. The proliferation of this visual rhetoric led to the enormous interests in research among consumer and communication researchers, gender experts and even academic institutions. Using sex to sell is a concept which was not conceived in just one day, this has been the philosophy that many advertisers and marketers embrace to market their products which was popularized by international brands like Calvin and Klein, Abercrombie & Fitch, Carl's Junior [7], Levis and Candies [8].

The proliferation of sexual information incorporated in advertisements proliferated in the 1960's with sex almost synonymous to nudity and other sexual information [9]. But the popularity of this technique can be traced back in the 1850s, when images of sex and gratification were utilized to sell various products and services [10]. Undeniably, the exploitation of human bodies in advertisements seemed to be a flesh-selling market or a syndicate of sex. Sexual information in advertising are presented in various forms of physical desirability, clothing and body exposure, sexual comportment, sexual interpretations and undertones [9]. Visual displays of human bodies both of men and women seemed to be a part of the pop culture [11]. Much have been investigated on the female, their bodies and representations but with less interest on its counterpart – the males.

Inquiries surrounding men are few since most researchers are focused on the subordinate and submissive, women. Many thought subsequently that masculinity does not encounter any gender issues because of its position in the society but its current condition as a tool in marketing does not make it exempted when it comes to evaluation of gender portrayals. What makes it more disturbing is that men appeared side by side the products or services being marketed: vulnerable, provocative and sexually objectified – demonstrating a masculinity in crisis. Sexual objectification is perceiving an individual as objects of erotic desire [12], focusing mainly on sexual and physical appeals [13]. It has been a staple topic among researchers, equally investigating the portrayal of men and women in various media platforms such as television, film, and print media among others. This is most commonly associated with femininity however imposes a serious threat on the masculine gender. Arguably, masculinity can be represented by either bodies or physical appearance [14], leading to the concept of objectification and sexualization [15]. Indeed, images of men started to flourish; appearing in printed and outdoor advertising such as but not limited to newspapers, posters, magazines and billboards.

3 Sexualized billboard advertising in the Philippines

Billboards are large structures usually situated for advertising purposes [16]. It is one of the most preferred categories among out of home advertising due to its ability to reach quickly its targeted audiences in certain locations [17]. The Philippines specifically Metro Manila is a hub for outdoor advertising, with roughly thousands of billboards situated in its various thoroughfares, making it one of the most visually distracting cities in Southeast Asia. But the proliferation of these advertising type is extremely distressing not just because of their large physical structures but its controversial and provocative contents resulting to moral outcry and criticisms among public, the government and ethical groups [18].

The emergence of male bodies as a marketing ploy is extremely visible in different billboards of products and services (perfumes, beverages, gadgets and fashion brands). One of the companies that resorted to this type of promotional technique is Bench which is owned by the Filipino-Chinese fashion business mogul, Mr. Ben Chan. Bench had pioneered in the country the incorporation of male models presented in highly provocative poses wearing skimpy or scantily-clad clothing with risqué captions and headings. Issues of sexual overtones and morality had stimulated the general public and deliberately outraging various sectors.

Among the other controversial outdoor advertisements that were criticized for its indecent images, sexual insinuations and erotic symbolisms are: Napoleon Quince's "Nakatikim ka na ba ng Kinse Anyos?" [Have you tasted a 15 year old?], Bayantel's Satisfaction Guarantee and Bench's underwear ads among others. These raunchy ads incorporate sex in its marketing formula making them disturbing and scandalous [19]. Of all these ads, Bench seemed to be the top notch in sexualizing gender imageries resulting to a market lead in the Philippine fashion brand industry. Bench's unceasing success was palpable on its massive, thought-provoking, and sensualized billboard advertisements that continuously strike the attention of every Filipino consumers.

4 Bench billboards and the realms of sexualized underwear ads

Bench is a Filipino global brand that started in 1987 as a single shirt outlet and later on had its reach in other countries like China, the Middle East [20], Singapore and even in the United States of America. From its humble roots, the company grew steadily fast; offering a wide array of clothing for men and women. Sooner, it ventured out to the field of beauty through its Bench Fix Salon bringing in hair and grooming products to Pinoy clientele. But to further its market lead, Bench overhauled its underclothing goods through its Bench Body, offering sleek, sexy, and trendy undergarments for both sexes.

With the Bench Body's reinvention, came a frantic move to maintain the lead by showcasing sensual men and women. Both presented in highly sexual poses with skimpy and scantily clothes, exposing some skin and body parts. But what makes it more divisive are humungous and imposing images of men in sexualized poses communicating sexual innuendos, wearing see-through clothes or silhouettes with their chiseled bodies [21] and crotch bulges noticeable from public's eyes. Bench Body's male models are often seen in a situation where one's masculinity should be ripped off in order to increase the sale of the undergarments. This was apparent in some of their billboard advertisements where good-looking guys show off their greased-muscular bodies wearing themed-costume from top Philippine fashion designers matching their scanty silhouette bikini briefs.

Nonetheless, this is not always the case, just like the billboard ads of the National Rugby Team of the Philippines (Philippine Volcanoes), which aims to promote rugby as a sport among Filipinos and raise funds for the team as they compete in various international competitions [22]. Philippine Volcanoes members posed individually and collectively but just like the common marketing ploy of Bench Body – where models wear measly undies showcasing their bodies and private parts as mere focal points of the ads; the photoshoots didn't fare well with the Filipinos, making these billboards a national issue with specific slants on morality and religion. It became a topic among debates and public outcry questioning its revealing sexual male imageries in too provocative poses. Others question the geographical locations of these out-of-home ads, as they were situated along a national road which may distract the focus and attention of daily commuters [23]. The Philippine Volcanoes' ad became a hit as everyone had their fair share of the issue, may it be in the academe, legislation, religious and moral sectors; lambasting the decency of the images presented and questioning the ethical standards of advertising in the country.

5 Emergence of sexualized billboard ads: a violation of Philippine advertising regulations

The themes of Bench Body billboard ads vary but the technique of presenting the product remains the same employing sexism or sexualization which makes it more appealing and sellable among Filipino consumers. This marketing ploy is indeed a violation of various Philippine advertising regulations as stressed in the Advertising Board of the Philippines (AdBoard) in its Advertising Code of Ethics (2000) [24] Article IV Section 1, where it is stressed that the presentation of human subjects in varied advertising forms should not be highly sexualized, vulgar, obscene and exploitative. Furthermore, Article IX stressed that, partly nude person is not allowed under the circumstances of: dressed in rickety or silhouette garments and dressed in skimpy or scantily-clad clothing exposing too much skin [female's body contours and male's bulging virilities].

But with the dissolution of the Ad Board, the Ad Standards Council (ASC) finally took over the role as the ultimate authority in the implementation of regulations surrounding ad contents [25]. Nevertheless, Bench Body's billboard ads are still in violation with the ASC's Revised Code of Ethics (2011) specifically its Article 1 [General Standards of Presentation] under Section 4 [Sex, Profanity, Obscenity and Vulgarity] provision where high profanity and extremely obscene advertising materials are prohibited. The employment of male models scarcely-dressed with Adonis-like physique and disturbing crotches, compromises these provisions. Particularly, Bench Body's billboard ads are presented in offensive nature and somehow contravene social and ethical standards. Moreover, these ads utilized obscene and vulgar poses, facial expressions, erotic situations, sensual props and setting and salacious taglines; suggesting sex as a way of selling the products. In addition, human subjects are deemed as sexual objects presented in stereotypical and violent themes, creating provocative presentations of sexual organs. Additionally, male subjects in the ads are displayed in critical poses simulating sexual intercourses and appeared in evocative physical contacts. Bench Body's billboard ads portray sexual acts and inhibitions which promote exploitative sexual tones that are of violation to Philippine society's morality standards.

6 Theoretical framework

This study is guided by the Reception Studies or popularly known as the Cultural Studies of Stuart Hall. The theory mainly discusses how audiences associate and interpret different media contents. Specifically, the study is anchored by the three media concepts of the theory: dominant reading, negotiated meaning and oppositional decoding. Dominant reading refers to the preferred meanings or interpretations that the producer of the message has for the content. Negotiated meaning on the other hand, pertains to somehow similar interpretation to the actual meaning of media texts but contests its philosophy and application while oppositional decoding literally means opposite association of meaning which is directly contradictory to those of the dominant and negotiated interpretations [26-29]. Also, the study is anchored on Suzanne Moore's Female Gaze Theory (1988) [30]. It argues that there is a female gaze in media which is far different from the male gaze. Furthermore, the theory explains that portrayals of men are patterned on traditional female representations such as nudity, scantily-clothed, sexual stares and sensual innuendos. Additionally, it stressed that this type of gaze is evident on the influx of male imageries in film and advertising.

The application of the concepts of Cultural Studies is focused on the interpretations or associations attached to the various billboard advertisements if they conform or compromise Bench Body's purpose of promoting the brand and its products. If these ads coincide with the brand's intention then they stand in dominant reading. But when these ads are in total opposite (demoralizing the image of the brand or downplaying its credibility) of the intentions of the undergarment brand then it is in oppositional reading. Negotiated reading is when these billboard advertisements concur with the preferred meaning (promotion of Bench Body's undergarments) but opposes its application (presence of sexualized male models or depiction of sexual objectification in the billboard ads). Female Gaze Theory on the other hand, explicated the idea that men same with its feminine counterpart can be subjected to sexual commodification in different parameters such as eye contact, posture, clothing and body language.

7 Methods

The study employed a qualitative approach since its goal is to describe the phenomenon of the emergence of sexualized male models in billboard advertisements. Furthermore, the study aims to determine the elements and the process involve in evaluating sexual objectification in the said out-of-home advertising. Specifically, content analysis is utilized which allowed the researcher to quantify the categories of various advertisements [33].

The researcher collected all possible materials for the study and a criteria was placed to filter the billboard ads to be used. Criteria in the selection of the ads are as follows: (1) should be a Bench Body billboard ad; (2) preferably under any Bench Body campaign; (3) could be in any theme or concept and (4) male model(s) in the ads should be Filipino [half breed or with Filipino blood]. After the selection process, the researcher came up with 27 billboard ads to analyze using a researcher-made coding sheet [based on various categories from previous literature and parameters presented in Visual Pleasure and Narrative Cinema (1975) [32] and Male Gaze (1975) [33]]. The coding sheet included various categories such as the theme of the billboard ad, underwear type and style, color, color intensity, gaze direction, gaze type, facial expression, body exposure, body type, body focus, body pose, advertising message and overall impression of the ad. The materials were examined, coded, classified and analyzed based on categories and occurrences, in order for the researcher to make inferences on their social and cultural contexts [34].

8 Findings and analysis

The samples were analyzed using various categories collated from previous researches and literature such as the theme, underwear type and style, color and intensity, gaze type and direction, facial expression, body type and focus, body exposure and pose, advertising message and the over-all presentation of the ad.

Content analysis revealed that the predominant theme of Bench Body billboard ads were sporty in nature, depicting male models posing with sports paraphernalia and equipment but there were also billboards with Victorian/Persian and Royalty themes. With regards to the underwear type and style, mostly bikini briefs that are low rise were used as main clothing of the male models. Scantily-clad or skimpy clothing constitute the idea of nudity [35], which is increasingly popular as an advertising technique. In terms of color and intensity, black and white seemed to be the top hue as shown through lighting and backgrounds of the advertisements. Most of the billboard ads looked dull and lack brightness and vividness, portraying a somber mood; allowing audiences to focus on the well-lighted parts of the billboard which are – the male bodies. In terms of gaze' types and directions, most models project a direct gaze to its audiences and predominantly are seductive or sensual in nature. Voyeuristic gaze implies a look towards the audience that is sexually arousing or objectifying towards same sex [36-37]. Facial expressions of the models were somehow mixed as some depict dominance, seduction and inviting vibes. With regards to body exposure and type, billboard ads displayed mostly whole bodies (head to knees) of the models and muscular bodies were usually the one strutting the various Bench Body billboards; exposing chiseled or hardly gym-ripped masculine physiques. This is a prototype of the New Man which is an eroticized representation of men often asserting certain body features, capturing audience attention [38, 15]. When it comes to body focus and pose, ads were dominantly showcasing male models standing with crotches or bulges which are quite obvious and noticeable from the audience eyes

Most of the Bench Body billboard ads conveyed indirect or underlying advertising message, not just promoting the company's product but also presented the idea of sexual objectification through the utilization of sexual imageries. Advertising unconsciously present subliminal messages aside from obvious visual metaphors [37]. Billboard ads were primarily presented in sensual, provocative and sexually- arousing manner.

9 Conclusions

The analysis results indicated that the idea of sex sells or sexual objectifications are highly exploited by fashion industry in order to promote their products and gain profit and revenue. And this is true in the case of Bench that tries to employ numerous marketing ploys ranging from sexualized male models to risqué headlines or captions; resulting to public scrutiny and outcry among consumers and moralists. Thus, the images of male models in the Bench Body billboards represented Filipinos as confident human beings but are portrayed as sexual objects with handsome looks and Adonis-like physiques. Indeed, Bench Body exploit human bodies as marketing ploys, compromising the standards of morality, religion and ethics of Philippine society.

10 Recommendations

Philippine government concerned agencies must impose firm regulations on advertising and its various types, specifically on its contents, themes, imageries and presentations. Furthermore, Metropolitan Manila Development Authority (MMDA) and Department of

Public Works and Highways (DPWH) must implement seriously the guidelines concerning placement of billboard advertisements in different public roads and thoroughfares as not to promote visual pollution and not to affect commuters' attention. Moreover, Ad Standard Council (ACS) and other advertising regulating bodies must be stricter in screening advertisements specifically its contents and implications to consumers. Additionally, advertising agencies and their creative think tankers should utilize human metaphors in lieu of actual human bodies. In addition, Bench and other fashion brands must resort to other advertising or promotional techniques or tactics in selling their products. These companies must retire from employing erotic and risqué models appearing side by side or incorporated with their products. For future research endeavors, the researcher recommends a holistic study dealing with all the billboard advertisements of Bench Body.

References

1. C. Chih-Chung, C. Chang, L. Wei-Chun Lin, Y. Nang, *Procedia-Social and Behavioral Sciences*, **57**, 352-359 (2012)
2. S.M. Busen, C.S Mustaffa, *Procedia-Social and Behavioral Sciences*, 155, 98-103 (2014)
3. P. Elena, *Procedia-Social and Behavioral Sciences*, 76, 686-689 (2013)
4. M. Muda, R. Musa, R. N. Mohamed, H. Borhan, *Procedia-Social and Behavioral Sciences*, **130**, 11-20 (2014)
5. GMA Network, retrieved from: http://www.gmanetwork.com/news/story/134812/money/penshoppe-bench-cite-strong-sales-but-profits-likely-to-get-hit (2008, November 21)
6. J. Sivulka, Soap, *Sex and Cigarettes: A Cultural History of American Advertising* (2012)
7. J. Speck, A. Guierrez, J. Sawayda, *Marketing Campaigns Use of Sex Appeal to Sell Products* (2011)
8. P. M. Lester, *Visual Communication: Images with Messages* (2014)
9. T. Reichert, J. Lambiase, *Sex in Advertising: Perspectives on the Erotic Appeal* (2010)
10. T. Hayri, *Procedia-Social Behavioral Sciences*, **62**, 1133-1137 (2012)
11. S.M. Alexander, *Sociological Perspectives*, **46** (4), 535-554 (2003)
12. M. Lövenberg, *The Elven Slave and Conan the Barbarian: Male Sexualization in Video Games* (2012)
13. J. Shewmaker, *Sexualized Media Messages and Our Children: Teaching Kids to Be Smart Critics and Consumers* (2015)
14. K. Milestone, A. Meyer, *Gender and Popular Culture* (2012)
15. S. Nixon, *Hard Looks: Masculinities, Spectatorship and Contemporary Consumption* (1996)
16. M. Dela Cueva, G. Gozar, C. Allingag, A. Salta, N. Enriquez, *Southeast Asian Ergonomics Conference* (2008)
17. E. B. Franch, C.B. Albiol, Online Journal of Communication and Media Technologies. **3** (2), 94-111 (2013)
18. J.E.A. Gomez Jr., *The Billboardization of Metro Manila.* Int. J. Urban Reg. Res. **37** (1), 186-214 (2013)
19. M.C. Hatol, Using Sex to Sell, retrieved from: http://www.mb.com.ph/using-sex-to-sell (2014, June 15)
20. Bench as a Global Brand, retrieved from : http://www.manilastandardtoday.com/lifestyle/107436/bench-as-a-global-brand.html (2013, June 25)

21. Spot.PH, retrieved from: www.spot.ph/peopleparties/48710/philippine-volcanoes-sizzle-in-bench- billboards?ref=cxsearch#photo-1370-1 (2011, June 30)
22. R. Gonzales, retrieved from: www.pep.ph/news/30082/national-rugby-team-philippine-volcanoes-pose- in-their-underwear-to-create-awareness (2011, July 6)
23. N. Almo, retrieved from: http://www.pep.ph/news/30109/mayor/requests-that-sexy-billboards-of-national-rugby-team-philippine-volcanoes-betaken-down (2011, July 8)
24. AdBoard, retrieved from: www.scribd.com/doc/71448729/AdBoard-Advertising-Code-of-Ethics- Manual#scribd (2000)
25. Ad Standard Council, retrieved from: www.asc.com.ph/our-standards/code-of-ethics/general-standards- of-presentation (2016)
26. McQuail, *Mass Communication Theory* (2010)
27. S. Littlejohn, K. Foss, *Encyclopedia of Communication Theory* (2009)
28. S.J. Baran, D.K. Davis, *Mass Communication Theory: Foundations, Ferment and Future* (2012)
29. E. Griffin, *A First Look at Communication Theory* (2011)
30. L. Gamman, M. Marshment, *The Female Gaze: Women as Viewers of Popular Culture* (1989)
31. G. G. Sparks, *Media Effects Research: A Basic Overview* (2013)
32. L. Mulvey, *Visual Pleasure and Narrative Cinema: Visual and Other Pleasures* (1975)
33. D. Chandler, retrieved from: http://visual-memory.co.uk/daniel/Documents/gaze/gaze11.html (2014, July 7)
34. E. Pernia, *Communication Research in the Philippines: Issues and Methods* (2004)
35. C.R. Wiles, J. A. Wiles, A. Tjernlund, J ADVERTISING RES, **36** (3), 57–66 (1996)
36. M. Patterson, R. Elliott, *Consumption, Markets and Culture*, **5** (3), 231–246 (2002)
37. U. Hakala, *Adam in Ads: A thirty-year look at mediated masculinities in advertising in Finland and the US* (2006)
38. F. Mort, *Cultures of Consumption. Masculinities and Social Space in Late Twentieth-Century Britain* (1996)

DOI: 10.1051/shsconf/20173300027

Experiencing the Digital News: The multiple dimensions of users' experience in news media use

Bahiyah Omar[1,*] and *Hasrina* Mustafa[1]

[1] School of Communication, Universiti Sains Malaysia, 11800 Penang, Malaysia

Abstract. The term media experience is seldom defined explicitly by communication and media scholars. Most often, it refers to media use and the feeling (such as enjoyment) resulted from the use. To fill this void, the present study turns to presence studies which espouse the idea of being there in mediated environment. As online news use involves a kind of mediated experience, this study adopts the existing measure of presence to determine the multiple dimensions of online news experience. A Principal Component Analysis (PCA) was used to establish the dimensions of the experience in this exploratory study. Four dimensions were extracted from the analysis: para-social interaction, transportation, absorption and social realism. The study found that online news users were experiencing beyond ordinary description; they interacted with the news, absorbed in the narrative world, and to certain extent felt that they were being transported into such world and also experienced the realism that the news stories tend to convey. Implications for understanding the experience in news media use from presence perspective are discussed.

1 Introduction

The adoption of the Internet as a mainstream news medium has become a global phenomenon. Since early 2000, the pattern of online news consumption across the globe has shown a strong growth of online news consumption especially in communities that have high Internet penetration and the developed segments of Asia-Pacific region [1]. Earlier studies indicate that the demographic characteristics of those who use the Internet regularly as a source of news have both increased and broadened [2] and a more recent report suggests the trend of using Facebook and twitter for news is on the rise [3]. The trend also suggests that the Internet has become the first choice for news and information among young people [4]. What leads towards this adoption is often linked to the attractive attributes of the Internet as the source of news. News research, however, seems to pay less attention to the experience of news use in online environment. Thus, examining the experience, and its possible dimensions, is central to the current study.

[*] Corresponding author: bahiyah@gmail.my

DOI: 10.1051/shsconf/20173300027

2 Literature Review

News researchers have attempted to understand the reasons for online news adoption. Examining online news use in Hong Kong, Chan and Fang [5] factor-analyzed 10 online news features and found that convenience, interactivity and multimedia were the key factors of adoption. Nguyen [6] found that an increase in online news attributes – in terms of 24/7 updates, customized news and no cost – predicts newspaper displacement. The attribute of convenience was found to be a strong predictor of the credibility of online news among politically interested news users [7]. Meanwhile, other study [8] found website usability and website attractiveness determined the selection of a particular news website over the other.

These studies shared one common theme; they examined the relationships between online news attributes and online news adoption within the innovation-diffusion framework. The attributes, to a certain extent, determine users' experience with the media. Yet, the two constructs are not identical. In consumer marketing research, for example, scholars define product attribute as users' perception of the product, not objective reality, which determine how it will be evaluated and purchased [9] while, product experience is defined as consumers' knowledge about and familiarity with a product or product category [10]. What constitutes attribute and experience in these studies signifies two different meanings.

The term 'experience' itself is loaded with various meanings. Most often, it refers to the real life, not the ideal or imaginary. It is also common to define experience as the knowledge or skill that results from experience. Another definition focuses on one's consciousness; "the conscious events that make up an individual life'[11]. This term, however, is seldom found in communication and media research. A closer examination of the studies in this field revealed that users' experience with media has been measured using various terms and in many ways; media use habit [12], media use [13] and media enjoyment [14]. These studies indicate that the frequency or pattern of media usage and the feelings resulted from the usage have been equated to the experience with the media.

It is also important to note that uses and gratifications approach has been applied to understand the experience of online news consumption [e.g.15]. Some studies [e.g.16, 17] have also examined different contexts, particularly online versus print environments, to demonstrate the effects of medium on news use. Notwithstanding, many studies especially in the field of computer sciences and cognitive psychology have longed been examining the experience of 'presence' which has been described as a sense of "being there" in a mediated environment. There was a growing interest among researchers [e.g.: 18, 19] to examine presence in order to understand the extent to which users find the virtual environment more engaging than the physical world. It is about the experiencing the "real" through media use.

Attempts have been made to conceptualize the concept of presence and provide theoretical evidence to establish its dimensions. Studies have established multiple dimensions of presence which include (a) social richness, (b) realism (c) immersion, (d) transportation, (e) social actor within the medium and (f) medium as social actor [20]. These dimensions have been further developed by Lee [21] who suggests (a) subjective or objective social richness; (b) perceptual or social realism; (c) transportation of self, place or other selves; (d) perceptual or psychological immersion; (e) social interaction with an entity within a medium and (f) social interaction with a medium itself as dimensions of presence. The concept of presence and its multiple dimensions demonstrate the experience of using a medium and how users perceive such experience.

Although presence is often examined in the immersive environment of virtual reality, it has also been tested in the context of different media. Previous studies have examined presence experience in video games [22] films [23] and even books [24]. Hence, the current study finds the logical appeal to adopt the concept of presence in explaining the experience of "being there" felt by audiences when they read a news story in an online environment. Such an experience refers to a direct relationship between the readers and the news story. In other words, readers feel that they are directly connected to the content and the story in such a way that they experience the shock, horror or pleasure conveyed by the news. Attempts to examine the experience of media use using this approach is almost non-existent. To fill the gap, this study is designed to explore dimensions of users' experience in online news consumption by using the measure of presence adopted from past research.

3 Methodology

A post-model experiment, exposing participants to an online newspaper, was conducted at one of the public universities in Malaysia. The study sample was a group of 147 undergraduate students who volunteered to participate in the experiment. On the day of the experiment, participants were directed to computer laboratories to read *Berita Harian* online without any time restriction imposed on them. After completing the reading task, they were asked to fill in a closed-ended questionnaire measuring their experience of reading the news during the exposure. The questionnaire consists of questions on presence dimensions, adopted from existing measures found in the literature, of which some of them were modified to suit the context of online news environment. All questions were written in both English and Bahasa Malaysia in one form (i.e: bi-lingual questionnaire). Principal component analysis (PCA) was used to establish the multiple dimension of users' experience in online news environment.

4 Results

Principal Component Analysis (PCA), with a varimax rotation, was used to determine the dimensions of online news experience. The analysis showed 10 components that fulfil Kaiser's criterion of eigenvalues greater than 1.0. Considering the study is concerned with the newly constructed scale for presence experience in the context of online news environment, factor loadings greater than 0.40 with secondary loadings above 0.20 difference were accepted. Scree plot analysis shows a break in the slope of eigenvalues after the fourth factor and hence the number of the dimensions were reduced to 4 factors. Some items were removed as they were conceptually inconsistent with the other items in the same factors. After elimination of these items, the final analysis produced 4 factors to represent presence experience in online news use. Table 1 presents the results of this analysis.

Table 1. Factor loadings using Principal Component Analysis with Varimax Rotation

Factors	**Indicators**	**1**	**2**	**3**	**4**
Para-social interaction	Showing non-verbal expressions to expressions/statements of people in news	**.74**			
	Reacting emotionally to expressions/statements of people in news	**.69**			

	Having the tendency to make verbal expressions to expressions of people in the news	**.69**			
	Ability to imagine the non-verbal expressions of people in news	**.66**			
	Ability to understood exactly what the people were going through emotionally	**.63**			
Transporta-tion	Ability to feel as if the self is at the location where the event occurred		**.77**		
	Being transported to the place where event occurred		**.76**		
	Picturing the self at the scene of news event		**.74**		
	Ability to feel that the events in the news is happening before the self		**.73**		
	Picturing the event as it is taking place		**.70**		
	Ability to feel as if the event was closer to the self than the real world		**.64**		
Absoption	Being pulled into the news story			**.76**	
	Ability to keep the mind focused on the news			**.75**	
	Ability to focus on the news; not thinking about other things			**.72**	
	Having the mind connected to the news; not wandering			**.66**	
	Being immersed into the news; not dragging			**.64**	
Social realism	Thinking that the events/incidents in the news could occur in real life				**.83**
	Believing that the events/incidents is a real life situation				**.71**
	Thinking that the news story sounds real				**.69**
	Believing that the events/incidents could occur to the self				**.53**
	Thinking that the news looks real				**.43**

Consequently, the current study interpreted the four factors accordingly. The first factor consisted of five items that reflect users' verbal and non-verbal engagement with the news. It was then labelled 'para-social interaction'. This factor explained 25.95% of variance in the experience of presence in online news use. The second factor that emerged from the analysis consisted of six items with 11.52% of variance explained. The factor refers to a sense of being transported to the location of the news event. Hence the second factor was labelled 'transportation'. The third factor which comprised of five items, was labelled the 'absorption' as the indicators suggest the extent to which users immerse themselves in the narrative world. All five items under this factor were reversed coded from distraction items to absorption items, and the revised statements representing the feeling of being absorbed in the news narrative were included in Table 2. This factor explained 7.65% of variance in the experience of online news use. Finally the last factor that emerged in the study is 'social

DOI: 10.1051/shsconf/20173300027

realism', which explained 5.73% of variance and comprised of five items. Social realism denotes the extent to which users feel that the news is real. The measure of all factors met the requirement of a reliable index as Cronbach alpha values for each factor was greater than 0.7.

5 Discussions

The aim of the current study is to understand media experience from different perspective. Using the existing measure of presence, this study explores how online users experience the news. The exploration leads to the discovery of four factors; para-social interaction, transportation, absorption and social realism, that represent the experience of presence in news consumption.

Firstly, this study found para-social interaction contributes the most in explaining the experience of news use. This concept is related to the fifth dimension, presence as social actor within the medium, as proposed by Lombard and Ditton [25].This dimension involves responding to social cues presented by a social actor within a medium. This study found that the para-social interaction, in the context of online news use, occurs when users involves in verbal and no-verbal expressions to the news stories that they read. The expressions could be in the form of emotional reaction to the stories and empathy towards the people in the news.

The second dimension is the sense of being transported to the news world. Lee [21] coins this dimension as transportation of self, place and other places which suggest two conditions; first is when users feel as if they were brought to a different time or place where the event occurred and second is when the objects or people from other places were brought to them through media use. In the context of online news use, users feel as though they are been transported to the news world, witnessing the events at the location where the events occurred. The finding could be implied by connecting this experience with the availability of multimedia contents in online news such as videos, picture galleries and interactive images. These features allow users to experience the transportation of self and place in online news media environment.

Absorption, the third dimension, is a concept closely related to transportation. It involves a sense of transitioning from the actual world to the narrative world [26]. The core to absorption is the experience with the narrative such as understanding and paying attention to it. In the context of online news use, this experience denotes the extent that users can focus and attentive to the news stories. Such an experience can also lead to the sense of being immersed in the news narratives.

Last but not least, social realism is the fourth factor found in the current study. It refers to a sensation of reality resulted from the real representations of objects, events or people produced by the medium [20]. The fact that news construct reality is often debated by journalism scholars. According to Lau [27], construction of reality happens when the reality is described in news and taken to be true by the public. The current study supports this notion as the finding suggest users find the news to be real and could affect their lives.

In conclusion, the current study provides a novel approach in understanding users' experience with the news and news media. This study contributes to the re-conceptualization of media experience as a multidimensional construct in media studies. On the practical side, uncovering users' experience in para-social interaction, transportation, absorption and social realism in online news consumption can help media practitioners to develop effective journalistic strategies to enhance news sites stickiness. Measurement validation for these dimensions is recommended for future research.

This project was funded by Exploratory Research Grant Scheme (ERGS), Ministry of Education Malaysia.

References

1. A. Nguyen. First Monday, **8** (2003)
2. A. Kohut, C. Doherty, A. Keeter. PEW (2004)
3. M. Barthel, E. Shearer, J. Gottfried, A. Mitchell. PEW (2015)
4. D.Code, World Association of Newspapers, (2007)
5. K. Chan, K, Fang, W. Research Paper, **8**, 4 (2007)
6. A. Nguyen, Journalism, **11**, 2 (2010)
7. T.J.Johnson, B.K.Kaye, Journ & Mass Comm Quarterly, **79**,2 (2002)
8. N.A. Zulkafli, B. Omar, and N.H. Hashim, SEARCH. **6**, 2 (2014)
9. G. Tom, T. Barnett, W. Lew, J. Consumer Marketing, **4** (1987)
10. J. W. Alba, J.W. Hutchinson, J. Consumer Research, **13**, 1 (1987)
11. "experience", in *Merriam-Webster Dictionary* (2015).
12. D.L. Lasorsa, W. Wanta, Journ & Mass Comm Quarterly, **67**, 4 (1990)
13. J. A. Brown, J. Comm, **48**, 1 (1998)
14. M.C. Green, T.C. Brock, G.F. Kaufman, Comm Theory, **14**, 4 (2004)
15. B. Omar, Procedia: Social & Beh Sci, **155** (2014)
16. B. Omar, *Proc of the 5th Int Conf on Computing & Informatics* (2015)
17. W.P. Eveland, S. Dunwoody, Comm Research, **28**, 1 (2001)
18. C. Heeter, Presence, **1**, 2 (1992)
19. H. Regenbrecht, T. Schubert, *Proc of the 5th Annual Int Workshop on Presence (*2002)
20. M. Lombard, T. Ditton, J. Comp Mediated Comm, **3**, 2 (1997)
21. K.M. Lee, Comm Theory, **14**, 1 (2009)
22. J.D. Ivory, S. Kalyanaraman, J. Comm, **57**, 3 (2007)
23. J. Lessiter, Presence, **10**, 3 (2001).
24. T. Schubert, J. Crusius, *Proc of the 5th Int Workshop on Presence* (2002)
25. M. Lombard, T. Ditton, *Proc* of the 3rd *Int Workshop on Presence* (2000)
26. R. Busselle, H. Bilandzic,. Media Psyc, **12**, 4 (2009)
27. R.W.K. Lau, Journalism, **13**, 7 (2012)

 DOI: 10.1051/shsconf/20173300028

Foreign Nationals as Offenders and Victims in Malaysian Crime News

Norealyna Misman[1,*], *Hamedi* Mohd Adnan[2], *Amira Sariyati* Firdaus[3], and *Che Mahzan Ahmad*[4]

[1,2,3]Department of Media Studies, Faculty of Arts & Social Sciences, University of Malaya, 50604 Kuala Lumpur, Malaysia
[4]Department of Communication, Kuliyyah of Islamic Revealed Knowledge and Human Sciences, International Islamic University Malaysia, 53100 Gombak, Malaysia

Abstract. Foreign nationals in Malaysia come from all corners of the world. They are here as migrant labour, highly skilled and professional migrants (expatriates), illegal migrants, refugees, asylum-seekers (Burmese asylum seekers with UNHCR card), forced migrants (human trafficking victims), students, and tourists. The influx of foreign nationals residing in Malaysia coincides with greater number of crime news featuring foreign nationals. This study explores the social construction of foreign nationals as the 'other' in the local crime news published by Malaysian newspapers. 94 news headlines and lead sentences of local crime news involving foreign nationals were identified and analysed for this study. Findings suggest that Malaysian newspapers magnify foreign nationals' migration status in each crime news.

1 Introduction

Malaysia has a long history on the immigrant for labour purposes where it started during the pre-independent era. The British colonization brought the Chinese from mainland China to work in the mining industry while the Indian from India to work in the plantation industry or estate. In the 1970s, Malaysia started to bring in foreign migrant from Indonesia to work in the construction industry [1]. In a meanwhile, due to Indo-China crisis, immigrants from Vietnam started to come to Malaysia as refugees and asylum seekers in the mid-1970s [2].

Currently, foreign nationals in Malaysia were from the various continent of the world. They were here as labour migrants, highly skilled and professional migrants (expatriates), illegal migrants, refugees, Asylum-seekers (Burmese asylum seekers with UNHCR card), forced migrants (human trafficking victims), students, and tourists. The Dewan Rakyat had announced that as at 2015, they were about 2.06 million foreign nationals working legally in Malaysia. The number constitutes of 6.9% the overall population in Malaysia [3].

* Corresponding author: elmisman@gmail.com

DOI: 10.1051/shsconf/20173300028

Crime issue is among the salient matter and highly concerned by Malaysian. An official report from the Royal Police Department stated that the crime rate in Malaysia declined by 17.7% in 2015[4]. Despite the decline, the crime news still grabs the attention of the readers and later become public anxiety.

Among the total of the crime index, about 11.2% of the total of crime cases were cases involving foreigners [5]. Statistically, the number is not much, but past studies show that the media is able to subtly construct foreign nationals as criminals although foreign nationals might be both offenders and victims in the news [6].

1.1 Media, crime, and foreign nationals

Due to the global diaspora, the studies on the representation of immigrants in the newspapers increase. As a country with high total number of foreign nationals, local researchers started to study on press coverage on illegal immigrants, the voices of people in power in the media about the immigrants and asylum seekers in Malaysia and the media reports on the influx of foreign workers in Malaysia [7,8,9]. Most of the researchers agreed that the media tend to report foreign nationals as criminals in the crime news. This could be described as minority signification [10]. Other than that, the public easily ponder a dramatic crime news such as murder, rape, and robbery cases [11]. For example, public understands easily a news on a Bangladeshi murdered his colleague than the financial significance of a corruption of public funds. Therefore, the media usually play the issue in this area and add foreign status to increase the news value.

1.2 Problem Statement

Foreign nationals that living in Malaysia are often considered as the 'other' as local citizen assumed themselves as 'us'. Therefore, the foreign nationals' status was magnified in the daily news especially the crime news. Mentioning nationality or foreign status in crime news increase the readability, drama and value of crime news. Hence, the purpose of this study is to explore the social construction of foreign nationals as the 'other' in the local crime news published by Malaysian newspapers. The analysis presented were based on the research question: What are the references or framed words that portray the foreign nationals as the other in the crime news of Malaysian newspaper?

2 Method

This study employs critical discourse analysis to analyse the data. This approach is used to trace in detail the construction of foreign nationals as the other/outsider/out-group in stereotype image of foreign criminals as in the Malaysian crime news [12].

Four main English language newspapers will be used for this study. They are *The Star, The New Straits Times, The Sun*, and *The Malay Mail. The Star* and *The New Straits Times* are mainstream newspapers and *The Sun* and *The Malay Mail* are known as alternative newspapers in Malaysia. These newspapers usually targeted people living in urbanized areas such as Klang Valley.

The analysis took place by examination of the headlines and the lead sentences [13]. The headline is important to see the tone of the article, the summary of the story and sometimes the only thing that the readers might read. In a meanwhile, the lead sentences serve the similar function to the headline and the news source to considerate from whose perspective the story being presented.

About 112 news articles were located randomly within the period of January to April 2016. However, only 94 articles were chosen for analysis. Other 18 news articles were eliminated because of redundancy of the topic reported. The time frame was chosen because, in the early of 2016, the Malaysian government announced to suspend the hiring of foreign workers in Malaysia. There was debate among various parties about the issue. Some people happily agreed with the decision because they think the influx of foreign nationals in Malaysia create uncomfortable environment but some are disagreeing stating that 'the country's economy is powered by foreign workforce [14,15].

3 Analysis

All the articles were coded from 1 to 94. *The Star* (n = 37), *New Straits Times* (n= 24), *The Sun* (n=22), and *The Malay Mail* (n=11). The analysis showed that (n=42) of the articles mentioned the specific nationality of the person involved in the headlines. (n=24) of the articles mentioned foreign national status in the headlines. However, all of the articles mentioned the foreign national status either at the headlines or the lead sentences of the articles.

The analysis presented were based on Wodak's discourse-historical approach [16,17]. The analysis below is to answer the research question on what are the discourses identified from the data. Two main discourses identified which are foreign nationals as offenders and foreign nationals as victims.

3.1 Discourse

Two main discourse identified from the analysis of the headlines and lead sentences of the data are foreign nationals as offenders and victims in the local crime news.

3.1.1 Foreign nationals as offenders

Five genres were identified under the discourse. First, the dominant genre under this discourse is foreign nationals were highly involving in illegal migration and human trafficking. (12,21,23,24,25,26,33,41,42,44,50,52,55,58,59,62,65,70,73,74,75,81,83,85,86,87,90). An example of the articles is '240 illegal immigrants jailed' (81). For human trafficking, such like 'Filipino charged for trafficking girl' (65). Second, the analysis identified foreign nationals as actively involved in violent crime events namely murder, assault and rape (3,7,9,19,30,47,56,72,88,91,93). An example of foreign nationals was framed in murder cases is 'Rohingyas charged with hiding body' (9). Other than that, 'Bangladeshi man arrested on suspicion of raping, molesting teen girls' (93) which portrayed foreign nationals as hunting local teenagers for bad purposes. Third, the data analysis addressed the construction of foreign nationals as involved in drug smuggling cases (2, 5,82,84,94). Among the articles depicting the genre are 'Thai national among three held in drug bust' (2) and 'Bukit Aman police plug Chinese drug pipeline, arrest 10 foreigners, 4 locals' (94). Fourth, another genre found is foreign national also involved in robbery cases (53) with the headline 'Gang Negro busted, and two held'. Fifth, the Malaysian newspaper also reported foreign national as involved in conning activities (6,8,10,17,44,46,66,69,76,79,80). Several categories of conning activities identified namely card–scam, 'Executive duped in card scam' (10). There was also car-cloning syndicate by Singaporeans, 'car-cloning syndicate busted' (17), and cases on foreign traders run businesses using fake business licenses, 'Foreign traders in control of local permits, premises' (80). The last genre is on the foreign nationals as a threat as there highly involved in terrorism activities such as kidnap for ransom (14,63), Sabah invasion (48,89), and Islamic States (IS) (13,51,68). The examples of headlines are 'Accused denies communicating with Lahad Datu intrusion leader' (89) and 'An Indonesian oil-and-gas consultant gets 9 months' jail for terrorism' (68). It can be seen that only Filipino were associated with abduction for ransom and Sabah intrusion while only Indonesians were portrayed as have a connection with IS.

3.1.2 Foreign nationals as victims

Apart from being depicted as criminals, foreign nationals also were reported as victims in the articles analysed for this study. 20 articles were portraying the foreign national status of the victims of the crime events. There were three genres identified which are foreign women as victims, exploited foreign nationals, and foreign-local relationship.

First, foreign women were portrayed as weak and usually be the victims of various crime incidents namely prostitution (21,65) robbery (29,60), murder (20,71), rape (45), and abuse (49). The headlines that can be highlighted are 'Three elderly men charged with trafficking women for prostitution' (21), 'Brazen Bangsar snatch theft - A Japanese housewife was dragged for several metres for her handbag' (60), and 'Man nabbed for molesting maid' (45).

Second, the status of foreign national might be the cause of them to be targeted for exploitation and manipulation. For example, 'Japanese man fleeced by a woman in Kota Kinabalu' (11), and 'Bangladeshi hurt in Mak Mandin flat robbery' (16). There were also be beaten up by the local because of minor misunderstanding event (27,28). Weaker foreign nationals also are the victim of murder (15,54,57,67).

Third, crime news involving foreign nationals also could be an indices of the changes of Malaysian social structure. The researcher found a genre of local-foreigner intimate relationship in the crime news. There are family conflicts that lead to crime cases. The husband or wife are foreign nationals (18,49,71,77). An example of the headlines are 'Wife admits doing a "Bobbit"' (18), Man held for abusing Uzbek wife' (49), and 'Family brawl turns fatal for Chinese national husband' (77).

4 Discussion

Thematically, the Malaysian newspapers socially constructed foreign nationals as both offenders and victims in the local crime news. The offenders were portrayed as cruel, horrific, and rogue especially in violent crime events. In contrast, victims were framed as weak, terrified, and pitiful in the crime news.

Despite all, through the examination, it is suggested that the newspapers revealed the nationality/ foreign status of the person involved in the crime events either in the headlines of the news or the lead sentence of the news. It is believed that the act is to add drama, readability, and value of the crime news [11]. Foreign national status such as Nepalese, Indian national, Indonesian woman, Bangladeshi, Myanmar nationals and others were among status mentioned in the data analysed [12].

News on illegal immigrants was reported the most to show that the illegal immigrants issue becoming one of the social ills in Malaysia. Moreover, the news represented the government's voice depicting that the responsible parties will combat illegal immigrant and human trafficking issues. Past studies on illegal immigrants in Malaysia also showed voices of authority are crucial to show they are actively curbing illegal immigration activities. Based on the data analysed, we can see that it is indeed lack of voices of foreign nationals themselves both as offenders and victims. The news analysed reported the voices of government, authority (Police and Immigration department officers), and also voices of the public (including reporters/editors of the newspapers). It is actually important to report the voices of the foreign nationals to see their views and judgments on this issue. Foreign nationals' views on both issue of 'othering' and 'criminalising' them are essential to avoid from over-generalisation, negative attribution, and prejudice. As a result, the study on this issue also will be more balance.

The study also portrayed the reality of Malaysian landscape presently. The increasing number of foreign nationals in Malaysia build a web of relationships between Malaysian and foreign nationals from various countries. Some are intimate relationships such as husband and wife or lovers. There is a relationship like supervisor and co-worker, and also the news depicted conflicts occurred among foreign nationals themselves. In some landscape, foreign nationals are accepted as part of the group, in contrast, Malaysian in general still not accepting the foreign nationals as part of them.

All of the newspapers' tone of reporting are quite similar regardless of their different ownership and status as mainstream and alternative newspapers. Therefore, it can be said that Malaysian media is in the same view on magnifying the foreign nationals' status as the other through crime news.

References

1. N. Widyawati, *Representations of migrant workers in Malaysian newspapers.* In The API Fellows. Are we up to the Challenge?: Current crises and the Asian intellectual community. (The Work of the 2005/2006 API Fellows (2006). Retrieved from http://www.apimal.org/blogcms/media/13/File/Nina.pdf.
2. Z. Mohd Don and C. Lee, *Representing immigrants as illegals, threats, and victims in Malaysia: Elite voices in the media.* Discourse & Society, **25**(6), 687-705 (2014).
3. W.S., Teh, Migrant workers: Invisible workforce. *Malaysian Digest* (2015). Retrieved from http://www.malaysiandigest.com/features/541277-migrant-workers-malaysia-s-invisible-workforce.html.
4. K. Karupiah, Crime rates are reducing and yet people feel unsafe. *The Ant Daily* (2015). Retrieved from http://www.theantdaily.com/Main/Crime-rates-are-reducing-and-yet-people-feel-unsafe.
5. N.a, 9,496 crime cases involving foreigners recorded from January to August. *The Sun Daily* (2015). Retrieved from http://www.thesundaily.my/news/1227187
6. H. Brookes, 'Suit, Tie and a Touch of Juju – The ideological construction of Africa: A discourse analysis of news on Africa in the British Press. *Discourse & Society,***6**(4) : 461- 94 (1995).
7. Abdul Rahim Abdul Hamid, Bachan Singh, & Muhammad Ammar Jamadi. Foreign labour employment in construction project. IJSRP ,**3**(11), 1-7 (2013).
8. H. Abdul-Rahman, C. Wang, L. Wood, & S. Low. Negative impact induced by foreign workers: Evidence in Malaysian construction sector. *Habitat International*, **36**(4), 433-443 (2012).
9. M. Idrus & I. S. Ismail, Petaling street almost Malaysian- A discourse analysis of news on illegal immigrants. *Procedia - Social and Behavioral Sciences,* **90**, 374 – 380 (2013).
10. R. Yamamoto, Bridging crime and immigration : Minority signification in Japanese newspaper reports of the 2003 Fukuoka family murder case. *Crime Media Culture,* **9**(2) 153 –170 (2012). DOI: 10.1177/1741659012461039
11. Y. Jewkes, *Media and Crime* (3rd ed.). London: Sage (2015).

12. M. Griffiths, The convergence of the criminal and the foreigner in the production of citizenship. In B. Anderson & V. Hughes, *Citizenship and its others* (1st ed., pp. 55-63). New York : Palgrave Macmillan (2015).
13. K.A. Lewig, &M.F. Dollard, Social Construction of work stress : Australian newsprint media portrayal of stress at work, 1997-98. *Work & Stress,* **15**(2), 179-190 (2001).
14. G. Dan, Not enough Malaysians to fill jobs done by foreigners. *The Malay Mail*, p. 11 (2016).
15. Z. Arifin, Country's economy is powered by foreign workforce. *New Straits Times*, p. 19 (2016).
16. P. Baker, C. Gabrielatos, M. Khosravinik, M. Krzyz˙anowski, T.Mcenery, & R. Wodak,. A useful methodological synergy? Combining critical discourse analysis and corpus linguistics to examine discourses of refugees and asylum seekers in the UK press. *Discourse & Society,* 19(3),273–306 (2008). DOI: 10.1177/0957926508088962.
17. M. Reisigl, M. & R.Wodak, The Discourse-Historical Approach. In R. Wodak & M. Meyer (Eds.), *Methods for Critical Discourse Analysis* (2nd ed.), pp. 39-45 (2009). London : Sage.

Conceptual view of Cyberloafing and Non-Work Domain

Patrick Chin-*Hooi Soh*[1,*], *Klan-Yeik* Koay[2] , and *Kok-Wai* Chew[3]

[1,2,3]Multimedia University, Cyberjaya 63100, Malaysia

Abstract. In an attempt to understand the reasons for employees' personal Internet use at work, known as cyberloafing, this paper attempts a new perspective to look at the phenomenon. As the barrier between the previously separate work and home domains reduces, employees are increasingly integrating their working and private world. This phenomenon has resulted in employees using their personal time for work-related tasks through technology in the privacy of their homes and private domain. Conversely, employees could be carrying out non-work related Internet use during office hours. This paper aims to explore the possibility of non-work domain as a factor for employees to cyberloaf. This paper proposed a conceptual model based on border theory and theory of interpersonal behaviour. The resultant augmented theory offers a richer explanation of how non-work domain affects employees' decision making process on cyberloafing.

1 Introduction

In a book named Work/Life Collision written by [1], the author describes about the collision between work and life with its subsequent impact to the society, families, and individuals in Australia. It was observed that there are drastic changes in household patterns with rising number of singlehood, single-parents households and dual-income earners households, more participation of women in labor force seeking for jobs and wealth, a redistribution of working hours and a greater emphasis on work-life-balance. These phenomena are even more conspicuous in today's world as compared to 12 years ago, when the book was written. Combined with the rapid advancement of technology that allows communications without geographical constrains, this has greatly minimized the barrier between the traditional boundary of work and family domains. These trends have driven employees to integrate their work and family domains. For example, employees in United States work at home twice a week [2]. Apart from that, utilizing the technology to perform work-related tasks such as checking office email, receiving instant messages from customers or colleagues and bringing office work home to do after official working hours has been a common practice. In today's global economy, it is prevalent for companies having businesses around the world. This leads to the need of employees of multinationals

* Corresponding author: chsoh@mmu.edu.my

meeting their counterparts who stay across continents with different time zones via teleconference late at nights or even in the early mornings for meetings.

To compensate the sacrifices of personal time for work-related purposes, employees may feel less guilty in and reciprocate by engaging non-work-related online activities during working hours for personal errands or stress relief. However, most of the organizations abhor employees from cyberloafing due to its negative ramifications such as clogging up networks bandwidth, degrading system performance [3], putting organizations at risk for legal liabilities (e.g., employees download illegal music or watch offensive videos) [4-7], and reducing job productivity. Surveys found that sixty one percent of American employees engage in non-work related activities during working hours and fifty-six percent of their time online, in fact, doing non-work related activities [8, 9] . About 20% to 30% of companies have terminated employees for cyberloafing activities like accessing pornographic sites, online gambling, and online shopping [10]. Given that there is no comprehensive framework that includes the potential impact of non-work domain in explaining employees' cyberloafing behaviour in the current literature, this conceptual paper seeks to:

1) Extends the theory of interpersonal behaviour by explaining the role of private demands and importance of non-work domain in affecting employees' cyberloafing behaviour

The resultant augmented theory provides a richer explanation of cyberloafing intention and behavior.

2 Theoretical background

This section explains Clark's work/family theory [11] and Triandis's theory of Interpersonal behavior [12]. Subsequently, we propose 6 propositions based on both theories in order to have a better understanding on how non-work domain affecting employees' decision making process on cyberloafing.

Border theory argues that people proactively try to attain work life balance and reduce role conflicts by constantly crossing the borders between work and non-work in order to satisfy both needs [11]. Changes in household patterns with rising number of dual-income and single-parent households have caused people holding multiple roles and responsibilities both at home and at work. As a result, in an attempt to manage the unique needs of each domain, people often crossing the border of work and family through communications and other actions. Following the logic and reasoning behind the border theory posits, this suggests that private demands can be a source of motivation for employees to use the Internet for non-work-related online activities. Private demands refer to family/social demands outside of the work domain. Employees may hold different roles such as being the breadwinner for the family or a part-time English teacher of tuition centre with various obligations and responsibilities in the society. Hence, employees with more private demands are more likely to fulfill those needs through using the Internet at work. For example, employees might use their Internet to buy movie tickets, pay home bill or arrange transportation to pick up their children from school due to working overtime. This is a concern in today's modern society where both parents are earning to support the family. As such, communication and coordination between family members are frequently done through during office hours.

A theory developed by Triandis to explain various types of human behavior named theory of interpersonal behavior (TIB) has been applied in the context of cyberloafing and proven to have a strong exploratory power on cyberloafing in previous empirical studies [12,13,14]. The theory postulates that a person's intention to act a certain behavior is best predicted by his or her perceived consequences, affect and social factors. On the other hand, the set of precursors that determine an individual's actual behavior are habit, intentions, and facilitating conditions.

Perceived consequences in this paper refers to "the degree to which cyberloafing behavior is valued positively or negatively by the individual; it is an assessment of the total set of outcomes that result from using the Internet for non-work related reasons during office hours" [14]. TIB posits that any behavior will generate a form of positive or negative outcomes which are evaluated subjectively by the individual based on the perceived degree of benefits and penalty. Affect refers to feelings of like and dislike felt by an individual when performing cyberloafing behavior [12]. The emotion linked to cyberloafing vary from exciting, happy, disgusting, dull or hatred, depending on how the individual perceive of cyberloafing [14]. When cyberloafing is seen as an enjoyable activity by employees, it is more likely they will form the intention to employ the Internet connection for cyberloafing. In this paper, social factors refer to the degree to which the perceived social pressure from important referent groups on the use Internet for non-work-related purposes in the workplace [12]. For example, referent groups include the employee's family, friends, co-worker, supervisor, IT department and top management. The influence of each social group on employees' cyberloafing very much relies on the perceived importance of the particular group by employees. When cyberloafing is perceived to be an acceptable behavior by others such as colleagues, top management or friends, the individual will be more likely to form intention to cyberloafing. Conversely, when people perceive their social networks disapprove of cyberloafing activities, the intention to cyberloaf will not be formed.

Facilitating conditions refer to factors within the individual's environment which are perceived to facilitate or impede the performance of certain behavior. Facilitating conditions can be classified into two dimensions [14]. First dimension refers to situational factor consisting of external conditions and access to the resources. Second dimension refers to internal factor referring to a person's self-efficacy to use the Internet to cyberloaf. For example, organizations that implement tight Internet monitoring system and follow strictly to the Internet usage policy, these impediments will reduce the likelihood of employees from cyberloafing [16]. In the same vein, employees who have the intention to cyberloaf without the access to the Internet access may not translate into actual behavior. Furthermore, the capability to use the Internet for personal reasons is crucial as ones must at least know how to technically utilize the technology given.

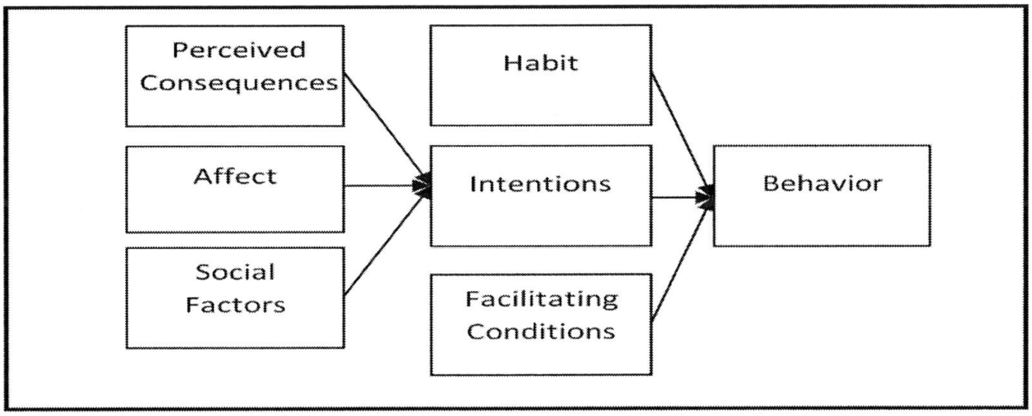

Fig 1. Theory of Interpersonal Behavior

Although theories and previous studies have provided indications on the possible impact of non-work domain in affecting employees' cyberloafing behavior, surprisingly the extant theories of cyberloafing in current literature have not explicitly considered them, including

DOI: 10.1051/shsconf/20173300029

TIB. To date, there is only a single study by [17] reporting that private demands have a positive relationship with cyberloafing. However, the underlying mechanism that private demands drives cyberloafing remains unknown for researchers. In this paper, we propose 6 prepositions to extend the existing theory of Interpersonal behavior, so that future researchers can take into consideration of the impact of non-work domain on employees' decision making process on cyberloafing.

P1: Private demands have a positive relationship with Perceived Consequences for Cyberloafing.

When employees utilize their private time (usually occurs at home) for work-related reasons, they are likely to feel justified to use their office hours to use the Internet for non-work related reasons. For example, an employee who spend hours at nights in the home doing work-related research would probably have little qualms to use part of his working hours in the office for online entertainment, relaxation or to maintain friendship through social networking sites. Another example is that an employee should have paid his monthly installment for his loan via online payment after working hours at night. However, due to he is so busy to finish his works until late night, he has forgotten to pay his bill. As a result, the needs to pay the bill may convert to (higher positive) perceived consequences for cyberloafing during work hours in view of the penalties of charges for late payments.

P2: Private Demands has a positive relationship with Affect for Cyberloafing.

To the extent that cyberloafing can be used to resolve private demands, employees will want to seek satisfaction in their work/life balance. For example, employees may seek to communicate with their loved ones on what time he/she is returning home for dinner, or perhaps after an argument with their spouse or turn to online gaming for escape from marital stress.

P3: Private Demands have a positive relationship with Social Factors for Cyberloafing.

In this current technological age, family, friends and social contacts regularly communicate with each other real-time and online through instant messaging, Facebook, Twitter etcetera. It follow then that people will perceive their friends and colleagues to be less disapproving (and might even be expecting) the use of the Internet for personal matters during working hours.

P4: Importance of non-work domain moderates the relationship between Private demands and perceived consequences for cyberloafing

P5: Importance of non-work domain moderates the relationship between Private demands and perceived consequences for cyberloafing

P6: Importance of non-work domain moderates the relationship between Private demands and perceived consequences for cyberloafing

Home/family, friends, community are the constituents of an individual's non-work domain. The priority people place on their work, home, or leisure is not the same and largely depends on cultural context and a person's life stage [18]. Work-oriented people may be dedicated to his or her job and thus the likelihood of them cyberloafing due to the influence of non-work domain will be lower as compared to nonwork-oriented people. For work-oriented people, cyberloafing may not be perceived as providing significant positive consequences such as maintaining relationship with family or friends or saving time for personal commitments, and generating positive affect such as happiness, loving or caring. Furthermore, young people who are at the beginning of their career might want to put more emphasis on their work rather than home or leisure. Consistent with this reasoning, the impact of private demands on perceived consequences, affect and social factors for cyberloafing is very much determined by the weights given to non-work domain by the employees.

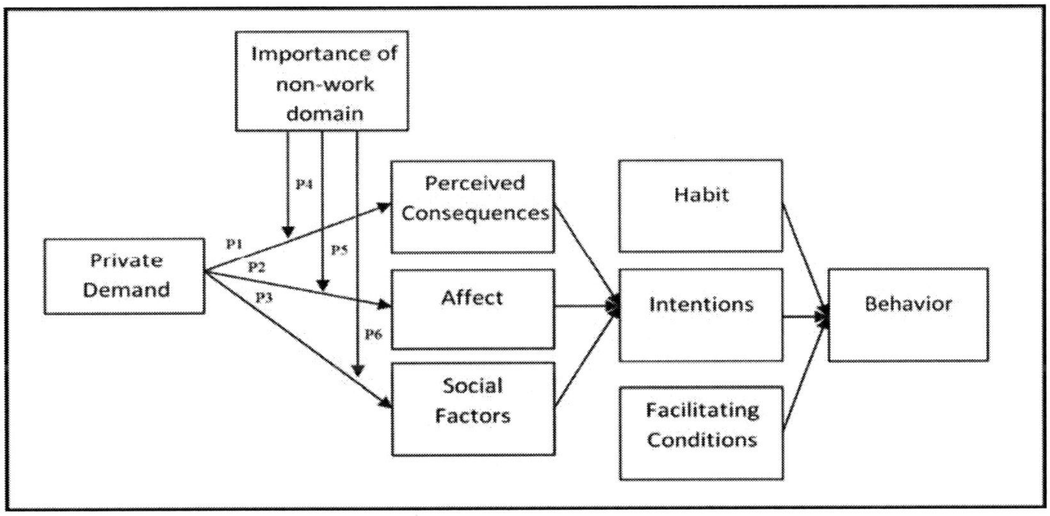

Fig. 2. Augmented Theory of Interpersonal Behavior

3 Conclusion

This conceptual paper attempts to explain the role of private demand and the moderating role of the importance of non-work domain on cyberloafing behavior through an extension of the theory of interpersonal behavior (TIB). As the boundary between work and private lives is becoming more intertwined, people need to balance their work and private domains. One chief means of balancing work and private domain is cyberloafing during office hours.

Future studies should empirically test the proposed conceptual framework to find out whether employees' non-work domain plays a substantial role in affecting cyberloafing behaviour.

References

1. B. Pocock, B. *The Work/life Collision: What Work is Doing to Australians and what to Do about it*: The Federation Press (2003)
2. Gallup News Service. Gallup Polll Social Series: Work and Education. In J. Jones & L. Saad (Eds.): Gallup News Service. (2005)
3. J. C. Sipior, B. T. Ward, *A Strategic Response to the Broad Spectrum of Internet Abuse*. Inform Syst Magage, **19,** 71-79 (2002)
4. A. E. Lichtash, *Inappropriate use of e-mail and the Internet in the workplace: The arbitration picture*. Dispute Resolot J, **59**, 26-36 (2004)
5. J. E. Mills, B. Hu, S. Beldona, J. Clay, *Cyberslacking! A liability Issue for Wired Workplaces*. Cornell Hotel RestauR Adm, **42**, 34-47 (2001)
6. R. R. Panko, H. G. Beh, *Monitoring for pornography and sexual harassment*. Commun ACM, **45**, 84-87 (2002)
7. L. R. Scheuermann, H. P. Langford, H. P, *Perceptions of Internet abuse, liability, and fair use*. Percepl and Motor Skill, **85**, 847-850 (1997)
8. I. Websense, Web@Work Employee Computing Trends Survey (2006)

9. MySammy, Cyberloafing Infographic: Personal Internet Use at Work Infographic. (2013)
10. C .J. Case, K. S. Young, *Employee Internet Management: Current Business Practices and Outcomes*. CyberPsychology & Behaviour, **5**, 355-361 (2002)
11. S. C. Clark, Work/family border theory: A new theory of work/family balance. *Hum Relat,* **53**, 747-770 (2000)
12. H. C. Triandis, Values, Attitudes, and Interpersonal Behaviour.In H. Howe & M. Page (Eds.), *Nebraska Symposium on Motivation,* 1979, 195–295. Lincoln, NE: University of Nebraska Press (1980)
13. G. D. Moody, M. Siponen, Using the *Theory of Interpersonal Behaviour to Explain Non-Work-Related Personal Use of the Internet at Work.* Inform Manage, **50**, 322–335 (2013)
14. L. G. Pee, I. Woon, A. Kankaanhalli, A, Explaining non-work related computing in the workplace: A comparison of alternative models. Inform Manage, **45**, 123-130 (2008)
15. T. K. Betts, A. J. Setterstrom, J. M. Pearson, S. Totty, *Explaining Cyberloafing through a Theoretical Integration of Theory of Interpersonal Behaviour and Theory of Organizational Justice.* J Organ End User Comput , **26,** 23-42 (2014)
16. J. C. Ugrin, M. J. Pearson, *The Effects of Sanctions and Stigmas on Cyberloafing*. Hum Behav, **29**, 812–820 (2013)
17. C. J. König, M. E. Caner de la Guardia, *Exploring the Positive Side of Personal Internet Use at Work: Does It Help in Managing the Border between Work and* Nonwork? Computer Human Behavior, **30**, 355-360 (2014)
18. E. R. Eddy, C. P. D'Abate, P. W. Thurston Jr, *Explaining engagement in personal activities on company time.* PERS REV, **39**, 639 – 654 (2010)

SHS Web of Conferences **33**, 00030 (2017) DOI: 10.1051/shsconf/20173300030
i-COME'16

Framing and Identity: How Mainland and Taiwan Media Represent Terrorist Attacks in China?

Yan Wang[1*]

[1] College of Humanity, 310023 Zhejiang University of Technology, Hangzhou

Abstract. Loss of identity is a leading cause of terrorism and a common bewilderment of Mainland China and Taiwan. Cutting into the research by the news representation of Chinese terrorism incidents, this paper discusses a special identity issue in a special identity history. This paper adopts mixed methods, which are constructed by content analysis, framing discourses and frequency statistics. 111 pieces of related news are selected and doubly-coded into both "identity package" and "news framing". The results show that the news representations inter-connect with the coexisting identity and confrontation between Mainland China and Taiwan.

1 Introduction

The occurrence of terrorism incidents usually reinforces the complexity of the interaction between media and society. What media frame agrees with or disagrees with closely relates to what the media finally presents. Likewise, terrorism is also the outcome of construction. The nature of terrorist incidents rests on a country system's recognition of the relationship between itself and the violence. News representation of terrorism incidents and its effect of publicity are inseparable from media's and audiences' recognitions on target countries respectively. Therefore, media representation of terrorism deserves dissecting and the relation between identity and terrorism deserves questioning.The pursuit of identity, a historical problem, also disturbs Mainland China and Taiwan. ---- Where do Chinese terrorists come from? How is terrorism represented on news? And how does media's stance on ethnic countries influence the news representation of terrorism incidents? How does news representation in different Chinese-speaking regions construct and attribute terrorism? From which perspective can we view the questions, for instance, "who are we", "who is the other", and "who are terrorists?" Taking mainland China and Taiwan's media representation of Chinese terrorism incidents as its core, this study tends to probe into the relation between terrorism and identity.

* Corresponding author: talktowangyan@gmail.com

2 Method

This study employs a mixture of method. Firstly, we select various representative media of different styles from both Mainland and Taiwan. (a) *People's Daily* and *the People's Daily Overseas Edition* are chosen from all the central-level media of Mainland China so internally and externally, but also reflect the distinct identity preference of ethnic Chinese people inside and outside China. (b) *Xinjiang Daily, Kunming Daily* and *Hangzhou Daily* are selected as representative provincial media, because of their geological closeness towards the incidents. (c) *China Times* and *Liberty Times* are collected in this study as Taiwan local media. The former one is famous for its conservatism while the later one for its strong appeal for Taiwan nationalism, "Taiwan first" and "Taiwan priority", as a result the combination of these two media can offer a diverse opinions of Taiwan society.

Secondly, according to the survey of KCIS, systematically sampled from all the 18 terrorism incidents happened in Mainland China in 2014 (*http://www.kcis.cn/8440*) in chronological order, taking a determined terrorism incident every three attacks, there appear all together six terrorist attack incidents. Taking as analytic targets the related news covered within three days right after the incidents happen. The sample totaled 111 news texts which were then coded and registered.

Thirdly, according to the preliminary observation before coding, the news contents can be categorized to nine news-coding categories: reporting date, media, page distribution, news narrating model, news source, event modality, event cause, news photograph and photography source.

At last, this study is dedicated to researching the relationship between the media framing and the national identity, so all news samples will be double-analyzed from both the respects of identity framing package and news framing.

3 Findings

The part of identity package refers to the theories "hegemonic domination"[1], "compromise discourse" [2] and "protest discourse" [3], which were developed by the author into three packages: "unconditional approval", "critical identity" and "critical disagreement". As the news framing part, the study reviews the contrasting framings employed by Chen [4], namely "critical & disagree", "identity & segregation", "propaganda & mobilization", "compatriots & sympathy", "critique & governance", "violence & Conspiracy" and "critique & dialogue" (Table 1)

Table 1 Identity package and news framing of this study

Identity Package	News Framing	Main Idea
Unconditional Approval	stability & struggle	Restoring social order, against violent terrorism, stability is prior to everything
	identity & segregation	The event's nature of terrorism incidents, prevent separatism
	propaganda & mobilization	Encourage people to throw oneself into restoration
Critical Identity	compatriots & sympathy	The security of the victim, the perpetrator and other party involved
	critique & governance	Improving governance, criticize the existing ethnic policy
Critical Disagreement	violence & conspiracy	Suspect that the authority withholds information

	critique & dialogue	Deny the terrorist nature of the incident, compare a wide range of news sources
Others	Unclassifiable Framing	

3.1 "Unconditional approval" package

The media in Mainland China, all taking Xinhua News Agency, the official news agency of the Communist Party, as their source of news, have been presenting their unshakable loyalty by "unconditional approval". The scenes, murderers, victims, causes of the event and countermeasures of terrorist attacks are also shelved and obscured. The reaction to crises and the scale of reports of local media in Mainland China has slightly progressed compared with the time before, but generally they are still positioned as a loudspeaker, following the order of officials and demonstrating a high level of national identity. In the regard of constructing the national identity, the role that the national media play in the "stability & struggle" and "identity & segregation" framing is much more influential than local media who pay more attention to the "propaganda & mobilization" framing.

3.1.1 "Stability & Struggle" Framing

First, it is the preemption that sets the basic tone for public opinion through authoritative sources. On March 2nd, *People's Daily* and *Xinjiang Daily* published on their own front page the press release of Xinhua News Agency, Beijing, *Xi Jinping's crucial instructions on the violent incident in Kunming Railway Station in Yunnan Province* (People's Daily, 2014.3.2:1). The first paragraph directly offer us the plot (main incident), and the next three paragraphs inform us of the instructions and dispositions on this incident from Chairman Xi Jinping, Premier Li Keqiang and other related figures in Yunnan provincial party committee and government in sequence, showing the will and the determination of taking resolute and effective measures to solve this complicated problem.

Second, the news agencies strengthen the accusative cognition and crackdown based on the source's preference. Also, the opinion of their own commentator is attached on the same page. The critique of *People's Daily*, *Punish Terrorism Crime, Secure Public Safety*, puts the terrorist attack in the context of "preparing for the Two Sessions and getting ready for comprehensively deepening reform", accuses the terrorism of breaching peace and stability, urges "1.3 billion people to be united to crack down violent crime", and finally summarizes "the terrorists' creating this bloody and horrible scene has woken up the justice and power in people's mind and strengthened people's conviction of defending social stability", and at the same time, "the party and the government has shown their solid determination to crack down violent terrorism, further boosting people's confidence of maintaining peace and harmony".

3.1.2 "Identity & segregation" framing

Compared with People's Daily which mainly targets at domestic readers, People's Daily Overseas Edition targets at Chinese people who live in Taiwan, Hong Kong or other foreign countries where information system is more open and national identity is more complicated. Therefore, the latter one tends to adopt a strategy of gathering people who agree with China and separating people who disagree when covering news. It also frequently quotes international voices that are coincident with Chinese authorities in terms of attribution and crack down those sources that take an opposite point of view, thus creating the binary framing, "identity & segregation", so as to establish a consensus.

In the basic mode of secondary incidents of Kunming 301 incident, the debate over determining the nature of terrorism took place unexpectedly. The specific cause was that CNN used quotation marks on the word "terrorist" in related coverage and claimed wickedly that it was not the first time that a knife attack had taken place and incidents like this happened in campuses in 2010 and 2012, however, without any political relevance. In AP's coverage, the prefix "what the official called" was added to the word "terrorist". In addition, instead of "terrorists", *New York Times*, *Washington Post*, etc. called them "attackers" (*People's Daily*, 2014.3.3:3). As for the assault on policemen in Aksu on May 8, 2014, Xinjiang which hasn't been reported by *People's Daily*, *People's Daily Overseas Edition* directly quoted Xinhua News Agency, emphasizing the efficiency of police – killed one and arrested one, and the stable and normal situation in Aksu in just one simple and concise sentence (*People's Daily Overseas Edition*, 2014.5.8).

3.1.3 "Propaganda & Mobilization" Framing

The "propaganda & mobilization" framing is an important means of gaining national identity. By conducting propaganda and mobilizing, Mainland China urges all Chinese people to stick together and crack down terrorism and intensifies work in reconstruction and reassurance in the disaster areas. The most intensively-covered basic modes are "sincere condolence" in "Maola's Murder in Kashgar on July 30" and "medical treatment" in "Bus Arson in Hangzhou on July 5". In all basic modes of the coverage of "Kunming 301" incident from People's Daily, People's Daily Overseas Edition and Xinjiang Daily, the content concentrates mainly on medical treatment, sincere condolence, initiation of social security system, recovery of normal order and other plot (secondary incident) basic modes. The Maola's special identities as a social celebrity, a patriot and a religious leader make the incident more than influential, thus reach an effect of "propaganda & mobilization". The Beijing authorities expressed condolence via the spokesman of State Administration for Religious Affairs of P.R.C. All 107 figures in the Xinjiang religious circle gather to hold a symposium on the 7.30 violent case, denouncing the violent and horrible act. There was no need for these leaders to show up and initiate the anti-violence and rescue systems due to the quick solution to the case, the Maola's death on the spot and no more casualties or site control.

3.2 "Critical identity" package

3.2.1. "Compatriots & Sympathy" Framing

Contrary to the "plot" basic mode of media in Mainland China, *China Times* attaches much importance to describing the whole process of the incident, as well as the scene, the murders and the victims, fully demonstrating its deep sympathy for compatriots in Mainland China. The sympathy for victims comes first. Taiwan's China Times covered the "Kunming 301" incident at large, including 3 basic modes of incidents - the crime scene, the background analysis, and the responses in society (*China Times*, ROC 103.3.3:A1). Second, it mentioned the sympathy for the attackers. In "East Turkistan Terrorists Resort to Violence under the High Pressure of Maintaining Stability" (*China Times*, ROC 103.3.3: A3), it was pointed out that Uighurs in Xinjiang were under high pressure of maintaining stability. "With 32 measures of maintaining stability, a network of prevention and control with complicated structure was set up in Xinjiang. Urumqi alone was divided into more than 2500 regional units and the communities were monitored 24 hours a day. The fact that delitescent East Turkistan separatists "started to become violent and horrible" was closely

DOI: 10.1051/shsconf/20173300030

related with the high pressure. Third, it agreed with the Beijing authorities, defining the incident as a terrorist attack. When Zhang Jinhua, a professor in the Graduate Institute of Journalism National Taiwan University, criticized that *China Times*' describing separatists in Xinjiang as terrorists was a partial quotation from Mainland China. Then China Times fought back with a question: "if this is not a terrorist attack, then what is?" (*China Times*, ROC 103.3.3: A3), further denounced that "*Terrorism Acts are Intolerable*" (*China Times*, ROC 103.3.4: A15), and reclaimed its position through a commentary from a civil servant "*Unconditional Hatred of Separatists*" (*China Times*, ROC 103.3.4: A14). In response to CNN's adding quotation marks to the word "terrorists", China Times published "*CNN's double standards on 'terrorists'*" and denounced that "the US wanted to separate our China" (*China Times*, ROC 103.3.4: A14). The phrase "our China" reflected a sympathetic emotion rooted from the close relationship.

3.2.2."Critique & Governance" Framing

First, it didn't quote any source from media in Mainland China; instead, it sent its own reporters to conduct an interview. The photos of the crime scene also take a major part, mostly quoted from the AP, the AFP and other foreign media. Second, it didn't quite agree with the Beijing authorities who called the attackers "terrorists" before confirmation and "Separatists in Xinjiang" or "Terrorists in Xinjiang" after confirmation. *China Times* quoted the official microblog of Kashgar's Communist Party Propaganda Office, justifying that not all the people in Xinjiang participated in the terrorist attack and it was more suitable to call the terrorists "East Turkistan Separatists". Third, *China Times* considered it as merely a fight between Uighurs in Urumqi and Han People, neither an incident caused by East Turkistan Separatists, nor a terrorist attack, which was different from the definition that the Beijing authorities had done to the July 5th incident in 2009. To its surprise, the political leaders in Beijing changed the district secretary in Xinjiang for this reason and enhanced the "high pressure" control (*China Times*, ROC 103.3.3:A3).Fourth, it reflected on the high-pressure policies that the Beijing authorities had put on Xinjiang.

3.3. "Critical Disagreement" Package

Though they are both Taiwan local media, Liberty Times name in its reports the cross-strait mainland as "China". In its "international edition", it even reports "Kunming 301" incident from a foreign perspective. This shows a strong disapproval on national identity.

3.3.1. "Violence & conspiracy" framing

First, underscore violence through representing the crime scene. Though Liberty Times and China Times have both covered the story of "the miserable throat-slit mother of one pseudonymous as Wang Yu", Liberty Times employed a calm and objective tone with little mention of subjective words, such as "lament" and "chocked", as China Times used in its reports. (Liberty Times, 2014.3.3 : A5) Second, the report <Kunming Bloodshed, Specialists: Unrelated with International Terrorist Organizations> adopts purposefully vague sources to create a conspiracy theory saying the top of CCP refuse to react. As the following citation from the report shows: "According to an informed source, some major general level PLA officers have been informed to keep indoors one or two hours before the incident, because 'the perpetrators are going to kill'." (Liberty Times, 2014.3.5 : A10)

DOI: 10.1051/shsconf/20173300030

3.3.2. "Critique & dialogue" framing

First, critiques on ethnic policies. This report quoted the human rights lawyer saying though the indiscriminate-killing "terrorists should be reproached, the fuse of such bloodsheds is "China's brutal ethnic policy", so "the Chinese government should reflect on its failed ethnic policy". (Liberty Times, 2014.3.3 : A5)Second, critiques on the Chinese government's lack of counter-terrorism ability and media's lack of quality. Thus, Liberty Times penetrates deeper than China Times by attributing terrorism incidents to historical reasons. However, they both ascribe terrorism incidents happened in China to observer's internal attribution not the actor's external attribution.Fourth, build a stage for dialog between Mainland China media and western media. Before quoted by Liberty Times, the Mainland China media can only use China Daily and its Overseas Edition to megaphone to western media for their "putting quotation marks on terrorists". Meanwhile, the named western media also appear as sources on Liberty Times. As a result, the two contesting groups of media meet and share the same status of Third-party sources on Liberty Times.

4 Conclusion

This study focuses on the news representation of terrorism incidents on both side of the Taiwan-Strait attempting to explore the relationships between terrorism, identity and media. The three major innovations are as followed:

First, the thesis detects that there is no fundamental improvement in the stalemate across the Taiwan Straits as well as the polarization of identity within Taiwan, which is manifested in the following aspects. (1) The confrontation between Mainland and Taiwan is embodied in an identity package, including "unconditional approval" (hegemonic domination), "critical identity" (compromise discourse) and "critical disagreement" (protest discourse). (2) There exist internal and external differences in the news reports on terrorism incidents. With the strategy of "one country, two faces", the separation and formation of identity are effectively prevented. (3) Though both taking a critical stance, *China Times* (Taiwan) and *Liberty Times* (Taiwan) are opposed in their identity.

Second, innovative research methods are adopted, which are constructed by "identity package" and "news framing". As a result, it might cause weak exclusiveness of the categorization, coding difficulty and low reliability and validity, finally leading to difficulties and limitations of the study. Therefore, the paper comes up with the following solutions. (1) By enhancing the integration of literature review and research perspectives, the paper refers to the solutions to some similar problems. (2) The test of mutual trust is expanded before the content analysis and the design of coding table is constantly optimized.

Another limitation of this study lies in its failure to carry out deep conceptualization and operationalization of "identity". The research of identity is quite complicated for it involves the fields of psychology, sociology, politics, etc. The study also fails to carry out an equivalent conversion among "the identity of news texts", "individual identity" and "collective identity". Besides, this paper seems to take advantage of reductionism by equating identity package with regional identity. Consequently, future researches can be done to study the identity of the audience for the news reports on Chinese terrorism incidents, and then serve as triangular cross-validation of my conclusions.

References

1. A. Gramsci, Selections from the Prison Notebooks, ed. and trans. Quintin Hoare and Geoffrey Nowell Smith. New York: International. (1971)

2. S. Hall, The West and the Rest: Discourse and power. *The Indigenous Experience: Global Perspectives*, 165-173. (1992)
3. S. Hall, *Cultural studies: Two paradigms*. (1986)
4. H. M. Chen. Framing and Attribution: a Comparative Study of the Reports about the Urumqi July 5 Riot, J.Commun 1 : 71-77. (2010)

A Case Study of Foreigner's Intercultural Identity in China via WeChat Moments

Ju Bei[1,*]

[1]International Business Faculty, Beijing Normal University Zhuhai Campus, China University of Macau, PhD student

Abstract. The study of communication and culture are interrelated with each other and with semiotics. This study attempts to explore the intercultural identity of foreigners via their use of social media in China. As the most popular social media used in China, a foreign teacher from America (named as Liz)'s posts via WeChat Moments from 2013 to 2015 were coded based upon photos and texts to indicate her identity in China. As a result of semiotic analyses, five main themes were identified in her posts, ranging from personal interests and needs, occupation, family, Chinese traits to American traits. Based upon these identified themes, Liz's intercultural identity as Chinese culture experiencer and promoter is reflected by her description of Chinese traditions, showing contrasts between American and Chinese culture, and using Chinese characters in writing. Overall, not much anxiety and stress during Liz's acculturation process are shown while most cultural posts reflect her active and ongoing adaption to Chinese culture.

1. INTRODUCTION

When new comers come in contact with people of different cultural backgrounds, identity variation takes place. Thus intercultural identity has often played a crucial role in intercultural communication. As [1] has pointed out "through face-to-face or mediated forms of communication, intercultural interfaces often present a multiple of challenges, including those that force people to confront and re-assess their own identity as well as the taken-for-granted practices of thinking, feeling and acting associated with identity (p.10)". As one of the most popular mediated forms of communication in China, WeChat Moments provides users with the platform to post what they believe and what they prefer. Meanwhile, the attitudes and well-being of WeChat Moments users are evaluated by their WeChat friends. Therefore, the present study targets at WeChat Moments of foreigners in China and tries to figure out what identities are hidden in their Moments posts. The better understanding of foreigners' identity in the host environment not only shows the ongoing cognitive and behavioral variations of foreigners but contributes to their successful acculturation.

[*] Corresponding author: jubei@bnuz.edu.cn

2. Literature Review

2.1 Identity in Intercultural Communications

In combination of personal and social factors in individual's identity, cultural identity has occupied the investigations of one's identity in intercultural communication. A new comer' cultural identity is initially perceived to be negotiable and manageable. Characterized as enduring and changing, [2] implied that cultural identities occur in daily discourse and social practices, and then are transmitted to other communicators. In [3,4] identity negotiation theory, an individual's developmental skills for competent identity negotiation are the center of effectively applying his/her knowledge of cultural and ethnic identity salience in diverse interactions. Drawn both from cultural identity theory and identity negotiation theory, [5] identity management theory stated that individuals are able to manage their cultural identities and facework when communicating with people of different cultural backgrounds. Moving beyond the flexibility of cultural identity, [7] employed the term "intercultural identity" to highlight the evolving and adapting nature of identity.

As stated by Kim [6,7] development of intercultural identity is characterized by two interlocked elements: individuation and universalization. As an individual rather a member of social category, one is able to clearly define oneself and others based upon unique personal traits. Along with individuation in the identity transformation process, individuals are able to identity common humanity and complementarity among differences and conflicts. Not limited to one's own group membership, individuals are better to view themselves as a part of a larger whole. With the continuing acculturation and enculturation, new comers experience their psychological conditions of stress, adaptation and growth in the changing environment. The existing studies [8-10] have provided the convincing evidence to this conceptualization of identity transformation in intercultural contacts.

2.2 Identity Studies in China

The highlight of identity studies [11–13] in China have focused on the empirical investigation of factors influencing social identity of migrant workers from rural to urban places and that of their children. Studies [14,15] targeting at foreigners including international students in China have mainly explored their acculturation situations. In [15]'s study, the good performances of foreigners in the level of economic situation and social interaction have been found. However, insufficient cultural and psychological integration of foreigners lead to their weak social interpenetration in overall. Certainly, Chinese scholars have attempted to discuss their own understanding of cultural identities. [16]'s explanation narrated the maturity development of intercultural identity involves three stages from unconscious, conscious to creative stage. Theoretically, starting from ethnocentrism, conscious intercultural identity characterized by reflective thinking and extended referencing is a threshold for new comes to go beyond before they finally accomplish self-transformation from ethnocentric to intercultural. Nevertheless, no studies exploring the foreigners' identity in China have yet been conducted. For this reason, the present study aims to investigate how foreigners perceive and develop their identity in intercultural contact with Chinese people. The better identity recognition will facilitate foreigners' acculturation into the new culture and establish their sense of belongs.

3. Research Questions

Launched by Tencent Company in 2011, WeChat has gained the greatest popularity on account of its interactive and personalized services. In its menu, four columns as Chats, Contacts, Discover and Me are displayed in navigation bar [17]. "Moments" (Peng you quan) means sharing the moments (posts about pictures, photos, texts and videos) with friends could be selected in the Discover. On the assumption that the usage of ethnic media contributes to the newcomers' acculturation in dominant culture, WeChat, as mediated form of communication, is of priority for foreigners to interact with Chinese people. Agreed upon [18]'s elaboration, symbols can serve as markers of identity ranging from national, occupational, religious to gender. Therefore, WeChat Moments is believed to be one form of symbols which include objects, behaviors, texts, ideas and people. In the present study, the semiotic analysis of foreign user's WeChat Moments is under discussion to answer the first question:

– What themes could be identified in WeChat Moments used by foreigners?

Based upon the description of thematic meanings shared in WeChat Moments, this study attempts to reveal the identities of foreign users via their use of host country media. Therefore, the following question is asked:

– How foreigners' intercultural identity is represented via WeChat Moments?

4. Methods

Defined by [18], culture is "a set of systems or codes of symbols and meanings, of communication as human symbolic activity". The study of communication and culture are interrelated with each other and with semiotics. Therefore, the semiotic analyses (based upon photos and texts) of foreigners' posts of WeChat Moments are conducted to uncover their intercultural identity when they are coming to China as newcomers. Liz (pseudonym) is from America and now is working as an IELTS Teacher for 2+2 program in one of universities in Zhuhai, Guangdong Province, China. Through her frequent updating posters in Moments, Liz shared lots of information about her work and life in Zhuhai with her WeChat friends including me. She was chosen to be the representative of foreigners in Zhuhai. And a case study (analysis of Liz's WeChat Moments) was conducted to discover her identities represented via WeChat.

Meanwhile, I have obtained Liz's consent for this study. Over January 15, 2013 to December 31, 2015, Liz has totally posted 138 posts and 1 short-video post via her WeChat Moments. Most posts have the photos with the texts in the bottom or the pictures with words, except for only three photo posts without text. Deleting three posts with no reference of words, 135 posts consequently were under analysis to identify themes which manifest a foreigner's identity in host environment. Moreover, if several photos are grouped in one post, only one screen shot is taken out in the present study due to the same topic shared. Then all 135 posts are coded based upon the photos and text illustrated in WeChat Moments. With regards to the photos whose meanings are vague, the texts act as complement to decide what topics are talking about. The post is not exclusive to each other. Some posts express multiple themes so they are repeatedly counted in each category. Take a post as an example, the screen shot primarily conveys the theme about Chinese culture by showing the red Chinese characters (*xin nian kuai le* meaning happy New Year) and the Chinese zodiac which indicates 2014 is the year of horse. In the bottom, the text acts as a reminder of class cancellation and exam. Therefore, this screen shot is counted as the post expressing both themes of Chinese culture and occupation.

More specifically, the posts describing her hobbies such as travelling (via food, scenery), hiking, gardening are labeled as personal interest. The posts with raised question like "Anyone know a good, trusted camera repair shop in Zhuhai to Guangzhou?" is classified as turning for help when Liz is living in host environment as foreigner.

DOI: 10.1051/shsconf/20173300031

Additionally, Liz is working as English language teacher in Zhuhai, all her posts involving students, assignments, examinations and classroom are categorized into the type relevant to her job. When Liz is posting photos with family members like her husband, parents, and relatives or writing the texts as "Waiting for my parents to arrive! First time I get to see my father on Father's Day in 3 years!" "Camping with fam", the themes of posts are coded as family life. It is noted that Liz's travelling experience with her husband are frequently posted in WeChat Moments. Those posts are not coded as family life, but as her personal interest in travelling instead. Other posts concerning typical festival, food, ceremony, language (refer to Chinese character), art, places in China and America are correspondingly coded as characteristics in each culture. Finally, the posts with no clear designated meanings are put as others, with no shared theme.

5. Results

Over period of the year 2013-2015, total 135 posts shared by Liz via WeChat Moments are in descending order, the number decreased from 63 in 2013, to 44 in 2014 and 28 in 2015 (Table 1). The posts shared in 2013 are nearly the sum of posts in 2014 and 2015. Five main themes are identified in all posts as personal interests & information needs; occupation; family; Chinese traits and American traits. The total number and percentage of posts in each category are shown in the following Table 1. The posts about personal interests & information needs (33.3%) and occupation (20.4%) occupy a half of all posts, while those on culture are up to 23.7%.

Table 1. Number of Moments' Posts in 2013/2014/2015

Themes \ Years	2013	2014	2015	Total (%)
Personal interest & needs	22	12	14	48 (33.3%)
Occupation	14	9	7	30 (20.4%)
Family (life)	10	10	3	23 (15.6%)
American traits	9	4	3	16 (10.8%)
Chinese traits	8	9	2	19 (12.9%)
Others	6	3	1	10 (6.8)
Total	63 (6 repeated)	44 (3 repeated)	28 (2 repeated)	

5.1 Theme one: Personal interests and information seeking

The posts relevant to personal hobbies reach to 39, with only 9 posts are about seeking information. More specifically, the interests can be further divided into subgroups like traveling (36) and gardening (3 posts). Within the subgroup of traveling, 31 posts are photos of scenery and individuals, and 5 posts are about food (as shown in Figure 1). These posts vividly narrate Liz's rich traveling experiences in places like Cambodia, Hong Kong, Korea, India, Indonesia, Japan, Nepal, Thailand and places in China. Her photos and text definitely convey the message that she is keen on traveling thus traveling lover is appropriate to describe her unique personal quality.

From her posts, Liz is believed to be a person fond of taking risks. She not only enjoys sightseeing but likes adventure travel by hiking, camping and climbing mountains.

Fig. 1. Traveling: food & Scener

This feature is particularly shown in the text of one screen shot, "Even the British Army turned back…too dangerous and impossible…deep snow and cold, but we did it". In terms of gardening, the theme focuses on Liz's preference to planting, enjoying self-planted food and fertilizer. Then Liz as a foreigner certainly needs helpful information when in the new environment. Therefore, she posts her needs in the Moments intending to get help from local people.

5.2 Theme 2: Occupation

Liz's posts (totally 30) apparently reflect her role as English speaking and writing teacher. These messages range from assignment, examination to in-class teaching (Figure 2). The first screen shot expresses Liz's hard work when grading students' journals.

Fig. 2. The role of teacher

The second aims to remind students of their lost personal belongingness in the class and the third acts as the notice announcing the examination time. All the posts incorporate the role of students when Liz plays the role of teacher. Liz also shares her posts to assist students who forgot to write down their names on assignment paper and who are eager for overseas learning. In addition, she takes photos with students as memory when one-semester class is ended. In this sense, Liz is thought to be a caring and responsible teacher and her relationship with students is very nice.

5.3 Theme three: Family

23 out of 135 posts talk about Liz's family life. All the family posts convey Liz's personal information. She got married in 2010 and now is living in Zhuhai with her husband. She "tells" us via posts that she plays a role of daughter as well as wife in family. She gets the Easter gift from her mother and takes her parents traveling in Beijing when they came to visit her in 2014. On the other hand, her married status could be easily identified by taking photos together with her husband. But this is not the criterion to categorize the posts into the theme of married life. The posts which Liz uses the words as "husband", "hubby", her husband's name, or the personal photos symbolize the wedding and anniversary are counted. It can be inferred that her happy and satisfying family life in Zhuhai.

5.4 Theme four: Chinese culture

When Liz from America comes to work in Zhuhai, how does she respond to the new culture? Primarily, differences like food, language, and social practice are illustrated Figure 3. Hot pot, especially in Sichuan province, is of great popularity all over China. As shown in the first screen shot, the spicy hot pot, chopsticks and drinks "王老吉" (*Wang Laoji*, a Chinese herbal tea, and perfect match for hot pot) all reflect being a part of Chinese culture. In the second and third screen shots, Liz is interested in traditions of Chinese wedding. The Chinese character "囍" (*Xi,* indicating double happiness) is particularly used in wedding such as printed on invitation card, pasted on windows and doors, and embroidered in wedding dress. And the red umbrella opened up in the wedding ceremony implies auspiciousness, joyfulness and good luck.

Fig. 3. Chinese traits (1)

Based on her experience of Chinese cultural, Liz makes a comparison between American and Chinese traits, and then attempts to use Chinese language.

As an example in Figure 4 points out, the "American" coffee sold in China is only the instant coffee and costs too much than that in America. The other interesting case is food difference shown in "Last meal in America (chips, tomatoes & beef)" and "First meal in China (Changfen, steamed rice rolls)". Though differences exist, Liz loves both of them based upon her use of emoji, smiling and loving faces. These comparisons of Chinese and American cultural traits, to some extent, indicate Liz's cultural sensitivity.

Apart from awareness of cultural differences, Liz also attempts to display Chinese culture and use Chinese character to express her ideas.

Fig.4. Chinese traits (2)

In one of the screenshots concerning Chinese culture, Chuanju (a type of Chinese opera originating in China's Sichuan province) is exhibited from the perspective of Liz. Correspondingly, Putuo temple (the largest temple of Buddhism in Zuhai) is also shown in another. Sitting the middle, Liz uses Chinese text "再見廣西" (*Zai jian Guang xi*, Farewell to Guangxi), expressing her enjoyment and her Chinese language learning.

Furthermore, Liz points out the error of information expressed in Chinese. For instance, in one screenshot, Chinese characters are written in the back-package of Heinz Ketchup, meaning that this product can be used in Chinese dish, pizza, bread and so on. But Liz makes her comment at the bottom of the text *"Do not trust Heinz! Ketchup does NOT belong on spaghetti, enchiladas, pizza, or croissants."* And in the other post, Liz further clarifies that "*...Ketchup is a condiment for French fries, hot dogs, hamburgers, and sandwiches and .... It should not be used as a sauce to cover noodles. Some things don't belong together. Noodles and ketchup are two of those things...*" In this case, Liz creatively uses her knowledge about American and Chinese food and recommends the correct way to do the cooking.

5.5 Theme five: American traits

When adapting to the Chinese culture, Liz's ethnic culture doesn't disappear when in host environment. She maintains the features rooted in American culture while experiencing Chinese culture. The themes about American traits could be identified by through Liz's "narration" of America as home. In the posts, she shared her hometown, graduated university, American holidays and foods. These messages, originated from being an American, transcend the geographical boundary, from locally to nationwide. Locally, Liz's hometown, Western North Carolina is posted as well as food in her hometown. Then it is the university where she studies. To a larger scale, as an America, she also posts photos about food and holidays celebrated in western country, like Easter, Thanksgiving and Christmas. In her family in Zhuhai, Liz often shares her homemade western food in WeChat Moments.

6 DISCUSSION AND FINDINGS

The five main themes have already been identified and described in Liz's 135 posts over last three years. Intercultural identity is defined by [6] as "an acquired identity constructed after the early children childhood enculturation process through the individual's communicative interactions with a new cultural environment". With regards to foreigners living and working in China, their identity shaped by the original culture and their identification to that group will be expanded beyond the conditioning of the original culture. Based up [7] and [16] explanation of intercultural personhood, the intercultural identity could be understood in the three dynamic phases of stress-adaptation-growth instead of and either/or state.

Among Liz's 135 posts, the number of posts concerning culture amounts to 35 (23.7%). Her identity as Chinese culture experiencer is revealed by her descriptions of Chinese traditions, contrasts between American and Chinese cultures and usage of Chinese language. There clues indicate that Liz is conscious of cultural differences and gradually "transform" herself. This is supported by [18] as "A different cause for transformation in the voluntary integration of a new sign into an old system, not because a particular old sign is no longer available, but just because the new one becomes available". To step further, these posts as signs are described by Leeds-Hurwitz as assimilation, the relatively straightforward acquisition of a sign by one culture from another.

Generally, not much anxiety and stress during Liz's acculturation process are shown by her posts, which may because Zhuhai is not her first destination in China. Most cultural posts reflect Liz's active and ongoing adaption to Chinese culture. Some posts even indicate her creative use of cultural knowledge especially Chinese language knowledge to express herself. Therefore, her posts indeed show the influence of culture on acculturation and thus ascertain the likelihood to further explore the foreigners' identity in Chinese culture. Though the universalized identity is out of possibility to investigate in the present case study, an individualized identity defined by [6] is salient reflected by Liz's acceptance of Chinese culture and seeking of intimate personal relationship with people from host environment.

REFERENCES

1. Y. Y. Kim, ICS, **4,** 1 (1994)
2. M. J. Collier, M. Thomas, Theor. ICC, **99**, 122, (1988)
3. S.Ting-Toomey, In R. Wiseman & J. Koester (Eds.), *Intercultural communication competence*, 72–111, (1993)
4. S.Ting-Toomey, In W. Gudykunst (Ed.), Theorizing about intercultural communication, 195–210, (2005)
5. T. Imahori, W. Cupach, In W. Gudykunst (Ed.), *Theorizing about intercultural communication,* 195–210, (2005)
6. Y. Y. Kim, *Becoming intercultural: An integrative theory of communication and cross-cultural adaptation*, (2001)
7. Y. Y. Kim, IJIR, **32**, 359, (2008)
8. T. J. Milstein, IJIR, **29** (2), 217, (2005)
9. M. J.Pitts, Paper presented at the annual conference of ICA, (2007)

10. R.Suro, *Strangers among us: How Latino immigration is transforming America,* (1998)
11. C. Rongwei, X. Yihan, Z. Yi, Chin. J.Sociol, **34**(4), 25, (2014)
12. Z.Wenhong, L. Kaichun, Sociol.Stud, **4**, 61 (2009)
13. W.Zhonghui, Z.Pan, G. Jin, Chin. J. Spec.Edu, **174**, 86, (2014)
14. H. Jiong-mei, Y. Xue-ling, J. Xinjiang NU, **35** (2), 129, (2014)
15. S. Ye, Master dissertation, ECNU, (2010)
16. X. D.Dai, CMR, **5** (2), 1, (2009)
17. T. L. Sandel, B. Ju, In T. Milburn (Ed.). *Communicating user experience: Applying local strategies to digital media design*, 103-126, (2015)
18. W. Leeds-Hurwitz, *Semiotics and communication: Signs, codes, cultures,* (1993)

Re-place-ing space in Libyan women's language learning and culture mapping

Melicent Jalova[1,*]

[1]Mindanao State University-Iligan Institute of Technology, Philippines

Abstract. Libya embarks on a new era in world relations where learning English has become a necessity. This study investigated how Libyan women managed to study abroad, found their place and learned a language despite the country's cultural subjection against women. It explored what motivated them to study English, how they felt about the learning environment, and the process of engaging themselves in intercultural exchange. Twelve female Libyan were subjects of this study. This study utilized a combination of the descriptive and analytical research designs. Since this is a qualitative study, the researcher used thematic analysis in analyzing and interpreting gathered data. It is founded in Milton Rokeach's Theory which says that each person has a highly organized system of beliefs, attitudes, and values, which guides behavior. Personal motivation and the desire to live through another culture without compromising its own concluded this study. The cultural construct that Libyan women can't decide and can't do a lot is unacceptable to these women who believe that they could play a very significant presence in the new government.

1 Introduction

Development encompasses many dimensions, levels, and processes. More so, various levels of development needed in any society necessitates a complex interweaving of strategies and programs to be able to effectively and efficiently deliver services to its citizens. Hence the dramatic effect of globalization today does not impede any country, culture or any person from learning a foreign language to help improve what it has in the past and present.

Libya, a country which suffers from political turmoil in the recent past is not exempted from this. After over four decades of oppression, Libyans are excited about living a good life and working to improve their educational system. And for a long time, they are now convinced of the great importance of improving the inadequate standard of English language in their country. As Libya embarks on a new era in world relations and under these circumstances, learning language, especially in English, has become a necessity and a key to success in business and study, and learning other culture. For more than a decade, Libya is one of the Arab countries where its government has invested heavily in English language learning by sending their teachers abroad to study.

* Corresponding author: melicent09@yahoo.com

The goal on learning English is to enable Libya to catch up in the development of its economy and to promote international exchange [1].

The government's mandate to send scholars abroad encouraged many women to continue their studies in countries like UK, US, Canada, Asia and many other countries. According to a September 2013 report by the International Foundation for Electoral Systems (IFES) Libyan women are highly educated. The majority of college graduates in the country are women, with 77% pursuing higher education courses, compared to 63% of men [2].

It is inspiring to discover the number of Libyans who want to learn English; the number is large and it is increasing as Libya embarks on a new era in world relations and opening up to foreign investment. Under these circumstances, learning languages, especially English, is becoming a necessity and a key to success in business and study, let alone facilitating the search for a job, not only in academia, but also in government and private sectors.

1.1 Embracing Language and Culture

Language and culture are interrelated. First, language is the vehicle of culture. There is not any isolated language that can exist without culture. Language is rooted deeply in national culture, and like a mirror, it reflects a nation's beliefs, values, customs, and others Rivers, cited in Kadyskyzy, [3]. When one learns a language, one has to embrace its culture as well. One's ability to learn and understand a particular language depends on her deep understanding of culture otherwise this causes nonstandard linguistic error. (Krasner 1999, as cited in Peterson and Coltrane, 2003) [4] stated that linguists and anthropologists have identified that forms and uses of a particular language reflect the cultural value of society in which that language is spoken.

1.2 Understanding Language and Culture in a Lexical Level

One can learn much about the connection between language and culture by studying the availability and use of words in language communities. Sapir as cited by Corbett [5] commented that vocabulary is a very sensitive index of the culture of a group of people. Even at the lexical or vocabulary level, one can often get a sense of how individuals perceive themselves as members of society and members of specific cultures. The differences in word usage between languages can have serious implications for those who attempt to converse in a language not their own if they assume the same words in both languages carry equivalent meanings [6].

1.3 Female Libyan English Teachers

Libyan English teachers are the fountain of knowledge among Libyan students. They are the source of ideas, skills and knowledge of learners. But often, female teachers are so limited in sharing their expertise because of the cultural constrains of the country. They have issues and concerns that are not acceptable in the society. This wall of culture restricts them to use appropriate techniques in language teaching and limit their ability to teach the language.

Since people do not learn a language in isolation, they engage in numerous activities and become active members to experience, observe, understand, explore, participate, meet and use the language. Wilson, [7] said that what a person learns and how they make sense of that knowledge depends on where and when they are learning, such as the social context.

Libyan teachers going abroad to study English become English Libyan teachers in a particular social context, using a particular kind of knowledge at a particular time, therefore, their practices have become socially constructed. These practices can be seen as constructed from their cultural background, views on learning, and the kind of education they have received.

2 General Objectives

To find out what motivated Libyan women to study English abroad, how they felt about the learning environment, and the process of engaging themselves in intercultural exchange.
Specific Objectives:
1. To find out how learning the English language abroad embody that cultural reality among Libyan women when they teach in Libya
2. To determine how Libyan women negotiate their cultural status in learning and teaching English
3. To determine the attitude of Libyan women towards learning English
3 Study Framework

How people behave-- how they greet one another, what and how they eat, what their manners among others – is motivated or shaped by their dominant attitudes, values, and beliefs. This idea is founded in Milton Rokeach's Theory which says that each person has a highly organized system of beliefs, attitudes, and values, which guides behavior.

An excellent way to illustrate the relationship among behavior, beliefs, attitudes and values is by making an analogy between culture and an iceberg, as suggested by Edward Hall [8] following Sigmund Freud's analogy of the unconscious mind with an iceberg. The tip of the iceberg, which he termed as the external culture, is really the smallest part. This refers to the explicit and conscious behavioral aspects such as language or verbal communication, customs, and typical food or dress. These can be acquired through observation, education or training. The largest part of culture is inside an individual's head and unconscious—beneath the water level of awareness. This part, which he called the internal culture, is what motivates behavior. It includes beliefs which are learned implicitly or unconsciously through interaction with others within a particular culture. They include likes or dislikes (attitudes) and things that are important to us which we want or desire (values). The largest and the most hidden part of any culture includes basic values, ways of thinking and perceiving, and non-verbal communication. They are implicitly learned before adolescence and are difficult to change.

3 Methodology

This study utilized a combination of the descriptive and analytical research designs. Since this is a qualitative study, the researcher used thematic analysis in analyzing and interpreting gathered data.

Participants. Twelve Libyan women participated in the study. Two studied at University of Malaysia, four studied at Essex University in UK, four at University of Manchester, one at University of Birmingham, and one from University of Glasgow, Scotland. Seven have Masteral degrees; while five had finished their Doctoral degrees.

Four of them are teaching at Sirte University, four in Tripoli University, and the other four are teaching at Gariounis University in Bengazi, Libya.

Administration. The researcher traveled in Tripoli and Bengazi to conduct the interview. Focus interview and FGD were used to get the information needed. These were done in a field setting, more specifically in the campus where the teachers were based.

A key informant was tapped to help and provide the necessary contact of the participants.

Meanwhile, the participants from Sirte University were interviewed in a more relaxed setting because the researcher also taught in the same university.

Method. The researcher formulated a set of guide questions which formed the core questions that was asked. The questions reflected the specific concepts needed to achieve the answers of research questions as well as to provide information on certain aspects of the study.

Data Analysis. To analyze the gathered data, the researcher separately coded the responses of the participants. Commonalities and peculiarities of the responses were noted.

Based on the framework of the study and the data collected, analyses and conclusions were drawn using the thematic approach. The themes used were: Libyan culture, other culture, negotiation and language learning.

4 Findings and Discussion

Motivation to deeply learn the language is the key factor in learning a language. Littlewood [9] explained that motivation is the crucial force which determines whether a learner embarks on a task at all, how much energy he devotes to it, and how long he perseveres.

Just like any other scholars who are thirsty to gain more knowledge and are passionate to learn and teach English, Libyan women are also equally empowered to do so. After a tedious application in the Ministry of Education in Libya, being accepted at the universities abroad, and a grueling argument who among in the family members can join her (husband if she's married; father or brother if she's single), at her desired country of choice, she travels and enjoy her scholarship. As one discussant quipped, "It took me a long time to study because no one could come with me. I had to wait for my brother to finish his studies so he could come with me. There is no way I could go and study without my brother or father. Sadly, others could not study abroad because no one could go with them and stay for two or more years. Finding a husband to accompany to study (sometimes) was the right thing to do."

The country's cultural subjection against women was the biggest challenge among the participants. But the motivation to learn and be immersed in an English speaking country was the biggest inspiration to push the goal. Different researches have been investigated on the effects of attitudes on motivation and proficiency in a large number of studies. Clearly, the women are motivated by a mixture of integrative and instrumental motivation, that is; the former is learning the language in order to communicate more satisfactorily and to gain closer contact to people and their culture and the latter was to further another goal.

Individual's drive to achieve a goal, curiosity of the new country to live, be successful, desire for new experience and be a true scholar are the factors that motivated them to study abroad. "I wanted to improve my English and teach English to Libyans, that's my goal." I wanted to be fluent and develop confidence when I speak English especially with my students, and "I wanted to feel how it was to speak with native speakers and speak with confidence when I was with them" are among the goals of Libyan women that drive them to get a Master's degree or PhD in another country.

Given the favorable factors that motivate them to study English language there are also risks in learning the language. Littlewood [9] stressed that there is a close link between the way a person speaks and the way the person perceives her identity and the world. He further stressed that when one tries to adopt new speech patterns, at some extent, one is giving up markers of her own identity in order to adopt those of another cultural group. In some respects, one is accepting another culture's way of perceiving the world. If she is

DOI: 10.1051/shsconf/20173300032

agreeable to this process, it can enrich and liberate her. If not, it can be a source of resentment and insecurity.

But Libyan women are very strongly connected to their roots. They very well define and assert their identity in learning English. "I studied ELT without compromising my culture or anything. "Yes, it was good to meet new friends, in a new and different world, but I always carry my roots, my religion ALWAYS. There's no compromise." "I was there to study only, I get to learn their culture, the importance of studying language and culture but at the end of the day, I am still a Libyan in words, in thoughts, and in deeds."

It is worthy to note how Libyans could learn the English language without really embracing the personality of the proponent culture. If Libyan women remain dogged in their ethnocentric stance and are reluctant in acculturating or assimilating themselves into a given culture, their learning of English maybe a difficult process. From the perspective of a native speaker their English may be badly spoken and hardly understood.

On the other hand, their intense motivation to learn a language transformed their learning process into a willful determination to surpass all odds.

The researcher believes that despite the process of getting acculturated into English-speaking cultures, Libyan women have developed a sense of cultural resilience or the ability to keep their core value of keeping their culture intact. Based on the participant's disclosure, "It's there. Wherever you go, you keep your own culture with you especially us women, it's precious in Islam,"

Although Libya's image is very fragile at this point, Libyan women's love of their country, knowing who they are is very evident. It cannot be snatched by external forces or agencies. It is deeply ingrained. Also, the perennial psychological link to one's country is wrought by human instinct to always go back to one's origin; in a way, "coming home."

Moreover, Islam is found to be a significant factor in the participant's emotional and deep sentiment about her being in other country. The teachings of Quor'an along with their family are the significant shaper of their children's mindset, also paved the way for the participant's "high context" mindset [8] - a very strictly defined orientation- characterized by adherence to the teachings of Allah. Everything outside of Quor'an, in their belief, is not worth their while when it comes to the realm of faith and values.

The Role of Libyan language instructors in teaching English

The role played by the female instructors is negotiated in certain ways that they are always "careful" when they teach the language without compromising their culture. Students' embarrassment may rise when conflicting issues in their culture get in the way. It emerged during the focus group discussion that the activities like role play, or even reporting are very difficult to do. "There are boys who would not recite or report in class if the girls are there. "We have to let the girls go home and then the boys would have their oral report." Or at times, "girls would not participate if there are boys." There was one instance that one female student stopped attending classes because the subject required to speak in front of the class and the teacher was called."

Being Libyan themselves, they understand the feelings of students. "That is why we give some subjects to foreigners because they know how to deal with students without embarrassing and offending them. Examples of these subjects are Literature and Spoken English classes where students are required to share, talk and do group activities. As Libyans ourselves, we cannot impose those activities to students."

There are profound consequences when an L2 learner participates in an L1 culture [5]. Since language learning involves oral activities to achieve fluency, one has to participate in class activities. But if a certain culture inhibits some students to participate because of the presence of the opposite gender then negotiations have to be made at certain extent. The teacher has to adjust her activities between the interests of the learners and the demands of the culture [5].

Bonvillain, [10] says that language fulfills diverse functions either separately or simultaneously. No matter how universal these functions of language are, methods of achieving them differ both in linguistic form and cultural style.

For a female Libyan teacher, teaching the language is teaching the culture itself. "When I open my mouth in English, am I not teaching foreign culture?" one participant asked. According to Stern, [11] a language teacher is the teacher of culture. No matter how and what strategies, techniques, and motivations she does in the classroom in teaching the language, she embodies the culture of other countries. She just have to adjust her language and decide what class activities that best suit the need of the learners without offending their own culture.

In Kramsch [12] notes of E. Sapir's, 'The Relationship of Language and Culture', he said that language is a guide to "social reality." It strongly affected our thinking about social problem and processes.

Libyan women whose ropes have loosened to educate themselves abroad have something in them to share to the world. That is -- without loosely giving in to the target language culture, they are able to survive learning the language abroad, teaching the language at home, and holding on to a positive attitude of elevating the standards of English language teaching in Libya.

5 Conclusion

It is inspiring to discover the number of Libyans who want to learn English; the number is large and it is increasing as Libya embarks on a new era in world relations and opening up to foreign investment. Under these circumstances, learning languages, especially English, is becoming a necessity and a key to success in business and study, let alone facilitating the search for a job whether in academia, government or private sector.

Libyan women are go-getters waiting for opportunities to come. The cultural construct that they can't decide and can't do a lot is unacceptable to these women who believe that they could play a very significant presence in the new government.

It takes only one good reason to learn English. Libyans have many reasons and a clear awareness of the significance of English. These female scholars who continue to pursue their dreams including their determination to secure their place in shaping Libya's future through their students are called for. It is up to the new government to make sure that the level of education in Libya now is at par with the very best in the world thus ensuring a secure and peaceful environment, where children and students could come and learn. It is also the government's responsibility to ensure that all the schools will meet a certain criterion so that students are equipped with the modern knowledge along with a better educational system to face the world head-on.

References

1. M. Sinosi, *Linguistic and textual analysis of classroom English interaction at Altahadi University*. (2010)
2. J. Sarhan, *Libyan women struggle to join the workforce* (2014)
3. A. Kadyskyzy, SINO-US E.T, **10**, 23-31 (2013)
4. E. Peterson and B. Coltrane, ERIC D. (2003)
5. J. Corbett, *An intercultural approach to English language teaching* (2007).
6. P. N. Chrosniak, *Encyclopedia of Anthropology* (2005)
7. K. Wilson, JOCRAL, **29**, 166-179 (1999)
8. E.T. Hall. *The silent language* (1998)
9. W. Littlewood, *Foreign and second language learning* (1998)

10. N. Bonvillain, *Language, culture and communication. The meaning of messages* (1998)
11. H.H. Stern. *Fundamental concepts of language teaching* (1983)
12. C. Kramsch, *Language and culture* (1998)

E-Queue Mobile Application

Yeen Lai Khong[1,*], *Bee Chen* Ooi[1], *Kok Eng* Tan [1], *Salizatul Aizah* Binti Ibrahim[1], and *Peck Ling* Tee[1]

Faculty Accountancy Management, University Tunku Abdul Rahman,Malaysia

Abstract. Waiting in line is a common experience in daily life, whether for a table at a popular restaurant or for the service at a bank. This experience is not always pleasant for most of people because they always have to wait for a long time to be serviced. The ability to interact with waiting customers is highly desirable because it allows businesses the opportunity to optimize their existing services and offer new services to waiting customers. However, interacting with individuals waiting in a queue has been inefficient and costly because employees must either talk with each waiting customer on an ongoing basis or the business must provide high tech devices that interact with each waiting customer. Agile methodology which will be used to develop this application, it incorporates the SDLC phases starting from the Planning phase up to the Maintenance phase. End of the research, we found that majority of respondents are prefer to use the proposed system compared with current method.

1 Introduction

Current system available for businesses to manage waiting customers, queue them for service, and notify them when service is ready for them have no means of interacting with each waiting customer. These system include physical queues, customers gather in a small lobby or in an area roped off from other customers, which awkward unappealing, and uncomfortable for customers. Intercom paging systems, customer's names are called out over a loud speaker to a crowd of waiting customers. This system can be cumbersome to manage because some guests may not hear the page the first time their names are called. This could also worsen the atmosphere for customers who are currently receiving service. "Take a number" system, each customer is given a number and an elevated display shows the "Now Serving" number to a crowd of customers. This system has similar problems as the intercom system with regards to crowd management. So, businesses not only interact with waiting individuals, they also cannot verify that waiting individuals have received notification that services is available, provide alternative services, or allow individuals to control their place in queue. The proposal aimed at developing a new Smartphone application, called "E-Queue mobile application" for businesses. Nowadays, more people use portable communication devices like Smartphone as their primary means of communications, these device and proposed system use to solve the queuing problem. This proposed system can reduce the costs of business to hire staff or purchase high tech devices

* Corresponding author: khongyl@utar.edu.my

to interact with waiting customers therefore customers also no need to spend long time to wait at waiting area.

1.1 Problem Statement

During peak period, customers must spend long time at waiting area to wait for service. They not dare to leave the waiting area to do others thing because they can't estimate the waiting time. With this proposed system, customers can view the estimate time by system and customers will received a notification when their queue number is nearing. In today's market "queuing" system, most of the business centres are using the number ticketing system which a number will be written on a piece of small paper when you press the button on the "queuing number system". However, based on researcher observation to the said system, its do have lots of disadvantages. At first, this method is not environmental friendly as they have to use so much paper in ticketing printing (especially some of the customer never turn up after took the number and this might create another environmental issue which is paper wasting).Customers will be given a queue number via this application without using a piece of paper. Businesses need to cost much money to hire employees and purchase high tech devices to satisfy waiting customers' unlimited needs. With this proposed system, customers can direct interact with system by using their own Smartphone to satisfy their needs.

1.2 Aim

The aim of this proposed system is allow customers monitoring the waiting situation and can interact with system through their Smartphone. With the proposed system or application, it enables the users to check the "on queuing waiting list". If the users wish to add themselves to the waiting list, user can just log in to the application with their login ID and password and get a queue number from the proposed application. This proposed system will show the estimate waiting time to customers, so they can go wherever mobile internet service is available while they wait. This proposed system also will automatically update current queue number. When customer's queue number is nearing; this application will send a notification to remind customers. However, the bigger advantages of this proposed system is enables the application system to manage their "wish" anytime after they received the notification or queue number. For example: application users can simply cancel the service request if they found they have something more urgent things to be done. Therefore the system will notify the number behind you to come early. Customers can have some "freedom" while waiting by using it. The proposed system is mainly focus on get a queue number, view queue list, view estimate time and user response to system server. The reason for proposing this E-Queue mobile application is to solve the problem that mentioned in the problem statement.

1.3 Objective

The purpose of developing this mobile application is mainly to research the current system to managing waiting customers, to find out the problem faced by customers and businesses and also to provide a better overall system to them. The objectives of this research would be To provide "freedom" to waiting customers by show estimate waiting time and sent notification when their number is nearing, allow customers interact with proposed system to modify their waiting status like delay or cancel the services, provide a proposed system that more environmentally friendly than current system and to provide a proposed system to solve queuing problem for Android 4 user.

1.4 Justification

This proposed system could help customers to get their queue number by using "E-Queue mobile application". This free application is able to download from Android apps store to their Smartphone. Customers can provide their feedback for the application, and those feedbacks can make as improvement in future. Customers must scan the kiosk by using their Smartphone to get their queue number and estimate waiting time will be show on their Smartphone. They can go anywhere that mobile internet service is available while their waiting. When customer's queue number is nearing, customer will receive a notification as remind. Customer can direct go counter to show their queue number and get services. Based on their queue number, counter staff can get customer's details and needs and provide services in shortest time. This proposed system also able to enhance confident of customers towards business because the function of the system.

2 Literature Review

The chapter reviews on topics of Android, Related works, SDLC methods, Requirement Gathering Techniques, Quick Response code(QR code). Proposed system should be assessed in detail from the each of the application functionalities to the application advantages and disadvantages. According to Haughey (2009) Google which acquire Android Inc. in 2013 has been constantly researching and updating the versions of Android to fix bugs faced by previous versions besides adding in more specifications and functionalities [1]. The latest stable version of Android which can be found in market is Android 4.3, Jelly Bean which was released August of 2013. Among the improved functionalities implemented on this latest version are restricted profiles limit access to apps and content, Bluetooth smart support makes Android ready for a whole new class of mobile apps that connect to fitness sensors, high performance graphics games, virtual surround sound, additional language support, location detection through Wi-Fi and auto complete (Developer.android.com, 2013). There are many strengths of Android which are worth to be applauded for. For instance, developers may most likely prefer creating apps for Android devices since Android offers an open development platform which allowing them to use third-party tools to create the particular apps. With the implementation of open source platform, anyone with the right skills and knowledge can create an apps specially tailored for Android devices without tight restrictions or approvals. Besides, this strength of Android also enhances flexibility since it allows developers to update their apps anytime to accommodate a range of different devices and play around with many features and functions of their apps during the development process [2] Developing Android apps may also be easy especially for Java programmers since all applications under Android are mainly use Java programming language [3] also commented that since Android implements the open source platform, available code samples and supports from fellow developers can be easily accessed thus, attracting more new potential developers to learn and create apps either with basic or complex functionalities. Meanwhile, an excellent testing environment also provided by Android that encourages developers to test and debug apps properly before presenting them to the Google Play (Android market). Android Inc. has also make developers' life easier by providing downloadable System Development Kit (SDK) in its website which comes with the API libraries and developer tools necessary to build, test and debug apps. [4] In contrast with its major competitor's application market, Apple's App Store which rejects application submissions that are deemed too simple or are similar to existing apps, Android's marketplace does not censor its apps. In other words, apps developed by any developers will have equal chances to enter and to be published in the market but of course, the best one will only succeed. Meanwhile, since Android is an open platform, manufacturers have the flexibility to pair the operating system with any hardware they wish, thus providing end users a higher number of device selections in terms of specifications, design, screen and other physical appearances [5]. However, Android does have its cons too. Though open source platform is part of Android major strengths, this technique has also raised security and quality control concerns to apps developers. Since anyone can develop and present new apps in the market without much restrictions and approvals, end users may purchase low-quality apps numerous times thus wasting their resources. With the open source concept, developers may also face piracy threats and loss of potentials profits since anyone could easily access to the same learning resources and produce similar apps. Meanwhile, hackers could also easily mimic popular apps and trick end users into using the fake apps which may contain malicious elements [6]. Since Android is being licensed to multiple phone manufacturers such as Samsung, LG, Huawei and HTC, apps developers also face problem in successfully presenting their apps across all Android devices since each manufacturer has its own configurations and specifications (especially processing power) being implemented on its respective Smartphone. In other words, there may be an incident in which some devices installed with certain versions of Android will not be able to run applications that work fine either on the same device or another device of different Android version. With such problem, fixing bugs and updating apps will also cause headache to developers as they have to make sure that the apps could be supported by all Android devices as soon as an update is performed [7].

3 Research Methodology

Agile method will be the main methodology used in this research paper. The reason of choosing this method due to agile method enables to enhance the collaboration between the system developer and application users. This benefit of using this agile method are it allow the developer retrieve the immediate feedback from the application user and notify the application user after the changing requirements has been changed within a minutes.

This research paper has been targeted completed within a year. System creators suggest 3 weeks times at least to be used to set up the scope of the system and agile system is one of the methods in assisting the above requirements as one of the pros of this agile method is it can easily to refine any of the development phases anytime if some things goes wrong (As this methodology covers up the whole system development learning cycle from planning stage to the maintenance stage)

4 Finding

In Order to gather required information for this research paper, survey is the best approach to be used. Survey method enable researchers gather feedback from students easily about the proposed system and also understand the "thought" of the targeted respondent to the "new proposed system as well. Targeted respondents for this survey proposed to be 800 students from UTAR. The survey is divided into three sections; Section A - Smartphone Users Among UTAR Students, Section B – Study on Existing System and Section C - Proposed System Expectations and Requirements. Each section consists of close-ended questions, students are more likely to spend their time to participate the survey and statistical data can be easily tabulated. We found that majorities of UTAR students are the user of Smartphone either of Androids or IOS, Thus, developing proposed system as an approach for Android- based ticketing system is rational move. Result shown more than 60% of participants are using Smartphone app more 6 to 10 times per day and 95% of them agree mobile application make their life easier. Besides that, result shown only small part of participants are always went to UTAR admission office because they don't want to take long time to wait at waiting area and this was supported by 35% of participants wait for more than 30 minutes during tuition fees payment period and it can be seen large part of students wait for more than 10 minutes at waiting area. 73% of participations feedback to researcher they are not satisfying the current service that provided and most of participations are agree with the currently method to distribute number is not environmentally friendly. No freedom while waiting will be the biggest weakness because participants dislike waiting long time at specific waiting area and unfortunate we have more than half participants are having such thought in their mind. When come to the last part of Proposed System Expectations and Requirements survey, 56% of respondents strongly agree should having a mobile ticketing apps and most of the respondents are satisfy for the user interface of current mobile ticketing app. More than 50% participants' claim that time estimation is important for the new mobile ticketing app and from the result above, time estimation is most important for the new mobile ticketing app.

Table 1. Results of Respondents feedback on The Current and Expected Application System

% of participants are using Smartphone app more 6 to 10 times per day	60%
% of participants are agree mobile application make their life easier	95%
% of participants are willing to take long time to wait at UTAR admission waiting area for more than 30 minutes during tuition fees payment period	35%
% of participants are not satisfying the current service that provided and most of participations are agree with the currently method to distribute number is not environmentally friendly	73%
% of participants are strongly agree should having a mobile ticketing apps	56%
% of participants are claim that time estimation is important for the new mobile ticketing app	50%

5 Conclusion and recommandations

With the availability of this proposed system name E-Queue, problem faced by UTAR students in waiting for service could be solved in a more effective and efficient way when compared to existing system. Plus, students can wait for service at anywhere as long as their Android phone is capable of retrieving Internet connection, thus ruling out chances for them to miss out their turn as the developed application is targeted to students and their collaboration in the testing phases is highly essential, the agile method is selected. As mentioned in the methodology part above, Agile methodology covers up the whole system development learning cycle therefore if anything goes wrong it enable the system creator back to the "wrong system stage" to rectify the system problem, improve and enhance the system according to the feedback and requirements from the system users. Interview session with one of the president of the student association, he could aid in getting a clearer view on what problem which is faced by students. In the implementation process, Java programming language will be used since Android development supports the mentioned language.As a smart phone user, we notice that system and application upgraded will be a never ending story issue, moreover more new and advanced system will be introduced. Therefore even though our proposed system able to accept by majorities at the launching stage but system creator should be always enhance the said system (for example: enhance and improve the application while found any system bugs or large group of users requirements to certain application function to be added) in order to retain the current application users and to grab new users. Thus, when the proposed system becomes stable and an upgrade of version is needed, developer should consider in adding on student's

DOI: 10.1051/shsconf/20173300033

requirements function into the proposed system. The addition of this function from time to time will eventually add in value and usefulness to the proposed itself, gaining more new users and maintaining existing ones. As this is a very new "queuing" system to be proposed to the public and still consider it a "newbie" system therefore we do not add must multimedia elements in the application to avoid heavily loading and downloading loading lacking due to these issue might cause the system instability therefore currently we keep the application design as simple as possible. With the current improvement in telecommunication industry, we hope that internet speedy would not be an issue if in the later stage multimedia elements added in the application. However, this is the suggestions and recommendations that we propose to the next researches in the future to have a look in this research look hoop.

References

1. D. Haughey, Waterfall v Agile: How Should I Approach My Software Development Project? . [online] Available at:http://www.projectsmart.com/articles/waterfall-v-agile-howshould-i-approach-my-software-development-project.php (2009)
2. V. Priya Android OS Vs. Apple iOS – Which is Better for Developers?.[online] Available at:http://mobiledevices.about.com/od/kindattentiondevelopers/tp/Android-Os-Vs-Apple-Ios-Which-Is-Better-For-Developers.htm (n.d.))
3. J. Hassell, Developing for the iPhone and Android: The pros and cons - Computerworld. [online] Available at:http://www.computerworld.com/s/article/9178684/Developing_for_the_iPhone_and_Android_The_pros_and_cons?taxonomyId=18&pageNumber=1 (2010))
4. Developer.android.com Android 4.3 for Developers | Android Developers. [online] Available at: http://developer.android.com/about/versions/jelly-bean.html (2013))
5. Escallier, P. Maximum PC | 10 Things Android Does Better Than iPhone OS - Page 1. [online] Available at:http://www.maximumpc.com/article/features/10_things_android_does_better_iphone? page=0,0 (2010))
6. J. Dempsey, *Android loses to Apple when it comes to security*. [online] Available at: http://www.cvbizjournal.com/columns/james-dempsey/android-loses-to-apple-when-itcomes-to-security.html (2013))
7. B. Rodriguez, Android vs. iOS: the Pros and Cons of Two App Platforms. [online] Available at: http://www.lunawebs.com/blog/2012/03/18/android-vs-ios-the-pros-and consof-two-app-platforms-825/ (2012))

Further explication of mega-crisis concept and feasible responses

Victor Yew-Cho Yen[1,*], and *Charles* T. Salmon[2]

[1,2] Wee Kim Wee School of Communication and Information, Nanyang Technological University, Singapore

Abstract. The contemporary world is in an era of mega-crises, caused by a host of physical, economic, political, social, and cultural factors. Through a step-by-step explication of the root concepts of *problem*, *mess,* and *crisis*, a mega-crisis is posited to be a set of interacting crises that are severe in impact, complex in nature, and global in fallout, with no seeming end in sight. Compared to a crisis, a mega-crisis is higher in severity and deeper in complexity. The paper argues for a stakeholder perspective in mega-crisis response, as relying on the organizational standpoint is inadequate for reaching long-term resolutions.

1 Introduction

In spite of its many innovations, opportunities and resources, the contemporary era can be viewed as one of a series of mega-crises, caused by a complex interplay of physical, economic, political, social, and cultural factors [1]. Due to its scale, complexities, and impact on many publics, the African famine can be considered a mega-crisis, i.e., "a complex constellation of political, religious, cultural, and commercial crises involving not only the media, governments, ambassadors, religious leaders, and citizens from many countries, but also private companies…" [2]. The issues are multi-faceted and related.

At the crux of the famine mega-crisis is a series of interacting problems, messes, and crises. These are some of the contributing factors [3]: Poverty; increasingly unpredictable weather which has the potential to reduce annual crop yields; drought; groundwater contaminated by pollution and deadly microorganisms; diminishing grain reserves; differential access to global commodities markets; distribution problems in allocating food resources; corruption; the variable price of oil; increasing meat consumption and concomitant diversion of grain to livestock rather than to humans; and increasing proliferation of agro-fuels, which similarly diverts grain away from human consumption.

Despite awareness of these various factors, responses by most international agencies remain predictable with every cycle of the famine problem in Africa: Provide food aid, provide loans to buy food (with interest), and provide technologies for more productive and efficient farming. Such seemingly reasonable measures, predominantly driven by authorities, are limited in their effectiveness because they fail to address the multiplicity of factors involved. Long-term solutions lie within a much broader web of social, political, and economical entanglements.

The paper endeavors to further the explication on *mega-crisis*, as guided by the *logic of concept explication* process [4]. This includes describing what a mega-crisis is and why it is different from conventional definitions of a crisis; examining empirical properties of a

* Corresponding author: yyen001@e.ntu.edu.sg

DOI: 10.1051/shsconf/20173300034

mega-crisis; and developing a tentative conceptual definition. The paper also discusses feasible responses to a mega-crisis.

2 Significance and opportunities

Currently, the application of crisis communication theory in a mega-crisis is inadequate. This is primarily because most seminal research responses are grounded in the perspectives and interests of an organization or perhaps an industry, which is insufficient in understanding a mega-crisis. Second, current definitions of a mega-crisis require further explication, as there are more attempts to define a mega-crisis in terms of what it is not rather than what it is. The paper begins by tracing the root concepts of problems, messes, and crises.

3 Literature review

3.1 Root concept: Problem

The explication process starts with tracing the root concepts of crisis and examining the critical similarities and differences in their escalating properties. At the most basic level is a *problem*. A problem must satisfy three conditions: "First, a decision-making individual or group has alternative courses of action available; second, the choice made can have a significant effect; and third, the decision maker has some doubt as to which alternative should be selected" [5]. A problem tends to be defined at a micro level, i.e., involving an individual or organization facing a decision.

There are three generally accepted ways to manage a problem: Resolve, solve, or dissolve [5]. First, the clinical approach would be to resolve a problem by choosing the method that yields a good-enough outcome, based on past experience and consultation of best practices. Second, the research approach would be to solve a problem by achieving the best results based on empirical evidence from primary research or statistical modeling and simulations. Third, the design approach would be to dissolve a problem by changing the environment in order to remove the problem [6]. It demands a different way of approaching the issue, by seeing the problem and possible solutions from a structural perspective.

3.2 Escalated concept 2: Mess

The next escalated concept from a problem is referred to as a *mess* [5], which is a large and complex set of interacting problems. Problems constituting a mess rarely can be clinically separated and analyzed for solutions [7]. In this regard, any single solution that professes to effectively solve or resolve a mess can only mean a misreading of the situation, i.e., it would not have been a mess, much less a crisis. A mess is the outcome of interacting problems [8] and is more complex than a single problem. Problems and messes also typically do not reach the threshold of media attention, which is a key factor to distinguish problems and messes from the more severe stages of crises and mega-crises. An example of a mess would be to discourage anti-littering behavior. Interacting problems may include a lack of litter bins, inconspicuous litter bins that are badly designed and placed in less-accessible spots, poor enforcement of anti-littering penalties, and dim lighting of the area etc. Resolving any one of these problems may not resolve the mess, as problems are intertwined. It is possible that all related problems have to be tackled holistically in order to yield positive results from the mess.

3.3 Escalated concept 3: Crisis

A *crisis* can have severely negative consequences for an organization. It is not a normal occurrence in the everyday flow of business. Defined as "a turning point for the organizational life" [9], it usually changes an environment drastically and irreversibly. A crisis is also a disruption that can threaten the existence of an organization [10]. The severity can "damage – even terminally – a company and individual careers, and can produce interminable legal suits and government investigations that can disrupt operations and smooth running of a company for years" [11]. This impact has largely to do with the presence of media as a *multiplier* in amplifying the crisis [12], thus exposing the reputational assets of an organization to risk.

A crisis goes through five common stages: detection, prevention or preparation, containment, recovery, and learning [10]. Detection involves the noting of warning signs that can lead to a crisis if unabated. Prevention involves either averting the crisis, lessening the impact of the crisis, or limiting the duration of the crisis. Crisis preparedness, which is driven by perceived seriousness and perceived controllability, can result in an organization's willingness to plan for a crisis scenario [13]. Containment refers to efforts that limit the duration or the spread of the crisis. Recovery happens when the organization adapts to an altered environment and set of practices precipitated by the crisis. Lastly, learning is the process of studying the crisis and taking stock of losses and gains. Learning will also improve detection of warning signs in the future.

The way crisis stages are managed is determined by an organization, as it both faces adverse consequences and possesses resources to ameliorate the crisis. The organization can adopt either a managerial perspective or a political perspective [14]. The former involves crisis managers devising a comprehensive strategy for crisis response. Most crisis theories have been derived from a managerial perspective: Apologia theory (to defend and protect image), image restoration theory (what is threatening image and which publics must be managed to redress image loss), decision theory (issues management and crisis prevention strategies), diffusion theory (how new procedures are accepted by the organization), and excellence theory (how organizations communicate with stakeholders) [10]. These theories consider organizational interests as a start point.

The latter – political perspective – is less concerned with functional responses to a crisis, and is purposed towards understanding its "social significance and political management that is partly determined by [its] psychological impact on the communities affected and the political choices that play up or down the importance, unacceptability and urgency of the events" [14]. In consulting the perspectives of affected stakeholders, the organization is able to assess the crisis comprehensively. The political perspective accepts a crisis as an ill-structured mess, which is an interacting set of ill-structured problems [8]. An ill-structured crisis, in turn, means that stakeholders have differing views, according to different backgrounds, values and interests. A political perspective posits that "every crisis is an existential crisis in meaning" [7].

3.4 Escalated concept 4: Mega-crisis

A *mega-crisis* is more severe and complex than a crisis, where both managerial and political perspectives are required to manage the situation. This is because a mega-crisis creates "deep uncertainty, and evoke(s) an extreme sense of urgency" [15]. Together, these two components are seemingly incompatible because a problem with *deep uncertainty* requires time to unravel, yet time is not available due to *extreme urgency*.

It is the *extreme urgency* of a mega-crisis that demands a managerial perspective. Due to media pressure, an organization must provide immediate relief to save lives and preserve property. However, as in the case of the African famine, quick and necessary responses in providing food aid, for example, do not address the core issues of a mega-crisis. Urgency can be *a red herring* to the real issues of a mega-crisis, where stopgap measures masquerade as long-term solutions due to public pressure.

Proposition 1a: A managerial perspective, adopted by the organization in a mega-crisis, is likely to meet urgent stakeholder needs.

Proposition 1b: A managerial perspective, adopted by the organization in a mega-crisis, is less likely to provide long-term resolutions.

Although the root word *mega* connotes *very large* or *out-of-the-norm* values, a mega-crisis is not *simply* a bigger crisis [16]. As mega crises evolve due to environmental changes, new emerging challenges require paradigm shifts in order for managers to cope. Consider the multifold complexities invariably linked with the concept of a *mega-crisis* – mega-disasters, mega-threats, mega-risks, mega-responses, mega-reactions and mega-opportunities – and one would begin to see the extensive work required in elevating crisis literature to mega-crisis levels. Thus, a mega-crisis has been defined as the "embedded engine of a chaotic world that evolves and mutates through global dynamics whose texture is made up of complex, unstable webs of constant, global, major dislocations" [16]. Globalization is a critical component in the identification of a mega-crisis [2, 17-18].

Also referred to as a *modern crisis* [8], a mega-crisis has an element of perpetuity, i.e., it is without a clear beginning or end. Characteristics of a mega crisis keep mutating [19] and its severity is not limited to the loss of lives and property, but potentially the legitimacy of the state as well. These challenge the five stages of a crisis because mega-crises remain in *prolonged limbo* [20], unresolved over time because of overwhelming complexities and waning public attention. Optimism bias, which refers to a perception that others are more susceptible to the risk of negative events, can happen to the organization [21]. This implies that mega-crisis managers may avoid or deny a scenario that is deemed too severe to comprehend. While a mega-crisis can be contained, recovery may not be in sight.

Beyond immediate responses in meeting extremely urgent needs, the political perspective is needed to address the component of *deep uncertainty*. Such is the level of uncertainty in a mega-crisis that past crises have limited value for understanding tomorrow's crises [22].

To truly understand a mega-crisis requires an appreciation of a state of constant flux, and *black boxes* of various disciplines and their tensions. It is an *ill-structured mess* [8], where different groups of stakeholders will define its problems based on their "values, interests, education, personal history, and the organizations for which they work". Every *modern crisis* is inter-disciplinary and even trans-disciplinary in nature, requiring a diverse community of professionals in many different fields for cooperation and integration in order to design a viable solution [7].

Proposition 2: A political perspective, adopted by the organization in a mega-crisis, is likely to provide long-term resolutions.

DOI: 10.1051/shsconf/20173300034

4 Discussions: Empirical Properties of a Mega-Crisis

The empirical properties of a mega-crisis are as follows. First, the mega-crisis is a complex set of interacting crises [5], in which there is "deep uncertainty" [15], and the absence of a simple or single resolution. An inter-disciplinary and even trans-disciplinary approach is needed to comprehend and manage this uncertainty [8]. Second, a mega-crisis has global implications [2]. Mega-crisis managers will need to grapple with the diverse cultural backgrounds and other differences of various publics. Third, a mega-crisis evokes a sense of perpetuity, with no distinct start and endpoint. Although intense media attention may heighten urgency in dealing with a mega-crisis in the short-term [19], media and public interest will decline when complexities unfold over time [20].

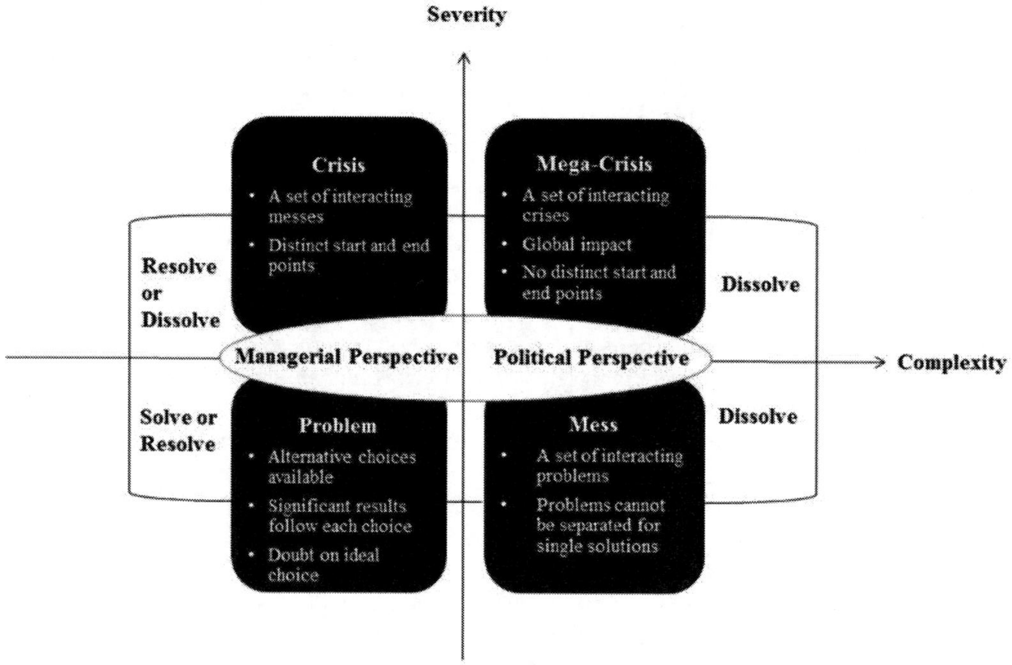

Fig 1. Mega-crisis mapping model

The mega-crisis mapping model is thus proposed. The x-axis illustrates the complexity and the y-axis illustrates the severity of a mega-crisis. First, a problem is low in complexity and severity, which makes it possible to solve it categorically or resolve it from a managerial perspective (organizational discretion). Second, a mess is high in complexity but low in severity. As problems are interacting, it is not possible to isolate them for resolution. Third, a crisis is low in complexity but high in severity. It is possible to prepare for and prevent a crisis from happening. It is also possible to contain and recover from a crisis, based how others have done so. As such, a managerial perspective can be applied. Lastly, a mega-crisis is the highest entity of the four in terms of complexity and severity. Mega-crisis managers have to deal with extremely urgent demands from stakeholders but yet be able to consider a long-term dissolution of the situation. Wide consultation of stakeholders will be required, through experts, in order to fully understand the situation, propose socially legitimate responses, and work with salient stakeholders to dissolve the mega-crisis in micro- and meso-levels.

 DOI: 10.1051/shsconf/20173300034

4.1 Responding to a Mega-crisis

The design approach, which proposes changing the environment to dissolve a problem [5], is the most plausible approach. The main challenge remains an accurate identification of the current environment, which calls for a *motivational framing* of the issue [6]. Motivational framing examines the situation for its intolerable state and determines why people should care. In order to do this, a political perspective that incorporates the views of multiple stakeholders should be adopted.

A multivocal approach has been argued to be a feasible process for the gathering of stakeholder inputs in a mega-crisis [2]. Within this approach lies the consideration of a macro-rhetorical arena and the micro-rhetorical arena [2]. The macro-rhetorical arena examines dynamics between different publics and their relationships with each other, taking into account cultural and social contexts. The micro-rhetorical arena is an in-depth study into each communicative process, which consists of three elements in "crisis communication, sender and receiver" and four parameters of "context, media, genre and text" [2]. A multivocal approach in mega-crisis communication focuses on the connections between stakeholders and how they are forged.

Stakeholder salience can inform these connections in the macro-rhetorical arena. The stakeholder salience model [23], which singles out three attributes in power, legitimacy, and urgency, illustrates what captures the attention of the organization. Power refers to the ability to carry out one's will to the extent of access to coercive (physical force), utilitarian (material or finances), or normative (symbols such as prestige, love and acceptance) resources. Legitimacy refers to socially accepted and expected structures or behaviors that may be interpreted differently at different stages of social organization [24]. And urgency refers to an attribute that has a time-sensitive claim that it is critical to the stakeholder. The more attributes a stakeholder group can muster or project to the organization, the higher the salience. The stakeholder salience model was later refined in a crisis setting and proposed another model with urgency as a key driving force [25].

However, in a mega-crisis situation that seeks to find long-term resolutions to a complex situation, urgency cannot determine salience. An alternative attribute can be one's ability to adapt to environmental changes over time. In adaptability, there are signs of resilience, which also enhances salience in a mega-crisis.

The dissolution of a mega-crisis (changing the environment to remove a problem), can be found in the micro-rhetorical arena. There are conceptual clues that point to how rituals can constitute a legitimate mega-crisis response. Design-oriented planning suggests that to *dissolve* a problem, it is "to be carried out as participatory as possible, ideally providing all stakeholders – not just their representatives – with an opportunity to take part" [5]. Unlike a routine, which consists of habitual actions without meaning, a ritual involves symbolic communication [26]. In a ritual, the situation is regarded to be more significant than its participants, and its metaphoric interpretation from reality is always purposive [27]. Components of a ritual include "group assembly (physical proximity), barrier to outsiders, mutual focus of attention, and a shared mood or emotional experience" [28]. Through rituals, stakeholders will be able to organize and cope in various groups, take on different roles on multiple platforms, and experience a common reality. In summary, the practice of multiple rituals by multiple stakeholder groups contributes towards a collective resolution of a mega-crisis. Rather than rely on a single resolution, a mega-crisis should be tackled in smaller parts that organizations can work with stakeholders on.

DOI: 10.1051/shsconf/20173300034

5 Refining the definition of mega-crisis as a concept

Based on previous definitions [2, 16], the following conceptual definition for a mega-crisis is posited: "A set of interacting crises that is severe in impact, complex in nature and global in fallout, with no distinct start and end points."

6 Conclusion

The model offered in this paper provides an analytic framework for comparing similarities and differences among problems, messes, crises and mega-crises. It can also be used to identify evolutionary stages of crisis development and targets for crisis management. That is, a well-managed problem does not have to evolve into a set of inter-related problems (a mess); a well-managed mess does not have to evolve into a set of interrelated messes (a crisis); and a well-managed crisis does not have to evolve into a set of interrelated crises (mega-crisis). A multivocal approach, which includes extensive stakeholder involvement, design-oriented planning, and symbolic communication, is a promising approach. Analytic generalization, which involves making projections about the possible transferability of findings from an evaluation based on research [29], is another. For crisis managers to confront new mega-crises that await, further research in the form of case studies is needed to trace the origins of mega-crises and to construct analytic generalizations.

References

1. J. Chomsri, P. Sherer, International Journal of Behavioral Science, **8**, 1 (2013)
2. F. Frandsen, W. Johansen, Hoboken, New Jersey: John Wiley & Sons Ltd (2010)
3. C. Chase-Dunn & P. Grimes, Annu. Rev. Sociol., **21**, 1 (1995)
4. J. McLeod, Z. Pan, S. Dunwoody, L. Becker, Concept explication and theory construction, in S. Dunwoody, L. Becker, D. McLeod, & G. Kosicki (Eds.), Creskill, New Jersey, Hampton Press (2005)
5. R. L. Ackoff, Interfaces, **11**, 1 (1981)
6. D. R. Loseke, J. Best, New York: Aldine de Gruyter (2003)
7. I. Mitroff, C. Alpaslan, Palo Alto: Stanford University Press (2011)
8. I. Mitroff, M. C. Alpaslan, S. E. Green, Int. Stud. Rev., **6**, 1 (2004)
9. M. Regester, London: Business Books (1989)
10. K. Fearn-Banks, New York: Taylor & Francis (2011)
11. M. K. Pinsodf, Lexington, Massachusetts: D. C. Heath and Company (1987)
12. O. Lerbinger, New York: Routledge (2012)
13. B. J. Wrigley, C. T. Salmon, H. S. Park, Public Relat Rev, **29**, 3 (2003)
14. A. L. Hughes, L. Palen, International Journal of Emergency Management, **6**, 3-4, (2009)
15. I. Helsloot, A. Boin, B. Jacobs, L. K. Comfort, Springfield, Illinois: Charles C Thomas Publisher (2012)
16. P. Lagadec, The Megacrisis unknown territory – In search of conceptual and strategic breakthroughs, in I. Helsloot, A. Boin, B. Jacobs, & L. K. Comfort (Eds.), 12-24, Springfield, Illinois: Charles C Thomas Publisher (2012)
17. M. J. Canel, K. Sanders, *The Handbook of Crisis Communication*, 449-466, Hoboken, New Jersey: John Wiley & Sons Ltd (2010)
18. P. A. Curtin, Negotiating global citizenship: Mattel's 2007 recall crisis, in W. T. Coombs & S. J. Holladay, (Eds.), *The Handbook of Crisis Communication*, 449-466, Hoboken, New Jersey: John Wiley & Sons Ltd (2010)

19. P. t Hart, & A. Boin, Between crisis and normalcy: The long shadow of post-crisis politics, In U. Rosenthal, A. Boin, & L. Comfort (Eds.), , 28-46, Springfield, Illinois: Charles C. Thomas Publisher (2001)
20. A. Downs, The Public Interest, **28**, 38-50 (1972)
21. C. T. Salmon, H. S. Park, & B. J. Wrigley, J Health Commun, **8**, 130-143 (2003)
22. U. Rosenthal, A. Boin, & L. K. Comfort, *Managing crises: Threats, dilemmas, opportunities*, Springfield, Illinois: Charles C Thomas Publisher (2001)
23. R. K. Mitchell, B. R. Aglc, & D. J. Wood, Academy of Management Review, **22**, 4 (1997)
24. M. C. Suchman, Academy of Management Review, **20**, 3 (1995)
25. C. M. Alpaslan, S. E. Green, & I. I. Mitroff, JCCM, **17**, 1(2009)
26. M. Spagnola, & B. H. Friese, Infants & Young Children, **20**, 4 (2007)
27. C. Bell, *Ritual theory, ritual practice*, United Kingdom: Oxford University Press (1992)
28. R. Collins, *Interaction ritual chains*, Princeton, New Jersey: Princeton University Press (2004)
29. R. K. Yin, *Case study research: Design and methods,* Thousand Oaks, California: Sage Publications (2009)

Framing crisis response messages on Facebook: a second level agenda analysis of MH370

Dayang Aizza Maisha Abang Ahmad[1,*], *Noratikah* Mohamad Ashari[1], and *Mus Chairil* Samani[1]

[1]Department of Communication, Faculty of Language Studies and Communication Studies, Universiti Malaysia Sarawak, 94300 Kota Samarahan, Malaysia

Abstract. A crisis can certainly bring great threats to an organisation. In time of crisis, what the organisation says and does may impose significant effect on the organisation's effort to survive its reputational damages. Although crisis responses are considered a common topic, this rapidly growing field of research is however vital to be critically explored. In light of Situational Crisis Communication Theory (SCCT) [1] and the Second-level Agenda Setting approach [2], this study seeks to analyse the message frames used by Malaysia Airlines System (MAS) in the formation of MH370 crisis response messages that that were disseminated directly to the organisation's stakeholders on Facebook. The convergence of framing and second level agenda setting has made this study significant as it advances the explication of potential crisis communication effects by underscoring the distinct importance held by certain attributes and frames, in the content of a crisis response message. This paper also includes discussion on the directions for future research on crisis response strategies particularly in the local context.

1 Introduction

Crises are taken as threat to every organisation's reputation. When a crisis strikes, the organisation's reputation can, to a great extent, be lost. Crisis that is sudden and unexpected can, in any form, powerfully disrupts an organisation's operations and poses both economic and reputational threat. In Malaysia, there were few notable organisational crises that can be reviewed throughout the first half to end of 2014 alone. Among Malaysia's most high profiled ones was the tragedy of Malaysia Airlines' flight MH370. Malaysian Airline System (MAS) Berhad or currently known as Malaysia Airlines Berhad (MAB) is a major airline, operating flights domestically and internationally. Malaysia Airlines is the flag carrier of Malaysia and a member of the Oneworld airline alliance. For the purpose of this paper, the studied organisation is still referred to by its previous name – MAS, taking into consideration that it was how the organisation was known and addressed as, during the hit

* Corresponding author: aadamaisha@unimas.my

of the crisis. MH370 was a scheduled international passenger flight that disappeared on 8 March 2014, while flying from Kuala Lumpur International Airport near Kuala Lumpur, Malaysia, to Beijing Capital International Airport in Beijing, China. The flight last made voice contact with air traffic control at 01:19 MYT (17:19 UTC, 7 March) when it was over the South China Sea, less than an hour after take-off. The tragic disappearance of Flight MH370 with a total of 239 passengers and crew members would be the second deadliest incident involving a Boeing 777 and the second deadliest incident in MAS history of establishment [3].

Crises are indeed inevitable for organisations regardless of their size, reputation or the industry they operate in. Any crisis situation, whether it is accidental, legal, financial or otherwise, will in turn forces organisations to establish a successful crisis response strategy that can help to control the damage and scope of crisis at hand. In surviving such damage, an organisation at an immediate effect, should implement crises response strategies (CRS). The scope of this study is to examine the framing of the actual messages that was disseminated by MAS to the public on Facebook in response to the mysterious disappearance of flight MH370 on the 8th March 2014. The findings reported in this paper is a preliminary result of a longitudinal study conducted to investigate how crisis response messages are framed and to determine the suitability of the strategies applied by the organisation when responding to the crisis. Parallel to this objective, this research content analyses a total of 7 statements that were released on the MAS official Facebook page on the day of the crisis. From the perspective of Second Level Agenda Setting Theory and Situational Crisis Communication Theory [1], this study analyses the two dominant message frames that were used in the formation of Malaysia Airline's MH370 crisis response messages – Information and Compassion.

2 Facebook in disseminating crisis response messages

According to Nel [3], 'the first data should, if at all possible, be provided within 1 hour after the event happened'. This is when communication platforms like the social media like Facebook can become strategically important. Facebook allows active communities to form online, which then increases the responsibility of organisations to manage and respond to a crisis effectively. With its real-time messaging and message linking to networks of its millions of wide range users, Facebook if utilised effectively can to a great extent help organisations to avoid potential crises situations [4]. It presents opportunities for organisations to tell its own account of a crisis without being subjected to journalistic gatekeepers and one-sided view of the crisis as normally practiced in the press media [5].

However, if used inappropriately and ineffectively, social media like Facebook can to a great extent worsen and intensify a crisis [4]. With the interactivity facilitated by Facebook, crisis response messages that were disseminated by the organisation can be forwarded and 'shared' among stakeholders and countless people. This could bring about the possibility of the stakeholders or other Facebook users to create and spread their own version of the crisis and thus create a likelihood of a boomerang effect to the initial crisis communication effort. The spread of false or incomplete information nevertheless could also form a risk, particularly due to the inability of social media users to distinguish between facts and rumours [6] which consequently could exponentially affect the organisation's image and reputation. As the recipients of the information, the reactions of the stakeholders would differ depending on the types of frames and attributes used to construct those messages. The attention given to the frames presented in the message could then influence their responses and determine how much responsibility they should assign to the crisis and the organisation involved [1].

3 Message Frames in Crisis Response Messages: Information and Compassion

Information and Compassion are the two primary message factors that are also viewed as message frames that are used as basis to the construction of organisation's crisis response messages [7].

Information Framing is grounded by two types of information that is essential to building a strong foundation crisis response effort - Instructing Information and Adapting Information. Instructing Information are terms used to describe the information required and expected by the stakeholders to receive from the organisations in time of crises. Therefore should be communicated by the organisation directly to stakeholders, informing them on what they need to do, mainly to protect them from the physical threat of a crisis. Instructing Information is considered critical to a crisis response due to the many expectations that stakeholders build towards an organisation when a crisis strikes. This expectation is primarily in form of the amount or type of information that organisation is expected to have and provide in response to the crisis. Organisation's ability to provide the needed information will in turn cultivates the stakeholder's perception of the organisation's degree of control on the crisis. Stakeholders are believed to likely perceive positively and respond favourably towards organisations that knows what happened, and able to advice on the next course of action and provide solution [7]. For the purpose of the analysis of this study, Instruction Information is conceptualised and operationalised based on three variants: (a) the what, why, when, where, and how of information about a crisis; (b) any precautionary actions stakeholders need to take; (c) actions taken by the organisation to correct and/or end the crisis and/or to prevent a repeat of the crisis [7]. On the other hand, Adapting Information functions to help stakeholders to cope with psychological threats from the crisis (e.g psychological stress that could be caused by the situation revolving around the crisis). To cope with this, stakeholders should be supplied with information about the 'what' and 'when' of the crisis, as well as the corrective actions taken by the organisation to convince stakeholders that they are safe thus reducing their psychological threat. The final aspect of Adapting Information that organisation needs to convey is the expressions of concern to the stakeholders; however not to the extent that can indicate admissions of guilt. Only after this part is established can the organisation focus on the crisis response strategies to protecting and restoring its reputation [1,7].

Expressing concern builds credibility with victims and other stakeholders. Therefore, the organisation should acknowledge and show concern for stakeholders' needs (especially the direct victims of the crisis). The compassion suggests the organisation credibility in turn, should facilitate the development of a favourable organisational reputation. Hence, to a certain extent may overlap with the final part of Adapting Information. Crisis communications strategies in relation to compassion framing could be arrayed from helping victims (i.e. accommodative) to defending organisational (i.e. defensive). Accommodation reflects compassion and concern for victims, whereas defensive lacks compassion by denying victims' needs. As such, the term Emotional Framing (and appeal) is found to have been used interchangeably with term the Compassion Framing. Emotional Framing refers to attributes used to appeal to the readers' emotions in form of words that expresses feelings (e.g. love and sad) [8]. Emotional Framing is usually executed by using drama as well as subjective and evaluative properties [9]. However, parallel to Coombs [7] the analysis will be conducted using the term Compassion Framing.

DOI: 10.1051/shsconf/20173300035

4 Analysing Crisis Response Message Frames: A Second Level Agenda Setting Approach

The explication of a second level agenda setting links closely with the process of framing. From the perspective of second level agenda setting, a frame is conceptualised as 'the central organizing idea for news content that supplies a context and suggests what the issue is through the use of selection, emphasis, exclusion and elaboration [2]. The two types of message frames used in crisis communication – Information and Compassion Framing comprises of particular attributes and different emphasis to the construction of a message [1, 7]. In other words, each frame is built upon the certain facts or values in form of attributes that are made salient and emphasised in the crisis response message by the organisation to the stakeholders [10].

Studies on framing usually further examine the attention of readers on the emphasized attributes and the opinions made in result to the attention given to the message [11]. Attributes as defined in the agenda setting theory, are 'characteristics and traits that fill out the picture of the object'; and thus at this level, the attributes emphasised over other elements that is believed could consequently influence the reader's perception of the message presented [12]. A handful of studies even have applied the second level agenda setting theory in the context of business by using the organisation as the "object" of study [13]. Hereby, attributes of corporations in the media content were evaluated and used to examine the effect of images transmitted in the media on the public's attitudes and opinions towards business firms. The theory was further pursued to measure the visibility and distinctiveness between one organisation to another, as well as in investigating issues and attributes in the study of corporate reputation [14, 15]. In similar vein, the current study applies second level agenda setting to content analyse the attributes used in the formation of messages by the crisis communication managers of MAS on Facebook in response to the disappearance of MH370.

Only a few studies to date have explored social media from a crisis communication perspective and specifically analysed the actual usage of social media by organisations for crisis management [4]. This study therefore fills the gap in literature by content analysing an actual crisis response messages in form of media statements that were released by MAS crisis communication team on Facebook to stakeholders on the day of the crisis, 8th March 2014. This research selected Facebook as the focus of the study as it is one of the prominent social media platform used by MAS in facilitating its communication effort in response to the MH370 crisis. From the perspective of Second-Level Agenda Setting Theory [2], this study conducted a thematic analysis in examining how the organisation frames crisis response messages on Facebook. The analysis performed for the purpose of identifying attributes used in the forming of Information and Compassion message frames (or interchangeably referred to as message factors) comprised of all 7 statements that were disseminated on MAS official Facebook page on 8th March 2014; hence the population of this study. This is considered sufficient taking into account the preliminary nature of this longitudinal study and the immense amount of messages released by MAS to the stakeholders throughout the conduct of the investigation since the disappearance of MH370 to the current state of the case.

Only the main page that is directly established by MAS in response to the crisis of MH370 was used for the purpose of this analysis. Other Facebook pages that were established by stakeholders or third party users or unknown sources (e.g PrayForMH370) are not considered relevant for this research. Stakeholders in the context of this analysis refer to the audience or recipient of the crisis response messages on Facebook, that probably range from victims' families, shareholders, media watchdogs, as well as clients (i.e. existing and potential passengers) throughout the globe.

5 Results

The analysis considered all the attributes used in the message, whether in form of a word or words; a sentence or sentences; as long as it matches or closest to falling in the category of Information and Compassion message frames. Following Coombs [7], this study operationalised Information Frame as the what, why, when, where, and how the information about a crisis, precautionary actions that stakeholders need to take, actions taken by MAS to correct the problem, as well as straightforward and objective information; while Compassion Frame as the denial, confessing of guilt, begging victims for forgiveness, emphasis of customer welfare including compensation, as well as feelings.

Due to the interactive and personalized nature of Facebook where exchange of information activities among users are mostly conversational and casual in approach, this research considers each sentence that makes up the crisis response messages as a form of conversation established by MAS to the stakeholders. Following Bales [16] in his analysis of conversation, single sentence conveyed in each media statement is considered a single unit of analysis. This single unit of analysis will be treated as a single unit of attribute that are communicated by the organisation to form frames, in which indicates the way that an information is presented and emphasised [1] in the organisation's crisis response message. The way a message is framed can to a great extent shapes how people define a crisis, assign attributions of the responsibility to the organisation in crisis and to know the solutions to the crisis [1]. Parallel to the objective of this study, a thematic analysis of the attributes was conducted for each of the 7 statements to determine the type of framing applied in the crisis response messages released by MAS on 8th March 2014. Results of the analysis showed that the attributes used in constructing the crisis response messages on the disappearance of MH370 on that date were mainly Information. The information framing was most prominent on one particular statement (i,e 3rd statement). Prominent herein refers to being emphasised to an extent that it could draw readers' attentions to focus on the prominent attributes when forming their perception and assigning responsibility towards the organisation.

A crisis creates a need for information. Communicating the 'required' information is part of the ethical responsibilities of the organisation and as part of their effort to address their concerns to the stakeholders. Based on the analysis, the Information Framing that was applied by MAS in its crisis response messages comprised of both types of information that is essential in establishing a strong foundation to an organisation's crisis response effort - Instructing Information and Adapting Information. Based on the analysis, the Instructing Information was highlighted in every statement except for the 6th Statement, while Adapting Information was only highlighted in 1st, 2nd, 3rd and the 6th Statement. Instructing Information encompassed of 18, while Adapting Information comprised of 13 from the total of 35 attributes. This makes Instructing Information as the most highlighted frame in the overall statements released by MAS on the 8th March 2014.

The information that the organisation supplied to stakeholders were found to be in line with Coombs [1,7] characteristics to defining Instructing Information. That is, the organisation was committed to providing stakeholders with information on 1) the basic press conference details that was conducted from time to time by the Group Chief Executive Officer of MAS (i.e. date and time of info release, and link to info update in form of hashtags - #MASalert). This information was found to be the most highlighted among other information under the Instructing Information frame (i.e. mentioned in 1st, 2nd, 3rd, 4th, 5th and 7th Statement). Following that information is the 2) details of the missing flight (i.e. route, location of departure and the supposed location of arrival, expected time of departure and arrival, number of passengers); as well as 3) the current status of the missing flight (i.e establishment of contact). More Instructing Information

DOI: 10.1051/shsconf/20173300035

were added by the hour about the details of the missing flight (i.e. nationalities of the passengers and its respective amount; the name of the pilot and first officer, their flying hours, the year that they started working with MAS; mentioned of the flight's code share with China Southern Airlines); and status of the flight at that current time (i.e speculation that the aircraft has landed at Nanming).

Meanwhile, the Adapting Information was found to be mostly in form of information that helps to explain the organisation's current effort to handle the crisis. The most highlighted information under this frame was on MAS's action in working with the authorities who have activated their Search and Rescue team in locating the missing aircraft. This information was consistently mentioned in 1st, 2nd, 3rd and the 6th Statement, making it the most highlighted among other 13 attributes that encompassed of the Adapting Information in the released statements. The organisation also alerted the stakeholders on their effort at that time in contacting the next of kin of the passengers and crews onboard MH370; a mention of their commitment to provide regular updates on the situation, and the effort to establish contact lines and website specifically to assist the public and the media on their queries. Immediately after the Instructing Information on the speculation that the flight has landed at Nanming was released, MAS provided further elaboration on the organisation's action in response to the speculation. Within the same statement (i.e 3rd Statement) following the information pertaining on the speculation, the organisation informed the stakeholders on their effort to verify the authenticity of the report at that time. Also in the 3rd Statement, stakeholders were further updated on the organisation's action focus to work with the emergency responders and authorities as well to as mobilise its full support as part of the effort to handle the crisis. The organisation nevertheless put into emphasis the Adapting Information that needs to be supplied to the victims' next of kin as guideline for their further action (i.e. direction to the support facility building at KLIA and contact number for enquiry of the direction)

When a crisis hits an organisation, any group of stakeholders can basically be victims who could suffer financially, mentally, and physically. As such, every crisis response should consider expressing compassion to victims. This is because; expressing compassion builds credibility with victims and other stakeholders and in turn facilitates the development of a favourable organisational reputation [7]. Past studies suggested few crisis response strategies that could be used as guideline for organisations in expressing compassion to victims. The strategy includes Mortification (i.e to confess guilt and beg victims for forgiveness), Super Effort (i.e. to focus on customer welfare including compensation) and Accommodative (i.e to reflect compassion and concern for victims) [7].

There were only 2 attributes of Compassion Framing that were detected throughout the 7 statements. That is, the same attribute (i.e. our thoughts and prayers are with all affected passengers and crew and their family members) was mentioned twice in two separate statements (i.e. 2nd and 3rd Statement). This attribute reflected precisely Coombs [8] 'Accommodative' strategy to expressing compassion to stakeholders as part of the effort to assist the building of favourable organisational reputation.

However, a single attribute of Compassion Framing in the 6th Statement demonstrated the similar criteria to Adapting Information. Similar to Compassion Framing, the final aspect of Adapting Information also emphasises on the organisation's needs to express concern and compassion to the stakeholders (especially to the victims); however not in a way that can imply admissions of guilt [1, 7]. Following Bales [16] this analysis considers the 2 sentences (i.e. We understand everyone's concern on MH370 pax & crew. We're accelerating every effort with all relevant authorities to locate the aircraft) that constitutes both Compassion and Adapting Information Framing as a single thought; hence a single attribute. A similar pattern of framing was also detected in the 1st statement (i.e. We deeply regret that we have lost all contacts with flight MH370 which departed Kuala

Lumpur at 12.41 am earlier this morning bound for Beijing) where Compassion Framing overlaps with Instructing Information due to the usage of 'deeply regret' in the sentence.

6 Conclusion and Discussion

A crisis represents a distinct and an inevitable source of reputational threat for an organisation. Hence, it is vital for an organisation in crisis to take considerable amount of attention and effort to ensure that the stakeholders' physical and psychological needs are put to priority. This includes making sure that the stakeholders are updated with the latest information and alerts via a fast and accessible means of communication like the social media. At the same time, to adhere to the responsibility and ethics of providing accurate and consistent crisis response messages to the stakeholders, especially during the initial stage of a crisis.

The findings to the analysis of MAS's crisis response messages on the disappearance of flight MH370 on 8th March 2014 has verified that the organisation has responded in accordance to Coombs [1, 7] recommendation of crisis response strategy. The thematic analysis result of the attributes found in all the crisis response messages were mostly Information type of message frames; consisting of both Instructing Information and Adapting Information. To supply stakeholders with sufficient Instructing Information is critical to a crisis response effort. The result of this study also showed that the proportions of attributes that constructed the crisis response messages on that day encompassed of mostly Instructing Information. Based on the attributes used and type of framing applied by the organisation in responding to the crisis, it can be concluded that MAS has displayed a great sense of responsibility and ethics to the stakeholders by not making the organisation's reputation as the focus to their initial crisis communication effort. But instead, the organisation's attention was solely devoted to serving the stakeholders' needs and interest by providing the important crisis-related information as well as keeping the stakeholders updated on the immediate action taken by the organisation in their struggle to stay in control of the crisis. According to Coombs [1, 7], this should in turn cultivate positive perceptions among the stakeholders toward the organisation which could consequently contribute to the development of a favourable organisational reputation at the later stage of the crisis. Result of the analysis additionally indicated that the organisation was also expressive of the concern they have towards the stakeholders, especially the direct victims of the crisis (i.e. next of kin of the passengers and crews on board). Although low in count, the amount of attributes in the Compassion Framing, were deemed sufficient to allow the organisation to show the required degree of empathy and care towards the victims' needs. Expressing too much of compassion attributes is not encouraged especially at an early stage of crisis as it could potentially do harm to the organisation. It could be misinterpreted as organisation's admission of guilt and organisation's degree of responsibility for a crisis [1, 7].

Based on the presented findings, this research concludes that framing plays a significant role in crisis evolution and the attribution of meaning to a crisis. The way a message is framed; and the facts or values of a crisis that the frame stresses can affect an individual's interpretation, opinion and even decision towards a crisis [1]. Like a crisis response message, crisis types can also be communicated in a form of frame. Crisis types are framed by having to feature certain aspects and cues of the crisis (e.g whether or not some external agent or force caused the crisis, whether the crisis was a result of accidental or intentional actions, whether the crisis was caused by technical or human error) [17]. The cues and features of the crisis that are made salient in the organisation's crisis response message will indicate how stakeholders should interpret, respond to a crisis, and even determine how much stakeholders assign attribute responsibility for the crisis to the organisation. In the case of the current analysis, the crisis response messages released by MAS on 8 March 2014 have not only supplied stakeholders with the information of the crisis, but have also somehow highlighted key features or cues that could allow stakeholders to have an indication on the type of crisis that they were encountering. The efficiency of SCCT guidelines concerning crisis response strategy and its suitability highly depends on the crisis type that the organisation is facing. Matching the crisis response strategy to the crisis type and framing type is highly important to ensure that the crisis communication effort to produce beneficial outcome for the organisation.

Taking into account the manifold effect of framing, this research strongly suggest for an organisation in crisis to be mindful on the way that a crisis response messages are presented especially in the interactive social media platform like Facebook. The advent of the social media in the field of crisis communication has not only provided organisations direct access to the public in disseminating information about the crisis, but it could also probably turned into a platform for the public, to collaborate in the building of a crisis frame which could even potentially create an adverse effect to the initial crisis communication effort. However, this could only be further verified with thorough analysis of the crisis response messages, its suitability as a strategy to protect the organisation's reputation, as well as stakeholders' responses toward the messages disseminated. Expansion of this study is therefore required, particularly in respect to the evolvement of the MH370 crisis and its status to date; as well as the immense amount of crisis communication messages disseminated and effort that were established by MAS through multiple media platforms.

References

1. W. T. Coombs, CRR, **10**, 163-176 (2007)
2. M. McCombs, *Setting the Agenda: The Mass Media and Public Opinion*, (2004)
3. P. Nel, *World Business, Finance and Management Conference*, (2014)
4. K. Freberg, PRR, **38**, 416–421. (2012)
5. B. F. Liu, Y. Jin, L. L Austin, JPRR, **25**, 51–67 (2013)
6. J. J. Hyun, H. H. Dae, JFP **44**, 95-102, (2014)
7. W. T. Coombs, JPRR, **10**, 3 (1998)
8. W. T. Coombs, PRR, **11**, 2 (1999)
9. A. –S. Claeys, V. Cauberghe, JBR, **67**, **2** (2014)
10. J. J. Hyun, H. H. Dae, JFP **44**, 95-102, (2014)
11. J. N. Druckman, PB, **23**, 3 (2001)
12. G. Lee, ICG, **72**, 8 (2010)
13. C. Carroll, M. McCombs, CRR, **16**, 36-46, (2003)
14. C. B. M Van Riel, C. Fombrun, CRR, **4**, 296-302, (2002)
15. M. Meijer, J. Kleinnijenhuis, JC, **56**, 543-559, (2006)
16. R. F. Bales, in K. Krippendorff and M.A Block (Ed.), *Content Analysis Reader* (2009)
17. W. T. Coombs, S. J. Holladay, MCQ, **16**, 165-186, (2002)

Stego Keys Performance on Feature Based Coding Method in Text Domain

Roshidi Din[1,*], *Azman* Samsudin[2], and *Sunariya* Utama[1]

[1]School of Computing, CAS, Universiti Utara Malaysia, Sintok, Kedah, Malaysia
[2]School of Computer Sciences, SoCS, Universiti Sains Malaysia, Pulau Pinang

Abstract. A main critical factor on embedding process in any text steganography method is a key used known as stego key. This factor will be influenced the success of the embedding process of text steganography method to hide a message from third party or any adversary. One of the important aspects on embedding process in text steganography method is the fitness performance of the stego key. Three parameters of the fitness performance of the stego key have been identified such as capacity ratio, embedded fitness ratio and saving space ratio. It is because a better as capacity ratio, embedded fitness ratio and saving space ratio offers of any stego key; a more message can be hidden. Therefore, main objective of this paper is to analyze three features coding based namely CALP, VERT and QUAD of stego keys in text steganography on their capacity ratio, embedded fitness ratio and saving space ratio. It is found that CALP method give a good effort performance compared to VERT and QUAD methods.

1 Introduction

With the rapid growth of Internet communication, information protection needs have become a critical issue in order to secure information [1]. One of the advanced promising researches on information protections for the next generation through untrusted communication channels is a steganology field. Unlike cryptology, which utilizes the encrypted and decrypted information of secret message which rendering the cover messages completely meaningless, steganology keeps the cover messages perceptually unchanged after concealing and detecting of the covered message. In speaking of steganology, there are two main branches [2]. Steganography is concerned with avoiding the suspicion of hidden messages in manipulated information [3]. Meanwhile, steganalysis concerned is discovering and rendering covert messages in given information [4].

Steganography is usually mislabeled with cryptography, because both are aimed at protecting valuable information. Whereas cryptography does not conceal the communication but only scrambles the data to prevent eavesdroppers understanding the content. Meanwhile steganography is study of hidden information [5] which is concealing

* Corresponding author: roshidi@uum.edu.my

the existing messages. There are two aspects of steganography, namely technical steganography and text steganography. Technical steganography concentrates on channel capacity which is concerned hiding information via a medium such as image, audio, video or other digitally invisible code, while text steganography concentrates on using written text to conceal secret messages [6].

2 Related Works on Text Steganography

Text steganography is the art of using natural language to conceal secret messages based on manipulating structure in the text [7]. There are two types of text steganography methods which are word-shift coding and feature coding.

Word-shift coding alters the text document by horizontally shifting the locations of words within the text lines to encode the document in a unique manner. Huang and Yan [8] developed a word-shift method based on character coding that modifies the inter-word spaces. Yang and Kot [9] have proposed an integrated character and word spaces which are based on facts that inter character spaces .Then, Liu et al. [10] proposed a method of word shift coding that can use in online to cover hidden message in plain text.

Next method, feature coding that alters a unique feature characteristic in text based on code word [11]. Rabah [12] has suggested some of the text features that can be altered in a large volume without making the reader aware of the existence information in the text. In Chinese letter, Sun et al. [13] has proposed a component-based method which is a novel text watermarking algorithm based on the mathematical expression. Besides, In Arabic letter, Shahreza and Shahreza [14] suggested that the information can be hidden in a text by changing the place of point in the Arabic alphabets. In Microsoft Word document, Stutsman et al. [15] has suggested a hidden message method by using generated sentences in multiple translations in term of Microsoft Word. Moreover, several methods of feature coding that manipulated pattern letter is English grammar such Changing in Alphabet Letter Patterns method known as CALP is trying to utilize English letters by mapping the binary sequence of the hidden for embedding process of the cover text [16]. Then, based multiple character-based [17] such as curve-based, vertical-based (VERT) and quadruple-based (QUAD).

Therefore, the objective of this study is to measure the fitness performance of stego keys using feature coding method in order to hide hidden text in the cover text.

3 Experimental Designs of Research

There are several stages that involved in the experimental design which is shown in **Fig. 1.**

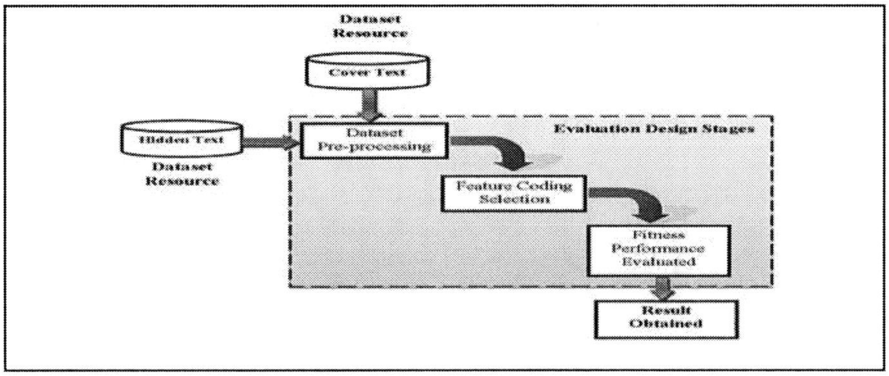

Fig. 1. Visualization of experimental design

This study consist of dataset such as cover text and hidden text that shown in **Fig. 2**. Then, three techniques are considered as feature coding methods used as the stego keys namely Changing Alphabet Letter Patterns called CALP [16], Vertical Straight Line called VERT and Quadruple Categorization called QUAD [17].

4 Fitness Performances Evaluated

The objective of this stage is to measure the fitness performance of the stego keys in text steganography domain from several researchers' works [18-21] such as capacity ratio, embedded fitness ratio and saving space ratio. Capacity ratio (CR) is used to determine the total of a hidden text that can be embedded in the cover text that can formulate as;

$$CR = \left[\frac{Total\ Bits\ of\ Hidden\ Text}{Total\ Bits\ of\ Cover\ Text} \right] \times 100\% \tag{1}$$

$$= \frac{\sum_{i=1}^{1<m<n} a_i}{\sum_{i=1}^{1<m<n} b_i} \times 100\% \tag{2}$$

where

a = *Total bits of hidden text*
b = *Total bits of cover text*

Meanwhile, embedded fitness ratio (ER) is used to determine the total fitness of a hidden text can be embedded in a cover text. The ER of the stego keys can be calculated as;

$$ER = \left[\frac{Total\ Bits\ of\ Stego\ Text - Total\ Bits\ of\ Cover\ Text}{Total\ Bits\ of\ Cover\ Text + Total\ Bits\ of\ Hidden\ Text} \right] \times 100\% \tag{3}$$

$$= \frac{\sum_{i=1}^{1<m<n} a_i}{\sum_{i=1}^{1<m<n} b_i} \times 100\% \tag{4}$$

where

a = *Total number of embedded bits*
b = *Total bits of expected stego text*

Finally, saving space ratio (SSR) is used to determine the total space of text that can be saved during embedding process in a cover text that can be calculated as;

$$SSR = \left[\frac{Total\ Bits\ of\ Expected\ Stego\ Text - Total\ Bits\ of\ Stego\ Text}{Total\ Bits\ of\ Expected\ Stego\ text} \right] \times 100\% \tag{5}$$

$$= \frac{\sum_{i=1}^{1<m<n} a_i}{\sum_{i=1}^{1<m<n} b_i} \times 100\% \tag{6}$$

where

a = *Total number of saving space bits*
b = *Total bits of expected stego text*

DOI: 10.1051/shsconf/20173300036

5 Fitness Performance Analysis

The fitness performance analysis of stego keys are investigated using 20 different types of cover text. The result of evaluated the stego keys are elaborated in the following discussion.

5.1. Fitness performance of CALP

This section explains the measurement used to measure the fitness performance of CALP on an analyzed text. Thus, the score values of capacity ratio (CR), embedded fitness ratio (ER) and saving space ratio (SSR) for CALP technique is shown in **Fig. 2**.

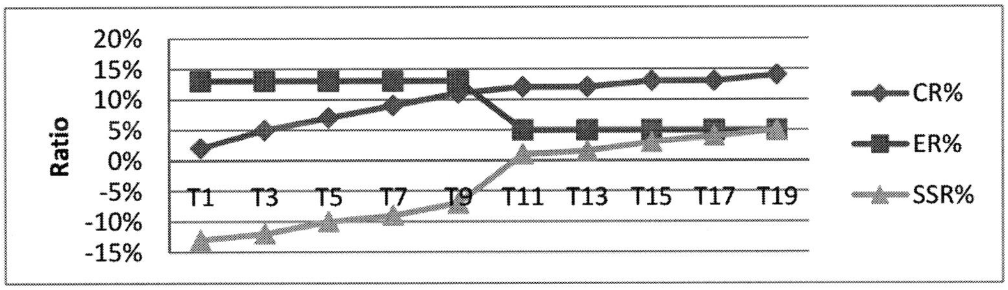

Fig. 2. Score values of fitness performance for CALP

Fig. 2 shows the number value of fitness performance for CALP. The result shows that the CR has a stable increase from 2% until closely 15% throughout the analyzed text. As for ER, the percentage ratio maintains between 14% and 13% throughout T1 to T9 and suddenly declines from 13% to 5% between T9 and T10. The percentage remains unchanged at 5% between T10 to T20. As for SSR, the percentage ratio starts at almost -11% at T1. Increase is seen at T1 when the percentage moves from -6% to 1% at T10 and gradually increases from 1% to 5% between T10 and T20.

5.2. Fitness performance of VERT

Fig. 3 shows the score value of fitness performance for VERT. The result shows that the CR has a steady increase from almost 1% to nearly 12% between T1 and T17 and slightly maintains at almost 12% from T17 to T20. As for ER, the percentage ratio has a gradual decrease from at nearly 13% to 12% between T1 and T9. However, there is a sudden spike closely at 12% to almost 16% of the percentage from T9 to T20. Meanwhile for SSR, the percentage ratio starts at less than -12% at T1 and sharply increases to -5% at T10. The trend remains stable at -5% from T10 to T17 before it falls slightly to -6% at T20.

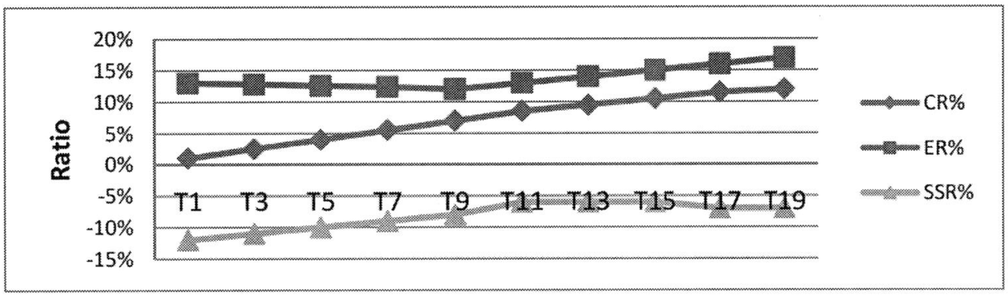

Fig. 3. Score values of fitness performance for VERT

5.3. Fitness performance of QUAD

Fig. 4 shows the score value of fitness performance for QUAD. The result shows that the CR has a steady increase from almost 1% to closely 12% from T1 to T17 and remains stable at 12% at T18. As for ER, the percentage ratio gradually decreases from nearly 13% to 12% between T1 and T10. Then, there is a sharp increase from T10 to T20 with a percentage of 12% and 17% respectively. Meanwhile, for SSR, the percentage ratio starts with less than -12% and steadily increases to almost -5% from T1 to T10. Then, the trend shows a constant movement from T10 to T17 with a percentage of -5% before it has a slight decrease to -6% at T20.

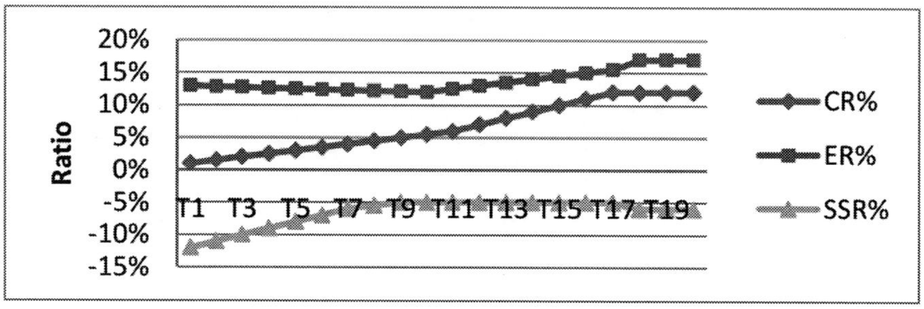

Fig. 4. Score values of fitness performance for QUAD

6 Conclusions

This study evaluated the three methods in implementation of fitness measurements of the stego keys based on capacity ratio, embedded fitness ratio, and saving space ratio. The three categories of text steganography techniques have been evaluated. It has been identified that CALP technique gives a better effort performance compared to VERT and QUAD technique. However, VERT and QUAD techniques give a quite similar result for embedding ratio and saving space performance compare to CALP. The results demonstrate that all of the methods are able to perform consistently with the utilization of hidden text and cover text. In future, this study proposes to evaluate a robustness of each technique in order to find a strength capability on text steganography from steganalysis activities.

This work was supported by 2016 Non-PI University Grant under RIMC Grant, Universiti Utara Malaysia (UUM).

References

1. J. M. Ahmed Z. M. Ali., IJCSNS **11,** 4 (2011)
2. R. Chandramouli, N. D. Memon, Proceeding SPIE Security and Watermarking of Multimedia Contents 5020,173 – 177 (2003)
3. C. Y. Chang, S. Clark, The 2010 Annual Conference of the North American Chapter of the Association for Computational Linguistics, 591 – 599 (2010)
4. H. A. Aboalsamh, H. Mathkour, M. Mursi, G. M. Assassa, Proceedings of the 12th WSEAS International Conference on Computers, 1011-1018 (2008)
5. R. Chandramouli, N. D. Memon, SPIE Security and Watermarking of Multimedia Contents 5020,173 – 177 (2003)
6. M. Chapman, G. I. Davida, M. Rennhard, Proceedings of the Information Security Conference (ISC '01), Malaga, Spain, 156 -165. (2001).

7. M. Chapman, G. I. Davida, M. Rennhard, Proceedings of the Information Security Conference (ISC '01), Malaga, Spain, 156 -165 (2001)
8. D. Huang, H. Yan, IEEE Trans. Circuits and Systems for Video Technology, **11**, 1237-1245 (2001)
9. H. Yang, A. C. Kot, IEEE International Conference on Multimedia and Expo, Taipei, Taiwan, **2**, 955 – 958 (2004)
10. M. Liu, Y. Guo, L. Zhou, Fifth International Conference on Intelligent Information Hiding and Multimedia Signal Processing, 807-811 (2009)
11. M. V. Nasab, B. M. Shafiei, Aust. j. basic appl. sci, **5**, 12 (2011)
12. K. Rabah, Information Technology Journal, **3**, 3 (2004)
13. X. Sun, G. Luo, H. Huang, Proceedings of the 3rd ACM International Conference on Information Security, Shanghai, China, **85**, 76-81 (2004)
14. M. H. S. Shahreza, M. H. Shahreza, International Conference on Intelligent Information Hiding and Multimedia Signal Processing, 1524-1526 (2008)
15. R. Stutsman, M.J. Atallah, C. Grothoff, K. Grothoff, Proceeding of the 21st Annual ACM Symposium on Applied computing (SAC 2006), Dijon, France, 338-345 (2006)
16. S. Bhattacharya, P. Indu, Duta, S. A. Biswas, G. Sanyal, Journal of Global Research in Computer Science, **2**, 3 (2011)
17. S. Dulera, D. Jinwala, A. Dasgupta, IJNSA, **3**, 6 (2011)
18. M. Agarwal, Computer Networks and Communications (NetCom). Springer, New York, 477-488 (2013)
19. L. Y. Por, D. Beh, T. F. Ang, S. Y. Ong, IAJIT **10**, 1 (2013)
20. C. F. Lee, and H. L. Chen, Springer Berlin Heidelberg, 155-179 (2013)
21. P. Akhilandeswari, J. G. George, Proceedings of International Conference on Internet Computing and Information Communications, Springer, India, 1-7 (2014)

Crisis Response Strategy and Crisis Types Suitability: A Preliminary Study on MH370

Noratikah Mohamad Ashari [1,*] , Dayang *Aizza Maisha* Abang Ahmad [1], and *Mus Chairil* Samani [1]

[1]Department of Communication, Faculty of Language and Communication Studies, Universiti Malaysia Sarawak, 94300 Kota Samarahan, Malaysia

Abstract. In time of crisis, the organisation's tactic in responding to the crisis according to its type/s may significantly affect the organisation's effort to survive its reputational damages. The study of crisis response strategy (CRS) requires further exploration within the Malaysian context, moreso with the greater and apparent use of social media as a platform for the organisation in crisis to reach out to its stakeholders. This preliminary paper studies the MH370 crisis in light of Coombs [1] Situational Crisis Communication Theory (SCCT). It aims to discuss the suitability of the CRS applied by Malaysia Airlines System (MAS) with the crisis type/s that was/were experienced by the organisation. Hence, the need for this paper to investigate the crisis type/s of MH370 and examines the CRS applied by MAS by content analysing media statements that were disseminated directly to the organisation's stakeholders on Facebook during the first day of the crisis. Results ultimately show an interesting analysis to the crisis type of MH370, and the suitability of the CRS applied by MAS for that matter.

1 Introduction

Crises implicate an organisation's reputation in ways that can lead to both economic and reputational challenges to the organisation. A sudden, unexpected crisis can damage or to great extent, lose the organisation's reputation if not managed carefully by the organisation in crisis. Regardless of the size, reputation or industry, there is always a possibility for every organisation to face any type of crisis situations.

This paper positions Situational Crisis Communication Theory (SCCT) [1] as the foundation of this study, particularly in regards to linking the crisis type and the application of crisis response strategies (CRS). At a glance, SCCT suggests that an organisation's reputation is a valued resource that is threatened by crises [2]. Moreover, SCCT functions to predict the reputational threat presented by a crisis and to prescribe CRS designed to protect reputational assets [1]. One can predict reputational threat with two steps [1]: (1) by assessing the initial crisis responsibility attached to a crisis (that is the crisis type) and then (2) assessing the other two intensifying factors of reputational threat – crisis history and prior relationship reputation. These crisis types, increased by factors of crisis history and

* Corresponding author: monaratikah@unimas.my

prior relationship reputation, allow the organisation in crisis to indicate the crisis responsibility that is posed upon it and thus anticipate its stakeholders' perception and reaction towards the crisis and the organisation itself.

Responsibility requires accountability and therefore the organisation must answer for its action [3]. After a crisis strikes, it is crucial that the organisation in crisis responds to its stakeholders immediately. This is where crisis responsibility connects with CRS. CRS are what the organisation in crisis says and does after a crisis. They affect how stakeholders perceive the organisation as well as the crisis itself. Based on the stakeholders' perception, the organisation in crisis can determine which CRS is or are more suitable to protect its reputation. Hence, in order to repair the reputation, reduce negative affect, and prevent negative behavioural intentions [1], useful CRS are designed in SCCT relative to the crisis types. Past studies confirms that as CRS becomes more accommodative and show greater concern for victims, stakeholders perceive the organisation as taking greater responsibility for the crisis [4, 5]. Therefore it is logical that organisations should initiate in preparing and establishing a successful CRS depending on the crisis type [1] to minimise the damage and scope of crises that are or will be faced upon.

The online environment through social media, websites, and even emails is one of the vital platforms used by many organisations to provide fast response to audiences in ways that is most appropriate to minimize damage and restore organisational reputation [1, 6] Plus, research indicates that a crisis website is deemed the best practice of internet usage during a crisis because it is another means for the organisation in crisis present its side of the story [7]. Apart from websites, the benefits of social media should not be ignored. CRS that are implemented through social media in disseminating information about the crisis to its key stakeholders is strategically important due to its efficiency, timely, and feedback-oriented nature. Social media present opportunities for organisations to tell its own account of a crisis without being subjected to journalistic gatekeepers and one-sided view of the crises as practiced in the press media [8]. In addition, social media are mostly used in sending cautions, to conduct situational wakefulness, and even to initiate actions [9]. Its networking abilities and the promise of a user-generated content therefore have made social media like Facebook an appropriate tool for resolving crisis communication. Despite its benefits, crisis communication in social media however provides organisations with little control of the audience response towards the response messages disseminated. Nevertheless, this points to the greater need for the organisation in crisis to strategise its response so that the CRS tallies with the crisis type that is faced upon.

2 MH370: Malaysian Airline System (MAS) as an Organisation in Crisis

There is still shortage of theoretical studies on orgnisational crises in the context of Malaysian organisations, despite the few notable organisational crises that can be reviewed for the study of crisis type and CRS suitability. Among Malaysia's most high profiled ones is the tragedy of Malaysia Airlines flight MH370. The tragic disappearance of Flight MH370 from Kuala Lumpur, Malaysia to Beijing, China with a total of 239 passengers and crewmembers had implicated the organisation with the utmost reputational threat in the history of its establishment. Conversely, the MH370 crisis had affected MAS stakeholders at such large scale, which includes the Malaysian government itself.

Coming from that, it is understandable that MH370 is fitting to the study of crisis types and CRS, given MH370 crisis is also well known and well speculated amongst various audiences at an international setting. Furthermore, as to the researchers' knowledge, there are no known empirical studies pertaining to the application of SCCT in a real organisation, especially in Malaysia. Although MAS is currently known as Malaysian Airline Berhad

(MAB), the organisation in crisis is to be addressed as MAS rather than MAB, relevant to the timeline of the study particularly when MH370 crisis is concerned.

This study considers selecting the social media that is prominent in facilitating organisation's CRS. That is to determine the official social media page of the MAS. Only main pages directly established by the organisation will be used for the purpose of this study. As this study intends to examine how the organisation employs social media for CRS, pages established by stakeholders or third party users or unknown sources (eg. PrayForMH370) are not considered relevant for this study. During the crisis, MAS had responded to its stakeholders through various platforms from the start, Facebook namely one of them. On the first day of MH370, seven media statements were disseminated in MAS's official Facebook page. Due to the preliminary nature of this study, the scope is limited to the first day when the crisis occurred to eliminate the possibility of the crisis be influenced by other external factors such as news media reports, blogs, hearsay, and other opinions made by other parties. Thus, the scope of this study is strictly revolved within the contents of official media statements from MAS within 8th March 2014. The researchers justifies that the first day in which the crisis occurred is relevant in looking for the authenticity of how the organisation responds to the crisis to its stakeholders, as well as the organisation's initial control over information.

In that respect, the main aim of this study is to discuss the suitability between the crisis types and the CRS applied during a crisis as suggested in SCCT. This research analyses a total of seven media statements that were released on MAS official Facebook page on the day of the crisis, 8th March 2014 hence the population of the study. Using thematic content analysis [10] as a method of study by adapting Bales [11] in his analysis of conversation, this study refers that a single unit of analysis is considered as a single unit of attribute that will form a frame. The way these frames are formed help the researchers determine the type/s of crisis and CRS applied by MAS in its media statements. The researchers considers all forms of word/s and sentence/s as attributes to the framing that are consistent and relevant to the categories (the 12 crisis types and eight CRS) and the propositions of SCCT. The interpretation of the coding is based on the collective frequency count of each category (ie, for crisis types: victim cluster – rumours; for CRS: Primary – diminish – excuse) for all seven media statements.

This study is significant as it analyses an actual CRS that was disseminated by MAS to the public upon the mysterious disappearance of flight MH370 on the 8th March 2014. Moreover, this study somewhat brings light out of the mystery crisis by looking at which crisis cluster it falls into and as to whether or not SCCT potentially needs revisiting to suit the current crisis conditions in the age of new media.

3 Identifying Crisis Types and Determining Suitable CRS

SCCT suggests that the initial crisis assessment is based upon the crisis type. The crisis type is how the crisis is being framed. Frames include the way (words, phrases, images, etc) that information is presented in a message [12]. How stakeholders define problems, causes of problems, attributions of responsibility and solutions to problems are shaped by the way a message is framed [13]. The framing effect happens when the organisation chooses certain elements to focus on, which for the purpose of this study, are texts and attributes contained in the seven media statements. The stakeholders who receive the message will center their perception on those attributes while forming their opinions and making judgments towards the crisis [12]. Crisis types are a form of framing, in which each type highlights certain aspects of the crisis. These cues show how stakeholders ought to translate a crisis [2]. The organisation tries to develop or shape the crisis frame by focusing on certain cues or attributes to provide information as to what the causes of the crisis –

DOI: 10.1051/shsconf/20173300037

whether the crisis is an accident, a force of nature, or intentional. It does make a difference if stakeholders see the crisis as a mishap, sabotage or criminal negligence because it determines the responsibility attributed to the organisation associated to the crisis by its stakeholders. There are 12 crisis types [2] identifed in which three crisis clusters are formulated based upon attributions of crisis responsibility [1] as shown and explained in Table 1.

Table 1. SCCT Crisis Types by Crisis Clusters (adopted from [1])

Victim cluster: In these crisis types, the organisation is also a victim of the crisis. (Weak attributions of crisis responsibility = Mild reputational threat)	
Types:	*Natural disaster*, Rumor, Workplace violence, Product Tampering/Malevolence
Accidental cluster: In these crisis types, the organisational actions leading to the crisis were unintentional. (Minimal attributions of crisis responsibility = Moderate reputational threat)	
Types:	*Challenges, Technical-error accidents, Technical-error product harm*
Preventable cluster: In these crisis types, the organisation knowingly placed people at risk, took inappropriate actions or violated a law/regulation. (Strong attributions of crisis responsibility = Severe reputational threat)	
Types:	*Human-error accidents, Human-error product harm, Organisational misdeed with no injuries, Organisational misdeed management misconduct, Organisational misdeed with injuries*

In this study, the crisis type is determined by analysing the attributes (word/s and sentence/s) that are relevant with the crisis types description, under the umbrella of the above three clusters.

Once the type of crisis or cluster is determined, the organisation in crisis can predict how much attribution will be placed on the organisation and how much reputational threat it is facing, and therefore, SCCT provides the organisation with a theoretical base to their CRS. Similar to the case of MH370, SCCT's list for responding to crises assumes that the organisation has accepted some level of responsibility for the crisis. The first type of CRS (primary) consists of three groups (denial, diminish, and rebuild) based upon perceptions of accepting responsibility for a crisis [1]. The secondary type of CRS (bolstering) functions as a supplemental strategy in too boost the organisation's existing reputation. The list is shown and defined in Table 2.

Table 2. SCCT Crisis Response Strategies (adopted from [1])

Primary CRS		
	Deny CRS	*Attack the accuser, Denial, Scapegoat*
	Diminish CRS	*Excuse, Justification*
	Rebuild CRS	*Compensation, Apology*
Secondary CRS		
	Bolstering CRS	*Reminder, Ingratiation, Victimage*

There are three objectives of CRS [1]: (1) to shape attributions of the crisis, (2) to change perceptions of the organisation in crisis, and (3) to reduce the negative effect generated by the crisis. In order to achieve such, the CRS must be at par with the attributed responsibility for the crisis – the crisis type or cluster. Each of the CRS assists the organisation in getting its side of the story into the media. Ultimately, the media's frames become the stakeholders' frames. For the purpose of this study, the media is MAS official Facebook page. A general guideline of suitable CRS is suggested [1] in order to suit each crisis cluster. The suitability can be adopted according to Table 3.

Table 3. SSCT Crisis Response Strategy Guidelines (adopted from [1])

1.	Informing and adjusting information alone can be enough when crises have minimal attributions of crisis responsibility (victim crises), no history of similar crises and a neutral or positive prior relationship reputation.
2.	Victimage can be used as part of the response for workplace violence, product tampering, natural disasters and rumors.
3.	Diminish crisis response strategies should be used for crises with minimal attributions of crisis responsibility (victim crises) coupled with a history of similar crises and/or negative prior relationship reputation.
4.	Diminish crisis response strategies should be used for crises with low attributions of crisis responsibility (accident crises), which have no history of similar crises, and a neutral or positive prior relationship reputation.
5.	Rebuild crisis response strategies should be used for crises with low attributions of crisis responsibility (accident crises), coupled with a history of similar crises and/or negative prior relationship reputation.
6.	Rebuild crisis response strategies should be used for crises with strong attributions of crisis responsibility (preventable crises) regardless of crisis history or prior relationship reputation.
7.	The deny posture crisis response strategies should be used for rumor and challenge crises, when possible.
8.	Maintain consistency in crisis response strategies. Mixing deny crisis response strategies with either the diminish or rebuild strategies will erode the effectiveness of the overall response.

4 Results, Analysis and Discussion

4.1 Crisis Clusters and Types Experienced

Based on timeline of this study, there were no mentions in the media statements as to the actual cause of the crisis. Therefore, the researchers interpret the contents of the media statements by analysing the framing cues that can help the researchers distinguish the type of crisis that MH370 fits into. Based on the result, MH370 interestingly belongs to all three crisis clusters (i.e. victim, accidental, and preventable).

The lowest count for type of crisis involving MH370 is in the victim cluster, specifically the Rumour type by two counts (i.e *There has been speculation that the aircraft has landed at Nanming).* Therefore, MAS did not emphasise that it is in a highly victimised crisis during the first day of the crisis. MH370 is also analysed as an accidental crisis. Herein, MH370 is framed as a Technical-error accident (i.e *Malaysia Airlines confirms that flight MH370 has lost contact with Subang Air Traffic Control*) due to the phrase "lost contact". This technical-error could consequently indicate what MH370 should have been if the flight has not lost contact with the air traffic control (i.e *MH370 was expected to land in Beijing at 6.30am the same day*). Accidental crisis also highlights the organisation's concerns on human lives, hence usually involves emergency plans (i.e *We are working with authorities who have activated their Search and Rescue team to locate the aircraft*). Under the same accidental cluster, there is only one count for Challenges type indicating that MAS has received stakeholders' claims about the lack of information given by MAS (i.e *We understand/ everyone's concern/ on MH370 pax & crew*).

As a preventable crisis, the statements repeatedly mention that the organisation itself is causing the incident (i.e *...that we have lost all contacts with flight MH370*) by taking note of the word, "we". However, by placing the word "we" (Human-error accident) rather than "the flight has lost contact with Subang Air Traffic Control" (technical-error accident) as mentioned in the accidental cluster has made the framing inconsistent in explaining the cause and nature of the crisis. The most prominent frame in determining the crisis type of MH370 is the preventable crisis cluster by 44 counts, in which Organisational Misdeed without Injuries and Organisational Misdeed with Injuries analysed concurrently; therefore possess equal number of counts. The researchers interpret that the word/s or sentence/s that mentions the passengers and crews affected in the flight (i.e *The flight was carrying a total number of 239 passengers and crew; 1. China – 152 plus 1 infant; 2. Malaysia - 38*) in which stakeholders are put at risk either physically and/or psychologically by the organisation. It is challenging to determine as to whether or not there were any physical injuries involved within that timeframe, despite the organisation's repeated mentions of its regret towards the stakeholders (i.e *Our thoughts/ and prayers/ are with all affected passengers/ and crew/ and their family members*). On the other hand, the statements highlighting the effort in finding MH370 (i.e *We are working with authorities/ who have activated their Search and Rescue team/ to locate the aircraft*), thus indicate that the organisation has put these stakeholders at risk in terms of psychological and emotional uncertainty. Nevertheless, in the aspect of Organisational Misdeed without Injuries, it can be interpreted as such due to the lack of information or evidence that indicates that there were injuries occurring within that timeframe, hence a pretense to stakeholders in the organisation not being able to provide the actual cause of the crisis in the statement. In other words, the framing of the statements were overlapping and vague in describing the crisis type.

Therefore, the researchers find a loophole in the crisis type category of SCCT. Based on the crisis type framed in the statements, there is no accurate indication as to the physical human injuries of the stakeholders on board in MH370, although the organisation has listed down the detailed information of the affected passengers and crew. Because there is no accurate description in describing a crisis that is 'uncertain' in terms of actual physical human injuries, perhaps another crisis type that describes events that are as such can be added to the theory.

4.2 CRS Applied and Suitability with Crisis Types

The lowest count of CRS applied by MAS in regards to MH370 is the deny strategy, specifically Scapegoat strategy (i.e *There has been speculation that the aircraft has landed at Nanming; We are working to verify/ the authenticity of the report and other.*). Deny strategies seek to establish a crisis frame by attempting to remove any linkage between the organisation and the crisis [1]. This is to point that if MAS is not involved in a crisis, it will not suffer any damage from the event. Therefore, based on the Rumour and Challenge crisis types that are framed in the statement, this CRS applied is deemed suitable.

Result show that MAS utilises a large number of primary CRS, specifically the diminish strategy (Justification) for 55 times. Diminish crisis responses attempt to assure MAS stakeholders that MH370 crisis is not as bad as it seems or that MAS is not responsible for the crisis occurrence [1]. In the aspect of Excuse response strategy, MAS hints its inability to control the event by using the phrase "lost contact" (i.e *Malaysia Airlines confirms/ that flight MH370 had lost contact/ with Subang Air Traffic Control at 2.40am, today*). Furthermore, these strategies help lessen the connection between the MAS and the crisis and help MAS stakeholders see the crisis in a less negative light. In the media statements, MAS generally uses Justification to show its credibility and minimise negative impressions of the organisation. Credible evidence have been detected in the Justification strategy (i.e *Flight MH370 was operated on a Boeing 777-200 aircraft*; *The flight was piloted by Captain Zaharie Ahmad Shah,/ a Malaysian aged 53./ He has a total flying hours of 18,365hours./ He joined Malaysia Airlines in 1981*). On the grounds that MH370 being an accidental crisis and that MAS has never had such a crisis before, the diminish strategy applied is deemed suitable, as explained in Table 3, item no. 4.

Rebuild strategies help alter perceptions of an organisation during a crisis [1] by offering real or symbolic forms of aid, that is Compensation (i.e *The airline will provide regular updates on the situation*) to victims and asking their forgiveness, that is a form of Apology (i.e *We deeply regret/ that we have lost all contacts with flight MH370).* These strategies attempt to take the focus off the crisis by taking positive action. Rebuild crisis response strategies should be used for crises with strong attributions of crisis responsibility (preventable crises) [1] regardless of crisis history or prior relationship reputation. In this study, although suitable in terms of technicality, the low count for rebuild strategies does not suit the high counts of preventable crisis type framing.

A secondary type of response, Bolstering is also used by only three counts in the media statements analysed, namely victimage in which MAS emphasises on triggering phrases that can make stakeholders sympathise with MAS by showing empathy towards the stakeholders at the same time (i.e *We understand/ everyone's concern/ on MH370 pax & crew./ We're accelerating/ every effort/ with all relevant authorities/ to locate the aircraft*). True enough, this strategy does apply suitably with the Rumour and Challenge crisis types that MH370 is framed as mentioned earlier, and that it does supplement the primary responses in the media statements.

In the aspect of overall suitability of CRS according to the crisis type, findings indicate that the highest count for CRS applied (diminish strategies) is not suitable with the highest count of crisis type that is framed in the media statements (preventable cluster). If the CRS weighs more on primary CRS, particularly on diminish response therefore MH370 should be associated more with victim or accidental cluster [1]. The media statements have framed MH370 as a preventable cluster, and therefore should apply higher count for rebuild strategies, which comprises only a total of 15 counts compared to 58 counts for diminish strategies.

5 Conclusion and Directions

At a glance, result shows that MH370 is a crisis that comprises of all three crisis clusters, mainly preventable in nature. Based on the frames portrayed in the seven statements, the researchers identify attributes that make up the frame of all three crisis clusters. Nevertheless, analysis of the crisis types denotes that there is a loophole in describing a crisis that is 'uncertain' in terms of the cause of crisis. Given that it is a preliminary study in which the statements are content analysed according to textual lines (word/s and sentence/s) as separate frames, the cause of the crisis is yet to be determined as a major frame. As far as the researches' observation based on the findings, one can assume that as a major frame, the type of crisis for MH370 is considered 'uncertain'. This is due to the repetition of the phrase 'lost contact' as the main indicator in determining the type of crisis. SCCT has yet able to cover other possibilities of crisis types. In this case, the researchers propose two directions for future studies: 1) analysis of major message framing in MH370 CRS to determine its ultimate type of crisis based on an extended timeline of media statements; and 2) expand and examine another crisis type that is similar to MH370.

Also, this study finds that MAS utilises the diminish strategy most prominently as the CRS, mainly in terms of justification. As far as the suitability is concerned, the researchers see it as a complex relationship, in which the guidelines overlaps and lack in describing the suitability of each crisis type concerning the case of MH370. Results show that when MH370 is analysed based on its respective crisis type as a whole, every CRS applied is deemed suitable. However, when the crisis is analysed based on the count of attributes, the CRS applied is found to be not suitable. For instance, although MH370 was framed more as a preventable crisis, the CRS applied however is more inclined towards accidental or victim crisis.

 DOI: 10.1051/shsconf/20173300037

As far as the researchers' knowledge, past research regarding SCCT does not study the timeframe aspect as part of their consideration in applying the right CRS to the stakeholders. Taking into account the timeframe aspect will aid a depth to the current guidelines of CRS. Apart from the timeline, the messages and attributes framed in CRS could also further explore the overall tone and effectiveness of how the crisis is represented by the organisation. Since SCCT is rooted from Attribution Theory [1] that suggests stakeholders tend to search for the causes of events that are negative and unexpected [14, 15, 3], stakeholders are also likely to react emotionally towards the event. Coming from that, this study can be expanded by analysing the message frames of the statements by looking at the statements' rational or emotional components (i.e information vs compassion). This can consequently bring light to the effects of CRS towards stakeholders' rational or emotional reactions. It is also proposed that the effects of CRS are studied by looking at the stakeholders' perspectives. This can be done by analysing comments by stakeholders in each of the statements disseminated by MAS, perhaps by examining the CRS effectiveness through the three objectives of CRS [1]: (1) shaping attributions to the crisis, (2) changing perceptions of the organisation in crisis, and (3) reducing the negative effect generated by the crisis, as indicators of effectiveness.

References

1. W. T. Coombs, CRR, **10**, 163-176 (2007).
2. W. T. Coombs, S. J. Holladay, MCQ, **16**, 165-186 (2006).
3. B. Weiner, *Social Motivation, Justice, and the Moral Emotions: A Attributional Approach,* (2006).
4. W. T. Coombs, S. J. Holladay, *Responding to a Crisis: A Rhetorical Approach to Crisis Communication*, (2004).
5. W. T. Coombs, S. J. Holladay, *Research on Emotion in Organizations: Volume 1: The Effect of Affect in Organizational Settings,* (2005).
6. A. Pang, APME, **23**, 2 (2014).
7. M. Taylor, M. L. Kent, PRR, **33**, 140-146 (2007).
8. B. F. Liu, Y. Jin, L. L Austin, JPRR, **25**, 51–67 (2013).
9. S. Mukherjee, IJMIBS, **4**, 2 (2014).
10. K. Krippendorff, *Content Analysis: An Introduction to Its Methodology*, (1980).
11. R. F. Bales. *Content Analysis Reader* (2009).
12. J. N. Druckman, PB, **23**, 3 (2001).
13. A. H. Cooper, EJPR, **41**, 37-80 (2002).
14. B. Weiner, PR, **92**, 548-573, (1985).
15. B. Weiner, *An Attributional Theory of Motivation and Emotion,* (1986).

Perceived Influence of Opposition Political Campaign Materials on Voters

Syed Arabi Idid [1] , and *Rizwanah* Souket[1,*]
[1]Department of Communication, International Islamic University, Malaysia

Abstract. This article focuses on how political attitudes of voters are shaped and affected during election campaigns. The study is constructed on the third person effect paradigm that posits that negative messages will have a greater impact not on 'me' or 'you' but on 'them'. Concurrently, perceived positive messages will have a greater influence on themselves compared to others, often referred to as the first-person effect or reverse-third person effect (Duck, Terry & Hogg, 1995; Perloff 1999). The present study was undertaken to understand how political communication messages can have a positive effect on own party supporters while the same message can be perceived to be biased and partisan to opposing party supporters. The current study is grounded on the research done by Idid & Souket (2014) that investigated the effects of Malaysia's largest political party, Barisan Nasional (BN) political communication literature on two Malaysian voter groups (BN voters and opposition voters), one of which regarded the message as partisan and the other as congenial. The present study attempts to investigate the effects of each of the opposition party political communication literature on two Malaysian voter groups (BN party supporters and the particular opposition party supporter).

1 Introduction

The Third Person Effect theory associated with Phillips Davison [1] postulated that audience members would deny media's direct effect on themselves but would perceive media's effect on others [1,2 & 3]. Several studies have focussed on voter perception and behaviour using the third person paradigm [4, 5 & 6]. It can be said that the classic third person effects are more pronounced when media or message content is perceived to be negative on the respondents. On the other hand, the third person effect can be reversed, called the first-person effect, when the desired message is intended to be positive on the respondents [7 & 8]. The third-person effect theory simply states that a message, perceived to be not welcome, can exert a greater influence on others than on the self. Thus, the message will have the least impact on 'me' (first person), you (members closest to the first person) and the greatest impact on them (others, the third person). Given the premise, the present study attempts to investigate the third person effect ideology among the Malaysian voters in the country's political settings.

[*] Corresponding author: rizwanahsouket@gmail.com

The Barisan Nasional (BN), the country's dominant party, has captured two-third majority seats in parliament since Malaysia got its independence from the British in 1957. However during the 2008 general election, BN lost its two-thirds majority parliamentary seats by winning140 seats while the opposition (Pakatan Rakyat) won 82 seats. Likewise, in the 2013 general elections, BN again lost its two third majority parliamentary seats to the opposition group. This time BN won 133 seats while the opposition party (Pakatan Rakyat) won 89 seats. This raises a concern over the uprising of a more well-informed Malaysian voter group. With the highest voter turnout in Malaysia's history (84.4%), all parties used their own strategies and political campaign materials (brochures, pamphlets, publications, leaflets and speeches) aimed at influencing the voters.

Based on the 2013 elections study, Idid and Souket [9] had found the BN respondents to have a 'first' person effect, indicating that the BN political communication literature had a greater positive impact on themselves (first person i.e. BN voters), their family and other BN supporters and a least or no positive influence on "them" (the opposition supporters and Undecided voters). The present study is keen to investigate further the influence of each of the separate opposition political communication literature on their respective opposition party voter (first person i.e. PAS, DAP or PKR voters), their family and other opposition party supporters and on "them" (the BN supporters and Undecided voters). By studying the campaign materials and their perceived influence, this is a reverse situation from the 2013 study.

The opposition coalition called Pakatan Rakyat consists of three parties: Parti Keadilan Rakyat (The People's Justice Party, PKR), the Democratic Action Party (DAP) and Islam Se-Malaysia Party (Pan Malaysian Islamic Party, PAS). The Parti Keadilan Nasional fused with the Malaysian People's Party in 2003 and called themselves People's Justice Party (Parti Keadilan Rakyat, PKR) advocating social-liberal values. The DAP was founded in 1965 and advocates for secular values. Its voters mostly consist of urban ethnic Chinese. The PAS an Islamic party established in 1955 advocates Islamic values and laws. Based on the third person effect hypothesis, this study attempts to investigate the views of the respective opposition party supporters on the influence of their own party campaign materials on themselves, their family and their own party friends. Likewise the study attempts to investigate the influence of each of the opposition party political literature on the BN supporters, their family and their friends.

Idid & Souket [9] found BN voters to have been positively influenced by the BN political campaign literature and they perceived a lesser level of influence on the opposition voters. They claim that the BN party voters are persuaded and convinced by the message projected by the BN political communication devices. As a result, the first person or BN respondents feel confident and comfortable when exposed to their own party communication devices and they feel that the other voters should also share a similar sentiment. In brief, the BN supporters view the BN political communication materials as a desirable message. On the other hand, when voters who had voted for the opposition parties were asked, they (Pakatan Rakyat) viewed the BN political communication message as undesirable and ultimately displayed a classic third person effect when investigating the influence of the BN campaign political literature on themselves, their family, and like party voters. They viewed the message to have the least effect on their own selves. Based on this finding, the current study hypothesizes the individual opposition political communication literature to have a greater level of positive influence on the respective opposition party voters and a lesser level of influence on the BN voters. Thus the respective opposition party voters (PKR, PAS, DAP) will view the message projected by their individual party as desirable and the BN voters would claim an undesirable effect on themselves.

2 The current study

The 2013th Malaysian elections was held on 5th May 2013. The study was conducted in July 2013 and sought the opinion of the registered voters drawn nation-wide on several matters concerning their behaviour during the elections. About 1463 voters were selected nationwide using quota sampling from two parliamentary constituencies in each state. A total of 100 trained enumerators and 12 supervisors were involved in the data collection. A 45 minute face-to-face interviews were conducted on the respondents. The respondents were asked whether the components of Pakatan Raykat (PAS, DAP, and PKR) political campaign materials such as brochures, pamphlets, publications, and leaflets and speeches had an influence on themselves, their families, BN supporters, opposition supporters, and uncertain voters. The 'Self' category included the respondent's perception of the political party communication devices influence on themselves, their family and other similar party supporters. While the 'others' category included the influences of the political communication devices on other party supporters and uncertain voters.

3 Findings of the study

The total number of 1463 respondents included 695 BN supporters and 768 opposition party supporters. Out of the total opposition supporters, about 22% of the respondents were PAS supporters (N= 169), about 48% were DAP supporters (N=365) and about 30% of the respondents were PKR supporters (N= 234). The study asked supporters to state the influence of opposition party materials on own supporters which included themselves, their family, and like party supporters. They were also to state what they perceived to be the influence on voters who were not their party supporters which could be the main party voters. Likewise, BN supporters were asked to state what they perceived to be the effect of the opposition party campaign materials on themselves, their family and their own supporters. They also estimated the influence of the materials on others than BN supporters which would likely be the opposition supporters.

The study first tested the influence of PAS party campaign materials on the BN voter and on the PAS voter (table 1). It was found that the BN voters perceived that the PAS campaign materials had a moderately positive level of influence on themselves (47%), on their family (48%) and other BN supporters (68%). Likewise, the BN voter perceived a positive influence of the same communication materials on the other party supporters (opposition supporters, 49% & uncertain voters, 48%). Hence a total of 54 percent of BN self-voter claimed to be influenced by the PAS communication materials on their self, family and like party supporters (other BN voters). To the BN supporters, the perceived influence of the PAS party materials on "the others" (besides the BN voter) was 48%.

When PAS supporters were asked about how they perceived the influence of their own party materials, they admitted a greater influence on self (97%). A total of 98 percent said the materials had an influence on their own selves, 97 percent said it had an influence on their family, and 96 percent claimed they had an influence on the party supporters. The influence of the PAS party materials on the others (beside the PAS voters like BN voters & unlikely voters) was 68%. A greater level of influence on self indicates a greater probability of a "reverse" third person effect or the 'first person' effect (table 1).

Table 1 suggests that BN supporters acknowledged the influence of PAS political materials on themselves and also on others. In a classic third person effect, the BN supporters would claim that others would be more affected by the undesirable PAS campaign materials.

DOI: 10.1051/shsconf/20173300038

Table 1. Perceived influence of PAS Political Communication Devices on SELF and OTHERS

Respondents (n=1463)	**Perceived influence (%)**			**Total on Self**	**Total On Other**
	On self	**On family**	**On like party supporters**		
BN supporters (n= 695)	47	48	68	54	48
PAS supporters (n= 169)	98	97	96	97	68

The study looked at the influence of the DAP political campaign materials on the BN voters and the DAP voters. The BN supporters indicated a slightly higher influence of the DAP political communication devices on themselves (50%) as compared to others (46%) (table 2). The BN voters indicated a comparatively low level of influence on self (41%) and on family (42%), however the voter seems to indicate the possibility of a greater level of influence on other BN supporters (68%).The influence of the DAP political campaign materials on the DAP voter themselves indicated a greater influence on self (95%) that comprised of self (96%), family (94%) and like party supporters (95%). The influence predicted to be on others indicated a positive influence of 88%, thus indicating a greater influence on the self of the voter (95%) as compared to the perceived others (88%). This finding confirms the 'first person' effect. The summary of the findings is presented in table 2.

Table 2: Perceived influence of DAP Political Communication Devices on SELF and OTHERS

Respondents (n=1463)	**Perceived influence (%)**			**Total on Self**	**Total on Other**
	On Self	**On family**	**On like party supporters**		
BN supporters (n= 695)	41	42	68	50	46
DAP supporters(n= 365)	96	94	95	95	88

The study then investigated the influence of PKR political campaign materials on PKR and on BN supporters. This study found that the BN supporters indicated a moderately high influence of the PKR political communication devices on themselves (52%) as compared to others (45%) (table 3). Likewise, the BN voters indicated more confidence in the perceived influence of the PKR campaign materials on the other BN voters (67%) than on themselves (44%) or their close family (44%). Congruently, the PKR voter indicated the greatest influence of the PKR materials on themselves (93%) as compared to the others (86%) reflecting the 'first person' effect.

Table 3: Perceived influence of PKR Political Communication Devices on SELF and OTHERS

Respondents (n=1463)	**Perceived influence (%)**			**Total on Self**	**Total on Other**
	On self	**On family**	**On like party supporters**		
BN supporters (n= 695)	44	44	67	52	45
PKR supporters (n=234)	94	92	93	93	86

4 Discussions

The current study focuses on the influence of each of the individual opposition party campaign materials on Malaysian voters. The study looks into testing the vulnerability of the Malaysian voters towards the political campaign messages projected by different political parties. This study develops from the third person premise that when a message has a low likelihood or perceived negative influence (undesirable) then the third person effect is more obvious. Based on this ideology, the opposition party political message is projected to be looked upon as an undesirable message for the BN voters and supporters, thus assuming a classic third-person effect. The opposition party political messages are viewed as desirable messages on their own opposition parties and are seen to be positive thus resulting in a more first person effect or reverse-third person effect. People will acknowledge the effects of communications that are usually regarded as socially desirable, healthy or good for self [8] to be positive. Reflecting on the current study, voters receiving messages from their own parties are perceived to have a higher likability of acknowledging the influence of desirable messages and at the same time they hold the belief that others are also strongly liable to be influenced by the same messages. The findings of this study yielded an interesting result. When each of the opposition political campaign messages were tested on their own respective supporters, it was found that the respective opposition party voters displayed a high level of confidence in their own political communication message and indicated a reverse-third person effect. The findings show that the opposition voters perceive to have a higher likability and confidence in acknowledging the influence of their own party messages (desirable) and at the same time they hold the belief that the BN voters are also strongly liable to be influenced by the same messages. Hence positive messages are perceived to be well received by themselves and by others.

The influence of the PAS campaign materials on the PAS voters (98%), on their family (97%) and the perceived influence on other like opposition supporters (96%) was comparatively higher compared to the perceived influence of the PAS campaign materials on BN supporters and other unlikely voters (68%). This clearly indicated that although the message to them was positive, yet they claim the influence on others was not as high. Similar to an earlier finding [9], the BN respondents displayed an extraordinary level of confidence in meeting the messages from the opposition party. The BN voters claimed a greater influence of the PAS campaign messages with 54% claiming the opposition party materials to have an effect on themselves but saying only 48 percent of non BN party supporters would be influenced by them. This could be associated with the self-enhancement postulate, whereby the BN voters considered the effects of the PAS messages to be positive on themselves and are liable to be influenced but it was not so with the other non BN supporters. On the other hand, the influence of the DAP campaign materials on the DAP voters indicated a high level of confidence on themselves (96%), their family (94%) and on other similar opposition supporters (95%) and a lesser level of influence on their perceived other voters and unlikely voters (88%).This finding could be attributed to the social distance between the self and distant others or the level of message desirability and subsequently the voters feel "I am smart enough to recognize the value of the message" [10]. The influence of the DAP campaign materials on the BN voter indicated again a contrary result whereby the BN voters indicated a greater influence on self (50%) compared to the perceived influence on others (46%). Finally on looking into the perceived influence of the PKR campaign materials on the PKR voter and BN voter it was found that the PKR voters displayed a strong influence of the PKR campaign materials on self (94%), their family (92%) and the other similar opposition party supporters (93%). The perceived influence of the PKR campaign materials on others and unlikely voters was again a high influence of 86%. This again confirms the confidence in the influence of the PKR campaign materials. The BN voter again displayed more confidence of being influenced by the PKR campaign materials on themselves (52%) but claimed it would not be a factor on non BN supporters (45%). Overall, the study has shown that people will likely accept congenial message as positive on themselves perceiving it to be in tune with their own cognitive needs [9]. Therefore, the opposition voters would claim that their campaign materials would not only have an effect on their own self, on those closest to them (family) but would also affect the BN party supporters. The earlier study [9] found that the BN supporters displayed a positive confidence on the influence of BN materials on self and on others. This study added that the BN supporters also found the opposition materials to have a reasonable influence on themselves more than the influence on others, denying therefore the third person effect. The opposition party materials were not seen to be unreasonable or negative to the BN supporters. Though they did not perceive the influence to be that big yet BN supporters perceived the opposition party materials to have lesser effects on non BN supporters.

The main objective of the study was to explore the effect of messages from the third-person effect perspective, but the findings have suggested a variation. The BN supporters have accepted the influence of their opponents' messages but they have rejected the influence on non BN supporters. BN supporters therefore had claimed the positive messages from the opposition parties but have rejected it to be of influence on other voters. It was also similar to the acceptance on the influence of the opposition messages on themselves and on other voters but acknowledging that they had lesser influence on other voters. In conclusion, we find each set of voters to display a positive confidence in the influence of their respective party communication messages on themselves, family and like supporters. Acceptance of a positive message does not mean acknowledging a similar level of acceptance on others. On the other hand while a negative message is accepted it does not mean that a party cannot acknowledge that the message has little influence on others.

References

1. P. Davison, Public. Opin. Quart. **47**, 1 (1983)
2. B. Salwen, Commun. Res. **25**, 259-285 (1998)
3. J. Brubaker, Am. Commun. J. **13**, 2 (2011)
4. S. Idid, Arabi. Wok. S, Analysing political campaign effects: a test of the third person effects theory. Paper prepared for ICOMC'10, 18-20 (2010)
5. A. Banning, Journal. Mass Comm. Q. **83**, 4 (2006).
6. F. Rojas, D. Shah, R. Faber, Int. J. Public. Opin. R. **8**, 2 (1996).
7. J. Duck, D.J, Terry, M.A. Hogg, Eur. J. Soc. Psychol. **25**, 2 (1995).
8. R, Perloff, Media. Psychol. **1**, 353-378 (1999).
9. S. Idid, Arabi R. Souket, AJPOR. **1**, 4 (2014)
10. J. Huh, D. DeLorme L. Reid, Journal. Mass Comm. Q. **81**. 4 (2004).

DOI: 10.1051/shsconf/20173300039

Discourse of transformation in organizational change management

Christiantine Della[1,*]

[1]Infrastructure University Kuala Lumpur, Malaysia

Abstract. This study examines the discursive construction of ideological change and identity within the practice of organisational control in organisational change management. The focus of the study was to examine how the organisation through its large-scale reengineering process to implement organisational change initiatives appropriated discourse of transformation to effect change among its organisational members. The organisation's focus is to change mindsets and persuade members to embrace characteristics, traits, attitudes and behaviour that are deemed to be beneficial to the organisation. Discourse of transformation is used as an object of discursive construction of reality in the construction of an 'ideal' member identity and ideological change. The theoretical framework for the study is informed by theories of identity and ideology in discourse, theories of power and language as articulated in the field of critical discourse analysis. The data consist of transcripts of 'Sharing Sessions' which were transcribed verbatim. The analytical framework for the textual analysis of identity and ideology is developed on a basis of a combination of concepts and methods namely, [1] analysis, intertextual analysis, Antaki and Widdicombe's principles for analysing identity in talk and [2] modes of identity regulation.

1 Introduction

This study seeks to investigate how discourse is used by management in the construction of identity and shaping of ideology as a focus for examining organisational control in an institution of higher learning. It does so via a case study of change in an institution of higher learning. It focuses on the various discursive means the organisation employs to exert influence over its members. It investigates the discursive processes by which organisational members are enjoined or influenced to define themselves, to think and to act or to adopt work orientations that are deemed congruent with managerially defined objectives and to make sense of the processes that constitute organisational change. The discursive strategies introduced at the institution during the time of the study appeal to the

*Corresponding author: chris@iukl.edu.my

DOI: 10.1051/shsconf/20173300039

notions of increasing efficiency within the organisation to enable it to be a profitable, sustainable and innovative organisation. This practice of managerial action to unobtrusively develop appropriate identities of organisational members is known as identity regulation by [2]. It is a form of ideological control or social engineering and is a preferred way to mould members as it is subtle and unobtrusive.

1.1 Organisational control

Organisational control is part and parcel of management. Management in seeking to address issues such as productivity, efficiency and survival needs to have control over its employees. Managerial control has not vanished in the postmodern organisation. It is only the means and methods which are employed to manage control processes which have changed. " ... employees develop multiple identities and roles at work, employing multiple sense making perspectives, and occupying positions of authority and subordination simultaneously" (p. 89). Employees do more multitasking and thus assume different roles and identities [3].

Deetz [4] expounds that managerial approach to organisational control is about "managing the 'insides' – the hopes, fears and aspirations of workers, rather than their behaviours directly" (p. 87). The aim is to win over the 'hearts and minds' of employees and direct their roles and activities towards the fulfilment or accomplishment of organisational goals. Definitions of reality as espoused by the management are thus established in an organisation and organisational members are exhorted to adopt this reality.

Flamholtz define organisational control as the process of influencing the behaviour of people as members of a formal organisation [7]. Similarly, [8,9,10] have interpreted control in terms of the influence on the subordinates to seek their compliance with organisational goals . Thus, organisations engage in various discourses to impart the 'reality' or ideology that they want to promote so that organisational members can be allied with it.

This study seeks to examine how control among organisational members is achieved through the discursive construction of ideology and identity in organisational change management initiatives. It specifically looks at the use of imperatives and allusions to pain as a strategy.

2 Discourse of imperatives

Imperatives refer to the use of verbs to instruct or direct someone in order for something to be done. Within the discourse of transformation is evident the discourse of imperatives. It is evident in most of the texts of the 'Sharing Sessions'.

2.1 Directives

Imperatives are used as directives to direct or instruct people to do something. Modals can also be used for the same purpose. Lines 95-140 (Excerpt 1) of Text 1 are extracted to illustrate the use of imperatives and modals. Due to the length of this excerpt, imperatives and modals are extracted and listed in Table 1.1. A directive is a command. By listing directives, the speaker in Text 1 tries to control the behaviour of the marketing personnel as well as the other members of the institution who attended this 'Sharing Session'. It is a top-

DOI: 10.1051/shsconf/20173300039

down directive to influence members about what they are supposed to do and what not to do. Members are in a subject position as they have to be told what to do and what not to do. Power is used to create and privilege beliefs that serve the interests of the marketing division in the institution.

Table 1.1.1 Excerpt 1: Text 1: line 95-140 List of directives

line		Imperatives	Modals
95-96	on how to increase our success rate for our walk in customers (.) I think we need to try our very best to satisfy our customers' needs		we need to try our very best to satisfy our customers' needs
96-98	when it comes to waiting time we need to reduce the waiting time (..) yah (.) I mean to provide good service put customer first	put customer first	we need to reduce the waiting time
98	try not to ask customers to come back and ah	try not to ask	
99	(..) they must go back with something at least (.)		must
100	I think and it's a clear cut case, then we should issue it (.)		should
101	so don't let them go without anything (..)	don't let them go	
102	try our very best to tell them about credit exemptions (…)	try our very best	
106-108	I think we need to understand today there are so many institutions around in Malaysia (.) so I think we have no choice but to treat the customer as special (.)	treat the customer as special	we need to understand
108-110	(.) the system must be friendly and staff MUST be friendly enough err (…) when it comes to err (.) err (…) when it comes to customer service (.)		must/MUST
110	(.) information provided must be accurate (.)		must
111-113	if you are not too sure about the information you're providing, always check with the relevant person in charge	check with the relevant person in charge	
122-124	when a phone call is coming in I think we need to observe our phone manners. (…) phone manners must be exercised here to ensure that they can create a good impression to the public.		we need to / must
124-125	*always* take time with customers' interests (...)	take time with customers' interests	
125-126	we must be able to provide general information and refer to relevant officer-in-charge		must
126-129	(...) try not to put the customer on hold because sometimes I also notice that we put customers on hold without any reason (...) when we put customers on hold we can be forgetful that the customer is there waiting for our answer	try not to	
129	also *never* ask the customer to call back	never ask the customer to call back	
131	don't ask customers to call back (...)	don't ask	
134-135	I think that we need to increase our turnaround time (..)		we need to

DOI: 10.1051/shsconf/20173300039

The excerpt serves to show the use of imperatives and modals as commands. There are numerous other examples found in this text.

Imperatives are used to command or instruct members on how to deal correctly with customers (students and parents). They are used as directives to direct correct behaviour or action. Firstly, there is reference to walk in customers and their needs, then there is reference to call-ins by customers requesting for information. The speaker directs the attention of the members on how to deal with customers properly. The use of imperatives is to direct members. Even though it seems commonsense, the speaker nevertheless lists the directives. Members are thus enjoined to appropriate this discourse of imperatives as something they must adhere to or follow.

The use of modals in lines 95, 97, 99, 100, 108, 110, 115 and 134 also emphasise the importance of treating the customer right. The speaker says in line 107-108; "I think we have no choice but to treat the customer as special (.) customer is changing". The obligation to treat the customer right is due to stiff competition among private institutions of higher learning in Malaysia. The modal "need to" occurs frequently in the excerpt to stress an impassioned plea for members to make the effort to understand their role in marketing. Marketing is the responsibility of everyone, including the academics. It is not entirely dependent on the marketing department as seen in the discourse of marketisation.

The use of "*always*" and "*never*" (italicised in the excerpt) in "*always* take time" and "*never* ask the customer to call back" serve to emphasise that it must be done at all times (one hundred percent of the time) without exception. They are strong commands and the speaker speaks with authority. It therefore does not allow for refute or rebuttal.

The use of directives is also seen in Text 2. The speaker in Text 2 (Excerpt 2) uses the modal "must" in "YOU MUST CHANGE (.) you must change with the times (.) with the circumstances (...)..." (line 52-53) and 'have to' in "you have to change" (line 55). These are strong commands and provide no alternative to members. It is an impassioned plea to change. He goes on further to say "... if you cease to change (...) if you REFUSE to change you will be staying where you are and you won't move on (..) you won't even have the chance to enjoy the fruits of change. (.)..." He uses 'if' twice to emphasise the consequences of not changing – "... you will be staying where you are...", "you won't move on", and "you won't even have the chance to enjoy the fruits of change. (.)...".

The words "cease" and "refuse" are negative connotations implying that if one has stopped to change or rejects change, one would face eventual failure or even death.

> YOU MUST CHANGE (.) you must change with the times (.) with the circumstances (...) change your business environment in order for you to adapt to the new situations and to survive (..) you have to change (…) if you cease to change (...) if you REFUSE to change you will be staying where you are and you won't move on (..) you won't even have the chance to enjoy the fruits of change. (.) so that's why I believe that I have to initiate this move

Excerpt 2: Text 2: line 53-58

In line 58, "that's why I believe that I have to initiate this move ..." serves to illustrate that members are not positioned to make changes and so may not 'move' or 'enjoy the fruits of change', so the managing director has to set the course of action for the organisation. This shows direct control over members. Members are subjected to changes as the management sees fit. This is in line with definition of ideology, "to study ideology is to study the way in which meaning (symbolic forms) serves to establish and sustain relations of domination" [5]. Since members are incapable of making changes, management has to step in and bring about changes.

Discourse of imperatives as seen in the excerpts serves to stress not only the urgency of action, but also can be seen as exerting power over members to conform. It sets out to exhort managers and members of the organisation to take concrete action to expedite change initiatives. The directives reflect the use of power or authority to 'direct' actions that are deemed to be beneficial to the organisation. There is therefore an unequal relationship between the speakers and the organisational members. Wodak defines power as; "Power is about relations of difference, and particularly about the effects of differences in social structures [5]. The constant unity of language and other social matters ensures that language is entwined in social power in a number of ways: language indexes power, expresses power ..." (p. 11).

Thus, power is wielded through the discourse of imperatives to get organisational members to conform to what is expected of them and as what Wodak says; "Language provides a finely articulated vehicle for differences in power within hierarchical social structures" (p. 11).

3 Discourse of pain

Discourse of pain is also appropriated in the construction of ideology among organisational members. One of the 'Sharing Sessions' in particular, shows evidence of the use of discourse of pain. In Text 3, 'Motivation', the speaker draws upon discourse of pain to emphasise the inevitability and need for change.

3.1 Change as a painful process

Discourse of pain is appropriated in the context of change. Change is a process and is often associated with discomfort or pain. The speaker in Text 3 effectively appropriates discourse of pain in highlighting the need for change. He draws an analogy between the process of change and the painful process of change that eagles go through in order to live longer.

3.1.1 Use of analogy

> this is really wonderful (...) we've got to learn from the eagles (...) this is VERY VERY incredible (....) ok (...) so just now I was talking about attitude (...) we have to have the right attitude (...) the right mental attitude and we must learn from the eagles (...) the eagle has the longest life span of the bird species (.) they can live up to seventy years (...) they can live up to seventy years (...) incredible, isn't it? (..) but in order for the eagle to reach this age (.) seventy years (..) the eagle must make a very hard decision (...) a very hard decision (.) not an easy decision but a hard decision (.) to reach the age of seventy (...) so what happens? (...) in its forties its nails become very long and flexible (...) its talons become very long and it cannot grab its prey anymore (...) normally how does an eagle catch its prey? (.) they fly and then they go down and grab the prey (...) but when they are in their forties they cannot grab the prey and then their beaks are no longer sharp and it becomes bent (.) no longer sharp but becomes bent (...) this is when it's in its forties (....) even the feathers become thick and become glued to the body (...)

because of that it becomes difficult to fly (...) so what happens? (...) the eagle is left with two options (.) two choices (...) one is to die (.) or to go through a very painful process (...) a process that takes one hundred and fifty days (…) how many months is that? (.) five months (…) so the eagle has to decide (...) it has to think (…) do I want to die or do I want to live? (…) if it chooses to live it goes through a painful process in a hundred and fifty days (…) the process requires that the eagle flies to a mountain top (…) it goes to a mountain top and there it knocks its beak against the rock until the beak cracks and breaks (…) the bent beak breaks (….) ok (…) then after that the beak grows back in a few months (.) then it will pull out its old nails (...) it will pull out the nails with its new beak (...) this is a very painful process (....) after some time the nails grow back (...) that is not the end of the process (...) with its new beak and new nails it uses its beak and nails to pull out its feathers which have become thick and glued to its body (...) it pulls out the feathers (...) the old, aged feathers (.) after five months (..) one hundred and fifty days the eagle takes its famous flight of rebirth and lives for another thirty years (...) err (...) another thirty years (.) what can we learn from this? (...) we learn about change (.) change is inevitable (.) it is necessary but sometimes we have to go through processes of pain (...) sometimes processes of change are painful

Excerpt 3: Text 3: line 145-176

The eagle has the longest life span among the species of birds. They can live up to seventy years, but in order to live that long, an eagle has to make a very difficult decision as the speaker reiterates, "in order for the eagle to reach this age (.) seventy years (..) the eagle must make a very hard decision (...) a very hard decision (.) not an easy decision but a hard decision (.) to reach the age of seventy (...)...". At the age of forty, an eagle experiences changes such as the talons becoming long, the beak becoming bent so that it cannot catch its prey as the speaker narrates, "(...) in its forties its nails become very long and flexible (...) its talons become very long and it cannot grab its prey anymore (...) normally how does an eagle catch its prey? (.) they fly and then they go down and grab the prey (...) but when they are in their forties they cannot grab the prey and then their beaks are no longer sharp and it becomes bent (.) no longer sharp but becomes bent (...) this is when it's in its forties (....) even the feathers become thick and become glued to the body (...) because of that it becomes difficult to fly (...)".

At this point, an eagle has two options, to die or to go through a "very painful process". The process of change takes approximately one hundred and fifty days. The eagle has to go through a process of rebirth or rejuvenation if it wishes to live. It has to first of all go to a mountain top and isolate itself for one hundred and fifty days (five months). The speaker explains the process of change the eagle undergoes (line 164-172). The eagle in undergoing this painful process lives another thirty years. It is a question of survival.

The speaker equates this process of change that an eagle goes through with that of an organisation, "what can we learn from this? (...) we learn about change (.) change is inevitable (.) it is necessary but sometimes we have to go through processes of pain (...) sometimes processes of change are painful (...)" (line 174-176). The speaker draws a parallelism between the processes of change that an eagle goes through to that of organisational change. Organisational change is often met with resistance, and is likely to give rise to misgivings, and discomfort but is necessary for survival.

Change is inevitable but necessary. It is often accompanied by pain, but the gain that one makes after that is indescribable. Thus, the speaker in using the discourse of pain shows the necessity and inevitability of change. After the process of pain, comes gain as in the maxim, 'There is no gain without pain."

Besides the use of analogy, the use of the present tense to express facts about eagles presupposes that it is the truth and therefore cannot be disputed. The statements of facts show a high level of speaker commitment. Members are positioned to think that although change is painful, it must be done. The speaker uses existential as well as value assumptions to inform and persuade the organisational members about change.

He draws the conclusion that "change is necessary for survival (.) just as in the case of the eagle (...)". Although the process is painful, it cannot be avoided. He rationalises change as changing "for the better (...) we don't change for bad (...) today is different from tomorrow (.) tomorrow should be better than today (...) so we got to change" (lines 177-178, Excerpt 4). Change is positive and not negative according to the speaker.

> sometimes we have to go through processes of pain (...) sometimes processes of change are painful (...) we change for the better (...) we don't change for bad (...) today is different from tomorrow (.) tomorrow should be better than today (...) so we got to change(.) change is inevitable (....) change is necessary for survival (.) just as in the case of the eagle (...)

Excerpt 4: Text 3: line 175-180

The modal "have to" in "we have to go through processes of change" gives a sense of definiteness – there is no option. The use of the modal 'should' in "tomorrow should be better than today" is a presupposition that change would bring about a better tomorrow. The emphasis of the repetition "change is inevitable (....) change is necessary for survival" also lends weight to what the speaker shares with the audience and the use of present tense makes it factual.

3.2 Discomfort and awkwardness

Discourse of pain also involves feelings of discomfort and awkwardness. The speaker in Text 3 conducts an exercise with the organisational members to illustrate this point. He draws a parallelism as a rhetorical strategy to influence his audience.

Drawing a parallelism

The speaker in Text 3 conducts a simple exercise with the members (Excerpt 5). He requests those wearing watches to remove their watches and wear them on the hand that is not often used to wear a watch, " (…) if you normally wear your watch on your right hand, change it to the left hand (….) if you wear it on your left hand (..) wear it on your right hand". He then asks the members how they feel. Some of the responses are; "awkward, isn't it? don't like it? (…) not used to it".

> change is not easy (...) I believe we all know that (...) right? (..) last week when I was talking to your colleagues (…) I asked them to do a very simple exercise (...) let's do it now (.) let's see all hands that are wearing watches (…..) ok (..) now I want you to change (…) if you normally wear your watch on your right hand, change it to the left hand (….) if you wear it on your left hand (..) wear it on your right hand (...) do it now (...) tell me how it feels like. awkward, isn't it? don't like

it? (…) not used to it? (..) these are all the responses (…..) [elicits responses from the audience] you say it's ok? (..) some more? gentleman at the back, how do you feel, sir? (…) do you feel comfortable? how many years have you been wearing your watch on that hand? (…) twenty years? wah! it's very simple (...) you've been wearing your watch on that same hand for twenty years (...) that is your comfort zone (….) suddenly today I asked you to change (.) of course you feel uncomfortable, awkward (…) err (…) all those negative feelings (...) believe me it will only take one or two days to get used to it (...) that's the difficult part, the initial step (...) that's why in life when there's something new, something different people feel so sceptical (..) so negative (….) these are things you hear when something new is introduced (...) why change? (..) we're ok so far (…) that's why some people can't move forward (.) because of that negative feeling, that thinking (…) no need lah! difficult! (..) let's do it like last time (..) we've survived so far etc. etc.

Excerpt 5: Text 3: line 190-208

He then comments, "that is your comfort zone (….) suddenly today I asked you to change (.) of course you feel uncomfortable, awkward (…) err (…) all those negative feelings (...)" (line 199-202). He assures the audience that the feelings of discomfort will last for just a few days after which it will become comfortable.

The speaker equates change to feelings of discomfort when change is initially implemented (line 203) "that's the difficult part, the initial step". It is natural for people to feel sceptical and have negative feelings about change. He cites questions normally posed by sceptics or cynics "why change? (..) we're ok so far (…) that's why some people can't move forward (.) because of that negative feeling, that thinking (…) no need lah! difficult! (..) let's do it like last time (..) we've survived so far etc. etc.". These examples of misgivings might possibly be what the organisational members are having. By mentioning these he is in a way making members feel ashamed of themselves as being not open to embrace change. It makes them reflect on their own personal responses to change. The speaker's skilful use of language serves to influence members to change their mindsets about change.

By drawing the parallelism between the discomfort of wearing their watches on their wrists which they do not normally wear their watches on and the association with the discomfort of embracing change, the speaker effectively influences members to adopt a more open attitude towards change.

Appropriating the discourse of pain is an effective way to change the mindsets of organisational members. By equating pain to eventual gain or survival helps members to adopt a more positive mindset. The effective use of the analogy about eagles and the use of parallelism of the exercise on the wearing of wristwatches serve to concretise in the minds of members that although change is a painful process, it brings with it tremendous benefits. The feelings of pain or discomfort in the initial phase are transient. Members therefore are positioned to adopt this line of thinking and have to negotiate their negative thoughts and misgivings about change.

4 Conclusion

The appropriation of the discourses of imperatives and pain set out to do ideological work. Discourse of imperatives serves to exert control over organisational members. Lists of 'dos' and 'don'ts' are replete. It is as if members need to be guided or taught to do the right thing in the execution of their responsibilities. This puts the employees in a subject position receiving orders or directives.

Discourse of pain serves to instil the idea that although change brings pain, it is worthwhile. Members are positioned to think that they need to endure pain and discomfort for the sake of the organisation. The analogy of the eagle which undergoes a painful process of change in order to survive is a strong case to persuade members to embrace change.

In conclusion, the discourses of imperatives and pain under the umbrella of the discourse of transformation serve to do ideological work in changing mindsets and work orientations of organisational members. They are used to garner support for the organisational change initiatives and persuade members to look upon them as sense making, rational, and reasonable and therefore commonsensical. The discourses set out to create an ideological commonsense. Fairclough defines ideological common sense as "common sense in the service of sustaining unequal relations of power" (p. 70) and in establishing and consolidating solidarity relations among members of a particular social grouping [6]. Ideologies are embedded in features of discourse which are taken for granted. In this way, the organisation is able to exercise unobtrusive control over its members.

References

1. N. Fairclough, *Analysing discourse: Textual analysis for social research* (2004)
2. M. Alvesson. H. Willmott, JMS. **39,** 619-644 (2002)
3. D. Jaffe, *Organisation theory: tensions and change.* (2001)
4. S. Deetz, *Transforming communication, transforming business: building responsive and responsible workplaces.* (1995)
5. R. Wodak, *Methods of critical discourse analysis*, 63-94 (2001)
6. N. Fairclough, *Language and power.* 2nd edition. (2001)
7. EG Flamholtz, TK Das, AS Tsui, *ACCOUNT ORG SOC,* 35-50 (1985)
8. JC Henderson, S Lee, *Manag. Sci.*, 757-777 (1992)
9. D. Kirsh, *Adapt. Behav,* 415-452 (1996)
10. WJ Orlikowski, D Robey, *INFORM SYST RES*, 143-169 (1991)

Public service broadcasting (PSB) regulation in Indonesia: Between market and public interest

Masduki[1,*]

[1]Senior Lecturer, Department of Communication, Universitas Islam Indonesia, Yogyakarta, Indonesia

Abstract: The reform of public service broadcasting (PSB) faces many obstacles in countries with political transition such as Indonesia. After 1998-political change, the arrival of ideas to establish PSB in Indonesia in two decades lacks of appropriate policies. As in other transitional states in Eastern Europe, there is the lack of regulatory design as well as its implementation. This paper examines process of formulating PSB law in 2002, particularly pays attention to industry capture over the making process as a critical tool to observe a long-controvercial of broadcasting law revision process in Indonesia (2012-2016). By use public interest and capture theories in regulation, this paper answers why the outcome of PSB regulation is weak by observing how the law was made. This study provides new analysis on PSB media law that still rare in Indonesia. As independent and non-profit body, Indonesian PSB represented by RRI (Radio of the Republic Indonesia) and TVRI (Television of the Republic Indonesia) is protected by Broadcasting Act.32/2002. It was enacted in favor of two actors: authoritarian officials with interest to use PSB as their mouthpiece in one side and the industry groups with interest to apply free market policy on the other.

1 Introduction and Research Design

As a public policy, redesigning public service broadcasting (PSB) regulation in the world is a complex negotiation process, influenced by political system and the strength of economic and social culture of a country. For instance, change in political system from authoritarianism to liberal in transitional countries in Europe and Asia has a direct effect on the process of formulating new media laws, actors involved and the result. It is mostly indicated by contradictory policy outcome between public and private interest.

Historically, follow the change of political system in Eastern Europe (late 1980s) and in Asia (late 1990s) from authoritarian and communist to liberal democratic system, PSB experienced significant regulatory reform. In line with the above radical reform, in the last 20 years, there was a trend of policy changes that move PSB ownership from state into independent public service media [1]. Furthermore, the changing PSB policy is a direct response to the emergence of new technology [2].

[*] Corresponding Author: masduki@hotmail.com

The political reform 1998 which marked by the end of Suharto power (1966-1998) has triggered media system reform in Indonesia [3]. For example, there was public pressure to reform RRI and TVRI amidst rapid growth of commercial media outlets. Prevously (1996-1998), media system was authoritarian in nature which both RRI and TVRI was state owned media and operated as a mouthpiece of the regime.

One of the important regulations born after the reform 1998 was Broadcasting Law no. 32/2002 which recognizes Public Service Broadcasting (PSB) and replaces the old Broadcasting Act no. 24/1997. The Act has formally changes the ownership status of RRI and TVRI from state to public. However, compare to similar policy in developed countries such as Germany and the UK, the Act is not clearly explaining structure, funding mechanism, content criteria and the ways that guide transformation from old to the new PSB system. Why?

In article 'Researching Public Service Broadcasting`, Moe and Trine explains four types of PSB studies, namely policy, institutional, role of PSB in democratic life and post modern research. Since PSB is dependent on decisions within political realism, much research has focused on policy type, mainly in the regulatory governance [4]. I classify this paper as study of PSB governance policy in Indonesia.

This paper examines process of formulating PSB policies in Broadcasting Act of 2002 and pays attention to industry capture as critical tool for the similar process in the recent years. This study proposes as best practice for media advocates in managing their advocacy in the 2016-Broadcasting Law revision. I will look at PSB regulatory outcome in the Broadcasting Act as starting point and to limit this study, I will select four debates: legal status of PSB, structure, service area and funding system among actors and it will be explained in these two periods: agenda setting/first discussion (1998-2000) and negotiation/decision making phase (2000-2002).

This paper answers two questions: First, how are debates among stakeholders on PSB and to what extent it was formulated in the Broadcasting Law? Second, how and why the existing media industry captured the PSB regulation? Broadcasting Act no. 32/2002 will be a primary document to analized, followed with other documents such as articles in journals, books and other literatures. My last observation in years 2000-2002 would also enrichs while other PSB policy model especially in Europe and the US will be used for comparison.

2 Theoretical Frameworks

2.1 PSB Regulatory Governance

PSB is widely understood as broadcasting made, financed and controlled by the public, for the public. According to UNESCO researchers, it is neither commercial nor state-owned and free of political interference and pressure of commercial forces [5]. In order to ensure above condition, Smith stated PSB regulatory governance should meet the followings: (a) legislative regulatory framework for its sustainable development; (b) clear mandate for public service broadcasters; (c) exact remit and standard of content; (d) corporate and financing structure that ensure editorial independence; (e) mechanism for public accountability and (f) role to meet global challenges: digital and transnational broadcasting [1].

In legal status, most of PSB regulated as independent entity to guarantee as public asset in safeguarding the interests of cultural, political democracy and to balance the increasing commercial media. For instance, BBC and ARD (Germany) are independent Corporations. Furthermore, to ensure the importance of this issue, the European Union imposes

Amsterdam Protocol of 1997. This regulation forced former Eastern Europe states that joined the Union to change their regulations and old-authoritarian form of PSB [6].

In Indonesia, political and economic crisis bring PSB regulatory issue to more complex one. Debates on PSB policy occur between both, the desire to create new rules to protect public interest by law in one side with desire to avoid strick rules resulted from historical and political trauma of authoritarian sytem in the past [7]. Amid the strengthening capitalist broadcasting industry, debates on PSB legal status, structure and funding which involved government, industry and civil society took long time. Recently, commercial radio and TV which controlled by monopolistic corporation [8] became the main enemy to establish public service media. To protect their power, commercial sectors actively try to limit PSB regulatory governance.

2.2 Public Interest in Broadcasting Sector

Normally, two issues: frequency scarcity and cultural porposes is classical rationales for PSB regulation. As a public domain, frequency is technically limited. For example, if it is used by any radio or TV company, the other can not use it at the same time. Hence, in the name of public and its limitation, frequency should be allocated for majority of the public, and the licensing process based on competitive roles, involves independent regulator (9). There is a general view that frequency should not be utilized by only government and industry players, but also for the benefit of citizen. Although the frequency scarcity doctrine is no longer valid at this time, its regulatory legacy still lives on the so-called "public interest" requirements over broadcasting contents. This commanded to satisfy public interest, convenience, and necessity [9].

Moreover, cultural protection over social impact from television/radio content is the other reason why broadcasting is strictly regulated [10]. Broadcasting that is free to air has entered private sphere of citizen at all social and economic levels. All over the world, radio and television are two major media which people can access news and entertainment, and are often regarded as the powerful media.

Basically, the term `Public Service` is the central point of necessary regulation for PSB at national and international level. Public service is used to describe a kind of provision deemed vital for every member of society, regardless of spending power, educational, social geographical location, or gender. At different times in different polities, varied services have belonged to this category, from public infrastructure like roads or telephone lines, via health care, to education as well as radio and television broadcasting [4]. In the liberal states such as the UK, in addition to market failure in distributing public resourse, PSB policy is a form of state's obligation to provide a public sphere [11]. Four aspects: legal status, structure, service area and the provision of PSB fund are the central policy debates that reflect degree of adoption to public interest in the PSB law.

2.3 Capture Theory of Regulation

As stipulated in Law no. 12/2011 on law making mechanism, formulation of national act in Indonesia is an absolute authority of both parliament and government officials. Hence, the legislator or government is policy maker. Unlike in the EU, Indonesia officials does not establish a permanent external agency or gives mandate to outside body to draft of a policy. Nevertheles, parliament may invite some experts as temporary consultant. By this power, issuing law draft formally should come from both government and parliament, through various steps. Civil society as well as industry may propose a draft to both institutions. In the formulation process, parliament formed a special team consist of

DOI: 10.1051/shsconf/20173300040

parliament members that represent political parties. The team intensively organizes both internal meetings and public hearings prior to the enactment of a law.

According to Barry M. Mitnick, problem in the regulatory making process arises when a policy maker (for Indonesia: Parliament or Government officials) acts in the industry side to damage public interest. It means, their policy outcome captured by industry as private-regulated party [12]. For instance, the agency's main mandate to regulate PSB to maximize benefit of society is captured and the agency acts in favour of the industry interest due to their lack of knowledge on the broadcasting system in the policy making process. Indeed, to regulate PSB system, policy makers require specific knowledge such as frequency allocation; remit, and funding system. Mitnick describes six forms of capture: constitutional; systemic; relational (governance); individual; functional and chronical [12]. For this study, I will explain two forms: individual and functional.

First, individual form: incentives shape decision making between legislators and the industry [12]. In the electoral system with high economic cost, capture occurs as a form of `political barter` among political parties and their politicians in one side and corporate donors on the other. In this model, regulatory members cum politicians need campaign funds to encrease their voters. The industry can easily provide `political investment` fund for them in order to have room to influence legislative policy outcomes of the policy makers in the future. Second, functional. It happens when regulated parties control and filter information essential to policy decision making, actively suply information that manipulate public opinion in order to keep their interest [12]. For instance, present on hearings intensively, active responses to the need of data, opinion spin in media and the involvement of industry expert as a temporary consultant of parliament.

In the long history of media reform, government, politician and independent commission of broadcasting are vulnerable bodies to intervere. For example, the scandal of internet neutrality regulation applied by Federal Communication Commission, the United States independent body [19]. In case of Indonesia, before ratification of the Broadcasting Act of 2002, the discussion and encatment of the Telecommunication Act of 1999 was also influenced by national and international telecommunication industries. They intended particularly to remove barriers for foreign investment in telecommunication market [13].

3 Debates on Indonesian PSB

In this part, I will explain result of document analysis, mainly Broadcasting Act of 2002 and its related documents enriched by my field observation in the past (2000-2002). During 1998-2002, there is strong debate among government, industry and public in the process of formulating the Act. Government interest represents by Ministry of Communication and Information Technology (MCIT), while Association of Indonesian Private Television (ATVSI) represents industry group. Meanwhile, public interest was voiced by various parties: academics of Indonesian University and some non-government organizations such as Indonesian Press and Broadcasting Society or MPPI [14]. In this situation, Parliament members tried to accommodate all opinions equally, but sometime falled in to one party interest. The table below desribes various proposals from all above parties over four issues of PSB governance: legal status, structure, service area and funding sources.

Table 1. Proposals on PSB Governances

No	Issues	Government	Industry	Public
1	Legal status	State agency under MCIT. A state agency under ministrial office and it established by	State owned corporation, a commercial entity under ministry of	State agency, independent, public owned, non-profit and it established by state

		government	state own corps	in the name of public
2	Governing structure	Board of supervisory and board of executive under MCIT, elected by MCIT	Board of commissary and executives as in the private corporation	Board of supervisory and board of executive as in the BBC, elected by parliament
3	Service area	Covers the entire Indonesian archilepago with proportional allocation of frequency between public, private and community broadcasting	Covers specifically for area that is not covered by commercial service, use only the existing frequency allocation	Covers the entire Indonesian archilepago accompanied by local public service broadcasting, supported with highest portion of frequencies
4	Funding sources	State budget, lisence fee, social donation, commercial advertising and other appropriate incomes	State budget, lisence fee, social donation, strick limitation of commercial advertising	Lisence fee, state budget, social donation, commercial advertising and other appropriate incomes

Source: [7, 14]

All debates influence over the making period of Broadcasting Act that spends more than 6 years from its agenda setting and drafting in the last 1997 up to the formal enactment in the parliament by November 28, 2002. On the PSB side, the result of these debates was contradicts to public interest. From all 64 articles in the Law, PSB sector is regulated only in 5 articles. First, article 13 that states PSB as one of four types of broadcasting institution in Indonesia alongside private, community and pay model.

Second, article 14 that explain legal status, PSB mandate and its structure. This article describes PSB as law entity established by state, operated as independent, neutral and not for commercial interest, and its function to serve society. This article also states clearly that RRI and TVRI are choosen for PSB in one side and provides room for establishment of local PSB outside both institutions on the other.

Third, article 15 explains funding sources (ranging from license fee, state annual budget, social donation, advertising and other appropriate incomes connected to broadcasting) and role of finance accountability (finance report should be audited by public accountant, published on media). Fourth, the two articles that regulate PSB's content: PSB to broadcast programs from internal country for 60 % at minimum a day; time allocation of business advertising (maximum 15 %) and social advertising (minimum 30 %) of the entire duration a day.

The enactment of these 5 articles indicates three conditions in favour for public interest. First, adoption of BBC-like model of independent and public owned PSB, changing RRI and TVRI's legal status from state to public entity. Second, introduction of lisence fee as source of funding alongside state budget is a sign for public based PSB and public participation. Third, establishment of supervisory board elected by parliament is a symbol for formal public access to the PSB operation that was not applicable in the past.

However, policy of service area, frequency allocation and advertising are in favour of market interest. As described above, the Act limits PSB to broadcast commercial advertising up to 15 % of the entire time a day. This followed by two government decrees of 2005 (translation of Broadcasting Act) that limit frequency for PSB up to 20 %, compare to private that receive 75 %. Limitation of commercial Ads as funding source will potentially disturbs PSB's sustainability, especially in the crisis of state budget and license fee as ideal fundings. Also, limitation of frequency reduces opportunity to develop new PSB channels in order to wider public services.

Based on these weaknesses, I argue there will be no significant reform of RRI and TVRI as PSB to reach audiences as citizen and create public service. To compare, before enactment of the Law, RRI uses only about 15 % of frequency allocation. The rest is in power of private radio stations. Since year 1989, commercial televisions utilize about 95 % of the entire TV frequencies while TVRI only use 5% [7]. Overall, the application of two articles of the Law will place PSB as a minor player in the audiences compare to private firms.

Furthermore, as discussed above, from the 5 articles, there is no detail explanation of PSB institutional design compare to specific PSB law as in the developed country. According to eight indicators developed by Smith [1] it only adopts 4 indicators and needs more articles to adopt the rest in the same level of regulation. For instance, more regulation needed to exact remit and standard of PSB content, finance that ensures editorial independence as well as the mechanisms to enhance public monitoring over the entire operation.

4 Captured by Industry

This section describes captures of industry power over formulation of PSB articles in the two periods: agenda setting (1998-2000) and negotiation/decision making phase (2000-2002). To define, agenda is the list of issues in a polity, consists of both political (the list of issues from decision-makers) and public agenda (the list of issues from general public) [15]. Decision making is a final choice, an outcome of cognitive processes leading to the selection of action among several alternatives [16]. I will start by explaining actors involved in the policy making period and the dynamics of their interaction.

During agenda setting and negotiation phases (1998-2002), Arifuddin [17] described enthusiasm of three actors: parliament members; officials under two presidents: Habiebie and Gus Dur and civil society to birth a new law which replaces old-Broadcasting Act no. 24/1997. In the ideological perspective, there are three kinds of actor: pro democratic broadcasting (CSO), authoritarian broadcasting system (Government) and pro liberal media system represented by them [18]. Furthermore, Irma [14] explained three different groups: (1) private firm that intend to protect their domination in the market. At agenda setting process, introduction of PSB in the draft was prevented by industry, (2) civil society such as the Alliance of Indpendent Journalist and academics from Indonesian University which proposed PSB institution in broader context than only as state agency. In their view, PSB may be a community based stations, (3) government (via MCIT) that proposed PSB system under MCIT. It was contrary to general idea of independent PSB in the parliament and CSO [14].

Compare to civil society, the industry's lobby to policy maker is more intensive involved a wide range of collaboration among ATVSI, advertising agencies, production houses and broadcaster associations. It includes regular opinion publication in their own media; propose law draft, and personal approaches to politicians. In contrast, civil society with limited support from RRI and TVRI broadcasters and international agencies focuses only their move to public opinion through community media, grassroot level of discussion and mass rally.

During 2000-2002, in summary, capture by the industry over PSB law took place in two steps: (1) agenda setting step, ability to setting up the central issue, (2) negotiation phase, ability to influence policy maker to accept policy in favour of their interest. From media coverages and notes of my field investigation in the past (years: 2000-2002), refer to Mitnick [12]. I found at least two kinds of capture below.

In the agenda setting, the industry applied functional model. With their highest resources of knowledge, fund and media outlets, private TV and radio owners with their professionals develop massive opinion to promote the importance to protect private industry as key actor in economic growth. Via more than 7 national TV channels such as

DOI: 10.1051/shsconf/20173300040

RCTI, Indosiar and ANTV, ATVSI published regular news, talkshow and advertisements. At the same time, ATVSI provides data and invite experts to set a law draft favouring the goal, and actively sends experts to public hearings in the parliament or discuss with government officials.

The long history of close-relation between private TV owners and political leaders obviously established common goal and influence outcomes of policy maker which depend on the direction from their party leaders. In Indonesia, political party needs good image and politicians need their safety in the parliament. The industry provides it by managing politician's images in the public though massive news coverages. This can be identified as individual incentive model [12] although more evidences are needed to ensure this practice.

Analyzing series of statements from ATVSI and PRSSNI (association of private radio owners), I conclude the main goal of their capture is protection of their companies, in the name of millon workers, high investment of broadcasting equipment and their claim of public utilily over contents. Hence, they criticize intiative to regulate PSB in the form of state intervention. Improvement of PSB was seen as 'serious danger' for their two sectors: huge area of audience service and advertising market incomes. Having the widest network, RRI and TVRI as PSB will be a strong competitor for the industry [7].

In negotiation and decision making phase, the firms adopt similar model with the agenda setting with more intensive pressures. Aside of actively attend public hearings and gathering mass rally, they influence both parliament and MCIT officials through direct meetings outside offices. Finally, in respond to the enactment of Act on 28 November 2002 which recognize independent broadcasting commission as their new regulator, the Industry raise protest by airing 'dark colour' in their TV stations which symbolizes a death of press freedom.

5 Conclusion

This paper shows that PSB debate in the policy making process in Indonesia during 1998-2002 has been intense. Actors separate into three groups: government, industry and civil society. There is unequal position which the industry has more resources to create pressure over policy makers in favour of their dominant in the market place than civil society. Formally, Broadcasting Act 32/ 2002 recognizes PSB as independent and public owned body. It is followed by introduction of public access to the supervisory body and parliament right to its selection process and to supervise PSB operation in the interest of public. However, there is no appropriate policy to ensure sustainable improvement of PSB management. The two policy sectors: limitation of Ads and frequency are obviously in favour of the industry. Hence, in terms of public interest, the regulation is very weak.

In my observation, this was a result of industry capture over the law making process. The goal of this action was mainly to keep PSB as a minor broadcasting actor in the field and maintain capitalist players that were already dominant. To sum up, I would like to address two possible solutions: First, Broadcasting Act no. 32/2002 needs to revise. Concerning the huge sectors involve in the revision process which will spend long time and energy, the 2015-proposal from parliament to draft specific law of PSB can be a faster way to accelerate PSB policy reform. Second, improvement of knowledge of policy makers on PSB governance is a central in managing public interest in the process of both the law revision and addressing new specific PSB law.

Finally, to prevent industry's capture in upcoming PSB law revision, strong alliance is needed amog local and international advocates in the process of regulatory making and its implementation. A model of supranational intervention such as European Union to the national policy of PSB can be adopted by ASEAN, a similar regional association of nations in South East Asia. However, in culture of secrecy, it is quite difficult to investigate deep-coalition among Indonesian politicians as policy maker with industry in the long period of policy making and its impact to policy outcomes. More research is needed on this particular issue in the future.

References

1. E. Smith, *A Road Map to Public Service Broadcasting* (UNESCO and AIBD, Kuala Lumpur) (2012)
2. I. Banerjee, *Public Service Broadcasting, A Best Practices Sources* (AMIC-UNESCO, Singapore) (2005)
3. Kristiawan, *Media Liberalization, Political Economy of Democracy and Media Industrialization in Indonesia,* (Indonesian University Publication, Jakarta) (2012)
4. H. Moe, T. Syvertsen, *Researching Public Service Broadcasting*, In K. Wahl, T. Hanitzsch (ed.), The Handbook of Journalism Studies (Routledge, London) (2009)
5. T. Mendel, *Public Service Broadcasting: A Comparative Legal Survey* (The Asia Pacific Broadcasting Development and UNESCO, Kuala Lumpur) (2000)
6. K. Jakobuwicz, *Finding the Right Place on the Map, Central and Eastern European Media Change in a Global Perspective* (Intellect Books, Bristol) (2008)
7. Masduki, *Broadcasting Regulation, From Authoritarian to Liberal* (Institute of Islamic and Social Studies, Yogyakarta, Indonesia) (2007)
8. M. Lim, *@crossroads, Democratization and Corporatization of Media in Indonesia (*Partisipatory MediaLab Arizona State University and The Ford Foundation, USA) (2011)
9. A. Thierer, *Why Regulate Broadcasting? Toward A Consistent First Amendment Standard for Information Age* (Communication Law Conspectus, Washington) (2005)
10. J. Dominick, *Broadcasting, Cable, the Internet and beyond, an Introduction to Modern Electronic Media* (McGrawHill Publishing, Boston) (2004)
11. J. Habermas, *The Structural Transformation of the Public Sphere* (Massachusetts Institute of Technology Press, Cambridge) (1991)
12. B. Mitnick, *Capturing Capture: Definition and Mechanisms*, In D. Levi-Four (ed.), Handbook of the Politics of Regulation (Edward Elgar Publishing, Cheltenham) (2011)
13. P. Rianto, *National Private TV Domination, The Crisis of Ownership and Content Diversity* (Pemantau Regulasi dan Regulator Media and Tifa Foundation, Jakarta) (2012)
14. W. Maimun Irma, *Social Construction of Public Broadcasting, Political Economy Analysis of Broadcasting Act, 32/2002*, (Indonesian University Publication, Jakarta) (2003)
15. J. W. Kingdon, *Agendas, Alternatives, and Public Policies* (Harper Collin Publication, New York) (1995)
16. F. Meyerhöfer, *Agenda Setting and Decision Making in the European Union: The Case of Galileo* (The Hertie School of Governance, Berlin) (2009)
17. Arifuddin, *Finding Undisputed Truth, Broadcasting Reform in Indonesia* (Nagoya University, Japan) (2014)
18. G. Heryanto, *Power Relation within Legal Entity Changes of TVRI: A Political Economy Study* (Indonesian University Publication, Jakarta) (2006)
19. R. Esguerra, *The FCC and Regulatory Capture,* https://www.eff.org, accessed on February 29, 2016

Law Documents:
1. Broadcasting Act, No. 32/2002
2. Mechanism of law making Act, No. 12/2011
3. Government Decree on RRI and TVRI as Public Service Broadcasting No. 11 and 12/2005

Social media usage by teen segmented commercial private radio in Bandung

Martha Tri Lestari[1,*], and *Hadi* Purnama[1]
[1]Communication Science Department, Communication& Business Faculty, Telkom University

Abstract. Radio is one of mass media which functions to inform, to educate, to entertain and to influence. In Indonesia, private commercial broadcast radios are the biggest in numbers with more than 700 radio stations throughout indonesia. Some of the private commmercial radio stations select audience segmentation of young people or teenagers. This is due to fact that numbers of young people dominate the whole population. Social media nowadays has become a primary need for most of its users, especially young people. The usage of social media as a tool in virtual social interaction, has also been used by many radio stations, especially whose target audience are teenagers. The usage of social media by teen segmented radio stations is not only for on-air activities, but also to support off-air activities. How the use of social media by teen segmented radio stations in Bandung, to support on-air and off-air activities, is the focus of this research. Through qualitative research with descriptive method, it is hoped to describe more comprehensively about the usage of social media by teen segmented private radio stations in Bandung.

1 Introduction

Media is not as traditional as before, conventional mass media also has limited time release or certain schedule which makes it difficult for the audience to access information at any time. The weakness of media such as limitation of place, time, or pages, can be covered by online media which is accessible anytime, anywhere [1]. This time, media has transformed into a 'new media' technology, known as social media.

Social media is the media we use to be social. The story is in the tactics of each of the hundreds of technologies, all of the tools that are avilable for you to connect with your customers and prospects, and the strategies necessary to use these tactics and tools effectively [2].

Industries which has developed or are developing nowadays rely on publication as one way to publish their companies, likewise media industry such as radio. Publication is one element of *Public Relations. Public Relations* itself is part of *Integrated Marketing Communication*. Rhenald Kasali in manajemen Public Relations, Konsep dan Aplikasinya di Indonesia in [3] said:

* Corresponding author: martha.djamil@gmail.com

"management function in public relations concept is aimed to create and develop best perception of an institution, organisation,company, or their products to people segmented,which activities have instant or uninstant effect towareds the future of organization, company, or product."

Public Relations activities is closely related to public opinion building and behaviour change of the society [4]. Furthermore, according to [3] function and task of *Public Relations* are *publicity*, *publication*, *advertising*, *promotion*, *press relations, etc* . through description above, writer can conclude that public relations function in an organization or company are important because it can help to support company's activity in building good relations and communication with its internal and external publics.

Radio is a creation of technology development which voice is transmitted together through radio wave in the air. [6] Bandung has many developing and developed commercial radio stations. On this opportunity writer intended to do a research to commercial radio stations which are Zora Radio, 99'ers Radio, OZ Radio and Ardan Radio. Writer selected these four teen-segmented radio stations based on several considerations such as internet penetration scale for Indonesian citizen is now more than 30 persen of total Indonesian. A survey conducted by Nielsen with respondents aged over 13 year old between March 2013 through May 2013 showed the smartphones penetration in developing countries in Asia Pasific has increased. The focus of this research is based on writer's curiosity in how the usage of social media by teenage-segmented private commercial radio stations in Bandung in supporting Public Relations activities (Promotion and Publication) as well as *on-air or offair radio programs.*

2 Research methodology

This research used qualitative research methodology. Jane Richie in [7] explained that qualitative research is an effort to display social world, and its perspective in world, from concept, behaviour, perception, and problems of studied humans studied. Qualitative research is specifically targeted in the use of facts. This research studies problems in society and how to behave socially in certain society or conditions, including relationship, activities, attitudes, point of views, and ongoing process and influence in a phenomenon.

2.1 Population and sampling

The population about this research are informans who know about social media function in their radio station. In this research we invited informans from 4 (four) teen-segmented radio stations which is most popular radio station in Bandung-Indonesia. There are Director&Marketing Manager from Zora Radio, Station Manager 99'ers Radio, Digital Communication unit from Ardan and OZ Radio.

3 Result and discussion

In result of this research, which 4 (four) teen-segmented radio stations in Bandung were the research's subject, researcher has summarized, studied according to theory used and analized results from questions in Focus Group Discussion (FGD) and the answers packaged based on questions list divided into 3 (three) discussion stages and answers form each informan key as representative of each radio stations. The result of FGD data processing:

3.1 Question stage I

On segment I in this research questions given are still on the basic stage such as since when do they use social media, the reasons or considerations in using social media, obstacles in using social media ,which social media used and which are the top priorities to be used by the radio stations. The results are:

3.1.1 Zora Radio

The answered that represented by Director and Marketing Manager Radio Zora , it is known that the using of social media in Zora FM started to be active when Zora radio switched its segment into teenagers listeners. They have used the internet for quite a long time, but communication using media social internally and externally has resulted in excalating interactions with listeners, users and even with internal staff of zora. The background why Zora radio uses social media is the need of client/costumers. The population of society using gadgets especially smartphones is so high,they even use multi gadgets. Media convergent era was one of the main reasons why social media is needed by teen-segmented radio stations. Obstacle found in earlier time in using social media was the human resources. Back then, it was difficult to find a rightful human resource to be responsible in media social division. Every social media has its own different characteristics because social media is something fluid and flowing, and social media age shorter time by time, social media content could be an obstacle in using social media.

3.1.2 99' ers Radio

99'ers Radio started to maintain its streaming application in early 2000. Nevertheless, he could not specify when exactly they used social media. At first, they used Facebook, but not for long, only lasted around one year from 2009-2010. They use Twitter more than other social medias. The website was not functioned in rising interactions until 2011, since they still interacted by sms, on air broadcast, etc. The background of using social media was based on the need from bottom to the top, meaning the need of market to the need of top management of the radio itself. Listeners' trends, the need of interactions between users/clients, listeners, and also radio employees have become one of the reasons in using social media, based on survey conducted by 99ers radio on social media usage, it can be said that social media usage was more advantageous than conventional media usage because it was more 'low cost', even 'no cost', then market needs will be fulfilled. The usage of social media today has become costumers behaviour and also can be a tool to support 99ers in conducting public relations activity. 99'ers Radio has a goal to "follow on to device" which means that 70% of the consumers using gadgets can use 99ers mobile application, and the last one is client demand; so the usage of social media is relevamt to the clients commercial needs. Obstacles in social media usage is almost the same with others, which is on finding the rightful human resource. At this time, 99'ers radio only has 2 staffs (webmaster and pic socmed) where they both have different objectives towards socialo media which made lack of coordination between the two.

3.1.3 Oz Radio

OZ Radio began to use the internet around the year 2000, then continued in 2005 by establishing a website. The usage of social media started with the released of facebook in 2006. In the beginning, they used personal account, not a fanpage. In last 2006 or early 2007 OZ Radio created official facebook and twitter accounts baring the radio's brand

DOI: 10.1051/shsconf/20173300041

name. According to OZ Radio's social media writer and content media editor, the backgorund of using social media was because it was the demand of their consumers (listeners). For example, listeners can find out about the song's title or the singer by twitter. Beside that consumers need, the usage of social media has become a main point asked by clients (advertiser, agency, and listeners). The obstacle in using social media according to informan of OZ Radio is the lack of human resources. They only have 2 persons responsible for social media usage and it is not adequate. He stated that ideally, 1 account should be managed by 1 person. This lack of human resources resulted in non-optimum usage of social media by the radio.

3.1.4 Ardan Radio

Ardan Radio has used the internet since the 2000's by using a website without online interaction with its listeners. Around 2005 through 2008, Ardan has used social media such as Facebook, Twitter but not optimally. The background of using the social media as communication and interaction tool was after seeing the market has moved towards digital, consumers, buyers, users , have become visual users and could also be called "gadget freak". It could be concluded that this digital era has tunerd into costumers trend. The obstacle in using social media focused in human resources. Ardan Radio was lack of human resources to maintain the websites. At this time, there is only 1 person who handles 4 websites of Ardan Group. Now Ardan Radio has its own division in handling social media usage which is Digital Communication Division while it is still difficult to find competent person for the position in the division. According to head digital communication Ardan radio who is the informan in this research, ideally this division should have around 6 persons / human resourcers.

3.2 Question Stage II

On stage II of the FGD, questions given to the informan were more specific. They were asked about social media usages such as the strength, weakness, and opprtunity in using social media. Special strategy in using social media, which unit handles the authority to use the social media, the advantages of using social media and how the social media function in building the radio's brand image. The results are:

3.2.1 Zora Radio

According to informan of Zora Radio, he stated that the strength of social media has to have visual power in each account. Just like in twitter and instagram, consumers will be more attracted to see them if the accounts are rich in visualization. PR and Promotion division is the one responsibles for the media social usage. The usage of social media in zora radio is to socialise, podcast, lowcost broadcast media which can also be used to analyze Zora's listeners' characteristics such as listeners profiling, which is advantageous in program or content selling to advertisers.

3.2.2 99' ers Radio

According to informan of 99'ers Radio who is the station manager, there are different strategies in using social media. Social media is one functional tool to support sales marketing and public relations programs as well as on air and off air programs (sales oriented, program oriented). Monitoring the usage of social media is really needed because it can give advantage in rising numbers of listeners by maintaining the social media

DOI: 10.1051/shsconf/20173300041

content. Participation from the radio internal is also important, for example in creating more lively creative campaign. The power of social media has to cope with the listeners since nowadays they are the visual generation where social media has becoma the main commodity in interaction. This division is under PR and Promotions unit.

3.2.3 Oz Radio

At this time, division responsible in handling social media usage is On Air program division. Previously, it was handled under PR and Promotions. The strength of social media in OZ radio is because it is always put forward on air content such as commercial-like content, built in between on-air programs' content. The opportunity in using social media related is so big. That is the reason why the usage has to be put into deep attention. For instance, the content in social media has to include a graphic designer in creating the posters. Furthermore, social media can become a media shorcut to interact with talents needed.

3.2.4 Ardan Radio

Ardan Radio has just established new division/unit related to media social handling, i.e Digital Communication Unit. According to Head of Digital Communication it is extremely important that the existing units integrate in supporting the usage of social media. It is because as we can see from the weakness of not having adequate competent personnels, the integration of units would be a good solution to the problem in social media usage. The opportunity lies ahead in social media usage can be seen from on air and off air sides, for examples the rising numbers of advertising clients. The special strategy is the "do and goals" of the social media have to be balanced. Besides that, Ardan is trying to collaborate between analog radio with the development nowadays which is 4 Os (On Air, Off Air, On Mobile and On Light) while the integration of all those services is being conducted so it is overlapping.

3.3 Question Stage III

On this stage III of FGD the questions is more focused to the outside of the usage of social media like to develop On Air or Off Air program, human resuources allocation, and financing allocation to manage social media, pengelolaan,influence towards society's loyalty and media prediction habit of each radio station.

3.3.1 Zora Radio

Social media has an important contribution to radio programs improvements, one way is by profiling study of its listeners. How the social behaviour ipof the listeners is another important element to support the improvements of On Air or Off Air programs.
From the profiling study, the media habit of Zora Radio can be mapped. But at this time, media habit analysis has not been done because it is still considered a short period of time since zora Radio changed its segmentation towards teenage listeners.

3.3.2 99 'ers Radio

Media Habit can be predicted but prediction will not be too far in the future because changes will influence listeners' media habit. Consumers who actually are listeners have a vision that radio is a conventional media. With the vast changing of social media in the last

4 years, prediction culd not be made too far, so 99ers would only be playing in what's up to date.

3.3.3 Oz Radio

Social media has an important contribution with maximum usage. Media habit prediction for OZ Radio besides Ac Nielsen which has online research, OZ has another way to do the analysis by conducting depth interview.

3.3.4 Ardan Radio

Social media has a high contribution, especially in changing media habit of the listeners. There are changes from previous radio activities into these days activities when technology has developed highly. For instance, high interests in podcast or selfie pics which are now trending today, were some of several behaviours of the listeners that should be supported by the right contents served by Ardan radio. Media habit of Ardan listeners are more about how to change the behaviour of old time era listeners into digital era or convergent media era like today.

4 Conclusion

The four of radio stations have and use active social media in communicating and interacting to conduct activities related to promotions or publications activities which are one of the elements of *Public Relations* activities. Strategy and usage in each radio station is different, but basically each station focuses on content. Visualization dimension is one of the strategies in using the social media. And they actually used the social media for doing some PR Activities such as cyber PR, sponsorship, companies identity, direct communications, publicity, CSR, special events and media inggriya [5]. Some of the radio stations use media habit analysis with indepth interview research on listeners/clients and studi profiling as well. As for media habit prediction, it can not be forseen too far in the future due to the rapid development of technology which will influence the usage of the social media itself. Suggestions which can be offered to the four stations is to allocate financial fund provided by the company, relating to the responsible unit of social media usage. Therefore the main key is to search for human resources who are creative, innovative and technology literate that can also cope with segmentation, targeting and also positioning of each station in accordance to keep on existing as radio publicity media for their audience/listeners which are not only in bandung area but also has spread to other urban cities through streaming application on internet media.

References

1. M. Romli, Asep Syamsul. *Jurnalistik Online : Panduan Praktis Mengelola Media Online*. Bandung : Nuansa Cendikia.**34**.(2012)
2. L. Safko. *The Social Media Bible. Tactics, Tools, And Strategies For Business Success. John Wiley&Sons Inc 2nd Edition. 3.* (2010)
3. R. Rosady. *Manajemen Public Relations & Media Komunikasi.Jakarta : PT Rajagrafindo Persada.**11**.* (2008)
4. F. Nova. *Crisis Public Relations (Bagaimana PR Menangani Krisis Perusahaan).*Jakarta : Grasindo. **38**. (2009).

DOI: 10.1051/shsconf/20173300041

5. P. Wahyuni . *Special Event. Alternatif Jitu Membidik Pasar.xxxi* (2013)
6. I. Astuti Santi. . *Jurnalisme Radio. Simbiosa Rekatama Media.* (2008)
7. J. Moleong, Lexy. *Metode Penelitian Kualitatif.6.* (2002)

A changing paradigm of interpersonal communication in divorce family

Lucy Pujasari Supratman[1,*]

[1] Faculty of Communication and Business, Communication Department, Telkom University, Telekomunikasi Street Number 1, Bandung-West Java, Indonesia.

Abstract. Indonesia is a country that embraces the Eastern Culture. The Eastern Culture identity attached to each family along with their values and norms. In the modernization era, there is a changing paradigm on looking at the value of family unity. The stigma in divorced families is no longer viewed negatively. It has been considered as nowadays lifestyle. According to statistics data from the Religious Affairs Ministry, the divorce rate in Indonesia is increasing. The results of this study are the changing paradigm of agreeing the parent's divorced decision to teenagers and the changing social interaction in the family. It has changed the interpersonal communication between the single parents and their young teenagers. The teenagers experience confusion to communicate in the beginning of the divorce with both parents. They traumatized, isolated themselves, and breakout. Effective interpersonal communication just started to shape after the single parents succeeded to adapt to a new social family pattern. Teenagers begin to shape their positive identity after it. While the negative identity is shaped to the teenagers who fail to communicate with their single parents.

1 Introduction

Family is the smallest social community formation. It is the smallest organization in society and the first container for psychological development for children. They will receive their first education on various orders of life in the family. All members of family consist of father, mother and children who obtain different roles. In general, father acts to protect and provide family needs. While mother role is to take care of the household and educate the children. Children role in the family is to accomplish the education. Parents must give good education to children, so they will have a bright future. Each role is executed according to the function in the family. The role of the family should be built together to foster harmonious family.

Parents will introduce the children to religious values, norms and community life as the foundation for them to socialize in social environment. In addition, parents also become a major encouragement when children need the support in life. Communication of father,

*Corresponding author: doktorlucysupratman@gmail.com/lucysupratman@telkomuniversity.ac.id

mother and children will be interconnected at any time in social interaction. Social interaction in the family occurs because they need each other.

Communication within a family is an important thing to be loved. When parents can no longer create good communication for children, the family atmosphere will not be conducive. The estrangement will create a communication gap. Personal ego of parents will sharpen into unsolved conflict to result in divorce. Parents have formed a failure in maintaining the balance of the family because they are not able to resolve the existing conflicts. The family on parents who choose the divorce solution to end the conflict has changed the social family hierarchy. It only consists of a father and the children, or mother and the children. Segrin [1] gave the termination of binuclear family, which is father or mother turned into a single parent. Parents no longer live together in the house. They should separate into non-intact families. A change in family form makes the gathering intensity of parents and children getting reduced.

This study is aimed to describe the changing paradigm of social interaction in divorce family. I would like to explain the formation of interpersonal communication between teenagers and their single parents after experiencing the divorce in globalization era.

2 Literature Review

There are many research about the interpersonal communication in divorce family have been conducted earlier. Most of all were published by using quantitative research method. As Fagan [2] stated on his research that the divorce brought a great impact to the prominent environment, such as family, worship place, school, marketplace and government. Those things were the main institution to create harmonization and to eliminate the side effect of the divorce to the children. Children were considered to be the victim in this research. It suggested that the academician, religion leader and government put the responsibility to the development of children from divorce family on their future education, morality, emotional, and physical. Whereas [3] Belinda et al elaborated their research about the impact of divorce to children in United States of America. Their research subject was teenagers from divorce family. Mostly, they refuse their parents separation. Moreover, teenagers from divorce family tend to follow the steps of their parent divorce rather than those parents who remain together. It was caused by teenagers' difficulty in adapting new family atmosphere. Their experiences as teenagers from divorce family effected family member health, outcomes and teenagers welfare. While [4] Kasoman stated that the student's education performances were influenced by divorce family situation. Students' performances were not able to get good marks in the school subjects. The research was located in some Mbala Disctrict schools. This research recommended five atmospheres to have the responsible in reducing the divorce rate to the edge. The more strengthen family bonding will be increased human resource capacity in the environment.

3 Methodology

The methodology of this research was qualitative research with phenomenological approach. It was a subjective perspective based on the informant's experiences. The data collection was obtained through the informant's frame of mind and their field of experiences as teenagers from divorce family. The writer collected the data through depth interviewing, informants observing and literature reviewing.

DOI: 10.1051/shsconf/20173300042

4 Data Analysis

In collecting the data, the writer used snowball sampling technique. There were ten teenagers who had become key informants in this research. Many teenagers had the experience of their parents' divorce since they were in kindergarten and elementary school. These are the informants' profile:

Table 1. The Informants Profile Analysis

PSEUDONYM OF MALE TEENS	AGE		THEIR PARENTS' DIVORCE AT THE AGE	REASONS TO DIVORCE	AFTER THE DIVORCE FOLLOWING
Teak	19		15 years old	Abandonment	Mother
Pine	19		16 years old	Financial Issue	Mother
Palm	19		15 years old	Disharmony	Father
Mahogany	19		17 years old	Betrayal	Mother
Damar	20		16 years old	Financial Issue	Mother
PSEUDONYM OF FEMALE TEENS	**AGE**		**THEIR PARENTS' DIVORCE AT THE AGE**	**REASONS TO DIVORCE**	**AFTER THE DIVORCE FOLLOWING**
Rose	19		16 years old	Betrayal	Mother
Jasmine	19		15 years old	Disharmony	Mother
Orchid	19		17 years old	Betrayal	Father
Ivy	19		17 years old	Disharmony	Mother
Lotus	19		17 years old	Violence	Mother

Source: Research Result (2015 to 2016)

5 Finding and Discussion

Teenagers really need both parents to develop healthy communication family formation. As Simanowitz [5] told about how difficult for teenagers in facing and building their new sense of identity without the help of family. They could have the interpersonal communication with their single parent to communicate each other and build teenagers identity. The support of their single parents in having good interpersonal communication made the teenagers learn to be brave by accepting the new family structure situation.

Without good interpersonal communication, teenagers will be confused to interpret their identity (self-concept). Based on my research result, the 8 teenagers can receive their parent's breakup after the interpersonal communication between single parent and the teenagers run well. They can even digest the reasons of their parents' divorce. But the 2 teenagers in this study keep committing negative actions after parental divorce. The negative actions are done by them to ask for more attention and their nonverbal symbol for the disagreement of parents' divorce.

Teenagers will observe how parents respond divorce. Negative behaviour can occur in teenagers as the impact of parental divorce. They feel victimized by divorce and the cause of their parents' separation. They display self-injurious actions in order to get the attention of parents. In addition to expect parent's attention, they do bad action in response to their grief over the divorce. These are the categorizations that faced by the teenagers who didn't do effective interpersonal communication with their parents.

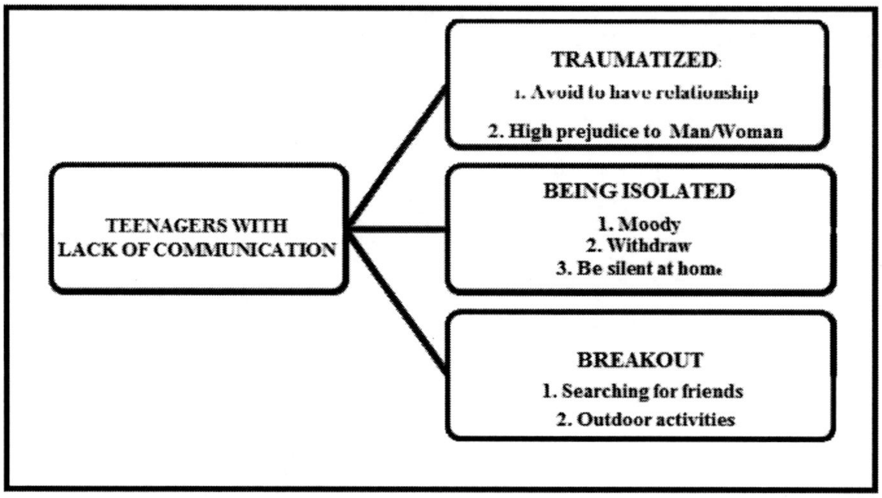

Fig. 1. Negative Teenagers Self Concept from Divorce Family

The role of communication in divorce issues that has been faced by the teenagers are in a prime position. Parents should talk about the reason of the divorce to the teenagers. Actually, they are able to use their logic properly to understand the reasons for the divorce of their parents. The interpersonal communication in the family must be established gradually. The support of their single parents in the form of good communication will make the teenagers learn to be brave by accepting their parents' decision to get divorce.

Effective conversation which occurred in daily talk will shape teenagers confidence. It is established gradually to tighten the relation of each member of the family. At the end, it built positive self-concept to the teenagers by the form of new identity value. This is not only for the teenager, but also for the single parents too.

The interpersonal communication has transformed teenager's perception about having the divorce parents. When having an interpersonal communication, the single parents should focus on the content quality to dive into teenagers mind and heart. There were different functions of having communication skill for the teenager in the family. Teenage girl had personal talk to strengthen the relationship, while teenage boy had the dimension to break the relations up with one of the single parent.

Eventually, the interpersonal communication between single parents and teenagers were transformed to transactional conversation. Teenagers would not feel lonely anymore because their existences were admitted by their single parents. They enter a brand new situation where they finally found their new identity as teenagers from divorce family. This was called the emergence of having an effective interpersonal communication in order to strengthen and support each member of the family.

The teenagers got their new vision on how to face their parents' divorce. They could encounter themselves from this problematic situation. This vision had constantly shaped their self-concept. In the earlier of divorce term, teenagers found it difficult to adapt into a new situation of having single parent. They tend to present negative self-concept. As the years went by, single parents kept making good interpersonal communication. At the end, teenagers' self-concept had finally transformed into constructive self-concept. While single parents who never had any talked during and after their divorce with their own teenagers, seemed to have negative self-concept.

6 Conclusion

Teenagers need the support of their father or mother in building good interpersonal communication. Interpersonal communication takes an important part to build the gap through losing parents togetherness. It can actually transform them to have good personality in order to avoid the negative effects of divorce. This form of communication is important in order to create a harmonious atmosphere for the establishment of teenagers. They will feel comfortable in communicating their fulfilment of self-actualization to their parents. Good interpersonal communication in divorced family which is done by the single parent will help teenagers to avoid deviant behaviour and destructive behaviours. They really need the figures of a mother and a father as their role model in shaping their self-identity. If it has not been done, they will lose motivation and the support to master their unstable emotions as teenagers.

References

1. Segrin, Chris, Flora, Jeanne. *Family Communication*. (London: Lawrence Erlbraum Associates Publishers) (2005)
2. Fagan, Patrick and Aaron Churchill, Journal of Marriage and Religion Research Institute. Washington DC. **202.393.2100.** (January 11, 2012).
3. Hewitt, Belinda, Zlatko Skribs, Mark Western, Journal of Australian Sociological Association & Sociological Association of Aotearoa New Zealand Joint Conference. University of Auckland, New Zealand. **20** (4-7 December 2007).
4. Kasoman, Frank, *The Influence of Broken homes on Pupil Academic Performance in Selected Schools in Mbala District: Lessons for School Managers.* (Dissertation: University of Zambia) (2012).
5. Simanowitz, Valerie, Peter Pearce. *Personality Development.* (London: Copyright Licensing Agency Ltd., 2003)

DOI: 10.1051/shsconf/20173300043

Ad-Creativity, Internal Consumer Response, and Purchase Intention: An Evaluation upon Ardiles "Flash Mom" TV Ad

Hatane Samuel[1,*], and *Janice* Aurellia[2]

[1] Marketing Professor at Faculty of Economy, Petra Christian University, Indonesia
[2] Marketing Program Alumny, Faculty of Economy, Petra Christian University, Indonesia

Abstract This research aim to analyse the effects Ad-creativity, Internal Consumer Response, and Purchase Intention as an evaluation upon Ardiles' "Flash Mom" TV ad. The research used 100 housewives as the sample respondents. The respondents were chosen when they were taking their children to go to some primary schools in Surabaya. Each of them were shown the ardiles' advertisement entitled "flash mom" twice and then they were asked to answer the questionare. The research variable Ad-Creativity as the independent variable consist of Divergence dimention which is measured by five indicators and Relevance dimention which is measured by three indicators. Internal Consumer Response as the mediation variable consist of cognitive and affective response dimentions, which each of them is measured by three indicators. Three indicators measure the Purchase Intention as the dependent variable. This research use Partial Least Square (PLS) as the quantitative analysis tool. The results prove that there are positive effect from Ad-Creativity to Internal Consumer Response, there are positive effect on Internal Consumer Response to Purchase Intention, and there are positive effect on Ad-Creativity to Purchase Intention.

1 Introduction

Domestic footwear production in Indonesia has an enormous sales value and can significantly influence Indonesian economic. Not only producing the footwear for the domestic residents, this production can also greatly contribute the export goods. The data from Indonesia trading department showed that the footwear industry growth in 2009 from 1.7 billion USD to 3.86 billion USD in 2013. There are lot of footwear companies in Indonesia that compete to get the maximum market share both in export trading and domestic. Thus, it is essential to create a good relationship among the customers.

The company's achievement in building the company image is made by creating the consumer perception upon their brand through promotion, [1]. Advertising is a marketing communication in the form of billboard, newspaper or tv ads, and signboard. Advertising should be able to provide value that can make the goal of success, which is able to affect a person to buy the branded products offered. Advertisement has exposure value that has to

*Corresponding author: samy@petra.ac.id

be done intensely by the company. Creativity is the main key to create the advertisement that can communicate the company goal, [2]. PT Wangta Agung Ardiles is one of the footware companies, particularly in school shoes. Ardiles uses persuasive media communication through tv ads that keep updating every year in order to approach the consumer. One of the tv ad from Ardiles in 2015 entitled "Flash Mom". Television as one of the media electronics is the universal media that is capable to inform audio visual.

In Indonesia, television is still the main medium for the society to consume the information (95%), followed by internet (33%), radio (20%), newspaper (12%), tabloids (6%), and magazines (5%), [3]. In majority, the household will turn on their tv for seven hours a day and it becomes an important thing in social environment, [4]. The company uses tv ad in order to create the awareness of a product or a new brand, inform the fitur and product excellence, create the taste upon the product or brand or even persuade the consumer to buy the product or brand that has been advertised, so that the prospective consumert might have persuasive and emotional interest, [5].

Attentional affect of an advertisement can be through the cognitive and affective response, which is called consumer internal response, [5]. It is important to know how the consumerts are able to show their response during or after watching the tv ads, and whether the advertisement is able to meet the communication goal, [6].

2 Literature Review

Advertisement is the form of nonpersonal mass media communication that can inform a fact about a product or remind the consumer to buy the favourite brand and it is done by the sponsorship or company, [6]. Advertisement is one of the promotion tools in the form of communication which is commonly used by the organization to deliver or introduce the company identity such as brand or product, [7]. A creative advertisement might help to attract the consumer attention. Divergence and relevance are two dimentions that can be used in order to enrich the creativity in the advertisement, [5]. Divergences refers to in what extent the ads contains stroryline, which is out of the ordinary. Divergences is considered as the main element in creativity. However, an advertisement must be relevance, which is meaningful, useful, and worth to watch, [7].

Many researches proved that the steps of ads exposure are attention to the ads, get the message, like the ads, and then purchase intention, [8]. The consumer response towards ads directly affect the cognitive or affective response, which determine the purchase intention, [9]. Cognitive response theory summarizes that an ad should have capability to influence consumer and attract the consumer upon the product offered that lead to purchase intention, [10]. Affective response theory is the emotional effect of cognitive response that occur to consumer when they see the ad, such as happiness or other emotion that occur from the message or the produc characteristic on the ad [11]. In general, the purchase intention is based on the effort to match between purchase motive and the attributes or the brand characteristics that are being considered by involving psychological aspect, [12].

Purchase intention is determined by: (1) perception, which is a process when someone is chosing, arranging, and interpreting the stimultant, so it becomes a complete picture and has meaning. (2) Attitude, which is a tendency where someone behaves consistently when he responses on an object or stimultant through either pleasant or unpleasant reactions. (3) Motivation, which is something that come from someone that push him to perform an action. (4) Integrase, which is the unity between attitude and action, [13].

3 Research Hypothesis

H-1: Ad-Creativity affect Internal Consumer Response
H-2: Internal Consumer Response affect Purchase Intention
H-3:Ad-Creativity affect Purchase Intention

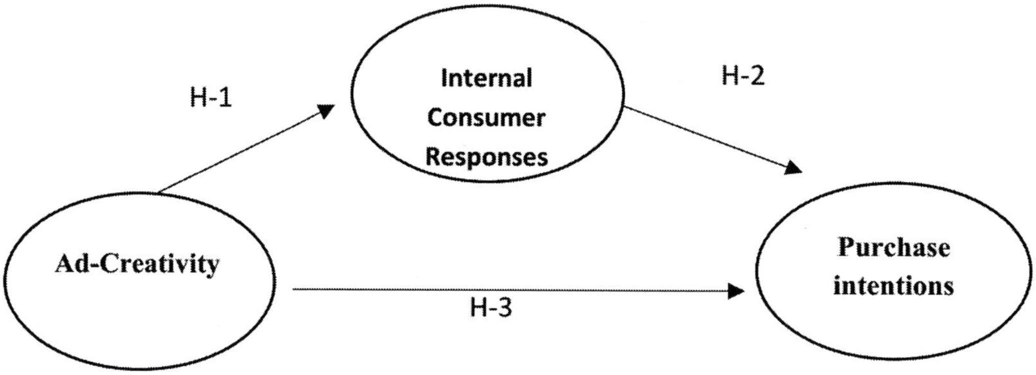

Fig. 1. Hypothesis Framework

4 Research method

Population in this research is all the housewifes who have children on the primary school and with the middle economic status. The sample were taken by purposive sampling, there were 100 housewives who have children in the primary school, belong to middle class economic status, had ever watched the Ardiles tv ad "Flash Mom", if they had never watched the ad, then would be shown twice, and then they would be asked to answer the prepared questionare. The 100 housewives were chosen when they were taking their children to school in some primary schools in Surabaya. The research variables used in this research are;

Independent Variable: Ad-Creativity (X) which is constructed by the divergence element (5 indicators) and relevance (3 indicators). Divergences refers to the extraordinary storyline or something new and difference from the others. Relevance focus on emotional aspect of the ad that is meaningful for the consumer. Those two aspect are able to make a successful ad.

Intervening Variable: *internal consumer response* (Z) summarizes that an ad should have capability to influence the consumer and attract them to the product offered, which is constructed by cognitive responses (3 indicators): that show consumer's responses in receiving information or message on the ad; and affective response (3 indicators): when the consumer seems to attract and like the ad.

Dependent Variable, (Y) *purchase intention* (3 indicators) is the responses on the creativity of tv ad that lead to an intention or even action to purchase the product.

DOI: 10.1051/shsconf/20173300043

5 Data analysis

The data analysis in this research used Partial Least Square model with the steps as follow: (1) reliability which is divided into two steps, *indicator reliability* evaluation and *internal consistency reliability* evaluation. (2) Data validity evaluation through *convergent validity* and *discriminant validity*. (3) *Path coefficient* and *coefficient of determination* assessment. (4) Hypothesis assessment through *t-Statistic*.

6 Discussion

From the data collected on 100 respondents who have answered the questionare, the researchers analyzed based on the steps as explained before which is first reliability evaluation through *indicator reliability* evaluation and *internal consistency reliability* evaluation. As shown in table 1, for the Ad-Creativity variable (X) which is constructed by divergence dimention (indicator x11-x15) and relevance dimention (indicator x21-x23) all the loading factors are above 0,7 so it can be said that both dimentions have good *indicator reliability*. In addition to that, the Cronbach alfa coefficient, which is 0.923 is supported with the composite reliability value that also above 0,70. It shows that this construction variable also has a good *internal consistency reliability*. Thus, the analysis can be continued.

The same thing happens to *Internal Consumer Response* (Z) which is constructed by cognitive responses (indicator z11-z13) and affective responses (z21-z23). All the indicators have the loading factor above 0,7 which mean they have good *indicator reliability*. Besides, the Cronbach alfa coefficient, which is 0.872, is supported with the 0.903 composite reliability value. Again, it is above 0.7 that means the conduction variable of *Internal Consumer Response* has good *internal consistency reliability*. Therefore further analysis can be done. For the purchase intention variable with its three indicators (y1-y3), it also has the loading factor above 0.70 so it means that it has good indicator reliability. The Cronbach alfa coefficient, which is 0.896 and supported with 0.936 composite reliability also shows that it has good *internal consistency reliability*. The AVE value or *communality* of those three variable shows that all of them are above 0.50 so it has possible *convergent validity* to be analyzed further.

Table 1. The Research Instruments Validity and Reliability Evaluation

Variabel	Indicator	Loading	AVE	Cronbach Alfa	Composite Reliability
Ad-Creativity (X)	X1.1	0.807	0.649	0.923	0.937
	X1.2	0.844			
	X1.3	0.759			
	X1.4	0.809			
	X1.5	0.813			
	X2.1	0.851			
	X2.2	0.803			
	X2.3	0.758			
Internal Consumer Response (Z)	Z1.1	0.818	0.609	0.872	0.903
	Z1.2	0.770			
	Z1.3	0.822			
	Z2.1	0.731			
	Z2.2	0.752			
	Z2.3	0.787			
Purchase Intention (Y)	Y1	0.850	0.829	0.896	0.936
	Y2	0.947			
	Y3	0.932			

Path coefficient evaluation is used to show how strong the effect or influence of each variable within the models. Meanwhile, the *coefficient determination* is used to measure how many endogen variables are affected by the other variables. Table 2 shows that the *path coefficient* of each variables are as follows: *Path coefficient* between *Ad-Creativity* and *Internal Consumer Response* is 0.819. It means that these two variables have positive effect relationship. This result is consistent as what shown in the previous study [5]. Positive effect of *Ad-creativity* upon *Internal Consumer Response* has 0,000 significat value. Statistically, it shows that this relationship is very significant and if it compares to the relation effect between variable within models, can be seen that these two variables are the strongest. It means that the *Ad-creativity* in Ardiles "Flash Mom" that is constructed by divergence and relevance dimentions has decent exposure value to *Internal Consumer Response.* It shows that H-1 is accepted.

The next one is *path coefficient* evaluation between the *Internal Consumer Response* and *purchase intention.* It shows that the coefficient is 0.312 with the 0.0049 significant value, which mean that there is positive effect between *Internal Consumer Response and purchase intention.* This result support the previous study [14], which find out that there is positive *Internal Consumer Response* upon purchase intention. By looking at the significancy shown above, it ccan be said that the H-2 is also accepted. Based on the finding, can be concluded that there is indirect effect between *Ad-creativity* upon *Purchase Intention* through *Internal Consumer Response* that is 0,819 x 0.312 = 0.255, it proves that *Internal Consumer Response* is a perfect mediation to *Ad-Creativity* upon *Purchase Intention.*

Table 2. Hypothesis Evaluation

Hubungan Pengaruh	Path Coefficient	Standard Erorr	T Statistic	Sig.
Ad-Creativity -- Internal Consumer Response	0.819	0.041	19.614	0.0000
Ad-Creativity – Purchase Intention	0.484	0.108	4.532	0.0000
Internal Consumer Response -- Purchase Intention	0.312	0.107	2.885	0.0049

By looking at the direct effect between *Ad-Creativity* and *Purchase Intention,* it shows that the path coefficient is 0.484 with 0.000 significant value. It can be said that *Ad-creativity* has strong positive direct effect upon Purchase Intention. It shows that H-3 is also accepted. Looking at the direct and indirect effects shown before, thus the total effect of Ad-Creativity upon Purchase Intention is 0.484 + 0.255 = 0.739. The coefficient value which is close to one proves that the exposure ability created by *Ad-creativity* on Ardiles "Flash Mom" is able to create Purchase Intention properly.

7 Conclusion and Research Limitation

The findings in this research has empirically demonstrated that there is a direct as well as indirect (through internal consumer response) effect of Ad-Creativity to Purchase Intention. These findings also show that the advertising design in Ardiles "Flash Mom" fits the target consumers. This research has a limitation in its generalisability due to the non-random sample. Besides, there is the possibility that the information obtained from screening ads had bias responses due to the diverged educational backgrounds of the respondents.

References

1. S.E. Ezekiel, S. A Kenechukwu., O.N. Leo Edegoh, International Journal of Asian Science **3**, 4 (2013)
2. G.M. Zinkhan, J.Advertising **22,** 1-4 (1993)
3. Nielsen Media Research. *Konsumsi Media Lebih Tinggi di Luar Jawa.* http://www.nielsen.com/id/en/press-room/2014/nielsen-konsumsi-media-lebih-tinggi-di-luar-jawa.html. (2014)
4. R.D. Joseph, *The Dynamic of Mass Communication* (**12th** ed). New York: McGraw-Hill. (2013)
5. R.E. Smith, X. Chen, Yang, J. Advertising **4 (1/2)** ,31-58 (2004)
6. M.R. Solomon, Marshall, Stuart. *Marketing* (**5th** ed). New Jersey: Prantice Hall. (2008).
7. S.B. Smith, MacKenzie, Xiaojing Yang, Laura Buchholz, William K. Darley, Marketing Sci **26**, 6 (2007)
8. Smith, J. Chen, X. Yang, J. Advertising **37**, 4 (2008)
9. D.J. MacInnis, D.J. Jaworski, J.Marketing **53**, 4 (1989)
10. E. Thorson, J. Moore, *Integrated Communication: Synergy of Persuasive Voices*. Mahwah. NJ:Lawrence Erlbaum. (1996)
11. M.P. Gardner, "Mood States and Consumer Behavior: A Critical Review". Journal of Consumer Research 12. (1985).
12. Morrison. *Periklanan dan Komunikasi Pemasaran Terpadu*. Jakarta: Ramdina Perkasa. (2007)
13. L.G. Schiffman, L.L. Kanuk, *Consumer Behavior* (9th ed). New Jersey: Prantice Hall. (2008)
14. Smith, _J. Chen, J. Marketing. **28**,5 (2009)

SHS Web of Conferences **33**, 00044 (2017) DOI: 10.1051/shsconf/20173300044
i-COME'16

Breaking The Traditional Communication Flow: Exploration of Social Media Opinion Leaders in Health

Emma Mohamad[1,*] , *Abdul Latiff* Ahmad[1] , *Sabariah* Mohamed Salleh[1] and *Ummu Khadijah* Wan Sulaiman[1]

[1]School of Media and Communication, Faculty of Social Sciences and Humanities, Universiti Kebangsaan Malaysia, 43600 UKM Bangi Malaysia

Abstract. Opinion leaders play an important role in mitigating health issues and are able to influence attitudes and health behaviors to a certain extent. However, in the world we are living today, individual health decision making no longer rely fully on doctor's advice, but often persuaded by people or organisations deemed reliable and trustworthy. Interestingly, one does not have to have a medical degree to be seen as "reliable" or "trustworthy" in giving out medical advices. Computer-mediated-communication changes the way opinion leaders emerge in virtual communities, specifically in the context of health. Through increased access to both traditional and online media, people have a fair access to communication technology and therefore, our definition of opinion leaders have shifted from how we used to understand it through Lazarsfeld's two-steps flow theory. In fact, the way opinion leaders operate via social media platform is very different from how it used to be through the traditional mass media. This paper will discuss on the characteristics of new media opinion leaders, particularly in the context of health, in hopes to learn new ways to promote health in future communication campaigns.

1 Introduction

Communications via the new media platform plays an important role in different aspects of our day-to-day life. Different patterns of interactions on the web and on social media challenged traditional theories of understanding how the circuit of mass communication works. Computer-mediated-communication changes the way opinion leaders emerge in the virtual community and how they operate is different from how we used to understand it through Lazarsfeld's two-steps flow theory.

Lazarsfeld et. al. [1] argued that opinion leaders are the first ones exposed to (traditional) media content, forms opinions on the issues that was discussed and infuse their opinions to the public. This infiltrated opinion will then later shape public opinion. In the present scenario however, this model is rendered problematic. In this day and age, we have

* Corresponding author: emmamohamad@ukm.edu.my

increased access to both traditional and online media, and even though digital divide is still present, people have a fair access to communication technology. Therefore, our definition of opinion leaders have shifted and is no longer limited to those who have access to the media. In fact, the way opinion leaders operate via social media platform is very different from how it used to be through the traditional mass media. This paper intend to explore how the definition and conceptualization of opinion leaders facilitated by communication technology have evolved particularly in health.

2 Changes in the way opinion leaders operate : Rethinking Two Steps Flow theory

First introduced almost seven decades ago, the Two Steps Flow Theory has served as a significant model in communication and is extensively utilised in many studies. Lazarsfeld et al [1] found that opinion leaders have certain influence towards the society through interpersonal communication. It is suggested that the more frequent and effective personal contact between opinion leaders and society, the more they can influence people's perception, motivation, attitude and actions.

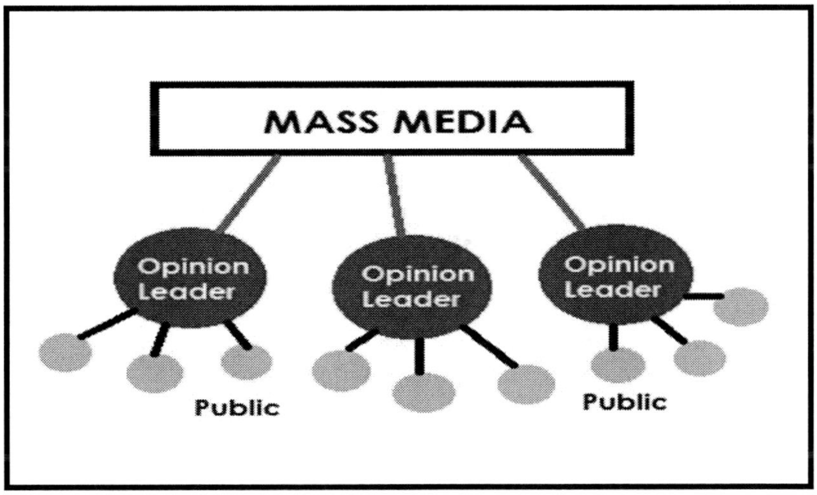

Fig. 1. Lazarsfeld et. al. Two-steps flow theory

However, the two steps flow theory only discussed how opinion leaders influence society through interpersonal communication, which is not entirely suitable anymore in our current environment today. This is mainly because mass media content is now accessible directly to society, thus opinion leaders might not able to control public opinion entirely.

Media scholars continue to debate on the relevance of opinion leaders in communication processes as well as the flow of communication itself. The idea of having 'multi steps', rather than just 'two steps' in communication flow from mass media to the public began to evolve and was accepted as the new way of looking at information dissemination. The multi steps flow suggests that communication can be multi directional and not limited to a one way street connected by opinion leaders. In fact, information flow can reverse from the public to the mass media through audience feedback process. Not only the multi steps flow better explains communication processes in current media landscape, it also acknowledges audiences as active media users. The process of forming public opinions also becomes more complex to understand but also very dynamic, whereby different groups of publics can influence other sets of publics when they share insights with each other.

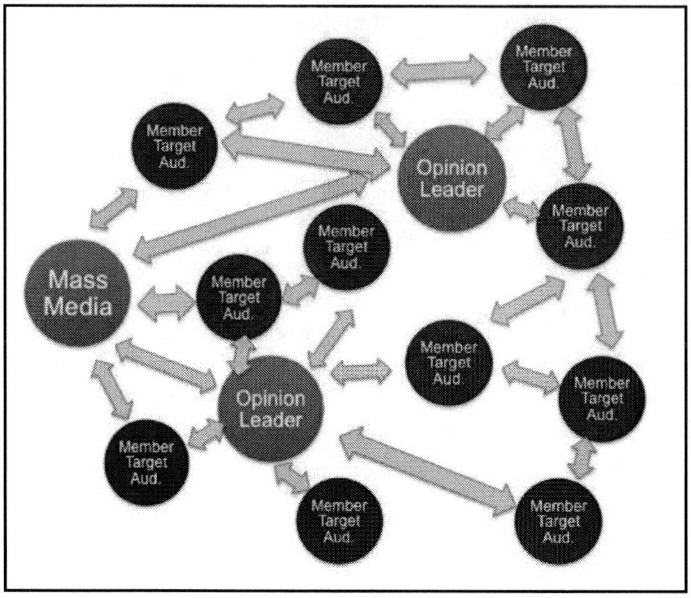

Fig. 2. Multi-steps flow.

The multi steps flow model also suggests that mass media messages can differ considerably from the original message. This is due to the exchange of meanings and interpretations by individuals who will then alter the original message according to his or her own understanding, hence giving it a new meaning. This new meaning will then be passed on and shared to other individuals, who then may also add on another layer to recontextualise the message.

Yet opinion leaders remained significant, but shifted its role and influence in a different direction. Secondly, the development of contemporary media creates a new breed of opinion leaders. These parasocial opinion leaders are more reachable virtually and easier to interact with rather than the traditional opinion leaders. The way they work to influence society's perception, motivation, attitude and actions through social media is also different than the traditional way. The multi steps flow is also known as diffusion of innovation theory developed by Everett Rogers in the 1960s. This theory specifically studies how innovation envelops through four stages i.e : invention, diffusion, time and consequences, whereby opinion leaders are seen as playing an important role in facilitating people's acceptance of new innovation.

With the increase of public's access to mass media content, one could argue that opinion leaders may lose its pressence and importance. After all, any individuals can form their own opinions based on how they understand media messages, and do not rely on opinion leaders to help them form judgement towards an issue. However, on the contrary, new media opinion leaders are stronger in pressence and increased persuasiveness in forming public opinion. Social media phenomenon in particular have witnessed the rise of opinion leaders in many areas, for example politics, fashion and health. Therefore it is important that researchers re-examines the structure and communication processes around digital/virtual opinion leaders to better understand why opinion leaders remain strong (if not stronger) in influencing public's opinion even when society nowadays are considered as active audience and media users.

DOI: 10.1051/shsconf/20173300044

3 Health opinion leaders : Local and international examples

In health communication, opinion leaders play an important role in persuading heath actions. A systematic review published in Cochrane by Flodgren et al. [2] found that opinion leaders alone or in combination with other interventions may successfully promote evidence-based practice. However the review found that in most studies, the role of opinion leader was not thoroughly examined and it is therefore unclear as to what is the best method to optimize the effectiveness of opinion leaders, and whether social media can play a role in influencing behaviours. It is timely that we re-examine the traditional way of defining opinion leaders and relook at its strategy in health communication.

Opinion leaders play a role in mitigating health issues and able to influence attitudes and health behavior to a certain extent. However, in this 21st century, individual health decision making no longer rely fully on doctor's advice, but often persuaded by people or organisations deemed reliable and trustworthy. Interestingly, one does not have to have a medical degree to be regarded as "reliable" or "trustworthy". In fact, thanks to the Internet, health opinion leaders may even transpire from small and marginalized group or people.

This paper will generally explore 5 Malaysian instagram accounts (considered opinion leaders in health). Investigation on these instagram content serves as an early exploration to find out what these health opinion leaders have in common. This preliminary observation on characteristics health opinion leaders have may help guide / give directions to further research on social media opinion leaders, particularly in relations to the communication flow and process.

Selection of these instagram users are random, based on several categories i.e: lifestyle, exercise, diet, sports and illness/health issue. We observed five latest instagram posts related to health posted by these five selected users. The five chosen opinion leaders are as listed in Table 1 below:

Table. 1. Instagram Opinion Leaders in Health.

Numb.	Instagram opinion leader	Instagram name	Category	Number of Followers
1.	Khairy Jamaluddin (Minister of Youth and Sports)	Khairykj	Sports	390K
2.	Kevin Zahri	Kevinzahri	Fitness and Weight Loss	155K
3.	Meameao	Meameao	Healthy Diet	9189
4.	Nana Al- Haleq	Nanaalhaleq	Fitness	14.1K
5.	Sharifah Sakinah	Sharifah_Sakinah	Exercise	966K

Generally all these five opinion leaders have several characteristics in common through observations of their health-related postings. Firstly, *they walk the talk.* All of them do not just preach about the importance of health, they become the example of what they believe in. For example, Khairy Jamaluddin is not just popular because of his status as Malaysia's Minister of Youth and Sports, he also shows that he is involved and trying to push himself to try new sports/activities. His latest post for example showed him crossing the finish line for Ironman 70.3 in Putrajaya. In his caption, he emphasised on the importance of having a positive mind and to be successful. He also shared how hard it was for him to complete the challenge and that showed his vulnerable side, something that could appeal to a lot of people.

Secondly, all these opinion leaders publish instagram posts that are aimed to inspire people to become like them. This *inspiration and aspiration* content can be seen in their achievements, for example through their weight loss and toned bodies, Sharifah Sakinah is an example who puts up pictures showing off her lean muscles, and are often pictured in the gym, working out. She is also an example who lost weight considerably and her 'before and after' pictures are a type of inspiration to others.

Thirdly, all these opinion leaders *breathe and live health*. They are not pretentious in their postings, often seen wanting others to achieve what they have achieved and willing to share their 'secret' to success. Meameao for example, do not only posts delicious looking healthy cuisines, she shares all her recipes in her instagram captions. This shows that she cares about her followers and that help to build a community of followers and build social support. Social media allows for great growth in online community and this is apparent in these opinion leaders' instagram account.

These three characters observed are seen consistent across all the five opinion leaders instagram pictures. While these characters are not exhaustive, they provide some ideas on how changes can occur in communication flow of social media opinion leaders in health. Therefore more indepth research should be carried out to better understand the changes in new media opinion leaders.

4 The way forward for health campaigners

Katz [3] highlights the debates on whether opinion leadership is formed by generalised trait or by issues (which then produces specialised opinion leaders). Indeed as Merton [4] had emphasized very much earlier on that opinion leaders on "local" issues are different than the "cosmopolitan" opinion leaders. After analysing examples in both local and international context, this paper would like to suggest that key players in healthcare to consider "specialised" opinion leaders through promoting or campaigning on health.

Social media platform in particular provides a big opportunity for health practitioners and health promoters to tap into via opinion leaders and social media celebrities. It is high time that we move from relying fully on the traditional forms of campaigning on health, by utilizing ready-made resources, or by grooming new online opinion leaders. By way of understanding both the new conceptualization and communication processes of these opinion leaders, health messages can be disseminated in a more effective manner, despite maintaining its nature of mass communication. Persuasion via online opinion leaders is impactful because of the specific-ness of health issue discussed, but also more importantly is the persona of these opinion leaders, leaving feelings of inspiration and aspiration to its audience/followers.

References

1. P. F. Lazarsfeld, B. Berelson, H. Gaudet. *The People's Choice*. (1944)
2. G. Flodgren, E. Parmelli, G. Doumit, M. Gattellari, M.A. O'Brien, J. Grimshaw, & M.P. Eccles. Cochrane Database of Systematic Reviews, **8**. (2011).
3. E. Katz, Int. J. Comm., **9**, 1023-1028. (2015) [http://repository.upenn.edu/cgi/viewcontent.cgi?article=1407&context=asc_papers]
4. R.K. Merton. *Communications research* (1949)

Probing intercultural competence in Malaysia: A Relational Framework

Syarizan Dalib[1,*], *Minah* Harun[2], *Norhafezah* Yusof[1] and *Mohd Khairie* Ahmad[1]

[1] School of Multimedia Technology and Communication, Universiti Utara Malaysia
[2] School of Education and Modern Languages, Universiti Utara Malaysia

Abstract. Studies in intercultural competence are quite numerous but they were mainly derived from Eurocentric experiences. Since Eurocentric scholars may become oblivious to certain elements or issues of intercultural communication that are pertinent to Asian people, the Western conception of intercultural competence have been argued for its relevance in the Asian world. This paper aims to revisit the current (Eurocentric) perspective of intercultural competence and probes an alternative perspective of intercultural competence by reviewing current Asian literature. Our review suggests that the conception intercultural competence must consider relational aspects when it is situated within Asian experiences. Since relational aspects were a noted gap in the existing Eurocentric definitions, this paper proposes a relational framework in probing intercultural competence in Malaysia.

1 Introduction

The need for higher learning institutions to produce interculturally competent graduates has emerged in this globalized era [1]. Intercultural competence is important to help university graduates to not only solve problems in a globalized society, but also for employability, job performance and personal development. If university graduates are required to possess intercultural competence, then higher learning institutions must play an important role to educate students on what it takes to become competent.

There are numerous developed models of intercultural competence that can be used to educate students. However, most models were completely defined from the perspective of Western scholars particularly the Euro-American community [1,7, 9, 13]. Accordingly, such Euro-American centric definition has led to delineation of theories and perspectives that were Eurocentric in origin [6, 13]. Recent development in the studies of intercultural competence indicated a challenge to the application of Eurocentric theories in other cultural settings [1, 3, 8, 9, 10]. Thus, calls have been made to re-examine whether the Eurocentric perspectives can be translated into other cultural contexts. Such re-examination is useful for scholars to offer comprehensive explanations on viewing communication that goes beyond the Eurocentric perspective.

[*] Corresponding author: syarizan@uum.edu.my

2 Defining intercultural competence: A Eurocentric bias?

The term "intercultural competence" is not only diverse in its definitions by Eurocentric scholars; they are also highly diverse in terminologies (see Bennett, 2009). Regardless of such various terminologies and definitions, the notions of appropriateness and effectiveness form the fundamental conception of intercultural competence [3]. Effectiveness refers to successful goal achievement and it is closely related to satisfaction of attaining desired outcomes [4]. Appropriateness refers to the ability to identify the rules of a given situation that provides a sense of what is acceptable and unacceptable to perform behaviours [4-6]

Given the fact that current conception of intercultural competence is dominated by the Eurocentric orientation, it is worthy to explain how competency is viewed from such orientation. Scholars noted that the existing Eurocentric definitions tend to place a primary focus on the individual as the locus of intercultural competence [2]. Although many Eurocentric models assume a partner, competence is still largely viewed as an individual concept [7]. Accordingly, most Eurocentric models define skills and knowledge as a function of the individual, thereby viewing intercultural competence as possessed by the individual. In addition to the notions of effectiveness and appropriateness, intercultural competence also includes three important components: motivation, knowledge, and skills. Based on a review of current models of intercultural competence, Spitzberg (2012) proposed that motivation refers to the many positive and negative factors that encourage a person to approach or avoid members of other cultural groups. Knowledge represents the possession of cognitive information about rules for appropriateness that helps in the performance of skills in a given context. Skills are the behavioural aspect of competence that is directed toward goal achievement.

Taking all the aforementioned components, it can be conjectured that the Eurocentric orientation of intercultural competence tends to focus on the ability of an individual to achieve effective and appropriate interaction in intercultural situations based on her or his knowledge, motivation, and skills [9]. This definition certainly provides fundamental understandings on intercultural competence. However, given that current ideas about intercultural competence are vastly drawn from Eurocentric scholars' experiences, do such definitions resonate with Asian experiences? Scholars attested that if the Western scholarship is used to analyze communication experiences in the West, it is a legitimate framework for such an analysis [10]. However, the problem arises when the single perspective is often presumed for its universality in many intercultural studies. Scholars contended that placing personal control at the heart of communication competence strongly reflect Western bias in which it may fall short to explain the workings of competency in other cultures [4]. Given that the intercultural competence field has developed world wide, the Eurocentric perspective has been challenged and questioned by both mainstream and other non-Eurocentric cholars [1, 11, 12].

3 A Relational Framework

One paradigmatic assumption that was conceived as offering an alternative answer to the dominant Eurocentric perspective is the notion of Asiacentric paradigm that provides the meta-theorical lens to view Asian communication practices [10]. Asiacentric paradigm provides a theoretical explanation that emphasizes Asian values at the centre of inquiry for viewing Asian phenomenon [10]. Based on Asian cultural and communication practices, the Asiacentric paradigm offers philosophical assumptions that include the ontological, epistemological, axiological stance of human nature [10]. The ontological assumption of Asiacentric paradigm rests in the form of relationality that views everyone and everything as interrelated [10]. Many Asian traditional ways of thinking is based on the fact of humans as interdependent and interrelated beings [3, 10, 11]. This ontological assumption is much more explicitly recognized in Asian cultures than in Eurocentric cultures that are traditionally dominated by the theme of individualism and independent self [3,7]. The foregoing ontological assumption leads to the epistemological assumption that since everyone and everything is viewed as interrelated, they can be meaningfully understood in relation to one another [3, 10]. The axiological assumption in Asiacentric paradigm is intertwined with the previous assumption in that, the valued goal of harmony is crucial in Asian communicative practices [10]. This Asian axiology of harmony is in mark contrast to the Western axiology that places freedom and control as the ultimate goal for communication [4].

Based on previous writings, the Asiacentric paradigm proposes five important aspects : (1) circularity, (2) harmony, (3) other-directedness, (4) reciprocity, and (5) relationality that gives implication on communicative practices [10]. These propositions essentially reiterated the themes of interconnectedness and mutual dependence that forms an integral worldview of Asian communicative practices [3, 10,]. Nonetheless, it is important to be cautious that Asian nations are plural societies that constitute remarkable varieties of communities that are divided by language, religion, case, and ethnicity. Accordingly, the Asiacentric propositions do not necessarily reflect real-life communication among Asians but they can be used as theoretical lenses to see an Asian version of communication [3,10].

The Asiacentric paradigm provides a meta-theoretical lens for us to re-examine the relevancy of Eurocentric perspective in the Asian contexts. However, it is important to note that such a re-examination does not suggest it is a tool that can be used to exclude the existing Eurocentric perspective. Rather, it provides the opportunity for finding convergent and divergent points between the Eurocentric and Asian perspectives. This attempt does not only contribute into providing an alternative outlook to the conception of intercultural competence, it also offers multicultural sharing that acknowledge cultural differences.

4 Probing intercultural competence in Malaysia

A recent phenomenological study in Malaysia indicated that intercultural competence is viewed as a co-created process between culturally different members that occurs through interpersonal relationship [1, 8]. Drawing from such study, a new conceptual model of intercultural competence has been delineated [1]. Accordingly, we propose a new conceptual model (figure 1) that re-defines the individual focused conception of intercultural competence into a framework that highlights a relational process between two culturally different members to achieve effective and appropriate outcomes. It is important to note at that, to echo Deardorff, although there are distinctions between the Western and non-Western perspectives in viewing intercultural competence, there are also certain elements that may be similar between Western and non-Western cultures [7]. As such, in the spirit of Chen's call to finding co-existence with Eurocentric perspective by accepting foreign elements and integrating them into one's cultural milieu [12], we contend that it is useful to re-conceptualize Deardorff's work [2]. Such a re-conceptualization does not only sensitive Asian perspective in describing competency in Asian context, but also provides a rethinking on the application of Eurocentric scholarship.

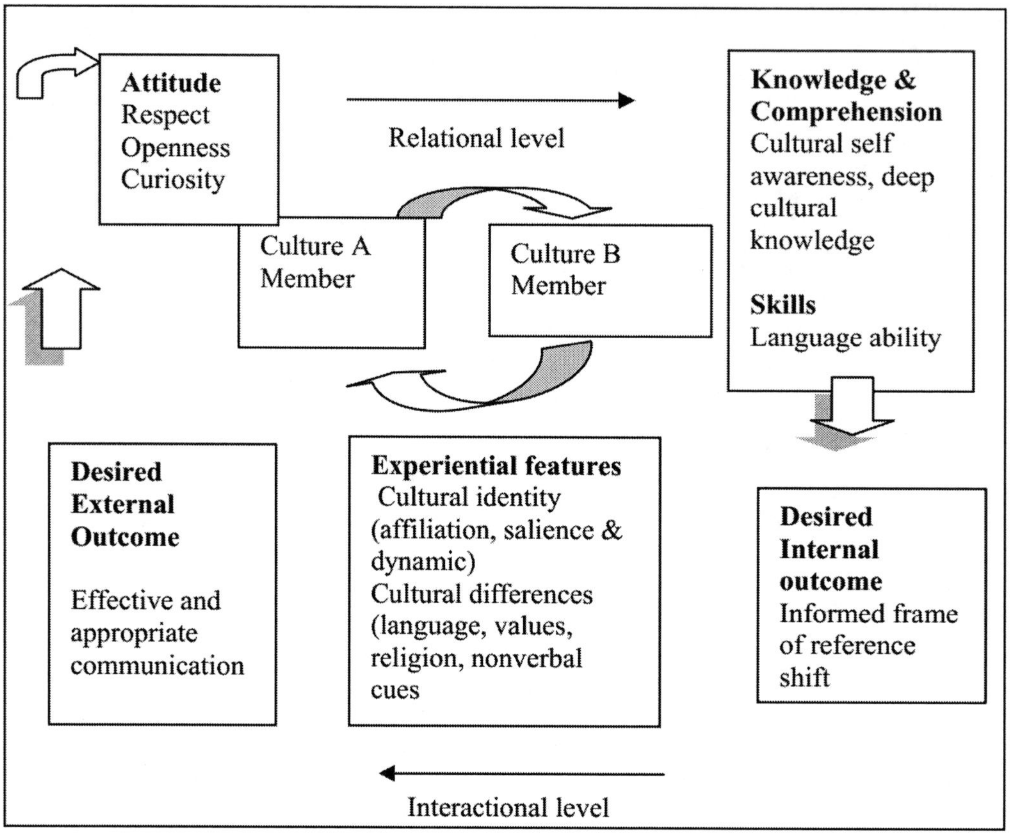

Fig. 1. Conceptual Model of Intercultural Competence

The model suggests the following perspectives on intercultural competence:

(1) Intercultural competence is a complex process to which it occurs through the movement and process orientation of various elements from the relational level into the interactional level.

(2) Culture A and Culture B members are interlocutors with different cultural identities. Intercultural competence is a mutual function of attitude, knowledge and skills of both interlocutors (Culture A and Culture B individuals). Such function occurs through interpersonal relationship.

(3) Attitude is viewed as the fundamental starting point that moves both interlocutors to acquire knowledge and skills, and to achieve internal and external outcomes.

(4) The internal outcome which involves the internal shift in frame of reference enhances the external outcome. The internal shift in frame of reference indicates the ability of interlocutors to adapt to different communication behaviors in their interaction with one another. The external outcome can be described as the ability of interlocutors to behave and communicate effectively and appropriately in their interaction.

(5) Experiential features indicate the level of interaction in which the identity of both interlocutors (affiliation to a certain cultural group, identity salience and identity dynamic) interacts with one another resulting in the experience of cultural differences (e.g. language, religion, values and nonverbal cues).

While this new model contributes into enriching the field of intercultural competence, it needs to be further explored to gain extensive insights. As such, we suggest a further exploration of this model as a theoretical framework for Malaysian researchers. We concur that a plausible setting to probe intercultural competence is within Malaysian universities. The influx of international students has produced an increased level of diversity in Malaysia's campuses. This development indicates that it is possible for Malaysian students to experience interaction with people from diverse national backgrounds in the campuses. Since previous studies indicated that students noted living in a multicultural campus necessitates them to develop competency [1], this suggests unique opportunity to investigate how the recipient of higher education (Malaysian students) experience interaction with cultural others (those holding different national memberships) and how they define intercultural competence out of their experiences. The findings of such research may not only enrich current understanding of intercultural competence. It also helps in educating intercultural competence among our students that resonates well with the Malaysian context.

References

1. S. Dalib, Unpublished doctoral thesis. Universiti Utara Malaysia (2014)
2. D. K. Deardorff, JSIE. **10,** 3 (2006)
3. S.Liu, Journal of Multicultural Discourses, **7,** 3 (2012)
4. M.R.Parks, Communication competence and interpersonal control. In M. L. Knapp & G. R. Miller (Eds.), Handbook of interpersonal communication (2nd ed., pp. 589-618). Thousand Oaks, CA: Sage (1994)
5. B.H. Spitzberg, W.R. Cupach, Interpersonal communication competence. Beverly Hills, CA: Sage.Lustig, Koester (1984)
6. B.H. Spitzberg, G. Changnon, (2009). Conceptualizing intercultural competence. In D. K. Deardorff (Ed.), The SAGE handbook of intercultural competence (pp. 2-52). Thousand Oaks, CA: Sage.Spitzberg (2012)
7. D. K. Deadorff (2009). Synthesizing conceptualizations of intercultural competence. In D. K. Deardorff (Ed.), The SAGE handbook of intercultural competence (pp. 264-269). Thousand Oaks, CA: Sage.Miike (2012)
8. S. Dalib, M. Harun, N. Yusof, Reconceptualizing intercultural competence : Aphenomenological investigation of students' intercultural experiences. *Procedia*, 13-135 (2014)
9. G.A. Yep, Encounters with the "Other" : Personal notes for a reconceptualization of intercultural communication competence. In M. K. Asante, Y. Miike & J. Yin (Eds.), The global intercultural communication reader (2nd ed., pp. 339-356). New York: Routledge.Chen & Starosta (2014)
10. Y. Miike, Harmony without uniformity: An Asiacentric worldview and its communicative implications. In L. A. Samovar, R. E. Porter & E. R. McDaniel (Eds.), Intercultural communication: A reader (pp. 65-80). Boston, MA: Wadsworth. (2012)
11. G. M. Chen, The two faces of Chinese communication. In Y. Miike, M. K. Asante & J. Yin (Eds.), The global intercultural communication reader (2nd ed., pp. 273-282). New York: Routledge. (2014)
12. G.M. Chen, Asian J. Commun., **19,** 4 (2009)

DOI: 10.1051/shsconf/20173300046

Designing and Implication of Pictograms in Universities of Tehran (IRAN)

Firouzeh Keshavarzi[1, 2, *], *Hamedi* Mohd Adnana[1,**]

[1]Department of Media Study, Faculty of Arts and Social Sciences, University of Malaya, 50603 Kuala Lumpur, Malaysia.
[2] Department of Graphic Design, College of human science, Islamshahr Branch, Islamic Azad University, Tehran, Iran.

Abstract. Pictograms have many implications nowadays and are used for simple and fluent pictorial interpretations. These concepts are usually informative, conductive and warning. These kinds of icons are used in order to simplify the communication in an environment as are used in airports, terminals, railroad stations, parks, zoo, hospitals, drug stores and so on. To provide an efficient contrast, pictograms are often in black and white. Nowadays in different public places and landscape design pictograms are used for perfect communication. By the ways these icons are related to the civilization and tribal culture. Designing pictograms need a permanent search for better plan for better information transfer. In this project researcher designed and executed pictograms for better conduction in some selected Universities in Iran at 2015. The project was done for internal and external places of selected universities. Absence of informative and pictorial icons in universities environments causes wandering about for novices and they are not able to find different places like libraries, restaurant, and offices and so on. This study resulted in better communication in campus environments. Researcher hope pictograms will find their real and proper role in Iranian art area.

1. Introduction

Pictograms are a group of signs which are used greatly nowadays in public places. In fact, pictography is a kind of writing ideas with drawings [1]. 5000 years B.C. Chinese used pictogram to communicate. These signs are used by Indians, African and etc... [9]. In 1937 London underground used pictograms. Nowadays these are commonly used for restrooms, restaurant and railway stations [2]. In early 1980's pictograms were used to communicate with low educated persons in public places. A form of worldwide pictograms is related to cloths (washing, ironing and so on). Pictograms about chemical hazards are another form of these signs [10].

Point pictograms are composed of 50 shapes which are mainly used by travellers through public places. These kinds of signs are mainly used in airports, hotels, stations and

* Corresponding author: keshavarzi@um.edu.my
** hamedi@um.edu.my

other public places. In fact, one of the main aims of pictograms is an efficient conduction. If this icon works perfectly there will be no need to employ some persons in information centres [4].

Pictograms are simplified icons which are somehow related to the main idea. There are some difficulties and disadvantages for pictograms too, for example pictograms on medicine are effective for medium- educated persons [8].

Communication is mainly based on person's visual intellect. Recent researches show that visual intellect is related to ecological, cultural and environmental factors. In designing pictograms, the target group is very critical [7] Pictograms by conducting a concept in a small pictorial form cause beauty of environment. Lemmom & Hyman 2006 believed that pictograms should be standardized to lower the dangerous results of some pictorial especially in medical usages [6].

Davies at al 1998 found that some pictograms are not totally understood by target group in Europe [3]. Katri et al. 2004 studied groups of children as target group for concepts which were presented by Pictograms on Medicines [5].

They found most kids new them perfectly. No research has been done in educational environments as schools and universities about pictograms. In this research, researcher tries to make pictograms for a campus environment in order to lesser the problems of newly arrived students while searching for different parts of our university [7].

2. Materials and Methods

Projects were done in a campus environment about 18000 m2 from 2014 to 2015. For the first step researcher draws many design with different colours. Some of these plans were composed of Iranian design. Researcher test many materials and colours to find the best one. Some of these plans were composed of Iranian design. Researcher test many materials and colours to find the best one.

This study uses mixed methods for collecting and analysing data; qualitative and quantitative methods can help the study complete its aims. Searching documents, the study surveys some previous researches in library and databases about designing and implication of pictograms in Tehran (Iran). Observation is another way of quantitative method for collecting data which study applies it. In this way study deals with looking at current designing and implication of pictograms and seeking their universities design. Interview, observation and questionnaire are other ways which this study uses.

3. Results

Pictograms are an effective way to communicate. These should be easily understood too. Researchers design some pictograms for open places and landscapes of universities. Researchers consider the education level of my target group and try to make pictograms which are concordant with their educational level. Researcher try to simplify pictograms while no part of essential information omitted. Designed pictograms were used for each faculty. Stories pictograms were homogenized (in form and colour) with open environment pictograms.

For construction of pictograms different materials were examined. Plexiglas is affected by light, cold and climatic conditions. These cause Plexiglas to be darken. So researcher use galvanized iron plate which has the most resistant material.

Pictograms were designed due to the theme of each parts of university. Designed elements were composed with the Icon of University Logo to find a beautiful and pleasant form.

Colour was very important in this project. The colours were chosen in accordance with the university logo so researcher tries to make different composition between different forms and colour to find an efficient composition.

4. Discussion

After using these pictograms, the university environment changed to be very pleasant. There was no need for information both. In the beginning of the semester, students easily find their ways and their questions were answered indirectly. Other university can use our patterns or their own ones to communicate with their target groups. In this project pictogram is used and proposed in universities for the first time it was previously used for public places. These pictograms were used for university restaurant, telephone booth, Mosque, Library, security, entrance and so on. Without these universities should employ some officer to help people find their way. The university should pay a lot for these people. Novice has many problems when they enter a university for the first time. Pictogram are efficient way to find way and have a better understanding of environment. Graphic designs are efficient way to make the environment more pleasant. Typical design of pictograms for Islamic Azad University in Tehran (Figure1).

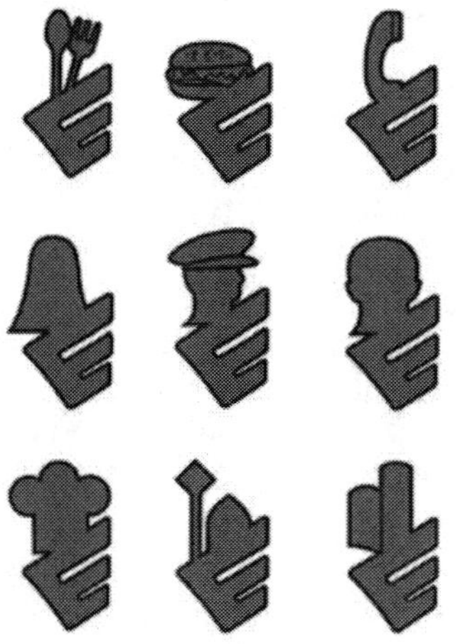

Fig. 1. Pictograms designed for different parts of the Islamic Azad University in Tehran.

References

1. A. Mosavilar, *Pictograms*. Jelve Honar (8 &9): (in Persian, 1997)
2. Anonymus. The Professional Association for Design for the U.S. Department of Transportation. *Symbol signs*, 2nd Ed. New York: American Institute of Graphic Arts (1993)
3. S. Davies, H. Haines, B. Norris and J. R. Wilson. *Safety pictograms: are they getting the message across?,* Appl. ergonomics **29**, 1 (1998)
4. E.Lupton,J.A. Miller,*Design Writing Research: Writing about Graphic Design*. New York: Kiosk (1996)
5. H.Katri, K.Kati,E.Hannes, J. Patricia,M. Airaksinen, *Do pictograms improve children's understanding of medicine leaflet information?* Patient Educ. & Couns. **55**, 3 (2004)
6. H.Lemmon, W.Hyman,*Pictowhat? Problems with the Use of Pictograms on Medical Devices*. J.Clin. Eng., **31**, 4 (2006)

DOI: 10.1051/shsconf/20173300046

7. M.S.Ehlers, R.Dowse, *Guidelines for the design of culture-specific pharmaceutical pictograms for use in a South African population.* Int J Pharm Pract, in press (2007)
8. R.Dowse,M.S.Ehlers ,*The evaluation of pharmaceutical pictograms in a low-literate South African population*. Faculty of Pharmacy, Rhodes University, Grahamstown 6140, South Africa, **45**, 2 (2001)
9. R.Dowse, M.S. Ehlers, *The development and evaluation of pharmaceutical pictograms.* In: Scott PA, Bridger RS, J. Charteris, editors. Global ergonomics. Proceedings of the Ergonomics Conference, Cape Town. Amsterdam: (1998)
10. R.Dowse, M.S. Ehlers, *Pictograms in pharmacy*. Int. J. Pharm. Pract. **6**, (1998)

Exploratory study of a crisis commander's perspectives on the role and value of public relations in crisis management

Saidathul Nizah Mat Tazin[1,*], and *Kiranjit* Kaur [2]

[1]Public Relations Department, Faculty of Communication and Media Studies, Universiti Teknologi MARA, 40450, Shah Alam, Malaysia
[2]Public Relations Department, Faculty of Communication and Media Studies, Universiti Teknologi MARA, 40450, Shah Alam, Malaysia

Abstract. This is an exploratory study into understanding the dominant coalition perspectives on role and values of public relations in crisis management in an energy company in Malaysia. In this study, the senior public relations of the energy company reports directly to the crisis commander (CC) in times of crisis thus, the CC was chosen as a sample where an in-depth interview was conducted. Notably, the CC perceptions represents the dominant coalition with the power to shape organisation policies and provide directions to public relations practice in managing media and communication in crisis management. Theory of excellence within the outlook of public relations empowerment was examined. The findings of this study have significantly confirmed that the patterns of the CC perceptions on role and values of public relations in crisis management are similar to other dominant coalition literature across the globe. This phenomenon needs changes to drive public relations to strive harder to be a problem solver and a strategic thinker in crisis management. This study is a significant attempt to understand the perception of the dominant coalition on public relations in crisis management and can be replicated in other sectors in Malaysia.

1 Introduction

The public relations role as a boundary spanner becomes significant in crisis management for trust building when there is constant communication with the society during crisis phases (pre, during and post). Public relations counsel to the dominant coalition contributes to the intention and action of accountability, transparency and integrity of organisational response in crisis management. Nevertheless, public relations dilemma in crisis management is prevalent when they became less influential with the dominant coalition. This study looks into the value of public relations from the viewpoint of a crisis commander (CC) in an energy sector in Malaysia. It is one of few attempts to examine public relations significance from the perspective of a member of a dominant coalition in crisis management in Malaysia.

*Corresponding author: saidathul06@yahoo.com

2 Literature review

Public relations literature [1, 2] emphasizes the public relations managerial function to be influential in decision-making in crisis management. Public relations as a managerial function plays a crucial role as a problem-solving facilitator engaging in environmental scanning and monitoring to identify issues and then adjust to the organization's mission to avoid conflict. However, public relations role engaging in a symmetrical communication model was critised by scholars for being unrealistic [3] and has limits especially in times of crisis [4]. Notably, public relations is based on ethical worldview that describes the normative standard of the practice.

Research is central to public relations practice [5,6,7] before having the credibility to advise dominant coalition to provide an ethical response in times of crisis. Public relations constructive role in crisis management also relies on communication planning [8], public relations function within organization and experiences managing crisis [9] and being practical, realistic and pragmatic in times of crisis [10]

Regardless of the multiple factors for public relations to be influential in crisis management, public relations significance in crisis management is dependent on the dominant coalition control and understanding over public relations value in crisis management. In other words, public relations empowerment is dependent on the power of dominant coalition who gives direction to public relations practice in organisations [11]. Relying on technical communication skills alone does not promise a successfully managed crisis without communication autonomy granted to public relations [12]. Grunig believes when the senior public relations is a member of dominant coalition, it helps organization manage its strategic public and achieve goals to become effective [13]. However, public relations has to extend its expertise as boundary spanners in various skills, resources, strategies, tactics and experiences to be accepted as member of the dominant coalition [14,15]. Public relations advises on ethical implications and maintenance of dialogic communication and relationships with publics is important to create ethical organizational behaviour [16] thus, manage public relationship for reputation.

Public relations role in managing crisis communication is growing in all parts of the world. Crisis communication practice focuses on public relations theoretical framework at large [17] where managing information at three different phases; pre, during and after a crisis is necessary [18]. In pre-crisis phases of previous organizational crisis experiences, it observed public relations involvement in strategic management managing communication in times of crisis [19].

The researchers employed the generic principles of Excellence theory and looked into public relations empowerment. This theory guides the researcher to understand the public relations role and value in crisis management in the dominant coalition perspectives. Besides, it is an added value research to study the excellence theory in terms of crisis management in Malaysia.

This paper asks two research questions. First, how does the dominant coalition perceive the public relations role in crisis management in the energy company in Malaysia? And, second, how does the dominant coalition perceive the value of public relations in the crisis management of the energy company in Malaysia?

3 Methodology

This study was conducted in a government-linked energy company in Malaysia due to its experiences managing multiple crises since 1980s till 2005. The global changing environment continues to threaten the energy company in Malaysia with multiple types of crises including financial, technological, natural, skewed management values and deception that may tarnish the organisational reputation and jeopardise the economy of the country. A single-case design was employed to understand the real context of the public relations practice in managing crisis communication in this particular energy company [20, 21].

An in-depth interview was conducted with the Crisis Commander (CC), who has the next highest position after the CEO in the hierarchy of ERP and chairs the Corporate Emergency Response Plan (CERP) steering committee in the energy company. The senior public relations reports directly to the CC. The interview was held for one hour with data being taped and transcribed. To avoid biases, the researcher uses *epoche* [22], a Greek word meaning to refrain from judgement [23]

4 Findings

The results of this study were discussed according to the research questions.

4.1 Dominant coalition's perception of the public relations role in crisis management in the energy company in Malaysia

According to the crisis commander (CC), the senior public relations manager plays a role as communication and media adviser in crisis management at two levels; in the strategic planning team, and the Emergency Response Plan (ERP) team. While on-going advice on communication is managed by the strategic planning team, specific communication advice is conducted in the ERP team during a crisis. The communication processes and procedures during a crisis are explained to guide the ERP team to communicate effectively to stakeholders. The senior public relations manager advices the CEO on guidelines for communicating with public figures in times of crisis with regard to the nature of the crisis encountered. The senior public relations manager also determines the appropriate time for the CEO or CC to make a public appearance and to respond accurately in press conferences. The CC admitted that public relations advice in managing media in times of crisis is essential because failure to do so may lead the media to "pick one of you and make you look horrible."

4.2 Dominant coalition's perception on the value of public relations in the crisis management of the energy company in Malaysia

4.2.1 Value of public relations in perception management

The dominant coalition trusts public relations to manage stakeholder perceptions as on-going process of crisis communication. According to the CC, "the value of public relations in managing perceptions is equally crucial to making the business sound." Ineffective management of stakeholder perception leads to failure to communicate with the parties concerned and it is the main risk of the company. However, the CC believes effective communication is the best strategy to manage stakeholder perception. Thus, public relations is expected to disseminate information about energy businesses and objectives to avoid misunderstanding. The CC said communication is effective when public disregards negative news and perceives only favourable news despite the crisis. The dominant coalition understands that managing smouldering issues are among the challenges met in managing perceptions in the energy company. Nevertheless, the CC said, "you cannot expect praises from everybody, but at least, people would say, nobody wanted the crisis to happen, but this company managed it well. So this is what we want at the end of the day." The dominant coalition believes there is room for improvement for public relations in management of perceptions.

4.2.2 Value of public relations in trust building

The dominant coalition of the energy company believes that the company is trusted because of its brand. The basis of this confidence is generated by competent electricity supply and efficient programmes. While the technical feature engages the engineers for trust building through effective delivery of products and services, the value of public support is measured by the efforts of corporate responsibility, philanthropy, strategic communication and media relations to build trust with stakeholders. The dominant coalition appreciates the role of public relations in educating the public; from consumers to school children, contractors and developers on the issue of electricity safety awareness.

Public relations efforts in educating multiple publics have raised the level of trust among its stakeholders for the energy company. The dominant coalition acknowledged that the public relations efforts increases the public understanding of the nature of the energy business and minimises difficulties in managing organisational response at any point in time. Similarly, other effective programmes including media night, sports tournament and donation drives support the energy power to promote familiarity with its stakeholders.

5 Discussion

The findings of the first research question presented that the CC of the energy power recognises the senior public relations manager as a member of the crisis management team along with other core divisions of the energy company, advising on communication and media. The role of senior public relations as media and communication counsel is a continuous one in strategic planning team and recognised most only during a crisis in the ERP team. The senior public relations manager continues to perform more of a technical function especially as media relations specialist and communication liaison. The senior public relations role gain power only when the position has a hierarchical authority, control of the environment (research) and network centrality [24]

The findings of the second research question described the dominant coalition's perception of the value of public relations in crisis management in perception management and trust building strategies and it is restricted to media relations activities and communication programmes. The dominant coalition has a poor perception of public relations commitment in research activities, monitoring and scanning of the environment. Formative research is crucial to gather information about public opinion and issues focusing an organisation and prepares the organisation for managing crisis and to minimise reputation damages.

6 Limitation

This paper focuses only on a single case study with a limited sample size. Nevertheless, the findings obtained are significant and contribute to future research. Future studies can expand the type of organisation involved with multiple sample size.

7 Conclusion

Research on public relations practice in various sectors in Malaysia needs to be expanded to improve the perception of the dominant coalition towards the profession. Public relations should not be complacent and rely only on skills and technical functions in crisis management but to conduct more formative research activities and expand the problem-solving function.

References

1. Grunig, J. E. *Excellence in public relations and communication management.* (1992)
2. Grunig, Larissa. A., Grunig, J.E., & Dozier, D.M. *Excellent public relations and effective organizations: A study of communication management in threes countries.* (2002)
3. Van der Meiden. Int. PR Review. **16,** 3 (1993)
4. Leitchy. PR Review **23,** (1997)
5. Cutlip S.M., Center A.H., & Broom G.M. *Effective public relations.* (2006)
6. Vercic, D & Grunig, J.E. *Perspectives on public relations research.* (2000)
7. White, J., & Dozier, D.M. Public relations and management decision making. *Excellence in public relations and communication management* (1992)
8. Reber, B.H & Berger, B.K. J. Comm Mgt. ***10, 3,*** (2006)
9. Guth, D.W. PR Review. ***21,*** (1995)
10. David, P. J. PR Research. **16,** (2004)
11. Rhee, Y. *The employee-public-organization chain in relationship management: A case study of a government organization.* (2004).
12. Marra, F. J. PR Review. ***24 (4),*** (1998).
13. Grunig, Larissa. A., Grunig, J.E., & Ehling, W.P. What is an effective organization. *Excellence public relations and communication management.* (1992).
14. Berger, B.K. Public relations and organizational power. *The future of excellence in public relations and communication management.* (2007)
15. Pritchard, R.S., Davis, D & Filak, V.F. The impact of the dominant coalition on health care public relations practitioners. http://www.instituteforpr.org/wp-content/uploads/ImpactDominantCoalition.pdf
16. Bowen, Shannon. A. J. Comm Mgt. **10,** 4 (2006)
17. Fearn-Banks, K. *Crisis communications: A case approach.*(2011).
18. Coombs, W.T. & Holladay, S.J. *Handbook of crisis communication.* (2010)
19. Fearn-Banks, K. *Crisis communications: A review of some best practices.* In Heath, R.L (Ed) & Vasquez, G (Contr.Ed), *Handbook of public relations.* (2001).
20. Yin, R.K. *Qualitative research from start to finish.* (2011)
21. Yin, R.K. *Applications of case study research.* (2012)
22. Merriam, S.B. *Qualitative research: A guide to design and implementation.* (2010)
23. Wimmer & Dominick. *Mass media research: An introduction.* (2006)
24. Dozier, David. M & Grunig, L.A. The organizational roles of communications and public relations practitioners. *Excellence public relations and communication management.* (1992)

SHS Web of Conferences **33**, 00048 (2017) DOI: 10.1051/shsconf/20173300048
i-COME'16

The Myth of White Skin: A Postcolonial Review of Cosmetics Ads in Indonesia

Roro Retno Wulan[1,*]

[1]Telkom University, Jl. Telekomunikasi, Terusan Buah Batu, Bandung Indonesia

Abstract. Cosmetics industry created the beauty myth for women through advertising. A cosmetic ad in Indonesia has spread a new concept of white skin: East Asia beauty myth. The white concept of Asia white skin basically derived from colonial legacy. The purpose of the research was analyzing the beauty myth in Indonesia ads using postcolonial perspective. The principal result brought the discourse analysis and postcolonial perspective a new insight in communication research. Particularly on media and cultural studies. Major conclusions showed that the beauty myth since the Dutch colonial period never been change. The main concept is always in colonialism's idea: "white is better". The West is better than the East.

1 Introduction

The world's changes have brought women to the positions that are difficult and confusing. The beauty industry had created the myth of global beauty. Women have constructed Western ideal of beauty: a tall-white-skinny-blond hair. Beauty myth maintained and promoted on a large scale by the mass media through advertisements of cosmetics, apparel, and health. The beauty myth enjoyed broad audience through the image of actresses in films, video clips, commercials and billboards hit every household Indonesia. At its core, capitalistic world today have colonized women. In the book "The Beauty Myth: How Images of Beauty Are Used Against Women", Wolf [1] shows that during this time black women, brown women, and white women in America dealing with the myth of beauty to be the perfect person for example: have a tall body, slim, white skin, and blond hair. Their facial skin should not have the slightest flaw and they should be as small waist size. This colonial beauty myth attached to women in colony territory until today. In fact this nature of beauty is subjectivity highly. Each culture has its own construction of what is called a beautiful woman. For example, Karen Tribe in Thailand who sees a beautiful woman whose neck long, another case with the Dayaks in Indonesia which saw a beautiful woman with long-eared. The white women herself is idealizing tanning as beauty while Indonesian women wanted to have white skin.

This is the beauty myth constructed by the colonial settlers from the colonial period. McClintock [2] in her book "Imperial Leather: Race, Gender and Sexuality" describe the relationship between conception clean or white skin color with colonialism. The colonists regarded the natives as the wild and nasty because of their skin color were black (in Africa) and brown (in Asia). The facts of skin color differences become the cornerstone to the imperialists to create social stratification in the colonial and indigenous people were in the lowest layer. Especially women as the subaltern. Her position was never improved. In such conditions it is understood that capitalism has collaborated with patriarchy to construct the women so that never satisfied with her.

On the one hand, patriarchy constantly reproduces itself to perpetuate their power over women as subordinate. On the other side capitalism had lived from women income. The increase number of female workers has been multiplying consumption. At the end, patriarchy in many ways is still trying to subjugate women. Being marginalized is actually

* Corresponding author: rorowoelan28@gmail.com

women's problem across Asia and the world, therefore, the study sought to analyze the beauty myth in Indonesia in connection with the proliferation of ads about whiter and brighter skin for women in Indonesia.

2 Literature Review

Postcolonial studies are born from the difference between the colonizers and the colonized. This raises the binary opposition of third world women who are fighting against two hegemonies, patriarchy and capitalism as a form of modern imperialism as the colonial heritage. As a study of theory, postcolonial cultural studies departing from cultural studies. It shows how the methodological concepts of postcolonial been explored through the depiction of sexual identity, distinctive character, historically aspect, and flexible. Haris (1983) in Aschroft, Griffiths & Tiffin [3] explains "Postcolonial texts may deal with divisions of race and culture which are apparently obdurately determined, each text contains the seeds of 'community' which, as they germinate and grow in the mind of the reader, crack asunder the apparently inescapable dialectic of history". Madison [4] argues "Post colonialism refers to the multiple forms and locations of discourse, performance, politics, value, and the "everyday" – both past and present- that emanate from the history of colonialism". Construction of identity does not impartially and not necessary to be feared as long as women have knowledge and know themselves. Thus, self-confidence on the uniqueness of their selves will grow by itself. The terrible thing is a myth. Myths will only live if maintained and the audience agreed with it. This myth if not soon be forgotten, will turn all women into puppet with sameness and uniform identity. On the other hand the birth of postcolonial theory has been open our minds to the concept of white skin became an obsession mostly on Asian women, rooted in the colonial period. The concept of Asian women beauty is certainly not to be compared with Western concept of beauty. Especially the issue of skin color and cleanliness.

3 Method

This research used discourse analysis to uncover representation of white skin concept. This research saw white concept as colonial period legacy. The white skin myth viewed as representation of colonial communication process. Lasswell [5] in Titscher et.al [6] stated in content analysis we look upon statements and signs as raw materials to be summarized in order to bring out either (1) the impact of content upon audience, or (2) the influence of control upon content.

If in some sense before the word "discourse" has limit to the definition unit of language, statements, thoughts or grounding determination and understanding of the facts, in the concept of Foucault, discourse implies the existence of power and the power behind these statements. This understanding believes that the power relations in society influencing and shaping the ways we communicate and how the knowledge was created. This research analyzed body lotion advertising in the last five years. They have several products with the specification of white skin branding. These strategies were advertising capabilities in constructing reality and affect the perception of audience. It has brought various changes of social and cultural values. The beauty standard of women is part of the ideal values that have been successfully transformed by advertisements and has become a uniform system as a whole. However, we never aware about intervention of colonial ideology.

4 Findings and Analysis

Women in Indonesia are now experiencing the identity distortion in relation to economic autonomy, political, and cultural. The identity distortion as a female identity is defined to be a complement to support the men as part of the patriarchal which can take the form of individual men, government, and country. Patriarchy class is defined as the "father", officials and state apparatus that acts as a protective father.

The loss of women's right to choice their future, like occupation and education, is also an issue to concern. Women were led into the means of production in service industry, manufacturing, or tourism, all of which make sex and female sexuality as a commodity to encourage people to buy service and goods. The problems of women in feudal times as if comes back now with a companion featuring women as a symbol of a husband who acts as a "Prameswari", as complementary under her husband's powers. Women trying to look attractive to be declared as the best for her man. The emergence of the concept of "white" is a legacy of the colonial era, to be a part of the construction of femininity and female sexuality today.

The theme of "white" appeared in several beauty ads lately. Discourse becomes white be dominating. White is not only appearing on soap ads, body lotion, face cleanser cream, face moisturizer, and body powder but also for the product for certain body parts that very intimate. The obsession against white skin and everything is marked up as white in my opinion is not just an obsession with beauty, but more than that ... the obsession with white can be categorized as a colonial nostalgia or even colonial trauma [7]. When colonialism affecting Asia, the imperialists are carrying the culture of "clean". Clean analogous to a clean of uncivilized or civilizing people who are still natural. The Europeans looked black, red, and brown as a dirty form. Dirty is not just the body that naturally colored but further defined

DOI: 10.1051/shsconf/20173300048

uncultured. Cleanliness is not merely literally refers to a state free of impurities, since both clean and dirty has social significance, racial and colonial [7]. Desirable white also has the quality to be hyper-real, not just white, but white is soft, smooth, healthy, and radiant. And more than that, being white is to become a global subject. Being beautiful, as recognized by the global community [7].

Now there is a shifting paradigm in the concept of white. Beside western whiteness, there is another whiteness to sell to Indonesia women, Asian white, such as Japan (Shinsui) and China (Tse Fuk). Even Nivea's product sells the whitening product in Indonesia. White concept of East Asia is also preferred since the presence of popular culture from all three countries: Japan, Korea and China. The emerge of K-Pop, Harajuku style and beauty of Chinese women in Hong Kong movies and soap operas has influencing worldwide and colonizing Asian women ideology to becoming white. Indonesian women which born in brown skin seduced to be white, and these body lotions can help them to becoming white. The hyper real condition to change their skin color by using lotion. The advertising had promoted the Asia's Whiteness to make the ideology of white more logical for Indonesian women. In this research, we focused to analyze Citra Body Lotion Product as the one which using Indonesian name. Citra Body Lotion also launched "Rumah Cantik Citra" to help the customer improving their beauty with Citra's products. Citra has gain several recognition as a product with loyalty from its consumer since 2006. Citra claimed their product as the beauty product that knows about Indonesian beauty. Citra promoted how beautiful skin can improve women beauty. Below is how the product changes their ideology for marketing strategic:

Table 1. Citra Body Lotion's Evolution [8].

No.	The Ads	Description
1		In 1984 Citra had launched the beauty of Javanese woman (kuning langsat). It is not selling the whiteness for the first time. They used natural material from Indonesia.
2		In 2006 they launched three new variants and started to sell the whiteness: Citra Bengkoang White Lotion, Citra Teh Hijau Beauty Lotion and Citra Mangir Beauty Lotion
3		Since 2013 Citra has improved their product to be more Asian and using Asia whiteness like Japanese Wakame, Korean Pearl, and Indonesian bengkoang. There are others variant and using Asian beauty whiteness myths. Like Korean royal jelly and pearls, Japanese rice, Japanese green tea, Himalayan Gojiberry, India grapes oil and almonds. From ten variants, 6 of those using the words "white" to attract consumer buying. The myth of whiteness assumed as cleanness and smoothness came from western. McClintock on her book "Imperial Leather"showed us how the Western influenced the East about cleanness and whiteness.

As we seen from the table above, Citra used white representation to sell their products. From the ads we can see how whiteness becomes something good and idolized. White skin color is more than a fantasy; it is a concept to describe beauty to Indonesian women. The current reality is determined by the market. Formerly women flocked make ringlets like Marilyn Monroe, but now more and more women who wish to have straight hair as a symbol of beauty from East Asia. Beauty products ads at this time seemed to discredit women to sacrifice the potential of hers and exalt physical perfection.

5 Conclusion

The development of communication theories cannot be separated from the development of the civilization. Advances in technology and culture affect the progress of Communication Science. Communication theory is a symbiotic creation that is made to explain the phenomenon that happens and it becomes a pattern that we can observe. The main aim of the theory is to explain, predict, and control the phenomena that occurred or will occur. So, this research tries to give the new insight of Communication. As the function of the theory is organizing experience, develops knowledge, stimulating and directing further research, gave way to the researcher to anticipate events that may not be able to observe [9].The phenomenon of white skin myth on women is based on the colonial ideology should be realized by women in Asia. The struggle against western hegemony will stimulate an increase in political and economic alignments, and directing research on improving the condition of women in post-colonial societies that bear the double burden that is subordinated by colonialism and domination of indigenous men. However, the development of literacy in Gramscian ethics of feminism is part of counter-hegemony to the ideas of masculinity which dominate the world of material and symbolic communities worldwide. The concept of white skin as a beauty construction of an idealization that is not eternal. It is changing with the times and interests.

References

1. N. Wolf, The Beauty Myth: How Images of Beauty Are Used Against Women. London: Harper Collins. (2002)
2. A. McClintock, Imperial Leather: Race, Gender, & Sexuality in colonial contest. London: Routledge. (1995)
3. B. Ashcroft, G. Garreth T. Helen, The Empire Writes Back, Theory and Practice in Post-Colonial Literature 2nd Edition. Routledge. London, (2002)
4. S. Madison, Critical Ethnography Method, Ethics, and Performance 2nd Edition. Sage Publication. USA, (2012)
5. H. Lasswell, Describing the Contents of Communication. NJ: Princeton University Press, (1946)
6. Titscher, S., Meyer, M., Wodak, R., & Vetter, E. (2000). Methods of Text and Discourse Analysis. London: Sage Publications.
7. A. Prabasmoro, Becoming White: Representasi Ras, Kelas, Femininitas dan Globalitas Dalam Iklan Sabun. Bandung: Matahari, (2003)
8. Rumah cantik citra http://www.rumahcantikcitra.co.id
9. D. Infante, A. Rancer, D. Womack, Building Communication Theory 2nd Edition. Illinois: Waveland Press. Inc (1993)

DOI: 10.1051/shsconf/20173300049

Validation of intercultural sensitivity three-factor model in Malaysian context

Norzita Yunus[1,2,*], *Ezhar* Tamam[2], Jusang Bolong[2], *Nor Azura* Adzharuddin[2], and *Faridah* Ibrahim[1]

[1]Infrastructure University Kuala Lumpur, Jln. Ikram-Uniten, 43000 Kajang, Selangor, Malaysia
[2]Universiti Putra Malaysia, 43400 Serdang, Selangor, Malaysia

Abstract. A plethora of studies have indicated the importance of intercultural sensitivity (IS) in today's highly interconnected, global world. Despite that, in Malaysian context, very limited reliable instruments are available to measure intercultural sensitivity. A study using Malaysian dataset by Tamam (2010) found a three-factor model of the Chen and Starosta's five-factor model of Intercultural Sensitivity Scale (ISS); however, the model was exploratory and has yet to be validated. Therefore, the purpose of this study was to further validate the intercultural sensitivity three-factor model within Malaysian collectivistic and multicultural context. Using survey as a means of data collection, 1000 undergraduate students at three higher education institutions completed self-administered questionnaires. The three-factor model found earlier was subject to Confirmatory Factor Analysis using Analysis of Moments Structures (Amos) software. The results showed that the intercultural sensitivity three-factor model showed a good fit thus indicating that the three-factor model is a viable alternative to the original model. In conclusion, the three-factor model is a valid and reliable instrument to measure intercultural sensitivity within Malaysian collectivistic and multicultural context.

1 Background of study

Within today's globalised and highly interconnected world, the ability to communicate across cultures is no longer an option. Numerous studies have established the importance of intercultural sensitivity (IS) as an important requisite for communication across cultures e.g. [1-11]. Cultural diversity is apparent in almost every sector. Within workplace contexts, according to Llyod and Hartel [12], organisations are recruiting employees from diverse backgrounds in optimising organisational outcomes. Within the higher education industry, cultural diversity is apparent in the increasing number of students involved in cross-border education (CBE).

* Corresponding author: norzita@iukl.edu.my

According to the United Nations Educational, Scientific and Cultural Organisation [13], the number of students studying abroad has shown a tremendous growth. With less than 200,000 students studying abroad in 1950s, the number is projected to grow to more than seven million by the year 2020. Within Malaysian multicultural and multi-religious context, intercultural sensitivity is inarguably crucial thus creating the need for more studies on the role of IS within Malaysian context. In undertaking these studies, a valid and reliable instrument to measure intercultural sensitivity is required. However, there is very limited reliable instrument available to measure intercultural sensitivity within Malaysian context. One of the very few instruments available is Tamam's [14] intercultural sensitivity three-factor model which is a local version of the Chen and Starosta's five-factor Intercultural Sensitivity Scale (ISS).

Nonetheless, the three-factor model requires further validation due to the limitations of the study. First, the model is exploratory thus is subject to further validation. Second, the model was produced from a single site study thus limiting its generalisability. Next, the study measured intercultural sensitivity based on intra-national interactions, which refers to interactions that occur among the multi-ethnic Malaysian undergraduate students. According to Kamal and Maruyama [15], intra-national interactions may not be as complex as cross-national interactions as the members share to a certain extent the same meaning of common symbols such as language and some cultural practices. Cross-national interaction refers to interactions that occur among local and international students who are of different nationality. In today's globalised world, Malaysians are not only limited to intra-national interactions thus measuring intercultural sensitivity from a cross-national perspective would be highly beneficial. Further, ISS requires replication and psychometric testing particularly in a non-American context [16, 17].

Therefore, the purpose of the present study is to extend previous work on IS by validating the three-factor model within Malaysian collectivistic context. This is done by addressing the limitations mentioned earlier. First, the three-factor model which was produced based on exploratory factor analysis (EFA) was subject to Confirmatory Factor Analysis (CFA). Second, data collection was carried out at multiple sites thus extending the generalisability of the findings. Third, the sample was undergraduate students who engaged in cross-national interactions; which means that intercultural sensitivity was measured based on respondents' engagement in interactions with international students on campus.

2 Intercultural Sensitivity Scale

Intercultural sensitivity (IS) is the affective dimension of intercultural communication competence (ICC) [1, 2]. Affective dimension refers to one's emotion that may motivate him/ her to engage in cross-cultural interactions. Chen and Starosta argued that without the "feel" which motivates people to engage in intercultural interactions, then the actual act (behavior) of engaging in interactions would not occur. In developing and validating Intercultural Sensitivity Scale (ISS), Chen and Starosta [16] carried out three studies; the first one was to examine the dimensionality of the items using Exploratory Factor Analysis (EFA); the second one was undertaken to evaluate the concurrent validity of the instrument; and the third one was to evaluate the predictive validity of ISS. The sample used was 414 college students in the US. The EFA generated a 24-item ISS with 5 factors. The factors are interaction engagement, respect for cultural differences, interaction confidence, interaction enjoyment and interaction attentiveness. The Cronbach's alpha coefficients for the first two studies were .86 and .88 respectively.

A study by Fritz, Mollenberg and Chen [18] tested ISS in a German sample using confirmatory factor analysis. The five-factor structure was reproduced thus indicating that the instrument was a valid, culture free scale for measuring intercultural sensitivity. Nevertheless, the authors indicated some minor weaknesses in the operationalisation of IS.

However, a subsequent study using German and US-American data by Fritz et al. [19] failed to reproduce the five-factor structure. The data did not fit both the original and alternative model. This finding raised the question on the applicability of ISS within different contexts.

A CFA of ISS within Malaysian collectivistic setting by Tamam [14] failed to reproduce the five-factor structure; the model fit the data poorly. Hence EFA was applied to determine the dimensionality of the scale. The result of EFA was a three-factor model with 21 items. The three factors are interaction attentiveness and respect (Factor 1), interaction openness (Factor 2) and interaction confidence (Factor 3). Some of the items on the original factor loaded on other factor thus the factors were re-named accordingly. The Cronbach's alpha coefficients for the scale was .914 and the Cronbach's alpha value for the subscales were .858, .895 and .849 respectively.

Tamam's [14] three-factor model indicated that ISS is not culture-free. Nonetheless, with high reliability coefficients and construct validity, the three-factor model is a promising alternative IS model for Malaysian context. Tamam [14] however recommended for validation of the model through replication studies.

3 Sampling and Data Collection Procedure

The population of the study was undergraduate students at three higher education institutions (HEIs) in Klang Valley. Due to the heterogeneity of the population, stratified sampling was used. Based on the stratification, the sample size required was 984. A total of 1150 questionnaires were distributed and the response rate was 92%. The management of each HEI was contacted for a sampling frame. Based on the sampling frame, respondents were systematically selected. A total of 1000 questionnaires were usable after elimination of questionnaires with missing data on key constructs as well as non-response. Respondents were approached to obtain their consent to participate in the survey. They were given token upon completion of questionnaires.

A majority of the respondents (70.8%) were Malays and the rest comprised of other ethnicities in Malaysia which were Chinese, Indians, Bumiputera and others. Slightly more than half (54.6%) of the respondents were female while the male respondents made 42.5% of the total sample. Twenty-nine respondents, equivalent to 2.9%, did not indicate their gender. A majority (56.9%) of the respondents were in the age group of 19 to 21 ($M = 21.37$, $SD = 1.62$) suggesting that most of them were in their first and second year of study.

4 Measurement

The aim of this study was to further validate Tamam's [14] ISS three-factor model; therefore, the model was used. This alternative model consists of 21 items from three factors which were Interaction Openness (8 items), Interaction Attentiveness and Respect (7 items) and Interaction Confidence (5 items). The items were measured using Likert's five-point scale ranging from 1 (strongly disagree) to 5 (strongly agree). Sample items include "I respect the ways people from different cultures behave" and "I can be as sociable as I want to be when interacting with people from different cultures."

DOI: 10.1051/shsconf/20173300049

5 Analysis Strategy

To validate the three-factor model obtained from EFA, as proposed by Hair et al. [20], the three-factor model was subject to Confirmatory Factor Analysis (CFA) using Analysis of Moments Structures (Amos) software.

6 Results

Table 1 shows the items from Tamam's [14] EFA results of the ISS three-factor model. Items with factor loading of 0.40 were included. The alpha coefficients of the subscales were good. The items were subject to CFA.

Table 1. Principal components analysis of Tamam's [14] three-factor solution of IS

		Factor Loading		
Items	**Alpha value**	**Factor 1**	**Factor 2**	**Factor 3**
Interaction attentiveness and respect	.858			
I respect the ways people from different cultures behave.		.896		
I am open-minded to people from different cultures.		.729		
I am very observant when interacting with people from different cultures.		.688		
I respect the values of people from different cultures.		.586		
I often show my culturally-different peers my understanding through verbal or nonverbal cues.		.554		
I try to obtain as much information as I can when interacting with people from different cultures.		.541		
I often give positive responses to my culturally different counterpart during our interaction.		.456		
Interaction openness	.895			
I don't like to be with people from different cultures.			.804	
I often get discouraged when I am with people from different cultures.			.752	
I would not accept the opinions of people from different cultures.			.750	
I think people from other cultures are narrow-minded.			.741	
I find it very hard to talk in front of people from different cultures.			.696	
I often feel useless when interacting with people from different cultures.			.687	
I get upset easily when interacting with people from different cultures.			.682	
I avoid those situations where I will have to deal with culturally-different persons.			.663	
Interaction confidence	.849			
I enjoy interacting with people from different cultures.				.678
I am pretty sure of myself in interacting with people from different cultures.				.701
I always know what to say when interacting with people from different cultures.				.681
I can be as sociable as I want to be when interacting with people from different cultures.				.855

DOI: 10.1051/shsconf/20173300049

I feel confident when interacting with people from different cultures.				.651
I have a feeling of enjoyment toward differences between my culturally distinct counterpart and me.				.463

Table 2 shows the fit indices of the three-factor model, for both the initial and fit model. In achieving model fit, three items were dropped resulting in 18-item three factor model. The model had a good fit with the latent constructs (χ^2/df = 4.885, p = 0.000, RMSEA = 0.062, GFI = 0.976, AGFI = 0.911, CFI = 0.916, TLI = 0.901, NFI = 0.897). The Cronbach's alpha value of ISS was 0.856 and the Average Variance Extracted value was 0.655. The alpha coefficients of the subscales was 0.859, 0.782 and 0.769 respectively.

Table 2. Goodness-of-fit indices and factor loading for intercultural sensitivity

Fit Indices	**Initial Model (n=1000)**	**After Attribute Dropped (n=1000)**	**Reliability**	**AVE**	**Recommended Value**
Chi-Square	816.6	635.09			-
Degrees of freedom	149	130			-
χ^2/ df	5.48	4.885			Equal or below 5.00
RMSEA	0.067	0.062			Equal or below 0.08
GFI	0.921	0.932			Equal or above 0.90
CFI	0.892	0.916			Equal or above 0.90
Factor 1		0.604, 0.562, 0.693, 0.756, 0.7, 0.724, 0.648, 0.549	0.859	0.65	Equal or above 0.30
Factor 2		0.586, 0.68, 0.655, 0.689, 0.638	0.782	0.65	Equal or above 0.30
Factor 3		0.648, 0.684, 0.524, 0.613, 0.639	0.769	0.62	Equal or above 0.30

7 Discussions and Conclusion

The aim of this study was to further validate Tamam's [14] ISS three-factor model within Malaysian cultural context. Validation was done by subjecting the three-factor model produced using EFA to confirmatory factor analysis. Based on the indices, Tamam's [14] ISS three-factor model demonstrated unidimensionality, construct validity and good internal consistency. This validation lent support to Tamam [14] and Fritz et al.'s [19] findings which indicate that ISS is not culture-free. Therefore, it must be used with precaution within different cultural context. The present study contributed to the body of knowledge by validating an instrument to measure intercultural sensitivity within Malaysian context. ISS three-factor model shows potential and can be used as an instrument in relevant cultural-related research and training.

Acknowledgements

Findings in this study are part of the doctoral work by Norzita Yunus, a PhD candidate at Universiti Putra Malaysia.

References

1. G. M. Chen, ICS. **19**, (2010)
2. G. M. Chen, W. Starosta, HC. **1**,(1997)
3. J. Chocce, IJIRIS. **1**, 5-11 (2014)
4. D. R. E. Cotton, R. George, M. Joyner, IETI. **50**, (2013)
5. D. K. Deardoff, JSIE. **10**, (2006)
6. M. R. Hammer, M. J. Bennett, R. Wiseman, Int J Intercult Relat. **27**, (2003)
7. J. -E. Jon, JSIE. **17**, 455-470 (2013)
8. M. W. Lustig, J. Koester, *Intercultural competence : Interpersonal communication across cultures*. 6th ed. (2010)
9. G. S. Matkin, J. E. Barbuto, JLOS. **19**, (2012)
10. C. L. Olson, K. R. Kroeger, JSIE. **5**, (2001)
11. L. T. Tran, L. Pham, CCOM. **20**, (2015)
12. S. Llyod, C. Hartel, JMP. **25** (2010)
13. P. G. Altbach, L. Reisberg, L. E. Rumbley, *UNESCO Report for world conference on higher education* (2009)
14. E. Tamam, JICR. **39**, 173-183 (2010)
15. A. A. Kamal, G. Maruyama, IJIR. **14**, (1990)
16. G. M. Chen, W. Starosta, HC. **3**, (2000)
17. D. Matsumoto, H. C. Hwang, JCCP. **44**, (2013)
18. W. Fritz, A. Mollenberg, G. M. Chen, ICS. **11**, (2002)
19. W. Fritz, A. Graf, J. Hentze, A. Mollenberg, G. M. Chen, ICS. **14**, (2005)
20. J. F. Hair, W. C. Black, B. J. Babin, R. E. Anderson, *Multivariate data analysis : A global perspective*. 7th ed. (2010)

 DOI: 10.1051/shsconf/20173300050

Media Literacy in 'Big Kitchen (1952)' historical documentary

Mohd. Nor Shahizan Ali[1,**], *Dan* Fleming[2], *Ali* Salman[1] *Novell* Lyndon[3]

[1]School of Media and Communication Studies, Universiti Kebangsaan Malaysia
[2]School of Arts, University of Waikato
[3]School of Social Development and Environmental Studies, Universiti Kebangsaan Malaysia

Abstract. Gen Y as a whole have different abilities in interpretation on the text media. Therefore, each issue necessary media texts not to cause confusion in the interpretation of the message. In line with the rapid pace of information technology to make all information more easily accessible lead the audience (Gen Y) became more confused in making the interpretation of the meaning; depending on their literacy level. The importance of media literacy should be considered in the process of reading on historical documentary. This article explores the interpretation of Gen Y at the Institute of Higher Learning on the historical documentary 'Big Kitchen (1952)' work of Malay Film Unit. A focus group discussion was conducted on seven informants selected. The discussion focuses on the literacy aspect and the overall interpretation of the historical documentary narrative as a message. A historical documentary has given rise to different interpretations among them. The results supports that the Gen Y who totally exposed to the media environment in everyday life able to understand the narrative of the historical documentary. The results showed that the transformation of social technologies and the impact of extensive and open information have influenced the interpretation of Gen Y.

1 Introduction

Technology plays a role as a trigger to change the flow pattern in line with the development of delivery of information time and communication activities. Development is dependent on the ability of technology to deliver information and communication patterns of information between network administration [1,2]. The changing landscape of delivery of information by importance (whether public or individuals) from time to time in accordance with the appropriate context. Media organizations in the country changed in line with technological developments.

* Corresponding author: shahizan@ukm.edu.my

 DOI: 10.1051/shsconf/20173300050

Changes of the print media to the electronic media in providing information of great hikes changes to the public in the process of accessing information. Research and discussions conducted is focused on identifying the maximum possible use of technology in obtaining information [3,4].

Various medium of information appear to have resulted in line with the times has influenced changes in the form and nature of the information. These changes are very significant when the information has become more compact in the form of icons, audio, visual, graphics, and text summary in accordance with the appropriate medium. Potter [2] argued that the dissemination of information and through a variety of medium makes it difficult to interpret the meaning of information. According to Renee [3], the use of visual to convey as a medium of information available is most effective since ages. However, delivery of information in the form of visual approach changed in line with technological changes. Progressively expand of human civilization; patterns are also more advanced of visual presentation so that it can be accessed anywhere. Evolutionary development of the dissemination of information through visual visibly starting from the walls of the cave, turn on television and via computer and now to smartphones.

Today, smart phones have played a big role as a medium for the dissemination of information. Various applications that can be used through a smart phone have emerged on the importance of information and the target user information. The development of this technology creates the diversity of the types of information that can be accessed only by using a smart phone. Gen Y is the most numerous generation access the information using a smartphone, then it implies that they have been exposed to the diversity of information that is transparent and publicly [2].

The development of this technology make the user needs to choose the platform (medium) information coincides with the information that you want to search. This situation since there are channels of communication that uses any platform available to provide information to the audience. The audience mainly consisting of Gen Y bestowed with various information, particularly in the form of audio-visual to discuss certain issues. Most of the issues under discussion are based on historical facts. So, Gen Y need information savvy (high level of visual literacy); so that the information that they received was not misinterpreted by them [5]. This aspect will make a way to audience to react as a literate audience in a sense of decision making [6].

2. Research issues

The emergence of various alternative documentaries nowadays use a realistic approach and uses historical facts as narration narrative like 'The Last Communist (2004)' work of Amir Muhammad assume that the communists in Malaya, an equally fighters helped drive the British out of Japan and Malaya in a bid for alternative. Similarly, alternative documentary 'Ten Years Before Independence' (2007) work that tells the story of Fahmi Reza hartal (strike or protest) that discusses the racial tension in emergency. Alternative documentary such as 'Bo Lang Chai (Forgotten) (2007)' Ong Boon Keong's work that explores issues of racial in minority groups considered Chinese tribal Penang Jetty marginalized groups said that while they struggled to make Penang as an international port also fought against the Japanese during the emergency.

Thus, the work of historical fact in documentaries especially noteworthy because it includes symbols that implied meaning. Audience consisting of the Gen Y is exposed to inundation information from a variety of media to date. In fact, they are showered with historical information differently from what they learned formally in school. Among other things, the emergence of alternative documentary approach has caused confusion to this new Gen Y to balance all this with their understanding the historical fact [6]. So, this gap is filled with such studies in media literacy skills for different types of audio visual material shaped alternative documentary work for the construction of meaning among the Gen Y in making interpretations.

Most of the results of past studies found that Gen Y is the savviest internet. Thought majority of them are affected by information obtained from the Internet. They have a critical thinking approach and transparent [3]. So, with the abundance of information in the new media (including the historical documentary); there needs to be studies to identify how patterns (read: interpretation) of meaning construction of the Gen Y present on audio visual materials (such as historical documentary).

Lester [7] and Dyer [8] were of the view that the visual element is the strengthening of the stereotypical image of a device. It's come to the conclusion that, whether an individual is identified by gender, age, cultural heritage, economic status or physical disability; visual message communicated by individual can be misleading [7, 8]. This is because the ability of the visual impact of emotions is higher than the words. Visual can stay in the mind for a long time. However, this statement is only based on visual elements that focused on imaging rather than visual interpretation of a visual element. Not mentioned was the material in the form of audio-visual materials, how the construction of meaning through? This was not explained by them. So, this gap that was evident in the relationship between literacy skills of meaning construction process audio visual materials especially the historical issues to the present Gen Y.

3 Method of study

Focus group interview was conducted in three groups A, B and C with twenty one informants (final year undergraduate students from Malaysia, comprising nine males and twelve females from all races who are studying at Universiti Kebangsaan Malaysia) and has watched the historical documentary 'Big Kitchen (1952)' produced by Malay Film Unit, which premiered at the Library Tun Seri Lanang (PTSL), Universiti Kebangsaan Malaysia (UKM). The analysis is descriptive about some interesting aspects related to cognitive and narrative message from an historical documentary on the construction of their views on the realism in daily life. This article focuses on two key votes on the interpretation (reading) of the educated youth of the historical documentary 'Big Kitchen (1952)'. The evaluation is to the reaction of the Gen Y informant's to the narrative message (read: reading) delivered by the historical documentary.

4 Result of the study

There are some impact caused the audience after watching the historical documentary. Including the rejection of the notion that labeled terrorists by the government to communists. Overall informants stated that the communist also the freedom fighters. The assertion is supported as follows:

Male C: If I opted in my opinion personally, did I say a struggle ....

Female C: A global struggle. As this communist, not only seen in Malaya, but it appears in the world, and brings a unique political structure.

Male A: For me this is a communist ideology. So, when the call ideology or thought, we must have our vision. So, it is called also struggle, even as we said to the communist, the socialist, and remained to fight for anything...

Female A: Yes I agree. Only when in Malaya, the fight was founded by the Chinese, so that the community in Malaya consider it only for the Chinese, and the fact that the successful implementation of the subject in school history.

Female B: They have their own beliefs, but may not be accepted by most people because of their beliefs rather ... like we are... we have a religion, so we cannot accept the concept of ideology...we got the same ideology. To think that they can fight, but with a different ideology.

Male B: Haa ... because everyone wants something that they must fight ... How we are teaching, if you want to target that, we have to fight ... Just like them, they want alternative ... what does it mean...they fight for that...

Female D: Yes, and it is not just a struggle, but an idea designed for the same direction around the world. Until today they still remain in certain countries.

Text factor (visually explicit) does not affect the Gen Y on their interpretation of the text of historical documentary. This statement is in line with Baran (2013), which states that any media content can be considered as a text made up of signs [9]. These signs are structured and interconnected with each other in specific ways. In the interpretation of the text, the recipient must be able to interpret the signals and also the structure of the texts. For example, when you read one sentence, you not only have to turn to the words of its own, but must also interpret the overall structure of the verse to understand it thoroughly.

This opinion was also agreed to by Potter [2] and Mohd. Nor Shahizan et. al [6], which suggests that the individual should have strong knowledge structures and information obtained from the real world, rather than mass media information only. In addition, the individual also needs a strong knowledge of the structure themselves to become critical media users. To separate themselves from the world of media and personal perspective on the media, then people need to know more clearly their personality in the two types of consciousness style of personal knowledge and personal goals. Two awareness is required in each individual locus. Structure strong knowledge can help individuals make decisions more wisely and fairly without being influenced solely by the content contained in the mass media. This situation can be supported by the responses by the Gen Y informants as follows:

Female C: Yes ... but from different angles. If you do not learn history at school a bit, maybe I'll accept blindly what the documentary that.

Male A: Not really ... because it affects my readings compare with the resources and information available and sometimes history justify his false and falsified the truth ...

Female D: Yes. Sometimes history also correct ... Can we trust ... but depend on self...whether to believe it or not ... just in this TLC documentaries... really communist look great... supposedly.

Female B: No. If history was monopolized by the parties, may be biased in fact. We take the example of the history of our family, we said our family history ... If we hear from our father, our father might be cheating, when we hear from the mother of us, there may be otherwise story.

Male B: If you follow the TLC that, from the beginning to the end I saw I did not focus so much on communist, I think the very wide range of communist ... somehow, I do not weigh against them, against government ... of early government have seen a distinctive agenda, wide they want something ... If include thinking, we already can think, the more we learn, the more we can think. We see government agenda...to ask themselves ... How do we see good people... not really affect my prejudices ...

 DOI: 10.1051/shsconf/20173300050

Female A: Yes, influenced, because already our old stories being subject to this communist crimes at school. But, what happens in the real world today make us more prejudicial to the west, in this case referring to the British. For we see now America wants to attack anyone participating like, and vary with the existing communist countries. They never make a mess, look at China... okay no problem, good economy, Americans are distracted them. So, this documentary west affirm what the nature of truth, they are evil and not communist.

This statement is supported by Gen Y informants who told that all of them surf the Internet to obtain additional information other than the information provided by the mainstream media. This statement is supported by the opinion below:

All: ...Yes.

Based on the results of focus group discussions, Gen Y are very often surf the Internet and have a high media literacy skills to material posted on the Internet. They use the Internet to get information, the latest information, entertainment, and social relationships. Overall, the majority of Gen Y description of the acceptance and trust of materials derived from Internet indicate they have a high tendency to censor any material disclosed to them. Gen Y is also available every thoughtful authenticity of the information they received. Consequently, the results show that they are not only affected by the text of the historical documentary, but the knowledge and side information obtained from other sources in interpreting the meaning of these historical documentary. In fact, the level of this interpretation and point to see the text implicitly.

5 Conclusion

The audience actively based on their observation will interpret thoughts, events and situations. They will also share ideas, feelings and experiences with the use of symbols in assessing and responding [9,10,11]. The results from the focus group showed that represent combination of mutual understanding (in technological community groups) and education. Clearly, this study confirms the importance of both of these factors to the development of individual literacy.

A combination of mutual understanding; in technological community groups and education respondents' prolonged start of the information and environment in everyday live from 1993 up to the year 2016 (majority of informants). Learning, reading and experiences of others had developed informants' mutual understandings. Moreover, the technological community groups was learned by informants from various media materials either cyber media, electronic or smart phone application (such as WhatApp group, WeChat Group, Telegram Group, KakaoTalk Group). History subject studied by informants during school days also affected them in their interpretation of the nature substance audio visual ideology. Surrounding effect whether formal or informal, namely the unique factors conditioning individual cultures derived from community members, family and friends also provide information and lessons to them. Therefore, by relation, all these factors have been brought together in a box and thought they had become a natural process (mental configuration).

Potter [2] analysis shown that the construction of individual meaning is influenced by the style of personal knowledge and personal goals of each individual. In this case, the formula finds that the individuals affected by the combination of mutual understanding (in technological community groups) and education as well as the uniqueness factor prevalence of individual accessibility to a variety of information using a variety of approaches in interpreting the ideology (everyday life).

According to Livingstone [12] and Sturken & Cartwright [13], this condition is based on the viewer's priority of social and cultural. From another point, the audience response or the ability to make interpretations are not too dependent on the interpretive process, but there are many sources of extra text that may attract television viewers in the construction of meaning as faith or belief, creed, the emotional, and social knowledge. In short, all the elements are referring to a condition that affects the pattern of their lives.

Among other things, according to Baran [9] and Nichols [14] most of the material can be interpreted in several ways, but the message issuer is generally intended to influence an audience to make interpretations based on the view when they deliver the message [9,14]. But the fact that the audience can make a different interpretation without being influenced by the social status. This view supports the action of the informants admitted that the combination of mutual understanding (in technological community groups) and education in the construction of the meaning of the 'Big Kitchen (1952)'. They choose to evaluate the narrative based on the technological effects and knowledge that they collect all this without being influenced by the alternative interpretation.

In conclusion, the fast-moving technological changes created the Gen Y a more analytical understanding of the material from the point of explicit and implicit [3]. This group (read: the Gen Y) is very active in social media and has access to high technology. Their thinking is more realistic and open in discussing issues serious [3]. Effect transformation of their thinking (the Gen Y) should be taken seriously by the leaders of the State, especially in the process of enacting any new policy that has to do with the group (the Gen Y) is. They (the Gen Y) not only make interpretations based on the experience and the environment, but they (the Gen Y) use of information technology skills to find and assess the facts of each particular those relating to historical documentary.

Potter [2] states that the message of the media has many elements that affect the audience (read: the Gen Y) on the state of the real world, but it also has elements that cause the audience departed from the real world situation. Television and movies are not interesting visual waive such thinking for themselves, reading, walking and other activities that do a lot of static in everyday life. However, the producer potrays the visual that can attract the most viewers (read: the Gen Y) to watch their issues. Most producers produce programs that use non-visual reality. The situation is different with the visual presentation of reality. Visual that shows reality included actions that really happened and this requires knowledge and skills to interpret the meaning of the visual reality [9].

Acknowledgement

Thanks to Universiti Kebangsaan Malaysia for granting Research Fund (TD-2015-006-Dana Penyelidikan TopDown) to support this study.

References

1. G. Kress & T.V. Leeuwen, *Reading images: The grammar of visual design*, 2nd Edition, (Routledge, New York, 2006)
2. W.J. Potter, *Media literacy*, 5th Edition, (SAGE Publications Ltd, London, 2011)
3. H. Renee, *Digital and Media Literacy: A Plan of Action* (The Aspen Institute, New York, 2010)
4. R. Verser & R.H. Wicks. J. Commun., 56, (2006)
5. E. Thoman & T. Jolls, Am Behav Sci, **48**, 1,(2004)
6. M. Ali, M. Salleh, and M. Rahman. Jurnal Sains Sosial dan Kemanusiaan, **9** (1), (2014)
7. P.M. Lester, *Visual communication: Images with messages,* 2nd Edition (Thomson Wadsworth. Belmont, CA, 2003)

8. G. Dyer, *Advertising as communication* (Methuen & Co., London, 2006)
9. S.J. Baran, *Introduction to mass communication theory: Media literacy and culture.* 7th Edition (McGraw-Hill, CA: Wadsworth Publishing Company, 2013)
10. D. Buckingham, *Media education: Literacy, learning and contemporary culture* (Blackwell Publishing, Maiden, 2003)
11. I. Heywood & B. Sandywell, *Interpreting visual culture: Explorations in the hermeneutics of the visual*, 2nd Edition (Routledge, New York, 2005)
12. S.Livingstone, (Edt.), *Audiences and publics: When cultural engagement matters for the public sphere* (Intellect Press, Bristol, 2005)
13. M. Sturken, & L. Cartwright, *Practices of looking: An introduction to visual culture* (Oxford University Press. New York, 2003)
14. B. Nichols, *Representing reality* (Indiana University Press, Bloomington, 2004)

 DOI: 10.1051/shsconf/20173300051

An Appraisal of Farmers Internet Use for Sourcing Agricultural Information in North-Western Nigeria.

Abdul-Aziz Haruna[1,*] *and D.* Baba[1]

[1]NAERLS, Ahmadu Bello University Zaria-Nigeria

Abstract: This paper examined the level of farmers awareness and experiences in use of the Internet for sourcing agricultural information by farmers in the north-western Nigeria who are also privileged to have Internet services and cafes' available in their communities. The study adopted survey method using questionnaire as an instrument for data collection. The population sampled for the survey was farmers who as well leave in both semi-urban and rural areas of Kaduna and Kano State of the North West agro-ecological Zone, of Nigeria. A total of one hundred (100) farmers were randomly selected using purposive sampling technics to form the sample size. Findings revealed that farmers lack awareness for using the Internet as an alternative means of getting Agricultural Information easily even with the shortage of extension agents. The study further identifies the farmer's constraints in using the internet for information to low awareness, low access and lack skills. The study thus, recommended that agricultural agencies and other stakeholders in Nigeria should embark on advocacy and awareness campaign to farmer's cooperative societies/groups on the uses of Internet for sourcing agricultural information.

1 Introduction

Agricultural development is vital for national transformation and increased of food security and to achieve this, farmers, researchers/scientist, and extension workers need to be well informed. In view of Jensen [5], the right information will guarantee knowledgeable, active, engaged and informed farmers who can participate wisely in the country's economic production and consequently development.

Radio, Television, Community newspapers and extension workers have been identified as various sources of agricultural information to farmers. However this paper looks at the introduction of the internet services to harness communication and information need for agricultural development purposes. The medium is a relatively new channel of disseminating mass mediated information. Adeyanju and Mbibi [1] are of the view that farmers should be aware and knowledgeable of ICTs significant role and utilize it effectively in obtaining their information needs. This is important considering the

* Corresponding author: abdulharunas@yahoo.com

successful employment of ICTs in Agricultural production across the world. Okunola [2] posits that in the 21st century, information dissemination and retrieval depend largely on the ability of one to access and utilize the internet effectively through the use of various technological means. Various researchers/scientists state that there have been efforts to harness communication and information media for planned developments purposes. This is in recognition of the function which information and communication media can take part in in fostering and mobilizing people towards national development. The world has become a global village as predicted by Mcluhan [8], with its rapid transformation and increased knowledge transfer.

2 The Internet Services

According to food and Agriculture Organisation FAO [4], the Internet was developed in the early 1960's it was then intended as a decentralized computer system that was less susceptible to nuclear attack than a centrally strained system, a devised whereby computers sent packets of information from one computer to another across the United States. Today, Internet connects researchers located at various Universities and colleges, individuals, organizations, continents and the world at large. Various studies revealed the potential of internet to all aspect of human development. Similarly, the internet service has turn out to be a vast and growing global network system that people use to converse, debate, meet, teach, learn, buy and sell, and virtually every kind of information including agriculture. Internet use is not merely limited to shopping and contact with people, but it also provides the atmosphere for news sharing and endorse the people to be resourceful with the latest news. We can access the data easily wherever it may be across the world. Internet is also a very good form of communication for people who want to express their own view points and feelings. For youth, internet is not merely an avenue of knowledge but also provide an entertainment aspect such as online games, downloading movies, music, charts etc. But this is not the only basis the internet appeals to the younger age group. They also established that the internet provides ease to explore and transmit research result to the end-users.

Agricultural Research Institutes in Nigeria are facing challenges in the area of extension communication delivery due to a number of reasons among which is shortage of extension workers in Nigeria. Extension workers are agents for communicating new agricultural innovation to farmers. The current extension/farmer ratio according National Agricultural Extension and Research Liaison Services "NAERLS [7]" Agricultural Performers Survey (APS) report stood at 1: 3500 farm families, even with these the institutes are essential to respond efficiently to the demands of policy makers, private sector investors and donor agencies, farmers and other stakeholders in the agricultural business. They are being asked to convey research outputs that will improve agricultural productivity, food security, in order to increase their countries competitiveness in global markets, and contribute to sustainable development. Therefore, this calls for need of agricultural research institutes to prioritize and strategize their responses in-terms of agricultural extension services.

As such, the internet offers opportunities for collaborative and parallel communication and opening up new, non traditional communication channels for rural communities and development organizations. Most notably, it can sustain button-up enunciation of development needs and perceptions, food and Agriculture Organization FAO [4] Going by the shortage of extension agents who facilitates adoption of new farming innovations in Nigeria, it is therefore paramount for all agricultural stakeholders most especially farmers to effectively utilize the potential of Internet to enhance their activities in terms of information and communication dissemination. Therefore the o**bjectives of this study are:-**

1. To identified the extent of farmers experience and skills in the use of Internet for Agricultural information.

2. To examine the relationship between farmers Age, level of education and level of internet awareness as a source of agricultural information.
3. To assess farmer's constraints in use of Internet for obtaining agricultural information.

3 Methodology

The study adopts survey method using questionnaire as an apparatus for data gathering. The population sampled for the survey was farmers who as well leave in both semi-urban and rural areas of Kaduna and Kano State of the North West agro-ecological Zone, of Nigeria. The choice of the zone was because farming is the major activity of the people in the zone and represent major food supplier in Nigeria, it is characterised by mixed cropping farming and livestock production for meat. It has a very few large and medium scale farmers that are engage into mechanised farming with majority smallholder farmers. A total of one hundred (100) farmers were randomly selected using purposive sampling technics to form the sample size. Descriptive statistics using table and chart was used to present the result while, regression analysis was used to determine the relationship between Dependent variables ie Internet skill, level of Internet awareness with Independent variables such as age, level of education, years of farming experience, Internet access, using internet to source information. Ordinary regression is a popular tool in analyzing relationship between dependent and independent variable it provides best explanation on existence, strength, magnitude, significance as well as form of relationship in a model Gujarati [3].

4 Result Presentations

The presentation and analysis of data generated for this study are presented below in tables and charts. The result in the tables indicates measurement of association between the dependent and the independent variables in other to find out the relationship between the variables using regression;

Table 1. Relationship between farmers Internet skill, Level of Internet awareness with * Age

Measures of Relationship	R	R Squared	Eta	Eta Squared	F	Sig.
Assessing farmers Internet skill * Age	-.192	.037	.499	.249	3.519	.065
Level of Internet awareness * Age	-.129	.017	.555	.308	1.722	.194

*P<0.05

The result in Table 1 shows that farmer's age has no any significance to the farmer's level of internet awareness and skills. This means that farmer's age those not matter when using the internet to source for agricultural information, in other words, internet usage cut across all ages whether young or old can utilized its benefit for self development.

Table 2. Relationship between farmers Internet skill, Level of Internet awareness with * Years of Farming Experience

Measures of Relationship	R	R Squared	Eta	Eta Squared	F	Sig.
Assessing farmers Internet skill * Years of Farming Experience	-.111	.012	.576	.332	1.493	.225
Level of Internet awareness * Years of Farming Experience	-.017	.000	.512	.262	.032	.859

*p<0.05

Result in table 2 also reveals that farmers years of farming experience has no significant difference in terms of farmer's internet skills and awareness which means that years of farming experience those not matter too, any farmer can utilized the potential of internet as agricultural information data base even without the assistance of an extension agents. This result further reveals that the internet can harness the activity of extension agents if utilized.

Table 3. Relationship between farmers Internet skill, Level of Internet awareness * Educational level

Measures of Relationship	R	R Squared	Eta	Eta Squared	F	Sig.
Assessing farmers Internet skill * Educational level	.709	.503	.822	.676	148.972	.000
Level of Internet awareness * Educational level	.560	.314	.585	.342	45.771	.000

*p<0.05

Table 3 shows that farmer's educational level has a 1% significant level also farmer's internet skill and awareness. This means that respondent's educational background has a relationship and significant role in farmer's internet skills and awareness of the internet to source for agricultural information.

Table 4. Relationship between farmers Internet skill, Level of Internet awareness * Farmers Access to Internet

Measures of Relationship	R	R Squared	Eta	Eta Squared	F	Sig.
Assessing farmers Internet skill * Farmers Access to Internet	-.111	.012	.854	.730	4.395	.039
Level of Internet awareness * Farmers Access to Internet	-.201	.040	.551	.303	5.553	.020

***P<0.05**

Result in table 4 indicates 5% significance level, this shows that relationship exist between farmers internet skill, level of awareness and farmers access to internet, which means that farmers in the study area might have availability of internet services but could not utilized it to source for agricultural information despites the high rate of extension/farmer ratio in the study area.

Table 5: Assessing farmers Internet skill, Level of Internet awareness * Ever use the Internet to source Information

Measures of Relationship	R	R Squared	Eta	Eta Squared	F	Sig.
Assessing farmers Internet skill * Ever use the Internet to source Information	.115	.013	.815	.665	3.794	.054
Level of Internet awareness * Ever use the Internet to source Information	-.076	.006	.388	.151	.650	.422

***P<0.05**

Table 5 shows a significant relationship at 10% between internet skills and ever use the internet to source for agricultural information. This means that using the Internet to source for agricultural information requires some skills. However, there is no significant relationship between farmers level of internet awareness and ever uses the internet to source for agricultural information.

Table 6. Assessing farmers Internet skill, Level of Internet awareness * Internet as alternative source of agricultural information

Measures of Relationship	**R**	**R Squared**	**Eta**	**Eta Squared**	**F**	**Sig.**
Assessing farmers Internet skill * Internet as alternative source of agricultural information	-.102	.010	.668	.446	1.787	.184
Level of Internet awareness * Internet as alternative source of agricultural information	.241	.058	.839	.704	18.841	.000

*** P<0.05**

The result in table 6 shows a 1% significant level of relationship between farmer's level of internet awareness, and with Internet as alternative source of information aside other sources. However, there is no relationship between farmer's internet skills and agreeing with internet as alternative source of information aside other sources. This means farmers needs to required internet skills for them to be capable to use the internet to source for information, similarly, the result further revealed that with good awareness farmers can adopt the use of internet as alternative source of information rather than relying on extension agents alone.

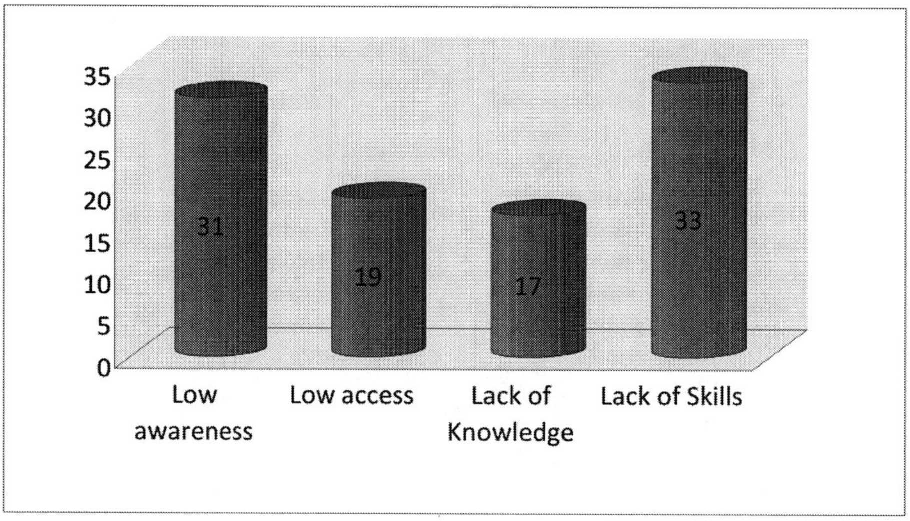

Fig. 1. Farmers constraints in using the Internet

Figure 1, identifies the farmers constraints on the use of Internet range from; low awareness, low internet access, lack of knowledge and skills to use the Internet.

5 Conclusions and Recommendations

The appraisal of farmers in internet use for sourcing agricultural information in the north-west Nigeria shows that the farmers lack awareness of using the Internet for sourcing agricultural information even with the shortage of extension agents whose duty is to facilitate adoption of new farming innovations. The study also shows that with proper awareness farmers could use the Internet as another source of agricultural information looking at the opportunities provides by the Internet if properly utilize. Michiels & Crowder [8] states that the introduction of ICT projects desires to be accompanied by support so that communities are conscious of the purposes and people have a clear considerate of their roles, and in particular how they will be element of decision making about objects, applications, content, etc. Extension agents can support and advocate for rural stakeholder participation in ICT project planning, implementation and evaluation. It was also discovered that low awareness, low access and lack of skills forms the constraints to internet usage by farmers in the area. According to Christian in web2for dev [9] the participatory network has open innovative conduct of interacting on the Internet, but certain constraints are identified: access, Cost, Time, Literacy and a certain level of media literacy. Particularly in developing countries, few people have Internet access or the means (literacy and media competence) to engage in such a conversation. Also just a few languages dominate and there are very few bridges between them. The majorities of online development debates are in English and exclude many groups from participating, a little of these constraints still remain or might even deepen web2forDev summary report [9]. Consequent to the conclusion of this study it is therefore recommended that:-

- The need for community enrollment mainly by NGOs or Public-Private-Community partnership for the need of using the Internet to source for agricultural information.
- In Nigeria agricultural extension information and knowledge services are usually from government and public sector organizations such like the National Agricultural Extension and Research Liaison Services (NAERLS), with other Agricultural Research Institute (NARIs) in Nigeria, to create awareness on the importance of using the internet as alternative source for Agricultural Information and to access their activities easily.
- The Nigerian government to address the problem of using Internet in agricultural and rural development via focusing on providing broadband connectivity and a content centric development approach particularly in the rural areas.

References

1. A.M. Adeyanju, J. Mbibi, "Knowledge, Awareness and Attitude of farming communities towards the use of ICT's for Rural Development". NJC. 5, 99 (2007)
2. A. Okunula EJSS. 11 (2) (2014).
3. D.N. Gujarati, Basic Econometrics: Special Indian Edition, (Fourth Edition), Tata McGraw Hill (New Delhi, India). 56-62 (2007).
4. FAO. The potentials of micro-computers in support of Agricultural Extension, education and training Rome: FAO. (1998)
5. J. Jensen. Journalism's Identity Crisis: What is the News for, www.reasononline.com (2003).

6. M. Mcluhan, The Gutenberg Galaxy: The making of Typograhic Man. Toronto: University of Toronto Press. Retrieved on 08/08/2011 from http://www.uky.edu/~drlane/capstone/mass/determinism.htm (1962).
7. NAERLS. Agricultural performance survey. Zaria: NAERLS press. (2011)
8. S. Michiels, L. Van crowder, Discovering the magic box. Local appropriation of information & communication technologies (ICTs).FAO, Rome. accessed: http://www.fao.org/sd/2001/KN0602a_en.htm (2001)
9. Web2 for Dev summary report, exploring the potentials of blogging for development. Htpp://natawilage.typepad: retrieved http://blogweb2fordev.net September, 2010.

 DOI: 10.1051/shsconf/20173300052

Counter-ideological forces of "Mr Gate" in online Journalism in Malaysia: A study of MalaysiaKini and The Malaysian Insider

Mohd Faizal Kasmani[1,*], *Rosidayu* Sabran[1], *Siti* Suriani Othman[1] and *Noor* Adzrah Ramle[1]

[1]Universiti Sains Islam Malaysia, Bandar Baru Nilai, 71800, Negeri Sembilan

Abstract: The online news media are said to have changed the understanding of the gatekeeping theory. Due to the virtually unlimited news space together with the interactivity in the online presentation of news, the Internet defies the whole conception of gatekeeping in journalism. This article seeks to examine the influence of gatekeeping forces in two influential Malaysian online news portals, MalaysiaKini and The Malaysian Insider (TMI), during the 2013 Malaysian general election. Based on the hierarchical model of news conception, this study assessed the influence of ideological gatekeeping forces on their news content. The reporting by both news portals was subjected to qualitative textual analysis. The results are triangualted with interviews with corresponding reporters and editors from MalaysiaKini and TMI. The findings show that although the reporting of MalaysiaKini and TMI is counter-ideological, the gatekeeping process remains intact. This is due to the speed factor and the requirement of the news portals to abide by the media-related laws in Malaysia.

1 Introduction

Gatekeeping theory is one of the most enduring theories in mass communication and has informed the study of journalism for more than 60 years. Derived from the channel metaphor by Lewin [15] , it was first introduced as a mass-communication concept by Manning [17]. Manning posited that 'Mr Gates', which mostly referred to the newspaper editor, decides which news is rejected and which will be published. At the same time, the gatekeeping practices also explain the power and influence of news editors over the news agenda. Editors basically maintain total control of the news agenda by deciding what should be published or rejected on behalf of news audiences[7].

Subsequent studies indicate that gatekeeping is actually a complex process in which news productions are filtered by a number of factors, such as the professional routines of journalists and the organisational structure [8,11,23] Gatekeeping, therefore, is seen as more of a group effort with multiple gates within the newsroom [8,25]. In 2000, Shoemaker and Reese [23] further refined the gatekeeping theory by introducing the hierarchical model of news influences, which allows for the study of gatekeeping on five levels of analysis: individual, routine, organisational characteristics, extra-media, and ideological.

The hierarchical aspect works in the way that each of the preceding forces may be thought to subsume the one before, suggesting that the ultimate level should be an ideological perspective [21,22]. The ideological level is concerned with how the media's symbolic content is connected with larger social interests and how meaning is constructed in the service of power. This necessarily leads us to consider that the recruitment of journalists, their attitudes, the routines they follow, their organisations' policy, and those organisations' positions in the larger social structure work to support the *status quo*. It means narrowing the range of social discourse by making the media agencies part of social control [22].

The Internet news media are said to have changed the understanding of the gatekeeping theory. Due to the Internet's virtually unlimited newshole, interactivity, and feedback available to readers of the online presentation of news, many scholars have written about how the online newspaper would challenge the gatekeeping concept in traditional media [10,24]. Interactivity in particular is said to have changed the power relations between journalists and audiences, where consumers have the ability to move at will through content and engage in computer-mediated conversations through the website [10].

For that reason, this article seeks to examine the influence of gatekeeping forces at ideological levels in online news portals. The reporting of two influential Malaysian online news portals, *MalaysiaKini* and *The Malaysian Insider* (*TMI*), during the 2013 Malaysian general election is used as a case study. The Malaysian Digital Association (MDA) and the comScore report for September 2015 showed that *MalaysiaKini* was the most visited news website and *TMI* the third

* Corresponding author: faizalkasmani@usim.edu.my

most popular news website in Malaysia. To investigate this subject, this article presents and analyses the reporting of the 2013 Malaysian general election together with interviews with the editors of *MalaysiaKini* and *TMI*.

2 Hierarchy of influence: the ideological forces

Reese and Shoemaker [23] argue that in understanding how the news is 'constructed', there are many forces at work that influence the news content. Their hierarchy of influences model suggests a framework for analysing media based on levels of analysis, from the most micro to the most macro — namely, individual, routines, organisational, extra-media, and ideological. It describes the multiple levels of influences that impinge on media simultaneously and suggests how each level interacts with the others [23] at different levels of strength in any shaping of media content.

The individual level considers the attitudes, training, and background of the journalist (or media worker, more generally) as influential. The routines level of analysis considers the journalistic routines that are structured, deeply naturalised rules, norms, and procedures that are embedded in newsmaking. The organisational level considers the imperatives that give rise to journalistic routines and how individuals are obliged to relate to others within that larger formal structure. The extra-media level looks at the influences from outside the media organisation to shape the news content, including the government, advertisers, public relations, influential news sources, interest groups, and even other media organisations [22]. The ideological force is seen as the highest level of the hierarchy of influences model. At this level, the media is said to construct common-sense understandings in media content that attach to certain interests of groups in power, which control significant social resources. This is made to appear natural with the meanings that people have become accustomed to [21]. Such 'common sense' meaning in the representation of news discourse could be seen as ideological. The concept of ideology in this context refers to a mediated ideology: it concerns how newsmakers represent the world in ways that make sense to the public, but the construction of the common sense is 'the view of those with most power in society' [19]. Fairclough [13] points out that the concept of ideology in media discourse is related to the nature of power relations between media and the existing power holders.

3 Methodology

This article analyses the reporting of *MalaysiaKini* and *TMI* during the 13th Malaysian general election. *MalaysiaKini* was formed by Steven Gan and Premesh Chandran in November 1999. Both were former reporters at the Kuala Lumpur newspaper *The Sun* [26]. *MalaysiaKini* started as a free service and initially received funding from press-freedom and democracy-advocacy groups, namely the Southeast Asian Press Alliance (SEAPA) and the Media Development Loan Fund (MDLF). In 2004 it became a subscription service [2, 26].

TMI began publishing in February 2008 and was owned by TMIDOTCOM Sdn Bhd until it was acquired by The Edge Media Group in June 2014. It describes itself as a news portal 'which offers an unvarnished take on events and personalities in Malaysia' [3]. It was reported that *TMI* was originally financed by a group of businessmen close to the then Prime Minister, Tun Abdullah Ahmad Badawi. Initially, it was supposed to act as a counterbalance to *MalaysiaKini*, which was seen as unfriendly to the Barisan Nasional [28]. Throughout the years, alongside *MalaysiaKini*, *TMI* has been regarded as an independent online news media in Malaysia giving voices by alternative views, filling a void left by pro-government traditional media.

Samples of news packages on the 2013 Malaysian general election from *MalaysiaKini* and *TMI* were accessed on each news portal. The timeframe for the sample selection covered one week between 1 May 2013 and 7 May 2013. The samples spanned four days before and two days after the election, which represented not only a critical time of active campaigning by both the opposition and the government but also the reactions to the election results in the aftermath. Samples were chosen from the NEWS section of *MalaysiaKini* and the HOME section of *TMI*. This section was known for including breaking news, and the choice of writing styles was of particular interest for the purpose of the study.

This article assesses the influence of ideological gatekeeping forces on the content of both *MalaysiaKini* and *TMI*. Their news reporting was subjected to qualitative analysis. Selected articles are analysed, and the analysis of textual features of the news was based on the analysis of the lexical features, sentence construction of texts, and analysis of direct and indirect quotations. The findings from the textual analysis are triangulated with personal interviews with reporters and editors of *MalaysiaKini* and *TMI* who produced news packages during the 2013 election. Four interviews were conducted with reporters and editors of *MalaysiaKini* and *TMI* around the state of Selangor and Kuala Lumpur between April and October 2015. The interviewees are Steven Gan, the editor-in-chief and the co-*founder* of Malaysiakini; Hafiz Yatim, a journalist and an assistant editor of the English desk at Malaysiakini; *Jahabar Sadiq, the* chief executive and *the* editor-in-chief of TMI and Zurairi Abd Rahman, a senior journalist of TMI. The interviews lasted between 30 minutes and one hour. The interviews were semi-structured, with follow-up questions added as the interviews unravelled. The questions focused mainly on the factors that shaped the news coverage of both news portals during the 2013 general election. The interviews were recorded and transcribed for analysis.

4 Analysis

There were 288 and 189 articles on the Malaysian general election produced by *MalaysiaKini* and *TMI*, respectively, during the period from 1 May 2013 to 7 May 2013. Of the 288 articles produced by *MalaysiaKini*, more than 23% (55 articles) comprised syndicated articles from a local news agency, BERNAMA, or news agencies such as Reuters and Bloomberg (10 articles). More than 10% (33 articles) were written by foreign columnists. Twenty-six of the articles were in the form of a review on readers' comments, collated and sorted by the editor according to the topics of the day; 6 of the articles documented chronological events or provided background information on the election. This means that during the 2013 general election, 156 of the articles, or about 55% of the overall articles, were produced by journalists from *MalaysiaKini* in the format of hard news.

Of the 189 articles produced by *TMI*, more than 30% (60 articles) were syndicated from BERNAMA, with less than 3% from other news agencies such as Reuters and Bloomberg. Almost 10% of the overall output (18 articles) was written by external columnists. Only 106, or 56%, of the overall articles of *TMI* during the 2013 Malaysian general election were written by its in-house journalists.

In terms of diversity of sources for the news reports, there were a total of 163 sources for *MalaysiaKini* and 100 sources for *TMI* (which excluded the editorial writing from the columnists as well as other types of articles). All of the sources were cited and attributed. Most of the sources for *MalaysiaKini* were representatives of Pakatan Rakyat, comprising 28% of the overall sources. Sources from the ruling party, Barisan Nasional, came in at second place with 21%. However, it is important to note that half of the sources from Barisan Nasional cited in *MalaysiaKini* were attributable to news syndicated from BERNAMA.

The majority of direct quotations in *TMI* were from Barisan Nasional. However, similarly to *MalaysiaKini*, the majority of the sources (24%) from Barisan Nasional were attributable to syndicated news from BERNAMA. This means that news written by *TMI* reporters represented only 12% of the sources from Barisan Nasional. As with *MalaysiaKini*, the majority of news sources provided by in-house journalists was from Pakatan Rakyat (19%).

Both news portals also gave voices to BERSIH, a coalition of non-governmental organisations (NGOs) that aims to reform the current electoral system in Malaysia to ensure free, transparent, and fair elections. Sources from the Malaysian Electoral Commission were also quoted by both news portals, although at less than 5% of the overall news sources. Both news portals seemed to have a similar strategy in terms of the representation of sources from Barisan Nasional and the opposition party Pakatan Rakyat. Although from the outset there was not much disparity in terms of the representation of sources from the ruling and the opposition parties – and for TMI it appears that there were more sources from Barisan Nasional – the majority of the sources from the ruling party were provided by BERNAMA. BERNAMA, a government-owned agency, was seen as the most important provider of news for mainstream news media in Malaysia. It was also found to function as a conduit between the government and the local mass media. There are two possible reasons why news syndicated from BERNAMA appeared prominently on both websites. Firstly, it made the news portals appear balanced in terms of representation of sources. The news portal primarily represented voices from the opposition groups, therefore, syndicating news from BERNAMA provided the voice of the government. Secondly, reporters of *MalaysiaKini* and TMI may have had difficulty gaining access to the sources from Barisan National because of their reputations as being pro-opposition party. Thus, receiving content from BERNAMA was a way to represent the story from the side of the ruling party.

The analysis of the selected articles from *MalaysiaKini* and TMI included lexical features and sentence construction in texts. Four articles were selected for textual analysis based on several key events occurring before, during, and after the election and which had been covered by both *MalaysiaKini* and TMI. The issues covered the campaign trails, the results of the election, and commentary on Najib's statement that the poor performance of the ruling parties was due to the "Chinese Tsunami". Consistently with the quantitative analysis, *MalaysiaKini* and TMI clearly gave a voice to the opposition party, which had generally been ignored by the mainstream media in Malaysia. However, there were disparities in the lexical structure of *MalaysiaKini* and TMI reports on the two political parties. Although both platforms relied heavily on direct quotations and indirect quotations from the opposition group, TMI adopted a less critical tone towards Barisan Nasional and its leader Najib Razak. This was best illustrated by reporting on the results of a survey by the Merdeka Centre for Opinion Research on 3 March 2013. *MalaysiaKini* adopted a more critical tone to coverage of the ruling party and its leader [18]. The report of the poll results opened with the lead "BN chief Najib Abdul Razak's approval rating sliding down a further 3% since the last poll". It then followed with a statement: "Pakatan Rakyat ahead of BN in terms of the favoured party to form the government". The report further explained that the statistic "reflect[ed] the slide in the 'feel good' factor that was previously generated by the large-scale distribution of Bantuan Rakyat's 1Malaysia (BRIM)".

TMI took a different angle when covering the story. The lead opened with "Barisan Nasional and Pakatan Rakyat are almost evenly tied to win Election 2013" [27]. It then explained that the polls showed that 42% had chosen Pakatan Rakyat to lead the government as against 41% for Barisan Nasional whereas 13% of voters remained undecided. The finding that the approval rating of Prime Minister, Datuk Seri Najib Razak, had slipped was mentioned in the second half of the story, reflecting a decision to give this fact less prominence than MalaysiaKini had done. *MalaysiaKini*'s attitude when reporting on the poll results with respect to Barisan Nasional's chances of winning the election was also reflected in the news headline. It wrote "Najib's rating slides, BR1M not working", which showed the use of negative verbs towards the subject. In contrast, TMI reported "BN, Pakatan in dead heat for Election 2013", which showed that the news portal preferred to refrain from predicting the poll results.

MalaysiaKini's [16] report on Anwar's campaign a day before the election further showed the opposition's indirect quotations in Paragraphs 1–5. Anwar's statements were paraphrased using the verbs "vowed", "urged", and "repeated

DOI: 10.1051/shsconf/20173300052

[his warning]", indicating, according to Barkho [5], "how the writer intrude[d] the event via what amount[d] to an evaluative comment". This demonstrated that *MalaysiaKini's* discursive patterns were closer to Anwar's viewpoints and his interpretation of events. In addition, the use of direct quotation at the end of the news report – "seize your destiny. Ini Kalilah!" – without any further attribution showed the reporter's support of Anwar's standing and position. TMI, on the other hand, gave a more detached reporting of the same event, as was shown in paragraphs 6–8 [20]. Except for one positive verb – "assured" – the rest of the coverage used the verbs "say" and "call", which carried little indication of the reporter's attitude towards Anwar's campaign and showed neither endorsement nor disparagement of his statements [9]

- … de facto PKR leader Anwar Ibrahim urged the country's 13 million voters to turn up and cast their ballots with "faith, courage and vision" [16]
- He also urged them not to get involved in the shredding or hiding of government documents if Pakatan were to seize Putrajaya from the BN tomorrow, after polling day [16]
- He also urged Malaysians not to fear [16]
- … he said in assurance [16]
- He repeated his warning to the Election Commission… [16]
- ...Pakatan Rakyat (PR) leader, Datuk Seri Anwar Ibrahim, assured the public that if his coalition wins Putrajaya, the transition will not only be historic but also peaceful [20]
- ... he said in a final press conference at Yayasan Aman here [20]
- Anwar called on the Election Commission to ensure a clean and fair election [20]

5 Discussion

Findings show that both *TMI* and *MalaysiaKini* provided alternative news reportage and analysis, which give competing discourses on political issues in Malaysia and challenge the version of the political reality as constructed by the mainstream media [14]. This is in contrast with media in Malaysia that was generally biased in favour of the ruling party [4]. During general elections, in particular, Malaysia's mainstream media were tasked with highlighting the Barisan Nasional government's successes, especially in the area of socioeconomic development. At the same time, adverting to the successes of other contesting political parties was restricted. If such parties appeared on the mainstream news, they were often depicted in a negative light [4]. This is in line with arguments by Steele [26] that specifically described *MalaysiaKini*, in particular, as legitimising 'alternative views of events, thus challenging the authoritarianism of the Barisan Nasional' (p. 91). Abbott [1] regarded *MalaysiaKini* as an example of the manner in which the Internet may have 'a democratising effect on the region's authoritarian and semi-authoritarian regimes by opening up a permissive online public sphere in which traditional hierarchies of power are challenged' (p. 333).

This counter-discourses coverage by *TMI* and *MalaysiaKini* is clearly part of journalistic ideology by both news editors. For *MalaysiaKini*'s Steven Gan and *TMI*'s Jahabar Sadiq, they view their news portal as breaking the deeply naturalised practice in Malaysian journalism of upholding only the official voices in the mainstream media. They also see themselves as breaking the 'common sense' meaning in the political news by presenting contending voices on political issues to ensure fair debates within Malaysian election discourse.

> *I feel very strongly that in a society like Malaysia where we are very divided, where people feel very strongly about things and have different viewpoints, we really need to come together and discuss them. We may at the end of the day agree to disagree, but what is important is that we know where we stand and where they stand on issues, and hopefully through that process, everyone comes a bit closer... (Steven Gan, personal interview, 20 May 2015)*

> *Why are Bernama, TV3, and RTM covering issues mainly from the perspective of the government or Barisan Nasional leaders and not so much from that of the opposition leaders? So we are a mirror reflection of the mainstream media...we try to balance the coverage ... (Jahabar Sadiq, personal interview, 26 October 2015)*

Both editors believe that in doing so they need to go back to the traditional practices of journalism of presenting facts to ensure that they report objectively and impartially. This also means the inverted pyramid technique — whereby journalists organise information from the most relevant at the top to the least relevant at the bottom — is the prioritised way of presenting news. It is in contrast with arguments that online news tends to incorporate a more conversational and narrative approach to news, indicating a more interpretative style of news writing that signifies a move away from the rigid news writing format.

> *...we report the facts of the day, issues of the day, as impartially as possible and as dispassionately as possible. This is the news. We clearly identify this is an opinion. We do not editorialise the news. The Malaysian Insider was set up in 2008 for the simple purpose of informing people so they can make informed choices and to give as much balance wherever possible in everything we cover. This is done in a very clinical way through journalism; this is news, and this is what is happening. This is a quote and this is what is happening. (Jahabar Sadiq, personal interview, 26 October 2015)*

> *...when it comes to reporting news, when we are writing news, doing interviews, etc., we stick by the book. We*

DOI: 10.1051/shsconf/20173300052

have journalistic ethics, and we stick to them. When we write news, of course, we cannot concoct whatever a person says in order to give a story our own slant. That would be breaking journalistic ethics. We need to be fair in the sense that we will get the other side to rebut and to respond. (Steven Gan, personal interview, 20 May 2015)

Going back to traditional journalism means that both online news portals very much stick to the gatekeeping practice where every news item needs to go through the editorial team. *MalaysiaKini*, however, as explained by Steven Gan and its journalist Abdul Hafiz Yatim, appears to have a more democratic style of management. This could also explain why the journalists' voices of *MalaysiaKini* occasionally appear in the reporting.

Most of the time we discuss an issue and then let Steven Gan make the decision. At our office, all of the officers, including the Chinese, Malay, and English, sit together without any segregation. In fact, Steven Gan also sits with us. We are a flat organisation. (Abdul Hafiz, personal interview, 20 May 2015)

There is communication between journalists and editor, and this is partly because there has to be checks and balances. A lot of our journalists are new, so they would require a bit of guidance. (Steven Gan, personal interview, 20 May 2015)

For *TMI*, the news process is more hierarchical and traditional. There is a clear role of Sadiq as the sole gatekeeper who makes the decision of what is in and what is out of the online newshole. This is further underscored by former *TMI* journalist, Zurairi Abdul Rahman, who was one of the news portal reporters during the 2013 election.

It is a pyramid, and I sit at the top of the pyramid. Every report that went out during the election campaign, during the timeline you specified, all had to go through me. If I think it is wrong, or it is off, I will say no, please rewrite it this way or I will rewrite it myself. (Jahabar Sadiq, personal interview, 26 October 2015)

The way we operate is that you should always just report, and the ones who will make the call will be the news editors in charge that day...they just send it in, and if the editor decides that it is not suitable, then you either re-write or kill the story. (Zurairi Abdul Rahman, personal interview, 20 October 2015)

6 Conclusion

Findings show that the notion of a 'gate' in the journalistic practice of two of the most popular news portals in Malaysia, *MalaysiaKini* and *TMI*, remains intact. Although the reporting of both news portals is contingent upon multiple explanatory factors as explained by the hierarchy of influence model, the editor-in-chief still holds immense power in deciding what information people can and cannot read. This is in line with previous studies concluding that journalists are retaining all or part of their gatekeeping roles as they move online, including the practice of story assignment and copy editing, reproducing those of their print counterparts [6,12,25]. The gatekeeping processes are more vital than ever in *MalaysiaKini* and *TMI*. In *TMI*, the editor acts as the sole 'Mr Gate' in deciding what news should be in or out.

Although the journalistic framework of both online news portals is still based upon traditional journalistic ideological practices such as fair, balanced, and objective reporting, the news content is counter-ideological. Since the gatekeeping process remains strong, this counter-ideological strategy is mostly decided at the news-editor level.

This article posits that the constant need to generate a counter-ideological flow of news while negotiating the *risk* of defamation, libel, and slander due to media laws in Malaysia means that it is imperative that both news portals retain all or part of their gatekeeping roles. In line with arguments by Singer [25], the sheer quantity of news that can be featured online, along with its wildly varying quality prepared particularly by rookie reporters, reinforces the need for 'Mr Gates' to sort it out.

References

1. J.Abbott. Democracy@ internet. org Revisited: analysing the socio-political impact of the internet and new social media in East Asia, Third World Quarterly, **33**, 2(2012)
2. J.Abbott. J. Contemp. Asia, **43**, 4, (2013)
3. About The Malaysian Insider (n.d.) Available from http://www.themalaysianinsider.com/about/, Accessed 1 December 2015.
4. M.K.Anuar. Politics and the Media in Malaysia, Philippine Journal of Third World Studies, **20**, 1, (2005)
5. L.Barkho. Unpacking the Discursive and Social Links in BBC, CNN and Al Jazeera's Middle East Reporting, Journal of Arab and Muslim, 1, 1, 11 (2007)
6. P.J.Boczkowski. J. Commun, **54**, 2, 197 (2004)
7. A.Bruns .BJR, **7**, 2, (2011)
8. W.P.Cassidy Gatekeeping similar for online, print journalists. Newspaper Research Journal, **27**,2 (2006)
9. L.Chen. Int Rev Appl Linguist Lang Teach, **43**, 33 (2005)
10. D.V.Dimitrova & M.Neznanski.. J Comput Mediat Commun, 12, 1, 248 (2006)

11. D.V.Dimitrova, C.Connolly-Ahern, A.P.Williams, L.L.Kaid, & A. Reid. Hyperlinking as gatekeeping: Online newspaper coverage of the execution of an American terrorist. Journalism Studies, 4,3, 401 (2003)
12. D.Domingo, T.Quandt, A.Heinonen, S.Paulussen, J.B.Singer & Vujnovic, M. Participatory journalism practices in the media and beyond: An international comparative study of initiatives in online newspapers. Journalism Practice, 2,3, 326 (2008)
13. N.Fairclough. *Language and power*. Pearson Education (2001)
14. S.Giersdorf & A.Croissant. a. Journal of Civil Society, **7**,1 (2011)
15. K.Lewin. Group decision and social change. Readings in social psychology, **3** (1947)
16. S.Loone. Anwar vows no retributive justice for political foes. Malaysiakini. (2013)
17. W.D.Manning. The Gate Keeper. Journalism and Mass Communication Quarterly, **27**, 4 (1950)
18. MalaysiaKini.Najib's ratings slide, BR1M not working. (2013)
19. D.Matheson. *Media discourses*. McGraw-Hill Education (2005)
20. O.Mok. Anwar says peaceful transition if Pakatan takes power. The Malaysian Insider. (2013)
21. S.D.Reese. Understanding the global journalist: A hierarchy-of-influences approach. Journalism Studies, **2,** 2 (2013)
22. S.D.Reese. BJR, **3**, 2 (2007).
23. P. J.Shoemaker & S.D. Reese . *Mediating the message*. White Plains: Longman. (1996).
24. J. B. Singer. Online journalists. *Journal of Computer-Mediated Communication*, **4**,1 (1998)
25. J B. Singer. Stepping back from the gate: Online newspaper editors and the co-production of content in campaign 2004. Journalism & Mass Communication Quarterly, **83**, 2 (2006)
26. J. Steele. The International Journal of Press/Politics, **14**, 1 (2009).
27. The Malaysian Insider. BN, Pakatan in dead heat for Election. (2013).
28. The Rakyat Post. Malaysian Insider joins the Edge Media Group. June 9. Available from http://www.therakyatpost.com/business/2014/06/09/malaysian-insider-joins-edge-media-group/, accessed 1 December 2015 (2014)
29. D. M. White. *Journalism and Mass Communication Quarterly*, **27**,4, (1950)

Accept or Reject? Predicting Ideation Outcomes through Enterprise Social Media

Nik Ahmad Rozaidi[1,*], *Jennifer* L. Gibbs[1], and *Julia* Eisenberg[2]

[1]Rutgers University, New Jersey, USA
[2]Pace University, New York, USA

Abstract. Implementing social media in the workplace may make it easier for employees to participate in knowledge sharing activities such as Q&A and ideation. However, vetting the quality of answers and ideas becomes more complex when anyone in the company can contribute. Research on the use of social media for Q&A has shown that certain characteristics and reputation algorithms can help determine the best answers. Less is known about the ideation process and the way it plays out in social media. This paper explores the use of enterprise social media (ESM) for ideation by employees in a large Russian organization distributed across nine time zones. In particular, we explore which characteristics of both ideas and their contributors predict whether ideas get accepted or rejected. Our analysis is based on logistic regression analysis of a sample of 488 ideas contributed in an ESM tool used in the organization as well as a content analysis of the types of ideas generated. Our results suggest that rather than being truly democratic and decentralized, ideation in ESM is driven by those in (or proximate to) positions of organizational power.

1 Introduction

Organizations need ideas to innovate. Ideas are solicited from a variety of internal and external sources. Employees are perhaps the most cost effective way to obtain ideas as they are experienced, knowledgeable, and accessible. For employees, the ability to participate in organizational decision-making is beneficial as it leads to greater employee satisfaction and productivity [1]. This is despite the paradox that the final decision remains in the hands of upper management [2]. The introduction of enterprise social media (ESM) and its increasing adoption within organizations may be expected to increase the level of employee participation in the ideation process, defined as the generation, development, and submission of ideas [3]. ESM has the potential to democratize participation by connecting employees without being constrained by geographical, functional, or hierarchical boundaries. Despite some resistance to ESM stemming from user perceptions and preferences that conflict with workplace norms and legitimacy [4.5.6], privacy and

* Corresponding author: nrozaidi@rutgers.edu

impression management concerns related to "context collapse" [7,8] and increased demands on one's time and attention [9,8], adopting organizations are finding that ESM affordances are enabling processes that were impossible or difficult before [10]. Employees can now connect more easily with co-workers and communicate across various organizational boundaries [11] as well as share and access knowledge from otherwise unknown colleagues [12,13]. This creates tremendous potential in terms of strengthening workplace ties and further enhancing the participatory climate in the organization. However, the extent to which ESM use is actually challenging traditional hierarchy, status, and power boundaries remains unclear.

In terms of ideation, ESM may provide unique affordances or possibilities for action [14,15] such that ideas can be more easily produced in greater numbers by almost anyone with access to the platform. The lack of credentials and expertise of contributors however may cause the idea quality to be questioned, and may undermine management's ability to evaluate and make a decision on which ideas to accept. As many organizations are focusing on innovation as part of their strategy and as adoption of ESM becomes more widespread, we believe it is important to understand the factors that can facilitate social media-based decision-making, including the process for ideation.

The aim of this study is to understand factors that may predict whether an idea generated in ESM gets accepted or rejected. To address this question, we borrow from the literature on answer quality from social Q&A sites, given the nascent use of social media for ideation. This paper contributes to the social media literature through its study of ideation as a consequence of the knowledge sharing aspect of the platform.

2 Background and related work

2.1 Ideation in Enterprise Social Media

We define ideation as a process of generation and development of ideas for problem solving. Ideation to us is a distinct concept, not to be confused with brainstorming or innovation. Brainstorming [16] is one of the techniques for ideation, while innovation is the broader process that includes ideation as a cornerstone activity and that which ends with the implementation of the chosen ideas [17]. Regardless, the goal of the ideation process is to produce as many good ideas as possible for subsequent selection and decision-making. The implementation of ESM in the workplace has created a novel forum for ideation to occur. While ESM is also a computer-mediated platform, there are a few key differences compared to traditional systems. Due to its broader reach to the whole organization and the likelihood that users are identifiable, ESM tools such as social network sites may afford greater and more egalitarian participation [18] that supports idea contribution, evaluation, and prioritization. ESM may also afford new connections (e.g. follower relationships) that transcend hierarchical, functional, and geographical boundaries [11] that further encourage participation in the ideation process. ESM also affords metavoicing [19] in which users can participate or express themselves in multiple ways in the ideation process, including commenting, liking, and following.

2.2 Answer Quality in Social Q&A Sites

Social media tools provide a unique venue for both enacting and observing the ideation process. Community Q&A sites like Yahoo Answers offer a way to publicly pose questions to a community of experts or knowledgeable users who are willing to spend time and effort to impart their knowledge on a subject. Even social network sites like Facebook are being used for Q&A albeit among the network of friends [20]. These are alternatives for finding

DOI: 10.1051/shsconf/20173300053

information online instead of using a search engine or emailing someone. For those providing the answers, there is often some benefit to be gained from providing a good answer, such as being recognized as an expert. However, there may also be an incentive to provide an answer, regardless of its quality, to earn points for participation. These incentives when combined with the openness of the platform mean that there is potential for spam, as providing feedback is not restricted to just those with the necessary expertise. To counter this, the one posting the inquiry can be requested to choose the best answer, or the community members can be asked to vote for it, but this is typically quite subjective.

As [21] found from their analysis of Yahoo Answers, the "best" answer in a community Q&A site varies depending on question topic (or context and nature of question) and can be more socio-emotional than factual-centric. For example, in support forums, the best answers tend to be the most emphatic or caring; and in discussion forums, the best answers tend to agree with askers' opinions. In predicting the best answers, [22] identified 20 features that strongly predict answer quality. Among them are: answer length; number of up and/or down votes ("thumbs-up" or "thumbs-down") an answer receives; the fraction of answers picked as best answers; and the average number of answers received by the question. [21] used answer characteristics and user attributes to predict answer quality. Performing a logistic regression on the answer outcomes across three categories (programming, marriage, wrestling), they found that reply length and total number of answers were strong positive predictors while thread length and number of user replies were significant negative predictors. Based on a review of related research, we theorize that answer quality depends on the properties of the questions, the answers, and the askers/answerers. Because ideation on ESM shares a similar problem as social Q&A in terms of finding the best contribution among many, it can be argued that the quality of ideas generated on social media is dependent on properties of both the ideas themselves as well as their contributors. We therefore ask: What factors predict whether an idea generated in ESM gets accepted or rejected?

3 Methodology

3.1 Data collection

A case study was performed on a company that had deployed an ESM application and used it for ideation. The company is a retail subsidiary of a multinational telecommunications company headquartered in Moscow, Russia that employed about 20,000 people in its corporate offices and retail outlets in approximately 1,000 cities throughout Russia. The company, TeleCom (a pseudonym), deployed an ESM application called TeleCom Life in Fall 2011 as part of a broader initiative to improve internal communication and knowledge sharing. The application enabled employees to create online profiles, find and articulate connections with co-workers by following them, and interact with others through the exchange of online messages or by "liking" them, among many other features. Employees' online activities were logged and stored in a file on a central server, which contained descriptive information about users (job title, department, and location) as well as data about when users posted messages, made comments, "liked" posts, and "followed" other users. The use of the ESM for ideation was undertaken through a section called Forum of Innovators. Employees (including management) were encouraged to visit this forum to post new ideas and/or comment on or "like" the ideas of others. Periodically, the management team would come in and review the ideas that had been posted and decide to flag them as either "accepted" or "rejected"; ideas carry the "new" flag by default. Winners were flown to Moscow for a meeting in which they received company recognition and some had the opportunity to be involved with further development of the idea.

 DOI: 10.1051/shsconf/20173300053

While we did not have direct access to the system, access was provided through a server log file containing metadata for activities from the entire first year of use of the system. The file was sanitized by removing personally identifiable information and texts of the posts, although the subject tags describing the posts, where provided, were retained. In order to facilitate the analysis of the log file, it was first converted into a MySQL database, its textual information machine-translated from Russian into English and verified by several native Russian speakers. Overall, the database contained 156,543 posts, 628,782 comments, 503,180 follower relationships, and 849,126 likes involving 17,192 users. Users were distributed across ten regions (Headquarters 11.0%, Chernozem 1.3%, Central 10.3%, Far East 5.5%, Moscow 9.0%, Northwest 10.2%, Siberia 9.3%, South 10.7%, Ural 14.2%, Volga 18.5%) and five hierarchical positions (top management 0.7%, middle management 14.9%, senior specialist 10.8%, specialist 67.1%, and support staff 6.6%).

3.2 Measures

Of the 156,543 posts in the database, 3,245 posts were explicitly labeled as "ideas". Of these, 155 were accepted ideas, 333 were rejected, while the rest were "new" ideas as of the time the log data were collected. There was no data regarding when decisions about the ideas were made. The ideas with decisions were organized into the dependent variable of *idea status* (accepted versus rejected), with "new" ideas being excluded from the dataset as they were still pending, for a total N of 488 ideation posts. The independent variables were comprised of two major dimensions: idea attributes and contributor attributes. *Idea attributes* included duration (number of minutes from first post to last comment), number of likes, and number of comments. *Contributor attributes* included employee position (recoded into five categories: top management, middle management, senior specialist, specialist, and support staff), location (ten regions that were recoded into headquarters versus non-headquarters), expertise points within the system (from their activities on the ESM and from ratings by other users), number of followers, number of people followed, number of ideas contributed, and how many ideas were new, accepted, and rejected. Due to the large degree of variance and number of outliers, the variables for duration, number of likes, and number of followers were log transformed to standardize them.

3.3 Data Analysis

We tested our ability to predict which idea characteristics and contributor characteristics were correlated with the outcome of the ideation evaluation process. Given that our dependent variable (idea acceptance versus rejection) was binary, we conducted a logistic regression analysis on the two sets of variables.

4 Findings

4.1 Inter-item correlation

An inter-item correlation matrix revealed moderate yet significant correlations among many of the study variables, indicating that they were associated but that multi-collinearity was not an issue. In particular, idea status was significantly positively associated with number of accepted ideas ($r = .51$, $p < .001$), number of followers ($r = .34$, $p < .001$), number of likes ($r = .32$, $p < .001$), duration of discussion ($r = .28$, $p < .001$), position ($r = .10$, $p < .05$), and expertise level ($r = .09$, $p < .05$), and it was negatively associated with number of ideas created ($r = -.15$, $p < .001$). Number of comments and region were non-significant. Table 1 shows the inter-item correlations.

 DOI: 10.1051/shsconf/20173300053

Table 1. Inter-item Correlations among Study Variables (N=488).

Variables	1	2	3	4	5	6	7	8	9
1. Idea Status	--								
2. Duration of Discussion	$.28^{***}$	--							
3. Number of Likes	$.32^{***}$	$.41^{***}$	--						
4. Number of Comments	.05	$.36^{***}$	$.51^{***}$	--					
5. Number of Followers	$.34^{***}$	$.11^{***}$	$.28^{***}$	$.09^{***}$	--				
6. Expertise Level	$.09^{*}$	.02	$.13^{***}$	$.06^{**}$	$.17^{***}$	--			
7. Number of Ideas Created	$-.15^{**}$	$-.10^{***}$	$-.05^{**}$	-.02	$.25^{***}$	$.13^{***}$	--		
8. Number of Ideas Accepted	$.51^{***}$	$.08^{***}$	$.14^{***}$	$.05^{**}$	$.36^{***}$	$.06^{***}$	$_{*}.24^{**}$		
9. Position	$.10^{*}$	$.09^{***}$	$.16^{***}$	$.04^{*}$	$.35^{***}$	.03	.03	.02	--
10. Region (HQ vs non-HQ)	.09	$-.05^{**}$	$.05^{**}$	.01	$.31^{***}$	.03	$_{*}.19^{**}$	$.18^{***}$	$.17^{***}$

* $p < .05$, ** $p < .01$, *** $p < .001$

4.2 Logistic regression

Table 2 summarizes the prediction results of the logistic regression, used to answer RQ1. Results showed that both idea attributes and contributor attributes predicted the likelihood of an idea being accepted. The overall model was quite good in predicting the degree of variability in the data. The significant chi-square indicated that the model fit significantly better than the empty model with just the constant. Further, the pseudo-R^2 (using Nagelkerke r-squared) was .538, indicating that the model explained about 54% of the variance in idea acceptance.

Table 2. Logistic Regression of Idea and Contributor Attributes on Idea Status.

	B	S.E.	Wald	df	Sig.	Exp(B)
Number of Likes	1.27	.41	9.63	1	.002**	3.55
Duration	.33	.13	6.67	1	.01**	1.40
Number of Comments	-.03	.01	8.34	1	.004**	.97
Number of Followers	.35	.26	1.76	1	.19	1.41
Expertise Level	.25	.67	.13	1	.72	1.28
Number of Ideas Created	-.09	.02	17.16	1	.00***	.91
Accepted Ideas	1.03	.13	60.37	1	.00***	2.80
Position	.00	.00	7.12	1	.008**	1.00
Region (HQ vs non-HQ)	.81	.37	4.88	1	.03*	.45
Constant	-5.32	2.61	4.16	1	.04*	.01

* $p < .05$, ** $p < .01$, *** $p < .001$

In terms of idea-related attributes, the most significant predictor was the number of likes an idea had. The coefficient of 1.27 indicates that for each additional like, the log odds of acceptance (versus rejection) of the idea increased by 1.27. Further, ideas with longer duration of development were also more likely to be accepted. Finally, number of comments was a significant but negative predictor, indicating that ideas with more comments were less likely to be accepted. In terms of contributor attributes, the strongest predictor was the number of prior accepted ideas a contributor had. The number of total ideas contributed also had a significant effect, although it was negative, indicating that those with more ideas were less likely to have an idea accepted. Position and region of the contributor were also significant predictors, although position (status) of the contributor had little effect. Being from headquarters significantly increased the odds of having an idea accepted. The number of followers a contributor had and his/her expertise level in the tool were not significant. To further explore the degree to which ideation crossed hierarchical boundaries, we examined the mean number of ideas contributed and accepted by region, position, and expertise level within the tool. Our comparison of means reveals that employees from headquarters contributed the most ideas overall, and that they had the highest mean number of accepted ideas by far (see Table 3). A one-way ANOVA confirmed that these differences by region were significant ($F = 30.364$, $p < .001$, $df = 9$ for total ideas; $F = 20.686$, $p < .001$, $df = 9$ for accepted ideas).

Table 3. Mean Number of Total Ideas and Accepted Ideas by Region.

Region	Mean Number of Total Ideas	Mean Number of Accepted Ideas	Region	Mean Number of Total Ideas	Mean Number of Accepted Ideas
HQ	13.13	.83	Northwest	6.96	.54
Chernozem	8.59	.00	Siberia	4.71	.21
Central	4.79	.16	South	10.05	.42
Far East	5.56	.10	Ural	10.34	.40
Moscow	9.07	.11	Volga	5.61	.21

Further, top management was not the biggest contributor in terms of number of ideas (which was highest for middle management and retail specialists), but top managers had a much higher number of accepted ideas overall than lower-level employees. A one-way ANOVA confirmed that differences by position were also significant ($F = 12.508$, $p < .001$, $df = 4$ for total ideas; $F = 9.064$, $p < .001$, $df = 4$ for accepted ideas). Interestingly, although expertise level within the tool was not a significant predictor of idea acceptance in the logistic regression when other variables were controlled for, those with the highest level of expertise contributed a much higher number of total ideas as well as accepted ideas than less active users. A one-way ANOVA confirmed that differences by expertise level were significant ($F = 20.426$, $p < .001$, $df = 3$ for total ideas; $F = 4.367$, $p < .01$, $df = 3$ for accepted ideas). This makes sense given that TeleCom Life was initially implemented by top management and then rolled out to lower-level employees in other locations (Table 4).

Table 4. Mean Number of Total Ideas and Accepted Ideas by Position and Expertise Level.

Position	Mean # Total Ideas	Mean # Accepted Ideas	Expertise Level	Mean # Total Ideas	Mean # Accepted Ideas
Top Mgmt	5.98	1.17	Level 4	8.52	.36
Middle Mgmt	9.69	.33	Level 3	1.32	.03
Senior Specialist	5.66	.27	Level 2	1.22	.00
Specialist	8.31	.35	Level 1	1.00	.00
Support Staff	7.28	.33			

5 Discussion

Social media is often heralded for its open architecture and resulting participatory culture. ESM offers new affordances for participatory decision-making in organizations, especially those that are large and geographically distributed, by providing a centralized forum to generate, evaluate, and prioritize new ideas that are contributed by employees from all levels and locations within the organization. Our findings shed light on which factors drive idea acceptance in a large Russian telecommunications company. We find that both idea attributes and contributor attributes predict which ideas are adopted.

Our findings suggest that idea attributes play a role in their likelihood of acceptance. We found that ideas were more likely to be accepted if they received more "likes" on the social media platform but fewer comments. This is consistent with how votes work to surface best answers in a community Q&A (e.g. [22]), and the lesser number of comments may indicate less convolution of the original idea. Put another way, ideas with more comments may be more controversial and require more discussion, generating more negative feedback that ultimately results in their rejection. Successful ideas did, however, have a longer duration from initial posting to final comment, than did unsuccessful ones, implying that more deliberation time increases the likelihood of idea acceptance.

Our findings also highlight the importance of contributor attributes. We found that ideas were more likely to be accepted if the contributor was from the head office and closer to key decision makers, alluding to the network proximity effect. In addition to location, position also played a role as top managers did not contribute as many ideas overall, but the ideas they did contribute were significantly more likely to be accepted as those of lower-level employees. This is likely to be explained by the fact that top managers and other employees who are located in proximity to management in the headquarters location may be more aware of the decision-making process and have more opportunity to advertise their ideas due to their access to key decision makers, thus ultimately having a greater chance of having their ideas accepted and adopted. This pattern seems to perpetuate itself, as those with a greater track record of having ideas accepted are much more likely to have a future idea accepted. These results are interesting in light of the democratic claims often made about social media, as they serve to temper this discussion. Our findings suggest that rather than being truly democratic and decentralized, ideation in ESM is driven by those in (or proximate to) organizational positions of power. This – as well as the fact that management made the final decisions about idea acceptance or rejection – suggests that a hierarchy of decision-making persists despite the fact that all employees have access and ability to contribute ideas, such that all ideas are not truly created equal in social media applications but are dependent on the status level of their contributors.

5.1 Practical implications

Our study highlights a new and arguably successful use of ESM for the important work activity that is ideation. Organizations that have been apprehensive about deploying ESM can take note of the benefits that may accrue from the unique affordances of ESM. Even the "socialness" of ESM can be helpful in building relationships and gaining trust among employees [7], which should translate into productive uses of the tool. For ideation, ESM can be a tool that enables greater idea production without requiring participants to be collocated, synchronous, or spontaneous. While a greater quantity should increase the chance of getting a good idea, the affordances of ESM, such as metavoicing and the embedded social network, are more likely to ensure that it happens. Decision makers can also leverage the metadata from ESM to assist them in deciding which ideas to accept or reject. Employees, having direct access to the system, should be able to know the types of ideas that are more or less likely to be accepted, and modify their contribution along that path.

5.2 Limitations and future work

The results, however, need to be read with certain limitations in mind, namely that this research is based on a single case study that may not generalize due to its boundary conditions of industry (telecommunications workers may be more technologically savvy), generation (primarily young retail employees in their 20s), and culture (workplace norms may be different in Russia than anywhere else). Future research should examine ESM-based ideation in other organizational and cultural contexts. Further, understanding of our data may be limited as we did not have complete textual content (only tags for a subset of posts) and had to contend with a language barrier. A second limitation is that this study is about ESM and less about public social media. While there may be overlap in the functionality of each, the circumstances and context surrounding the use of ESM are different than public tools such as Facebook and Twitter. Given that most research has focused on public tools, however, our study makes an important contribution to knowledge of ESM use for Q&A and ideation in the workplace, about which less is known. Moreover, these findings should be equally applicable and relevant for decision-makers in any social sites that ask for ideas, or those attempting to find the proverbial needle in the haystack, in trying to locate quality information within the vast amount that exists. As organizations are increasingly implementing ESM platforms, it is important to understand how knowledge is shared and more importantly, how it influences processes that facilitate idea generation, evaluation, and adoption within an ESM application.

References

1. K.I. Miller, P.R. Monge, AoM J, **29**, 727 (1986).
2. C. Stohl, G. Cheney, MCQ, **14**, 349 (2001).
3. T. Kelley, J. Littman, *The Art of Innovation: Lessons in Creativity from IDEO, America's Leading Design Firm* (2001).
4. M. Lüders, First Monday, **18**, 8 (2013).
5. M.M. Skeels, J. Grudin, *Proceedings of GROUP (*2009).
6. J.W. Treem, S.L. Dailey, C. Pierce, *Proc. of ICA Conference (*2013).
7. N.B. Ellison, J.L. Gibbs, M.S. Weber, Am. Beh. Sc. (2013).
8. C. Lampe, J. Vitak, N. Ellison, *Proceedings of CSCW* (2013).
9. J.L. Gibbs, N. Rozaidi, J. Eisenberg, JCMC, **19**, 102 (2013).
10. J.W. Treem, P.M. Leonardi, Comm. Yearbook, **36**, 143 (2012).
11. J. DiMicco, D.R. Millen, W. Geyer, C. Dugan, B. Brownholtz, M. Muller, *Proc. of CSCW* (2008).
12. M.J. Brzozowski, T. Sandholm, T. Hogg, *Proceedings of GROUP* (2009).
13. J. Huh, L. Jones, T. Erickson, W.A. Kellogg, R.K.E. Bellamy, J.C. Thomas, *Proc. of CHI* (2007).
14. J.J. Gibson, *The ecological approach to visual perception* (1979).
15. I. Hutchby, Sociology, **35**, 441 (2001).
16. A. Osborn, *Applied imagination* (1957).
17. R.A. Wolfe, J. Mgt. Studies, **31**, 405 (1998)
18. D. Halpern, J.L. Gibbs, Comp. in Human Beh., **29**, 1159 (2013).
19. A. Majchrzak, S. Faraj, G.C. Kane, B. Azad, JCMC, **19**, 38 (2013)
20. R. Gray, N. Ellison, J. Vitak, C. Lampe, *Proc. of CSCW* (2013).
21. L.A. Adamic, J. Zhang, E. Bakshy, M. Ackerman, *Proc. of WWW* (2008).
22. E. Agichtein, C. Castillo, D. Donato, A. Gionis, G. Mishne, *Proc. of WSDM* (2008).

An Analysis of Malaysian Political Communication and Its Role in Electoral Process

Suhaimee Saahar @ Saabar [1,*] *Ismail* Sualman[1] , and *Mohd Adnan* Hashim[2]

[1]Faculty of Communication and Media Studies, Universiti Teknologi MARA (UiTM), Shah Alam, Malaysia
[2] Faculty of Communication and Media Studies, Universiti Teknologi MARA, Melaka, Malaysia

Abstract. This paper attempts to discuss political communication in Malaysia especially after the 12th General Elections (GE12). The purpose of this study is to analyse the political efficacy or sense making process which involve analysing the techniques and creative execution that had been used by political parties and its effects on Malaysian voters. The results of GE 12 had fashioned a political tsunami that influenced many important aspects in the Malaysian political landscape, be it national and economic policies, social unity or campaign strategies. Therefore, the main focus of this paper is to discuss the elements and importance of political communication research in Malaysian political communication campaign by focusing on the aspects of creative execution, political information efficacy and sense making process and its impact on the choice of communication tools amongst Malaysian voters.

1 Introduction

In a democratic country, citizen's participation in the decision making process is important where the process of information exchange make them to exercise the inalienable right [1]. Election provides equal rights to the people to choose their representatives in the government system [2] which serve not only to select leaders acceptable to voters but also to make the leaders accountable for their performance during their tenure in office [3]. Therefore, election and campaign has become a vital platform for any contesting candidates and political parties to position their ideologies and values, through persuasive, systematic and strategic communication and media such as political advertising campaigns. On the other hand, it is the right channel for a citizen or voter to engage, participate and execute their right under the constitution to choose their leaders. In Malaysia, electioneering system is based on three principles, majority votes based on method first-past-the-post system; single member territorial representation; and multi-party electoral system.

*Corresponding author: aimeesavvy@gmail.com

Since independence in 1957, Malaysia has experienced several general elections mostly contested by political parties and a small number of independent non-party affiliated candidates and these elections occur at regular intervals. The role of political communication in election campaigns is important and less research about in outside the United State of America is more needed to understand its impact on the electoral process [17] on political advertising and political communication in general in Europe, Canada, Israel and East-Asian countries [15]. Even though Malaysia has conducted 13 general elections since the first one in 1955, very little attention was given on research relating to political communication and advertising. As mentioned by Kaid (2001), research on the role of political communication and advertising outside the United States of America is still in its infancy and need more study to understand this phenomenon on the voters. However, there are several important similarities and many interesting differences that warrant some discussion on this subject [16]. The limited number of political communication studies originating in Asia can also be explained by understanding differences in the political systems themselves, which make it difficult to apply theories of media effects developed in Western democracies. For example, the procedures of how political candidates are chosen and elected in Asia, the way people participate in politics and how political information is discussed in the media all differ significantly from political processes in Western nations [18].

2. Research Gap

The limited number of this study in this subject can be traced based on the number of research done by local scholars. Most of the past studies on the role of political communication in elections were focused on political media coverage and the effects of the mass media during election [7,5,11,20]. There are tendency to study the role and impact of media especially on the political messages such as the effect of political news to voters. Their main concern is to understand how mass media vehicles are used as the main political tools in influencing and persuading Malaysian voters. There is very limited study on political advertising in Malaysia. Even so, the researchers involved on this topic were not from the advertising field. Studies on the effects of media and political campaigns from the perspective of, for instance agenda setting, framing and priming were also few and far between.

It is in fact, difficult to identify precisely the beginnings of political communication research in Malaysia [10]. Among the earliest studies on election is the work of two experienced scholars in communication studies: Syed Arabi Idid and Safar Hashim. Their first book entitled *General Elections: A Political Communication Perspective* which was published in 1993 marked the birth of Political Communication Studies in Malaysia [10]. Their work were mainly focused on the role of media during election campaigns and the effects of agenda setting on issues, images of party leaders, information seeking and media use especially to the voters. Even though there are admittedly limited research work in this field, studies on Malaysian politics and election campaigns have gained interest among foreign media, especially from the west [11,24,18].

Therefore, this study will focus on the types of communication tools usually used for political electoral process in Malaysia. In particular, this study intends to answer the following research question: Are mainstream media more popular than social media for political electoral process in Malaysia? By achieving the aforementioned objective, it is expected that this study will help to advance current knowledge about the role of mainstream media and social media as the main political tools in influencing and persuading Malaysian voters. In addressing this question, the paper has the following structure. In the next section, the overview of political communication in Malaysia is

DOI: 10.1051/shsconf/20173300054

discussed. After that, the methods used are explained thoroughly. In the last part, results and discussions are presented in detail along with the conclusion.

3. A Review of Political Communication in Malaysia

According to Kaid and Holtz-Bacha [6], the history of political communication and advertising should begin with the United States of America (USA) because it is here that the techniques of the form were pioneered wherein they have reached their highest level of sophistication. The USA, having become the twentieth century's most successful capitalist power, went faster and further, in commodifying the political process by the utilizing advertising as a strategic approach compared to any other country in the world. It did not take long for the techniques developed in the USA to be exported to Britain and other countries [4]. In Malaysia, political advertising is a field that is still young [5]. As such it is difficult to say when political communication and advertising actually started in Malaysia. Nonetheless, the usage of placards, posters, banners, fliers, radio and television advertisements can be traced back since the election campaigns in the 1960s. It is therefore quite appropriate for researchers interested in political communication and advertising in Malaysia to embark on further research in this field.

Based on some early studies in the mid-1980s conducted by early researcher in this field, political advertisements had been in use is Malaysia since the 1960s. For instance, in the 1986 general election campaign, political parties in Malaysia in particular the National Front (*Barisan Nasional*) started using political advertisements in the form of caricatures. And during that time, these caricatures which appeared in the mainstream newspapers seemed to have attracted the attention of readers. Although in the 1995 general elections, political advertisements were quite noticeable [6], it was in the 1999 general elections that these advertisements were widely used [7]. Malaysia's 11th general elections campaign held in 2004 witnessed changes in political campaign with the widespread use of advertisements in television, newspapers and other media [8]. During this time, media vehicles were used as the main political communication tools to convey and disseminate political information to the eligible Malaysian voters [9].

For any contesting parties, political advertising is one of the best tools for them to disseminate their political messages. Another significant development in political advertising practices during this time was the entry of social media that had shaped and influenced political communication landscape in the country. However, as indicated earlier, the turning point in Malaysia's changing political campaigns was in the late 1960s where the use of newspapers as a subtle campaign tool for voters was apparent [10]. As a matter of fact, the contemporary political campaigns in Malaysia have evolved from mainly relying on the use video tapes, banners, and flyers to advertise their agenda to using mainstream media (electronic and printed) and the world wide web [6]. Before the entry of social media, in the 1999 general elections, the National Front (*Barisan Nasional*) being the ruling party which owns the leading mainstream television channels and newspapers took opportunity to utilise these media channels to the fullest to promote their party and aggressively hit out at the opposition parties and its candidates. This is not surprising as the National Front had been using the mainstream media as their campaign tool since the existence of these media vehicles in Malaysia [1,11,7,12,13,5,14].

In the 1995 general elections revealed that Malaysian society has become accustomed to advertising blitz especially via the media and other means of mass communication, [12]. In the most recent general elections held in 2013 where political advertising were utilised to the fullest, contesting parties spent millions of ringgit to reach their potential voters. To reclaim some lost seats in GE 12, the National Front had spent millions either on print advertisements, television and radio spots, billboard and other outdoor advertisements and also the social media channels to disseminate their political messages to Malaysians. This development clearly shows the increasing role of political advertisements in modern political campaigns. This high increase in expenditure to the marketing and advertising activities of the political campaigns could justify the attention political communication scholars gave to the study of political advertising [15]. Nevertheless, even though the National Front spent a million of ringgits in their campaigns, they still lost some important parliamentary seats to the opposition parties.

4. Methodology

The respondents employed in this study are the voters of the June 18, 2016 Sungai Besar by-election in Selangor. Convenience sampling is used in this study as respondents were selected because of their availability during the survey held on June 13, 2016 wherein data were collected on this date. The data of this study were generated from responses to questionnaires completed during the actual survey. A total of 360 questionnaires were provided. The trained enumerators assisted in the distribution of questionnaires to the voters of Sungai Besar. The questionnaire survey was employed because the survey method is a popular method used by researchers especially for political communication scholars to study election and voters. It is a method of data collection in which information is obtained directly from individual persons who are selected so as to provide a basis for making inferences about the larger population [1]. This method remains a dominant methodological tool for political communication analysis [22]. The SPSS 20 was employed in order to conduct frequency analysis. The analysis is used to extract the percentile of the profiles of respondents in terms of their gender and race, the percentage of selection of media for political issues during election campaigns and the choice of media vehicles by the respondents for information on political issues.

5. Findings

The present study involves a total of 360 respondents who are voters of Sungai Besar, Selangor by-election. The race distribution consisted of 64.3% Malay, 24.4% Chinese and 11.3% Indians. Meanwhile, gender distribution consisted of 51.0% male and 49% of female. Table 1 highlights the type of media that the respondents choose to get political information or political news during the election campaign. Based on the table below, it is understandable that most of the past research had indicated that mainstream media still play a significant role as the main political information source for voters in the rural and sub-urban areas. During the election campaigns, most of the political parties will work hard to highlight their presence inside the mainstream media to influence the political information efficacy and sense making process of the voters.

During an election campaign, mainstream media may compete with the social media. Based on the findings, early assumption can be highlighted to identify the information needs and information seeking activities among the voters. Traditional media such as television can be the main information needs for the voters to seek more information that can directly influence their political information efficacy.

 DOI: 10.1051/shsconf/20173300054

Table 1. The selection of media for political issues during by-election campaign

Media	frequency (*f*)	percent (%)
Mainstream media	265	73.6
Social media	95	26.4
Total	**360**	**100.0**

Source: Derived from current study

It is interesting to note that in Table 2, a major portion of the respondents who relied on television news consisted of 32.1%, followed by news on radio 20.8%, 11.3% of newspaper advertisements and outdoor advertisements, 9.4%. The response for social networking sites, though substantial was quite low with less than 6%.

Table 2. Exposure to media vehicles for political issues during by-election campaign

Media Vehicles	**frequency (*f*)**	**percent (%)**
Newspaper articles	18	6.8
Newspaper advertisements	30	11.3
News on television	85	32.1
Advertisements on television	20	7.5
News in Radio	55	20.8
Radio advertisements	17	6.4
Outdoor advertisements (*Banners/posters/billboards*)	25	9.4
Social networking sites (*political blogs, political online videos, party websites, Political ads on cell phones*)	15	5.7
Total	**265**	**100**

Clearly, the results show that the television medium is still a popular medium that voters relied on for information on political issues. This is not at all surprising since people in the sub-urban and rural areas tend to see this medium as dependable and trustworthy for information that they need [22].

6. Conclusion

The purpose of this study is to investigate the types of communication tools usually used for political electoral process in Malaysia. The results of this study suggest that mainstream media play a significant role as the main political information source for voters to get political information. Despite the increasing popularity of social media sites, mainstream media is still relevant as tools of communication in short-term political campaigns, particularly for by-elections in rural and sub-urban areas where there are more middle aged people who are either baby boomers or from the gen x group of people aged 40 and above.

This study provides important implications for political parties wanting to participate in by-elections held in rural and sub-urban areas with a campaign period of two weeks or less. It is undeniable that although the rapid changes in communication technology has attracted the younger generation to embrace the latest communication tools; it is quite obvious that this new technology has very little effect on the more elderly voters in non-urban areas. Nevertheless, this phenomenon still needs to be studied in detail, and not merely grounded on assumptions as the results of this study is based on an exploratory study. As this study has a number of limitations, among which respondents were chosen only in one location, the findings cannot be generalized in other locations. For future studies, we recommend that varied samples are included to increase the generalizability of findings. However, this study has certainly provided an improved understanding of the types of communication and advertising tools usually used for political electoral process in Malaysia.

References

1. Mustafa K. Anuar. APME, **24**, (2014).
2. Mohd Roslan, M. N., & Mahmud, A. MJDES, 1. (2013).
3. Idid, S. A., & Kee, C. P. ASS, 8. (2012).
4. McNair, B. (2011). *An Introduction to Political Communication* (fifth). Oxon: Taylor & Francis Group,
5. Syed Arabi Idid dan Safar Hasim, Pilihan raya umum: Satu perspektif komunikasi politik. Kuala Lumpur: Dewan Bahasa dan Pustaka.(1993)
6. Kaid, L.L & Holtz-Bacha, C. Handbook of Political Advertising. Thousand Oaks, CA: Sage. (2006)
7. Arfah Yusof, JK, 17 (2001)
8. Yaacob, R. A. I. R., Embong, A. M., Endut, M. N. A. A., & Amin, A.-R. M. IJSSH, 4. (2014).
9. Ratnam, K. J., & Milne, R. S. PA,43.(1970)
10. Ezhar Tamam and Manimaran Govindasamy. (2009). Political Communication Practices and Research in Malaysia. In L. W. Aw (Ed.), *Political Communication in Asia* (pp. 135-153). Oxon: Routledge.
11. Welsh, B. JD, 24 (2013)
12. Mustafa, K. Defining Democratic Discourses: The Mainstream Press. In F. L. Teik (Ed.), *Democracy in Malaysia : Discourses and Practices* (pp. 138-164). Surrey: Curzon. (2002)
13. Kuppuswamy, C. S. *10th General Election in Malaysia - an Analysis: It is now Malays vs Malays.* India: South Asia Analysis Group. (1999)
14. Weiss, M. Introduction: Patterns and Puzzles in Malaysian Electrol Dynamics. In M. Weiss (Ed.), *Electoral Dynamics in Malaysia : Findings from The Grassroots* (pp. 1-16). Petaling Jaya, Selangor: Strategic Information and Research Development Centre. (2014)
15. Alqaseer, J. *Political Advertising in Kuwait A Functional Discourse Analysis Florida State University 2008 For the Degree of Doctor in Philosophy in Accepted by :* University of South Carolina. (2013)
16. Kaid, L. L. (Ed.). *Handbook of Political communication research*. New Jersey: Lawrence Erlbaum Associates. (2004)
17. Lancendorfer, K. M.. *The role of attributions on voter response to political advertising*. Michigan State University.(2005)
18. Willnat, L and Annette J. Aw. Political Communication in Asia: Challenges and Opportunities. In L. K. Lynda, *Handbook of Political Communication Research* (pp. 479-505). New Jersey: Lawrence Erlbaum Associates, Publishers.(2004)
19. Case, W. MJDES, **1** (2013).
20. Sualman, I., Shariff, S. Z., Yaakop, S. H., & Mahmood, A. IPEDR, **64** (2013).
21. Mun, W., & Li, L. SEARCH, 3 (2011).
22. Syed Arabi Idid dan Safar Hasim. Pilihan raya umum: Satu perspektif komunikasi politik. Kuala Lumpur: Dewan Bahasa dan Pustaka (1993)

Between Ebola epidemics and Boko-Haram insurgency: Media agenda transition interplay

Lateef Adelakun[1,*], and *Hamedi* Adnan[2]

[1] PhD Student, Department of Media Studies, University of Malaya, Malaysia
[2] HOD, Department of Media Studies, University of Malaya, Malaysia

Abstract. The menace of Ebola outbreak coincidence with Boko haram insurgency left not only the policy formulators in disarray but also rocked the media with indecision on which to prioritize. Media set agenda on the most important issue or on an issue the media want to make important but the dissection of which suppresses the other when two issues are keenly considered of public interests fames discussions on agenda transition. Ebola epidemic rocked Nigeria at a time when media coverage of Boko Haram insurgency was at its climax, setting double battles for government to fight and for media to frame as public agenda. Content analyses of two Nigerian newspapers between June and December, 2014, explained the fade-in and fade-out strategy that the media employed to set, sustain and shuttle agenda front-and-back on the issues. The media interest to guide the audience decisions on the duo (health and security challenges) also explained the indispensability of media agenda role and the priority leverage among issues.

1 Introduction

Terrorist activities attributed to *Jamā'at Ahl as-Sunnah lid-Da'wah wa'l-Jihād* popularly known as 'Boko Haram" in 2009 just after the 'extrajudicial' killing of its leader, Mohammed Yusuf by the Nigerian police did not only make the ideology of the group known but also marked the beginning of the long record of insurgency Nigeria has ever fought. *Boko-Haram,* (often translated as "Western culture is heresy"), the name most often used to refer to the group, which captured one of its ideologies, has become a sobriquet for terrorism in the media across the globe. Religious antecedent of the group's ideology, fuelled by political aberration and rivalry upon which Nigerian government failed to nail it in the bud aggrandised terrorism in Nigeria. As terrorism started gathering momentum especially since 2011, it structurally became a common news focus fading away the public attention on other major concerned issues. Since 2011, issues on Boko Haram insurgency have been consistently enjoying priority on the media agenda. Mass media, particularly newspapers explored the many news angles through which the issues surface and resurface to keep the public interest aglow of the saga until Ebola Virus Disease (EVD) broke out in Nigeria in 2014 [1].

EVD came to Nigeria at the time two major issues (Political Campaign and rivalry that precipitated 2015 general election and the incessant and uncurtail Boko-Haram insurgency) had wrapped up both media and public attentions. EVD in Nigeria, though surfaced at the time the insurgency was at the peak of media agenda competed for media saliency. Then, 'how does a new issue creep into the height of media agenda and fade away or be at neck-on-neck with the one already enjoying media saliency?', which is the central question addressed by this work deepens explanation on saliency transfer among issues upon which media/public agenda are set.

Making an issue salient among the most important issues competing for limited audience attention though is regarded as sole responsibility of the media, the expectation is that gravity of the issues should remain the major criteria upon which the media decision to set agenda is based. Going by the trends of the topmost issues in the mid-2014, Boko Haram Insurgency and Ebola Attack topped the media agenda list. Media could have used other parameters to decide which of the issues should be more emphasised than the other upon which the agenda was set for the public. But our concern in this study is not only to verify which of the competing issues at the time was made salient in the media but also to explain which agenda-setting strategy did the newspapers employed to sustain the saliency of the prioritised between the issues.

* Corresponding author: delak058@yahoo.co.uk

2 Literature Review

Vinson [2], in his review of Boydstun [3] work related to media patterns of coverage that determine choices of issue upon which media agenda is built, emphasised on the two approaches itemised. The approaches, according to Vinson (pg. 761) – "alarm mode, where some event suddenly calls reporters' attention to an issue and the patrol mode, where the media regularly report on certain issue in a more traditional watchdog role" though do not totally explain reasons why media make some issue salient at the expense of others, both add up to criteria upon which the media decision on what issue it consider the most important for the public consumption. Having established that the media cannot give the same proportion of attention to all issues, the media decision on which should be made more significant is moderated by plethora of factors among which the need to meet the nature of market information demand (what issue interest the audience the most) speaks volumes. The consideration of the factors also tells much on how long an issue enjoy media saliency (how long it remains on the top rung of media agenda ladder) and at what point does a new or another issue fades away the one enjoying media saliency. Martin [4], while reviewing the same work, pays attention to discusses, which Boydstun described as momentum, skew and explosion as indices that determine while some issues easily getting on the media agenda and some others don't. Martin's satisfaction with the generalisation test of the Boydstun's alarm/patrol hybrid model that explain reasons some issue find a way into media agenda can explain while EVD, despite its short period in Nigeria competed with Boko-Haram issue in the race towards forming the media agenda.

This description in actual sense makes the theory relevant in addressing the central obligation of the media to manage news in favour of societal development. Mustapha [5], agrees that the application of agenda-setting theory cut across various issues. Coleman, McCombs [6], also confirm that researchers have adapted the theory to discuss and understand number of issues. But the concern is how layers of agenda setting influence the public agenda and decisions and how do the media and public agenda shift from an issue to another most especially when issues saliency are competing for audience attentions and public concerns. Whether first level or second level of agenda-setting or even ''agenda-building'' [7] and ''agenda melding'' [8], the central focus, which is salient shift is a confirmation of media involvement at all levels.

The media agenda-setting concern of 'What the audience think about' could be regarded as the primary stage of the media influential might on audience. As the inequitable treatment of myriad news items competing for audience attentions elevate audience interests in a news item above others, it paves ways for other media influential concepts such as framing and priming to germane. It saves and redefines media framing concern of 'how to think about all issue' to a more succinct and manageable concept, 'how to think about the 'most important issue'.

2.1 Agenda Setting Transition

At the time Ebola broke out in Nigeria, there were other issues of perhaps equal importance competing for audience attentions such as Boko-Haram insurgency, which was then at its tensed stage, and the issue of political transition, which was hanging loosely as hyper tension general election (first of its kind, in which the opposition unseat the ruling party) drawn near at the time. Making choice out of the competing news items upon which the media agenda is set is a media task that depends on certain number of influences ranging from media perception of reality and the nature of media environment including audience wariness to the happenings around them. Audience wariness (how active the audience) is an important factor that segregate between what the media considered important and what the audience assumed essential [9], a dichotomy that leads to 'transfer of salience'.

Shifting audience attentions from what they earlier considered most important in relation to the earlier media positioning of issues reported is another area where the essence of media agenda is felt [10]. This requires that the media proportionately reduce the saliency (the amount of attention and prominence given to issues) of one issue to give in to new one that requires public attention. Prior to EVD outbreak in Nigeria, *Boko Haram* insurgency was the main agenda (both media and public), which went as far as influencing international perception, assessment and definition of peace in Nigeria upon which the country foreign policies depend. Despite the worrisome nature of the insurgency, media repositioning of issues (Agenda-Shifting) subdue its saliency to allow EVD crept to the top of agenda [11].

However, shifting agenda from one issue to another can be discussed under two faces – full and partial. A full shift of media agenda (which could be regarded as Agenda-setting transfer) occurs when there is over domination of public attentions on new issue as a result of the total shift in the saliency that the previous issue enjoyed in the media. It results in the assumption that the old issue has rather been over-flogged or it has been concluded and there is no fresh angle to the issue, which may press for public attention or debate. In this kind scenario, new issue of media agenda benefits immensely as more attention is given to it from both media and public. EVD outbreak did not enjoy the kind of agenda shift as other issues of proportionate importance competed for the media saliency as well as public attentions. The case of EVD outbreak fits in to Partial agenda shifting (Agenda-setting shift) in which new issue competes with the old media agenda for media saliency as well as public attention. Though the main agenda may be shifted to the new issue, old issues are not totally wiped off but rather subjected to saliency reduction as well as shortfall in the public attentions. As EVD outbreak became the media and public agenda, the importance attached to *Boko Haram* insurgency and general election, which were at the time burning issues subsided.

Studies [12-14] have used Agenda-setting transfer to depict how media agenda translate to what the audience consider as the most important issue, a concept used to describe issue salience in what is considered first-level agenda-

DOI: 10.1051/shsconf/20173300055

setting. The concept mostly explains how "issue attributes salient in the media were functioning as significant dimensions of issue evaluation among audience member" [12]. The operational assumption of agenda-setting transfer in this work diversifies into the transition from one issue salience to another rather than movement of the salience within the agenda players (media and audience).

Most often, the rate at which an issue fade-in to allow the agenda set on another issue predominantly depends on the differences in the gravity of both issues as well as the longevity of the period the earlier issue remains on top of the media agenda based on the issue lifespan [15]. The lifespan of an issue includes the period from issue conception (perhaps, when first reported in the media) until salience is built around it to enjoy public attention and recognition (public agenda). Agenda climax sets in when such issue becomes the central focus of both the media and the public and remains the priority on the ladder of individual concerns. Fresh angles to the issue usually dictate how long it remains at the climax of media and public agenda ladder. Mass Media capitalise on the fresh angles to prolong the salient period an issue enjoys from the public. Until a new issue of predominant importance occurs to facilitate 'return to scale' of the previous issue, public attentions remained attached to the previous, which enjoys its agenda climax.

2.2 Influence of Agenda Sponsoring on Agenda Setting Transition

Studies have argued front [8, 13, 14, 16] and back [17] on 'where does agenda-setting of an issue originate'. Having established that the power to set agenda is not a monopoly of the media but rather a reflection of media-public reciprocal influence, it is necessary to justify how agenda are sometimes sponsored to cause agenda shift. In Nigeria, influential people including politicians, parties and groups grant the media audience whenever they want to set agenda on issue as a form of attention diversion from real issue at the moment. Since their views make news, and the media benefit from their economic dominance through advertising charges raise their voices in the media above others; their voices become louder and the issue they emphasise in the media becomes a new agenda.

Could this have constituted agenda shift between boko haram saga and Ebola case? No empirical justification to substantiate that the media attention on Ebola issue was apolitical. However, information surplus in relation to scarcity of attentions induces the struggle to influence the agenda-setting function of the media. The competition for the limited attention as a tenet for agenda-setting delineates the origin and the routes through which an issue become prioritised – whether through natural cause or artificial generation and sponsoring of issue for the purpose of attention diversion. Both routes use media to facilitate the spread of the issue in order to ensure that public interest and attention are won. The relationship between them encourages that an issue may be stripped off the saliency it enjoys as a result of the importance of another issue to the public or actions of influential individual and groups.

3 Research Methodology

Media reports on Boko-Haram insurgency and Ebola issues in two of the most popular daily newspapers in Nigeria (*The Sun* and *The Guardian*) were content analysed. The two newspapers selected for the study had extensive coverage of both the EVD outbreak and Boko-Haram insurgency. The analysis covered seven months; a month before and after Ebola rampage in Nigeria (June to December, 2014). Three major variables are used as tools to establish how the media saliency is shifted from an issue that enjoys media agenda to another. The variables are: a) Number of stories published on either of the issues at interval within the time lag; b) News page: what fraction of the publications on each issue makes front-page news since the front-page story is considered the most salient story of an edition; and c) News-hole: what fraction of the publication on each issue is given better treatment in term of space allotted. Each of the variables is quantitatively analysed and represented on line graphs to show the relationship in term of the saliency shift between the two competing issues.

4 Findings and discussion

Cross examination of how frequent the newspapers feature each of the issues gives impression on which of the issues received more media attention than the other at a time within the interval of the study. The frequency of an issue in newspapers in relation to the page in which it frequently appears in newspaper; and the proportion of space it often occupies, which dictated how in-depth the issue is treated are tantamount to the degree of saliency such issue enjoys and the pattern of saliency shift that is likely to occur when competing with other issues that compel lower, similar or higher media attention. The shift in the media saliency between Boko Haram insurgency and Ebola outbreak were measure at a time when both issue were burning to chronologically measure the change in the newspaper coverage of the issues.

DOI: 10.1051/shsconf/20173300055

Table 1. Frequency of Published Stories in each of the Issues per Month

	No of Stories Published by two newspapers per Month	
Month (2014)	**Boko Haram Stories**	**Ebola Stories**
June	78	5
July	64	78
August	48	199
September	46	194
October	52	171
November	73	134
December	82	59
Total	443	840

Despite that Boko Hara insurgency had been enjoying media attention long before Ebola broke out, the table above shows that there is inverse relationship between the media attention given to the issues; as the outbreak broke out, there was significant decrease in the proportion of news stories on the insurgency between June and August. The relationship changed to direct in September; both having less media coverage than before (perhaps some other issues compelled media attention). Return-to-scale set in and returned the relationship back to inverse proportion between October and December but rather in favour of Boko Haram insurgency.

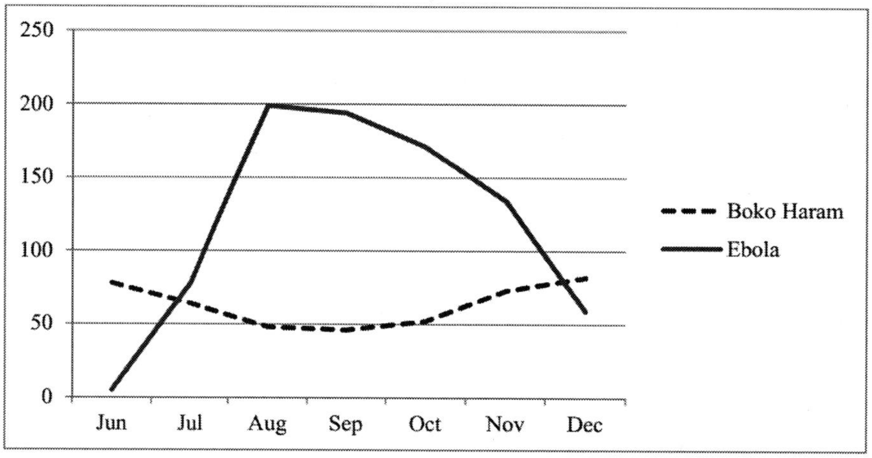

Fig. 1. Line Graph Showing Relationship between the Issues in Term of Number of Stories

Going by the overall attention given to the issues in both newspapers, more than three-quarter of 1103 stories on both issues was centred on Ebola. This shows that Ebola received more media coverage than Boko-Haram. The frequency of occurrence of Ebola news in the newspapers sent signal to the audience on how important the issue. It also justified what constitute public discussion since the issue media report frequently is relative to what becomes public agenda.

$$\frac{840\ (Total\ Ebola\ stories\ reported)}{7\ (Months) \times 2\ (Sampled\ newspapers)} = 60\ (Average\ Ebola\ News\ stories\ per\ newspaper\ per\ month) \quad (1)$$

This implies that an average of two news stories was published in each Nigerian newspaper per day within the period under study (See (1)). Contrarily, an average of a news story on Boko Haram insurgency was published in each of Nigerian newspaper per day (See (2)). This signifies the intensity of each of the issues and the equivalent attention that the media attached to each.

$$\frac{443\ (Total\ Boko\ Haram\ stories\ reported)}{7\ (Months) \times 2\ (Sampled\ newspapers)} \approx 32\ (Average\ Boko\ Haram\ news\ stories\ per\ newspaper\ per\ month) \quad (2)$$

 DOI: 10.1051/shsconf/20173300055

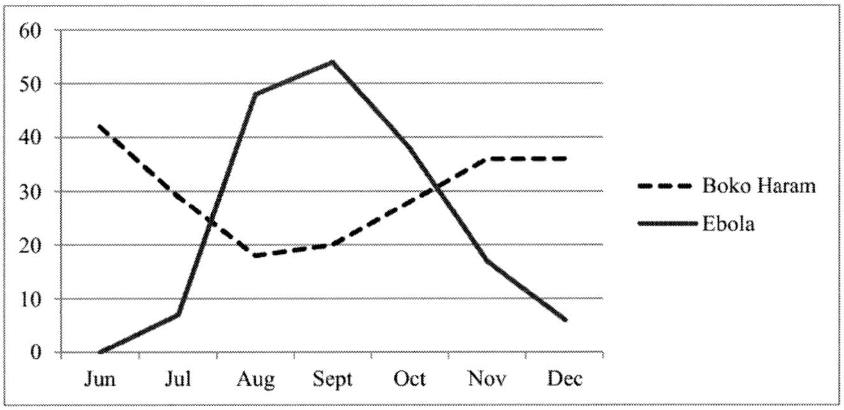

Fig. 2. Line Graph Showing Relationship between the Issues in Term of Front-page Advantage

The page on which a story appear and the amount of space it occupies in the newspaper are important criteria for measuring how important the story. Newspapers usually dedicate the front page to the most important news of the day while other stories of lesser importance are buried inside the newspapers. The front-page stories are regarded as attention catcher as well as Unique Selling Proposition for the edition in addition to its agenda setting function. As the newspaper started given Ebola stories attention on the front page in July, that of Boko-haram started falling until both get to equilibrium towards July ending. The fact that both enjoy return to scale at different periods (Boko Haram in mid-August and Ebola in mid-September) and also in opposite direction is an indication that the competitiveness for saliency between the issues is keen. It is also an indication that the shift in the saliency from Boko Haram to Ebola is neither absolute nor continuous (See Fig.2). The similar relationship was also established in the amount of space allotted to stories. Stories that are allotted a minimum of half page are considered detail and more important than those with less space (See Fig. 3)

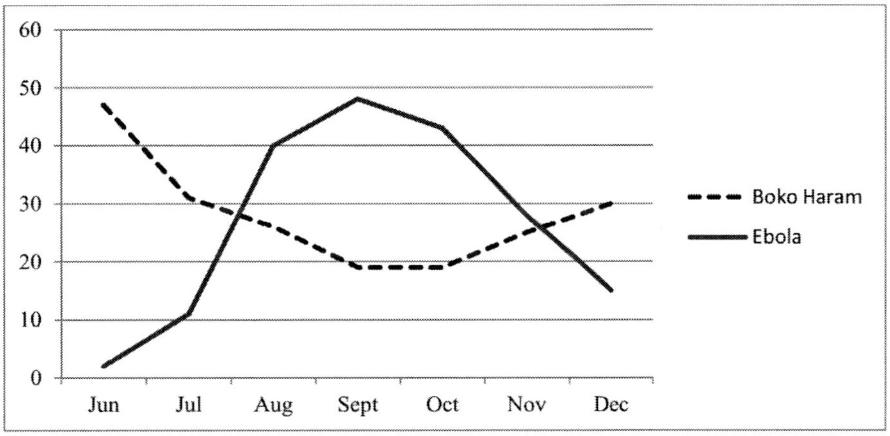

Fig. 3. Frequency of Issue of Focus in Relation to the Space Allotted in the Newspaper

5 Conclusion

Newspapers' attention was not shifted at once from Boko-Haram to Ebola despite that Ebola issue was then health emergency of international concern. The media agenda was gradually transferred from the issue that was enjoying the media attention to the new issue considered more important. EVD outbreak did not totally fade away *Boko Haram* insurgency; though its importance at first caused a major reduction in the salience the insurgency enjoyed and consequently brought EVD issue to limelight. It diverted higher proportion of public attentions away from *Boko Haram* insurgency, within its shot lifespan and the media agenda return to the latter, which had enjoyed long period of uninterrupted media agenda. The return of media saliency back to the insurgency explains the part play by issue prominence in the media preference of issue upon which agenda is set.

References

1. L.A. Adelakun, Mass Communication and Journalism, **4**, 193 (2014).
2. C.D. Vinson, *Making the News: Politics, the Media, and Agenda Setting*. Public Opinion Quarterly, **78**(3), 761 (2014).
3. A.E. Boydstun, *Making the news: Politics, the media, and agenda setting*. University of Chicago Press (2013).
4. J.A. Martin, *Making the news: Politics, the media, and agenda setting*. Journalism, **16**(2), 298-299 (2015).
5. L.K. Mustapha, *Agenda-setting Theory: A Reflective and Visionary Analysis*, in *Critique and Application of Communication Theories*, N.T. Ekeanyanwu, S.N. Ngoa, and I.A. Sobowale, Editors., Convenant University Press: Ota, Ogun State. 105-132(2012).
6. R. Coleman et al., *Agenda setting*. The handbook of journalism studies, 147-160 (2009).
7. P.A. Curtin,. Journal of Public Relations Research, **11**, 1 (1999)..
8. P.A. Shaw et al,. Int. J. Public Opin. Res, **11**, 1 (1999).
9. P. Bantimaroudis, S. Zyglidopoulos, and P.C. Symeou, J. com, **60**, 4 (2010).
10. M.W. Ragas, and M.S. Roberts, Journalism & Mass Communication Quarterly, **86**, 1 (2009).
11. M. Besiou, M.L. Hunter, and L.N. Van Wassenhove, *A Web of Watchdogs: Stakeholder Media Networks and Agenda-Setting in Response to Corporate Initiatives*. J. Bus. Ethics, **118**, 4 (2013).
12. S.H. Kim, D.A. Scheufele, and J. Shanahan,. Journalism & mass communication quarterly, **79**, 1 (2002).
13. S. Kiousis, and M. McCombs,. Communication Research, **31**, 1 (2004).
14. S.J. Moon., Communication Research, **40**, 5, (2013).
15. R. Vliegenthart, and S. Walgrave,. Party Politics, **17**, 3 (2011).
16. U. Nwanne,. Global Journal of Medical Research, **14**, (2014).
17. G. Thesen, *Political Agenda Setting as Mediatized Politics? Media-Politics Interactions from a Party and Issue Competition PerspectiveInt. J. Press/Polit*, **19**, 2 (2014).

DOI: 10.1051/shsconf/20173300056

The Intercultural Adaptation Experience of Afghan Students in Malaysia

Abdul Latiff Ahmad[1,*], *Hameedullah* Azimi[1], *Sabariah* Salleh[1], *Emma Mirza Wati* Mohamad[1] and *Shahrul Nazmi* Sannusi[1]

[1]School of Media and Communication Studies, Universiti Kebangsaan Malaysia

Abstract. Studying abroad is an experience which can benefit both students' original and host countries. The Afghan government in their quest to improve the human capital have decided to send their students to pursue their studies abroad. It is hoped that these students not only succeed in their academic endeavor but also to learn from the context of cultural and social experience of others, of the technological and globalising world; and bring the rich experience home to help develop the country. The research uses the U-Curve model of intercultural adjustment (Lysgaard, 1955) as the foundation which describes adaptation as a process that moves from a "honeymoon" period into "culture shock" and on to recovery or "adjustment" and "mastery" stages. It looks at the experiences of Afghan students in their adaptation process. Data were obtained from three focus groups discussion done at three different public universities in Malaysia that enrolled the Afghan students. Studies showed that though students went through a bit of culture shock earlier, they are able to adapt in the end which is similar as to the stages discussed in the model.

1 Introduction

Globalization has significantly contributed towards the movements of students for higher education. Moving into new countries, students are challenged with a new adaptation process, trying to familiarize themselves with new surroundings and new culture.

In this context, globalisation refers more to the sharing and transfer of knowledge between those moving between countries. The growth of knowledge-based economy has changed the level of competition where not only are future employers competitive in seeking the best talent but universities are also becoming more selective in training the future workforce. [1]. On the other hand, internationalisation refers to the awareness, understanding and acceptance on the cultural aspects between those in the higher education globally [2]. Globalisation clearly presents new opportunities, challenges and risks for higher education.

One of the most important subjects in terms of globalisation is migration and cross borders study and this study focuses on the movement of international students, specifically Afghanistan students to Malaysia. In general, globalisation refers to the flow of people,

* Corresponding author: alba@ukm.edu.my

technology, culture, values, ideas, knowledge, and economy across borders. There are a large number of definitions of the term "globalisation of higher education"[3].

2 International Student Experience

Kim (1988) defined Intercultural communication as "*a communication process of using signs and symbols that elicit meaning in another person or persons for whatever intent, or even without conscious intent, on the part of the person producing the symbols or signs*" [4]

From the above explanation we understand that intercultural environment are created when two or many different cultures of people are interacting with one another using various symbols as medium. The experience from an this unfamiliar setting can create anxiety, confusion and depression, because for the newcomers, they faced language problems , difficulties in finding accommodation, economic stress, loneliness and these conditions can lead to further problems such as anxiety, despression and various forms of illness.

A research was also conducted on the relationship between students' adjustment factors and their intercultural adaptation process, and they concluded that having travel experience, a high level of education and strong command of languages contributed significantly to the differences in cross-cultural adjustment among participants [5] . The research also found that there is a significant relationship between academic matters, personal, emotional and environmental factors and how an individual is able to adapt in a cross-cultural environment There is also a tendency for international students to stay within their respective cultural members, within a safe enclave [6]. This limits the opportunity for a greater exposure of cross-cultural interaction.

The shortage of higher education places and under-resourced universities in many developing countries has also contributed to many students going overseas to pursue their higher education, and statistics have shown a growing trend of international students from developing countries [7].

3 Malaysia as an International Education Hub

Malaysia is making its mark in the international education hub .A report prepared for the UNESCO 2009 World Conference on Higher Education entitled "Trends in global higher education: Tracking an academic revolution", mentioned that more than 2.5 million students are studying outside their home countries [8] . It further predicted that there will be an increase to 7 million international students by 2020. Malaysia was then placed at the 11th position in the list of countries with the most numbers of international students, where about 2% of international students in universities around the world chose Malaysia as their destination for higher education. The Ministry of Higher Education of Malaysia estimates that in the year 2020, they will be able to attract 200,000 international students [9].

4 The Afghan students

Afghanistan government sent their first batch of 196 students to Malaysia in 2012. This research will gather the experience of Afghan students' academic life in Malaysia, and focus on the challenges they face in their adaptation process. It looks at the intercultural experience that the students have had and to understand how they dealt with the adjustments.

According to the World Bank report entitled 2013 Higher Education in Afghanistan: An Emerging Mountainscape report, in Afghanistan "only 5 percent of the university academic

DOI: 10.1051/shsconf/20173300056

staffs are with Ph.D level of education. Most universities have either no staff with PhD or just one or two academic staffs with such qualification. Approximately 38 percent of university staffs have Masters, while 57 percent of university staffs have only Bachelor's degrees"[10]. Today, according to the statistic from the Embassy of Afghanistan in Malaysia, there are around 342 students in three public universities in Malaysia. Some of these are students are sponsored while some are not.

The objective of this research is to understand the adaptation process of the Afghans and to explore the challenges they faced while studying in Malaysia. The study was conducted using Lysgaard's (1955) U-Curve model known as intercultural adaptation or cross-cultural adjustment theory which focuses on the four phases of a) Honeymoon, b) Culture shock, c) Adaptation or Adjustment, and d) Mastery phase [11]. This study looks at whether the Afghan students experienced the same stages as those as described in the U-Curve model.

5 Methodology

This research was conducted in three public universities in Malaysia; *Universiti Kebangsaan Malaysia* (UKM), *Universiti Islam Antarabangsa Malaysia* (UIAM), and *Universiti Sains Islam Malaysia* (USIM). The participants are students who are fully sponsored by the Afghan government that are studying in both the undergraduate and postgraduate level. There are eleven males and three female Afghan students aged from 19 to 33 years old and the programs they are doing are based on the respective Government to Government agreement between Afghanistan and Malaysia. The first group consisted of four male students and one female, the second group consisted of four males while the third group has two males and two females.

6 Data Analysis

6.1 Upon arrival

The Afghan master students specifically from UKM had nice memories of the warm welcome by the International Relations office of UKM, especially the head of the department. They appreciate the effort and are very thankful to the IR office of UKM and also the Afghanistan embassy for the warm welcome. The undergraduate students were welcomed at the airport by the Afghanistan embassy representatives in Malaysia and also the Afghan students' community of UIAM.

Hamid, (G.1): "*I found Malaysia beautiful and peaceful country.... I really forgot all my worries and sadness that I had before coming to Malaysia.... I really appreciate university specially IR office staff of UKM. They help us during coming from the airport to campus and also they provide dinner and morning breakfast for all Afghanistan students.*"

Salma, (G.3): "*First time, when I arrived in Malaysia international airport, we welcomed by representatives of the Afghanistan embassy in Malaysia and also Afghan students who came before us to Malaysia. The first night was really interesting the Afghan student community provided us food and few days were really we don't know where are coming and the beauty of Malaysia keep us so busy.*"

6.2 Culture Shock and Homesickness

Being homesick is common for those who had to leave their home, family and friends when they traveled abroad, away from their comfort zone The Afghan students in UKM (G.1) felt

DOI: 10.1051/shsconf/20173300056

depressed as this is the first time he is being away from his family. It is tougher for the married students as they had to be separated from their families and they really missed their children and family.

During the focus group discussion, they cited the problem of living in an unfamiliar environment of the university. They had trouble finding the supermarket and also with the campus transportation system. They were unsure of who to ask and the transportation system was rather complex. As a result they had to walk from one place to another and most of them would just stay in their room due to their frustration with the transportation system.

Shafiq, (G.1): "*We walked many kilometers by foot. I remember that we walked from Za'Ba College to PPS it takes 30 minutes to arrive there. The weather was too hot. Because university and also the Afghanistan embassy in Malaysia didn't hold any program to help us understand and be familiar with a new environment.... I was also with my friends to go outside by foot more than 6 to 7 kilometers to change our money. And buy some necessary things.*"

One of the Afghan students recalled the three months being completely free with nothing to do. They just ate and slept which made them sick, and they became lazy day by day. The first three months was really depressing for some of them.

Jawed, (G.3): "*three months we completely free, there was nothing to do. We just eat and slept which made sick our mind, and we became lazy person. The first three months really put worse or negative effect on our mind, also on our knowledge and learning process becomes slow. One big problem we faced here we cannot sleep at night, in Afghanistan it was normal we go to our bed for sleeping at 10 or 11 pm at night. But here, I can't able to adjust my sleeping time. Here in the Malaysia 4 and 5 am we can't sleep. I also missed my family, the weather, my friends in Afghanistan and Afghanistan people.*"

The second problem that was mentioned in the group discussion concerns the married students. Prior to coming to Malaysia, they were promised that they could bring their family along, but after arriving in Malaysia, they faced problems in getting visas for their family. It took one year for the visas for their families to be sorted. Khalil (G.1), who was able to get visa for his family after one year, really missed his family in his first month in Malaysia.

Khalil, (G.1): "*For Afghanistan people are really difficult to get visas for family during study in Malaysia. I really miss them.*"

Shafiq, (G.1) came from a very critical war zone (Hilmand Province) in Afghanistan to Malaysia. He also faced the same problems, he was worried about his family in Afghanistan and at the same time he faced some problems in the new environment. All of these make him very stressful.

Shafiq, (G.1): "*After that I realized that I came from a far part of the world to Malaysia and feel worried about my family. They lived in a war zone. Also, we had problems of transportation, foods ....*"

6.3 Perception of Malaysians

Some of them find that the local Malay students are good and friendly in class. The Malay students have a lot of respect for the Afghan students. They seem to be friendly even in the first meeting and tend to smile a lot. In general, female local students are more open minded, and they talk nicely and have good behavior compared to the male locals students.

Khalid, (G.2): "*We are around 20 Afghan students in the same faculty and same class. I had some local friends also. We always exchange our ideas. Afghan students more they have the same way of thinking and the same idea. I am really happy to talk and have international friends.*"

Walid, (G.3): "*The first time, when I came to Malaysia, my roommates were all Afghans, after when I started my first semester, university makes some changes and I became roommates of the three locals. We had nice time together, many times they invite me to their house, I traveled with them to their hometown. They are so open mind boys;; there weren't any difference among us. We were so friendly, First days of my life in the Malaysia were limited to my Afghan group friends. Because before coming to Malaysia I don't know anyone of them. ..*"

6.4 Similarities of Culture

Afghan students in Malaysia agree the same religion can be one of the factors, which make it easy the adaptation process for them to fit in the local Malay society. The Afghan students were happy that their government chose a Muslim country as the destination for their higher education. They were able to adapt well to the local environment since the issues of halal food and religion which are the main concerns did not cause them any problems. Malaysia being a multi-cultural and multi-religious country can facilitate foreigners no matter whether they are Muslims or non Muslims.

Farid, (G.1): "*it has naturally a few effects, maybe it is negative shock or positive. But for me it was so easy to adapt to the Malaysian culture, maybe one of the reasons because Malaysia a Muslim country, we have access to halal food, Masjid, our study environment is really beautiful with all facilities and also more important Afghanistan government sponsor me a full scholarship.*"

After staying in Malaysia, the Afghan students now think differently from the past. They mentioned during the focus group discussion about the wrong perception that Afghan in general have about life overseas. They believe that the lack of communication and information on other countries formed their negative perceptions on foreign countries.

Walid, (G.3): "*I belong to close society. There in our ancient culture, people think if someone travels, he will lose his religion, and he will be completely changed. And some other negative ideas about overseas. But what I find when I came to Malaysia, it was different from my previous expectation. I was wrong, now I had communicated with different type of people. I learned lots of issues.*"

The Afghan students stated that once they adapt to the Malaysian environment, they do not face any problem being a foreign student in Malaysia. The only problem that most of them still consider a problem is the spicy Malay food, even though it is not a big problem as Afghani and Arab food are also available at cafes in their campus and also many restaurants nearby.

7 Discussion and Summary

The paper aims at presenting the Afghan student academic and cultural life experience in Malaysia. To better understand their life experience in Malaysia, the study presents in the various phases of the students journey in pursuing higher education in Malaysia. The students went through the honeymoon and culture shock stage. Most of them were excited and happy to be pursuing their studies in Malaysia and were touched by the welcome done by the respective universities. They went through the culture shock stage due to some differences between the cultures and when homesickness seems to sink in. Eventually they build more positive perceptions of Malaysians and discovered to focus more on the similarities between the two cultures rather than the differences. These assisted the students throughout their adaptation process in Malaysia.

 DOI: 10.1051/shsconf/20173300056

References

1. A.Arokiasamy,. Aust. J. of Bus. and Mngmnt Res.**1**,9 (2011)
2. R. Yang. Int. Edn. **13**,1 (2002)
3. M. Wende, High. Edu. Pol. **14**,3 (2002)
4. Y.Y. Kim. *Communication and cross-cultural adaptation*. (UK, Clevedon, Multilingual Matters, 1988)
5. C.S. Mustaffa and M. Ilias. Int. Comm. St. **22**,1 (2013)
6. Y. Chen, H. Chen. *Int. Conf. on Bus. Mgmnt and Info. Tech. App.***98**:233 (2009)
7. R. Yang. Int. Edn. **13**,1 (2002)
8. G.A, Philip,L.Reisberg,, L.E.. Rumbley, Trends in Global Higher Education: Tracking an Academic Revolution.(2009)
9. A.S. Olutokunbo, I.A. Ismail, T. Suandi. Int. J. Of Learning and Development. **3**,6 (2013)
10. *Higher Education in Afghanistan: An Emerging Mountainscape*. (The World Bank, Washington D.C, 2013)
11. S. Lysgaard. Int. Soc. Sc. Bulletin.**7**,1 (1955)

Permissive Attitude toward LGBT in Jakarta: Study of Differences in the Level of Permissiveness among individuals born in 1950-1970's and 1980-2000's

Endang Setiowati[1,*]

[1]Faculty Member, Communication Studies, Vocational Program, Universitas Indonesia

Abstract. Lesbian, Gay, Bisexual and Transgender (LGBT) movement is strongly resisted in Indonesia. This is not only that Indonesian people are still religious, but also the traditional norms and values are still very strong in daily life. Based on the Pew Research Centre in 2014, the homosexuality level of acceptance in Indonesia is only about 3%, while the rejection is 93%. LGBT has been started since the 1960s, but it became strength since the private television in early 1990s promoted the effeminate characters to be the top celebrities paid hundreds of millions of rupiah. People who born in the 1950s to 1970s that consume only one TV station will have different respond to LGBT with people who born in the 1980s to 2000s with some TV stations. Purpose: to look for the differences in the level of permissiveness toward homosexuality between those two groups of people. Method: using a mix method of qualitative and quantitative. Qualitative methods interviewed 6 informants. The quantitative will be a survey of 100 respondents. Result: The level of permissiveness on LGBT of the younger group is higher than the older group. According to acceptance factor both group mostly reject LGBT. But according to properness, the older group tend to say disturbing and disgusting, while for the younger group assume not disturbing.

1 Introduction

At the end of January 2016 appeared a poster that was distributed virally through social media. The poster mentioned if a student from the University of Indonesia (UI) offer counselling services to lesbian, gay, bisexual, and transgender (LGBT). The poster also emphasized that they are not alone and the institutions called themselves as the Support Group and Resource Centre on Sexuality Studies (SGRC) will help them cope with difficult times as a result of the identity crisis. In the posters also contained the photos and names of four students and alumni who pose as LGBT persons. Suddenly the poster became a trending topic on various social media as well as in the mass media.

Indeed the LGBT issue becomes widespread in the media and in the community chatter. The pros and cons opinion also appear. This is because LGBT is still a taboo. Results of research on Global Attitudes Project conducted by the Pew Research Centre in the year 2013 which was updated in 2014 stated that the level of acceptance of LGBT socially in Indonesia only reaches 3% only, while rejecting as much as 93%. Level of public acceptance is not just because of religious interpretation but also norms and values of Indonesian people who are not comfortable against LGBT presence [1].

* Corresponding author: setiowati@gmail.com

DOI: 10.1051/shsconf/20173300057

The LGBT movement became strength after it promoted by some private TV stations that established in the early of 1990s. Private TV stations compete with each other to show the effeminate figures in any program to attract an audience. At first the effeminate figures is only used as the butt or derision, but in line with the increase in the rating on each program featuring the effeminate figures, it became a new commodity for private TV stations. After the 90s came the new pansy celebrities such as the late Olga, the late Avi, Ivan Gunawan, or a feminine figure of a man like Indra Bekti, Irvan Hakim, Ruben Onsu and so forth.

Characterizations of gay and lesbian began to frequently show in soap operas or movies. It seems that homosexual began to be accepted and considered normal by society. For people born in the 1950s to the 1970s, of course characterizations of gay, lesbian and transvestite could be something disturbing. because although many effeminate figures like Sukardjo AC/DC, Tessy Srimulat, even the late Benjamin S and Bing Slamet, the two famous Indonesian comedian ever played a transvestite or effeminate figures in their movies.. But people know that they are actually a man in real life. Unlike the effeminate figures appearing on private TV era, where in real life they also show their love in the same sex. They also dress up as women in their real life. This could lead to them being born in the 1980s to the 2000s, which was the main target of the private TV, will become accustomed to seeing the effeminate figure in real life.

This research will focus on how those two groups of people in seeing the LGBT. Whether or not they have different opinion and attitude toward LGBT.

2 Literature Review

LGBT movement in Indonesia has existed since at least the 1960's which then began to develop in the 1980s along with the popularity of Fantastic Dolls dance group consisting of the transvestites. This movement began to explode in the new millennium until now. The first LGBT organization is the Association of transvestites Djakarta (HIWAD), that was protested by the Indonesian Ulema Council (MUI). In the 1980s established various organization focus on LGBT empowerment such as Lambda Indonesia for gay people, PERLESIN for lesbian and the most well-known is GAYa Nusantara that embodies the gay, lesbian, and transgender. [2] The dramatic changes occur in the political system and the government in 1998 in which freedom of opinion started to become a staple. Indonesian Women's Congress in December 1998 formally include representatives of lesbians, bisexual women and transgender men and include them in the sector XV in the organization of Women's Coalition for Justice and Democracy (KPI)

The rise of activities and campaigns in favour of the movement of Lesbian, Gay, Bisexual, and Transgender (LGBT), either directly or indirectly, have negative effects for society and especially the child's age. Moreover, LGBT campaigns conducted through information channels and social media users are target school age children and adolescents.

Apparently, had many children and adolescents are victims of age. Data compiled by the Ministry of Education and Culture in 2016, it seems LGBT children and adolescents in the world is quite serious. There are 3000 students in the city of Batam known as a predominantly gay. [3]

AusAID survey, as many as 700 or 22% of children aged 16-20 years in Tanjungpinang and Bintan behaves likes the same sex. Increased homosexual HIV sufferers of 6% (2008) to 8% (2010) and continues to be 12% (2014). Furthermore, there is a SD Gay community, Gay junior high school Gay number of followers on twitter with thousands. At the same time, there are 119 organizations that are concerned about LGBT issues in Indonesia and the numbers continue to rise. [3]

The research conducted in 3 cities, Shanghai, Hanoi, and Taipei on adolecents and young adults perception of Homosexuality showed that adolescents and young adults of higher economic status and educational level, whether they are students or not, and/or if from an urban area, are likely to have a more positive perception of homosexuality. The other findings showed that increased media viewing is correlated with a more favorable perception of homosexuality, and this relationship is most pronounced for those with the least amount of social contact with lesbians and gay men. Media may be able to influence the beliefs of adolescents and young adults, particularly, for those who have little direct social contact [4].

3 Methods

This study uses qualitative and quantitative approaches. [5] The qualitative research interviews 6 informants consisting of a woman born in 1955 (WM), a man born in 1961(DA), a man born in 1977 (AZ), a woman born in 1983(DK), a woman born in 1992 (KL) and a man born in 2001(AA). Questions about how their opinion on LGBT and whether they agrees that persons with LGBT that remains themselves or had to go and get back into a man or woman as nature.

Quantitative research is the development results of interviews with six informants in qualitative research. The research was conducted online, and the link of the questionnaires distributed via Twitter, Facebook, some chat groups in Blackberry, WhatsApp, and Telegram, to the people who were born in 1950 to 2001 or from 15 years old to 66 years old. Restrictions birth year to 2001 is because they were born in the year, is already 15 years old and went to high school, considered opinions given quite trustworthy.

Using close and open ended questionnaires, as many as 11 questions related to 3 major topics such as LGBT as abnormal behaviour, the role or characterizations of a LGBT in the TV program and also about the existence of the LGBT their environment. Time distribution of the questionnaire was for three days and got 187 responses.

From the 187 responses obtained, consisting of 63 men (34%) and 124 women (66%). Age is composed of 77 people (41%) aged 37 to 66 years and 110 (59%) aged 15 to 36 years. Their domicile is in South Jakarta as many as 51 people (27.3%), East Jakarta 22 people (11.8%), West Jakarta 18 (9.6%), Central Jakarta 3 (1.6%), North Jakarta 2 (1.1%) and those living around Jakarta (Depok, Bogor, Bekasi, and Tangerang) as many as 91 people (48.7%).

Religious affiliation of respondents is 131 people (70.1%) are Muslims, 29 people (15.5%) are Christians, 23 people (12.3%) are Catholic, 1 person (0.5%) are Hindu and 3 people (1, 6%) replace it with words that are not related to religion, and one person call himself, as a humanist.

4 Results

4.1. Result of Qualitative Research

They were born in the 1950s to the 1970s and 1980s to the 2000s birth assume that LGBT is an inappropriate behaviour as incompatible with religion and norms of Indonesian life. But their opinion is different when asked whether LGBT is a disruption to their lives. The older groups regarded as intrusive and should be eliminated from television shows and movies. Such as what WM said,

> *"…. presenter seperti Olga..maaf bukan mau ngungkit yang sudah almarhum ya.. atau Ivan Gunawan itu kenapa sih harus diekspos? Kan bisa cari orang lain. Kayaknya tuh kalo bencong pasti laku.."*

The older group also assume that homosexual behaviour is contagious behaviour that should be prohibited from developing in Indonesia. Such as what DA said,

> *"… homoseksual itu menular lho.. di kantor saya ada satu boss lain bagian sama saya yang nyata banget kalo dia itu hombreng, eh.. tahu-tahu anak buah dia berubah semua jadi gemulai… saya tahu banget ada seorang yang tadinya normal cowok banget sekarang kelihatan jadi berubah salon banget penampilannya, terus juga baju-bajunya kelihatan banget deh udah jadi hombreng sekarang… "*

However, about LGBT can be contagious is also stated by one informant who was born in 1983 know that DK as she said the following,

> *"...menurut saya seperti penyakit bisa menular karena temen saya pernah menjadi korbannya.. kalo keganggu atau ngga, sedikit terganggu karena dalam agama manapun ngga ada yang mendukung lgbt ini.. kalo buat berobat atau ke psikiater*

menurut saya kalo dalam dirinya merasa benar mau diobatin gimanapun bakalan susah..."

While the younger group was not bothered, because there are many LGBT in their peer groups. So if there is a rule should prohibit LGBT scene containing on television does not also have an impact on their environment which is already a LGBT. According to them, people who have LGBT indeed consciously choose their life. As far as the LGBT do not interfere with their lives and do not invite them to join their LGBT, they would not pay attention to their existence. Such as KL said,

"...gak masalah... banyak koq mahasiswa yang hombreng.. kalo lesbi belom pernah kenal... pokoknya asal mereka gak gangguin saya..gak ngusik hidup saya ya biar aja mereka toh punya komunitasnya sendiri kan, jadi meski di TV ada larangan penampilan LGBT kayaknya gak ngaruh deh... yang hombreng tetep aja hombreng.."

As well as the AA 15-year-old boy said,

"..selama mereka tidak menyinggung sesuatu yang bersangkutan dengan kesusilaan dan agama saya rasa sih fine2 aja, toh itu hanya hiburan..."

4.2. Result of Quantitative Research

The quantitative analysis is using SPSS Version 21. Regarding the responses to the question whether LGBT is an abnormal behaviour, the majority of respondents, 148 (79.1%) responded that LGBT as abnormal behaviour. When viewed from the age of 62 turned out to vote (80.5%) of respondents aged 37 to 66 years stated LGBT abnormal behaviours, while from age 15 to 36 years as many as 86 people (78.2%). If the views of gender turned out 85.7% of men (54 people) stated that LGBT abnormal behaviour, while women only 75.8% (94 people).

In terms of religion, the majority stated LGBT is an abnormal behaviour. The one and only Hindu respondent also stated that LGBT is an abnormal behaviour. On the contrary, they who are not state religion, the majority 66.7% (2 persons) states that LGBT is normal behaviour. From the majority of the states LGBT is an abnormal behaviour, those who are Catholic turns out the largest percentages that is 87% (20 people). While the Muslims respondents who are the majority of respondents only 78.6% (103 people).

Table 1. LGBT as an Abnormal Behavior by Age and Religion

AGE			RELIGION					TOTAL
			1	2	3	4	5	
37-66 y.o	Y	Count	35	15	11		1	62
		% within topic	56.5%	24.2%	17.7%		1.6%	100.0%
		% within religion	83.3%	83.3%	73.3%		50.0%	80.5%
	N	Count	7	3	4		1	15
		% within topic	46.7%	20.0%	26.7%		6.7%	100.0%
		% within religion	16.7%	16.7%	26.7%		50.0%	19.5%
15-35 y.o	Y	Count	68	5	12	1	0	86
		% within topic	79.1%	5.8%	14.0%	1.2%	0.0%	100.0%
		% within religion	76.4%	100.0%	85.7%	100.0%	0.0%	78.2%
	N	Count	21	0	2	0	1	24
		% within topic	87.5%	0.0%	8.3%	0.0%	4.2%	100.0%
		% within religion	23.6%	0.0%	14.3%	0.0%	100.0%	21.8%
TOTAL	Y	Count	103	20	23	1	1	148
		% within topic	69.6%	13.5%	15.5%	0.7%	0.7%	100.0%
		% within religion	78.6%	87.0%	79.3%	100.0%	33.3%	79.1%
	N	Count	28	3	6	0	2	39

DOI: 10.1051/shsconf/20173300057

AGE			RELIGION					TOTAL
			1	2	3	4	5	
		% within topic	71.8%	7.7%	15.4%	0.0%	5.1%	100.0%
		% within religion	21.4%	13.0%	20.7%	0.0%	66.7%	20.9%
GRAND TOTAL		Count	131	23	29	1	3	187
		% within topic	70.1%	12.3%	15.5%	0.5%	1.6%	100.0%
		% within religion	100.0%	100.0%	100.0%	100.0%	100.0%	100.0%

Note: 1. Muslim, 2.Catholic, 3. Christian, 4. Hindu, 5. Others

The reason why they consider LGBT is a most abnormal behaviour mostly is that LGBT violating the predetermined nature of Allah (50.8%), violates the religious and social norms (27.3%), and considers LGBT as a disease (7%). While people who consider LGBT is not an abnormal behaviour mostly gave reason that people has the right to choose their own will (9.6%), because LGBT is a destiny (2.7%) and the impact of social environment (1.6%). The interesting finding is even though there are some of them who say that LGBT violates the nature of God (28.2%) and the abuse of religion (20.5%) but they stated that LGBT is not an abnormal behaviour.

Regarding the response to the role or characterizations of a LGBT in the TV program, majority 75.4% (141 people) said that they often see the LGBT roles or characterizations exposed on TV programs. While when asked about whether they were bothered by the appearance of LGBT characters in TV program majority 68.4% (128people) said they are bothered by the LGBT appearance on TV. But strangely when compared of those who responded had rarely or never seen LGBT appearances on TV, with whether they interfere with the LGBT appearances on TV, there were 65.2% (30 people) who stated annoyed with the LGBT characters exposed on TV. Whereas previously they claimed rarely or never seen LGBT characters on TV. Judging from age who answered annoyed with LGBT appearances on TV, as much as 68.8% (53persons) of aged 37 to 66 years, while respondents aged 15 to 36 years have much more percentages, as much as 68.2% (75 people) was disturbed by the appearance of LGBT on TV.

Regarding the response of whether LGBT appearances on TV should be banned, 73.3% (137 people) said that LGBT appearance on TV should be banned. When viewed from age, there were 72.7% (56 persons) aged 37 to 66 years stated LGBT should be banned on TV, while for those aged 15-36 years are also in line with the other age groups, a majority of 73.6% (81 people) considers LGBT appearances on TV should be banned.

Regarding the response of what to do if there is LGBT persons in their neighbourhood, mostly (67.9%) stated that they will remain friend to the person, as long as they do not ask or persuade them to be LGBT. It is only 8% stated that they will not continue being friends with them. Relating to the age, 60 % of people who will discontinue the relationship with the LGBT are from the age of 37 to 66 years. Respondents who stated that they will remain friend to the person, as long as they do not ask or persuade them to be LGBT are mostly from the age 15 to 36 years (73.6%) while only 59.7% respondent of the age 37 to 66 years stated the same. Another interesting finding is respondent from the age of 37 to 66 years are more compromised to the LGBT as they stated that they will remain friend however their behaviour. It is 28.6% of those ages. While in the ages, is only 20,9% who stated that they will remain friend however their behaviour.

Regarding the question of what should be done if there is a person who induce or encourage others to be LGBT, mostly (72.7%) stated that the LGBT person should be reprimanded and advised. 13.9% stated that as long as the person who was induced is accepted the invitation, so that they will do nothing. 9.6% stated that the LGBT person should be severely punished. The rest of 3.2% of the respondent chose to answer in other reason, it was if the invited are children, then the LGBT should be punished, but if they are an adult, it is up to the invited. Relate to the age group, the age of 15 to 36 is more strictly because 11.9% from this group stated that they should be severely punishment, on the other hand only 6.5% of the age of 37 to 66 years stated the same.

Regarding the question what the LGBT persons should do for their life, there are three categories of answers, which almost have equal numbers of each. The highest number (35.3% or 66 persons) is the answer that nothing to be done for them as long as he does not invite or influence others to be LGBT. Then 34.2% or 64 persons responded that they should find psychiatrist to be back normal. While the rest (30.5% or 57 persons) tend to say that LGBT people should repent and strive to live a normal life. Relate to the age group most of the 37-66 years old tend to say that the LGBT persons should find medication to be back normal (39%), while the 15-36 years old tend to say nothing should they do as long as they do not affect others to behave the same (40%).

5. CONCLUSION

In Indonesia, program containing LGBT is categorized as a form of information and entertainment that are not in the same breath with Indonesian cultural values system. Norms prevailing in Indonesia still refuses LGBT as a normal thing. Indonesian Broadcasting Commission (KPI) insisted on retaining their decision to block LGBT from public media. But the appearance of sissy characters is not regulated in the Code of Conduct of Broadcasters and Standard od Broadcasting Program (P3SPS). Regarding the norm, there are quite a lot of positive law that do not accommodate the position of LGBT. Law Marriage Law 1 of 1974 does not recognize the term "mating type". Later, Law No. 23 Year 2006 concerning Population Administration does not recognize LGBT. Law No. 4 of 2008 on Pornography includes the term "deviant intercourse" as one element of pornography. In the pornography laws also included prohibited is "intercourse chatter" as "... oral sex, anal sex, lesbian and homosexual". In addition, the Government Regulation No. 54 of 2007 concerning the adoption, explicitly stipulates that adoptive parents cannot be a homosexual couple. Not only that, the Minister of Social Affairs Regulation of 2012 stipulated that the so-called social welfare issues, including those due to their sexual behaviour becomes obstructed in social life, such as; transvestites, gay men and lesbian women.

Based on the above, it appears that LGBT still be the pros and cons. The religious life that is still strong in Indonesia also affect permissiveness. Evidenced by the results of quantitative and qualitative research that both those aged 37 to 66 years and 15 to 36 years, quiet equal to refuse LGBT. The level of permissiveness is higher in younger age groups, especially on the answer about what to do the LGBT persons in his life, they answer as far as not trying to persuade or influence people there is nothing to do.

Therefore both qualitative and quantitative research found that there are no significant differences in opinion and attitude toward LGBT between the age of 15 to 36 years and the age of 37 to 66 years. They mostly said that the LGBT persons are abnormal, and considered them as people who are sick and should be treated by the experts to be back normal. Most of the respondents think that the role of LGBT in TV station should be reduced, because TV stations are considered to be the vehicle to spread the LGBT way of life as a normal life.

References

1. Wike, R., & Horowitz, J.M. The Global Divide on Homosexuality: Greater Acceptance in More Secular and Affluent Countries. New York: Pew Research Centre. (2014)
2. P, R. A. Menelisik Perjalanan LGBT di Indonesia. Retrieved March 2, 2016, from Republika Online: http://www.republika.co.id/berita/jurnalisme-warga/wacana/16/01/28/o1n41d336-menelisik-perjalanan-lgbt-di-indonesia (2016)
3. Susanto. Civic Intelligence dan Propaganda LGBT. Retrieved March 14, 2016, from Detik.com: http://news.detik.com/kolom/3164153/ltigtcivic-intelligenceltigt-dan-propaganda-lgbt (2016)
4. Yongliang F, PhD., Lou, C., Gao, E., Tu, X, Cheng, Y, Emerson, M. R., Zabine, L . *J. Adolesc. Health*, (2011)
5. Riduwan. Metode dan Teknik Menyusun Tesis. Bandung: Alfabeta. (2006)

DOI: 10.1051/shsconf/20173300058

Border Issue: Misperception between Indonesia and Malaysia

Irwansyah Irwansyah[1,*]

[1]Universitas Indonesia, Communication Department, Kampus UI Depok 16424, Indonesia

Abstract. Indonesia and Malaysia border has been disputed and led to misperception. By using Krauss and Morsella's misperception concept, this study described that the border issue was occurred only online community member who never existed at the border area. Online community members constructed the discourse about border issue based on the agenda of mass media. While online community members and mass media lead discussion on sovereignty and nationalistic ideology, local border people concerned how they maintain the mutual relationship. The misperception which leads to conflict in online community forum showed the extent of civil conflict mediated by mass and new media.

1 Introduction: Border Issue between Indonesia and Malaysia

Mass media has shown the Indonesian–Malaysian relationship to be one of conflict rather than one of cooperation [1]. Conflict between Indonesia and Malaysia date to 1960. At that time, Indonesia refused to recognize the emergence of the Malaysia Federal State, which was considered a colony under British imperialism. The conflict has continued and the border dispute between Indonesia and Malaysia has resulted in high tension between the two governments, including their military forces [2].

Since both countries proclaimed their liberty, at least four serious issues concerning border claims have fueled discourse in the mass media. First was the dispute over the Sipadan–Ligitan islands, which Malaysia won through the International Tribute to Justice (ITJ) on December, 2012. Second was Malaysia's claim on the Ambalat block, which was considered a threat to the political and military forces of Indonesia. Third was the dispute over the placement of the frontier pole on the border line between West, East, and North Borneo in Indonesia and Sarawak and Sabah in Malaysia. Fourth was Malaysia's claim on Tanjung Datu, West Borneo, secured by their construction of a lighthouse in Indonesian territory [3].

Meanwhile, the development of communication and information technology has stimulated the evolution of these issues from state conflicts to personal conflicts [4]. Border area conflicts grow in the virtual world as the online community dramatizes these conflicts [5]. As a result, potential conflicts between Indonesia and Malaysia are issues of public discourse not only in the offline world and mass media but also in community-based online forums such as Topix.com®.

* Corresponding author: irwansyah09@ui.ac.id

2 Misperception from Communication Perspective

Argumentation and disputation concerning border conflict can be analyzed according to misperception concepts [6]. Misperception is defined as the wrong assumption about how communicators send a message about something known by other people. This may occur when people estimate the knowledge, beliefs, and values of other people. Such estimations tend to be biased if they are based on subjective knowledge and belief. This specific situation may occur when each culture uses a different sign system. Indicators of misperception include (a) different perspectives; (b) misassumptions; (c) group differentiation; (d) different communication objectives; (e) lack of collaborative conversation or dialog; and (f) use of unfamiliar words. Misperception about border issues may exist when either senders or receivers underestimate any of these six indicators.

Misperception occurs in almost every process in the exchange of ideas and messages. People have different fields of experience and frames of reference, which affect their encoding and decoding [7]. In the case of online community conversations, different subcultures may expose and learn different beliefs and values to guide them [4] in interacting and communicating with each other in the same online forum. Although misperception has been found in comparisons of Indonesian and Malaysian users' blog, facebook, and twitter accounts [1], online forums that reflect the different character of new media have not yet been studied. Therefore, this study was an exploration of the specific online forum, Topix.com, which has 250,000 online community forum members and 750,000 users every day [8]. At the same time, this study also served to confirm and compare specific topics in the online community forum, including military activity, with the actual situations in the border area between Indonesia and Malaysia.

3 Discovering the Indonesian–Malaysian Border Conflict

This study used netnography [9] which is adapted from ethnography. This method is also referred to as web ethnography or virtual ethnography [10]. Netnography was used to identify conflict potential based on the different perceptions among online community forum users. Because the virtual world has a legitimate culture and meaning as if in the "actual or real world" [11], one can perceive the virtual world as having the capability of taking social action to construct a new culture that may be studied ethnographically [10].

There are five procedures on doing netnography: (1) making cultural entrée, (2) gathering and analyzing data, (3) ensuring trustworthy interpretation, (4) conducting ethical research, and (5) providing opportunities for culture member feedback [9]. This section would focus on making cultural entrée and how to gather data. Analyzing data and other procedures would be discussed in different sections of the article.

Cultural entrée of this study was two threads or conversations about Indonesian and Malaysian conflict: (a) Ekspedisi Khatulistiwa-GGK, Senoi Praaq, Sarawak Rangers (Ekspedisi Khatulistiwa) and (b) "Lucu!!! Perbatasan Indonesia-Malaysia, RI Andalkan Tank Scorpion" (Tank Scorpion). These threads were selected based on (a) border conflict issues and (b) the most number of comments. Each text that contained specific discourse on country borders and military forces in border areas was included.

Moreover, the study gathered the data from the registered Malaysia forum for Topix.com® conversations focused on the border conflict between Indonesia and Malaysia. Two main themes emerged: (a) military forces from both countries and (b) the frontier areas of both countries. The military force theme was categorized into (a) main devices in defense systems, such as weapons, tanks, and combat flights and (b) military personnel strength in the border area. The uncertainty of the location of the frontier in the border area triggered conflict concerning territorial claims between Indonesia and Malaysia. This uncertainty occurred because frontier poles were sometimes misplaced. The misplacing of frontier poles was ascribed to the Malaysian military as a part of confrontation. Several members of the online forum indicated that the Malaysian Army Forces (*Angkatan Tentara Malaysia*–ATM), the Royal Malaysian Police (*Polisi Diraja Malaysia*), or the Sarawak Rangers (*Senoi Praaq*) were responsible.

The Ekspedisi Khatulistiwa thread, created by Malaysian user Formalay, contained discourse on the issue of Indonesian military activity in the border area, with 594 comments posted between September 24, 2012, and March 20, 2013. The initiator of the thread wrote that special Malaysian troops, such as the Special Action Group (*Group Gerak Khas-GGK*), the Sarawak Rangers (*Senoi Praaq*), and the Air Force of Royal Malaysia (*Tentara Udara Diraja Malaysia*) oversaw the expedition to make sure that the Indonesia military did not cross the border and did not execute unexpected or illegal activities. In response to the Ekspedisi Khatulistiwa thread, an Indonesian user named Forindo created the Tank Scorpion thread to convey news from Indonesia regarding the placement of Scorpion tanks along the border between Indonesia and Malaysia. The thread had 1,002 comments posted between January 7, 2003, and March 28, 2013. These two threads led to different perceptions, understandings, and responses between Indonesian and Malaysian users and members of online forums who contributed their comments.

The user name of Formalay and Forindo could be identified as insider who have strong ties to the online group and to the consumption activity [9]. The user also tend to be long-standing and frequently referenced members about border issue between Indonesia and Malaysia. However, in regards to ethical issue on netnography, this study ensured the confidentiality and anonymity of any users including Formalay and Forindo.

4 Misperception about Border Issues on Online Forum

As a part of analyzing data from netnography, misperception was identified through the different perspectives found in members' arguments. For example, user named *Garnett* from Indonesia said, "Malaysia was over reacted on *Khatulistiwa* Expedition . . . in contrast when lighthouse construction at *Karang Unarang*, the Indonesian builders were interfered by Malaysian boat patrols by moving the wave to lighthouse site . . . until day light . . . Indonesia military were not over reacted.

Misperception was also identified through the misassumptions of both parties. In the *Ekspedisi Khatulistiwa* thread, for example, user named *Paskal* assumed the Indonesian military expedition was similar to a scouting expedition: "ATM [Royal Malaysian Force] has been playing by moving the border pole. TNI [Indonesia military] has just come to get the expedition which similar to scout expedition." User named *Pengamat* assumed that the Indonesia military expedition was military training and showed the strength of military devices at the border. *Pengamat* said that the Indonesian military expedition was a show of force for the Malaysian troops: "Indonesia has shown a challenge standpoint to Malaysia troops through the training of Indonesia military nearby the border. Moreover, the time is similar to Independence Day of Malaysia. Indonesia has instructed to Trimaran ship on the sea border but Malaysia is just quiet and afraid. Where do the Royal Malaysian Force hide? On pondering room?"

Group categorization also resulted in different perceptions. The use of the names *Formalay* or *Forindo* was part of group categorization that showed users or members' affiliations to specific states. The use of *Indon* for Indonesian people and *Malon* for Malaysian people triggered members of both groups to attack each other in the virtual world. Different perceptions that could cause conflict in both countries were identified through uncollaborated conversation. Such conversations were similar to undialogical statements from both Indonesian and Malaysian users. In the *Ekspedisi Khatulistiwa* thread,

 DOI: 10.1051/shsconf/20173300058

four users made undialogical conversation. Garnett commented on sea patrols of Malaysia at *Karang Unarang*. The next user, *Wes NgeXXX*, commented on Java Island's (central government of Indonesia) manipulation of news facts on the field. *Paskal* then responded by comparing Indonesian frogmen troops with Malaysian or Korean special forces. Finally, Malaysoak responded by asking about the lonely Ambalat zone and by informing members about the Malaysian Royal Navy breeding turtles.

Misperception could also be identified through the different objectives of communication. Online discussion labeled *Forindo* was identified as discussion making virtual war on Malaysia. Many Indonesian users or forum members followed this objective. The thread developed by *Forindo* on the Malaysian forum had the fifth highest of amount of discussion traffic on Topix.com®. The purpose of creating this thread was to construct negative images about the status of Malaysia in that country's forum. Comments on *Forindo* increased twice as fast as those in Formalay's thread, needing three fewer months than Formalay to increase conversation.

Finally, different perceptions that could cause specific conflict were identified through the way both parties used unfamiliar words. These unfamiliar words could lead users from both parties to bully and insult users from the opposite group. For example, *Imoes Gigo* did not understand the terminology *group gerak khas* (special action of military group). As a result, the user interpreted the terminology as a dance or clown group. Similarly, when Sergio Ramos used the word *lari bertaburan* ("escape randomly"), user *Satrio* said the interpretation of those words was strange and irrelevant. The user might have wanted to insult Sergio Ramos with the unfamiliar words *hanjut sanan sinun* ("go anywhere").

5 Media as a Trigger of Border Conflict Discourse

Border conflict between Indonesia and Malaysia existed in the online community forum because users did not see the real border context. Users responded to the various issues about the border based on information from mass media. Mass media reported any comments about the border conflict, including statements from sources or news makers outside the border area. These sources or news makers were representatives of government agencies [12] and used negative tone and confrontational and harsh words. Moreover, the media did not cover both sides of the story and did not confirm any of these statements with either people from the other country or the local people of both countries.

Any media news about the border conflict was a discourse between the government and people far from the border area. In the border area, the local people saw the economic gap between Indonesia and Malaysia. As long as their economic needs were covered by both markets, the locals never complained and never thought about conflict. Indonesian informants said that the real conflict was any negative news about the border. People from the House of Representatives (Dewan Perwakilan Rakyat/DPR) called for nationalism concerning the border conflict to attract the attention of their voters. The point of view of the central government concerning the border conflict was based on colonialism and current agreements about outstanding border problems. Because military elites thought border conflict to be a threat to the unity of ideology, they campaigned to enhance troop spirit to protect the border. Every dispute on media about the border area resulted in inconvenient interactions and tension between the local people and their relatives in other countries. Malaysian informants said that any inconvenient situation or tension resulting from the Indonesian media was a cause of concern not only for their family relationships but also for their economic sustainability.

Conflict in online community forums such as Topix.com® also showed the extent of civil conflict mediated by mass and new media. Seedlings of actual conflict had grown from a series of border conflicts between Indonesia and Malaysia. However, when the real

DOI: 10.1051/shsconf/20173300058

conflict about the border between the two countries subsided, people from both countries, especially people far from the border, who wanted the continuity of conflict created a cyberspace war. Not only could war in the virtual world be used as a propaganda tool, but such cyberspace wars could also prolong real conflict [13]. From the perspective of group communication, conflict became part of the group process [14], allowing members of the group who had similar backgrounds to communicate more openly. Group members who did not share such backgrounds constructed negative communications, suggesting misconstruction of perceptions [15].

6 Conclusion

Debates, disputes, and conflicts about the border area between Indonesia and Malaysia are mostly triggered by mass media and then extended into new media such as online community forums. Mass media quotes statements from any sources or news makers, including people or representatives of institutions that do not understand the real situation in the border area. Although many of these sources may think about sovereignty, they are often not available to listen to the local people and to assist with their needs. Similar to the members of the online community forums from both Indonesia and Malaysia, they express their voices, ideas, opinions, and comments without thinking about the voices of the people in the border area. Both mass media and members of online community forums only respond to the issues concerning the border, including border poles and the military located at the border, attacking and defending without any accurate knowledge of what is occurring in the real field of the border area. Therefore, misperception will become more complicated because the users have no real information about the situation and life in the border area.

As a result of this study, further research should be conducted concerning border conflict from the perspectives of politicians in DPR, news makers or sources, and government officers concerned and understanding of the needs of the local people. In addition, research should be conducted concerning the creation of balance and harmony between sovereignty or nationalism and economic development at the local community level in the border area.

References

1. Irwansyah, Jakarta, UI-Press, pp. 194-234 (2012).
2. E. Fitriani,Jakarta, UI-Press, pp. 11-66 (2012).
3. M. Muhyiddin, *Malaysia Hentikan Pembangunan Mercusuar di Tanjung Datu*, 21 May (2014). [Online]. Available: http://www.tempo.co/read/news/2014/05/21/078579426/Malaysia-Hentikan-Pembangunan-Mercusuar-di-Tanjung-Datu. [Accessed 17 April 2015].
4. G. M. Bowler, The Qual. Rep., **15,** 5 (2010).
5. J.-H. Shin and G. T. Cameron, Journ. Com. Quart., **80,** 3 (2003).
6. R. M. Krauss and E. Morsella, "Communication and Conflict," in *The Handbook of Constructive Conflict Resolution : Theory and Practice*, San Faransisco, Jossey-Bass, pp. 131-143 (2001).
7. J. T. Wood, *Interpersonal Communication : Everyday Encounters* 6th Edition, Boston: Wadsworth, (2010).
8. websitetrafficspy.com, 23, 4 (2013). [Online]. Available:

http://topix.com.websitetrafficspy.com/. [Accessed 23 4 2013].

9. R. V. Kozinets, , Jour. Mar. Res., **39** (2002).
10. J. Sumiala and M. Tikka., Com. Cult. Crit., **6,** 2 (2013).
11. T. Boellstorff, *Coming of Age in Second Life: An Anthropologist Explores the Virtually Human*, Princeton: Princeton University Press, (2008).
12. S. Z. M. Shariff, H. A. A. Aziz, N. Yusof and B. A. N. M. Thaheer, Int. Jour. Hum. Soc. Sci., **2,** 23 (2012).
13. P. Mitra, *Cyberspace War : The Kashmir Propaganda Battles Move to the Web*, Far East. Econ. Rev., p. 28, 5 November (1998).
14. B. A. Olaniran, Int. Jour. Conf. Man., **21,** 1 (2010).
15. O. B. Ayoko, An Int. Jour., **14**, 2 (2007).

Advent of The New Media: The influence of intrinsic and extrinsic motivation in online photo sharing behaviours on Facebook among young Malaysian adults

Wan Ching Lee[1], *Thinavan* Periyayya[2*], and *Nair* GV[3]

[1]Universiti Tunku Abdul Rahman, Faculty of Creative industries

Abstract. Social networking sites have become an important channel of communication in people's daily lives. Online photo sharing, a type of artefact sharing on Facebook has also grown in popularity. Thus, the purpose of this research is to study the relationship of intrinsic and extrinsic motivations in online photo sharing behaviours on Facebook. The study will also examine privacy concerns regarding online photo sharing. A non-random purposive sampling approach gathered a total of 422 Malaysian young adults aged between 18 and 34. Data was collected via a web based questionnaire distributed on Facebook. The results show that intrinsic and extrinsic motivations have positive correlations with online photo sharing behaviours on Facebook. Intrinsic motivations showed a stronger connection to online photo sharing behaviours. There was no significance found between privacy concerns and online photo sharing behaviours on Facebook. There was also no difference between genders in their online photo sharing behaviours. However, males showed higher privacy concerns in regards to online photo sharing on Facebook.

1 Introduction

Online social networking sites have gained tremendous popularity around the world over the past few years [1-2]. Facebook is the most popular social network, and has surpassed the 1 billion mark of users worldwide as of March 2015 [3-4]. Results from a recent study showed that Facebook is also the most popular social networking site in Malaysia [2]. Online photo sharing is one of the top applications used on Facebook [1], and in Malaysia, 88% of Malaysian online consumers upload their photos on social networks such as Facebook [5]. Majority of Facebook users worldwide, as well as Malaysia, are young adults in the age group of 18 to 34 years old [6]. Generally, reports show women are more active on social media as well as in online photo sharing [7].

In several studies on online sharing behaviours, theories such as the motivation theory of self-determination, social exchange theory, social capital and the social cognitive theories were drawn on to help explain individuals' motivations to share. Researchers have found that intrinsic and extrinsic motivations both act as a crucial determinant in knowledge and artefact sharing behaviour on online social networks [8]. This study also seeks to investigate how privacy concerns of users affect their online photo sharing behaviour.

2 Problem statements

A key question would be what motivates Malaysian netizens to be so heavily invested in such photo sharing behaviours. Sharing photos online is one of the most popular activities on Facebook [5], and is also rapidly growing among Malaysian Facebook users [2-6]. Research has also been limited to certain aspect of motivations and has so far only focused on the Flickr online community in the United States [9-10]. Thus far, there is no published research on what motivates users to share their photos on Facebook. The lack of data regarding online photo sharing behaviors on Facebook in the Malaysian context provides an opportunity to explore and contribute new information to this area of study, as geographical and cultural differences may result in different findings [11]. The impact of privacy concern and photo-sharing behaviour on Facebook has not been researched yet and as such this aspect will also be explored in this study.

* Corresponding author: thinavan@utar.edu.my

3 Research questions

Based on the problem statement the following research questions were formulated to guide this study:

- What is the relationship between intrinsic motivations and online photo sharing behaviours on
- Facebook among young Malaysian adults?
- What is the relationship between extrinsic motivations and online photo sharing behaviours on Facebook among young Malaysian adults?
- Which motivational factors (Intrinsic or extrinsic motivations) are stronger in users' online photo sharing behaviour on Facebook among young Malaysian adults?
- What is the relationship between online privacy concerns and online photo sharing behaviours on Facebook among young Malaysian adults?
- Is there a gender difference among young Malaysian adults' online photo sharing behaviours on Facebook?
- Is there a gender difference among young Malaysian adults' online privacy concern and online photo sharing behaviours on Facebook?

4 Literature review

In online social networks, the content is completely dependent on its creation by the users themselves [10], thus, without user content sharing, it would be meaningless to be a member of such sites. Sharing can have separated into two distinct forms namely, knowledge sharing, and artefact sharing [10]. Knowledge sharing is written information to be shared on one's network. This has been widely studied. Artefact sharing focuses on the sharing of personal content such as photos and videos where content created may or may not be shared. Studies on Artefact sharing is lacking and thus, this study aims to close this gap and contribute data that would continue to create a more comprehensive understanding of the motivating factors that drive users to share online photos on Facebook.

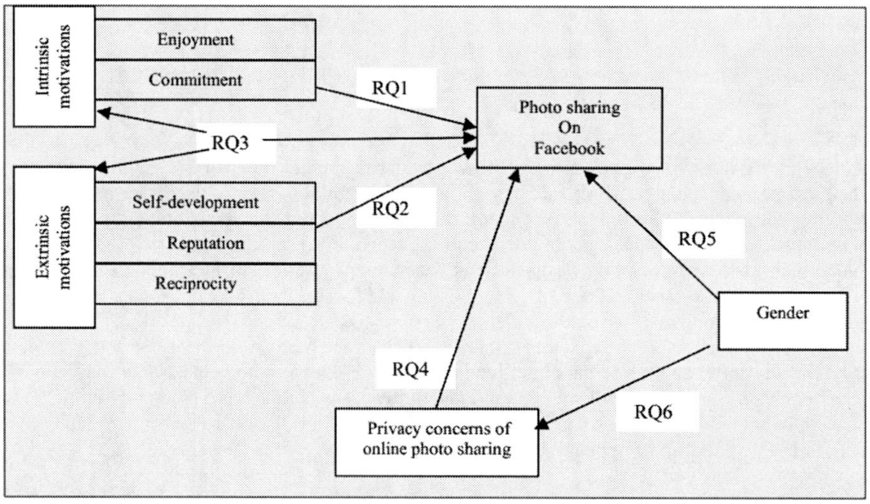

Fig.1. Conceptual Framework

The conceptual framework in Fig 1 is constructed based on various theoretical perspectives such as the self-determining theory of motivation [12], the social exchange theory [13], as well as the social capital theory [14]. These theoretical perspectives were applied to many studies that focused on the motivations behind online sharing behaviours of internet communities. Thus, this study also draws the strengths of these theoretical perspectives and focuses them on the research questions of this study.

Previous studies have shown that intrinsic motivations such as enjoyment and commitment play a crucial role in motivating users to share online. Enjoyment was found to have a positive correlation to share knowledge and content in online communities such as open source soft wares and online communities such as Wikipedia [15-16]. Artefact sharing communities such as Flickr was also noted to be influenced by the motivation of enjoyment [9,10,17]. The intrinsic motivation of commitment was positively correlated to both quality and quantity of photo sharing on an online community on Flickr [9-10]. Based on this it can be hypothesized that there is a significant relationship between intrinsic motivation and young Malaysian adults' online photo sharing behaviour on Facebook.

Several studies also looked into the relationship between extrinsic motivations (self-development, reputation, and reciprocity) and photo sharing behaviour. It was also found that social network users were more likely to continue using the social networks when other users reciprocated positive encouragement. Several studies similarly showed that software contributors placed greater emphases on self-development and reputation-gaining [15,18,19]. In one study it showed that users rated the motivation of reputation as a strong motivator for their friends to share information on Facebook [17]. Therefore, it can be hypothesized that there is a significant relationship between extrinsic motivation and young Malaysian adults' online photo sharing behaviour on Facebook.

Research question three aims to find out if intrinsic or extrinsic motivations play a stronger role in user's online photo sharing behaviour. One study showed that intrinsic motivations were more significant while in another [15] it leaned towards extrinsic motivations [19]. Given the mixed results it can be hypothesized that there is a significant difference in the influencing role of intrinsic and extrinsic motivations in young Malaysian adults' online photo sharing behaviour on Facebook.

Research question four aims to investigate the relationship between users' privacy concern and how it influences their online photo sharing behaviours on Facebook. Prior studies concluded that there is an inconsistency between users' privacy concern and their behaviour. Participants with profiles on social networking websites were less concerned with identity information disclosure and allowed anyone to view their profile without toggling their privacy setting [20]. Based on this Hypothesis four was formulated: that there is a significant relationship between privacy concern and young Malaysian adults' online photo sharing behaviour on Facebook.

Gender differences in online photo sharing behaviours are also evident in previous researches. In general, studies have shown that females posted more photos than males [7,21,22].On privacy concern, male participants showed greater risk taking attitudes and were less concerned with identity information disclosure where else women showed more concern over their privacy, had less friends but still were more likely to share [20].Based on this hypothesis five and six was formulated that is, there is a significant gender difference in young Malaysian adults' online photo sharing behaviour on Facebook; and that there is a significant difference between males and females on privacy concern and young Malaysian adults' online photo sharing behaviour on Facebook.

5 Method

The online questionnaire for this quantitative study targeted members of Facebook in Malaysia. Non-probability purposive, convenient sampling was used to obtain the participants for this study. The multi-step distribution procedure, also known as the snowball sampling method was used to collect the required data. The online survey with regards to the 2013 version of Facebook was conducted until enough participants were reached, and the duration of this exercise was from November 2013 to April 2014. The online questionnaire consisted of a total of 59 questions. All the scales used in this study were validated in the original studies, and were validated again in the present study. The variables of the study consist of the independent variables of intrinsic motivation (enjoyment and commitment), extrinsic motivation (self-development, reputation, and reciprocity), and the dependent variable is Facebook users' online photo sharing behaviours. Users' online privacy concern will be an independent variable to see if it impacts the dependent variable that is photo sharing behaviours on Facebook. SPSS was used to run statistical analysis and tests such as Pearson R, multiple regressions, and independent T test was used to test the hypotheses. The pilot test conducted showed that the reliability of the scales was all above 0.70 for the Cronbach's alpha. Out of 501 respondents, a total of 422 valid participants were used in this study.

6 Results

6.1 Demographic profile & Facebook usage

Out of 501 respondents, a total of 422 valid participants were used in this study. This was a response rate of 84%. The participants' age was between 18 and 34 years old (M=24, SD=3). Most of the participants' current residence was in Malaysia (94.1%). Female participants accounted for 59.2% while male participants accounted for 40.8%. A total of 77% of the participants were university students; 30.3 % young working adults; 2.8 % of the participants self-employed and 6.2 % unemployed.

Majority of the participants, 43.6 %, have been Facebook members between four to five years and another 40.3 % members of Facebook for six to seven years. A total of 7.6 % of them have been members of Facebook for eight to nine years and only 6.4 % were members for less than three years. In terms of friends the users' had on Facebook, 94.2% of the participants had more between 500-900 friends in their profile. On the question of usage of Facebook on a typical weekday, 73.2 % of participants spent between one to five hours on Facebook while 14.5% of the participants spent between five to ten hours on Facebook. A total of 3.8 % spent more than 10 hours on Facebook and only1.2 % of them spent less than an hour on Facebook.

Number of photos shared. Participants were asked on how many photos they have currently shared on Facebook. A total of 29.1% of the participants shared 100 to 499 photos on Facebook; 18.2% of them shared photos in the range of 500 to 999 photos;17.3% had shared less than a 100 photos on Facebook; 15.9% of the participants shared between 1000 and 1499 photos on Facebook;7.3% shared about 1500 to 1999 photos;2.8% responded that they have shared between 2000 and 2499 photos, and another 2.8% shared between 3000 and 3499 photos; 2.4% shared between 2500 and 2999 photos and 4.0% of them shared more than 3500 photos. Respondents were also asked on the frequency of the type of photos (Events, self-portrait, friends, family, pets/animals, scenery, and objects) they would usually share on

Facebook. For events, 37.7% of the respondents said they would often share photos of events on Facebook, 35.1% responded sometimes, 13.5% responded always, 11.4% responded rarely, and 2.4% responded never. Self-portrait photos were less frequently shared as 41.7% of them said they would rarely share photos featuring themselves only, 29.6% responded sometimes, 14% responded often, 10.7% responded never, and a small percentage of 4% responded always. The frequency of sharing photos on family, pets and animals, scenario and objects were not popular.

6.2 Results on Hypothesis Testing

Results for hypotheses one showed a significant positive correlation between intrinsic motivations and online photo sharing behaviour (*r=.60, p<.001)*. Both the intrinsic motivation of enjoyment (*r = .58, p<.001*) and commitment (*r = .47, p<.001)* showed a significant positive correlation with online photo sharing behaviours on Facebook. For Hypothesis two, there was significant positive correlation between online photo sharing behaviour and extrinsic motivations (*r=.40, p<.001)*. The extrinsic motivations namely self-development (*r = .34, p<.001)*, reputation (*r = .41, p<.001)*, and reciprocity (*r = .32, p<.001)*, showed significant positive correlation with online photo sharing behaviours on Facebook. Results for hypothesis three showed that the intrinsic motivation of enjoyment, β=.44, *t* (421) = 9.82, *p* < .001, and commitment (*B*=.19, *t* (421) = 3.64, *p* < .001) were stronger predictors compared to extrinsic motivations of self-development, β=.013, *t* (421) = .25, *p=.806*, reputation, β=.12, *t* (421) = 1.98, *p* = .048, and reciprocity, β= -.023, *t* (421) = -.43, *p* = .669. Hypothesis four showed that there was no significant correlation between photo sharing behaviour and privacy concerns (*r = .06, p=.181)*. The results for hypothesis five showed that there was no significant gender difference between males (*M* = 16.13, *SD* = 3.34) and females (*M*= 16.10, *SD*= 3.33) in their online photo sharing behaviour on Facebook. Results for hypothesis six showed that there was a significant difference between males and females in their online photo sharing behaviours and privacy concerns on Facebook, t (420) = -2.31, *p*= 0.21. Males (*M* = 60.98, *SD* = 12.78). The results are summarised below.

Table 1. Summary of results from Hypothesis testing

Hypothesis	Results
H1: There is a significant relationship between intrinsic motivation and young Malaysian adults' online photo sharing behaviour on Facebook.	Supported
H2: There is a significant relationship between extrinsic motivation and young Malaysian adults' online photo sharing behaviour on Facebook.	Supported
H3: There is a significant difference in the relationship strength between intrinsic and extrinsic motivation in young Malaysian adults' online photo sharing on Facebook.	Supported
H4: There is a significant relationship between privacy concern and young Malaysian adults' online photo sharing behaviour on Facebook.	Not Supported
H5: There is a significant gender difference in young Malaysian adults' online photo sharing behaviour on Facebook.	Not Supported
H6: There is a significant difference between males and females for privacy concern and young Malaysian adults' online photo sharing behaviour on Facebook.	Supported

7 Discussion

Results supported hypothesis one and the findings are consistent with previous study done on knowledge sharing behaviours as well as online photo sharing behaviours where enjoyment and commitment were found to be positive correlated to knowledge sharing in online communities [15-16]. This positive correlation with intrinsic motivations and online photo sharing behaviour suggests that young Malaysian adults are highly motivated intrinsically to share their photos and enjoy doing this; they also feel committed and obligated to share photos to keep their online community updated.

Hypothesis two on extrinsic motivation and online photo sharing behaviour was consistent with the findings from previous research. Findings in this research were similar to previous research that investigated artefact sharing in online communities such as Flickr and YouTube. Research findings also linked that the trade-off between quality and quantity in online photo sharing in which users who were motivated by self-development and reputation were more concerned about the quality of their photos [9]. Earlier research results showed that sharing YouTube videos were strongly correlated with attention received by others which could be translated to the intrinsic factors namely self-development, reciprocity and reputation [23].

For hypothesis three, intrinsic motivations were stronger predictors due to the nature of the online social network of Facebook. Different social networks cater to different types of users, as well as different types of users could be present in the online network itself and hence results may vary from different social networks [10]. The type of online social network that is social in nature and have known social connections on a more personal level, as is the case of Facebook, users in this network indulge in sharing photos for intrinsic rewards.

Hypothesis four was rejected as there was no significant relationship between the online privacy concerns and online photo sharing behaviours on Facebook. Higher privacy concerns did not lead to an increase of online photo sharing behaviours on Facebook, and neither did it deter them from sharing photos. The results are in line with prior research that has noted the inconsistency between users' privacy concern and their behaviour [17,24,25]. The lack of privacy concern of users was reasoned that the perceived benefits of using the online social networking site and how deeply Facebook has integrated into users' daily routine as an indispensable tool to maintain one's social capital and to maintain communication through technology outweighs the risks of sharing personal content, and users often regarded risk and negative consequences as a third person phenomenon [25].

Hypothesis five was rejected as there was no significant gender difference in regards to their online photo sharing behaviour on Facebook. Prior research noted that females posted more photos than males [7,21,22,26]. However, in another study it was reasoned that the similarity across genders in online behaviours may be because that social networking sites may be a place where the creation of mini-cultures, in which separate set of social norms in an online environment are formed with the effect of neutralizing any gender differences [8].

Hypothesis six is accepted as there is a significant difference between male and female in regards to their online privacy concerns on online photo sharing behaviour. In which, research results showed that males are more concerned in regards to their online privacy on Facebook compared to females. Prior research reported that generally there is an inconsistency between users' privacy concerns and their actions [17,20,24,25].

8 Limitations and Recommendations

This study used convenient sampling and therefore the results cannot be generalized. It is premature to generalize the findings of this study to other population groups unless the results can be replicated. For future research, the use of qualitative approach should be considered to further enhance the findings of this research. A qualitative research would be able to delve deeper into individual perspectives and motivations to indulge in photo sharing.

9 Implications of study

Researches and practitioners can use the data to better understand users' behaviours and how it affects users' personal life, their community, as well as provide further insights on how to improve human interaction, communication and the social bond that society strives on. Designers of such online social networking sites could use this data to explore ways to make the users photo sharing experience more enjoyable and feel more strongly attached to the community to encourage more online sharing behaviours.

1. ComScore,*Facebook passes Yahoo! Now the third most trafficked website worldwide*Retrieved from http://www.ditii.com(2011)
2. S.E.Mustafa, A.Hamzah,Int. J. of Soc. Science and Humanity, **1**, 2 (2011)
3. Statista,Global social networks ranked by number of users. Retrieved from http://www.statista.com(2015)
4. Statista,*Active social media penetration in Asian countries.* Retrieved from http://www.statista.com 2015)
5. TNS, Largest ever digital research project reveals major changes in online behaviour around the world. Retrieved from http://www.tnsglobal.com(2010)
6. Socialbakers, Heart of Facebook statistics: Facebook statistics by country. Retrieved from http://www.socialbakers.com(2013)
7. A.N Joinson,*'Looking at,' 'looking up' or 'keeping up with' people? Motives and uses of Facebook. Proceeding of twenty-sixth annual SIGCHI conference on Human factors in computing system*, Italy (2008)
8. N.J.Hum, P.E. Chamberlin, B.L. Hambright, A.C. Portwoord, A.C. Schat, J.L. Bevan, *Comp. in Human Behav.***27**, 5, (2011)
9. O. Nov, C Ye, *Why do people share photos online? Antecedents of photos' quality and quantity. Proceedings of the Americas Conference on Information Systems (AMCIS 2009),* San Francisco, (2009)
10. O.Nov, M, Naaman,C. Ye, C. *Proceedings of AAAI International Conference on Weblogs and Social Media (ICWSM, 2009),* San Jose, CA. (2009)
11. Y.Kim, D. Sohn, D, S.M.Choi,*Comp. in Human Behav.* **27**, 1 (2011)
12. E.L. Deci, R.M. *Ryan,Intrinsic motivation and self-determination in human behavior.* (Plenum, New York, 1985)
13. P.M. Blau, *Exchange and Power in Social Life.* (Wiley, New York, 1964)
14. J. Nahapiet, S. Ghoshal, S. *J Acad. Mgmt.***23**, 2 (1998)
15. K.Lakhani, R.Wolf, *Why hackers do what they do.* In J.Feller, B.Fitzgerald, S. Hissam, K. Lakhani, (Eds.), *Perspectives in free and open source software.* 3-22. (MIT Press, Cambridge, MA, 2005)
16. O. Nov, *Comm. ACM,* **50**, 11 (2007)
17. A. Acquisti, R. Gross, *Proceedings of 6th Workshop on Privacy Enhancing Technologies,*36–58, (Robinson College, Cambridge, UK, 2006)
18. S. Oreg, O. Nov,*Comp. in Human Behav.,* **24**, 5 (2008)

19. M Wasko, M. S. Faraj, *MIS Quarterly,* **29**, 1 (2005)
20. J.Fogel, E. Nehmad,*Comp. in Human Behav.* **25**, 1 (2009)
21. A.M. Kimbrough, R.E. Guadagno, N.L. Muscanell, J.Dill, *Comp. in Human Behav.* **29**, 3 (2013)
22. A.Mendelson, Z.Papacharissi,*Look at us: Collective narcissism in college student Facebook photo galleries.* In Z. Papacharissi, (Ed.), *The networked self: Identity, community and culture on social network sites,* 151-173, (Routledge, New York, 2010)
23. B.A. Huberman, D.M. Romero, F. Wu,J. *Info. Science,* **35**, 6 (2008)
24. S.Ahern, D. Eckles, N. Good, S,King, M. Naaman, M., R. Nair, *Proceedings of the SIGCHI conference on Human factors in computing systems*, **7** (ACM, USA, Press 2007)
25. B.Debatin, J.P.Lovejoy, A. Horn, B.N. Hughes, *J. Comp. Mediated. Comm.***15**, 1 (2009)
26. A. Lenhart, M. Madden, *Teens, privacy, & online social networks.* (Pew Internet and American Life Project Report, 2007)

SHS Web of Conferences **33**, 00060 (2017) DOI: 10.1051/shsconf/20173300060
i-COME'16

Tun Dr. Mahathir's Leadership Communication: The Confucian Perspective

Joyce Lynn-Sze Cheah[1,*], *Norhafezah* Yusof[1] and *Mohd Khairie* Ahmad[1]

[1]Communication Department, School of Multimedia Technology and Communication, College of Arts and Sciences, Universiti Utara Malaysia, 06010, Sintok, Kedah.

Abstract. Previous studies have provided valuable insights into the impact of culture on the concept of leadership communication. However, only a few studies that focused on exploring and understanding the values of Confucianism and its impact on the Chinese culture, especially in Malaysia. Thus, this study is among the first study to examine the leadership communication from the perspective of Confucianism towards the leadership of Dr. Mahathir from the perspective of the Chinese community in Malaysia. This study aims to assess the role of Confucian values in Dr. Mahathir's leadership communication. This study interviewed 15 Chinese leaders. The findings reveal that three main Confucian values are relevant to Dr. Mahathir's leadership communication. The Confucian values are *Zhi* (wisdom), *Xin* (trust) and *Xiao* (filial piety). This study provides a new perspective on leadership communication from the context of cultural diversity in Malaysia which emphasizes on Confucian values.

1 Introduction

Leadership is a set of behaviours that are intended to play a role in achieving the goals of the organization. Studies have shown that culture differences directly affect the effectiveness of leadership and influence the satisfaction and motivation of the followers [9]. Culture signifies the basic values and belief systems of individual that are expected to impact leadership processes.

Among the challenges faced by the leaders in Malaysia are the ways to maintain the development in light of rapid changes in the country. The priority is, therefore, to be able to sustain the impressive achievements made thus far in the rather challenging context of a mullti-racial and multi-religious society, especially the need to understand the cultural thinking of the Chinese community after the General Election (GE13). Therefore, it is important to understand the Chinese culture and values so to manage effectively.

Dr. Mahathir is the fourth Prime Minister of Malaysia, is said to be a charismatic leader [3]. Dr. Mahathir practiced a strong and decisive, both in his leadership and communication style. Malaysia is composed of a majority of Malays, but its business drive comes from the Chinese as the minority group. To keep the economy growing, Dr. Mahathir must allow the Chinese minority to prosper, thus he need to provide some rhetorical favors to the Chinese as well.

* Corresponding author: joycecheah@uum.edu.my

Thus, this paper seek to identify the influences of Confucian values in Dr. Mahathir's leadership communication. Through interview with fifteen Chinese leaders, we first explain the concept of leadership communication, then we introduce the core values of traditional Confucian, focusing on content relevant to leadership. We further aim to identify the influence of these values in the leadership communication practices of Dr. Mahathir. Finally, we discuss how these values could be a rich source of understanding for the future leadership research in Malaysia.

2 Leadership Communication in intercultural context

Leadership research has begun in the study of sociology and political science. Scholars said that leadership is basically rooted in the language and communication. [4, 5]. The importance of communication to leadership has been stressed by a number of scholars over the years [2, 7]. Research shows that communication is a powerful catalyst for establishing and sustaining trust. Furthermore, loyalty can be achieved through high level of leader communication.

Leadership practices that are successful in one culture are likely be unsuccessful in another if cultural differences are not considered [6, 8]. Similarly, it can be seen between the Malays and Chinese culture in Malaysia. To understand the ethnic relations between Malays and Chinese, we should identify the different cultural patterns and values that exist among the various ethnic groups in Malaysia.

The Malays believe strongly in religious value. Therefore, they expect their leaders to act as role models who are spiritually and religiously in tune. In communication, the Malays practiced tact and indirectness. The Malays also uphold the value of self-respect, politeness, sensitivity to feelings and value relationships [1].

Meanwhile, the Chinese are motivated by financial rewards [1]. They value hard work, wealth or prosperity, face, harmony and risk taking. It was also observed that the Chinese relationship between leader and follower in work and education was based on the Confucian ideal of filial loyalty.

Despite the cultural difference between the Chinese and Malays, they also possessed the same values on leadership and management principles. The cultural values that were common among the Malays and Chinese were filial piety, honesty, knowledgeable and trustworthiness [13]. Not surprisingly, the Malays shared the value of filial piety with the Chinese where emphasis was put on minding and respecting the parents.

This clearly shows that many challenges and difficulties arising from intercultural communication [9]. This paper suggested that a study on leadership communication from a Chinese perspective should be explored. Therefore, this study was undertaken to explore and understand the values of Confucianism and its impact on Chinese culture in Malaysia.

2.1 The influences of Confucian values

In the last two decades, Western theories have been increasingly focusing on the humanistic aspects of successful leadership that considered similar to Chinese values [6, 8, 10]. At the same time, Chinese leadership has adopted many Western leadership attributes. Many scholars recognize that some theories may not be applied by the current intercultural country [9, 14]. In other words, Malaysian leaders may be unable to effectively manage the country if they are unable to understand the practice of other cultures in Malaysia. Therefore, it is necessary for leaders to understand the culture of the Chinese community by considering the impact of Confucian values, to be effective in the leadership.

To maintain harmony through the relationships, Confucianism promotes five virtues: *ren* (benevolence), *yi* (righteousness), *li* (propriety) *zhi* (wisdom) and *xin* (trustworthiness) [12]. In Confucian cultures, leaders are expected to display *ren*, meaning benevolence or

DOI: 10.1051/shsconf/20173300060

humanism. *Ren* is sometimes translated as "goodwill" or "goodness" towards others. The Confucian leader is expected to be a good-natured leader and to manage with kindness [11,12].

Yi, or righteousness means that the leader is expected to uphold the highest standards of moral conduct [11]. Individual self-interest is to be sacrificed for the good of the organization. Meanwhile, Confucius was very concerned with relationships. Appropriate behavior or *li,* is dictated through Confucian thought in terms of one's relationship with superiors, parents, husband/wife, elders and friends. In a typical Chinese organization, decisions are made by the leaders and followers is expected to carry out the task without question [12]. Employees are expected to be loyal and devoted to their organization and in return, the organization is expected to take care of them.

The value of wisdom or *Zhi* has always been held in high esteem. Wisdom and age are closely associated in Chinese culture, and it is not surprising to find great deference paid to older members of society [12]. Finally, Confucian leaders are expected to possess *xin*, or trustworthiness. The Chinese leader is responsible for maintaining control and insuring that all subordinates follow policies consistent with the mission of the organization [12].

Malaysia is a multi-ethnic and multi-cultural society. The Chinese ethnic represent approximately 24% of the population. Bahasa Malaysia and English are the official languages, but there are many areas where Chinese is spoken. Malaysian Chinese put great emphasis on the education, sticking to the traditions of Chinese culture and teaching themselves Chinese literature. Chinese philosophies, especially Confucianism, greatly impact the Chinese Malaysian and have done so for many years [1]. Clearly, traditional philosophies are still part of the cultural fabric in Malaysia today.

2.2 Dr. Mahathir Mohamad Premiership

Under Dr. Mahathir's long Prime Ministership (1981-2003), Malaysia's political culture became increasingly centralized and authoritarian, due to his belief that the multi-ethnic Malaysia could only remain stable through controlled democracy. Dr. Mahathir is a visionary leader and at the same time he is pragmatic, dynamic and intelligent. As a leader, Dr. Mahathir does not only know what has to be done, but also he knows how to get it done [3].

Besides, Dr. Mahathir is one of those few leaders who has the ability to capture our mind when he speaks. His outstanding rhetorical abilities have placed him at the forefront of eloquent leaders in the eyes of his audiences. He speaks calmly, courageously and with magnanimity in words and thoughts [3]. A leading Malaysian politician like Dr. Mahathir with his charismatic character has created the impression that his success is not isolated from his persuasive abilities and has raised the curiosity to investigate his leadership communication particularly in this age which witnesses an interest among researchers to understand the application of traditional values in developing his leadership communication.

In general, charismatic leadership is more appropriate to describe Dr. Mahathir's leadership style in his communication. His speeches that were filled with powerful rhetorical devices had reinforced his credibility as a speaker. His discourse incorporated persuasive and compelling vision to support his main goal, to help unify the people. Apart from being characterized as a charismatic leader, his communication is also helpful in understanding Dr. Mahathir's leadership style. As a leader, power is of the utmost importance to ensure the reform efforts in the administration. It is well-known fact that communication was one of his personal skills as the Prime Minister, especially his skills necessary for interacting with other leaders and societies that would form an important power and his leadership style.

3 METHODOLOGY

This research is based on in-depth interviews with a total of fifteen Chinese leaders in Malaysia. These interviews were done during the period from July 2013 to January 2014.

The interviews were designed to showcase the leadership thoughts and their perception towards Dr. Mahathir's leadership communication. Beyond their leadership status, the leaders also recognized as role models for other leaders for at least two years. In order to collect a holistic perspective on leadership communication, informants were selected to represent three groups of leaders, the political leaders (n=7), leaders of business organizations (n=5) and reporters (n=3). The three group of leaders have been chosen because these groups are believed to give the maximum variation that contributes to the understanding of the concept of leadership communication.

Each leader answered a standard set of semi-structured questions, for a total of eighteen questions. The interviewed were audio-taped (with permission), transcribed, the text prepared, edited and verified by the interviewees before they were analyzed. To analyze the data, the researchers used NVivo 10 software program.

4 FINDINGS

The results showed that the Chinese leader highlighted the development of Confucianism as a philosophy of morality and personal ethics in relationships between leaders and followers. Specifically, in this study is their perception towards Dr. Mahathir's leadership communication. From the aspect of leadership, the basic values of Confucianism will produce good leaders, humane and fair. The interviewed leaders mentioned the top three values that leaders should serve as role models to the followers. The three values frequently mentioned are *Zhi* (wisdom), *Xin* (trust) and *Xiao* (filial piety).

4.1 Zhi 智(wisdom)

Wisdom or *zhi*, is defined as knowing, understanding and intelligence. *Zhi* is one of the main Confucian values to be possessed by a leader. Wisdom also means the ability to think creatively and rationally to solve problem [10]. Such leaders are constantly striving to enhance their knowledge and develop ideas and are always open to new ideas.

Most of the informants (nine informants) agreed that Dr. Mahathir has great wisdom. When talking about Dr. Mahathir, they argued that reading habit has made him a visionary, knowledgeable and capable leader so as to achieve remarkable development and progress in the country.

Informant 15 mentioned that one of the secrets why Tun is a wise leader is because he likes reading. In addition, the knowledge also helps him to make the right decision in different situations.

"A leader who have an extensive knowledge. Tun deserve respect and recognition. Although he was trained as a medical doctor, he was able to express quality ideas in various fields". (Informant 12)

In essence, what is stressed here is the fact that a successful leader must not only continue to learn, but needs to widen his or her visions and horizons. Confucius said a leader must not only have the expertise, talent, ability and diligence, but he also has to care about other people, knowing and willing to help out with [11].

4.2 Xin 信(Trust)

Trust, or *xin* addresses the issue of personal integrity with particular emphasis on faithfulness and trustworthiness. For Confucius, trust is very important in interpersonal communication, particularly in leadership communication (Han, 2013). Leaders must be honest and truthful in their words, consistent with their actions, be responsible and must learn from their mistakes. In this way, they will earn the trust of the people. Confucian ideology instructs that leaders need to gain trust of the society in order to mobilize and communicate with them [11].

Four informants have explained that trust links with Dr. Mahathir's leadership communication. Two of the 15 informants said: "As we can see, the past Prime Minister of Malaysia like Tunku Abdul Rahman and Tun Razak, have won the trust of the community. They make a good decision for the country. Similarly, because people believe in his abilities, Dr. Mahathir has proven to the people by transforming Malaysia from an agricultural to a modern state". (Informant 10)

"The Chinese have always been grateful to Tun. Since independence in 1957, the Chinese people have been sharing the wealth and power with the Malays. We can see a lot of Chinese people have succeeded and some have become millionaires during Tun's administration. For example, Robert Kuok, Tan Sri Quek Leng Chan and Tan Sri Vincent Tan. Thus, Malaysians, especially the Chinese community has confidence in Tun". (Informant 2).

4.3 Xiao 孝(Filial Piety)

Generally, the concept of *Xiao* (filial piety) is a key principle in Confucian ideology. The original version of this concept was limited to immediate family, but it evolved to include the relationships outside the family, such as leader-follower [11].

According to Confucian, obedience and respect for parents has developed a way of life that is acceptable [11]. For Informant 3, *Xiao* is a common practice and is a very important in Malaysia. "The most important thing a child should learn is obedience to his parents. This practice starts from home and extends to the society".

5. DISCUSSION AND CONCLUSION

This study found that the Malaysian Chinese emphasizes on Confucian values. The findings demonstrate that leaders with Confucian values are evaluated highly by their followers. It seems that people in the Chinese context would agree with the potential contribution of these values to leadership effectiveness. This study also provides insights into understanding leadership communication and the Chinese culture. Finally, the current study also has practical implications for issues of leadership in Malaysia. The values identified in this study can be applied to other political leaders in a multi-ethnic context to assess leadership behaviors.

As a conclusion, Malaysian leaders need to learn and apply the concepts of moral leadership to improve leadership practices in general. In cross-cultural situations, Malaysian leaders need to understand the importance of Chinese values. Dr. Mahathir's leadership communication behavior accords well against the modern-day Confucian leadership.

References

1. A. Asma. *Malaysian Management Review* **27**(1), 3–18 (1992)
2. H. A. Bakar, G. Jian, & G. Fairhurst. *Asian Business & Management*, **13**(2), 143-170 (2012)
3. A. Chong, & K. S. Balakrishnan *The Pacific Review*. DOI: 10.1080/09512748.2015.1013496. (2015)
4. J. A Conger. *Academy of Management Executive,* **5**(1), 31–45 (1991)
5. K. Czech, & G. L. *Communication Quarterly,* **58**(4), 431-457 (2010)
6. P. Dorfman, M. Lavidan, P. Hanges, A. Dastmalchian, & R. House. *Journal of World Business*, *47*, 504-518 (2012)
7. G. T. Fairhurst, & S. L. Connaughton. Leadership, 10(1), 7-35 (2014)
8. J. S. Gao, J. K. Arnulf, & H. Kristofferson. *Journal of Management*, **27**, 55-65 (2011)
9. D. I. Jung, & B. J. Avolio. *Academy of Management Journal*, **42**(2), 208-218 (1999)
10. P. King, & W. Zhang. *Journal of Management Research,* **6**(2), 1-21 (2014)
11. L. Ma, & A. S. Tsui. *The Leadership Quarterly*, **26**, 13-24 (2015)
12. G. M. Chen, & J. Paper presented at the Annual Meeting of the Speech Communication Association (79th, Miami Beach, FL, November 18-21, 1993)
13. D. Zawawi. *Int. Journal of Economics and Management,* **2**(2), 409-426 (2008)
14. J. L. S. Cheah, N. Yusof, & M. K. Ahmad. *Malaysian Journal of Communication*, **30** (Special Issue), 129-144 (2014)

DOI: 10.1051/shsconf/20173300061

Health Related Campaigns in Social Media and Its Practical Aspects for Youths in Malaysia

Suffian Hadi Ayub[1*], *Yesuselvi* Manickam[1], *Mohammed Rezal* Hamzah[2], *Julinawati* Suanda[3] and *Husna Afifi* Mohd Yusoff[2]

[1]DCLA, School of Arts, Sunway University, 46150 Petaling Jaya, Selangor, Malaysia
[2]IKOM, Universiti Malaysia Perlis, 01000 Kangar, Perlis, Malaysia
[3] PPIPT, Universiti Malaysia Perlis, 01000 Kangar, Perlis, Malaysia

Abstract. The paper discusses the importance of social media in today's society and review how health related campaign could penetrate the youths in Malaysia. In the internet age, the youths are divided into two; the digital natives (born after 1980) and digital immigrants (born before 1980). Further to that, the paper provides an insight on how past efforts by relevant stakeholders were utilised in creating awareness to Malaysian youths through the social media. Upon identifying the efforts through extensive literature review, the usage of social media in propagating behavioural changes in youths' were also discussed. Several meaningful impacts were discovered and must be carefully considered in terms of its practical implications to suit Malaysian youths.

1 Introduction

In the technological era, the Internet has been touted as the most effective way to reach the people. Its reachability and coverage surpasses other media; either traditional or electronic. Any service sector has now utilised the Internet to send messages and promote its product to users. The government and non-government alike too are moving towards strengthening its presence in the Internet. The advent of social media introduce flexibility to its users in managing information. The social media too have opened up more room for the industry to flex its muscle in engaging the publics and this include health industry. Although the government and private sector play an important role in communicating with the publics on health issues, the role that Non-government organisations (NGOs) carry too are equally important. It is the NGOs that are more close to the stakeholders and these groups too are responsible to strengthen whatever effort that the government is working on.

The health related campaign via the social media often targeting youths in particular, because this group spend more time online. This further illustrates the importance of the Internet to the authorities in reaching out to its intended/targeted public. Undoubtedly, NGOs have always been the backbone to both the government and the people. Its very existence elucidates the importance of the cause that it champion. In the health industry,

*Corresponding author: suffianay@sunway.edu.my

NGOs play a pivotal role in educating the public about current situation and early prevention. These activities, in return assist the government in reducing the treatment costs and other associated financial issues which has alarmingly becoming higher than before. In today's society where youths are the important investment for the country, lack of education and awareness on diseases and its prevention has put them at risk and even more so making it highly challenging to combat due to the rising cost of medication and treatment.

2 Methodology

The paper employed content analysis to probe further into the usage of social media among organisations and public figures pertaining to health related campaign aiming at youths in Malaysia. The primary goal of the content analysis was to investigate the role of social media usage in the internet and the types of information provided by the organisations and public figures with regards to health related campaign and its outreach to youths specifically and public in general.

The fact that youths largely contributed to the development of the nation signal the call for the government to play greater emphasis on youths' health issues and government alone could not achieve this. Subsequently, NGOs are expected to champion the health related campaign to create awareness among youths. The measurement of health related campaign and its effectiveness will ensure a comprehensive and sustained effort by the NGOs. This is vital as it deduce a meaningful impact in educating youths which later will create a holistic health conscious society which is detrimental to nation's building.

Youth constitutes half of the Malaysian population [4]. Statistics from Department of Statistical Malaysia shows that in 2008, from 27.7 million people in Malaysia, 11.26 million people (40.6%) is from youth group. Meanwhile, it is expected that in 2012, the number of youth in Malaysia will be increased to 11.65 million (40.2%). Youth represent a significant group in a society. Youth characteristics within society are able to either give hope or despair to the country. In the current era, most of the issues and social problems happened in Malaysia involved the youth [5]. As such, the notion to create a health conscious society in Malaysia must include youths. In producing a public health campaign aims at youths, the NGOs must be able to categorically define what appeal the most to this demographic group.

Generation Y actively contributes, shares, searches for and consumes content – plus works and plays – on social media platforms [8]. They are the first generation to have spent their entire lives in the digital environment; information technology profoundly affects how they live and work [2, 10]. Further to this, marketing categorised health campaign as one of social marketing where marketing plays a vital role in market penetration strategies. Marketing in its wider definition is suitable to be applied in assessing the social media usage through the differences of market segment such as the immigrants and natives digital users [10]. This provides a cautionary reminder that the insertion of public health campaign must be carefully constructed to ensure maximum exposure and behavioural changes. As a matter of fact, social norms and behaviour may be changing due to Generation Y's use of social media – affecting civic engagement, attitude towards privacy, nutrition, health care practices and public safety in the general population [8].

DOI: 10.1051/shsconf/20173300061

3 Findings

The results of the content analysis on the usage of social media among organisations and individuals (governments, NGOs, public figures) have indicated that the trend is increasingly positive with a steady growth in the numbers of followers / subscribers / likes. The content analysis was conducted on various organisations and individuals across the spectrum of social media platform such as Facebook, Instagram, Twitter and Youtube. Result of the content analysis are provided in Table I.

Table 1. The usage of social media among organisations in Malaysia

Institute / Organisation	No. of like / subscribers	Urls / link	Focus	Status	Type of social media
Ministry of Health (MoH)	628,194	https://www.facebook.com/kementerankesihatanmalaysia/?fref=nf	Health campaign, diseases, origin of sickness, treatment, symptoms	Government	Facebook
MoH	37.7k	Various health related information	Health campaign, diseases, origin of sickness, treatment, symptoms	Government	Instagram
Mercy	56,201	https://www.facebook.com/MERCYmalaysia/?fref=ts	Health campaign	NGO	Facebook
Makna	39,324	https://www.facebook.com/maknacancer/?fref=ts	Health campaign	NGO	Facebook
Dr. Sahit Arshat	236,311	https://www.facebook.com/hamidarshat45/?fref=ts	Diseases, origin of sickness, treatment, symptoms, health campaign	Individual / physician	Facebook
Prof. Dr. Muhaya	103,9812	https://www.facebook.com/profdrmuhaya/?fref=ts	Diseases, origin of sickness, treatment, symptoms, health campaign	Individual / physician	Facebook
Mmgazette	4,639	@MMGazette	Diseases, origin of sickness, treatment, symptoms, health campaign	NGO	Twitter
MedTweetMy	6,556	@MedTweetMYHQ	Health campaign	NGO	Twitter
Myhealthkkm	360 videos	https://www.youtube.com/MyHEALTHKKM	Diseases, origin of sickness, treatment, symptoms	Government	Youtube
Mediktv	638 videos	https://www.youtube.com/user/MedikTV	Diseases, origin of sickness, treatment, symptoms	Ngo	Youtube
Dr Zubaidi Ahmad	218,774	https://www.facebook.com/DrZubaidiHjAhmad	Diseases, origin of sickness, treatment, symptoms	Individual / physician	Facebook
Kevin Zahri	466,129	https://www.facebook.com/zahri.kevin/?fref=ts	Health management	Individual / health lifestyle	Facebook

				promoter	
Persatuan Autism Malaysia	22,337	https://www.facebook.com/groups/autisme.malaysia/?fref=ts	Autism	Support group	Facebook

In Malaysia, various organisations and government agencies has capitalised in propagating health related campaign and research indicated that social media bandwagon such as Facebook, Youtube, Instagram and Twitter are among the top choices as it resonates the public preferences when seeking information. MoH is the official government body to provide latest information of health to the public while for NGOs; Mercy and Makna have been the most popular site that the public referred to for information related to health. Apart from that, individuals too have gained momentum in assisting the government to spread awareness on health issue. Public figures such as Prof. Dr. Muhaya, Dr. Sahit Arshat and Kevin Zahri have been able to attract more than 100,000 followers and this show how the information on health flow from on source to many that it creates a ripple effect due to its speed, reachability and minimal cost to attain information. (Truthfully) an inexpensive media campaign using social media reaches a relatively large volume of potential consumers, at any time of day, and has a higher ROI than with traditional media. So it created higher value added to organisations [12].

The overarching focus on public health campaigns has been to utilize theoretical frameworks to plan, design, implement, and evaluate communication-based programmes that are aimed at initiating and sustaining changes in knowledge, beliefs, attitudes, and behaviours of target audience members with respect to positive health behaviours [8, 15]. Even more so in the technological advancement where the NGOs are expected to be IT savvy and possess the skills to engage online. Campaigns must be aimed at a particular, target audience for maximal impact [8, 11]. The social media has been recognised as the most suitable outlet for the NGOs to reach out to youths in Malaysia and evidently this demographic group populates the cyberspace compare to others.

4 Discussions

Research on social media has grown significantly in the last decades and contributed to understanding the nature of social media and how it works. It has also paved the way for behavioural studies of its users. The usage of social media and the Internet to promote ideas and products too are becoming relevant with the advent of technologies. As technologies become cheaper and readily available, the accessibility of information through social media too thrives exponentially. As it becomes more personal to individuals, social media provide social supports to those in need. People of all demographics are adopting these technologies whether on their computers or through mobile devices, and they are increasingly using these social media for health-related issues [7]. More often than not, it serves as the silent partner to those in need. It is in this nature that the NGOs partake in communicating to the youths. The role that the NGOs play in health related campaigns assist the government to better educate the people and social media is one of the platform suitable when approaching youths.

Social network sites (SNS) are permanently changing the ways in which we interact and make sense of ourselves. This is where social media is seen as an effective tool to reach the intended public; the youths. As youth are some of the most avid users of social media, the development and availability of tailored content for this age group provides an opportunity to extend health promotion efforts [2]. Current technological development in Malaysia allows its citizen to enjoy hassle free internet connectivity and this is where the social media flourish especially among youths.

Given this goal and the conditions of the health market, there is a need to create unique positioning for the campaigns in order to generate public awareness [1]. Service organisations, managers, researchers and public policy makers are interested in Generation Y's use of social media because it affects people's behaviour in many domain – with positives and negative outcomes for customers, firms and their employees, and society [11]. Taking into account the influences of social media towards youths are different in each country, further study into how Malaysian youths use social media could help to determine the most suitable health campaigns that would attract in tandem with their nature, behaviour and upbringing [12]. Evidently, this clearly signals that a collective effort is much needed within and among the government, private sector and the NGOs. Whilst the government could provide up to date information and related legislations, the private sector could contribute financially and the NGOs through its widespread influence could solidify these messages through social media bandwagon.

There is a visible trend in public health literature, which emphasizes the need to effectively market public health by exploring alternative, innovative strategies [1]. The importance to carefully organize and execute a public health campaign is paramount as it involves cost and causal effects [14]. The governing bodies such as government, private sectors and NGOs are responsible to ensure the message transmit through the campaign reach and influence its target audience. In addition, public health campaign involves a systematic communicative process that aims to persuade a relatively large collection of individuals to adopt certain prescribed health messages and/or behaviours. Similarly, the NGOs championing the health related campaign must ensure that the publics are able to differentiate it so that the campaign will allow them to initiate behavioural changes in their lifestyle while at the same time initiates understanding the importance of health.

References

1. A. Basu & J. Wang "The Role of Branding in Public Health Campaigns". *Journal of Communication Management*, Vol. 13 Issue 1, 1995, ISSN 1363-254X
2. B. L. Joan, et al. "A Social Media Approach to Inform Youth About Breast Cancer and Smoking: An Exploratory Descriptive Study". *Collegian* **21.2**. pp. 159-168. (2014)
3. S. Bennett, K. Maton, & L. Kervin, L. *British Journal of Educational Technology*, Vol. 39 No. 5, pp. 775-786. (2008)
4. C. Luth, G. Ranzini, & M. Meckel "Stress 2.0: social media overload among Swiss teenagers". Communication and Information Technologies Annual Doing and Being Digital: *Mediated Childhood Studies in Media and Communications*, **8** pp. 3-24. (2014)
5. D'Silva, J.L., H.A.M. Shaffril, J. Uli & B.A. Samah. *Am. J. Applied, Sci.*, 7. (2010)
6. D. Dzuhailmi, A. Haslinda, H. Azimi, A.I Ismi, T. Ezha, A. Abdul-Lateef, S.G. Sarjit, A. Nobaya & S. Turiman. *American Journal of Applied Sciences* 9 (7): 974-978, 2012, ISSN 1546-9239
7. J. Palfrey & U. Gasser. *New York, NY: Basic Books*. (2008)
8. K. Holly & I. Zena "Harnessing Social Media For Health Promotion And Behavior Change". *Health Promotion Practice,* **14.1**. pp. 15-23. (2013)
9. K.N. Kline & M. Mattson. *Health Communication*, **12**, pp. 1-21. (2000)
10. L.B. Snyder "Health communication campaigns and their impact on behaviour" *Journal of Nutrition Education and Behaviour,* **39**, pp. S32-40. (2007)
11. M.S. Wesner, and T. Miller. Organizational Development, **26** No. 3, pp. 89-96. (2008)

12. N.B. Ruth, A. Parasuraman, A. Hoefnagels, N. Migchels, S. Kadabayi, T. Gruber, Y.K. Loureiro, & D. Sornet. *Journal of Service Management*, **24** No. 3, pp. 245-267. (2013)
13. P. Farwina, J. N. Ismawati & S. Ainin. *Journal of Telematics & Informatics, Elsevier,* **32**, Iss 1, pp. 67-78. (2014)
14. S. Julinawati, D. Cawley, M. Brenner, & N.J. Rowan "Exploring health marketing strategies by the Ministry of Health Malaysia in preventing cervical cancer: A preliminary review". *The 2nd International Malaysia-Ireland Joint Symposium on* Engineering, Science and Business (2012)
15. S.M. Noar "A 10 year retrospective of research in health mass media campaigns: where do we go from here?". *Journal of Health Communication*, **11**, pp. 21-42. (2006)
16. W.J. Paisley. In Rice, R.E. and Atkins, C.K. (Eds), Public Communication Campaigns, 3rd ed., *Sage, Thousand Oaks:* CA, pp. 3-21. (2001)

DOI: 10.1051/shsconf/20173300062

Metaphor of 'Adaptation': A Metaphoric Rhetorical criticism on Aminuddin Baki's Torch Movement Speech 3

Dil Froz Jan Sayed Halem Shah[1*], *Abdul* Mua'ti @ *Zamri bin* Ahmad, *PhD*[2], *Hamisah Zaharah* Hassan, *PhD*[2], *Rosya Izyanie* Shamshudeen, *PhD*[2] and *Faridah* Ibrahim, *PhD*[2]

[1] Faculty of Arts, Communication and Education, Infrastructure University Kuala Lumpur,
[2]Faculty of Modern languages and Communication, University Putra Malaysia

Abstract. This paper looks at the Aminuddin Baki's Torch Movement Speech 3 (*Ceramah Gerakan Obor 3).* It's is to identify the type of metaphors used and how the metaphors are conveyed in the Torch Movement Speech 3 on 'Adaptation' of the Malays. The study hopes to identify the direction of the 'Adaptation' metaphor which is liken banana tree and its offsprings. Many studies have been carried out on the biographical portion of Aminuddin Baki but none on all his Torch Movement Speeches that he had made on the progression and development of the Malays. A qualitative method was used through a rhetorical analysis. A Neo-Aristotelian Criticism is used in the analysis of speech of which this study demonstrates the relevance in the application of ancient Malay metaphors in the modern day communication interaction among the Malays.

1 Introduction

Aminuddin Baki was a rhetor and committed to seeing the advancement of Malaysians by communicating educational advancement. He made considerable contributions to the revolution of education in Malaysia by giving speeches on how the nation can pace proactively into the future without falling behind. He was aspired to see his country prospers whereby the natives would be given the respect and equal rights that they deserve.

1.1 Research Context

Aminuddin Baki was a revolutionary rhetor who tried to impart change into the phlegmatic Malay culture. He disagreed with most of the standpoint maintained by the Malay heritage [5] with regard to progress. The rhetor was capable of being disagreeable to certain deeply rooted cultural issues without inciting rebellion among the audience [3]. A rhetorical methodology was carefully designed by the rhetor to achieve a particular communicational goal which is yet to be analyzed within the paradigms of contemporary theories. Metaphors ensure a familiar but forceful connection to be made between the listeners and the rhetoric message. It requires a soigne concoction however, since the wrong metaphor would fail to

* Corresponding author: dilfrozjan@iukl.edu.my

invoke the desired response from the audience, and consequently falter in creating a singular impact on the community of interest [22]. It is therefore highly critical for the purpose of this research to solve the aforementioned problem.

1.2 Research Objectives

This research focuses solely on the use of metaphors in attaining the desired response from the audience. This is congruent with the level of education possessed by the audience at that time. This research aims to complete the following objectives:

a. To identify the metaphors employed within a certain context
b To determine the structure of the metaphor within the certain context of the speech

2 Literature Review

Metaphor is a form of figurative speech that enables indirect communication to take place. Metaphor imparts meaning by adopting external concepts. As such, it requires a flexible interpretation of dual ideas. This is examined under the Contemporary Theory of Metaphor [21], which deals with the prospect of employing metaphors in mapping, reasoning and problem solving.

2.1. Rhetoric

If speech is inextricably linked to a purpose, then rhetoric is the art of communicating this purpose [23].

2.2. Metaphor

Development, as the term suggests, would challenge the community to leave their comfort zone into a new horizon. In doing so, it is vital for them to adapt to change. This can be a perilous process. Adaptation [31], requires the community to be flexible to the dynamic nature of the environment.

Kalaulah usaha (penyesuaian) ini gagal, akan berbuahlah pisang dua kali [2]
If this attempt (for adaptation) fails, then the banana would bear fruit twice

2.3. Speech

The speeches given by leaders of the largest organization in America was studied by [5], and three general corporate strategies were introduced to eliminate problems. They were – influencing government policy, education on economic for the public and increasing social responsibility of the business part.

3 Methodology

This is a content analysis research that uses the neo-Aristotelian criticism. Metaphoric criticism is an approach that emphasizes mostly on a very specific strategy implemented within rhetoric. By employing metaphors, the meaning attached to a concept can be

DOI: 10.1051/shsconf/20173300062

dynamically changed [26]. The ability of metaphor to alter the perception of a message makes it very alluring [6, 3] and compel the audience to be more receptive. This study analyses the metaphoric content of Aminuddin Baki's Torch Movement Speech 3 which emphasizes on 'Adaptation.'

4 Analysis and Discussion

Metaphorical meaning is to combine the meaning of both tenor and vehicle to derive the mapping that exists between the two [34]. Both meanings must complement each other for the metaphor to work. Once the meaning is clarified, its impact is probed more closely. Given the unavailability of audience and exigence from 1960s, only the desired impact of the rhetorician is scrutinized. Metaphors detected in Baki's speech are interpreted based on several themes.

4.1. *Kalaulah usaha (penyesuaian ini gagal, akan berbuahlah pisang dua kali* [2]
If this attempt (for adaption) fails, then the banana would bear fruit twice

Since the obstacle varies with time, so must the community. Being static is a dangerous alternative. The force of nature [15] often displays the extinction of a species that does not progress with its environment.

Deciding the course of action in responding to change is a serious issue that needs to be addressed [8]. Without the right supervision, adaptation can disintegrate the community. People can make the wrong decisions and act poorly in the face of calamity.

4.2. Metaphorical Element

Banana is a tropical fruit in Malaysia coming from the banana tree of genus Musa. It is usually yellowish in color. The structure of the fruit is long and curved. Its content is soft and slightly mellow in taste. The fruits of the banana exist in a cluster and hang and the top of the banana tree. Monocarpic [20]. By nature, the banana tree bears fruit only once in its entire lifetime. Before passing, the banana tree would naturally propagate new offspring at the bottom of the tree. The life cycle of a banana tree is conceived simplistically as a five state process. It begins with the propagation of seedling, which will develop into an offspring. As time passes, the offspring grows into a banana tree. The tree experiences fruiting where it produces the banana. Ineluctably, the final state is the death or demise of the tree. A notable trait of the banana tree is its fruiting repertoire.

The banana tree metaphor ideates the construct of mistake in a rather queer manner. Strange in the sense that it paints the fruiting process of a tree in a deplorable limelight, when it should have been a positive one [35].

Now, it is compelling to have a closer look at mistake [13] before bridging it to the metaphor in question. To note, mistake is defined as the course of action which causes a negative effect as compared to the intended goal.

The banana tree metaphor ideates the construct of mistake in a rather queer manner. Strange in the sense that it paints the fruiting process of a tree in a deplorable limelight, when it should have been a positive one [35].

In this respect, the construct of mistake [29] can be decomposed into four main components – original intention, actual occurrence, destructive action and unforeseen penalty. The original intention is the true motivation behind the act of the individual. It is usually tied to a course of action.

A possible explanation can be found by the inability to make the right estimation in adapting [7]. For the sake of argument, assume that a general guideline has been established with regard to the acceptable amount of parasite to be sprayed on the crops. Without loss of generality, assume further that for one acre of land, two liters of pesticide is recommended.

Unfortunately, the logic behind the safety of pesticide may not be mathematically based [27]. There might be other factors, such as accumulation of pesticide or environmental ramification. Here, the mistake is caused by the failure to adapt to the change. The capacity to adapt to change is essential in making the right decision and avoiding mistakes.

4.3. Metaphorical Meaning

The metaphor connects the banana tree to the action performed by an individual. It is a truism that a banana tree grows with time. A similar conception can be said on a series of actions that mount through time. When more actions are deployed on a particular endeavor, the amplitude of strain is greater [16].

Ironically, the metaphor associates the bearing of a fruit as an instigation of mistake. This is quite paradoxical, given the fact that the fruit of a tree that can be eaten, should distinctly be perceived as the emergence of good tiding, instead of an ominous one.

From the error making point of view, it can be inferred from this portrayal that it is forgivable to commit a mistake for the first time [28], as it would serve as an initial orientation to the learning experience.

In retrospect, the metaphor indirectly conveys a subtle lesson that failure is a natural phenomenon in life [1]. No one is free from fault.

4.4. Metaphorical Impact

The desired impact of using the banana tree metaphor is to promote the importance of learning from past mistake such that it will not recur. This can be quite challenging, given the fact that a fierce competition always exists between right and wrong, from the way a person defines his own values [11].

Learning from a past mistake can happen in two ways. The most common is to learn from one's own experience [4] in order to change.

Previously, a detailed analysis is given on the metaphor that represents mistake as the second fruiting of the banana tree. For comparison it is contrasted with another metaphor that insinuates 'mistake as music.' Logically, this too is a quaint metaphor, given that music is said to be therapeutic in the present time [10].

Face the music
Mistake is like music
If this attempt (for adaptation) fails, then the banana would bear fruit twice
MISTAKE IS LIKE THE SECOND FRUITING OF THE BANANA TREE

Understanding the music metaphor requires a slight explanation to its rationale. It is hypothesized that the metaphor had originated from the suspense experienced by a performer before appearing on stage [30]. As the curtain is lifted and music is being played for the sake of introducing the play, the performer must face his fears and the audience.

In term of its goal, the music metaphor encourages the audience to embrace accountability [32] in dealing with the consequence of a mistake.

5 Conclusion

For the Malays to be successful in any vocation, they must learn to adapt to the ever changing needs and challenges of the environment [2]. In doing so, learning from their past mistakes, as well as the mistakes of others, is one of the most critical requirements of success. An individual can only be progressive if he can honestly look at himself in the mirror and ascertain his own weakness without excuse.

Having the right set of beliefs to succeed, or self-efficacy as experts would say, is a cultural trait to a certain extent. Certain communities are highly deterministic. They relate nearly all forms of success to humanism i.e. how we act as whole. The Malay community may not share the same zest for humanism. With the influence of Islam, they are taught to be humble in their attempt for success.

A common saying among them is *"Effort must be made but the final decision comes from God ..."* Sadly, certain extremists propagate a false interpretation of this principle. They promote the belief that every failure is fated and in doing so, discounts the importance of making effort entirely.

This is a portion of doctoral research work undertaken by Dil Froz Jan at Universiti Putra Malaysia. The other three metaphors were also discussed in previous papers and journals. The methodologies used are similar to the previous papers.

References

1. E. Addessi, F. Bellagamba, A. Delfino, F. De Petrillo, V. Focaroli, L. Macchitella, V. Maggiorelli, B. Pace, G. Pecora, S. Rossi, A. Sbaffi, M. I. Tasselli, F. Paglieri. *Cognition*, Volume 130, Issue 3, March 2014, Pages 428-441. (2012)
2. Aminuddin Baki. "Cheramah 3 Siri Gerakan Obor" *Radio Malaysia.* (1964c)
3. Ibarretxe-Antunano, B.I (1999). Polysemy and Metaphor in Perception Verbs: A Cross Linguistic Study: PhD Thesis. University of Edinburgh, 1999.
4. J. Angouri, & M. A. Locher. *Journal of Pragmatics 44 (2012)* 1549-1553. (2012)
5. Myers, R.J, & Kessler, M S (1980). The Rhetorical Analysis of Business Speech: Unresolved Questions Journal of Business Communication, July 2010 47: pages 346-368
6. S. Billett. *Educational Research Review*, Volume 12, June 2014, Pages 1-13. (2014)
7. R. Biria, A. Mohammadi. *J Pragmatics 44 (2012)* 1290-1302. (2012)
8. L. Cirillo. *Heinz Werner and Developmental Science Path in Psychology 2005*, pp 109-119. (2005)
9. S. Dekker. *Appl Ergonomics,* Volume 34, Issue 3, May 2003, Pages 233-238. (2003)
10. P. Gentle, & T. N. Maraseni. "Climate change, poverty and livelihoods: adaptation practices by rural mountain communities in Nepal." (2012)
11. D. Fancourt, A. Ockelford, A. Belai, *Brain, Behavior, and Immunity,* Volume 36, February 2014, Pages 15-26. (2014)
12. V. V. Gouveia, T. L. Milfont, & V. M. Guerra. "Functional theory of human values: Testing its content and structure hypotheses." (2014)
13. C. Harteis, J. Bauer, H. Gruber. *International Journal of Educational Research,* Volume 47, Issue 4, 2008, Pages 223-231. (2008)
14 M. M. Islam, S. Sallu, K. Hubacek, J. Paavola. *Marine Policy*, Volume 43, January 2014, Pages 208-216. (2014)
15. F. Kadri, S. Chaabane, C. Tahon. *Simulation Modelling Practice and Theory,* Volume 42, March 2014, Pages 32-52. (2014)
16. K. Kitajima, K. A. Carol. *Ecology (jstor)* 70 (4): 1102–1114. (1989)

17. G. Lakoff. *Ortony, A. (Ed.), Metaphor and thought* (2nd ed.), Cambridge, UK: Cambridge University Press. (1993)
18. M. Marra. "Disagreeing without being disagreeable: Negotiating workplace." (2012)
19. E. Oishi. "Austin's Speech Act Theory and the Speech Situation." *Esercizi Filosofici* 1, 2006, pp. 1-14. (2006)
20. M. M. Rodriguez Marquez. "Patterns of Translation of Metaphor in Annual Reports" in *American English and Mexican Spanish*. PhD Thesis, University of Surrey, (2010)
21. A. Saini. *New Scientist*, Volume 204, Issue 2731, 21 October 2009, Pages 42-45. (2009)
22. R. Santagata. *Linguistics and Education,* Volume 15, Issues 1–2, spring 2004, Pages 141-164. (2004)
23. B. Scholnick, N. Massoud, A. Saunders. "*Journal of Financial Stability,* Volume 9, Issue 1, April 2013, Pages 26-37. (2013)
24. R. Studer, B. Danuser, H. Hildebrandt, M. Arial, P. Gomez. *Journal of Psychosomatic Research*, Volume 70, Issue 6, June 2011, Pages 557-564. (2011)
25. N. Suckall, E. Tompkins, L. Stringer. *Applied Geography,* Volume 46, January 2014, Pages 111-121. (2014)
26. P. E. Tetlock, F. M. Vieider, S. V. Patil, A. M. Grant. (2013)
27. R. Tourangeau, R. J. Sternberg. *Cognition*, Volume 11,Issue 3, May 1982, Pages 203-244. (1982)
28. L. W. I. Wairegi, P. J. A. van Asten, M. Tenywa, M. Bekunda. (Musa spp. AAA-EA). (2009)

Portrayal of Women's Images in Television Dramas: A Malaysian Case

Faridah Ibrahim, Phd[1,*] , *Norzita* Yunus[1], *Dil Froz Jan Sayed* Halem Shah[1], *Munirah* Ilias[1] *and Amli Hazlin* Ahmad Mokhtar[1]

[1]Faculty of Arts, Communication and Education, Infrastructure University Kuala Lumpur, Malaysia

Abstract. Current situations brought about by globalisation and sophisticated innovations in new technology not only have created a new understanding of the world around us but also conscientiously sensitised society, particularly the women towards a more active role in shaping their world. An important contributory factor that can determine the success of women is their centrality in the media, through media positions held or through media images portrayed. What this paper has endeavoured to show is the portrayal of women in the media particularly, television dramas. In today's scenario, the role of women keeps changing, so does the portrayal of women's images in television dramas. This paper, based on a qualitative content analysis study of five popular Malaysian television dramas, found that the stereotypical generalizations of women's images are still existing but to a lesser degree.

1 Introduction

In modern society, with the onset of globalization, the mass media are seen as a powerful tool that can shape behavior and capitalize on the potentials of women which can be used to their advantage. While on the other hand, media reinforces existing stereotypes, attitudes, expectations and values which could be a hindrance to women's advancement in organizations and societies [1]. For decades, we have seen that seminars on mass media at local and international levels have issues of gender as one of their focus.

But in spite of the orientations, emphasis and exposures, the women have not been treated fairly by the media. Except for very few selected personalities, on the overall, media treatment of women can best be described as shallow and narrow. It is contended that on films, drama and soap operas, the newspapers and broadcast media, traditional and new media, women's activities and interests typically go no further than the confines of home and family [2, 3, 4]. Women are rarely portrayed as rational, active and decisive merely because of their dependent, romantic, caring and submissive characteristics. Both as characters in fictional media programs, such as TV dramas or as decision makers portrayed on television news bulletins and newspapers coverage, women are under-represented in

* Corresponding Author: faridah@iukl.edu.my

terms of quantity. Such absence has marginalized women and reinforced their inferior status in many spheres.

1.1 Research Context

Television dramas are always considered to be a female-viewer-attracted industry. With the advantages of visuals, sound and movement, they have the potential to accentuate social values and popular images of personalities and culture [5]. The dialogues in television dramas have the tendencies to create the images of gender through the languages used and these would leave long lasting impressions in the minds of the audience. Where gender relationship is concerned, the content of the media via media languages and broadcast dialogues have the latent power to emphasize the prominence and dominance of masculinity and femininity. Language is indeed the most sophisticated symbol that mankind has to deal with. Language forms opinion and this becomes patterns and models of the way people perceive things [6]. Furthermore, the images portrayed in television dramas has a definite impact on the thought patterns of society [7]. Hence, positive words would foster good and healthy images for the women and likewise, negative words would conjure negativities. This would be downright harmful for the women and somehow weakens continuous effort to empower them. The paper aims to look at current trends in media portrayals of women via television dramas; what are the dominant images and what are the positive and negative orientations of women's images portrayed in five selected TV dramas in Malaysia.

1.2 Research Objectives

This research focuses on the portrayal of women's images in five selected Malaysian television dramas. This research aims to answer the following objectives:

- To identify the dominant themes in the portrayal of women's images
- To identify the positive and negative images of women through selected television dramas.

2 Literature Review

Although the situation is not as extreme as in the earlier years, the portrayal of women in television dramas remains divided, showing the dominant social value of masculinity versus femininity. Women are always associated with negativities either through words/dialogues or through actions such as submissive, sensitive, dependent, emotional, bossy, indecisive or fickle-minded. However, the men are usually pictured positively as very rational, independent, assertive and competitive. Two scholars from India [8] found in their study of Indian television serials that women are projected according to the stereotypical image existing in the society leading to reinforcement of stereotypical role. Sex stereotyping is also evident in television portrayal of men and women in their televised roles. Women in the world of television are presented in the role of domestic help, a wife, a mother and being portrayed as submissive and engrossed in common family affection and duties. Men are depicted as employed, spirited and combative.

In Malaysia, the question of female sexuality has proven to still be a critical issue today and was particularly so in post-war Malaya. Most of the studies that looked into the portrayal of women were based on Malaysian films. Several studies found that the image of women was seen from the two-value orientation for instance the dichotomous Virtuous versus Fallen/Passionate woman [9]. Other local studies [10, 11, 12] also demonstrate that media imagery of women are profound in Malay movies. Over the past decades, a great deal of movie messages via Hollywood, Bollywood and even 'Mollywood' (the label for Malaysian movie industry) has perpetuated the notion of women as property or commodities "available" for sexual and other use. These messages are played over and over again in movies which show images of women as the gentler sex, having common traits like passivity, weak, needing protection, subservient, scatter-brain, dependent, comforting, nurturing, caring as well as beautiful (so convenient as sex object). The men on the other hand, are pictured with positive traits such as forceful, strong, capable, aggressive, demanding, independent, successful, tycoon, a hero with the stoic macho-man and leadership qualities.

Basically, decisions by media practitioners or program producers regarding portrayal of gender are based on assumptions about the current state of social and cultural values. In most societies, such values incorporate certain definitions of 'masculinity' and 'femininity' which are replayed and reinforced by the media. A study on contemporary Malay movies by [13] found that the male dominance versus female submissiveness dichotomy is dominant. If the movies of the fifties portrayed women as objects of domesticity, contemporary movies portrayed them as objects of sexuality. These modern movies portrayed women in the public sphere as working professionals but once out of their homes, they are seen as objects to be admired, teased, used, desired and played upon by the opposite sex. Such a portrayal about women is not healthy and culturally detrimental.

An earlier study by [14] looked at the detrimental words used by the media, especially the global media, to picture women's images. When the media chose to allocate space on women issues during the 1995 Fourth World Conference on Women in Beijing, they used devilish terms such as 'hardcore feminists', 'burning bras', 'nude women running around the streets of Huairou' and other generalisations and stereotypes which generally eclipsed many positive accomplishments at the Conference.

A comparative study on women's images in Polish and Malaysia's print media [15, 16] found that although the study of gender images and representations are undertaken from different cultures, there are some universal similarities. Male dominance is still prevalent and the female gender is still involved in the long continuous struggle to make their presence felt. But on a positive note, the print media in both countries, both magazines and newspapers showed evidences that highlighted the continuous struggle and determination by women in both countries to empower themselves. These evidences were traced from several headlines in the Polish magazines that spell out these words 'A women does not have to..', 'a woman can..', 'a woman is able,' 'a woman conquers the world'. The Malaysian magazines are carrying healthy headlines regarding women such as 'The next generation entrepreneurs..' ' A woman of success…', 'The noble cause and the golden voices..' These evidences showed that more and more women who, inspired by the medial image of a free, liberated, self-determined woman, are trying to change their fate.

3 Methodology

This study utilizes the qualitative content analysis approach to collect data pertaining to the themes and dialogues that show reflections of women's images in the selected television dramas. This study is part of a bigger study on women and children's representation in television dramas [17]. For this paper, only a total of five popular Malaysian television dramas were selected for this study namely Sehangat Asmara on Astro, Ariana Rose on TV3, Salju Sakinah on TV Al-Hijrah, Ummi on TV3 and Keluarga Iskandar on Astro. All these TV drama serials were aired on several different television stations in the year 2013 and 2014 and several titles were adaptations from best seller novels in Malaysia such as Ariana Rose and Sehangat Asmara. The data were analysed using qualitative content analysis based on recurring themes [18].

4 Analysis and Discussion

Three types of genre were observed from the five television dramas that were selected for the study. *Ummi* and *Salju Sakinah* come under the drama genre, whilst *Sehangat Asmara* and *Ariana Rose* are categorized under the love and romance genre , and *Keluarga Iskandar* comes under the family genre. In all the drama serials, there are strong women representation in the leading roles as well as strong women characters in the supporting roles. Based on the qualitative content analysis, it is observed that the recurring themes of the dramas is about portraying women as a strong character, and women with great determination. Only one out of 5 dramas understudy showed the main character as a housewife in *Keluarga Iskandar* who is associated to domestic problems related to families, in-laws and children. There are dialogues in this drama that place the main woman character in a stereotypical generalizations. In *Keluarga Iskandar*, Kay who is the leading woman character thought that she is pregnant because she has wrongly read the pregnancy gauge. Hence, she goes around telling her in-laws that she is pregnant. Basically, it is supposed to be comical and hilarious but such ignorance is not good for an educated woman like Kay. Some of the dialogues which she uttered do not put her in a positive light. These are the words that have been used: *"Urrgh..Acu, tolong kak Ngah ambil air please. Sejak mengandung ni, tulang belakang Kak Ngah rasa macam nak patah." [Acu, please get me some water. Ever since I am pregnant, I feel that my back is so painful]; "Acu, tolong ambil majalah..yang ada exercise sikit tu"[Acu, please get me the magazine, the one with that little exercise]; "Excuse me! Excuse me! Tolong jangan kecoh boleh tak? Saya tengah sakit ni. Sejak mengandung baby ni, macam sakit-sakit sangat badan saya." [Excuse me!Excuse me! Please don't make so much havoc. I am in pain. Ever since I conceive this baby, my body is aching all over"]*

Nevertheless, on a positive note, the other four dramas do project a healthy image of the leading women characters. Positive images portrayed by these characters are loving and caring, confidence, patience, intelligent, active, well-mannered, strength, honest, full of spirit, skillful, possess leadership qualities and independent. The study showed high and dominant recurring themes emphasizing active, strength and independent in the leading women characters. This has shifted them from the usual stereotypical generalizations of women who are so dependent on their male partners or counterparts.

Ariana in *Ariana Rose* is portrayed as a beautiful and highly educated lady who had been raped but had managed to rise in her career as a corporate lady amidst her traumatic experience. She is also noted for her kindness with the orphans and is always thinking of her creator. Her daily prayers had strengthen her faith, *"Ampunlah aku Ya Allah. Aku tidak kuat menghadapi ujian mu Yang Allah" [Oh Allah Almighty. Please forgive me. I am not strong to face your test].* Umi in *Ummi* is a strong-willed lady who is suffering from cancer but is so resilient and spirited to raise her three adopted children. She earn her leaving as a hospital cleaner and selling recycle items.

Kay in *Sehangat Asmara* who is pregnant and left by her husband is able to stand up and finish her university education to become a successful lawyer. The same goes to Sumayyah in *Salju Sakinnah* who has to brave herself through hardship in a foreign land. These are good images constructed by the content or program producers. These are positive images that would empower the audience, especially the women.

However, the portrayal of women's images in Malay dramas are seen to be packaged with dual image orientations. On one hand, the audience are being fed with positive and good images of women, and on the other, it will be thwarted with some portrayal of bad and negative images of women. These characters also have dominant roles in the dramas but they will usually end up in disaster, mishap and death. They are also pictured as beautiful and successful but unlike the leading characters, they are evil. Mimi, who is Kay's best friend in *Sehangat Asmara*, died in a car accident after several attempts to destroy Kay's marriage. Puan Sri Syarifah Zaharah, Kay's mother, is not a representation of a good mother. She has schemed with Mimi to destroy her own daughter's marriage. Her evil heart is evidenced in these words: "*Nak kahwin, kahwin tapi jangan harap dapat restu dari mak dan ayah. Bila kahwin jangan berani jejak kaki di rumah mak. Mak tak terima Zainal sampai bila-bila, tahu tak?" [You want to marry, go ahead. But don't expect blessings from me or your Dad. And don't step into my house once you are married. I will never accept Zainal as my son-in-law, do you understand?]*

Liyana in Ariana Rose died a tragic death after several attempts of plotting to win back Tengku Adam from Ariana. Her evilness was found in these words: *"Kalau aku tak dapat miliki kau Adam, tak de siapa yang dapat miliki kau" [If I don't have you Adam, no one else will.]*. Beautiful and sexy Liyana even go to the extent of plotting to kill her own brother so that he does not come in between her and Adam.

This study found that although women's representations in the dramas transcend through time, the portrayal of images still remains the same which is dichotomous in nature that defines the images of women as either perfectly good or absolutely bad or evil. Towards a certain extent, these selected television dramas had some empowerment elements which are reflected through the dominant images of women that showed strong-willed women fighting to stand by their own beliefs and values. This study also found that the women's social class, job status, level of education and character traits are being positively depicted more than before. Nevertheless, the images of women holding supporting roles in dramas need not always be stereotypically represented.

5 Conclusion

Media content, especially television dramas that portray women and women issues need a reorientation. Present day coverage and programs on women tend to emphasize the 5 F's namely Fashion, Food, Family, Furnishing and Fun that spell out to be Femininity. Programs should devote more on people, lifestyles, cultures, attitudes and values and downplay trivialities about women. Television dramas should be more future oriented, rational, highly informative, educational and inspiring that could empower women.

There should be a serious re-look at the media content, both entertainment and news. The media is a powerful tool for education and social change. It has capacity to preserve, record and define human culture and history. Nevertheless, due to insensitivities on the part of content producers, it tend to project stereotypical generalizations of women's images. Although the images of most leading women roles are showing some positive orientations, the supporting women roles in TV dramas are always having negative orientations. Such insinuations are not improving the overall women's images. This calls for a gender-sensitized content producers. If the media can shape and structure people's perception of discriminating roles and values, the same institution can be tailored to give a more dynamic and real images of women as human beings in their own rights. Perhaps, it is appropriate to end this paper with a quote from a profound and truly a legendary figure of our time: "*Kind words can be short and easy to speak, but their echoes are truly endless"* – Mother Teresa

Acknowledgement

The findings in this study are extracted from a joint research project funded by IPPTAR-IUKL 2014-2015.

References

1. I.Faridah, AMIC Conference (2005)
2. L. Strumska-Cylwik, I. Faridah, IJAS, **7**, 5, 647-682 (2014)
3. H.B.K Fuziah, I. Faridah, ICLC Conference. (2002)
4. Faridah, H. Rahmah, JKMJC. **14** (1996)
5. H. Ingham, http://www.aber.ac.uk/mwdia/students/hzi940.html. (2007)
6. L. Strumska-Cylwik, I. Faridah, IJAS, **7**, 5, 647-682 (2014)
7. S.Kaul, S. Shahni, Stud Home Comm Sc. **4**, 1, 15-20 (2010)
8. S.Kaul, S. Shahni, Stud Home Comm Sc. **4**, 1, 15-20 (2010)
9. A.S.H.Kueh, MA thesis. Murdoch University. (1997)
10. H.B.K. Fuziah, I. Faridah, SAMA Conference. (2002)
11. H.B.K. Fuziah, I. Faridah, ICLC Conference. (2002)
12. H.B.K. Fuziah, I. Faridah, ICLC Conference. (2004)
13. H.B.K. Fuziah, I. Faridah, ICLC Conference. (2004)
14. Faridah, H. Rahmah, JKMJC, **14** (1996)
15. L. Strumska-Cylwik, I. Faridah, IJAS, **7**, 5, 647-682 (2014)
16. I.Faridah, L. Strumska-Cylwik, IICON-LANCOMME. (2014)
17. I.Faridah (ed), *Televisyen: Wanita dan kanak-kanak. Representasi dan pemaparan* (IPPTAR-IUKL, Kuala Lumpur, 2015)
18. M.B. Miles, A.M. Huberman. *An expanded sourcebook qualitative data analysis*, (Sage Publications, London, 1994)

SHS Web of Conferences 33, 00064 (2017) DOI: 10.1051/shsconf/20173300064
i-COME'16

Determinants of Community Involvement: A Review of Literature and New Hypotheses

Salisu Aliyu Abubakar[1,*], and *Bahtiar* Mohamad[1]

[1]School of Multimedia Technology and Communication, 06010 UUM Sintok, Kedah, Malaysia

Abstract. This article provides an understanding of what community relations is all about. From the numerous literatures consulted, it observe that relationship between companies, and their various host communities was characterized by conflict destruction and misunderstanding but sometimes positives. Against this background, a new framework on community involvement was proposed as a response to the failing attempts by several recommendations and model on how to improve community relations. Furthermore, this paper's attempt to unravel what need to be done in creating and enabling environment that will bring about peace, understanding and patronage. This article argues that, community involvement enhances effective participation of host community. However, based on the proposed model, the determinants of community involvement are: trust, control mutuality, commitment and satisfaction. This linkage of determinants to community involvement could only be relevant if the company agreed to work together, based on what this paper hypothesised in line with the assumption of social exchange theory.

1 Introduction

A community relation is an organizational strategic function of reaching out to various stakeholders, especially host communities with a view to fostering understanding [1]. Similarly, a community relation is a technique of monitoring the concern of host communities, and to building rapport to earn long-term support of community members. It is an organizational internal mechanism which reduces risks that might confront companies [2]. The variety of risk includes: host community's activism, boycott of company's product and even negative media coverage.

Companies utilize community relations strategy to promote mutual benefit with their stakeholders, especially their host communities. Globally, companies employed various programmes in order to improve their relationship with host community to improve understanding and support [3]. Managing relationship that can benefit both company and the host community is the ground for contemporary community relations.

Several decades, company community relationship has been characterized as unfriendly and hostile, but sometimes positive [4]. In recent times, the violent conduct of host communities' towards business firms, especially multi-national company is alarming,

* Corresponding author: mbahtiar@uum.edu.my

and the restiveness is prevailing [5], leading to several incidences of killings, kidnapping, boycott and vandalize of properties, and sometimes lead to temporal or permanent shutdown of a business site. However, this predicament has now compelled companies to be decisive in looking down an avenue to patch up their dispute with their respective host communities. Therefore, in order to address these negative and harmful effects causes to the environment and the inhabitants; companies ought to engage in developmental community activities, empower the inhabitant and address the imbalance in terms of resource allocation and utilization [6] by involving host community in the decision making and other activities. Genuine involvement with the community and transparency has been described as a panacea to improving community relations, but that could only be achieved through the help of relationship determinants: trust, control mutuality, commitment and satisfaction respectively.

Therefore, this paper intends to review literature and develop new hypotheses on the determinants of Community Involvement.

2 Review of Literature

A community relation is a multi-dimensional field, which various scholars viewed differently. For instance, it is aims at ensuring that stakeholders, particularly host community need to be integrated into developmental activities that will improve their lives, with the view to earn patronage [1]. The focus on this definition is on company's interest, because by doing well you will get better.

Furthermore, community relations aimed at resolving dispute or rancour that is capable of threatening harmonious coexistence between companies and their various stakeholders. For instance, the environmental effect caused by company's activities, is the basis of several conflicts between companies and host communities. Therefore, the best and suitable way to address it is through dialogue and appropriate consultation. Based on the above listed definitions of community relations, it is observed that three characteristics of community relations have been highlighted below thus:

Involvement: To achieve an enduring community relation, companies should have supposed to carry local community along in developmental programmes. Involving community through a developmental initiative comes in a different form such as educational support, youth empowerment, social welfare and even providing aid to the needy societies, and more importantly these new ways of involvement are blended in company's core area of proficiency for business strategy and marketing [7]. Lack of involving community members is largely the major causes of misunderstanding and rancour between companies and community.

Communication as a basis for dialogue: Communication serves as fundamental means of creating and enabling environment for understanding between company and host community. Communication is a veritable tool for resolving conflict and proper solution through consolation. Communication management entails effective supervision of free flow of information and organizational external relations, which consist of public issues and engagement with social groups, the media and other related agencies [8]. An effective two ways flow of communication gives room to monitor any change or variation in community attitude or anticipation, so as quickly take the necessary measures which could result to commitment and satisfaction between company and community.

Based on the identified characterization of community relations, this paper adopted community involvement as the most effective instrument of building harmonious coexistence between companies and their host communities. However, community involvement could only sustain only if it is built based on trust, control mutuality, commitment and satisfaction respectively [9].

3 Research Frameworks and Hypotheses Development

In line with the objective of this paper that is to review literature and develop new hypotheses on the determinants of Community Involvement. Conceptual model is designed to further signify the causative relationship within the various constructs, and other factors influence community involvement. Figure 1 indicates the key constructs.

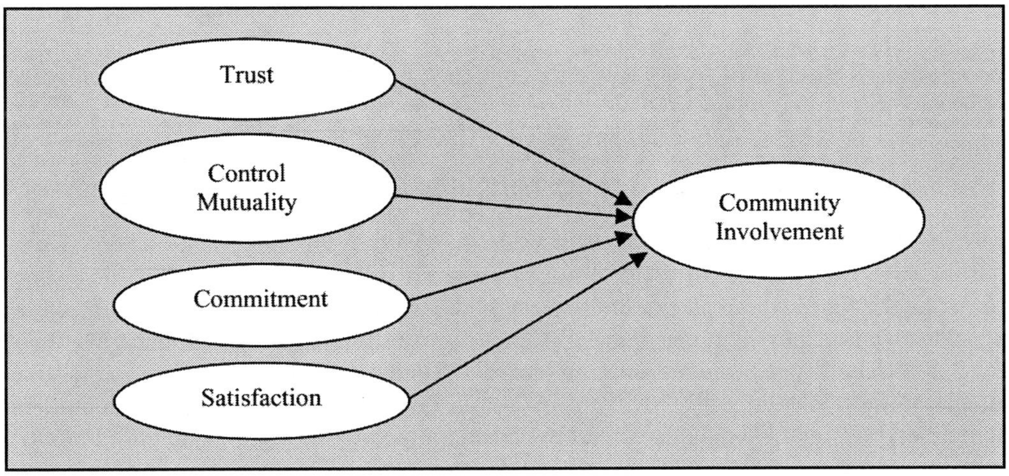

Fig. 1. The proposed model of the antecedents' Community Involvement

The antecedents of community involvement and relationship determinants are trust, commitment, satisfaction, and control mutuality.

3.1 The Relationship between Trust and Community Involvement

According to [10], trust is defined as one party's level of assurance in readiness to be submissive to transact with the other party. Trust, however, can bring about tranquillity with one another and strengthen the relationship between two or more individuals in relationship or between organization and community. Similarly, trust is a significant component which aids organisations or companies in actualizing their goals and aspiration, because is a cordial linkage that bond organizational system. In support of above assertion, [11] affirm that, trust depicts to extend of reliance and determination upon one party's readiness in fulfilling its commitment toward another.

Although these variable's trust and community involvement are different concepts, but they have a significant linked, because the degree to which local communities are involved in company in community development activity heavily relied on the understanding and relationship that have been built between them. In reality, [12], asserted that, in-depth consultation by the company to its various key stakeholders, especially local community base on clear and openness throughout the implementation activity are vital ingredients of a genuine community development. This according to [13] could only be attained if host community are deeply involved and inform about a certain key decision by either the company, organisation, or even government authority in taken decision that directly affects the community.

Previous researchers have identified trust as a veritable instrument for determining a relationship between companies and their various stakeholders, particularly host communities, but [14] found out in their study that, trust is not a determinant in an online business purchase relationship, but rather people heavily rely on interactivity.

There is no empirical confirmation so far in relationship between trust and company community involvement, moreover to support the position that there exists a link several studies attested to that [15]. Considering the above discussion, it is pertinent to propose that:

Proposition 1: The level of trust between a company and host community will positively affect involvement of community.

3.2 The Relationship between Control Mutuality and Community involvement

There exist a relationship between control mutuality and community involvement. Even though the two concepts are different, but they shared a link. Control mutuality according to [16] is the degree in which parties consent on who has the authority to influence the other party. While, it is a new adopted business imitative by the companies in engaging their local host community by giving them, a sense of belonging and seeks a competitive advantage. Corporate community involvement entails how companies engaged stakeholders in the deliberations and executions of various developmental programmes that are targeted at achieving effective community relations and development.

The aim of every business is to make significant profit and compete favourably among the competitors; therefore, this could not be achieved without effective community relations, in other words, the business would not prosper in tense environment, so a key to effective community relations is through involving the local populace, so that they could have a sense of belonging. There exist a link between control mutuality and community involvement is in the area of meeting stakeholder's expectation, especially host community. Every company that deliberately overlooks community expectation by refuse to involve them in the company's community developmental programmes or other related activities that concern them, then the company will have itself to blame. Taking into cognizance the above discussion so far, it therefore, propose that:

Proposition 2: There is a significant impact of control mutuality of a company with host community involvement.

3.3 The Relationship between Commitment and Community involvement

Commitment was defined as a degree to which relationship is sustained, even to the extent of compromising some immediate needs to secure long-term interest [17]. Commitment has also been defined as the extent to which each party's readiness and believe that the relationship is worth been sustained and promote to next level. Many scholars asserted that commitment signifies how partners look at target a successful and persistent relationship [18]. It is important to note that, the current debate among researchers on business and society, emphasize that, organization and companies competing in the market circles should strive to demonstrate their ethical commitment by providing their audience with satisfying information concerning product content and social services.

DOI: 10.1051/shsconf/20173300064

It is interesting to note that, many US giant companies and other reputable firms around the world are shifting their focus by devoting substantial time and resources towards community involvement initiative, which comprises of capacity development and skill acquisition, corporate support and other developmental project because gradually companies have realised that, for them to be faithfully accepted as neighbor of choice and have their license to operate, they must exhibit uttermost commitment by involving host community in their developmental activities [19].

Although scholars shared different views on how company's commitment shaped their relationship with host community. According to [20], companies lack of internal coordination in their developmental projects are the reasons behind failures of several community relations programmes, because it has been shown that in spite, the huge amount budgeted for community developmental programmes by multinationals companies, some host communities are not convinced with the level of company's commitment because they have not been involved throughout the process and execution. A study by [21] further re affirmed this position that huge investment alone without compliment it with internal commitment will not bring the desired result.

Commitment towards community developmental projects measured company's readiness in providing essential services in improving the general wellbeing of society and benefit company's business competitive advantage [22]. This form of initiative helps in fostering mutual trust and understanding, and subsequently brings an economic advantage to the company. Considering the above discussion, it is therefore proposed that:

Proposition 3: The level of a Company's commitment will positively influence host community involvement.

3.4 The Relationship between Satisfaction and Community involvement

Satisfaction is described as the degree to which both parties have the impression that the relationship is favourable to them because of real expectation of the outcome characterized satisfaction as enjoyable attainment [23]. Literature reveals that, significant number of researches conducted on satisfaction viewed it as either on product or service evaluation derived from. However, this study intended to identify the satisfaction as an evaluation of relationship outcome between company and host community.

A number of researches have established the relationship between satisfaction and community involvement. This is evidently clear on how people evaluate community satisfaction with an outcome of various levels of involvement within the community setting [24]. Thus, network structure within the community determines their satisfaction. The community with solid and formidable structure of involvement and networking are susceptible in bringing a diverse group of settlers such as investors together and reach an accepted goal.

Even though satisfaction has been described as the fulfilment derived from the service been rendered or delighted with the situation but in some cases brand reputation or organizational reputation plays a significant role in relationship building rather than the actual satisfaction from the service been rendered. However, [25], have contrary findings. Their study reveals that satisfaction is not a key a factor of shaping relationship. Considering on the above discussion; therefore, this study proposed that:

Proposition 4: There is a significant impact of host community's satisfaction and the level of community involvement.

4 Conclusions

Based on the above analysis, the proposed model explains how community involvement companies can provide companies with various benefits that can transform into quality and beneficial relationship, between companies and host communities. Another valuable expected contribution to be derive from this paper, is that, companies wish to embark on community relations activities to improve their harmonious coexistence with host communities, will find this article relevant. Furthermore, this article will be useful to managers and practitioners in suggesting to their respective companies as well as all their representatives as to what project is suitable and desirous of the host communities.

This paper deliberated on the research framework as well as the hypothesis development. The framework is designed as a mechanism to develop a causal effect of relationship between determinants and community involvement in the relations between company and host community. Furthermore, highlighted are the antecedents of community involvement, which are: Trust, satisfaction, commitment and control mutuality respectively. The paper ends with showing link between various construct and hypotheses as proposed accordingly. Based on this, we proposed this model.

References

1. B. W. Altman, Business and Society Review, 102(1), 43-51. (1999)
2. K. Rehbein, & D. A. Schuler, Business & Society, 54(6), 794-821. (2015)
3. C. Sengupta, International Education and Research Journal, 2(6).11-12 (2016)
4. A, Hassan, & R. Kouhy, R. International Journal of Energy Sector Management, 9(2), 204-226. (2015)
5. F. E. Nlerum, & M. D. Epepe, Open Access Library Journal, 2(12) 1-7. (2015)
6. C. P. Lin, Y. H. Tsai, S. W. Joe, C. K. Chiu, Journal of business ethics, 105(1), 83-93. (2012)
7. D. Hess, N. Rogovsky, T. W. Dunfee, California Management Review, 44(2), 110-125. (2002)
8. S. A. Waddock, & M. E. Boyle, California Management Review, 37(4), 125-140. (1995)
9. J. E. Grunig, & L. A. Grunig, A Report for the Department of Energy Office of Science. Department of Communication. University of Maryland. (2001)
10. Y. H. Huang, Y. H. Journal of public relations research, 13(1), 61-90. (2001)
11. A. Sharma, & R. Kiran, African Journal of Basic & Applied Sciences, 4(3), 95-105. (2012)
12. S. Du, & E T. Vieira Jr, Journal of Business Ethics, 110(4), 413-427. (2012)
13. A. M. Durko, & J. F. Petrick,. Journal of Travel Research, 1-15 (2015)
14. C. E. Ugouwanyi, Global Journal of Human-Social Science Research, 16(2). 31-45 (2016)
15. M. E. Clarkson, M. E. Academy of Management Review, 20(1), 92-117. (1995)
16. L. C. Hon, & J. E. Grunig, Florida Institute of public relations. (1999)
17. S. C. Ting, Journal of Service Science and Management, 9(03), 276.(2016)
18. C. P. Lin, Y. H. Tsai, S. W. Joe, C. K. Chiu, Journal of business ethics, 105(1), 83-93. (2012)
19. S. M. Loureiro, I. M. D. Sardinha, L. Reijnders, Journal of Cleaner Production, 37, 172-178. (2012)
20. E. E. Izogo, Journal of Retailing and Consumer Services, 23, 107-117. (2015)
21. A. M. Durko, & J. F. Petrick. Journal of Travel Research, 1-15 (2015)
22. L. Y. Do Thi Truc, Centria University of Applied Sciences, Kokkola; Thesis. (2016)

23. A. Sharma, & R. Kiran, African Journal of Basic & Applied Sciences, 4(3), 95-105 (2012)
24. J. Bernard, Sociologia Ruralis, 55(2), 205-226. (2015)
25. A. S. Sengupta, M. S. Balaji, B. C. Krishnan, Journal of Business Research, 68(3), 665-674. (2015)

Revisiting video game ratings: Shift from content-centric to parent-centric approach

Hee Jhee Jiow[1,*], *Rayvinder* Athwa[1], *Ling Ling* Chew[1], *Muhammad Helmi* Elias[1], *Nina* Lim[1], *Kenneth* Woo[1]

[1]Singapore Institute of Technology, 10 Dover Drive, Singapore 138683, Singapore

Abstract. The rapid adoption of video gaming among children has placed tremendous strain on parents' ability to manage their children's consumption. While parents refer online to video games ratings (VGR) information to support their mediation efforts, there are many difficulties associated with such practice. This paper explores the popular VGR sites, and highlights the inadequacies of VGRs to capture the parents' concerns, such as time displacement, social interactions, financial spending and various video game effects, beyond the widespread panics over content issues, that is subjective, ever-changing and irrelevant. As such, this paper argues for a shift from content-centric to a parent-centric approach in VGRs, that captures the evolving nature of video gaming, and support parents, the main users of VGRs, in their management of their young video gaming children. This paper proposes a Video Games Repository for Parents to represent that shift.

1 Introduction

The rapid adoption of video gaming among children has placed tremendous strain on parents' ability to manage their children's consumption [1]. To which parents rely on video games ratings (VGR) for information to support their mediation efforts [2]. Yet, there are many difficulties associated with such practice. This paper will highlight the challenges associated with current VGRs and argues for a shift from content-centric VGR approach to a parent-centric approach. This paper will begin by exploring several well renowned VGRs.

The Entertainment Software Rating Board (ESRB) lists a number of games and its ratings on their website, and even provides parental control guides to better manage their children's video gaming behavior [3]. Appropriate age warnings ranging from Early Childhood (EC) to Adults Only (AO) and brief content descriptors such as violence and sexuality were provided as well to better inform the users [3]. ESRB aims to provide information to empower parents in their video game management habits of their children.

The Pan European Game Information (PEGI) functions similarly to ESRB, with clear age restrictive warnings as shown from the numbers 3, 7, 12, 16 and 18 that corresponds with the respective age groups, and content labels in forms of pictures depicting violence (a

* Corresponding author: jhee.jiow@singaporetech.edu.sg

picture of a fist) and foul language (a picture of various symbols in a speech bubble) [4]. PEGI advocates parental supervision and video game gate keeping, and provides assistance for consumer decision-making during video game purchase [4].

It is evident that VGRs generally provides video game content information to support parents' mediation efforts. However, there are two major issues with VGRs' content-centric approach – its subjectivity and its relevance.

2 Subjectivity of Game Information

First, this paper argues that such information is highly subjective considering how video games have evolved [1]. The validity of game ratings provided by the industry such as those from ESRB and PEGI have frequently been called into question, especially from parents [5, 6]. Studies have shown that parents saw industry's game ratings to be too lenient and imprecise. While parents generally agree with industry when a video game is rated as unsuitable for children, those that were rated appropriate for children were mostly perceived differently by parents [6]. When parents were made to rate those games in comparison to ESRB's ratings, Walsh discovered that 13 percent of E-rated (for Everyone) video games were perceived as unsuitable for children aged 3 to 7; and 31 percent of T-rated games were given M ratings by parents, highlighting the more conservative opinions of parents versus the industry [7]. Furthermore, Thompson & Haninger's quantitative study of video games showed that more than half of E rated games contained intentional violence, while less than half of those games were not issued with content descriptor [8]. Thompson, Tepichin & Haninger found similar issues in their assessment of the accuracy of content descriptors in adult video games, where majority lacked content descriptors apart from violent content – thus questioning the validity and accuracy of many VGRs [5].

In addition, the exposure to the actual gameplay of some video games has caused parents to become more critical of industry's ratings, especially for violent content. A study by Becker-Olsen & Norberg discovered that upon revealing actual gameplay sequences, parents change their perceived rating pre-exposure (based on game titles or box art) to a harsher rating [9]. Children, however were less sensitive and rated violent games more leniently. Such observation can be attributed to the fact that parents are more sensitive to video game violence, that has evolved to appear more realistic with the help of technological advancements. However, such discrepancy presents a risk of miscommunicating information by game industries to parents – where box art and game advertisements indicate game ratings that might be seen as 'misleading' for parents [9]. Furthermore, most parents use the television or movies rating system as a point of reference when providing ratings for games [10, 11], which also raises the issue of confusion in the systems across the different media platforms.

Differing opinions of game content between industry and parents is not the only factor that raises the subjectivity of VGR information. Additionally, the perception of what is suitable and unsuitable for different parents vary – in accordance with their norms, values and experience which shapes the expectations they have of their child [9]. Some parents were most concerned with violent content, while others found that foul language in video games was their biggest issue [10]. Other differences were also noted, such as fathers who have experiences in the army perceive war games as 'acceptable' violence for their child, whereas those involving elements of street violence were deemed unsuitable [10]. Hence, every parent has a different way of applying and interpreting VGRs, which raises perceptual bias and subjectivity of VGRs. In addition, different cultures reveal how parents might have varying perspective on what is deemed as appropriate. For example, parents in U.S. were more concerned with violent content in video games, whereas parents in more conservative places, such as Singapore, worry more about sexual themes [12, 13].

DOI: 10.1051/shsconf/20173300065

While there are many calls for a universal rating system across all media types, scholars have argued against its feasibility because such approaches have overlooked some of the basic characteristics of video games as a medium - that they are impossible to be played in its entirety prior to the issuance of ratings [9, 14, 15, 16, 7]. Most importantly, video games have unique features such as "content variability, content modifiability and content meaning" [16, p.12] that differentiates itself beyond the static nature of other forms of media.

ESRB indicated how video games are "player-controlled and there are many different permutations of gameplay" [3, p.FAQ], which aptly describes content variability. Game content might change in accordance to how the user plays, hence players can choose from many different journeys in order to complete the various missions [16]. Free roaming games such as Grand Theft Auto (GTA), Bully and World of Warcraft (WoW) are examples of those that might provide multiple experiences in accordance to how the users decide to play them. In contrast, games such as Pac-Man and FIFA are more straightforward in terms of its content and the aim of the games, not allowing for much content variability. There are even games with no clear game narrative, such as Sims where players control the well-being of virtual families. Ultimately, the content experience of video games vary greatly which makes it difficult to convince consumers of a single rating or content indicator.

Moreover, gamers themselves contribute greatly to modification of the game content and gameplay experience [17]. ESRB specifies how "game experience may change during online play" [3, p.FAQ] and "online interactions not rated by the ESRB" [3, p.FAQ] on games that affords online engagements, in efforts to distance the industry from being responsible for how players interact in-games. Young gamers expose themselves to not just the content of the video game itself, but also the gaming 'culture' of others which might include the use of profanities in online voice chats [18, 19]. In this situation, the actual game content does not contain strong language, but the experience of the user was altered by other players. Additionally, content modifiability can be experienced through "mods" [16, p.15] where programmers and other users construct their own variation of the game to provide a different experience. Moreover, these mods can be shared online for other users to play, thereby perpetuating the subjectivity of game content. In the life simulation game of Sims, the game developer originally censored the genitals of the avatars when they use the toilet or showers to keep it appropriate for the young. However, on an occasion, a user-generated mod was made available online, which removed the censorship of the genitals and further created genitals for the avatars [16]. Another example would be the controversial mod known as Hot Coffee – an interactive sex scene in Grand Theft Auto: San Andreas that was rated M for Mature [11]. Although this particular content was not part of the original game, it demonstrates how modifications might greatly change the gameplay and experience of players, and thereby reduce the validity of game ratings. Unlike television programs and music, rating the content of video games by playing them in its entirety is impossible [20].

3 Relevance of VGRs in the Evolving Video Game Space

Second, this paper argues that content-centric approach adopted by VGRs is inadequate in this new media era. Jiow & Lim described how the affordances of video games - portability, sociability, perpetuity, accessibility, interactivity and identity multiplicity – have evolved and raised concerns, and challenged parents' monitoring and management of their children's video gaming habits [1]. These concerns that parents have about video gaming revolve around other dimensions beyond content, namely time, interactions and effects [3, p.FAQ]. Parents are mainly concerned that prolonged video gaming could result

DOI: 10.1051/shsconf/20173300065

in displacement of time intended for homework or studies [2, 21, 22, 23]. Parents are also concerned that their children will withdraw from healthy social activities as a result of spending excessive time on video gaming [22, 23]. Another concern is that children's academic grades will be compromised if they spend too much time playing video games [22]. Yet, some parents "sometimes used video games as a childcare management tool and reward, in hopes it would motivate the child to exhibit other positive behaviors" [2, p.123]. Parents are also concerned about the social interactions present within video gaming too. Concerns of social interactions goes beyond content modifications of the video game. Duggan & Rainie noted that "72% of parents of online teens are concerned about how their child interacts online with people they do not know, with some 53% of parents being 'very' concerned" [24, p.2]. Parents are typically worried that their child may be harassed, stalked, sexually exploited, cheated, mixed with the wrong company, or even subjected to unwanted advertising by online strangers [2, 25]. The effects of video gaming on the children's social and academic functioning, which is also commonly used as a measure of pathological video gaming, is also a pressing concern among parents [26, 27, 2, 28]. Jeong & Kim highlighted that excessive video gaming, particularly online gaming, has a strong correlation to the high absence of social relationships with family or peers – players indulge in the online world of social interaction while their social self-efficacy in reality are negatively affected [29]. They would spend more time fostering relationships in the online space and were reported to have more problematic offline relationships [29]. In addition, relationship problems with peers and teachers in school tend to result in poor grades as adolescents find it difficult to adapt to the school environment [29]. Hence, players use the online platform to escape from reality and into the virtual arena where they have that supposed 'freedom' to do anything. Internal issues such as loneliness, low self-confidence, anxiety and depression are some of the factors that may arise from excessive online video gaming and have the potential to affect the child's development both in the offline social arena as well as their academic aptitude. Parents are also concerned about the amount of money spent sustaining their children's video gaming habits – the effects of video gaming on finances – and typically frown upon video games that require payment [2].

4 Shifting to a Parent-Centric Approach

As such, a revisit of VGRs is required, and this study proposes an online VGR system, cogently named Video Games Repository for Parents (VGRP), that embodies a shift from content-centric approach to parent-centric approach. Doing so, would better support parents' efforts in mediating their children's video game use in the new media space. The next section describes and proposes that game duration, content, effect on players' occupational functioning and financial information be available in the VGRP.

As mentioned earlier, parents' main concern is about the displacement effect of time on video gaming children, specifically about effects of prolonged gaming, beyond the video gaming content [2]. VGRs generally do not provide data on the average amount of hours required to complete a particular game mainly because the amount of time required to play or complete it differs and varies from player to player, and game to game [30, 31, 16]. However, information highlighting the average time needed to complete a segment, quest or mission in the game would greatly aid parental mediation. Parents can then negotiate play time with their children either as a form of compromise or reward [2]. A major hurdle of parental concerns regarding time displacement is the lack of knowledge about new games. Traditional mediation efforts are largely based on a standard playing time given to their children however, these techniques operate on the assumption that all games operate on a standard unit of playtime. Yet, information regarding the average playtime of game is available from platforms that allow gamers to share their time taken to complete the game

[32, 33]. For example, games such as *DOTA*, where players are involved in skirmish battle mode, has each battle lasting for an hour on average, however for games such as *WOW*, players go on quests as long as 12 hours, as they progress further into the game [30, 34]. As such, this paper proposes that information on game duration be included in the VGRP since time is of paramount concern to parents, and such information is reasonably attainable.

While this paper argues a shift from content-centric approach, it does not call for an absence of game content information. Parents are concerned about the influences of game content on their children, especially violent, sexual or content that promotes values contrary to parents' beliefs. However, such game content information needs to be informative (allowing the user to judge for themselves) versus prescriptive (prescribing an age category to it). As such, this paper proposes that screenshots and/or gameplay sequences be shown in the VGRP.

Jiow found that parents were concerned about the effect of video game play on the occupational functioning of the children [2]. Studies have shown that prolonged video gaming has detrimental effects on the occupational functioning (mainly social and academic functioning) of young gamers [27, 35]. These effects may include "social withdrawal, self-neglect, family and marital problems as well as sleep deprivation and fatigue resulting in poor educational and/or work performance and job loss" [28, p.749]. While the presentation of this information may not be directly applicable to individual gamers (children), this paper proposes that Pathological Video Game Use data on specific games be included in VGRP.

It is paramount to note that whilst game content and duration information are causes of concern for parents, the financial cost of video games cannot be ignored. The financial cost of purchasing video games has always been a pertinent issue in the past, but then, it was restricted to the one-time purchase of the physical video games [36]. However, video games can now be played across various mediums and devices. And avails more opportunities for spending money without parental involvement. The financial concerns are more prevalent in the modern gaming industry as a variety of business tactics are used to encourage gamers to spend money on these video games. The evolution of business practices adopted by the gaming industry create a new demand for items that are deemed to be important in enhancing the gaming experience of players.

5 Limitations and Conclusion

This paper has explored the popular VGR sites, and highlighted the inadequacies of VGRs to capture the parents' concerns and to represent the evolving nature of video gameplay culture and consumption. This paper has argued for a shift from content-centric approach to a parent-centric approach in VGRs, and has also proposed an online VGRP to represent that shift. However, this paper based its analysis on existing literature, and has no concrete data on how parents might receive the proposed VGRP. As such, future studies may seek to qualitatively and quantitatively explore video game information that is desirable for parents to aid in their mediation efforts. The authors seek to do just that.

References

1. H.J. Jiow & S.S. Lim, B. Sci. Technol. Soc. **32**, 455 (2012)
2. H.J. Jiow, *Parental mediation of video games in Singapore* (Doctoral dissertation, National University of Singapore, Singapore 2014)
3. Electronics Software Rating Board, http://www.esrb.org/ratings/ratings_guide.aspx (2016)

4. Pan European Game Information, http://www.pegi.info/en/index/id/33/ (2016)
5. K.M. Thompson, K. Tepichin, K. Haninger, Arch. Pediat. Adol. Med. **160**, 402 (2006)
6. D.A. Walsh, D.A. Gentile, Pediatrics. **107**, 1302 (2001)
7. D.A. Walsh, *Sixth annual video and computer game report card* (Report 2001)
8. K.M. Thompson, K. Haninger, JAMA-J. Am. Med. Assoc. **286**, 591 (2001)
9. K.L. Becker-Olsen, P.A. Norberg, J. Advertising. **39**, 83 (2010)
10. A.D. Brekke, *Video game ratings: Does the system work for parents?* (Master's thesis, University of Tennessee, Knoxville 2006)
11. A. Brekke, C. Lepre, *Burned by Hot Coffee?: Parental trust and usage of the video game rating system* (International Communication Association 2007)
12. M.K. Miller, Polit. Pol. **38**, 705 (2010)
13. P. Nikken, J. Jansz, S. Schouwstra, Eur. J. Commun. **22**, 315 (2007)
14. A. Lenhart, J. Kahne, E. Middaugh, A. Rankin Macgill, C. Evans, J. Vitak, *Teens, Video Games, and Civics: Teen's gaming experiences are diverse and include significant social interaction and civic engagement: Prepared by Pew Research Center & the MacArthur Foundation* (Report 2008)
15. N.J. Stroud, A. Chernin, J. Child. Media. 2, 1 (2008)
16. J. Tocci, *Rated M for Moral Panic: Content considerations in video game legislation* (National Communication Association 2007)
17. M.D. Griffiths, Educ. Health. **28**, 65 (2010)
18. H.J. Jiow, S.S. Lim, J. Lin, *Game On: Refining the parental mediation framework for a video gaming landscape* (International Communication Association 2016)
19. S. Livingstone, E.J. Helsper, J. Broadcast. Electron. **52**, 581 (2008)
20. J.B. Funk, G. Flores, D.D. Bunchman, J.N. Germann, Youth. Soc. **30**, 283 (1999)
21. A. Khoo, Asia. Pac. J. Educ. **32**, 416 (2012)
22. L.A. Kutner, C.K. Olson, D.E. Warner, S.M. Hertzog, J. Adolescent. Res. **23**, 76 (2008)
23. M.M. Skoric, L.L.C. Teo, R.L. Neo, Cyberpyschol. Behav. **12**, 567 (2009)
24. M. Duggan, L. Rainie, *Cell Phone Activities 2012: Prepared by Pew Research Center & the MacArthur Foundation* (Report 2012)
25. A. Lenhart, O. Lewis, L. Rainie, *Teenage Life Online: Prepared by Pew Research Center & the MacArthur Foundation* (Report 2001)
26. D.A. Gentile, Psychol. Sci. **20**, 594 (2009)
27. D.A. Gentile, H. Choo, A. Liau, T. Sim, D. Li, D. Fung, A. Khoo, Pediatrics. **127**, 319 (2011)
28. T. Sim, D. Gentile, F. Bricolo, G. Serpelloni, F. Gulamoydeen, Int. J. Ment. Health. Addict. **10**, 748 (2012)
29. E.J. Jeong, D.H. Kim, Cyberpsychol. Behav. Soc. Netw. **14**, 213 (2011)
30. J. Billieux, M. Van Der Linden, S. Achab, Y. Khazaal, L. Paraskevopoulos, D. Zullino, G. Thorens, Comput. Human. Behav. **29**, 103 (2013)
31. Interactive Software Federation of Europe, *Videogames in Europe: Consumer Study* (Report 2012)
32. IGN, http://ap.ign.com/ (2016)
33. AskAboutGames, http://www.askaboutgames.com/ (2016)
34. F. Karlsen, Int. J. Game. Stud. **8** (2008)
35. D.A. Gentile, C.A. Anderson, *Handbook of Children, Culture and Violence* (Sage Publication 2010)
36. P. Zackariasson, T.L. Wilson, Com. For. **11**, 56 (2013)

Internet use among urban Malaysians: Network diversity effects

Antoon De Rycker[1,*], *Nicole* Yang Lai Fong[1] , *Catherine* Lee Cheng Ean[2], and *Ramachandran* Ponnan[1]

[1]Taylor's University, School of Communication, 47500 Subang Jaya, Malaysia
[2]Sunway University, Department of Communication and Liberal Arts, 46150 Subang Jaya, Malaysia

Abstract. This study examines social network diversity in urban areas relative to residents' usage of information and communication technologies (ICTs). Individual-level variation in social network diversity was measured using position generator data collected as part of a survey conducted in Malaysia's Klang Valley (N = 808). Regression analyses were performed to assess the extent to which network diversity is related to ICTs. We find that most ICTs have a negative effect on diversity. Only frequent use of the Internet at work, mobile access to the Internet, and reading online news or blogs contribute positively to diversity. Findings support both a tendency toward 'networked individualism' and the more recent 'glocalization' thesis that some ICTs may also afford participation within local space rather than only across distant space.

1 Introduction

The growth of social media platforms (e.g., Facebook) has renewed interest in the analysis of sociality through social networks, focussing on network diversification and social capital effects (e.g., [1-2]). The central question concerns the ways in which digital technologies influence social relations and Internet users' personal networks.

A personal or 'egocentric' network is generally defined as 'the network surrounding one node, known as the ego' [3]. For different egos, this set of socially relevant 'alters' may vary in diversity, or within-group homogeneity [4-5]. As a social construct, it corresponds to the notion of variety: the extent to which individuals differ from one another on certain sociodemographic, behavioral or intrapersonal attributes (e.g., occupation or education but also whether someone plays volleyball, likes gardening or has visited Paris) [6-7]. Given the wide range of attributes in which individuals can differ, various operationalizations of the same 'social network diversity' construct is possible.

Our main purpose is to identify the social network affordances of new media among urban Malaysians. Secondly, we would like to find out whether the Malaysian findings support any of the main arguments put forward in connection with the role of ICTs, new media, and digital culture in contemporary society. For example, relationships between the Internet and social network diversity have been found to be either direct or mediated, either

* Corresponding author: antoon.derycker@taylors.edu.my

positive or negative; considerable attention has been paid to so-called 'networked individuals' and their lack of participation in traditional local settings (e.g., public libraries and coffee shops) [8]. It is important to examine these and similar statements for different populations and to see whether findings can put some of the existing Western scholarship in a broader perspective. Given the relative paucity of social network research on Malaysia (e.g., [9-11]), our study also seeks to contribute to a better understanding of the social texture of urban Malaysian society.

2 The Internet and social network diversity

Researchers examining Internet use and social networks are in agreement that as a conceptual category, 'Internet use' is too undifferentiated to capture the unique 'social affordances' [12-14] of each digital media or device. Facebook, for example, can be used to meet new people, to find people to date, to keep in touch with old friends, to check out people a person has met socially, etc.; all of these utilities, however, have different social network implications [1]. By looking at the detail of social media, researchers are now able to make more precise, more qualified, and less sweeping statements than the either/or positions that characterized some of the earlier research and social commentary [15].

The detailed findings can be related to the more general arguments that have been developed in the literature. A survey of two decades of Internet research identifies three main theories: (1) the Internet helps form close and meaningful relationships around shared common interests, both globally and locally; (2) the Internet isolates and alienates people from their communities, negatively impacting social network diversity and social capital; (3) the Internet supplements other forms of interaction and communication, serving 'an additive role' [1]. The third perspective corresponds to a view of the Internet as enhancing community instead of weakening or transforming it [16]. However, different explanations of the use and impact of the Internet need not be mutually exclusive: 'they all may be happening simultaneously', and also, different ICTs 'may have contradictory relationships' [4]. Facebook and other social networking services, for example, contribute to social network diversity directly as well as indirectly through neighborhood ties; yet, while the former effect is positive (more diversity), the latter is negative (less diversity). These opposite findings are consistent, however, with earlier studies that relate social networking sites to higher aggregate levels of network diversity but reduced neighborhood involvement [17, 4].

The second theoretical perspective – isolation and alienation – can be related to early Internet scholarship such as the 'networked individualism' argument [18]. Under this view, the Internet provides a new social setting for the formation and maintenance of social ties. However, when more time is spent interacting online globally than in offline groups or social settings, this creates a transformational shift from so-called place-based to person-based sociality. Personal networks become dichotomous, i.e., 'online and global' versus 'offline and local.' Although 'networked individuals' may form many new ties online [19], those ties are not necessarily diverse. A plausible explanation is that the Internet may 'function more like the private home than a public space – building large homophilous networks at the expense of diverse ties' [4]. Thus, the replacement of 'densely-knit and tightly-bounded groups' by 'sparsely-knit and loosely-bounded networks' [20] would not only increase individualism and privatism but also decrease social network diversity overall.

Networked individualism is not the only theory that has been proposed when attempting to describe online behaviour. Within the long-standing debate about the effect of the Internet on personal networks, Hampton and his co-researchers have recently argued in favour of what they refer to as 'glocalization.' [4]. Originally, the term referred to 'the intermingling of intensive local and extensive long-distance relationships;' however, glocalization is now used within media and communication research to refer to 'the combination of global connectivity and local activity' [15]. Whenever the Internet is adopted by a critical mass of individuals repeatedly interacting within a shared local setting (e.g., the neighborhood or workplace), it will increase awareness that the Internet affords communication within local space, i.e., with social actors in the vicinity rather than across distant space [21-22]. In other words, glocalization describes the phenomenon that individuals can maintain multiple, partially interlocking and differentially diverse personal networks that are 'offline and local,' 'online and global' but also 'online and local.'

Note that the present study is not aimed at developing a new way of thinking about the relationship between Internet use and the diversity of social networks. Also, networked individualism, glocalization, and other 'substantive theories' are restricted in descriptive accuracy and explanatory scope to a particular social setting [23]. Still, and as observed in Section 1, our findings can be used to test or refine existing theories, with urban Malaysia as a comparative case [24].

3 Methodology

3.1 Data

The dataset for this analysis is based on the survey responses of 808 adults, aged 31 to 55, living in the Klang Valley. The Klang Valley is Malaysia's most densely populated, most heavily urbanized, and most ethnically diverse region [25]. The area includes the country's capital city, Kuala Lumpur, with a population of around 1.6 million; the total population of the Klang Valley is estimated at 7.5 million [25]. For more details about the sampling breakdown, see [10].

The questionnaire is based on a previously validated and published survey instrument [4, 26]. Minor adjustments were made to make the items and responses suitable for use in Malaysia (and thus maintain content validity). First, two occupations, i.e., 'Congressman' and 'middle-school teacher,' were replaced by respectively 'Member of Parliament' and 'secondary school teacher.' Second, when considering the response options for employment status, we separated 'other' into 'self-employed: professional' (e.g., architect) and 'self-employed: business owner' (e.g., shopkeeper). This distinction was introduced to reflect the Klang Valley's small-scale, entrepreneurial economy more accurately.

Note that neither of these changes consists in rewordings, reframing of the original questions or adopting a new scoring system. Phrases like 'Member of Parliament' do not introduce comprehension issues either; on the contrary, they are simple, straightforward, and appropriate for the reading level of the respondents [27]. Still, to minimize the risk that our localized version would affect the instrument's accuracy and validity, a pilot test was conducted among colleagues prior to administration.

3.2 Measures

The dependent (outcome) variable of interest is social network diversity. The independent (predictor) variable used to account for variability in the level of social network diversity is respondents' self-reported use of various ICTs. Selection and operationalization of the key variables were modeled after [4].

3.2.1 Social network diversity

As in many social network studies, a position generator instrument was used to measure the dependent variable (e.g. [28]). Respondents were asked to indicate whether or not they know someone in a specified occupation, and if so, to assess the closeness of the relationship (1 = no one, 2 = someone but not close at all, 3 = less close but still important, 4 = close, 5 = very close). The more ties someone has to other social positions, and the stronger those ties are, the more diverse (or heterogeneous) a person's egocentric network. The generator includes occupations that vary in socioeconomic standing; for the complete list of 22, see [4]. but also race and religion. In regard to race, the main reason is that Malaysia can be characterized as 'a society of multi-ethnicities' in which different cultures co-exist without necessarily blending into each other [29].

As in [4], responses to the generator were subsequently combined into a composite scale measure; note that this increases the reliability and validity of the instrument [24]. Specifically, an additive index was calculated based on 29 items: the 22 occupations (ranging in prestige from security guard to Chief Executive Officer), the three main races (Malay, Chinese, and Indian) and the four main religions (Muslim, Buddhist, Hindu, and Christian). The mean social network diversity score was 2.57 (SD = .622).

3.2.2 Information and communication technologies

To measure Internet use, survey respondents were first asked how often they went online (1 = every day, 2 = several times a week, 3 = once or twice a week, 4 = once or twice a month, 5 = less than once a month). Next, they were presented with a series of dichotomous questions about whether or not they had Internet access at home and at work, and if so, what types of access were available (e.g., digital subscriber line). Finally, given this study's focus on the network diversity affordances of different ICTs, we asked respondents whether or not they engaged in a variety of online activities (a total of twelve) such as instant messaging, sharing digital photos, social networking, and banking.

The mean frequency of Internet use was 1.50 (i.e., going online several times a week up to at least once every day) with a standard deviation of 1.064. A majority of respondents uses the Internet from home (n = 532, 65.8%), compared with a slightly lower number for Internet use at work (n = 473, 58.5%); for both, wireless connection is the most common access method. Most respondents engage in five to six online activities (n = 227, 28.1%), with very few reporting no activities (n = 36, 4.5%) or all twelve (n = 38, 4.7%) (data not shown).

3.3 Analysis procedure

The data were analysed by running a series of ordinary least squares (OLS) regression analyses. Generally, when studying the multivariate nature of the social world, multiple regression analysis is the tool of choice but only when all variables are continuous; in the current study, however, only the outcome variable is continuous while each of the ICT-related predictor variables is either categorical or rank. For more justification, see also [24]. Following [4], we used unstandardized regression coefficients to capture the associations instead of the more usual standardized ones.

4 Findings and discussion

A first major finding is that different types of ICT contribute differently to social network diversity. The more specific results of the regressions have been summarized in Table 1.

DOI: 10.1051/shsconf/20173300066

Table 1. Ordinary Least Squares (OLS) regressions of social network diversity outcomes (N = 808).

Regression coefficients	**Network diversity**
Constant	2.639***
Frequency of Internet use	.045
Frequent Internet use at home	–.107*
Internet access at home	–.081**
Frequent Internet use at work	.003
Internet access at work	.019
Internet access through mobile devices	–.085
Instant messaging	.133
Sending emails	–.244***
Blogging	–.015
Posting messages (e.g., forum)	–.168**
Sharing digital photos	–.002
Exploring an interest or hobby	.091
Purchasing (e.g., music and books)	–.134*
Playing video games and MPGs	–.173**
Social networking (e.g., Facebook)	.038
Reading news or blog posts	.035
Banking	–.023
Stock transactions	.077

For the majority of ICTs and online activities, we find that the relationship to social network diversity is negative or that it has no statistical significance. Among the statistically significant relationships of diversity, one third are negative: social network diversity is lower among respondents who have Internet at home, access the Internet from home every day, send emails, post messages, buy over the Internet, and play online (multiplayer) video games. People who send emails, for example, have networks that are, on average, .244 SD less diverse than those who do not (or do so less frequently). As for the remaining online activities, no statistically significant linear dependence of diversity on these various regressors was detected. They include sending instant messages, blogging, sharing digital photos (e.g., Flickr), exploring an interest or hobby, banking, stock transactions, and perhaps unexpectedly – considering [4] – social networking services (e.g., Facebook or Weibo). Although ranked second after instant messaging (n = 635, 78.6%), it could not be established whether this type of social media use contributes to explaining the variability in personal network diversity.

Taken together, our ICT results suggest that urban Malaysians' online behaviour has created a new social setting that is in (partial) competition with traditional ones but that has a higher likelihood of reducing social network diversity. Within the theory of 'networked individualism' [18], this would then suggest that urban Malaysians function socially within dichotomous personal networks rather than 'glocalized' ones. Social network diversity among urban Malaysians is medium to low (M = 2.57, on a scale from one to five). When discounting racial/ethnic and religious diversity responses and after conversion into a binary categorical variable (0 = no one, 1 = at least one person), we arrive at a social network diversity index of M = 11.71 (SD = 5.59); this is only slightly higher than M = 9.25 (SD = 5.30) found in the US ([4]). This figure would be difficult to account for within a 'glocal' perspective: for one thing, percentages for Internet access and frequency of use in the Klang Valley are high (see Table 1), and for another, Malaysia's overall ratings for Internet affordability, skills, and individual usage are above average compared with other upper-middle-income countries ([30]). These results would thus support Putnam's original 'time displacement' hypothesis ([31]), at least, as applied to the Internet and today's digital technologies ([32]).

Despite these tendencies, the data do not point to a deeper, more structural shift toward 'networked individualism.' Within the set of negative relationships, coefficients are low, ranging from .081 SD (frequent Internet use at home) to .244 SD (sending emails). What is striking is that in urban Malaysia, affordances for higher network diversity are less in evidence than in the US (e.g., [4]). For example, in the US, the use of social networking services contributes directly to egocentric network diversity but in urban Malaysia, respondents who use social media (78.6% of the sample), the regression analysis does not allow us to draw any conclusions.

Explanations for the mixed results and the similarities with, and differences from, the US findings may be looked for in (1) the specificity of the individual ICTs examined in this study, (2) respondents' 'network practices' [33], as in, for example, the amount of active communication on Facebook [15], (3) properties of their egocentric social networks other than diversity such as size, density or centrality, and (4) structural aspects of contemporary Malaysian society. Given the large number of variables and combinations among them, we will not explore this further in the present study. The main purpose has been to gain insight into the 'bigger picture' relating digital media (and devices) and egocentric network diversity. Further examination of the many individual survey findings will allow a deeper, more nuanced understanding of social network diversity among urban Malaysians but this will be reported in a separate publication.

This study was supported by Grant FRGS/1/2012/SS09/TAYLOR/03/1 from the Malaysian Ministry of Higher Education, 2013–2015.

References

1. N. B. Ellison, C. Steinfeld, C. Lampe. Journal of Computer-Mediated Communication **12**, 3: 1143–1168 (2007)
2. A. L. Gonzales. Communication Research **42**: 1–20 (2015)
3. A. Marin, B. Wellman. The Sage handbook of social network analysis. Thousand Oaks, CA: Sage Publications, 11–25 (2011)
4. K. N. Hampton, C. J. Lee, E. J. New Media and Society **13**, 7: 1–19 (2011)
5. J. Porter, M. O. Emerson. Social Science Quarterly **94**, 3: 732–757 (2013)
6. D. A. Harrison K. J. Klein. Academy of Management Review **32**, 4: 1199–1228 (2007)
7. M. McPherson, L. Smith-Lovin, J. M. Cook. Annual Review of Sociology **27**, 1: 415–444 (2001)

8. L. Rainie, B. Wellman. Cambridge, MA: MIT Press. (2012)
9. H. M. Adnan, S. R. Mavi. Asian Social Science **11**, 15: 1–9 (2015)
10. A. De Rycker, L. F. Yang, C. E. Lee, R. Ponnan, V. S. Lokasundari. International Journal of Asia Pacific Studies **11**:61–92 (2015)
11. W. M. W. Jaafar. The Macrotheme Review **3**, 2: 170–187 (2014)
12. E. Bradner. Paper presented at the CHI 2001 Conference on Human Factors in Computing Systems, Seattle, WA. (2001)
13. B. Wellman. In: K Christensen, D Levinson (Eds.), Encyclopedia of community: From the village to the virtual world. Thousand Oaks, CA: Sage, 560–563. (2003)
14. S. Zhao. Journal of Computer-Mediated Communication **11**, 3: 844–862 (2006)
15. A. Bohn, C. Buchta, K. Hornik, P. Mair. Social Networks **37**: 29–41 (2014)
16. B. Wellman, J. Boase, W. H. Chen. IT & Society 1:151–65 (2002)
17. K. N. Hampton, L. F. Sessions, E. J. Her, L. Rainie. Pew Research Center. Washington, DC. (2009)
18. B. Wellman. International Journal for Urban and Regional Research **25**, 2: 227–252 (2001)
19. C. Haythornthwaite. Information, Communication & Society **8**, 2: 125–147 (2005).
20. B. Wellman, A. Quan-Haas, J. Boase, W. Chen, K. Hampton, I. Díaz, K. Miyata. Journal of Computer-Mediated Communication **8**.
21. K. N. Hampton. American Behavioral Scientist **53**, 8: 1111–1132 (2010)
22. K. N. Hampton KN, B. Wellman. City and Community **2**, 3: 277–311 (2003)
23. J. W. Creswell. Research design: Qualitative, quantitative and mixed methods approaches (4th ed). London: Sage (2014)
24. W. P. Vogt, E. R. Vogt, D. C. Gardner, L. M. Haeffele. Selecting the right analyses for your data: Quantitative, qualitative, and mixed methods. New York: The Guilford Press (2014)
25. National Census. Population distribution and basic demographic characteristic report 2010 (updated 5 August 2011). Department of Statistics Malaysia. Available at: http://www.statistics.gov.my (2010)
26. N. Lin, B. H. Erickson BH (Eds.). Social capital: An international research program. New York: Oxford University Press (2008)
27. L. A. Clark, D. Watson. Constructing validity: Basic issues in objective scale development. Psychological Assessment **7**, 3: 309–319 (1995)
28. N. Lin, Y. C. Fu, R. M. Hsung. The position generator: Measurement techniques for investigations of social capital. In: N. Lin, K. Cook, R. Burt (Eds.). Social capital: Theory and research. New York: Aldine de Gruyter, 57–81. (2001)
29. A. R. Embong. State-led modernization and the new middle class in Malaysia. Basingstoke: Palgrave. (2002)
30. S. Dutta, T. Geiger, B. Lanvin (Eds.). The global information technology report 2015: ICTS for inclusive growth. Geneva: World Economic Forum and INSEAD. (2015)
31. R. D. Putnam. Tuning in, tuning out: The strange disappearance of social capital in America. Political Science and Politics **28**, 4: 664–683 (1995)
32. M. Kestnbaum, J. P. Robinson, A. Neustadtl, A. S. Alvarez. IT and Society **1**, 1: 21–37 (2002)
33. B. Hollstein. Qualitative approaches. In: J. Scott, P. J. Carrington (Eds.), The Sage handbook of social network analysis. Thousand Oaks, CA: Sage Publications, 405–416. (2011)

Deathless after Death: Humanize the Tradition and Celebrate of Death in Java

Sumekar Tanjung[1,*]

[1]Department of Communications, Islamic University of Indonesia, Yogyakarta, 55584, Indonesia

Abstract. This paper aims to describe important side about death conception in Java, especially Bejikarto's people as one of Javanese in Yogyakarta. This research located in Yogyakarta city and take the informants purposively. Technique of collecting data doing through participate observation and in-depth interviews, while the data analyzed by ethnographic descriptive. For Javanese, death is not the end of the journey in this temporal world. Death as a manifestation of the body's extinction and the new life's genesis that is eternity. In Javanese culture, all of the life and death's process have concept and control. Death upheld to maintain harmonious relations with others, with nature occult or supernatural with God. To celebrate the death of the bodies considered as an appreciation, respect, and the act of a religious nature. In Javanese, human is the unity with macrocosm. Celebration of death is reflection of Javanese desire in order to reach *manunggaling kawula lan gusti* (unification with God).

1 Introduction

Bejikarto is located about 15 kilometers east of the main border city of Yogyakarta and 10 kilometers north of Prambanan Temple. The village is in the midst of large farms and lush, which is the source of village economy. Communication with the surrounding villages and the town is quite smooth and openness, with adequate infrastructure support.

In Javanese people, death is not seen as a problem to arbitrarily decide the common thread between the spiritual and the physical body. Death upheld and generate social realities in their traditions. Love of the Javanese in Javanese his identity cannot be dilunturkan dogma other teachings, even his own religion. An understanding of all conceptions of life and death can be obtained through deductive paradigms scripture. Meanwhile, local traditions departing from inductive paradigm, where the tradition of ancestral become the main handle. When religion sign in Java, both paradigms are met. Both push and pull each other in the form of acculturation and syncretization. Therefore, Javanese adherents of all religions, spiritualism has an attitude and thoughts are relatively common, and they are the true Javanese. This means more emphasis on those who carry out their religious beliefs mingled with Java.

Generally, Javanese form of belief in God are non-doctrinal, not stuffy, and openned. According Bejikarto people, *Gusti Pangeran* (God) is *Tan Kena Kinaya Ngapa* (immutable

* Corresponding author: sumekar.tanjung@yahoo.com

and cannot be instance). They allegorized *Gusti Pangeran* relationship with humans symbolically, *koyo kodhok kinemulan ing* leng (frogs swaddling burrows and frog swaddled burrows). The distance between God and creation are considered not measurable anymore, unlimited, and close but not touching (*adoh tanpa wangenan, cedhak tanpo senggolan*). That is why a lot of people trying to exploit the symbolism of Java on the song Macapat to achieve and get closer to the *Gusti Pangeran* (God).

"For Javanese people, the most important thing is to know ourself. And it also means, recognize and honor that God is free to move hearts of everyone. Not as if religion is a private matter, but because the person himself must feel for where came from, and where will go, God called him." (Mandi, September 25, 2014)

Those statement supports Javanese expressions that hate dogmatism, exclusivism, fanaticism, and arrogance as disclosed Mandi as doyen of Bejikarto. In Javanese culture, autonomy for people to find themselves at the bottom of his soul, the coordination of God is very respected.

According Ciptoprawiro in Prabowo [1], Javanese philosophy throughout its history has always conclude that God is *sangkan paraning dumadi* (origin and the return of all instance). When death comes, there are shapes and forms such as human self that comes out of the body and luminous white, Bejikarto people refers to it as *mayangga seta*. If the shape already seen, then silence, and silence is considered as an effort to unite themselves with God when it dies.

Javanese literature do not view death as the opposite of life. Therefore, the concept of death is seen as a result of their birth. In short, if there is no birth, then during which there will be no death. Meanwhile, for Bejikarto people, life always comes before and after birth. So death is often expressed as *mulih mulo mulanira* (back to origin) in Serat Wedhatama on one couplet of *Pucung* song [2]. This implies that death is not the end of the journey in the mortal world.

Bejikarto people believes that live just as the same as allusion, *urip natural neng donyo iku mung nunut ngombe* (live in this world just stopped for drink). The word "drink" interpret as transience, when drink is enough, then the journey to the final destination can be resumed. Man must remember that life journey will follow later. Given that there are three concepts of Javanese life cycle on great dharma. Javanese people often refer to *tigaan* concept. *Purwa* (beginning) as a preliminary step of man being born, *madya* (center) is a step in the life of man facing older adults, and *wasana* (end) is the final stage of an elderly man to dying [3].

By Mandi's statement, the three degrees of dead are distinguished in terms of procession or treatment when the remains will be interred.

"Death is loss of the life of body. Death has three degrees. The first is main death, the death of a man of honor. Intermediate die, natural death or ordinary death like most people. Third degree, nistha death, death from suicide or accidents." (September 25, 2014)

In Serat Asmaralaya described how man who is facing death's door. Asmaralaya are metaphors of life of every human being. Figuratively question is think of dying or death of human consciousness. That God is essentially intangible and enveloping light emanating universe. If the time of death has arrived, there are shapes and forms such as himself out of the body and luminous white (*mayangga seta*). If the kind of shape it hass been seen, then silence, and silence is an attempt to unite ourselves with God.

2 Method and Participants

Communications, rather than specific languages, provide the frame of reference for analyzing the place of language in any particular society or culture. The focus of studies in the field is on individual speech communities, which are clusters of people using common signs to communicate. Particularly, communication ethnography is interested in the way that communication within a speech community is organized into various systems of communication events and how they interact with every other system in the culture. It looks to answer the basic question of what a speaker needs to know to communicate correctly and appropriately within a given speech community and how a speaker learns to do this. Researcher analyze different speech situations, such as ceremonies, or speech events, greetings or compliments, to determine how their structure and content are culturally determined. Regardless of the topic of the study, researchers in the ethnography of communication focus on a speech community. The potential participants for these research were two local residents in Bejikarto. The actual participants at each meeting varied depending upon who showed up or asked to be present.

3 Results and Discussion

Among the causes of death are suffering. The nature of this illness as well as death, easily happen anytime. For Bejikarto's people, pain is a matter of comfort and discomfort. Dul Sadin, who is resident in this village. He is second child of five siblings. He had eight siblings before, three others were died just for one reason: *masuk angin* (catch a cold).

Masuk angin is a symptom of getting sick (usually the incubation period of getting flu or cold). The symptoms are feeling dizzy, body temperature is above normal (but feel cold, like a fever), feel sick inside stomach (wanting to throw up, similar to feeling seasick if we have ever been in that situation), and cold sweats. *Masuk angin* is the ordinary Indonesian's term for feeling unwell. Some peoples are so vulnerable to changing weather, from dry season to rainy season (we call this *pancaroba* period in changing weather or season), and they blame the wind as the cause. It is called *masuk angin* because the wind gets into body. If we hang out all night long (*begadang*), short of sleep or rest and get wind into our body (night wind), you could catch *masuk angin* and feel unwell when you wake up in the morning.

For Javanese people, *masuk angin* has emerged as the concept of pain in the form of discomfort, discomfort in the body, believed that caused by the influence of the wind. Actually, *masuk angin* is not a medical term, but people called it. Presenting symptoms include abdominal bloating, nausea, pain, plus could not fart and burp [4]. For Javanese, winds are said to fall into human body through the pores of human body as well as *babahan hawa sanga* (there are nine-holes in human body). As a result of the wind elements that enter to body of excess body imbalance. For Javanese people, health is being balance, while illness, injuries and misfortunes regarded as an imbalance.

Due to differences between Javanese concept of medical and then *masuk angin* was viewed differently by two sides. There is no medical terminology for entry into *masuk angin* that the wind is only considered as a collection of symptoms of a disease. The medical side is equal to mention catch a cold as same as influenza. *Masuk angin* is considered to be a symptom of different diseases such as influenza, hypertension, diabetes, hepatitis, or post-traumatic stress. Therefore, catch a cold will be treated by medics based on the symptomatically felt by the patient.

> "*When I was catch a cold, my eyes felt heavy while opened. My body felt lazy when I was used for work. If there is spare time, so can only be used to bedrest. Just when I smoke, it tasted was bad, bitter tongue, it was definitely I getting catch a cold,* " (Sud, September 25, 2014).

Generally, people with *masuk angin* feels weak, lethargic, and mouth is not delectable to eat and drink. Javanese distinguished cold weight basis mild symptoms, more or less the number of symptoms as well as the ease of the process of healing. Late *masuk angin* (*kasep*), are allowed by the sufferer so bring sudden death. As a result many sudden deaths are often preceded by malaise, shortness of breath or out a lot of sweat. The expression of traditional healer, Kelik confirmed that, "*Angin duduk it should start from ordinary cold, but not felt. It is dangerous because not doing any work just sat there, suddenly died instantly.*"

The doctor who practice medicine both in health centers, hospitals and private practice patient often find a common complaint that his body does not feel good. They told the doctor that he was prone to colds. Prior to their doctors already performed *kerokan* (drawing an oiled coin over skin repeatedly until skin turns bright red, it makes the wind exit from body and will get well), herbal drink, and take a pill *kecethit* (despite it was name pills back pain or twisted pain, but people in rural areas believe that this pill to cure various diseases). Therefore, they went to the doctor who is considered more knows because it has a complete medical equipment. Shortly, all the discomfort or discomfort is felt, they give the term as colds.

Medical systems in traditional societies have different concept of health and illness, as well as treatment efforts. In Javanese people, Foster and Anderson [5] classify the causes of disease ascribed personalistic system or naturalistic system. In personalistic concept, the disease is believed to be the intervention of an intermediary for the active form of magical or non-human. In Bejikarto society, they call it as unreasonable or unnatural disease. These types of diseases such as *kelebon*, *digawe wong*, *kualat*, and *kebendhu*. The healing is done through ceremony or offering to neutralize cause pain. Naturalistic concept is regarded as *lumrah* (proper) or natural disease. Usually affecting the health of the body is the weather, food, bacteria, and toxins.

Views of the health anthropologists divide the disease into three health trust model, the magico-religious models, biomedical models, and holistic models. Javanese society which has strong traditional system, all illness is described in a holistic manner. This model emphasizes the aspects of harmonization or balance the body, regarding environmental, social, cultural, and behavioral. A healthy body condition is seen as a state of mind or a balance between the physical and the mental. Javanese sighted, "*batin karep, raga nututi*", meaning if the mind wills, the body too. That the body, mind, and spirit or *sukma* (soul) is a unity of mutually influential.

Javanese people are understood that the body is maintained by *kakang kawah adhi ari-ari* (everyone has a twin in their life), so human life survive they must understand the universe as a symbol of God's power. It includes four elements, *kawah* (fetal membrane) for the east, blood fo the south, *tali pusar* (umbilical cord) for the west, and placenta for the north, which serves to maintain physical and mental health in order to keep in equilibrium, in the context of *memayu hayuning bawana* (man must be always maintain harmony and balance) Healing intended to restore the balance. Meanwhile, cured or dead, do not know when it happened.

4 Humanize the tradition

In Bejikarto and other region in Java, to celebrate death of the bodies considered as an appreciation and respect that should be perpetuated. Even been understood together that event *tahlilan* a ceremonial ritual is usually done as a generality society. And their relatives, *handai taulan* (friends) and neighbors who all are men (because that was involved in this ritual are male), do *dzikir*, while certain prayers to be sent to the corpse.

Rila or sincere, means let everything consciously, as a guideline though often difficult to obtain, but always endeavored everyone. In Javanese philosophy, bamboo philosophy adapted to central element of Javanese culture, namely *rila* (sincere), and *nrima* (grateful), and *sabar* (patient).

In Bejikarto, this ceremonial usually has done after burial ceremonial on afternoon before *maghrib* (sunset call to prayer). *Selamatan* on third days of death (*nelung dina*) is intended as an attempt heir to honor the dead. In this regard, they believe that the spirit of the deceased still inside the house. But the spirit was not in bed anymore. Spirit is already wandering to find a way and easily leave the house. After the spirit is considered out of the house, then held *mitung dina* (seventh days of death). Spirit that has come out to be paused in the courtyard. To ease the way, the guests held *tahlilan* to pray. *"Tahlilan means read tayibah sentences. The completion of the guests later given bancakan rice and side dishes."*

The term of bancakan on bancakan rice is likely to come from the Pungkur, made of woven bamboo distance tenuous. This webbing called a rack. Later developed and adjusted to *Bancak*. In addition to the mention of rice *bancakan*, some are calling it a blessing rice (from Arabic baraka).

On the fourteenth day of death, the spirit has been toward the grave. Given offerings also vary. It is intended as offerings to the spirits and bodies should be enhanced, such as blood, flesh, nails, hair, bones, and muscles [6]. To enhance all of them, do *nyatus dina* or one hundred days of death. Grain is almost the same, the difference is only in the stocks, sticky rice and compote. *Pasung* that made such a mountain (*payung*) of jackfruit and filled material from wheat. Its meaning is that the dead get an *payung* (protection). Because the person who died is like going through a long and hot, then for him made of sticky rice as lemek (*base*), so that the feet are not hot. *Ketan* also significantly *raketan*, it means to closer to God. Also equipped compote dish which comes from the creator of universe or *kolaq* (God is The Creator). With this kind of dish, expected dead will smoothly facing The Creator. It then lasts up to one year and two years death of death.

On the one thousand days (*nyewu dinane*), relatives and neighbors came to donate (give money donations). This tradition is expected to entertain the bereaved family as well as making the message that living also must be followed (death) at a later date. Is also inseparable presented their supper each event held. Model presentation of the dishes are usually varied, depending customary walk at the venue. But the entree menu was more than modest, or tends to resemble a festive menu of dishes that smell. If they were not taking food after *tahlil*, usually *selamatan nyewu* put a meal together. After eating together and held a feast. This event so impressed a little party and become a norm.

5 Conclusion

It should be noted, all the conceptions that has been described merely formal phenomenon. Melee core Javanese people to pick up and interpret the human life cycle process can not be described by anyone. In his search, they often hit a lot of things and it all comes down to the individual spiritualism. The point of the highest value, Javanese spiritualism hopes to reach *ngangkah ningrat nunggal Hyang* (reach the silence on the upper level one with God) [7]. A number of studies, including this paper, only slightly touched on the concept of death in Java.

References

1. P. D. Prabowo, Pengaruh Islam dalam Karya-karya R.Ng.Ranggawarsita (Narasi, Yogyakarta, 2003)
2. Ki Sabdacarakatama, Serat Wedhatama: Karya Sastra K.G.P.AA. Mangkunegara IV (Narasi, Yogyakarta, 2010)
3. S. Endraswara, Mistik Kejawen: Sinkretisme, Simbolisme dan Sufisme dalam Budaya Spiritual Jawa (Narasi, Yogyakarta, 2006)

4. A, Rustami, A–Z Tips Hidup Sehat Sehari-hari (Wahana Totalita Publisher, Yogyakarta, 2009)
5. M. Sudarma, Mengenal Variasi Layanan Pengobatan Alternatif. Sosiologi untuk Kesehatan (Salemba Medika, Jakarta, 2008)
6. T. W. Bratawidjaja, Thomas Wiyasa. Upacara Tradisional Masyarakat Jawa (Sinar Harapan, Jakarta, 1993)
7. H. Mulyani, J. K. J. **2,** 115 (2006)

Islam in the Indonesian religious soap opera: faithfulness for the sake of commodification

Muzayin Nazaruddin[1,*]

[1]Universitas Islam Indonesia, Communications Department, Jl. Kaliurang 14.5 Sleman, Yogyakarta, Indonesia

Abstract. This study discusses Islam representation in Indonesia, focusing on the adaptation of Islamic-mystic magazines into religious soap operas. Its aim is to reveal the norms underlying the adaptation processes. It analyses Hidayah magazine, one of the most popular Islamic-mystic magazine in Indonesia, as the source text, and Rahasia Ilahi, the highest-rating religious soap opera, as the target text. The study finds that the basic norm underlying the adaptation processes is market orientation. It determines two subsequent norms, namely the narrative standardization and faithfulness. The flexible narrative sequence in the magazine should be standardized into the fixed cause-effect narrative structure of soap opera. It creates a more specific norm, namely extreme binary opposition. Meanwhile, faithful norm has three aspects, namely adequacy, factual, and Islamic teaching conformity principles. This three-aspects of faithfulness is specific feature of the religious texts translation.

1 Introduction

Within the Indonesian society, in which religion, especially Islam, plays important roles in the socio-cultural dynamics, the representation of Islam in almost all types of mass media is everyday phenomena and has a long history.

This paper focuses on one recent phenomenon of Islam representation in Indonesia, namely the adaptation of Islamic-mystic magazines into religious soap operas. The aim of the study is to reveal the norms underlying the adaptation processes. The study analyses Hidayah magazine, one of the most popular Islamic-mystic magazine in Indonesia, as the source text, and Rahasia Ilahi (The Secret of God), the highest-rating religious soap opera, as the target text. Thus, the research questions of the study are: what kind of norms underlie the adaptation processes from Hidayah magazine into Rahasia Ilahi? How these norms work in such processes of adaptation?

2 Research method

The research is done through two stages. First, I conducted some general analysis dealing with the huge number of religious soap operas and Islamic-mystic magazines. This step was hand in hand with my daily activities as audience who had routinely watched and read them,

*Corresponding author: muzayin.nazaruddin@uii.ac.id, begawan.nazaruddin@gmail.com

since 2002 to 2006. Second, I specifically analysed nine editions of Hidayah magazine, randomly chosen from 2003 to 2006: Hidayah 28, November 2003; Hidayah 33, April 2004; Hidayah 45, April 2005; Hidayah 46, May 2005; Hidayah 49, August 2005; Hidayah 53, December 2005; Hidayah 54, January 2006; Hidayah 58, May 2006; and Hidayah 59, June 2006. In these nine editions, there were 41 mystical stories in *'Iktibar'* (The Real Experience) column, which always become the source text of the adaptation into religious soap operas. Besides, I also analysed 18 episodes of Rahasia Ilahi, an adaptation of *Iktibar* column in Hidayah magazine.

In this study, I follow the analysis model proposed by Cattrysse [2], consist of four steps: the selection policy of the source texts, the adaptation policy of the texts that have been selected, the function of the target text within its contexts, and the relations between selection–adaptation policies and the function of the target text within its context. But, due to the limited pages, this paper would only discuss the first and second steps. This model is chosen because of the similarity of the object study that includes large amounts of texts, not just a single-pairs text (from one source text to one target text), and the similarity of the study aim, that is revealing the norms behind the adaptation. Cattrysse has analysed American noir films made from 1940 to 1960. In this study, I analyse the religious soap operas aired from 2004 to 2007.

3 Selection policy

Religious soap opera first emerged in 2004. A year later, in May 2005, according to AC Nielsen survey, there were 35 titles of them [9]. Even, I found at least 44 titles of religious soap operas in all Indonesian television during 2004 to 2007 [16]. Of course, not all of them are the adaptation of certain source texts. Then, how many of them are the adaptation and how many are not, or purely based on the screenplay? The lack of adequate documentation of television programs in Indonesia makes a considerable difficulty to answer this question quantitatively. However, I found that some of them mentioned clearly the source text of their stories (Table 1).

Table 1. Religious soap opera and its source text

No	Religious soap opera	Television	Source text
1.	Rahasia Ilahi (The God Secret)	TPI	Hidayah magazine
2.	Takdir Ilahi (The God Destiny)	TPI	Hadith of Prophet Muhammad in the *Bukhari Muslim* book
3.	Astaghfirullah	SCTV	Ghoib magazine
4.	Iman (The Faith)	SCTV	Hidayah magazine
5.	Taubat (The Repentance)	TransTV	Insting magazine
6.	Hidayah (The Guidance)	TransTV	Hidayah magazine

Source: main data (collected, compiled and arranged by the researcher)

We may notice from the table that there are three magazines become the source texts, i.e. *Hidayah*, *Ghoib*, and *Insting*. These three magazines belong to the same genre, that is Islamic-mystic magazine. According to Irawanto [7], this genre is a new phenomenon emerged after the Indonesian political reform in 1998. The genre was pioneered by *Hidayah* magazine in 2001. Since the first edition, this magazine immediately gained success, sold 10,000 copies, and constantly raised their printed copies every year. In 2004, according to AC Nielsen survey, *Hidayah* magazine becomes the most widely read magazine in Indonesia [12]. Then, this success created a new trend, marked by the emergence of the epigones, such as *Hikayah, Ghoib, Insting, Alkisah, Kisah Nyata, Posmo,* and others. They

DOI: 10.1051/shsconf/20173300068

have common identity signifier, i.e. having specific column for mystical or irrational stories, interpreted with Islamic teachings, generally as the evidence of God's power.

Obviously, this market success initially drives the emergence of religious soap opera. Having a great success, more than ten production houses offered cooperation with *Hidayah* magazine in terms of adapting and broadcasting its mystical stories into television program [13]. In 2004, Kusuma Eka Perkasa (KEP), a production house in Indonesia, produced *Rahasia Ilahi* (The God Secret) as an adaptation of mystical stories in *Hidayah* magazine, and broadcasted it on TPI (*Televisi Pendidikan Indonesia,* now known as MNC TV), one of Indonesian national televisions. This first religious soap opera in Indonesia gained a great success. Then, TPI launched the second religious soap opera, titled as *Takdir Ilahi* (The God Destiny). It has reached the top-ten rating of Indonesian television programs only in two months after its launching. According to AC Nielsen survey, during March and April 2005, *Rahasia Ilahi* got 14.9 rating score and 40.29% share score, it was on the top of the 50 programs in all Indonesian televisions. Besides, *Takdir Ilahi* was in the third place with 9.8 rating score and 22.8% share score. The success of these two religious soap operas had placed TPI as the highest rating television at that period [19, 22].

Motivated by this success story, almost all Indonesian televisions produced similar soap opera, in which the mystical or irrational stories and Islamic symbols become the distinctive identity. Then, this genre rapidly grew from 2005 to 2007; some of them were very popular and reached 10 highest-rating television programs [19, 4, 8]. Thus, regarding its history, the rating or market orientation seems to be the basic motivation of the adaptation of Islamic-mystic magazines into religious soap operas. Hence, the media looks at religion, especially Islam, as a marketable commodity. It also means, this is not a matter of Islamic propagation. This conclusion may lead to the thesis of "marketization or commodification of Islam" that has pointed out by many Indonesian experts, such as Sasono [17], Masduki [14], or Setiawan [18].

4 Adaptation policy

In the adaptation policy level, there are several points should be analysed. The most fundamental one is the shift in the genre, from the factual genre towards the fictional one. *Hidayah* magazine claims that their story is not just a fiction. It is news that is packaged in a feature style, and is written based on the report conducted by their journalists. Another Islamic-mystic magazines followed this factual claim; most of them claim that the stories presented are real experiences. The factual claim is supported by the presence of informants or witnesses. In addition, the witnesses statements are often written separately from the main text, in special additional columns, in order to emphasize its importance.

Actually, many columns in *Hidayah* magazine which are claimed as journalistic reports were full of writer's interpretations. There was no clear differentiation between facts and opinions. This is actually a proof of the general phenomena of the journalism weakness within Islamic media in Indonesia, as has been shown by Muhammad [15] and Irawanto [7]. But, in general, though very interpretive, *Iktibar* column in *Hidayah* magazine could still be classified as a journalistic report.

However, in the adaptation level, the model chosen is soap opera, which is generally considered as fictional genre, not factual. Eventhough, religious soap operas also make similar factual claim. In the opening shot of *Rahasia Ilahi*, we will find a big note on the television screen: "This is a real story taken from Hidayah Magazine – name and character have been changed". Some other religious soap operas also claim that their stories are fact, gathered from various sources (see Table 1).

The problem also lies on the change of production mode, from magazine (print media) to soap opera (electronic media). Change in the production mode is complex in terms of

 DOI: 10.1051/shsconf/20173300068

interpretation and translation, so it always produces new meaning. In this case, the producer interprets and translates the content of magazine into visual and aural format of television show, and ultimately blur the boundary between fact and fiction. It has important consequences to the narrative structure of the text. My observation on *Iktibar* column finds five types of narrative structure, namely the story of *su'ul khotimah* (bad death), the story of *khusnul khotimah* (good death), the story of repentance, the story of God's punishment, and the story of God's mercy. Likewise, my observation on religious soap operas, especially *Rahasia Ilahi*, also finds those five stories. Each story has different narrative structure, but has some fundamental similarities. Table 2 describes in detail these five types of narratives.

Table 2. Five types of narrative structures

Narrative type and elements	Comparison	
	Hidayah Magazine	Rahasia Ilahi
The story of *su'ul khatimah,* generally tells the tragic death. a. Depiction of tragic death of the main character. b. Description of bad behaviour of the main character during his/her lifetime as the cause of tragic death. c. Reflection on the story presented, usually by citing the Quran or hadith.	No definite narrative sequence. However, the most used sequence: reflection (lead), tragic death, bad behaviour, and reflection.	A fixed narrative sequence: opening reflection, bad behaviour, tragic death, and closing reflection.
	The story of tragic death and bad behaviour are the main narrative portions, longer than the reflection.	The description of bad behaviour becomes the largest narrative portion.
	Informant testimonies in the text, especially in some separate additional columns.	No informant testimony.
The story of *khusnul khatimah*, generally tells the wonderful and memorable death. a. Description of beautiful death of the main character. b. Description of good behaviour as the cause of beautiful death. c. Reflection.	The most used narrative sequence: reflection (lead), beautiful death, good behaviour, and reflection.	A fixed narrative sequence: opening reflection, good behaviour, beautiful death, and closing reflection.
	The story of beautiful death and good behaviour are the main narrative portions.	The description of good behaviour becomes the largest narrative portion.
	Informant testimonies.	No informant testimony.
The story of repentance, tells the repentance event. a. Description of bad behaviour of the main character before repentance. b. Description of the repentance processes. c. Reflection.	The most used narrative sequence: reflection (lead), bad behaviour, repentance processes, and reflection.	A fixed narrative sequence: opening reflection, bad behaviour, repentance event, and closing reflection.
	The description of repentance processes and bad behaviour are the main narrative portions.	The depiction of bad behaviour becomes the largest narrative portion. The repentance event only depicted shortly in the end.
	Informant testimonies.	No informant testimony.
The story of *azab*, tells the punishment from God. a. Description of bad behaviour of the main character. b. Description of the God's punishment for the bad behaviour. c. Reflection.	The most used narrative sequence: reflection (lead), bad behaviour, God's punishment, and reflection.	A fixed narrative sequence: opening reflection, bad behaviour, God's punishment, and closing reflection.
	The description of God's punishment and bad behaviour have same narrative portion, longer than reflection.	The description of bad behaviour becomes the largest narrative portion. The punishment from God only depicted shortly.
	Informant testimonies.	No informant testimony.

 DOI: 10.1051/shsconf/20173300068

The story of God's mercy, tells the beautiful and bizarre events as evidence of God's love. a. Description of the good behaviour of the main character. b. Description of bizarre events. c. Reflection.	The most used narrative sequence: reflection (lead), good behaviour, God's mercy, and reflection.	A fixed narrative sequence: opening reflection, good behaviour, bizarre event, and closing reflection.
	The description of God's mercy and good behaviour have same narrative portion, longer than the reflection.	The depiction of good behaviour is the largest narrative portion. The bizarre event depicted shortly in the end.
	Informant testimonies.	No informant testimony.

Source: main data (collected, compiled and arranged by the researcher)

We may see that there are fundamental similarities between the narrative structures of *Rahasia Ilahi* and *Hidayah* magazine. Nevertheless, there are also some differences between them indicating important changes in the adaptation processes. First, there is a standardization of narrative structure in the adaptation process. In *Hidayah* magazine, the stories do have constant narrative parts, but the order of the parts is flexible. Meanwhile, in *Rahasia Ilahi* the narrative sequence is standardized with the pattern of cause-effect: started with the causes and ended with the effects. Considering the production aspect of religious soap opera, the standardization of the narrative structure is understandable since its processes is much more complex than the production processes of printed magazine. Moreover, since the production intensity is very high, one episode in a week on average, the standardization of the narrative structure therefore is a needed strategy.

The other dimension of adaptation policy should be analysed is the difference between the character patterns of the source and target text. In *Hidayah* magazine, the story usually focuses on the main character; while other characters, such as families, friends or neighbours usually appear as witnesses or informants who talk about mystical event experienced by the main character. Secondary characters exist, but their role is very limited. In *Rahasia Ilahi*, secondary characters are usually more prominent. In fact, *Rahasia Ilahi* often adds other characters that do not exist in the source text. According to Greimas's actantial model [6], the most important additional character is the opponent. It creates an extreme binary opposition between the main character and the opponent. If the main character is a bad person, the opponent will be a very good person, who always persecuted by the main character. Conversely, if the main character is a very kind person, the opponent will be a very cruel person, who always disturbs the main character. Hence, there is an extreme binary opposition: extreme-good person and extreme-bad person. It already exists implicitly in the magazines, but it presents explicitly in a very strong way in the religious soap operas. It confirms Lotman's idea that the binary opposition principle is the basis for every literature text and work of art, and even for every sphere of human life [11].

The change in the narrative portions also strengthened such binary opposition. In *Hidayah* magazine, description of the causes and effects are generally in equal quantity; while in *Rahasia Ilahi*, description of cause dominates the show. For example, in the story of *su'ul khotimah* (bad death) and *azab* (punishment from God), the overall duration of the soap opera is dominated with the depictions of bad behaviour of main character. Instead, in the *khusnul khotimah* (good death) and God's mercy stories, the show is filled with the depictions of good behaviour of main character.

Religious soap operas effectuate an over exaggerated of generalization that the bad person will forever be bad, otherwise the good individual will forever behave good actions. The processes of changes, to be good from bad condition, or otherwise, or to be more mature, never shown. In the repentance story, it is more obvious. Typically, from the beginning to the middle scenes, the show is dominated by many bad behaviours of the main character, just to show that he or she is a sinner. Then, in the end of the show, suddenly the main character repents. The struggle and contemplation for repentance is not shown at all. Repentance is depicted as something happen suddenly and easily.

Besides the addition of the opponent, *Rahasia Ilahi* usually adds Islamic cleric (*ustad* or *kyai*) character as the helper. I find at least four representation types of *ustad's* roles: praying for the dead man being punished by God; fighting against the demon or witch; giving advise someone who has just converted; and preventing people who will do anarchists against the criminal. This representation is an excessive reduction. Various studies have demonstrated strong evidences of *kyai's* roles; especially through their *pesantren* (Islamic boarding school), in the development of education, economic empowerment, preservation of local literature and culture, as well as development of democratization [1, 3, 5, 21, 25]. These various roles do not exist in religious soap operas. They reduce the role of *kyai* in such a way: only in the 'mystical' and 'superstition' world. One of the very common closing scene in the religious soap opera is the fight between *kyai* and shaman, represented as the fight between 'goodness' and 'wickedness' [26]. It produces another binary opposition, between *kyai* as 'extreme angel' versus shaman as 'extreme evil'.

The last important dimension is the adaptation of 'the reflection' as a part of the narrative structure. In *Hidayah* magazine, the reflection is written by the journalist, as integral part of the story, in the beginning as the lead, and then reaffirmed in the end of the story. The same pattern is adapted in *Rahasia Ilahi*, there are opening and closing reflection. But, in *Rahasia Ilahi*, the reflections are actually not included as part of the soap opera itself, but done by Arifin Ilham, a young *ustad* who is very famous in Indonesia. Arifin Ilham is featured not to be an actor in the soap opera, but to open and to close the religious soap opera with Islamic teachings. He always appears in the opening and closing, quotes the verses of Quran or the hadiths of Prophet Muhammad, and dress as an *ustad*. It suggest that Arifin Ilham is giving an Islamic sermon, yet the medium is television, not mosque or Islamic school. Thus, it confirms Sisley [20] who has emphasized the notion of faithfulness in the translation of written bible to audio-visual media. The complexity of the translation, from one medium to another, is not merely about producing the new meaning, but also the problem of 'expected reception' when the producer expect the audience to belief in their products.

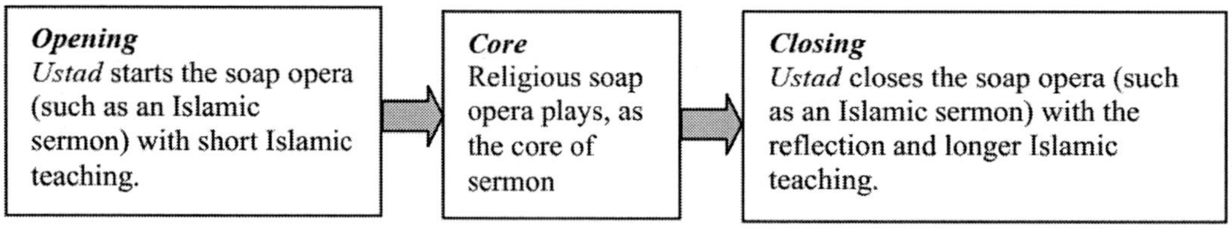

Fig. 1. The Islamic sermon as the narrative pattern of religious soap opera

The similar strategy, presenting popular *kyai* or *ustad* to give Islamic teaching, is also performed by another religious soap operas, as shown in Table 3.

Table 3. Popular *ustad* in the religious soap opera

No	Soap Opera	Television	Indonesian popular *ustad*
1.	*Rahasia Ilahi*	TPI	Arifin Ilham
2.	*Takdir Ilahi*	TPI	Ali Mustafa Yaqub
3.	*Astaghfirullah*	SCTV	Jefri al Buchory & Yusuf Mansur
4.	*Kuasa Ilahi*	SCTV	Jefri al Buchory
5.	*Pintu Hidayah*	RCTI	Luthfiah Sungkar

Source: main data (collected, compiled and arranged by the researcher)

Carefully observing this norm, we will find slightly different notion of the faithfulness with common understanding in the translation studies, which usually understood as 'correct' translation of source text. In the correct translation, the principle of adequacy or equivalence between the adaptation and its original determines the quality of translation [2, 23, 10]. In the case of religious soap opera, the adequacy principle also seems important, by which they tend to state clearly the source text and prefer to follow the narrative structure of the source [24]. However, in the religious soap opera there are another more important dimensions of this faithfulness, i.e. the conformity with the teachings of Islam, based on a true story, or combination of both. Hence, there are some aspects of the faithful norm: adequacy, factual, and Islamic teaching conformity principles. I suppose this double meaning of faithfulness, refers both to the truthfulness adaptation of the source, and the religious teaching and real experience conformity, is the specific feature of the translation of religious text.

5 Conclusion

This study finds that there are some norms underlying the adaptation of Islamic-mystic magazines into religious soap operas, which are related to one another in certain hierarchical relation. Figure 1 illustrates the hierarchical relationship of these norms.

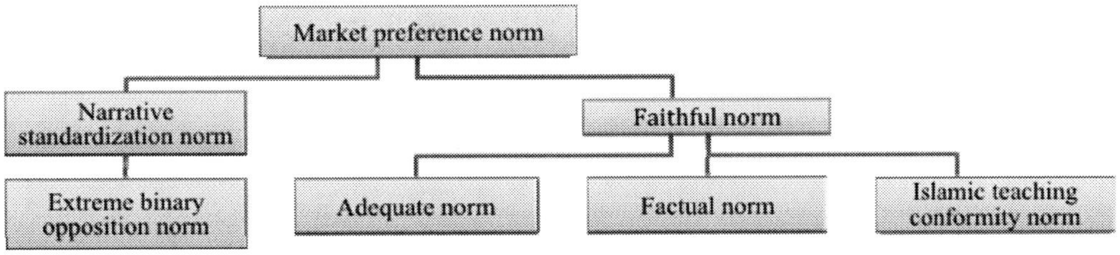

Fig. 2. The norms of the adaptation processes

The most fundamental norm is the popularity or market preference, determining the choice of adaptation from popular Islamic-mystic magazines, into soap operas that are also extremely popular. Thus, it is not a matter of Islamic propagation, but commodification of Islam. This economic motivation determines two subsequent norms, i.e. the narrative standardization and faithfulness. The flexible narrative sequence in the magazine should be standardized into fixed cause-effect narrative structure of soap opera. It creates a more specific norm, namely extreme binary opposition, which are already exist in the source text but amplified in the target text. Meanwhile, faithful norm has three aspects, namely adequacy, factual, and Islamic teaching conformity principles. This three-aspects of faithfulness is specific feature of the translation of religious texts. It is obvious in the explicit signifier in the religious soap operas mentioning the story as a real experience, also in the appearance of popular *ustad* to open and close the soap opera with Islamic teachings.

Acknowledgements

This article is part of the author's study at the Master Programme of Semiotics, University of Tartu, Estonia, sponsored by Directorate General of Higher Education (DGHE), Ministry of National Education, Indonesia. The author is very grateful to Elin Sütiste (Department of Semiotics, University of Tartu) for insightful comments on the draft of this article.

References

1. L. Binder, Islamic tradition and politics: the kyai and the alim, Compr. S. in Soc. and Hist. **2** (1959).
2. P. Cattrysse, Film (adaptation) as translation: some methodological proposals, Target, **4.1**, 53-70 (1992).
3. Z. Dhofier, *The pesantren tradition: a study of the role of the kyai in the maintenance of the traditional ideology of Islam in Java* (ANU, Canberra, 1980).
4. Fathuri, et al., Menjual kuburan di layar kaca, Majalah Syirah, July (2015).
5. C. Geertz, The javanese kyai: the changing role of a cultural broker, Compr. S. in Soc. and Hist. **2** (1959).
6. A.J. Greimas, *Structural semantics* (Lincoln & London, Univ. of Nebraska Press, 1983).
7. B. Irawanto, Transformasi pers Islam di era reformasi, J. I. Sos. dan I. Pol. **9.3**, 297-315 (2006).
8. S. Ivvaty, Tahun horor di republik hantu, Kompas, 30 December (2007).
9. A. Khudori & A.R. Pitakasari, Sinetron keagamaan melorot setelah banjir, Gatra, 13 June (2005).
10. J. Lambert & C. Robyns, Translation, In R. Posner, K. Robering, T.A. Sebeok (eds.), *Semiotics: a handbook of nature and culture* (Berlin & New York, Walter de Gruiter, 2004).
11. Y.M. Lotman, Notes on the structure of a literary text, Semiotica, **15.3**, 199-205 (1975).
12. Majalah Hidayah, Hidayah: empat tahun menorehkan sejarah, **49**, August, 98-99 (2005).
13. Majalah Hidayah, Menengok perjalanan Majalah Hidayah di layar kaca, **49**, August, 106-107 (2005).
14. Masduki, Religiotainment di era "budaya pop" televisi, J. IPTEK-KOM, **8.2** (2006).
15. A. Muhammad, Jihad lewat tulisan: kisah sukses Majalah Sabili dengan beragam ironi, J. Pantau, **2**, 15 (2001).
16. M. Nazaruddin, Sinetron religius: sinetron islami?, J. Kom. **2.2**, 315-330 (2008).
17. E. Sasono, Religiusitas layar kaca, Tempo, 3-9 October (2005).
18. I. Setiawan, Komodifikasi dan representasi nilai-nilai keislaman dalam tayangan televisi, J. IPTEK-KOM, **9.1** (2007).
19. Sinar Harapan, TPI melanjutkan perjuangan untuk tetap di atas, 8 April (2005).
20. J. Sisley, *Translating from one medium to another: explorations in the referential power of translation* (Univ. of Warwick, 2000).
21. K.A. Steenbrink, *Pesantren, madrasah, sekolah* (Jakarta, LP3ES, 1974).
22. Suara Merdeka, TPI andalkan sinetron religius, 22 February (2005).
23. P. Torop, Intersemiosis and intersemiotic translation, Eur. J. for Semiotic S., New Tartu Semiotics, **12.1** (2000).
24. G. Toury, A cultural-semiotic perspective, In T.A. Sebeok (ed.), *Encyclopedic dictionary of semiotics* (Berlin & New York, Mouton de Gruyter, 1994).
25. E. Turmudi, Religion and politics: a study on political attitudes of devout muslims and the role of the kiai in contemporary Java, Southeast Asian J. of Soc. Sci. **23.2**, 18–41 (1995).
26. K. Van Heeren, Cruelty, ghosts, and verses of love, ISIM Review **22**, Autumn, 20-21 (2008).

Commodification broadcast television program on local television in Bandung, West Java

Aep Wahyudin[1,*]

[1]Faculty of Da'wa and Communication, Islamic State University of Sunan Gunung Djati Bandung, Indonesia

Abstract. The media has now become a tool that performs many functions, beyond its function as a conduit of information to the public. As an industry, television programming impressions always count with the economic calculations that led to the result of profit. Media with strengths and opportunities will see any loopholes that benefit the industry into a viable commodity. The commodification in a critical view of the media sees the commodities and commodification with two things that connect the object and process. It is one indicator of global capitalism that is now taking place. Commodification is the transformation of the shape of the relationship, which is initially trafficked into things that are free of the commercial nature of the relationship.

1 Introduction

In less than two decades, the world of television and entertainment in Indonesia experienced a remarkable development rapidly. The proliferation of private television in Indonesia, making human beings can survive television personality must exist for their own companies, because they differ with subsidized government television. This fact is the private television demanding to search for life and a large profit. Given the media industry is one of the promising businesses. But in fact the capital creativity they can take heart in the wider community to become loyal viewers. What they presented turned out favored, even though there are so many forms of commodification made by the personnel of private television.

The media is far different with BBC. BBC has deliberately oriented specific channels to certain age groups such as youngsters and children in order to attract viewers who are drifting away in the competition. Media has now become a tool that performs many functions, beyond its function as a conduit of information to the public. Media are in triangulation state, market, civil society. Media that linking these three components, the reform era triangulation relationship is dominated by the market, whereas formerly dominated by the state. Media liaison these three components, and must not be located or in favor of just one component [1].

As time becomes business promising media. The media industry can be used as a propaganda tool and for commercialization. Especially with the proliferation of private

* Corresponding author: fwahyu2000@yahoo.com

television which certainly does not get subsidies from the government, so they have to rack my brain to be able to survive and continue to exist in the television world, is to create programs that dance that the public likes what the television present, so the profits (profit) can they. One of the commodification of land except entertainment, talk shows, reality shows and more prevalent on television is a religious impression. Religion with a variety of formats into a land that always never dry to be explored at once exploited in various forms impressions. Modern society is hungry for spiritual cleansing would make religion a mooring end to solve life's problems. The rise *tausia'h* business since the early 2000s became one of the clear evidence of how religion is now a basic human need. Especially in the moments of the annual ritual of some sort of Ramadan comes, religion turned into a business area that inspired many creative people to produce events that can suck the audience, because it is not uncommon that ultimately it was stuck into the game symbols away from religious values pure.

In the context of popular culture, mystique soap opera is just a phase of the ongoing commodification of culture. The television station made programs for mystical scented soap opera, and certainly not just a passion born of mere propaganda. More than that, such broadcasts are also laden trade policy because it provides enough profit from the business side. Absorb enough ad impressions lot plus persistence models thus give concrete evidence of ulterior motives are behind view program.

2 Literature review

Local private television media in West Java, could not be separated with the business, the business interests that will enable local television as an industry have to chase profits. In this aspect, the television will make and create programs to gain commercial advantages. In fact, Supratman said that local television function is to present local genius television programs. In the media industry are found examples of symptoms of commoditization [2]. The media content is made an extent as to attract the market so-called commodity markets could be sold then distributed audience then also be impact on the standardization of commodity goods. By the existence of feature, platform, and the newest function, the way the audience access the media is changing [3]. The broader impact is when the symptoms of the modification and standardization have experienced modification means that these commodities are scattered and spread in a massive people's lifestyles. Public or community can easily reflect his personal life with what is in the media. Commodification according vocabulary in Marxist terms is a form of transformation of the relationship, which was originally free from the things that are trafficked into the commercial nature relationship. Commodification described as a way of capitalism by bringing accumulated his capital purposes, or simply can be described as a change in the value of the function or to become an exchange. There are two dimensions of the relationship commodification in communication: 1. The process of communication and technology contribute to the general process of commodification in the overall economy; 2. The process of commodification at work in society as a whole penetrate communication processes and institutions, so that development and contradictions in the process of commodification affect communication as a social practice.

3 Data analysis

The following overview shows the composition of the local TV broadcast:

DOI: 10.1051/shsconf/20173300069

Table 1. Format Composition Program Local TV Broadcast

TV Station	Information Program		Education Program		Entertainment Program		Others		Amount	
	F (Hour)	%	F (Hour)	%	F (Hour)	%	F	%	F (Total Hour)	% (total percentage)
BTV	*49*	*46.7*	*5*	*4.7*	*52*	*48.6*	*0*	*0*	*106*	*100*
STV	*34*	*31.3*	*4*	*3.7*	*47*	*43.3*	*23*	*21.6*	*108*	*100*
PJTV	*36*	*28.2*	*12*	*14*	*58*	*46*	*15*	*12*	*127*	*100*
Amount	*119*	*34,90*	*27*	*7,91*	*157*	*46,05*	*38*	*11,14*	*341*	*100*

4 Discussion

Analysis of the table above, should programs broadcast content that is inserted into the program classification factual and non-factual is the encouragement of provision of broadcast according to the law of broadcasting and regulatory programs broadcast in the P3SPS, which can encourage and be inspirations program based factual and non-factual. However, these programs are for the benefit of the market and the commodification of advertisers only. This is the commodification of culture in the Program of Bandung Raya (BTV). Bandung Raya TV is a local news program that presents the news accurately packaged and populist from the coverage of events and happenings that occur around Bandung or West Java. Contain actual news and event program is aired every day.

Fig. 1. Regarding Snapshot Image on Bandung Raya TV

There are also the commodification of Religion and Health (Special Treatment) on three Program Local Television (BTV, PJTV, STV). Program therapies and alternative medicine is an interactive talk show that discussed the treatment alternative. In this event featured an informant (an expert in alternative medicine) which discusses how the treatment he did. "This event in addition to taking an informant as a therapist (but not a doctor), there is also a patient ever cured or healed by therapists and issued earlier testimony. In other contexts, this event sells the drug in the form of herbs and the like as well as the identity and address of where the drug can be obtained. After giving a presentation or explanation of what and how the medication, then opened the interactive dialogue, which allows viewers to ask questions or consultations by telephone. Usually the problem, when the therapist had to provide treatment that are "unseen" or paranormal, so it is feared falls on mystical impressions which are prohibited by P3 and SPS. Therapy and treatment program that took place in Bandung TV for instance in the Pearl Therapy program, Dr. Kita, Special Dialogue, the Alternative Healthy with Bandung TV, Totok Clinic Stomach Mega Power, Mayang Tea Clinic. This display program packaged in the form and format some program, then presented in several different packaging program title, treatment program was aired on average in the afternoon, at 16.00 up to 17.00 pm.

Treatment programs are impressions of therapy or treatment of various diseases complained and submitted by viewers either directly or through his phone. Almost all local television programs of alternative medicine impressions and some of them could be considered mystical. Most radio also broadcast ads this treatment.

Television is industry entertainment, and entertainment is the creativity of artistic freedom. Apparently, it is already an advanced adage for national television, including the local television in Bandung, West Java. Of course, this problem can be seen and studied from broadcasts the program. In the context of the commodification of entertainment this, Music Holic, SoundXplore, Piloks, 8Beat Euy, Momentoz De Paz, Sinemania make and make entertainment programs packaged in a show, concert, music, and fromat other programs, which are more likely perform aspects of sensuality, sex, gossip, vulgarity, pornography, and other displays that are inconsistent with public ethics.

Fig.2. Commodification Entertainment Program 'music' PJTV (2010)

Commodification entertainment of the show program "Music Holic", "SoundXplore", Piloks, "8Beat Euy" is, by showing clearly the visualization of scenes and action swings Nudity, exposing the sensitive for the public, perform movements erotic and challenging (or such movements having sex), singing a song with the sounds erotic, and also respiratory sigh-sigh sensual, performance dress in skimpy outfits and sexy, coupled with the lyrics and lyric songs conformity, sex and vulgar.

References

1. N. Yoshiko, Y. Ritsu. *Public Broadcasting and Changing Audiences in The Digital Era: Services and Social Mission.* (NHK Broadcasting Studies, 2008)

2. L. P. Supratman.. Observation Journal. *The representation of local wisdom at regional news program 'Kalawarta' on TVRI West Java*. Volume 14 Number 1 Year 2016. ISSN 1412-5900. Page 27. (2016)
3. R. Kubey. *Media literacy in the information age: current perspective, information and behavior Volume 6.* (New Jersey, USA: Transaction Publishers, 2001)

DOI: 10.1051/shsconf/20173300070

Local Roots, Global aspirations: Impact of culture on work environment and organizational culture in Malaysian Small and Medium Enterprises in the Information Technology Sector

Vandana Saxena[1,*], and *Anindita* Dasgupta[1]

[1]Taylor's University, Malaysia

Abstract. This paper investigates the role of culture in hiring, team formations and workplace interactions in Malaysian small and medium enterprises (SME) in the Information and Communication Technology (ICT) sector. This research used the case study approach, with multi-method data collecting instruments like observation, interviews, and analysis of the data available on the websites of the two ICT SMEs under study. The participants selected for the study were the owners, managers and senior employees of both firms. While both firms operated in similar fields, the workforce of one consisted largely of Malaysian employees, while that of second company consisted largely of foreigners. The findings revealed a considerable bias and preference towards cultural homogeneity.

1 Introduction

Does culture play a role in the hiring, team formations and workplace interactions of professionals in Malaysian ICT SMEs? If yes, then is it a peripheral factor, or does it exert a significant influence on these organization practices and thus shape the nationality and social demographics of the ICT sector in Malaysia? Existing studies are divided over the role culture plays in labour market stratification. [2,3] have argued that culture is peripheral to occupational sorting, while [10] contend that culture is an important basis on which symbolic rewards like access to desirable jobs, recognition and status —are distributed. Lauren Rivera, a prominent researcher in the field of cultural sociology and social stratification, argues that the decision to hire an employee, a crucial moment in labour market stratification, is actually a process of "cultural matching" between candidates, evaluators and firms. Shared culture often outweighs the concerns about absolute productivity of an employee [13].

This study is an attempt to unpack cultural dynamics that come into play during the crucial processes like hiring, team formation and workplace interactions in Malaysian ICT

* Corresponding author: Vandana.Saxena@taylors.edu.my

DOI: 10.1051/shsconf/20173300070

SMEs, a sector which is believed to employ a diverse workforce drawn from a number of foreign countries. The paper explores the perceptions of the top management vis-a-vis both the 'insider' (the Malaysian employees) the 'outsider' (the foreign employees).

ICT SMEs offer an interesting arena to test the theories of cultural matching. Skilled human capital can be drawn from a vast set of global resources. Hiring foreign workers is a common practice. Subsequent team formations and interactions shape the organizational culture of the said ICT. In Malaysia, Indians and Filipinos form the core ICT workforce. They create a unique work environment where the Malaysian companies and their local employees work alongside the expatriates who are 'experts' in their respective fields of technical knowledge but 'outsiders' to the 'Malaysian' cultural milieu. The dynamics of such multicultural workforce are neither simple nor direct. Several organizational studies have explored the complications that arise from the entry of the cultural outsiders into a relatively homogeneous field [1,5,12,11]. This paper adds another dimension to the debate by analyzing the role of culture in processes by which an outsider gains entry and becomes a part of an organization.

The organizations under consideration for this study are service oriented ICT SMEs that have been in business in Malaysia for almost 20 years. They deal with diverse software technologies like Systems, Applications and Products in Data Processing (SAP) and OpenSource Solutions. Providing the case study of the ICT firms, we explore the influence of cultural factors like tastes, language, and geography and so on, on important decisions like hiring, team formations and interactions in order to: -

- determine the extent to which cultural matching influences the entry of a newcomer into the organization,
- understand its influence on the interactions between the team members and the resultant shaping of the work environment.
- to investigate into the way in which the top management's perceptions of cultural diversity influence demographics of the given ICT enterprises.

The study builds on the theories of organizational culture including cultural sociology and social categorization. After a discussion of these theories, the paper highlights the methodology by which the two case studies were chosen, the tools of data collection and the research questions central to this study. A close consideration of the nature of SMEs in Malaysia and the features of the two case studies leads to the final discussion of hiring practices, team formations and group dynamics where culture has a major impact.

2 Literature Review

Apart from professional expertise and skills, several other factors influence the newcomers in an organization, right from their entry to their everyday workplace interactions. [13] contends that after taking into account the skill sorting, human capital and social capital during hiring, there is still a great degree of unexplained variance in the process. This, she contends, is because most studies on hiring concentrate on quantitative data available from employers and organizations. For a greater understanding one needs to delve into the process itself, into the perceptions and thought processes that affect the evaluation, comparison and selection of candidates. Impressions conveyed before and during the job interview influence the decision making; 'subjective impressions of candidates that employers develop through interviews are strong drivers of hiring decisions, often carrying more weight than candidate's résumé qualifications' [13]. During the length of the interview, the cultural similarities are subjectively assessed and they become a major factor in the selection of a candidate. Through these observations, Riviera concludes that hiring is 'fundamentally an interpersonal process.' Cultural similarities serve as emotional glue generating a feeling of trust and comfort [4,8,14,16], they also become basis to evaluate merit [9]. [7] coined the term 'cultural matchmaking' to highlight the role of culture in the hiring practices. Cultural matching to builds on the theories of social categorization and social identity [17] and the similarity attraction paradigm [3] adding on the dimension of culture and its impact.

While most of these studies talk about cultural aspects of hiring, that is, the process of entry into an organization, this study goes beyond hiring to explore the way culture plays a role in shaping the work environment, the intercultural dynamics and the organizational structure in Malaysia ICT SMEs.

3 Methodology

Our data collection was carried out in two ICT SMEs located in the financial hub of Klang Valley in Malaysia. The very nature of the research problem necessitated the use of an inductive qualitative approach. Culture in general is a challenging concept to discuss in a multicultural and fast globalizing context that is Malaysia. [6] have highlighted the importance of using the twin methods of observation and in-depth interviews in the course of researching culture. As culture is often perceived as being a deeply personal feeling, besides being a multi-layered and complex concept to express and understand, we believed that a series of in-depth interviews with the respondents- mostly on an individual basis- would be the best method to unpeel its many layers and interpretations. Following [15], we also placed a great deal of importance to observation as our second technique of data collection. Additionally, we carried out a detailed analysis of the data available on the websites of the SMEs under study.

While both firms operated in similar fields, the workforce of one consisted largely of Malaysian employees, while that of second company consisted largely of foreigners. This differentiation was the critical factor that guided the section of the firms as we were keen to understand the reasons for such divergent workforce demography.

The participants selected for the study were the owners (3), managers (2) and senior employees (5-10) of both firms. At the beginning of fieldwork, the owners of both SMEs were interviewed several times to understand the history of the respective firms. We then selected two team leaders- both of whom were hiring managers- from each SME in order to gather qualitative data that provided us with clues to cultural meanings and interpretations, and a deeper understanding of the background of the respondents themselves.

A small group of employees i.e. IT developers and delivery teams from both SMEs were also interviewed for feedback from 'within' the teams. Observation techniques were extensively used to gather additional and corroborative qualitative insights.

We collected the data over a period of three months during which time we went back to the SMEs several times. The data were analyzed by content analysis from the multi-methods used in the qualitative approach.

4 Key Features and Profile of ICT SME firm

According to the guidelines issued by the SME Corps, Malaysia in the year 2013, a service sector enterprise can be deemed as an SME if its sales turnover does not exceed RM20 million or the full-time employees are not more than 75. A detailed account of the organizational features of Malaysian ICT SMEs based on the websites of both SMEs and interviews are as follow:

- They follow an integrated set of human resource policies and practices which work in tandem with the policies of the government. Especially relevant are the policies regarding the hiring of foreign professionals and retaining their services.
- Extreme work pressures and great number of projects require a workforce that is focused on task at hand. For the management as well as the professionals, timely completion of the task is crucial.
- Since the ICT sector in Malaysia is in the growth phase and on the list of Government's high priority sector, the firms need workforce that is creative as well as innovative. Facing the lack of skilled local workforce within Malaysia, ICT firms shop for professionals from countries like India which are a pool of IT resources. In many firms, foreign professionals far outnumber the locals.
- Most of the expatriate population in the ICT sector is transient. Tight financial and budget constraints of an SME affect hiring, contracts, wages and so on. An oft-cited complaint against expatriate ICT professionals in Malaysia is that they see Malaysia as a stepping stone to greener pastures like Singapore. On the other side, the expatriate professionals argue that other places offer them pay commensurate with their expertise. This is an important factor which as we shall see, underpins the attitudes of the local workforce towards the foreigners and also of the foreign professionals towards work, organization and the cultural environment.

5 Themes

Despite falling in the broad category of ICT SMEs there are some key differences between two companies chosen for this study. OpenSource solution provider is a medium sized enterprise that caters to a niche market. It boasts of a flexible hierarchy with open channels of communication between different levels of staff. This enables the managers and technical personnel such as developers/consultants to provide comprehensive solutions to the customer's needs.

Started by a local entrepreneur, the company hires through references and the socio-professional networks of its employees within Malaysia. Majority of the employees are ethnic Malays. Though they move from one team to the other, the employees stay with the company for a long time. Foreign professionals in the company are rare and when hired, they work under Malay team leaders and managers. The company seems satisfied with the local staff but also showed an awareness of future growth trend whereby it would have to open its doors for the foreign professionals soon.

The second organization deals with Systems Applications and Products in Data Processing or SAP solutions, a fiercely competitive segment of ICT industry. Like the earlier company, it tailors its services to the needs and budgets of the clients and provides appropriate expertise to diverse organizations including mid-market companies to larger multinational companies (MNC) and government-linked companies (GLC). The focus is primarily on quality, planning and timely delivery. Most of the staff consists of ICT experts from India. There are also some professionals from Philippines and Malaysia.

Started by a Malaysian entrepreneur of Southern Indian origin, the company hires experts mostly from the south of India, a region which has emerged as a global ICT hub. The practice reflects the wide perception of Indians being the experts in IT industry. During interviews, a Malaysian ICT developer shared that 'Indians as young as 12 or 13 years old were becoming expert developers' whereas in Malaysia 'even ICT graduates possessed limited skills'. He related this to the serious lack of role-models and outstanding leaders in the Malaysian ICT sector. Another source of comparative advantage the Indians possess is the availability of virtual "networks of experts" that they can rely on for troubleshooting technical and computer matters on a daily basis. Hence despite the extra cost and effort involved in the hiring of expatriates, Indian IT professionals brought in the advantages that are not available to the local workforce.

However several factors beyond skill set and expertise become operational right from the moment of hiring. The managerial staff and team leaders showed preference for team members who came from the same socio-cultural background hence perpetuating homogeneity. The senior management too seemed aware of these interactional dynamics and the ensuing conflicts arising out of the dominance of one socio-cultural group.

Thus demographic constitution of the two companies offers an interesting case of contrasts and similarities. Though the staff of the two companies differ in their nationalities, both have essentially homogenous teams which are a product of the hiring process and also of the in-group and out-group dynamics which ensures who was hired, retained, and who stayed with the organization for a long term. Selection of 'fit' candidates operated in both the companies, at a conscious and an unconscious level.

The discourse of cultural similarity was found to be structured in the formal and the informal interactions among the employees. It manifested in a variety of interpersonal relationships like friendships, mentoring and techno-social networks. These overlapping themes were expressed largely within the discourse of a family.

6 Findings: Family, Techno-social Networks and Fitting into the organizational culture

The discourse of family and familial relationships surfaced repeatedly in the interactions between the employees in the two companies, albeit in different disguises. The seniors in both the organizations assumed the role of the company elders, familiarizing the new employees with the projects, organizational structure and generally helping them find their feet in the new work environment. This interpersonal relationship worked at professional and personal level. The seniors 'looked after' the newcomers not only during the working hours but also during the leisure time and after work hours. Eating together, meeting over the weekends, co-planning leisure activities - it seemed as crucial to initiate the newcomers in these in-group activities as it was to introduce them to the project.

In the company with the predominance of Malaysian employees, the perception of organization as a family existed alongside the perception of the foreigner as almost an intruder. It came across in the main complaint against the foreign employees. One of the senior managers told us that 'they (generally) don't have a strong sense of belonging.' 'It is because they don't socialize much (with locals),' a young Malay team leader added. 'They tend to work alone. It sometimes creates problems.'

At the professional level, the complaint was concerned with the noncompliance to the standard methods and ways of handling the projects, writing the programs and so on. 'This creates problems in the later audit and maintenance stage,' the team leader confided. 'We don't know how to fix a program because they haven't done it the way we do... wastes a lot of time.'

The tendency to 'work alone' trickled from the official to non-official interactions. While foreign professionals tend to focus on completing the projects, the local employees looked forward to social and personal interactions, eating together during lunch times and other such common activities. Religious and food taboos also emerged as points of contention. Foreign professionals and the local employees often ate separately, a tendency viewed with annoyance among the local employees. 'We are not hiring robots,' opined one of the local managers.

Beneath the complaint, there were noticeable undercurrents of the dominance of the majority. The onus of 'fitting -in' the organizational 'family' lay with the expatriate employees. They were expected to follow the codes of the majority. For instance, when a

DOI: 10.1051/shsconf/20173300070

Chinese employee brought pork for lunch, the manager had to tell him that he could eat pork but not mention it or eat with others.

Another issue of contention was the lack of cultural awareness among the newcomers. It was cited as a major problem of the expatriate professionals coming to Malaysia. 'A little example,' the team leader told the researchers, 'the foreigners are not aware that government clients cannot be called during the time of Maghreb or the evening prayer. A local employee knows these things automatically.'

Thus foreign professionals are seen as intruders in the "warm" "family-like" environment, and often stereotyped as being culturally insensitive or 'awkward'. The case of the sole woman employee in the staff offered an interesting insight. She is a fresher and a foreigner and she has been living in Malaysia for nearly twenty years. With her knowledge of the local language and customs, she 'fitted- in' the organization. She undertook the job of taking care of the visitors, the refreshments and other duties of hospitality. The senior manager advised her to marry a Malaysian. 'This,' she laughed 'was an everyday talk. I have a father here.''

Similar family discourse was evident in the second company which consisted primarily Indian professionals (and a small group from the Philippines). Working away from home, the expatriate managers and professionals tended to create a home-like atmosphere by finding people of their own kind and sticking with them.

The company was started by a Malaysian entrepreneur of South Indian (Kerala) origin in the late 1990s, approximately at the same time when unprecedented upsurge in the field of Information Technology began in India. Apart from technical expertise, one of the main criterion of hiring was extended family or close friends 'back home ' ('from my village in India") or people that he could personally vouch for or trust. Over the years, this expanded to include those similar in the regional, linguistic and cultural background.

The trend continues to dominate. As the organization has grown, the power of decision making has trickled down from the senior management to the hiring managers and project leaders. Yet the project managers prefer to work with team members who come from same region and culture. The young professionals hired from India also gravitate towards the seniors who hail from similar cultural background.

Like the earlier case study, the set-up echoed the structure of a close-knit family with the hiring managers and project leaders assuming a paternal role, expecting obedience and gratitude from their protégés and at the same time, 'looking after them' in the foreign land. Thus the mentoring extended beyond the professional sphere into the personal arena. Food, housing (most prefer to live in close proximity to each other) and leisure glued the team together.

The expatriate hiring managers revealed their positive perceptions of the new hires from their own ethnic groups. The Indian hiring managers and project managers found it easier to communicate effectively and easily with the employees of similar social, cultural and regional background. Their higher degree of technical expertise, similarities in working patterns, effective delivery -were cited as major advantages to the team. The individual work teams communicated in their own mother-tongue within and across teams, for example in Telugu or Malayalam, with scant regard for minority team members who belonged to other ethnic groups even among the Indians themselves. Such linguistic collusion was a common mode of creating in-groups and out-groups.

Though the teams worked in harmony, the dominance of one ethnic group sometimes created friction or competition within the organization. One of the employees from another nationality shared that the 'Indians wanted to form their own groups, and stick to it. They did not want to merge with the local population.' Other commonly held perceptions about Indians were that their sense of timing was different from Malaysians that they promised much more than they could actually deliver, that many of them spoke English with thick local accents which were often difficult for Malaysians to understand. According to the senior management, this friction among the staff members was a major complication since the company employees are foreign as well as local, based on market expertise and latest demands of the ICT industry.

Thus the companies face a new dilemma today. Whereas the senior level Malaysian staff has reservations about hiring expatriates from the same region, the expatriate hiring managers look deep into their own socio-ethnic network for a 'good fit' in the existing teams. Even when the management hired professionals from other nationalities and cultures, these new professionals found themselves excluded from the teams and its in-group networks, leading to problems like communication gap and employee dissatisfaction. Eventually, the employees who did not 'fit,' left.. One outgoing Malaysian professional (a minority in this SME) complained that his boss treated the Indians and non-Indians differently. He cited this as one of the major reasons for quitting the organization.

7 Implications: Similarities and distinctive characteristics

Despite the difference in the composition of the workforce, the two SMEs showed a remarkable degree of similarity in their day to day functions, employee perceptions as well as the problems they encounter in face of growing diversity in the ICT sector. Competing with the multinationals in the ICT field, the thrust was on the qualities of loyalty and sense of belonging. The right 'fit' into the work culture is as important as the professional expertise. Furthermore, hiring through their 'techno-social networks' with references from friends, family and hometown created value additional to the merit of the candidate [6]. Thus social and cultural similarities served as 'powerful emotional glue that facilitates trust and comfort, generates feelings of excitement and bonds individuals together' [13].

The model of a family emerged time and again, in overt discussions and also as underlying subtext in other situations and utterances. Instead of a clearly defined professional role, the employees played the part of family members. The older members and mentors undertook the role of family elders, helping, guiding and at times chastising the new employees. The new employees, in turn, fitted in the with the older member's perception by being loyal to their senior managers.

The notion of fitting together worked at organizational as well as the emotional level extending the sense of security to the employees but at the same time perpetuating homogeneity. Culturally similar employees were seen to strengthen the harmony in the organization and enhance team productivity by minimizing conflict of the team. On the individual level, the team members seemed secure, confident and at ease with each other. Cultural outsiders were perceived as intruders in the warm atmosphere of the company.

Apart from these similarities, the two SMEs showed some distinctive features due to the nature of the hiring process. While the company with predominantly local population hired long term employees and look forward to long lasting associations, the expatriates in the other company were hired on short term contracts. Most of them created short term relationships with their own kind, concentrating on the work and being more concerned about the project and money rather than fitting and adjusting to the new cultural environment. Loyalty when it surfaced was directed towards the project managers who belonged to same socio-cultural background instead of the company at large.

8 Conclusion

Two SMEs under consideration for this study stand at similar crossroads that the Malaysian ICT sector faces on a larger level. They are poised at the level of enterprise growth where looking beyond the network of family and friends for hiring is crucial to expand and compete within the market driven ICT sector. Despite its wariness, the company dominated by Malaysian employees is planning to expand and hire foreign employees. 'It is inevitable,' said one of managers recognizing the need of the hour. The other SME has already taken the plunge. Moving beyond Malaysia to enlarge its talent pool, it now faces issues that expatriate employment brings. At the moment it is dealing with the power struggles that diversity brings.

The cultural interactions within these SMEs provided an interesting study to test the theory of cultural matching. According to Rivera, cultural similarities affect the organizational processes, the cognitive processes, and the affective processes during the process of hiring. Considering the organizational processes in the Malay dominated SME, the thesis was found to be largely true since the organizational set up is small. The thrust is on finding the candidate who 'fits' into the organizational 'family.' In case of the second SME, the study revealed that as the organization grows beyond the local confines, skill set becomes as important as cultural matching. The regional dominance of Indian professionals is a result of their skills and expertise as much as their cultural background.

The study also revealed that the process of cultural matching is active before the process of hiring and remain so long after it. Experiences in the past, observations of the environment around, open and covert conflicts - all these create the perceptions through which the candidates are evaluated. These perceptions shape the idea of the right 'fit' even before the candidate or the vacancy materializes. These social and cultural factors also played a major role in creating a conducive work environment for some employees while excluding the others. Hence cultural matching had a determining influence in the power struggles within an organization and shaped its ethnic demographics.

Acknowledgements

This study was supported by a Fundamental Research Grant (FRGS/2/2013/SSI08/TAYLOR/02/1) of the Ministry of Higher Education, Malaysia.

References

1. L. Argote, D. H. Gruenfeld, C. Naquin. 'Group learning in organizations.' In M. Turner (Ed.), *Groups at work: Theory and Research* (369–411). Mahwah, NJ (2001).
2. Blau, Peter, Duncan, Otis. The American Occupational Structure. Free Press (1967)
3. D.Byrne. *The Attraction Paradigm.* Academic Press (1971).
4. R. Collins. *Interaction Ritual Chains.* Princeton University Press (2004)
5. T. H. Cox, S. A. Lobe, P, L McLeod. 'Effects of Ethnic Group Cultural Differences on Cooperative and Competitive Behavior on a Group Task'. AMJ, **34, 4** 827-847 (1991)
6. S. Selvadurai, A. Dasgupta. Culturally Diverse Teams and Social Interaction Influence on Team Effectiveness in Small and Medium Enterprises in the ICT Sector in Malaysia, *Millennial Asia* April 2016 7: 42-62, Sage Publications (2016)
7. P. DiMaggio. "Nadel's Paradox Revisited: Relational and Cultural Aspects of Social Structure." In Nohria and Eccles (Ed.) *Networks and Organizations: Structure, Form and Action.* 118–42 HBS Press. (1992)
8. B. Erickson. "Culture, Class, and Connections." AJS *102*, 217–51 (1996).
9. M. Lamont, M. Vira. "The Study of Boundaries in the Social Sciences." ARS **28**,167–95 (2002)

10. A. Lareau, W. Elliot. "Cultural Capital in Educational Research: A Critical Assessment." Theory and Society **32**,567–606 (2003).
11. F. Milliken, L. Martins. 'Searching for common threads: understanding the multiple effects of diversity in organizational groups.' AMR **21**, 402–433 (1996).
12. K. W. Phillips, A. L. Katie, A. N. Margaret. 'Is the Pain Worth the Gain? The Advantages and Liabilities of Agreeing with Socially Distinct Newcomers. *Personality and Social Psychology Bulletin*. **35**, 336-350 (2009)
13. L. A. Rivera. 'Hiring as Cultural Matching: The Case of Elite Professional Service Firms.' ASR **77**, 999-1022 (2012)
14. V. Roscigno. The Face of Discrimination: How Race and Gender Impact Work and Home Lives. Rowman (2007).
15. S. Selvadurai, D. Anindita 2016. "Culturally Diverse Teams and Social Interaction Influence on Team Effectiveness in Small and Medium Enterprises in the ICT Sector in Malaysia." Millenial Asia **7**(1), 1-12 (2016)
16. K. Stainback, T. D. Donald, S. Sheryl "Organizational Approaches to Inequality: Inertia, Relative Power, and Environments." ARS **36**, 225–47 (2010).
17. H. Tajfel, T. John. "The Social Identity Theory of Intergroup Behavior." In Worsheland and Austin (Ed.) Psychology of Intergroup Relations. Nelson-Hall,7-24 (1986)
18. C. Tilly, C. Tilly. Work under Capitalism. Westview Press (1998).

DOI: 10.1051/shsconf/20173300071

A Critical Discourse Analysis Study of Cyberbullying in LGBTQ's Instagram Account

Sukma Ari Ragil Putri[1,*] *and A.A.I Prihandari Satvikadewi*[1]

[1]University of 17 Agustus 1945 Surabaya, FISIP Untag Surabaya, Jl. Semolowaru No. 45 Surabaya Indonesia

Abstract. As the advancing of technology, bullying became online, it moved to social media, which was then called cyberbullying. Instagram is one of social media that contain a lot of cyberbullying, it usually happened to public figure. It became interesting when the public figure in the bully is LGBTQ, namely Dena Rachman (transgender). This study explored how discourse formed the logic of truth which naturalize the perpetrators to commit cyberbullying act without hesitation. This study used Critical Discourse Analysis (CDA) method and Teun A. van Dijk's analysis technique and data interpretation. This study found that the logic of truth that became the basis of cyberbullying is already a natural logic and became the perpetrator's mindset because it has been institutionalized in everyday life through education, religion, social life, and state regulations.

1 Background

Bullying is a negative action that is often aggressive and manipulative, performed by one person or more against another person or group of people for any given period of time, that contained with violence and imbalance of power. As the advancing of technology, bullying became online, it moved to social media. People often encounter conflicts caused by misunderstandings on social networking sites (SNS), one of which is in the form of cyberbullying.

Social media or social networking sites (SNS) was becoming increasingly rampant in recent years. Even Joost van Loon [1] stated that the term of networking is no longer representing the terminology in information technology but also in the terminology of anthropology, sociology, cultural, and social sciences fields. Networking does not just involve devices such as computers but also involves an individual or networking actor [1]. Social media can make communication between humans become borderless because people can communicate with anyone in any country at any time. Aside from being a medium to communicate with others directly, social media becomes a medium for communicating a person's identity.

* Corresponding author: sukmaariragilputri@gmail.com

DOI: 10.1051/shsconf/20173300071

Uploading photos on social media is one of the way a person communicating their identity to others. While the likes and comments are a form of interaction. Interaction is a concept that has been used to distinguish the traditional media with new media. According to Graham [1] the presence of communication technologies are basically provide the opportunity for anyone to interact easily, and connect to each other without the constraints of distance and time. Technology has became a medium of all human activity.

Manovich [1] revealed that in fact, the concept of interaction in the new media made boundaries of the physical and social interaction begin to blur. For example, resident of country A can easily interact with the resident of country B without having to meet before. Resident of country A can also praise or denounce the resident of country B without inhibitions, even if it turns out the resident of country B is a public figure. This is where the physical and social boundaries begin to blur. Social media eventually become a new container to the marginalization of the groups or people who are considered incompatible with the existing normative order in society, one of which is LGBTQ.

2 Teminologies to LGBTQ Understanding

LGBTQ identity has been regarded as something different. Before discussing about LGBTQ identity, we need to understand the most important terminology regarding human sexuality first. There are gender identity, sexual orientation, and sexual behavior. People tend to misunderstand those three terminologies as a single unit which caused confusion and wrong meaning.

Gender identity is often misunderstood as sex. However, those two things are like chalk and cheese, very different. Sex was categorized based on the physical characteristics or biological characteristics of human. For example, man was characterized by having penis, testicles, and sperm, while woman was characterized by having vagina, ovum, and uterus. Those biological characteristics are natural and cannot be changed, except for the genital parts that can be modified with the advancing of technology. However, the modification will never change its function [2].

In contrast to sex, gender identity was a social construction. It refers to a set of roles, functions, rights, responsibilities, and behaviors of man and woman that constructed by people based on culture, social, economy, and politic situation. For instance, man was constructed as a gender that has masculine traits such as strong, bold, rational, and tough. This world was kind of black and white, it always been applying binary opposition concept. Therefore, when there was strong, there was weak. That is why woman was constructed as a gender that has feminine traits such as weak, fearful, irrational, emotional, and fragile.

A fatal error occurs when masculine and feminine were regarded as natural as sex. Yet according to psychological study, every human being has masculine and feminine traits since they were born. However, as they grow up, the tendency of a person's masculinity and femininity are influenced by parenting, cultural values, education in the formal school system, and the interpretation of religion values. According to Judith Butler [3], the masculine gender is not always in man's body, as well as feminine gender is not always in woman's body. Furthermore, there is no reason to assume gender also categorized into two like sex.

Butler's opinion was based on Simone de Beauvoir's statement in The Second Sex that suggest that someone was not *born* a woman but rather *become* a woman. For her, gender was constructed. She explained that a person that *become* a woman was under culturally duress to be female with all of its feminine traits. And clearly for her, that compulsion does not come from biological sex. There is no a guarantee or provitions that a person who *become* a woman needs to be female [3].

DOI: 10.1051/shsconf/20173300071

The second most important terminology is sexual orientation. Sometimes people confuse with gender identity and sexual orientation. For instance, people assume that a man who has feminine traits also like man. Though not necesseraly the case. As described previously, gender identity is a social construction whereas sexual orientation is a natural thing and cannot be changed. Someone cannot choose to be born with a particular sexual orientation. Being heterosexual, homosexual, bisexual, and asexual is not an option, nor because of social construction [2].

Similar opinion was expressed by Suryakusuma [4]. In one of his paper, he said that the concept of gender identity is often confused with the concept of sexuality. According to him, gender identity is psychological and cultural expressions of gender that evolve into specific roles and behaviors that then identified with specific sex. The third terminology is sexual behavior. It can be understood as a way of expressing someone's sexual relationship. Sexual behavior is strongly influenced by social construction because it can be learned and not a natural conditions.

Those three terminologies, gender identity, sexual orientation, and sexual behavior, each of them has category that considered normal and abnormal by social convention. For example, normal gender identity would be masculine man and feminine woman. Normal sexual orientation would be heterosexual, and normal sexual behavior would be penetration of penis into vagina. Other identity, orientation, and behavior outside that normal category then categorized as LGBTQ which is a category that accommodate individuals who are considered abnormal by social convention.

3 LGBTQ and Instagram

On daily life, LGBTQ have been bullied all the time in many ways by people who thinks they are abnormal. It becomes interesting when LGBTQ begin to express their identity on social media, one of it, is Instagram. Social media as a medium for people to express their thought, feeling, expressions, basically everything, become a bullying medium for minority, which is LGBTQ. People with dominant ideology tend to marginalize minority by uploading comments that were rude and judgmental. They even considered those comments as something fair and reasonable to do.

One of LGBTQ that express her new identity on Instagram is Dena Rachman. Dena Rachman was known as an Indonesian child artist with the name Renaldy Denada and famous with various songs, as well as a host of various music shows. She also starred in some of TV series. Born as a man, Dena Rachman began to change her appearance to be a woman when she studied at University of Indonesia in 2005. Her changing invited many pros and cons from people in Indonesia.

People's assumption in Dena Rachman's gender identity were various. When assumption goes one step further, it becomes bullying. People tend to say or write their assumption on Dena Rachman's Instagram. They commented on any photos uploaded by Dena Rachman with indecent comments and sometimes had nothing to do with the photos itself, they intended to attack Dena Rachman's personality and identity. Of these problems that arise within the rising of new media, I want to find out how discourse formed the logic of truth which naturalize the people to commit cyberbullying act on Dena Rachman's Instagram without hesitation by using Teun A. van Dijk's Critical Discourse Analysis.

4 Teun A. van Dijk's CDA

Teun A. van Dijk stated that discourse in the media cannot be separated from the concept of power and access. Power is fundamental in the formation of a discourse. Power is a tool that is owned by social group, institution or organization that is used to control the actions and thoughts of others, with the aim of limiting the freedom of other groups. In this case a group that are limited are a minority, namely LGBTQ.

The power of dominant party distributed through public domains such as the realm of politics, media, law, education, and business, which then resulted in different centers of power in certain groups. Power is based on the ease of access of a group on the social resources such as wealth, jobs, status, or simply access to public communication space [5]. In general, the power of media is symbolic and persuasive, the media has the potential to control the audience thought, though not directly control their actions.

In relation to social media, social media only serve as a medium. Even it merely a medium, it has the most important role. It is used as a tool to carry out the exercise of power of the dominant groups. Social media is free to be used by anyone who has access to it. In the end, it still controlled by dominant groups that even before the existence of social media had been in charge of power in real life. In this study, the dominant group is the heterosexual community living with gender identity in accordance with their biological identity (sex). They use social media as access to perpetuate the practice of their power over subordinate groups which is LGBTQ. Social media becomes a medium for them to do an extension of marginalization in the form of cyberbullying.

Teun A. van Dijk's critical discourse analysis focused on the role of discourse in the reproduction and the challenge of dominance. Domination itself was interpreted as a social power of the elite, institution, or group, which produced social imbalance in various fields such as politics, culture, class, ethnicity, race, and gender imbalances. He thought that social inequality is not simply produced by the action of an individual, for example with the command. This imbalance was produced jointly by the parties that dominate people's thought so that their domination feels natural and legitimate [6].

This study used van Dijk's CDA based on several things. First, van Dijk [6] stated that social cognition is an important part of critical discourse analysis. Social cognition monitors discourse, communication, and all forms of action and interaction. The same thing applies to our understanding of social events or social institutions and power relations. It can be argued that social cognition becomes a medium between the micro and macro levels of society, between discourse and action, between individuals and groups. A social cognition itself has many forms including opinions, attitudes, ideologies, norms, and value. Second, this study used social media and made comments in social media as a unit of analysis. As we know it, the comments in social media were usually straightforward and comparatively brief. Therefore, it did not require very deep text analysis. This can be accommodated by van Dijk's CDA that did not require a detailed text analysis.

4.1 Text Analysis

Table 4.1. Text Analysis Step

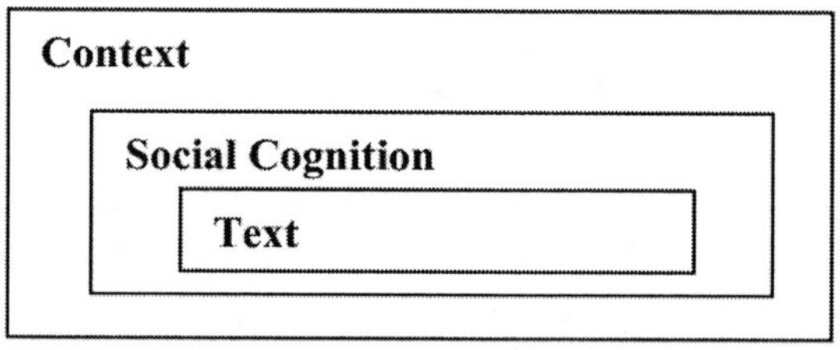

Based on the picture, van Dijk's CDA has three dimensions; the text, social cognition, and social context. At text level or referred as the microstructure, we can look at how the structure of text has been used to emphasize a particular theme. In this study, first, I selected several pictures of Dena Rachman in her Instagram account that showing the performance of her gender identity, sexual orientation , and sexual behavior. Then, from those pictures, I observed, selected, and categorized their comments. Based on initial observations, the comments that have been uploaded in Dena Rachman's Instagram account can be divided

DOI: 10.1051/shsconf/20173300071

in three main context; (1) physical context, the comments that discuss the body, (2) identity context, the comments that discuss the victim's identity, (3) religious context, the comments that brought religion issues. Those comments that consist of multiple context will be analyzed by looking for their meaning regarding to the choice of words and sentences by using some elements of discourse proposed by van Dijk.

4.2 Social Cognition Analysis

At the level of social cognition, I studied the text production that involves social cognition of the text producer. As I mentioned earlier, social cognition many forms such as ideology, values, attitudes, and norms that can be known by conducting in-depth interviews to determine the social cognition of text producer .In this study, I chose the most one commentator that uploaded negative comments all the time in Dena Rachman's Instagram account. To understand the social cognition, van Dijk used scheme. He called it as a model of mental structure that includes how we see other human being [5].

Table 4.2. Social Cognition Analysis Step

Person Schemas This scheme illustrates how one defines and view others.
Self Schemas This scheme relates to how someone sees, understand, and describe themselves.
Role Schemas This scheme relates to how someone sees and portrays role and position occupied by someone in the community.
Event Schemas This scheme is perhaps the most common scheme, because almost everyday we always experience events.

Model or the scheme is a central element in van Dijk's framework, because the interpretation of an event based on this model. Therefore , the model or the scheme is an important guide to this study.

4.3 Social Context Analysis

The third level to understand discourse construction is social context. Discourse is part of a growing discourse in society. In the study of discourse, to understand the discourse, we need to understand the social context that manufactured and constructed in the community. In this study, I analyzed how the LGBTQ discourse developed in the community by using scientific literature or history research. According to van Dijk, there are two important point, which are power and access.

Van Dijk defines power as the ownership that owned by a group (or members) for the control of the other group. His CDA gave considerable attention to the so-called domnination, wether domination of race, religion, social class, gender, etc. Domination reproduced by granting special access to one group than the other group. He also drew attention to the discourse production process through legitimate mind control. If controlling discourse is a first major form of power, controlling people's minds is the other fundamental way to reproduce dominance and hegemony. Mind control involves even more than just acquiring beliefs about the world through discourse and communication. Suggested below are ways that power and dominance are involved in mind control. First, recipients tend to accept beliefs, knowledge, and opinions (unless they are inconsistent with their personal beliefs and experiences) through discourse from what they see as authoritative, trustworthy, or credible sources, such as scholars, experts, professionals, or reliable media. Second, in some situations participants are obliged to be recipients of discourse, e.g. in education and in many job situations or in daily life [7].

5 Practice of Power and Access Affect LGBTQ Discourse

Based on comment analysis on Dena Rachman's Instagram, I found out that Dena Rachman who has a proportional body and dress up like 'real' woman accept bullying in the form of words which attacks the physical characteristics related to sex, such as the word breast (boobs, tits, etc), genitals (penis), and also deals with such physical authenticity of the results of surgery or not. In addition, Dena Rachman is a transgender, and now she is in the process of sex reassignment surgery. He was attacked in the form of comments by the words mentioning that transgender is a bad thing, like the word pansy, 'bencong', 'waria', etc.

Furthermore, from interviews with several comentators, I obtained a judgment that they position themselves on a higher position than Dena Rachman. Moreover, by looking at how they put forward the issue of ethics and religion to bully Dena Rachman, suggested that they position themselves as a party who have a better understanding of ethics and religion than Dena Rachman. It is happening because they feel that the correct identity according to the provisions of the religion is the identity of the 'normal', heterosexual, masculine men and feminine women. Identity outside those criterias are inappropriate. Associated with the widespread use of social media Instagram and many LGBTQ that express themselves in social media, the commentators felt uncomfortable with that event. Based on the results of in-depth interviews, all of them admitted that it is inappropriate that LGBTQ express their identity in the form of photos on Instagram.

Then I associated the findings above with van Dijk's two main concepts, practice of power and access. Practice of power that occurred in this study were happened when people with 'normal' gender identity (heterosexual, masculine male and female feminine) have domination of the minority group in Indonesia, identities which are considered 'weird' and abnormal (lesbian, gay, bisexual, transgender, queer). Domination by van Dijk [5] is understood as a form of social power abuse that is legally and morally a legitimate practice of control over other people's personal interests, which often leads to social inequality.

Based on the analysis of comments and analysis of social cognition in the previous stages, I saw that commentators' thought or logic about the identity of LGBTQ has been formed early on, because the education of parents and the surrounding environment, which then unknowingly been embedded into the mindset and even life principle then they passed to their descendants. Logic formed since childhood then generate attitudes, behavior, and positioning as reflected in the comments and social cognition of commentators. It becomes interesting when it appears one space, a new medium, for LGBTQ people to express their identity. The new medium is social media, which is Instagram. In practice, the logic of truth in the communities associated with LGBTQ, still on the same page as before. They carry wherever they are, even if it's in social media. This is why cyberbullying in Instagram against LGBTQ was arisen.

The second concept of van Dijk's CDA is access. According to van Dijk, access affects discourses. Teun A. van Dijk [6] put greater emphasis on how to access between each group in society. The group that dominates have greater access than the group who are not in power. Therefore, those in power have a greater chance to have access to media, and greater opportunities to influence public awareness. With their access to influence the discourse on Instagram, making them more free to express all forms of resentment and hatred.

DOI: 10.1051/shsconf/20173300071

Conclusion

'Manual' bullying requires face to face interaction between the offender and the victim, but now it moved to the virtual world which is then referred to as cyberbullying. Perpetrators do cyberbullying to the figures which are considered to have the identity out of the ordinary, in this case is transgender. They give negative comments and tend to be rough on transgender's Instagram account. The logic of truth raised by the commentators on Instagram is already become their mindset because it has been institutionalized in everyday life through education, social life, and state regulations. They feel threatened with something outside the 'normality'.

One thing we should understand is technology will continue to develop and it is very possible that LGBTQ people will express theirselves in other social media when people restrict their expression in Instagram. I concluded that the public itself is still overlap between accepting the existence of LGBT as a reality that must be faced or support their existence. Here, I hope the community, not stuck on the logic of normative truth, and although not yet to accept the existence of LGBT people, but at least they do not do rejection and attacks against LGBT.

References

1. R. Nasrullah, Teori dan Riset Media Siber (Cybermedia), (Kencana Prenadamedia Group, Jakarta, 2014)
2. Mulia, Siti Musdah, Islam dan Homoseksualitas: Membaca Ulang Pemahaman Islam, (Jurnal Gandrung 1(1): 9-31, 2010)
3. J. Butler, Gender Trouble, (Routledge, New York, 1990)
4. Suryakusuma, Julia I, Konstruksi Sosial Seksualitas: Sebuah Pengantar Teoritis, (Prisma: 3-14, 1991)
5. T. A. van Dijk, Discourse, Power and Access, (Routledge, London, 1996)
6. T. A. van Dijk, Principles of Discourse Analysis. Discourse & Society, J. E **4**, 2 (1993)
7. T. A. van Dijk, Power and The News Media, (Hampton Press, New Jersey, 1995)

DOI: 10.1051/shsconf/20173300072

Digital Democracy in Rural Indonesia

Ranggabumi Nuswantoro, *M.A*[1,*]. *Mario Antonius* Birowo, *Ph.D*[2], *Idha* Saraswatiand[3] *and Ferdy* Fachrudin[4]

[1,2]Lecturer of Communication Studies, Universitas Atma Jaya Yogyakarta
[3,4]Combine Resource Institution

Abstract. Citizen participation is the key to democracy. The development of information and communication technologies makes citizens more empowered, have broad access to information and able to influence policy making, especially at the local level. This paper will describe digital democracy in rural indonesia, when the villagers report their local isues and fight for their right to have a favorable policy. “Speaker Kampung” is an online community media that express the aspirations of citizens in Ketangga Village, East Lombok, West Nusa Tenggara, Indonesia. “Speaker Kampung” is a prototype of a digital democracy in Indonesia.

1 Preface

The information and communication technology has pushed the development of community media in the whole world. The internet is changing the pattern of media, from one-way, to two-way, or even multi-direction. Community media is media that their sources of information coming from many sides (crowd source). Internet, with its flexible format, is highly relevant for community media.

Internet highlight the advantages of communication media, which are hyperlocal and decentralized (non-centralized). Internet make community media can present information that has not been touched, even unthinkable, by the mainstream media. Community media can produce and reproduce the information in accordance with the aspirations or interests of the citizens.

Digital democracy arises when media started using the internet community. "Speaker Kampung" is an internet community media in Village Ketangga, Suela sub-district, East Lombok, West Nusa Tenggara, Indonesia. "Speaker Kampung" is a form of digital democracy, when the internet bridging gaps of space for people to express their aspirations to government officials, particularly at the local level.

2. Theoritical Study

* Corresponding author: ranggabumi_nuswantoro@mail.uajy.ac.id

DOI: 10.1051/shsconf/20173300072

The theory that used in this paper is the main theories about digital democracy, in particular the concept of deliberative democracy by [1]. This deliberative theory provides a perspective that democracy in "Speaker Kampung" is a democracy that leads to the fulfillment of the conditions of deliberative. The condition of deliberative is a condition which people consciously discuss with each other (community) about the important things in political life that affected them, including public policy.

2.1. Digital Democracy

The digital era is marked by the emergence of the Internet. As stated by [7], that people are now looking for the data is no longer in the form of an index in a book, or review in audio and video tapes, but in digital form. Today, people are easily typing in data search websites such as Google or Yahoo. In terms of politics and democracy, according to [7], there are at least four things that distinguish the digital world with the old world (2003: 135-136):

First, in this new age of reason. Digital communications will empower individuals at the expense of the elite, meaning the process of delivering information more politically independent and not dependent power elite (which may control the media).

Second, this new communication medium will move at a speed that is almost incomprehensible. Deadlines become irrelevant or, in fact, does not exist. The demand for real time news has increased dramatically.

Third, with the increasing speed of communication, the citizens in the digital world will demand more political information in accordance with their needs, both technical side (media) and content, to determine the pace in a general election. Convenience individual will become an integral part of the political conversation in the future.

Fourth, the public will experience an explosion of political information. Starting from the information shared by the conventional media to digital media. If not managed the information explosion it would be counterproductive for the development of democracy in the future.

Fourth novelty of politics and democracy in the context of the digital world over Franklin Roosevelt proved statement that "Democracy is not a static thing." Digital world politics and democracy would bring more dynamic, because the digital world provides an opportunity for people to express and discuss ideas. Individuals no longer depend on the elite. Democracy will be stronger because its base is the individual who has the freedom to determine their own destiny

2.2. The Condition of Deliberative

Digital democracy has changed the conventional democratic practices by representatives become participatory democracy through information and communication technologies. The capacity of technology has created a new political aggregation and articulation, when the citizens can delivery of their opinions directly and not depend on their political representatives [2]. Digital democracy became citizens' choice when the representative democracy are not able to deliver ideas, opinions, and interests of citizens related to public policy. At this point, digital democracy theoretically form a deliberative condition, which is a prerequisite for the realization of deliberative democracy.

The definition of deliberative condition is a condition where people consciously discuss with each other about the important things in political life that affected them, including public policies [1]. Citizens realize that representative democracy is less concerned about their interests, therefore people look for ways to be able to express their opinions directly.

 DOI: 10.1051/shsconf/20173300072

The general election the citizens have to position the new media as a tool of democracy, use it as much as possible for the creation of open societies and dialectical [8]. Citizens discuss about the things around the election through the new media.

In addition to [8], there is another theory expressed by [5] about the relationship between digital and conditions of deliberative democracy in the context of elections. According to [5], internet promote mutual understanding by developing some critical public space which guarantees freedom of expression. New media (social media) have the potential to reconfigure the power relations in the context of the communication. With social media, citizens no longer have to be passive consumers of propaganda of political parties, the government or the news media, but was actually given space to share alternative perspectives and publish their own opinions. Strengthen the theoretical relationship between digital democracy with deliberative conditions [5].

3. Description and Profile of "Speaker Kampung"

"Speaker Kampung" presents to express the aspirations of the citizens. During this time the activists of "Speaker Kampung" think that the government doesn't care to the real conditions around the community. "Speaker Kampung" dreams that citizens have stronger bargaining power and able to determine the direction of public policy. Besides that, "Speaker Kampung" encourage two-way communication between the government and the citizens. "Speaker Kampung" disseminating local information that has been rarely covered by major media, whether in form of text, photos, video, and audio.

3.1. Vision and Mision of "Speaker Kampung"

"Speaker Kampung" has a vision to be a community media that encourages freedom of expression villagers, particularly those in the Ketangga, District Suela. While its mission is to: 1) organize information management activities through the citizen journalism. 2) develop the capacity of journalism member so the information that is processed and disseminated has an ethical responsibility.

3.2. "Speaker Kampung" as a Digital Media

At the beginning, "Speaker Kampung" was a print newsletter that circulated in the village Ketangga. Due to cost constraints, then the caretaker choose to use blog because it is cheaper. Transformation process occurs at the level of personal caretaker. Hajad Guna Roasmadi as editor in chief then experienced three models of technology: bulletins, radio, and finally blog and facebook page (internet).

Blog and facebook ultimately used by "Speaker Kampung" as an information medium. News and local information quickly uploaded by members of the media "Speaker Kampung", which is only 6 people, in the facebook page because it is more responsive than a blog. After appearing on facebook, Hajad Guna Roasmadi will edit the text and uploaded it on the blog of "Speaker Kampung", http://speakerkampung.blogspot.co.id/.

4. Aspirations of Citizens in the "Speaker Kampung"

Members of the media "Speaker Kampung" become the eyes and ears in responsing the aspirations of citizens. In a blog of "Speaker village", there are many pages that have distinctive and solely intended to accommodate the many different interests of the citizens.

The various pages including "Village Environment", "Village Development", "Village Products" and "Village Self-help".

"Village Environment" contains a variety of information about events around the community, which is usually not covered by the major media. As on March 2, 2016, "Speaker village" published information about public facilities in the Perigi Village of Wells District of East Lombok, Suela, that are broke by the flood that occurred in the three-point watersheds. Once this information appears on the blog "Speaker Kampung", aid for Perigi Village residents came from all sides.

"Village Development" contains a variety of information about the regional development process that involving residents, including a variety of administrative matters of information / resident population. One of the information that published in 15 November 2015 loaded the complaints of citizens on the birth certificate free service that was held at the District Suela by the Department of Population and Civil Registration (Discapil) in East Lombok District. The people complained because they have to wait for a very long time and stand in a line in an overcrowd situation.

"Village Product" loads a variety of information about the ultimate products of the village. As reported on March 2, 2016, the village chief Sapit keep doing innovation for the sake of building a community. In 2015 he tested the cultivation of fruit crops. So that in the future, Sapit Village, District Suela, East Lombok become the central area of agro tourism.

"Village Self-help" loads a variety of information about the habits of citizens, or citizens distinctive culture that is positive for their daily lives. On February 13, 2016, "Speaker Kampung" published an information about besiru. Besiru is mutual helping activities in the planting season that is still ongoing and sustained by the Sasak tribe community in Lombok. Besiru usually carried out during the rainy season, where the farmers help each other to plant their fields.

5. Analysis and Conclusions

Community media "Speaker Kampung" has become an alternative media in delivering the aspirations of citizens. During this time, while lots of community media using radio or television, "Speaker Kampung" using the internet (blogs and facebook) to accommodate the aspirations of the citizens. Motto of "Speaker Kampung" which is from citizens, by citizens, and for citizens becomes their principles in running digital democracy. Through the "Speaker Kampung", citizens are fully aware that they can express their opinions and aspirations directly.

References

1. J. S. Fishkin. *When The People Speak: Deliberative Democracy & Public Consultation,* New York: Oxford University Press. (2009)
2. B. N. Hague, D. L. Brian D.Loader. *Digital Democracy: Discourse and The Decision Making in The Information Age*, New York: Routledge. (1999)
3. Lievrouw, Leah A. & Sonia Livingstone. *Handbook of New Media: Social Shaping and Social Consequences of ICTs*, London: Sage Publications. (2006)
4. Lim, M. "*The Internet, Social Network and Reform in Indonesia*". Contesting Media Power: Alternative Media in A Networked World. Ed. N. Couldry and J. Curran. Rowan & Littlefield. 273–288. (2003)

5. S. Prokhorov. *Social Media and Democracy: Facebook as a Tool for the Establishment of Democracy in Egypt*, unpublished thesis, School of Arts and Communication Malmo University. (2012)
6. Wilhelm, Anthony G., 2003, *Demokrasi di Era Digital: Tantangan Kehidupan Politik di Ruang Cyber* (terjemahan), Yogyakarta: Pustaka Pelajar.
7. D. Winston. Digital Democracy and the New Age of Reason dalam *Democracy and New Media*, Henry Jenkins dan David Thorburn (Eds), Cambridge: The MIT Press. (2003)
8. Y. Zhao. New Media and Democracy: 3 Competing Visions from Cyber-Optimism and Cyber Pessimism, *Journal of Political Sciences & Public Affairs*, Volume 2: 114, http://www.omicsonline.org. Diakses tanggal 23 Mei 2014.(2014)

Transparency and communication can improve wildlife welfare outcomes: A case of kangaroos

Peter Simmons[1,*]
[1]School of Communication and Creative Industries, Institute for Land, Water and Society, Charles Sturt University, Bathurst, Australia. The author is a member of the Bathurst Kangaroo Project.

Abstract. All countries manage human and wildlife coexistence. Where traditionally humans may have killed animals perceived to be a problem, this is often no longer legal or socially acceptable. Decision-makers tend to feel less strongly about coexistence issues than the people who attempt to influence them on behalf of human or wildlife interests. It has been argued that links between human interests and decisions affecting wildlife should be transparent, and that open decision making processes involving a range of local stakeholders will improve outcomes for humans and wildlife. This paper examines one case incident in an ongoing conflict between an international car racing track and kangaroos that have occasionally been found on the track during a race, causing danger to themselves and race participants. A secret local government report and plan to cull kangaroos was obtained using Freedom of Information legislation. When released to the media the subsequent public discussion showed a much greater concern for kangaroo stress, harm and right to live than the official report, and called for consideration of a range of alternatives to culling. This led to postponement of culling plans, and commitment to a more open community discussion of options. The case clearly supports claims that greater transparency and local stakeholder participation in management decision processes can improve welfare outcomes for non-human animals.

1 Introduction – culling kangaroos, a contentious issue

The kangaroo is one of the world's iconic non-human animals alongside lions, elephants and tigers. Yet at home in Australia discourses concerning kangaroos vary between reverence for a national symbol on crests and currency, and contempt as a pest to be eradicated. Decisions about whether or not to cull kangaroos are ongoing and highly contentious. It has been argued that officials who work for authorities and make decisions affecting wildlife often feel less strongly about issues of coexistence than people who lobby or otherwise attempt to influence them on behalf of wildlife or human interests [1,2]. Communicators have an important role to play in the wellbeing of non-human animals because they can explore and influence attitudes and understanding of publics, stakeholders, and the humans who make important decisions affecting kangaroos. Recent research has called for greater transparency in decision processes affecting wildlife, as a way to improve outcomes for kangaroos as well as humans [3-5].

* Corresponding author: psimmons@csu.edu.au

In every country humans make decisions about how to coexist with non-human animals. This paper presents a recent incident in the case of a long-running conflict between lucrative high technology human recreation in the form of a motor race, and an ancient species, the kangaroo. The paper daws on recent research into human-wildlife coexistence to analyse a confidential document recommending kangaroo culling obtained under Freedom of Information legislation in March 2016, and newspaper accounts which appeared in subsequent days and weeks. Irrespective of the rights and wrongs of the claims of different human parties involved, the case illustrates the negative consequences that can arise for wildlife when decisions affecting non-human animals are made without transparency and accountability to a range of stakeholders.

The paper provides a background to kangaroos and the issue of culling, and communication and decisions affecting wildlife. It then presents the case, illustrating key themes concerning the importance of communication and transparency, followed by a discussion of insights from this case.

2 Culling, justifications and calls for more transparency

Kangaroos are a protected species in Australia and it is necessary to obtain a licence to kill them. Recent research has called for increased transparency in processes and enhanced accountability in justification for killing kangaroos. Licenced killing of kangaroos is the world's largest commercial kill of wild animals on land. Each year licences are approved to kill more than 1 million [6]. They are mostly killed for their meat and fur [3]. The annual value to Australia has recently been estimated at $88.8 million AUD ($63.1 USD), and 880 full time equivalent jobs [3]. Although a small part of Australia's agricultural economy, the jobs are significant to those workers and their families. When communicating about culling the kangaroo industry tend to emphasise themes of pest control, harvesting an abundant resource, and adherence to a strict national code concerning the requirement to kill instantly with a single shot to the head with approved firearms [7]. Critics of culling processes have stressed cruelty [4] and the lack of enforcement and compliance with the provisions of the code in the remote sites where killing tends to take place [6]. Some have also questioned the science and other claims used to justify killing, such as over abundance and damage to environments. In a recent newspaper story about culling kangaroos, Dr Daniel Ramp suggested confusion between attitudes to culling and land use:

> "The problem with this notion of over abundance is that it often gets muddied with attitudes of land use...In actual fact from purely an ecosystem point-of-view, I have never seen a situation where kangaroos on their own without anything else happening are causing problems where they're out of balance or their not being driven by anything else in the environment. It just doesn't happen." [8].

To improve kangaroo welfare Boom et al (2013) called for a range of improvements in processes including improved training in humane killing, increased penalties for violations, and more effective screening and scrutiny in the granting of culling licences [6]. Officials who make important decisions concerning the coexistence of humans and wildlife often find themselves between interest groups with polarised views, and in situations where some people are bound not to be pleased, whatever the decision [5]. According to Boronyak-Vasco & Perry [3], economic arguments tend to dominate policy decisions affecting kangaroos, thus favouring industry interests, while 'animal protection advocates represent a group of stakeholders who have legitimate concerns over the commercial kangaroo industry', but 'currently have no economic mechanism to voice their interests' [3]. Researchers have reported that the values of the wider public and government agencies responsible for wildlife management are frequently quite different [9] and that the views of large numbers of people are either ignored or not heard [10], raising the possibility that large numbers of animals are killed or otherwise harmed in ways that would not be approved by the citizens that decision-makers are representing. Studies in Canada reported that the public and experts agreed on potential harm levels, but management strategies that involved killing animals often lacked broad public support [11].

An online study of the reasons people give for their attitudes to culling kangaroos found that attitudes to culling were nuanced, complex and often contradictory, even within individuals [5]. There were small minorities who held extreme attitudes either for, such as killing for enjoyment or against culling under any ciorumstance. More commonly, attitudes were conditional. Some accepted culling when the meat or fur was not wasted, others said culling was unacceptable for economic gain or human convenience, and would only kill if the individual kangaroo or mob were in danger from disease or starvation. Importantly, they reported that attitudes often conflated with other interests such as cost, inconvenience or perceived attitides of family and social groups.

> 'An urban-rural divide promulgated by some culling proponents may fuse cultural/political allegiances with attitudes to culling in ways that undermine kangaroo welfare. Humans charged with wildlife management and decision-making should aim to make assessments that benefit wildlife as well as humans' [5].

The researchers called for research into the circumstances influencing acceptance of different approaches to wildlife management. They also called for research into identifying and making relationships between human interests and decisions affecting animal welfare more transparent [5]. Ramp (2016) also called for greater transparency, arguing that many decisions are made without consultation or understanding of the priorities and preferences of the wider publics represented by decision-makers [8]:

> 'We live in a democracy and we should always have open discourse about the decisions we take, particularly when they include killing animals. Governments really should be open to listening to good management and good science, and also different attitudes within the community. I think you would find that most people don't feel comfortable about the killing of kangaroos. And governments should be listening to that and finding alternate ways to share space.' [8]

Authorities in democratic and even many non-democratic countries are under increasing pressure to listen to stakeholders and enable them to participate in decision-making. Calls for transparency in treatment of animals used in research have led to a range of written commitments and agreements and ongoing work towards alignment of industry practices and public expectations of openness about suffering, procedures used to enhance animal welfare, and possible alternatives [12, 25]. In the case of wildlife, traditional responses to kill problem wildlife are no longer legal or socially acceptable [13] and killing today should be treated as a last resort [10]. Decisions need to be carefully justified, and familiarity with stakeholder and public beliefs about wildlife management aids good communication [14]). Environmental researchers have argued that communicators and social scientists have an important role to play in understanding attitudes to human and wildlife interaction [15,13]. Better understanding of perspectives and options [2] and involvement of a range of local stakeholders in decision processes [13] can improve outcomes for humans and wildlife.

This section overviewed the culling of kangaroos and perspectives of proponents and opponents. It has also suggested that economic arguments tend to dominate policies affecting kangaroos, that attitudes to culling often conflate with social and economic interests, that culling is today more likely to be popularly regarded as a last option, and that kangaroo welfare may be improved with more transparency concerning the relationship between human interests and decisions to cull. The next section presents the case of a proposal to cull kangaroos to improve the safety of motor racing drivers.

3 Culling kangaroos for a car race

3.1 Background – 2009 cull leads to community complaints

Bathurst in NSW Australia is the home of an internationally renowned motor car race. The circuit on Wahluu/Mt Panorama is 6.2 kilometres, and just 3 kilometres from the city centre. Total attendance over 4 days at the main annual race event in 2012 was 207,205, bringing $25.28 million and the equivalent of 255 jobs into Bathurst [16]. Car racing thus brings significant income to the city and region.

The winding track is usually a public road with a 60kmh speed limit. During races cars average around 170kmh on the circuit and achieve top speeds of nearly 300kmh. At each meet there are several classes of vehicles racing over several days, with many millions of kilometres covered over the years.

Wahluu/Mt Panorama is also home to several families of mostly eastern grey kangaroos, totalling about 1000. Occasionally kangaroos get onto the circuit during a race. Around roads Kangaroos are unpredictable and dangerous to themselves and motorists at even the best of times. During a high speed race they are potentially lethal. Since 2004 there have been 9 incidents involving kangaroos on the track, 5 where a car collided with a kangaroo. No humans have been injured in these incidents involving kangaroos [17].

In 2009 the public found out through the media that Bathurst Regional Council (BRC) had secretly killed 140 eastern grey kangaroos on Mount Panorama [18]. The media reported that the decision had been approved by senior council management and correctly licenced, although it emerged that the full committee of elected councillors had not been consulted about the decision to cull:

> 'I'm in a very difficult position as a councillor because council staff took it on board as an internal management issue and didn't make the councillors privy to their decision to cull the kangaroos. And I'm very concerned for the huge outcry there has been for the killing of the kangaroos and I can't account for how it was done, where it was done, whether it was necessary, because I haven't been privy to the information ... I think it would have been very different had there been a consultation with both the elected representatives and the fauna management people in the region of how to deal with this issue. So I think it's been a bit of a disaster for the council... From the advice I've had from the fauna experts it [culling] is kind of the last tool that they resort to. There are other measures that they would try to introduce to manage kangaroos, other than a direct cull.' [19].

Once word got out to the public, the council received considerable negative feedback locally, nationally and internationally about the culling, through various media, and also from direct public complaints. In following years the issue has continued to surface periodically, and has attracted media attention globally. The Bathurst motor-race circuit / kangaroo conflict features on news pages in many countries including leading media sites such as the BBC [20] and in several languages. The race is televised internationally and earns considerable broadcast revenue, and links between the race and killing kangaroos are undesirable for commercial interests. The 2009 cull and subsequent threats to cull kangaroos led to some organised local movements to find alternative solutions. In 2014 the Bathurst Kangaroo Project (BKP) was formed as 'a collaborative science & research project supporting informed solutions for coexistence between kangaroos & community'

[21]. This community-led initiative attracted the interest of conservation biologists at the University of Technology in Sydney and led to a series of studies of the Wahluu/Mount Panorama kangaroos and their habitat. The Bathurst Regional Council contributed funds and other resources to support several research projects.

3.2 The 'secret plan for a cull' in 2016

On 31 March 2016 Bathurst's newspaper the Western Advocate ran a story headed 'Bathurst Regional Council's secret plan for a cull on the Mount', see Figure 1 [22]. Using Freedom of Information (FOI) laws the Australian Society for Kangaroos (ASK) had obtained and passed onto the media a BRC report from December 2015 revealing plans for a cull of kangaroos after the 2016 Easter races. A report marked 'CONFIDENTIAL' from the Director Corporate Services & Finances recommended application to cull kangaroos posing '.. a serious threat to motorsport participants competing at Mount Panorama during scheduled motor racing meetings' [17]. It listed 9 incidents involving kangaroos since 2004. 'Luckily to date these encroachments have only resulted in vehicular damage and no injury to persons' [17]. The report did not mention harm to kangaroos.

The report says the population of kangaroos appears stable, and that there is increasing race activity each year, particularly prior to sunrise when kangaroos are most active. The report does not discuss the possibility of racing at other times when kangaroos are less active. The report refers to research conducted by UTS, and improved understanding of kangaroo movements and habits. 'To date no real solution has been presented' [17].

> 'Given the high potential for conflict between kangaroos and Motorsport events, the time is now opportune to create an impact on kangaroo numbers in the Mount Panorama precinct by way of a further cull' [17].

The report then says that relocation of the kangaroos is not viable, and that current fencing is 'not providing a sufficient deterrent' [17] to kangaroos getting on the track during racing. A memorandum from the Director of Corp Services and Finance on 10/12/015 to the Director of Engineering Services, also obtained under FOI, notes that council has resolved to approve application for a cull of kangaroos. Another memorandum to elected councillors on 14/12/2015 and signed by the General Manager says that no application will be made until after the Easter car race event, and that before an application is made 'a report will come back to Council in open session' [17].

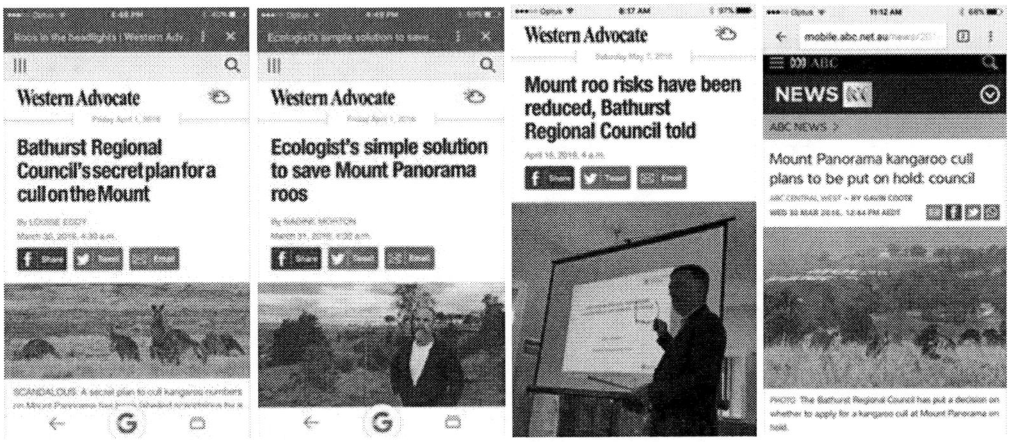

Figures 1, 2, 3, 4. Plans to cull changed quickly once stakeholders were aware and involved

In days following the story of the planned cull another media story appeared: 'Ecologist's simple solution to save Mount Panorama Roos', see Figure 2 [23]. A local Ecologist with a long experience of kangaroos on Wahluu/Mount Panorama was reported saying that the problem could be fixed by using knowledge of kangaroo movements to install appropriate fencing.

> 'It's not a numbers thing, it's an attitude thing … there's just a mindset that culls have to happen' [23]

A founder of the Bathurst Kangaroo Project was also quoted saying that council needed to take advice from experts, and fix holes in fences and their 'track clearance protocols' [23]. A community-led 'roundtable' was organised where local wildlife experts and landholders, and biologists, ecologists and social scientists from three universities, suggested alternative perspectives and approaches to managing coexistence of kangaroos and car racing.

When the story of a 'secret plan for a cull' broke in the media the council were quick to downplay the inevitability of a cull [Figure 4]. The mayor was quoted in interviews across media [see Figure 3] saying that following recent race events held without kangaroo incident there was optimism that fencing and other strategies were working [24]. Further, that no decision about a cull would be taken until after 'effectiveness of current coexistence programs could be measured' and an 'all-embracing discussion' [24].

The president of the Australian Society of Kangaroos, the organisation that obtained the report and made the process more transparent by sharing it with the media and other stakeholders was also quoted. She expressed concern that a cull had been considered at all by the Council, but that it was '… good that they're taking on board our concerns and concerns of the public ...' [24].

4 Conclusions

The important insight we glean from this case is that the welfare prospects of animals, who cannot ever speak for themselves, were clearly enhanced when a closed local authority decision process was opened to a wider group of stakeholders. In this instance, deliberations over the wildlife management strategy were made transparent only when the authority was required by Freedom of Information request. It may transpire that a future, open and 'all-embracing discussion' [24] process in the community results in a decision to cull Wahluu/Mount Panorama kangaroos, to make racing safer for drivers who visit several times a year. At the very least, the kangaroos around Wahluu/Mount Panorama that would have been culled, and their families, have received a stay of execution.

The case supports recent suggestions from Ramp [8], Mehmet and Simmons [5] and Bergstrom et al [10] that a lack of transparency in coexistence decisions can lead to adverse outcomes for wildlife welfare. Behind the scenes the official correspondence emphasised human concerns for recreation safety and economic interests [3], and was dismissive of alternative solutions to culling, when compared to the subsequent public discourse in the media. The inclusion of media and other stakeholders - local ecology and fauna experts, community and animal protection groups – led to much greater recognition of the welfare interests of the kangaroos including stress, injury and their rights to live around Wahluu/ Mount Panorama. A broader set of non-lethal options for problem-solving were canvassed than the traditional cull. The case suggests support for Treves et al's assertion that incorporation of local stakeholders as partners in planning and management can improve outcome for wildlife as well as humans [13], and the claim that publics are often more likely than authorities to prefer non-lethal solutions in wildlife management [11].

'Transparency involves *communicating meaningful information* (eg, data or details of decision-making processes) to audiences, openly and honestly, with the intention of *informing*, enabling *understanding* and meeting responsibilities of *accountability*' [25].

Studies in the UK indicate popular support for transparency about how non-human animals used in research are used, how many suffer and how much, care shown for welfare, and whether alternatives to suffering have been considered [12]. There is increasing evidence of popular support for similar transparency and accountability concerning the treatment and welfare of non-human animals in the wild. Decision processes will need to include meaningful participation by community members [11] and communicators may play increasingly important roles in ensuring the feasibility and quality of such processes.

The authority to make life and death decisions comes with great responsibility. Increasingly that responsibility includes meeting community expectations of transparency and sensitivity to the needs of non-human animals. This paper supports and recommends further research into processes where humans manage life, death and welfare for non-human animals. Non-human animals will benefit when processes involve participation of, and accountability to, a wide range of stakeholders.

1. E. S. Paul. *Society & Animals*, **3(1)**, 1-21 (1995)
2. S. Dubois, H.W. *Human Dimensions of Wildlife*, **18(1)**, 1-19 (2013)
3. L. Boronyak-Vasco, N. Perry. *Ecological Economics*, **114**, 159-167 (2015)
4. D. Ramp. *Journal of bioethical inquiry*, **10(2),** 267-272 (2013)
5. M.Mehmet, P. Simmons. *Environmental Communication* (to be published)
6. K. Boom, D. Ben-Ami, L. Boronyak and S. Riley. *International Journal of Rural Law and Policy*, **2** (2013).
7. J. Kelly. Kangaroo Industry Background. Kangaroo Industry Association of Australia (KIAA) (2013) . Retrieved 23 April 2016 from http://www.kangaroo-industry.asn.au/morinfo/kangaroo_industry_background.pdf
8. D. Ramp in J. Hall. Doubts over kangaroo culling science. *The Age newspaper*, 23 April, 2013. Retrieved online on 24 April, 2016 from http://www.theage.com.au/act-news/doubts-over-kangaroo-culling-science-20160408-go1u5k
9. T.L.Teel, M.J. Manfredo. Understanding the diversity of public interests in wildlife conservation. *Conservation Biology*, **24**(1), 128-139 (2010).
10. B. J. Bergstrom, L. C. Arias, A. D. Davidson, A. W. Ferguson, L. A. Randa, & S. R. Sheffield. *Conservation Letters*, *7*(2), 131-142 (2014). http://onlinelibrary.wiley.com/doi/10.1111/conl.12045/full
11. S. D. Dubois (2014). Understanding humane expectations: public and expert attitudes towards human-wildlife interactions. Unpublished Phd dissertation retrieved on 13/5/2016 from: https://open.library.ubc.ca/cIRcle/collections/ubctheses/24/items/1.0165855
12. Ipsos MORI. Openness in Animal Research—the public's views on openness and transparency in animal research. Ipsos MORI, 2013. Retrieved on 20 May, 2016 from : http://www.understandinganimalresearch.org.uk/policy/concordatopenness-animal-research/openness-in-animalresearch-public-dialogue/
13. A.Treves, R.B. Wallace, L.Naughton-Treves, A. Morales. *Human Dimensions of Wildlife*, **11(6)**, 383-396 (2006)
14. S.L.Hanisch-Kirkbride, J.P.Burroughs,S.J. Riley. *Human Dimensions of Wildlife*, **19(3),** 253-266 (2014)
15. M. J. Manfredo, T.L. Teel, K.L. Henry. *Social Science Quarterly*, **90(2)**, 407-427 (2009)
16. Western Research Institute (WRI). *2012 Supercheap Auto Bathurst 1000. Economic Impact Report*. Retrieved on 22 April 2016 from http://www.wri.org.au/images/docs/annual_reports/Final_2012_Supercheap_Auto_Bathurst_1000_Economic_Impact_Report.pdf

17. Bathurst Regional Council (BRC). *Mount Panorama Fauna*. Bathurst Regional Council report and proposal to cull kangaroos. Dated 9 December 2015. Copy held by the author and available on request.
18. B. Cubby. Kangaroos are the big losers in this year's Bathurst 1000. *Sydney Morning Herald newspaper*, 2 October, 2009. Retrieved on 23 April 2016 from http://www.smh.com.au/environment/kangaroos-are-the-big-losers-in-this-years-bathurst-1000-20091001-ger3.html
19. T. Carpenter. Bathurst Kangaroo Cull: Cr Tracey Carpenter. *Youtube video.* Retrieved on 23 April 2016 from https://www.youtube.com/watch?v=BwCaX8jXtI0
20. F. Molloy. Finding kinder ways to live with Australia's wildlife. *BBC News*, 5 November 2014. Retrieved on 21 April, 2016 from http://www.bbc.com/news/world-australia-29644228
21. *Bathurst Kangaroo Project (BKP) Facebook page*. Retrieved on 23 April, 2016 from https://www.facebook.com/bathurstkangarooproject/
22. L. Eddy. The *Western Advocate* newspaper, 30 March 2016. Bathurst Regional Council's secret plan for a cull on the Mount. Retrieved online on 23 April, 2016 from http://www.westernadvocate.com.au/story/3818137/roos-in-the-headlights/
23. N. Morton The *Western Advocate* newspaper, 31 March 2016. Ecologist's simple solution to save Mount Panorama roos. Retrieved on 21 April, 2016 from http://www.westernadvocate.com.au/story/3821474/ecologists-simple-solution-to-save-mount-panorama-roos/?cs=115
24. G. Coote. Mount Panorama cull plans to be put on hold: council. *ABC News website.* Retrieved on 23 April 2016 from http://mobile.abc.net.au/news/2016-03-30/concern-about-proposed-kangaroo-cull-at-bathurst/7284296
25. J.W. Yeates, B. Reed. Animal research through a lens: transparency on animal research. Journal of Medical Ethics, **41(7)**, 504-5 (2015).

DOI: 10.1051/shsconf/20173300074

The Emerging Web 2.0 Social Media as a Tool to Develop Engaged Audiences for Behaviour Change in Health Promotion: An Overview

Hui-Nee AuYong[1,*] and *Guan-Gim* Yeoh[2]

[1]Faculty of Business and Finance, Universiti Tunku Abdul Rahman, 31900 Kampar, Perak, Malaysia.
[2]Faculty of Information and Communication Technology, Universiti Tunku Abdul Rahman, 31900 Kampar, Perak, Malaysia.

Abstract. Sharing of health information which is related to health becomes easier with the introduction of new technologies such as Web 2.0 social media networks. Social media networks represent a rather innovative way of online information and knowledge repositories. On the other hand, many non-communicable diseases are caused by unhealthy urban lifestyle. Hence, the practice of the healthy lifestyle behaviors should be promoted for sustainability. Besides, in order to cut down the chronic diseases, workplace is one of the key channels for the delivery of interventions among adult populations. Latest development in information and communication technologies (ICT) has given positive influences in health communication. The new technology is believed to be more engaged because it involves visual, audio and animation simultaneously, thus it is more reachable and collaborative. This paper reviews strategies to manage health promotion. Nevertheless, improvement is still needed to achieve successful health promotion. This paper also provides insights of issues and poses future topics of health promotion to consider.

1 Introduction

With the Economic Transformation Program (ETP), Malaysia aims to transform to a high-income economy. Tee [1] suggested that these changes were coupled with changes to sedentary lifestyle while the diets could also change to high fat and low fiber which contributed to the high-energy content. As a result, the disease pattern is now more inclined to non-communicable diseases. People with metabolic syndrome (MetS) are getting more due to overweight. The practice of unhealthy lifestyle practices may make a person at risk of developing coronary heart disease (CHD) which can be tested such as by a cardiac catheter test [2]. The Malaysian Ministry of Health (MoH) through the Health Education Unit (infosihat) uses ICT dissemination of information in a timely, equitable and innovative manner. Infosihat also utilizes five (5) health promotion strategies [3]. The MoH has also established Malaysian Health Promotion Board (MySihat) in 2006. One of top five

*Corresponding author: auyonghn@utar.edu.my

challenges faced by Malaysian Health Promotion Board is new technologies [4]. Both provision of access to quality healthcare and promotion of healthy lifestyle are implemented for the quality of life improvement [5]. Health information can include following instructions after a doctor's visit, treating a chronic illness, or taking a medication properly [6]. According to [7], tailored messages that succeed in making information relevant appear to stimulate greater cognitive activity. Health promotion is a combination of educational, organizational, economic and environmental supports [8] Studies include [9] and Karan [10] suggest that health campaign are able to change behaviors to a healthy lifestyle. On the other hand, the concept of workplace health promotion lies in preventive strategy.

The process of enabling people to improve their health and increase control over, as defined as health promotion [3]. It was reported that the WHP provided a high return on investment. [11] identified workplace health promotion program return on investment (ROI) at US$3–$10 per Dollar invested, which was higher than ROI estimated by the European Agency for Safety and Health at Work [12] which was ROI at 2.5–4.8 Euros per Euro invested. Therefore, these studies show that health promotion is an effective means for improving the health of workers. Wellness at work is important also for influencing positive behavior of the workers. Nevertheless, this concept is still to be understood and implemented by many workplaces. Besides, media campaign with the aim of increasing physical activity has been one of the major health education and promotion activities by the Ministry of Health [13]. The aim of this paper is to review the influence of social media in health promotion.

2 Review of Past Research

2.1 Workplace Health Promotion (WHP)

Healthcare promotion does not limit to the public healthcare provider only as it gives impacts to the private sectors as well as to the social in general. McLeroy, Bibeau, Steckler and Glanz [14] propose an ecological model for health promotion which focuses attention on both individual and social environmental factors. In monetary terms, [15], found that medical costs fall by about US$3.27 for every dollar spent on WHP and that absenteeism costs fall by about US$2.73 for every dollar spent. National Health and Morbidity Survey 2015 [16], indicated that 29.2% of adults in Malaysia experienced some form of psychological distress. Nevertheless, the study from [17], indicated that there are limited workplace health promotion (WHP) that showed a significant cut down in mean total cholesterol levels and cigarettes smoked. [18] concluded that the wellness programs help in reducing the impact of stress at the workplace, improve overall employee satisfaction, as well as help in reducing absenteeism.

2.2 Health Communication: Behavioral Models

In this field, [19] identified that the most frequently quoted theories were the Health Belief Model (HBM), Social Learning Theory, and the Theory of Reasoned Action (TRA). HBM used in health behavior explained health behaviors and actions [20]. On the other hand, Deci and Ryan [21] suggest that Self-Determination Theory (SDT) is a needs-based motivation theory. [22] explains that self-efficacy measures predict what people will do under more specific circumstances in Self-Efficacy Theory (SET).

Social Learning Theory states that to motivate individuals they must be made to evaluate their vulnerability and susceptibility to diseases that have severe outcomes.

People's beliefs in their collective efficacy to accomplish social change, therefore, play a key role in the policy and public health perspective to health promotion and disease prevention [23]. In this case, people with stronger self efficacy correlate more to healthcare knowledge behaviour. Information, Motivation and Behavioral Skills Model (IMB Model) describes the attitude of prevention, and interprets information that can be translated into action and eventually able to make up their attitudes and behaviors [24].

Besides, there are also emerging health behavioral models reported in the last decade. For instance, [25] reported that the diffusion theory has inspired work on dissemination of public health. [26] anlyzed the individually-centered theories used in the healthcare communication research, which include Goals-Plans-Action Theory, Uncertainty Theories i.e. Uncertainty Reduction Theory and Uncertainty Management Theory, and Action Assembly Theory. Pietromonaco, [27] used the Attachment Theory, while [28] report on other models used in healthcare such as precaution adoption model (PAM), goal setting theories, the elaboration likelihood model, and behavioral self-regulation in a meta-analysis.

2.3 Potentials of Social Media in Health Communication

Though social media does not eliminate disparities between groups, these social media tools have lowered barriers in almost all corners of the world. [29] suggest the Internet can be used for persuasive health communication. [30] stresses the importance to ensure a new digital divide are not created. Those who are not actively seeking public health information through the internet have to be identified [31].

Recently, Facebook CEO Mark Zuckerberg launched the company's 10-year technology road map with ways connecting and bringing people together for causes including fighting global disease [32]. The emerging internet has become a main medium to enable health information sharing in various forms. A consumer survey by Ed Bennett Consulting (ebennett.org) quoted by [33] says that 72 % of patients searched for online information before or after a doctor visit. It clearly shows that larger hospitals are early adopters of Social Media. Channel wise, both Facebook (30%) and Twitter (28%) are equally popular, with You Tube (19%) and Linkedin (18%) coming in second. Blogs (5%) are a distant third. Consequently, the widely available "mobile" health (m-health) can play an important role in healthcare [34]. According to [35], the definition of Health 2.0 is the use of social software to promote the collaborations among patients, medical professionals and those who are involved in healthcare [35].

Appropriate use of social media can improve user access and satisfaction in obtaining healthcare information. According to [36], World Health Organization has listed that patient use of social media recorded higher satisfaction. [33] also suggest that healthcare organizations to grow business and visibility using the social media. Social media is a leveling of the health playing field through its removal of geographical boundaries and opportunities and connectedness [37]. Social media and Facebook in particular, can benefit more than just a platform for the dissemination of information. Particularly, in a survey conducted by Timian et al. [38], healthcare explored Facebook as an evaluation tool for hospital services through social media.

In contrast, application of social media for public health communication or private health promotion has to be used with care particularly among healthcare professionals. For instance, [39] and [40] warned for the potential harm associated with unsafe social media content, as people seeking health information has to recognize that the authorship is difficult to determine, sources are rarely provided, and users may post their personal opinions in social marketing [41] and cause concerns of inequalities [42], privacy, misinformation and lack of evaluation [43]. For instance, there is a scandal involved

 DOI: 10.1051/shsconf/20173300074

China's top internet search engine, Baidu, of which the company's search engine provides medical information based on bidding process that caused the death of a university student [44].

With a significant social media presence, health professionals can no longer assume that their authority and social position will afford them greater influence [45]. Hence, one of the health communication research focus suggested by [46] is developing e-health literacy programs to train patients in health information gathering, evaluating, and use. [47] further encourages researchers to work with healthcare professionals and educators for social media educational strategies. This has to be done to overcome challenges encountered by physicians i.e. confidentiality, lack of participation / time / trust, workplace support, and information anarchy [48]. Public health agencies too must develop a strategic communication plan that incorporates best practices [49]. Health promotion must move to high engagement for enhanced communication [50]. They can get people involved in creating environmental and social conditions. In other words, partnerships and participation are viewed as the culmination of social media [51].

3 Conclusion

There is a tenacious need to improve the behavioral outcomes of health communication interventions. Public health promotion agency has to embrace use of social media networking which is also important in promoting its trust by the public. Furthermore, promotion of healthy lifestyle should also be extended to the private sector. This is due to the fact that an unhealthy workplace may bring about unsatisfactory / complaint and ultimately turnover of the employees. Hence, Employers should also consider savings against illness related costs as a result of WHP. Other benefits are increased production and improved morale. Intervention programs target the individual, and the work environment also needs consideration. In short, social media has changed communication and furthermore to transforming healthcare. The use of social media in health communication also helps in providing a better understanding to health information.

References

1. E.S. Tee, *Obesity in Asia: prevalence and issues in assessment methodologies*, Asia Pacific Journal of Clinical Nutrition, **11**, 3 (2002)
2. National Health Service, NHK UK, *Coronary Heart Disease – Diagnosis*. http://www.nhs.uk/Conditions/Coronary-heart-disease/Pages/Diagnosis.aspx (2016)
3. World Health Organization (WHO), *Ottawa Charter for Health Promotion,* Geneva (Switzerland) (1986)
4. E. Arps, *The Use of Internet and Mobile Phone Based Health Promotion Interventions in Youth Populations,* Health Promotion Forum (HPF), Newmarket, Auckland, (2014)
5. Ministry of Health (MOH) Malaysia, *Country Health Plan 2011-2015*, Putrajaya (Malaysia), (2011)
   F.M. Moy, A.A.B.
6. Pfizer (2016). *Health Literacy*. http://www.pfizer.com/health/literacy
7. M.W. Kreuter and R.J. Wray Tailored and Targeted Health Communication: *Strategies for Enhancing Information Relevance*, Am. J. Health Beh. (**Suppl. 3**), (2003)
8. L. Green, and M. Kreuter, *Health Promotion Planning: An Educational and Environmental Approach.* Mayfield, Mountain View, CA, (1991)
9. L.B. Snyder, *Health Communication Campaigns and Their Impact on Behavior*, J. Nutr. Educ. Behav., **39**, (2007)

10. K. Karan, *Impact of Health Communication Campaigns on Health Behaviors in Singapore*, Soc. Marketing Q., **14**(3), (2008)
11. Goetzel RZ, Ozminkowski RJ. *The health and cost benefits of WHP programs*. Annu. Rev. Public Health, Cross Ref. Medline, **29**(1), (2008)
12. European Agency for Safety and Health at Work (EU-OSHA), *Workplace health promotion improves productivity and wellbeing*, Bilbao (Spain), (2010)
13. G. Rampal, L. Rampal and S.H. Lee, *Health promotion in Malaysia. In: Health Promotion Theory and Practices*. CUHK, Hong Kong, China, (2012)
14. K.R. McLeroy, D. Bibeau, A. Steckler & K. Glanz, *An Ecological Perspective on Health Promotion Programs*. Health Educ Behav. **15**, 4 (1988)
15. K. Baicker, D. Cutler, and Z.R. Song, *Workplace Wellness Programs Can Generate Savings*. Health Affairs. **29**, 2, (2010)
16. Institute for Public Health (IPH), *The Third National Health and Morbidity Survey 2015*. Kuala Lumpur, Malaysia: Ministry of Health (MOH), (2015)
17. F.M. Moy, A.A.B. Sallam, & M.L. Wong, *The results of a worksite health promotion Program in KL, Malaysia*. Health Promo. Int. **21**,4, (2006)
18. D.N.M.A. Abdullah, A. Ali, and Y.L. Oon, *A Study on the Relationship bet. Coporate Wellness Programs and Employees' Stress, Job Satisfaction & Absenteeism*, (2008)
19. K. Glanz, F.M. Lewis, and B.K. Rimer, *Health Behavior And Health education: Theory, research, and practice*, San Francisco, CA: Jossey-Bass, (1990)
20. M.H. Becker, R.H. Drachman, and J.P. Kirscht, *A new approach to explaining sick-role behavior in low-income populations*. Am J. Public Health. **64**, (1974)
21. E. L. Deci, and R. M. Ryan, *Intrinsic Motivation and Self-Determination in Human Behavior*. New York: Plenum Press, (1985)
22. A. Bandura, *Self-Efficacy: The exercise of control*. New York: Freeman, (1997)
23. A. Bandura, *Health promotion from the perspective of social cognitive theory*. Psychol. and Health, **13**, (1998)
24. J.D. Fisher & W.A. *Fisher Changing AIDS risk behavior*. Psychological Bulletin, **111**, 3 (1992)
25. L.W. Green, J.M. Ottoson, C. Garcıa, and R.A. Hiatt, *Diffusion Theory and Knowledge Dissemination, Utilization, and Integration in Public Health*, Annu. Rev. Public Health, **30**, (2009)
26. C.L. Bylund, E.B. Peterson, K.A. Cameron, *A practitioner's guide to interpersonal communication theory: An overview and exploration of selected theories*, Patient Educ. Couns., **87**, 3 (2012)
27. P.R. Pietromonaco, B.Uchino, and C.D. Schetter, *Close Relationship Processes and Health: Implications of Attachment Theory for Health and Disease*, Health Psychol., **32**(5), (2013)
28. E. DeLeon, L.W. Fuentes and J.E. Cohen, *Characterizing Periodic Messaging Interventions Across Health Behaviors and Media: Systematic Review*, J. Med. Internet Res., **16**, 3, (2014)
29. M.M. Cassell, C. Jackson, B. Cheuvront, Health *Communication on the Internet: An Effective Channel for Health Behavior Change?* J. Health Comm., **3** (1998)
30. E. Arps, *The Use of Internet and Mobile Phone Based Health Promotion Interventions in Youth Populations,* Health Promotion Forum (HPF), Newmarket, Auckland, (2014)
31. O. Washington, *Using Social Media Platforms to Amplify Public Health Messages*, Georgetown University (2010)
32. Facebook, *Technology to Connect the World: F8 Day One Roundup*, 12th April 2016, http://newsroom.fb.com/news/2016/04/f8-2016-day-1/ (2016)
33. Wipro (www.wipro.com), *Impact of Social Media in Healthcare*, (2010)

34. J.G. Kahn, J.S. Yang, and J.S. Kahn, *'Mobile' Health Needs and Opportunities in Developing Countries,* Health Aff., **29**, 2, (2010)
35. J. Sarasohn-Kahn, *The Wisdom of Patients: Healthcare Meets Online Social Media*, California Healthcare Foundation, (2008)
36. C. Hawn, *Take two aspirin and tweet me in the morning: How Twitter, Facebook, and other social media are reshaping healthcare*, Health Aff., **28**, 2 (2009)
37. R. Gorham, L. Carter, B. Nowrouzi, N. McLean, & M. Guimond, *Social Media & Health Edu: What the Early Literature Says*, Intl. J. E-Learning & Edu, **26**, 2 (2012)
38. A. Timian, S. Rupcic, S. Kachnowski & P. Luisi, *Do patients "like" good care? Measuring hospital quality via Facebook, Am J Med Qual.,* **28**, 5 (2013).
39. G. Eysenbach, and C. Köhler, *Does the Internet harm health? Database of adverse events related to the internet has been set up,* British Med. J., **324**, (2002)
40. A.Y.S. Lau, E. Gabarron, L. Fernandez-Luque and M. Armayones, *Social media in health*, Health Info Mgmt J, **41**, 2 (2012)
41. K. Vance, W. Howe, R.P. Dellavalle, *Social Internet Sites as a Source of Public Health Information*, Dermatol Clin., **27**, 2 (2009)
42. F. Khatun, A.E. Heywood, P.K. Ray, S.M.A. Hanifi, A. Bhuiya, S.T. Liaw, *Determinants of readiness to adopt mHealth in a rural community of Bangladesh*, Int. J. Med. Inform., **84,** (2015)
43. A.F. McGloin and S. Eslami, *Digital and social media opportunities for dietary behavior change*, Proc. of the Nutr. Soc., **74**, (2015)
44. S. Jiang, Student death triggers probe of China's Baidu, *CNN Money*, http://money.cnn.com/2016/05/03/technology/baidu-china-student-death/ (2016)
45. G. Bodie, M.J. Dutta, *Understanding Health Literacy for Strategic Health Marketing: E-Health Literacy, Health Disparities, and the Digital Divide*. Health Mkt Q., **25**, 1, (2008)
46. M.J. Dutta, *Health Communication: Trends and Future Directions*, in J.C. Parker and E. Thorson (Eds.) Health Commun. in the New Media Landscape, 59-92, (Springer Publishing Co., New York, 2009)
47. D. Currie, *Public health leaders using social media to convey emergencies: New tools a boon*, Nations Health, **39**, 6, (2009)
48. S. Panahi, J. Watson and H. Partridge, *Social media and physicians: Exploring the benefits and challenges*, Health Inform. J., **22**, 2, (2016)
49. R. Thackeray, B.L. Neiger, A.K Smith and S.B. Van Wagenen, *Adoption and use of social media among public health departments*, BMC Public Health, **12**, 1, (2012)
50. B.L. Neiger, R. Thackeray, S.A. Van Wagenen, C.L. Hanson, J.H. West, M.D. Barnes, M.C. Fagen, *Use of Social Media in Health Promotion: Purposes, Key Performance Indicators, and Evaluation Metrics*, Health Promo. Prac., **13**, 2, (2012)
51. B.L. Neiger, R, Thackeray, S.H. Burton, C.G. Giraud-Carrier, M.C. Fagen, Health Promo. Prac., **14**, 2, (2013)

Factor of Awareness in Searching and Sharing of Halal Food Product among Muslim Families in Malaysia

Siti Zanariah Yusoff [1], and *Nor Azura* Adzharuddin[2,*]

[1]Department of Modern Languages and Communication, Faculty of Languages and Communication, Universiti Sultan Zainal Abidin, Gong Badak Campus, 21300 Kuala Terengganu, Terengganu.
[2]Department of Communication, Faculty of Modern Languages and Communication, Universiti Putra Malaysia. 43400 UPM Serdang, Selangor.

Abstract. Information search activities are fundamental in information sharing, especially for the context of information confusion in the market for halal products. This research paper will focus on awareness factor in information searching and sharing of halal products in the Muslim families. Recently, the halal issue has become the hot topic as it involves community religious tenets. This study aims to enrich communications literature in terms of information seeking in halal food products. The sample of the research consisted of 340 Muslim families in Bangi, Selangor. The data are collected through questionnaire and analyzed using descriptive analysis and inferential analysis such as correlation. The result of the study revealed that awareness factor is significantly correlated with information seeking behavior among Muslim family.

1. INTRODUCTION

The growth of Muslim population is increasing over time. Muslim population taking up 30 per cent from the total population and it is forecasted that in 2025 the Muslim population will be increasing up to 35 per cent [1]. Due to this, the Halal product consumption is increasing and so does the demand for Halal products. Halal and its related industries have become an important market and lucrative business, same goes to its great potential of investment. Halal industry covers food, non-food and services area [2, 3]. Muslim group is seen as representing the entire population of Malaysia with an average of 60.4 per cent which makes halal issue as an issue of interest to the user [4]. In general, consumers' purchase intentions towards food are highly associated with individual and environmental factors such as marketing information, situation and food specific properties which include product [5]. This is because the roles of information, knowledge and Sharia are very important for Muslims in making their food choice. Search activity information is fundamental to the achievement of an individual as it is the key to the level of competence and to achieve information needs. Information seeking is one of the learning tools which

* Corresponding author: drazura318@gmail.com

involve search process, gather and use the information that had been found for certain purposes in the future.

1.1 Problem Statement

The exposure of halal issue becomes the hot topic to debate among society as it is vital for religion issue. Report shows that The Islamic Development Department (Jakim) withdrew halal certification from Secret Recipe Manufacturing Sdn. Bhd. due to violations involving hygiene offense under the Malaysian Halal Certification Procedure Manual and Good Manufacturing Practice [6]. This kind of issue becomes polemic in Muslim society. They have expressed their mixed responses on the matter since the issue has sparked concern among Malaysians, particularly the Muslim community. It can be seen through hot debate in media social, as it provides easier way to publish people thoughts and reach people all over the world [6]. 'Halal' is a simple concept but the current scenarios in the industry make it difficult and led to consumer confusion [7]. This matter becomes more complicated with the development of technology and also the diversity of resources obtained in the production of food including other daily products The issue that arises is attracting the attention of Muslim consumers to new processing methods of foods and other products [8]. The halal issue that hit our country is getting worse and it should been taken care by all parties, especially consumers. This is because although various measures have been taken by the authorities concerned with the enforcement of halal logo, halal certification and other matters related to halal issues, however it is not exhaustive. The proof is, more industries or premises that actively selling daily use products, especially foods that are allegedly halal. The actual threats regarding halal issue is the lack of awareness in individuals itself. Perhaps consumers fail to build their own awareness because of the internal factor within themselves such as value, religiosity commitment and attitude towards halal products, information processing or external factor like environmental influence including culture, opinion leader and community [9]. The Muslim community in this country still does not have confidence against products manufactured by Muslims, even though the products are in choice. Moreover, no one can prevent people from using the product, but as an attempt to deal with it, an awareness campaign can be implemented through non-governmental organizations or Islamic Consumer Association [10]. Even though there are many studies have been done related to halal issue, this study, however, are focusing on the information seeking of halal food products. Most individuals avoid and ignored any information related to health especially things related to cancer. They purposely avoid the information because they are afraid knowing the result. The ignorance of information happens when individual knows about the problem yet chooses not to face it. Therefore, this study was carried out to achieve the following research objectives: 1). to identify the factor of awareness in searching for halal food products among Muslim families, 2). to identify the relationship between factor of awareness with information seeking behavior in Muslim families and 3). To identify the relationship between information seeking behavior and information sharing behavior.

2. Literature Review

2.1 Information seeking

Every human being wanted to know the feeling naturally [11]. Based on this reason, there are many researches has been conducted to investigate human information seeking behavior from all aspects of life. Most of the researches that have been done contributed to literature development in psychology, computer science, business administration, religion studies, medical, genetic counselling, engineering and plantation. This situation shows that information seeking is vital in maximizing human cognitive needs and as a communication tool in human life. Information seeking action sparked by curiosity and awareness of the existence about the knowledge gaps. Usually, women dominated this kind of behavior in certain family. Woman tends to be the most dominant person in seeking information regarding their child and the whole family, especially for themselves [12]. This situation happens everywhere and it clarifies their role as a very caring person could not be argued anymore. Information searching behavior can be defined as the ways of how individuals collect information for their personal purposes, to add existing knowledge and for self-development [13]. In order to generate a better generation with a pure heart, information seeking on halal products is a must.

2.2 Awareness

Muslim's awareness is a great power in order to educate the manufacturer of halal food products in the market. As a Muslim, understanding and alert of halal and haram concept is essential, we cannot just simply depend on act and the authority. For example, Muslim consumers should take note about the content and ingredient of their food or any products when they shop. They also should be aware of the processing method whether it follows *shariah* compliance or not. There are nine categories of halal and haram related to food products. *Mashbuh* and *makruh* are in between halal and haram law [14]. *Mashbuh* means something that lead to curiosity, meanwhile, *makruh* means something that not encourages being used [15]. It is vital for all Muslim consumers to have knowledge about any products they want to buy falls under which category. In Malaysia, there are a few researches related to halal product such as [16]. Even though factor of awareness among Malaysian has been acknowledged to be increase, yet their level of awareness are in low category compared to other countries such as Indonesia and Thailand. The main reason that leads to this situation is the limited sources of product information in the market. This kind of issues is at worst when media reported about the misuse of halal logo by irresponsible person almost every day whether in print media and electronic media. Thus, to overcome this issue, researcher wants to identify the level of consumer awareness, next to suggest a suitable concept in promoting halal products.

2.3 Halal food

Food is a basic necessity in human life. In that case, Islam is very concerned about the issue of halal food. A food will be categorized as halal when the production of raw materials, material components, additives and processing based on the guidelines set out by legislation [17]. Therefore, it includes the preparation, processing, storage, packaging, handling and transportation that meet Sharia. Generally, Islam requires consuming all the foods that are fine and forbid eating the foods that are disgusting and dirty *(khabaith)*. Standards of good nutrition for Muslims expressed as halalan toyyiban. In the course of Facebook Sahabat Halal Jakim, the food is fulfilling the concept of *halalan toyyiban* if it is free from Islamic banned substances, free from the substances that can cause harm, not filthy and also clean and safe. Cases like halal logo abuse, material fraud or content used in a product and also the management of premises that do not comply with *halalan toyyiban* standards commonly reported. The existence of such cases may raise questions about the validity of halal [18].

2.4 Conceptual Framework

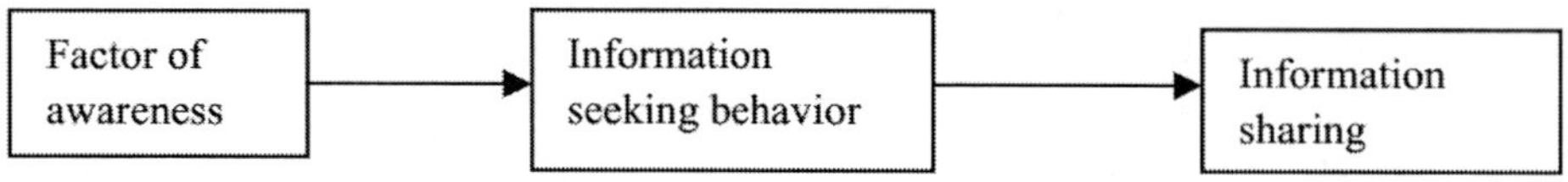

2.5 Hypothesis

H1: There is a significant relationship between the levels of awareness towards halal products with information seeking behavior among parents.

H2: There is a significant relationship between the levels of awareness towards halal products with information seeking behavior among children.

3. Methodology

The questionnaire seeks to identify the level of awareness in information seeking behavior among Muslim families, and consist of five sections on demographic, exposure towards halal products, level of awareness, information seeking behavior and information sharing behavior. Pre-tested data among 30 families showed significant result with Cronbach Alpha's value equals to 0.950. 340 set of questionnaires have been distributed to the respondents who live in Bangi, Selangor. Researcher decides to select Bangi as research location because majority of Muslim population live there [19]. Other than that, Bangi is an urban area that exposed to process food compared to rural area. Empirical data was collected through survey by using questionnaire. Data was analyzed using (i) descriptive statistic and (ii) Pearson correlation. Data were analyzed using SPSS and were presented using correlation.

4. Analysis and Findings

This study analyzed the relationship between the levels of awareness with the information seeking behavior among Muslim families. As noted in literature review, level of awareness among Muslim consumer is still low. Statistical analysis in Table 1 showed both parents and children has high level of awareness towards halal food products with 92.9 percent and 86.5 percent. This shows that parents are more concerned and aware of halal food products compared to children. Parents' high awareness exist because they attended a lot of awareness campaign held either by the government or non-governmental organizations (NGOs) to help Muslim consumer increases awareness in them about halal food products. Furthermore, Muslim consumer awareness of the issue concerning dubious counterfeiting halal logo also increased due to the campaigns held. According to them, an awareness campaign held was so relevant when there are dumping of fake halal logo and also variety of tactics used by traders to confuse consumer. This research finding is consistent with the findings from [20] which states that the factor of Muslim consumer awareness is high when it comes to the issue of the selection of halal food products. However, these findings contradict the issues raised by a local newspaper regarding people today are not compassionate about the matter and consider whatever sold is halal to them. This proves that Muslim society today no longer take for granted about the halal and haram food because realizing its importance to the development of an outstanding personality.

DOI: 10.1051/shsconf/20173300075

Table 1. Respondent Distribution towards the Level of Awareness of Halal Food Product

Awareness Level	Frequency		Percentage (%)	
	Parent (N=340)	Children (N=340)	Parent (N=340)	Children (N=340)
High (36-50)	316	294	92.9	86.5
Medium (23-35)	24	46	7.1	13.5
Low (10-22)	0	0	0.0	0.0

Table 2. Distribution of Respondents towards Halal Products Information Seeking Behavior

Information Seeking Behavior	Frequency		Percentage (%)	
	Parents (N=340)	Children (N=340)	Parents (N=340)	Children (N=340)
Active	321	314	94.4	92.4
Passive	19	26	5.6	7.6

According to Table 2, it was found that parents and children are active in seeking information. A total of 321 parents (94.4%) and 314 children (92.4%) said they actively seek information about halal food products. This means that, everyone in the family being active in seeking information of halal food products among Muslim family and not putting the responsibility on the shoulders of the parents alone, but the children also play a part thereof. This situation may be because they are exposed to the influence of the surrounding communities as well as a wide exposure in the media. Ease of access to the internet is one of the factors the children easily get information about halal issues, for example through social sites being visited, such as Facebook, Twitter and also BlogSpot. However, there are a small part of the respondents were passive in seeking information on halal product that is 5.6 per cent for parents and 7.6 per cent for children. They were passive in seeking this information because they were likely to rely on others in the family. In addition, it may be due to indifference and lack of self-awareness about the importance in searching the information of halal products in the family.

The information sharing is important in the family to ensure the harmony of the family institution. This includes the sharing of information of halal food products. According to Table 3, it was found that many occurrences of sharing in the Muslim family of 250 parents (73.5%) said they preferred to share information with families about halal food products. Similarly, the children, they said a lot of sharing in the family occurs in related of halal food products, but its percentage is quite low compared to parents with 68.5%. To enable the sharing of information occurs in a family, each member should have knowledge about what they want to share. This knowledge is usually acquired through the activity of information searching.

Table 3. Distribution of Respondents towards Information Sharing of Halal Food Products (N=340)

Information Sharing	Frequency		Percentage (%)	
	Parents (N=340)	Children (N=340)	Parents (N=340)	Children (N=340)
High	250	233	73.5	68.5
Medium	83	89	24.4	26.2
Low	7	18	2.1	5.3

4.1 Relationships between levels of awareness with information seeking behavior

H3: There is significant relationship between the levels of awareness towards halal products with information seeking behavior among parents.
H4: There is significant relationship between the levels of awareness towards halal products with information seeking behavior among children.

For the awareness factor and its relation to information-seeking behavior, it is found that coefficient of correlation and significant value for parents is ($r = 0.472$, $p = 0.000$). The correlation coefficient value is dissimilar to the correlation coefficient and significant value for children, ($r = 0.477$, $p = 0.000$). This study has found that there was a weak positive correlation between the level of awareness and information seeking behavior of halal food products for parents and children. Significantly smaller value of significance level also suggests that the second hypothesis (H3 and H4) of this study is accepted. This means that the higher level of awareness, the more active Muslim families seek for information. Muslim consumers are more concerned and realize the importance of finding and using halal products through inquiries to the responsible party whether they can trust the halal logo alone before buying a food product [21]. Accordingly, it can be concluded that consumer awareness on halal food products affect their information seeking behavior. In addition to this, the awareness aspect of halal food products can also be associated with exposure to halal issues. High awareness exists when Muslim families are exposed to halal issues featured by media or other sources, thus encourage their information seeking behavior.

5 Conclusion

The study shows that Muslim families have high level of awareness towards halal food products. Apparently, they no longer depend on information given by JAKIM only, yet they are highly active in information seeking of halal food products by their own. As a family, they communicate frequently in sharing of halal food information. This small study drips some information that factor of awareness are vital in information seeking of halal food products among Muslim families [22].

References

1. Esfahani, A.N, Shahnazari, A. J. Basic. Appl. Sci. Res., **3**, 2 (2013)
2. Ahmad, H., Fazullah, A., Borham, A. J., Hashim, H. & Razak, M. J. Arts and Sci. Conf., Austria (2011)
3. Hanzae, K.H. & Ramezani, M.R. Interdisciplinary J. Res.in Bus., **1**, 5 (2011)
4. Abdul Aziz, Y. & Chok, N.V. 3rd International Conference on Business and Economic Research Proceeding, (2012)
5. Golnaz, R., Zainalabidin, M. & Mad Nasir. S. *Non-Muslim consumers' understanding of Halal principles in Malaysia*. J. Islamic Mark., **3**, 1, (2012)
6. Astro Awani. Secret Recipe halal certificate revoked. Retrieved on 19th October 2015 from http://english.astroawani.com/malaysia-news/secret-recipe-halal-certificate-revoked-60809 (2015)
7. Abd Hafiz, K. Dalam Prosiding 2nd Global Islamic Marketing Conference. Organized by: The United Arab Emirates University and the International Islamic Marketing Association (2012)

8. Suddin L, Geoffrey H. T, Amin H. Int. J. Islamic and Middle Eastern Finance and Manage. **2**, 1, (2009)
9. Dun Panchor. Kesedaran penggunaan produk halal masih rendah. Retrieved on April 27, 2012from http://www.dunpanchor.com/v2/index.php?option=com_content&view=article&id=104:kesedaran-penggunaan-produk-halal-masih-rendah&catid=40:program-kerajaan-negeri-kelantan&Itemid=108(2010)
10. Case DO., Andrews James E., Johnson J. David and Allard Suzzane L. Avoiding versus seeking: the relationship of information seeking to avoidance, blunting, coping, dissonance, and related concepts. **93**, 3 (2005)
11. Aristotle. Metaphysics. Apostle HG, translator. Grinnel, IA: Peripatetic Press (1979)
12. Warner, D. & Procaccino, J. D. J. Am. Soc. Inform. Sci. and Tech., **55**, 8, (2004)
13. Kakai, J.M., Ikoja-Odongo, R., & Kigongo-Bukenya, I.M.N. A study of the information seeking behavior of undergraduate students of Makerere University,Uganda. World Libraries 14 (1).Available: http://www.worlib.org/vol14no1/print/kakai_print.html. (2004)
14. Eliasi, J. R. & Dwyer, J.T. J. Am. Dietic Assoc. **102,** (2002)
15. Jamal, A.N.S. and Mohd Ariffin, N.S. *Dimension of Halal Purchase Intention: A Preliminary Study*. Universiti Malaysia Sarawak (2008)
16. Mohd Dali, NRS, Nooh MN, Nawai, N & Mohammad, H. J. Manage. & Muamalat, **2**, (2009)
17. zizan, A., Yaacob, M., Khairudin, N., Saad, N.L., Mohammad, N. *Kesedaran terhadap pemakanan halal dalam kalangan pelajar Islam peringkat sekolah rendah di negeri Perak*. Prosiding WCIT2012 (2012)
18. Azhar, A, Abdul Wahab, H, Ahmad Shariff, N & Badarulzaman, MH. *Perlindungan pengguna terhadap produk makanan Halal: Tinjauan hukum dan undang-undang di Malaysia*. Prosidings WCIT2012 (2012)
19. Alam S.S, Mohd R, Badrul Hisham. J. Islamic Mark. **2**, 1 (2011)
20. Endang, S.S.J. Indonesian Soc. Sci.and Humanities. **3**, (2010)
21. Ahmadi, M. *Factors Influencing the Muslim consumer's level of confidence on halal logo issued by Jakim: An Empirical Study*. Tesis Sarjana yang tidak diterbitkan. Universiti Sains Malaysia (2005
22. Yusoff, S. Z., Adzharuddin, N. A., & Bolong, J. Faktor pencetus perilaku pencari maklumat produk halal dalam kalangan keluarga Islam. *GMJME*, **3**(2), (2014)

Communication and nurturing to sustain collaborative partnership

Suhaini Muda[1,*]

[1]School of Multimedia Technology and Communication, Universiti Utara Malaysia, 06010 UUM Sintok, Kedah

Abstract. One of the major challenges of sustaining collaborative partnership in a longstanding community service organisation in the study involves ambiguous communication and cultural constrains among the stakeholders which due to the lack of communication and nurturing in the relationships. Based on the responses of multiple stakeholders, this paper brings forward the discussion on the need for communication and nurturing, the aspects which importance is commonly known but often being disregarded in the collaborative settings.

1. INTRODUCTION

This qualitative case study was done in Malaysia within the context of a particular community service organisation (CyberCare), a non-governmental organisation (NGO), aiming to improve the life of the children in orphanages through the mechanism of collaborative partnership between various stakeholders. Numerous programmes and activities have been carried out since 1998. The findings of this study bring together the views of multiple stakeholders concerning the challenges of sustaining collaborative partnership in a longstanding community service organisation. This paper will discuss one of the major challenges which involve the issue of ambiguous communication and cultural constrains from the perspective of the selected stakeholders which includes five CyberCare, four corporate, two other NGO, nine volunteer, and nine orphanage administrator representatives. The paper flows from the review of the literature to the brief of findings, and discussion and conclusion.

2. Communication and nurturing

The sustainability of collaborative partnership has been connected by a number of authors to transparent communication and members continuously nurturing the relationship [1-5]. The importance of constant communication and nurturing in maintaining and sustaining collaborative partnership has been discussed extensively in the literature. This is especially obvious in that in which the scholars emphasise collaboration as an interactive

* Corresponding author: suhaini@uum.edu.my

DOI: 10.1051/shsconf/20173300076

process [1,3] and as emerging processes [6-8]. These writers argue that collaboration evolves as participants interact throughout the course of collaboration. They also emphasise the importance of collaborators being involved in constant communication and nurturing. Huxham6 describes the nurturing in collaborative partnership in terms of a gardening metaphor. In this context, fragile plants need gentle care, and an overgrown garden needs more decisive tactics like chopping down excess growth, pulling up weeds, and caring for the plants that have been overpowered by others to nurture the plants and garden back to health [6].

The reasons why so many writers stress the need for constant communication is clarified by Wildavsky [8]. He argues that when collaborators come from different fields or bring widely varying perspectives and backgrounds, there is often a gap between expectations and understandings. He claims that the only way to bridge the gap is through communication over time among all the participants. In another study, the authors11 suggest that established informal relationships and communication links are one of the factors in the success of community collaborations. In their study, the community collaboration used both formal and informal communication which gave the mixture of formal and informal ways of building and sustaining collaborative relationships. The study showed that the members of the collaboration were able to balance the informal nature of communication with their capability to communicate about formal and professional issues in an open and effective way. Consequently, the personal connections the members had to each other made them more willing to make efforts to resolve any arising issues together because they shared personal relationships.

The significant role of communication in sustaining a successful partnership is further discussed in reports by the multi-stakeholder partnership group, Global Knowledge Partnership (GKP) [9]. This work claims that it is important for the partners to maintain regular contact with each other, thereby preventing differences from becoming conflicts. Rather, through regular communication, issues can be jointly solved. GKP [9] suggests:

> *If the original set of agreements governing the partnership has been properly structured, ensuring on-going communication and transparency should be no more than a question of implementing the agreed procedures*" (p 25).

In this claim, the GKP also strongly relates constant communication with collaborative partnership agreement where it is believed that a well-structured agreement will help in the smooth running of the implementation process through communication and transparency. However, as has been discussed, it is a very difficult task because the process of reaching agreement itself is time-consuming, and partners come with different perspectives and expectations.

Huxham and Vangen [2,10] and Vangen and Huxham [5,11] emphasise the need for nurturing to address the various challenges. They specifically stress the importance of constant nurturing in coping with the challenges of a multiplicity of aims, power imbalances, sustaining trust, multifaceted membership structures, the dynamic of collaborative structures, and supporting leadership activities and encouraging community involvement in collaborative partnership [2,11]. In terms of managing the barriers of time and energy, Wildavsky [8] also suggests careful nurturing as a route to renewing enthusiasm as the partnership goes along.

It has been argued that continuous nurturing of the relationship is essential to address these challenges even in situations where collaboration is working well with a good degree of trust [5]. This continuous effort is needed to ensure that a sufficient level of trust can be maintained [5,11]. Despite the importance of nurturing, undertaking it is not easy in practice. Huxham and Vangen [2] claim that the main challenge in managing membership structures and dynamism is "learning how to identify, live with and progress despite ambiguity and complexity" [2] with constant nurturing [2,12] like through constant communication via email and social media networks. Many other authors also agree with the need to nurture the collaborative or partnership process and the challenges in practice. For example, Prins [13] who shares her experience as a process consultant or facilitator for multiparty collaboration in the context of foster care admits that there is a tension between the great need to take time to nurture the process and the need to produce results as expected by the collaborative managers of the collaborative project. The reviewed of the literature show that communication and nurturing are critical concerns in sustaining collaborative partnership

3. AMBIGUOUS COMMUNICATION AND CULTURAL CONSTRAINS

The responses from the selected stakeholders reveal that part of the challenges in sustaining the collaborative partnerships in this study concerns ambiguous communication and cultural issues among the stakeholders. Even though all orphanages signed a memorandum of understanding (MOU) to join the partnership, they claimed a lack of clear communication as a main challenge. Where there was no communication or follow up of the programmes, the orphanage administrators were most likely to make their own assumption. The challenge of communication was also voiced by the volunteers and the service organisation. Although the participants seemed to share similar cultural values, and the partnership practised flexible working, the findings reveal that there were some related cultural and structural constraints. In these terms, the service organisation, corporate, NGO, and volunteer participants discussed the difficulties of partners understanding different approaches, and the difficulties of volunteers adapting to cultural differences. The discussion on ambiguous communication and cultural constraints in a range of contexts comprise the challenge for all stakeholders, and the specific challenge between the service organisation and the Lions Club.

Table 1 provides the brief of the responses from the participants concerning the communication and cultural constraints among the stakeholders. While Table 2 presents the brief of responses from participants regarding the communication and cultural constraints between CyberCare and the Lions Club in particular.

Table 1. Communication and cultural constraints among the stakeholders

Issues	**Brief of the responses**
Miscommunication with the corporation	Poor communication with the corporate partner led to a change in agreement with the corporation.
Assumptions about discontinued programmes	Lack of communication, and follow-up programmes led the orphanage administrators to assume that the programmes were discontinued.
Imprecise collaborative partnership's benefits	Ineffective communication of the programme benefits to the partners can lead them to become suspicious
Lack of direction in moving forward	Unsatisfactory communication from the service organisation left partners without direction
Difficulties in adapting to cultural and religious differences	Cultural differences between volunteers and coaches made it hard for volunteers to adapt the coaches' ideas.

DOI: 10.1051/shsconf/20173300076

Table 2. Communication and cultural constraints between CyberCare and the Lions Club

Issues	Brief of the responses
Adhering to the protocols of the Lions Club	The impression is created that in order for club members to be accepted and gain support from other club members, they had to attend all the club meetings and functions where not all members were in tune with the practice.
Different understandings of sponsorship promotion	CyberCare valued the right of those who provide financial support to CyberCare to be promoted more but the Lions Club may see that they should be given greater priority as a patron to the organisation.
Difficulties in understanding different organisational approaches	The difficulties in understanding each other's ways of carrying out programmes, due to different cultural approaches within the organisations.
Disputes over shared resources	The disputes over the resources collected under the "Lions Club of CyberCare Kuala Lumpur" arise when CyberCare decided to withdraw from the Lions Club network

The issues dealing with communication and culture between stakeholders in this collaborative partnership setting showed that the stakeholders (especially from the group of orphanage administrators and volunteers) strongly valued continuing commitment in the programme, and constant and clear communication with the partners in sustaining collaborative partnership. However, these stakeholders only emphasised the need for constant communication with the service organisation, rather than with all of the stakeholders involved. This further explains the dyadic relationships that exist between the two stakeholders in this collaborative partnership. The challenge of communication and culture in specific to CyberCare and the Lions Club dealt more with organisational working cultures.

4. THE NEED OF COMMUNICATION AND NURTURING TO SUSTAIN THE COLLABORATIVE PARTNERSHIP

The lack of clarity and transparency in communication led to negative assumptions which may threaten relationships and implementation. For example, one of the orphanage administrators suspected that CyberCare had a hidden agenda to attract sponsors in asking for the participation of children from his orphanage. This was because CyberCare did not clearly explain to the administrator the details of the programme and the way the orphanage could benefit from participation.

The findings support previous studies that emphasise the importance of constant communication and nurturing in sustaining collaborative efforts [1-2], [14], [5, 13]. At first glance, it might seem that my study suggests that this collaborative partnership was lacking the aspects of nurturing and communication which are important to sustain the collaborative efforts and relationships. However, some of these conflicts are structural (such as the timescale of interns from university, and the anxieties of orphanage administrators about the internet's influence on children), in which case they are not easily resolved via communication.

Even in situations where collaboration is running well with a relative degree of trust, Vangen and Huxham [5] emphasise the need for continuous nurturing to sustain satisfactory levels of trust, and Wildavsky [8] suggests careful nurturing to renew partners' enthusiasm during the process. The lack of such constant communication and nurturing as revealed in this study has impacted the collaborative partnership's structure and implementation of its programmes. The most significant impact was seen in relationships with the orphanages as perceived by the orphanage administrators. Despite the agreement at the beginning of the partnership, due to the gap in communications some of the orphanage administrators were still uncertain of their partnership status: whether to consider their orphanages as partners or to regard CyberCare as a partner. On top of this, not all of them viewed the programmes as sustained. It is not just constant communication that is required to carry the collaborative partnership forward, but more than that, it is important for communication to be clear and transparent.

While the worldwide network of innovators of multi-stakeholder partnerships in the area of ICT for development, GKP [9] regarded implementation of a properly structured partnership agreement as able to ensure continuing communication and transparency, this thesis has argued that correct implementation based on initial agreement cannot resolve communication problems in all partnership settings. For example, in this study, there are a few cases of miscommunication and lack of transparency even after the partners have formally agreed on the related procedures and what they wanted to work on collaboratively. This shows that having a well-structured agreement is not the only factor ensuring on-going communication and transparency. In addition, some scholars [1] take the view that "*communication among partners is based more on interdependent relationships than on contractual agreements*" as presented in my study. This is due to the complexity in implementing collaboration, the voluntary participation, autonomous actors, and less visible traditional management systems like "*hierarchy, standardisation, and routinisation*"[1]. This may explain why some participants in my study expressed the need to have a more organised management structure in CyberCare.

Besides the issues mentioned, there were also some advantages to the styles of communication practised in this collaborative setting. This was portrayed in this study where some decisions to join or continue with the collaborative efforts made by individuals within the partnerships were because of the political or social links between them which McQuaid [15] refers as informal structure. It was the informal structure of the partnership [15] through the established informal relationship and communication which drove this collaborative partnership forward. This finding also supported Perrault et al [14] view that established informal relationships and communication can contribute to successful community collaboration. For example, the main representative of CyberCare (YW-SO) used to help the representative from the Lions Club (SL-NGO) with the Lions Club programmes before CyberCare was formed as a Lions Club. The relationship built through the programmes led YW-SO to approach SL-NGO, and SL-NGO to agree to collaborate. This partnership used both informal ways of communicating, like text messages, and more formal means such as meetings of the board members. Established informal relationships provided flexibility for CyberCare members to achieve collaborative decisions like discussing collaborative plans and programmes in a member's home or in a cafe while at the same time reaching professional/formal decisions. This informal structure did not just contribute to the smooth running of the decision-making process especially when the project was conceived, but also led the stakeholders to develop the formal structure with general agreement to set up and carry out the collaborative project.

The ability of this collaborative partnership to survive through transformation was consistent with the conclusions of Melaville and others' [16] research, namely that a long-term collaborative partnership was enabled by a series of interrelated activities designed to solve the shared problems and create a new system of services for children and families. This process prompted changes in the system including integrating and restructuring services [16]. This research supports the view that change within collaborative partnerships is a positive outcome rather than a threat. Through the collaborative programmes, in the long run, some of the close partners in this collaborative partnership were aiming to change society's acceptance of the child-centric programmes and perception of the community service work.

I would like to express my heartfelt gratitude to my principal supervisor, Dr Nicole Matthews for her continuous support throughout my enduring PhD journey.

REFERENCES

1. A. M. Thomson, J. L. Perry, Public Adm. Rev. (**Special Issue**), 20-32, (2006)
2. C. Huxham, S. Vangen, *Collaborative Governance: a new era of public policy in Australia?* 29-44, (2008)
3. D. J. Wood, B. Gray, J. App. Behave.Sci.,**27**(2), (1991)
4. Overseas Development Institute and Foundation for Development Cooperation, *Multi-stakeholder partnerships: issue paper*, (2003)
5. S. Vangen, C. Huxham, J. Appl. Behave. Sci., **39**(1), (2003)
6. B. Gray, *Collaborating: finding common ground for multiparty problems*, (1989)
7. B. Gray, D. J. Wood, J. Appl. Behav. Sci., **27**(1), (1991)
8. A. Wildavsky, *PS,* **19**(2), (1986)
9. Overseas Development Institute and Foundation for Development Cooperation, *Multi-stakeholder partnerships: issue paper*, (2003)
10. C. Huxham, S. Vangen, *Managing to collaborate : the theory and practice of collaborative advantage*, (2005)
11. S. Vangen, C. Huxham, *Network Theory in the Public Sector: Building New Theoretical Frameworks*, (2014)
12. C. Huxham, Public Ma. Rev., **5** (3), (2003)
13. S. Prins, J. Appl. Behave. Sci., **46**(3), (2010)
14. E. Perrault, R. McClelland, C. Austin, J. Sieppert, *Administration in Social Work,* **35**(3), (2011)
15. R. W. McQuaid, *Public private partnerships : theory and practice in international perspective*, (2000)
16. A. I. Melaville, M. J. Blank, G. Asayesh, *Together we can: a guide for crafting a profamily system of education and human services*, (1996)

SHS Web of Conferences **33**, 00077 (2017)
i-COME'16 DOI: 10.1051/shsconf/20173300077

Effects of TV Crime Shows on Behavioural Development of Children

Mudassar Abdullah[1,*], *and Nik Adzrieman B. Abd Rahman*[2]

[1] Ph.D. Scholar, School of Multimedia Technology and Communication , University Utara Malaysia 06010 Sintok Kedah, Malaysia
[2] Senior Lecturer, School of Multimedia Technology and Communication, University Utara Malaysia 06010 Sintok Kedah, Malaysia

Abstract. Television crime dramas and shows are very popular all over the world. This popularity is not bound to a certain age group, rather all the TV viewers like these shows very much. Like other countries, dozens of TV channels are telecasting these crime shows in Pakistan. Furthermore, few of the channels telecast crime shows at prime time which attests the popularity of such genre. Some of the media contents behave in morally disputed ways. The crime depictions as re-enactments of TV crime shows are questionable in the field of research signifying diverse cultural contexts. A large number of people are habitual to watch these shows, which may probably come out with negative behavioural outcomes. Especially the children who are at their behavioural developmental phase; are more susceptible to adopt negative behavioural leanings. In this research effort, introduction and detail of TV crime shows in Pakistan are provided, the literature concerning "media as risk factor" in children development is discussed, and relevant theories inferences are deliberated.it was found that media has powerful role in behaviour formulating of children and violence media portrayal (TV crime shows) may appear with grave concerns. Previous scientific literature was reviewed to find and discuss the problem in hand. In the research effort, the literature review provides research propositions to explore further dimensions to TV crime shows' effects and possible negative or positive behavioural outcomes in children behaviour.

1. INTRODUCTION

Some programs of media entertainment industry are generally perceived to be "good programs", but some other programs are considered as immoral at least some of the time. However, these immoral programs are often liked and watched by a large number of

* Corresponding Author: mudassar80d@yahoo.com

viewers. Crime shows are comparatively a new genre in Pakistani television industry. This type of programs was introduced five or six years before in Pakistan, since then very popular among all segments of viewers. In a simple definition, crime show/drama is "A story or drama about the investigation of a crime by the police" [1]. In Pakistan, the crime shows are produced with few modifications in which a real crime event is presented by re-enactments. The primary focus of these shows is to describe the fare of criminal acts taken from real life events. However, there may be some other effects of these crime shows having negative behavioural tendencies in viewers, especially children.

Last five decades of media research attest to the potential power of media to influence virtually all concerns that societies have about young people, ranges from aggressive behaviour, drugs, obesity, school performance, depression and suicide [1]. It is also argued that media cannot be accused being the main player on such behavioural problems, but it can make a considerable contribution. At one hand we have good behaviour impacts of different TV programs on children (e.g. Sesame Street in increase school readiness) [2], on the other hand, some other programs increase the risk of negative behaviour development in children [4–6]. In this complex behavioural implications, television programs can be powerfully beneficial in the lives of children; but, much more research is needed in this area to recognize better usage of modern media and act accordingly [6].

2. Purpose

The purpose of this paper is to detect the need for research in media (television) effects. From general to specific, only one type of TV content (television crime shows) is in focus. Research on the effects of television crime shows on children in their behaviour development may provide new comprehensions. Moreover, this paper is intended to the social context of Pakistan, where the TV crime shows are very popular for last few years.

3. Methodology

The existing literature is used to explore and fulfil the purpose of this paper. Previous literature provides the findings of the topic chosen as well as the further dimensions to conduct more research. The major intention of the literature review is to provide, evaluate, analyse and then synthesize the knowledge shared by other researchers focusing the problem under study. For this paper, the literature review on the *behaviour formulation of children by media* and the *theoretical inferences* is made to provide dimensions for future research.

4. TV Crime Shows in Pakistan

In television format, the technical term to identify the crime shows is "procedural". Procedural shows represent a type of shows in which a problem (mostly crime related) is introduced, its complete procedure is elaborated, then solved. All this process follows a process beginning with the climax of crime. Later, the different stages of the problem or issue are particularised by showing past events. Mostly, the criminal is caught by police as these are based on real solved police cases. All the issue is portrayed like drama show with the help of re-enactments. Few of the examples of these crime shows at international lever are *CSI*, *Law & Order* and *NCIS*. Few of such genre programs in Pakistani news channels are "*Shabbir tou Dekhay Ga*" (Express News), "*Meri Kahani Meri Zabani*" (Samaa), "*Wardaat*" (Samaa), and "*Aisa Karega Tou Marega*" (Express News). The above procedural dramas (crime shows) of different private TV channels are very famous in the country due to its dramatic nature, lack of long storylines, and true events depictions.

5. Behaviour Development of Children and Media as a Risk Factor

Child psychologists observed that most of the children experience a single psychological or physical risk factor suffered little if any continuing harm, but multiple risk factors can lead to psychological disorder [7]. From these multiple risk factors, media influence can be a significant factor to investigate. Recent research evidence raises concern about media effects of on children aggression, substance usage, sexual behaviour, eating disorders and academic difficulties [8]. Excessive watching of television during childhood appears significantly in a criminal conviction, personality disorder with aggressive behaviour tenets [9]. Primarily, the psychological processes of behaviour development of children rely on learning with observation, and with frequent exposure to specific media content, the cognitive , emotional and perceptual responses can be predicted [10].

Few longitudinal studies, many static observational studies, and numerous experimental studies indicate that TV exposure (violent content) is related to antisocial behaviour in children [11-12]. Huesmann, Moise-Titus, Podolski, and Eron [13] found that children's TV viewing containing violence is significantly correlated with later aggression. In sum, children spend more time watching TV and other media than doing any other routine activity [14], [15], and some of the content may have effects on key behaviours including antisocial behaviours and aggression. Furthermore, it is also witnessed that children might use media to facilitate the key developmental process in their social context. For further research, it is significant to examine media use through a developmental lens in studying behaviour development of children. Research on children viewership of TV crime shows may help to understand why they watch and like these shows and what influences they adopt from.

6. Theoretical Inferences

A significant body of social theory has developed in explaining the processes through media exposure have effects on viewers. According to Coyne and colleagues [14] there is two theoretical bases in terms of the role of media in childhood and adolescence. One rationale includes the theories of media effects, such as *Social Learning Theory* [16], *Cultivation Theory* [17], Information *Processing Theory* [18], and the *General Aggression Model* [19]. These theories are based on the notion that media have some effects on children and adolescents, whether good or bad, either on attitudes or behaviour. On the other hand, another way of investigating media role provided by *Used and Gratifications Theory* [20]. This theory rational is that adolescents/children gravitated toward media as they have specific needs and media help them to fulfil these needs [21]. A detailed investigation of previous literature reveals that children use media to satisfy their specific psychological needs, however, they are influenced by media in different aspects.

Research on controversial television contents combining social learning theory, reveals that violence content increases the likelihood of antisocial, aggressive and violent behaviour in long-term as well immediate contexts [22]. Furthermore, social leaning theory explains that repeated observation of real life antisocial/aggressive models portrayed in media like TV, children develop a behaviour that aggression is proper, and then they obtain scripts of how to behave aggressively [23]. In this context, social learning theory has become one of the important theories to investigate behaviour development of children

 DOI: 10.1051/shsconf/20173300077

denoting the research on TV effects. On the other hand, cultivation theory postulate that television is the most powerful storyteller in a culture, which repeats the ideologies and myths, facts and pattern of relationships that elucidate the social order [24]. Repeated dose of TV, over time, acts like the pull of gravity toward an imagined centre which results in a shared set of concepts and expectations about reality among diverse viewers [25]. As the proposed area of interest is TV crime shows' effects of children, so cultivation analysis has its worth to study behaviour development through a specific TV show.

The social information processing intent attribute is being considered as one optional mediator in media effects research relating violence. The previous professional literature argues that social instability and hostile attribution are quite associated and has well demonstrated with adults as well as with children [26-27]. Crick [28] further investigated and found that hostile acknowledgement is theorized to contribute to the development of aggressive behaviours in children. In this reference, the social information processing theory has its own significant to investigate behaviour development of children with reference to TV crime shows. The General Aggression Model (GAM) [19] is another theoretical perspective to explain the development of aggressive cognitions, attitudes, and behaviours. This model states that "exposure to media violence may contribute to enhancing or mitigating the aggressive constructs"[29]. Anderson and his colleagues GAM describes that exposure to violent media content can affect aggressive thoughts and feelings then arousal [30], which may lead to some negative outcomes. Television viewing, especially watching a specific content may be a meaningful way to investigate behaviour development of children [31]using uses and gratification [32] approach . Previous literature attests that viewers seek out various type of media content to fulfil their certain needs [33], so the TV viewing by children to a specific content (TV Crime Shows) may be investigated through uses and gratifications theory.

7. Propositions

The main focus of this paper is to analyse the possible negative and positive effects of TV crime shows in Pakistan. Due to the content sensitivity of TV crime shows and fewer research in this area, it is scientific need to investigate further. The following propositions may be important to consider further research in behaviour development of children in the Pakistani context.

- Television Crime Shows develop negative behavioural leanings in children
- Television crime shows provide a procedural script to children who intend to commit a crime
- There are few positive inferences of TV crime shows on the behaviour of children
- Children watch TV crime shows to satisfy their previous behaviour formulations

8. Conclusion

Research on behaviour development of children with reference to antisocial behavioural tendencies has been nourishing for last few decades. What children learn, how children learn, these are two burning questions for child psychologists and social scientists. Different contents of media have always been in research focus to measure its effects on viewers. These contents are also linked with behaviour development of viewers. TV crime shows are popular in Pakistan among different segments of viewers. The purpose of these shows is to convey hatred with crime, and depict the distressing outcomes of crime, but the way of presentation of this message need to be investigated. As these programs show a criminal happening via re-enactments, the background of this event, and the relevant

DOI: 10.1051/shsconf/20173300077

proceedings later on, which may cause harmful behavioural implications in children. Different dimensions of behavioural implications on children by the crime shows need to be investigated in media effects and usage research.

References

1. V. Strasburger, B. Wilson, and A. Jordan, *Children, adolescents, and the media.* 2nd edition. Thousand Oaks (CA): Sage, (2009)
2. S. Fisch, R. T. Truglio, and C. F. Cole, *The impact of Sesame Street on preschool children: A review and synthesis of 30 years' research,* Media Psychol., **1**, 2 (1999)
3. B. A. Dennison, T. A. Erb, and P. L. Jenkins, *Television viewing and television in bedroom associated with overweight risk among low-income preschool children, Pediatrics*, **109**, 6 (2002)
4. C. Jackson, J. D. Brown, and K. L. L'Engle, *R-rated movies, bedroom televisions, and initiation of smoking by white and black adolescents, Arch.* Pediatr. Adolesc. Med., **161**, 3 (2007)
5. M. M. Garrison, K. Liekweg, and D. a Christakis, *Media use and child sleep: the impact of content, timing, and environment,* Pediatrics, **128**, 1 (2011)
6. V. C. Strasburger, A. B. Jordan, and E. Donnerstein, *Children, Adolescents, and the Media:. Health Effects,* Pediatr. Clin. North Am., **59**, 3 (2012)
7. M. Rutter, *Stress, coping and development: Some issues and some questions.,* J. Child Psychol. Psychiatry., **22**, 4 (1981)
8. V. et al. Strasburger, *Health effects of media on children and adolescents,* Pediatrics, (2010)
9. L. a Robertson, H. M. Mcanally, R. J. Hancox, A. L. a Robertson, and M. Helena, *Childhood and Adolescent Television Viewing and Antisocial Behavior in Early Adulthood*, Pediatrics, **131**, 3 (2013)
10. J. Maier and D. Gentile, *Learning aggression through the media: Comparing psychological and communication approaches*, Psychol. Entertain. Media, (2012)
11. L. Huesmann, *The effects of childhood aggression and exposure to media violence on adult behaviors, attitudes, and mood: Evidence from a 15-year cross-national longitudinal study.*, Aggress. Behav., **25**, (1999)
12. W. L. J. Pers. Soc. Psychol., **53**, 5 (1987)
13. L. R. Huesmann, J. Moise-Titus, C.-L. Podolski, and L. D. Eron, *Longitudinal relations between children's exposure to TV violence and their aggressive and violent behavior in young adulthood: 1977-1992.,* Dev. Psychol., **39**, 2 (2003)
14. S. M. Coyne, L. M. Padilla-Walker, and E. Howard, *Emerging in a Digital World: A Decade Review of Media Use, Effects, and Gratifications in Emerging Adulthood*, Emerg. Adulthood, **1**, 2 (2013)
15. Alloy and M. & Marketing, "Alloy Media & Marketing," *http://www.marketingcharts.com*, (2016)
16. A. Bandura, D. Ross, and S. a. Ross, J. Abnorm. Soc. Psychol., **66**, 1(1963)
17. G. Gerbner and L. Gross. J. Commun., **26** (1976)
18. L. R. [Ed] Huesmann and L. D. [Ed] Eron, *Television and the aggressive child: A cross-national comparison*, Telev. Aggress. child A cross-national Comp, (1986)
19. C. A. Anderson and B. J. Bushman, *Human aggression,* Annu. Rev. Psychol., **53**, (2002)
20. E. Katz, J. G. Blumler, and M. Gurevitch, *Utilization of mass communication by the individual*, Crit. readings media audiences, **5578** LNCS, 19–31, (1974)

21. A. M. Rubin, *The uses-and-gratifications perspective of media effects.*, Media Eff. Adv. theory reserach,165–184, (2002)
22. C. a. Anderson, L. Berkowitz, E. Donnerstein, L. R. Huesmann, J. D. Johnson, D. Linz, N. M. Malamuth, and E. Wartella, *The influence of media violence on youth*, Psychol. Sci. Public Interes., **4**, 3 (2003)
23. B. J. Bushman and L. R. Huesmann, *Short-term and long-term effects of violent media on aggression in children and adults.*, Arch. Pediatr. Adolesc. Med., **160**, 4 (2006)
24. J. D. Brown. J. Sex Res., **39**, 1 (2002)
25. G. Gerbner, L. Gross, M. Morgan, and N. Signorelli, *Growing up with television: the cultivation perspective*, (1994)
26. K. A. Dodge and C. L. Frame, *Social cognitive biases and deficits in aggressive boys.*, Child Dev., **53**, 3 (1982)
27. [27] N. R. Crick and K. a. Dodge, *A review and reformulation of social information-processing mechanisms in children's social adjustment.*, Psychol. Bull., **115**, 1 (1994)
28. N. R. Crick, *Relational aggression: The role of intent attributions, feelings of distress, and provocation type*, Dev. Psychopathol., **7**, 2 (1995)
29. D. A. Gentile, S. Coyne, and D. A. Walsh, *Media violence, physical aggression, and relational aggression in school age children: A short-term longitudinal study*, Aggress. Behav. **37**, 2 (2011)
30. C. a Anderson and K. E. Dill, Anderson, C. A., & Dill, K. E. J. Pers. Soc. Psychol., **78**, 4 (2000)
31. A. M. Rubin. J. Broadcast., **27**, 1 (1983)
32. W. J. Severin and J. W. Tankard, *Communication Theories: Origins, Methods, and Uses in the Mass Media.* (2010)
33. S. M. Coyne, L. M. Padilla-Walker, and E. Howard, *Emerging in a Digital World: A Decade Review of Media Use, Effects, and Gratifications in Emerging Adulthood*, Emerg. Adulthood, **1**, 2 (2013).

DOI: 10.1051/shsconf/20173300078

Exploring the Pattern of Internal Communication in Total Quality Management Implementation in Manufacturing Companies

Sharina Samsudin[1,*], *Nor Hayati Mohd* Jalil[1] , *and Marzura* Ibrahim[1]

[1]Universiti Utara Malaysia, 06010 UUM Sintok, Kedah, Malaysia

Abstract. Communication is an important aspect of organizational dynamic process and it is irrefutable one of the critical success factors of the Total Quality Management (TQM) implementation. Although there are increasing awareness concerning the importance of internal organizational communication, knowledge appears to have rarely translated into practice or there is lack of guidance to effectively practicing it. Therefore, a study was proposed to investigate the pattern of internal communication practice in TQM implementation. This paper concerns on focusing into internal communication practices in TQM implementation factors and the communication patterns in terms of message forms and mediums of communication to convey TQM messages. Reliability and Factor analysis were carried out to confirm items that form each factor in internal communication in TQM implementation. Based on the analysis of survey questionnaires distributed to 104 large manufacturing companies implementing TQM in Malaysia, it was found that there are certain communication patterns practice by the management when conveying and disseminating TQM messages, but they have no specific guidelines. Therefore, the internal communication practices are not well planned. Finally, it is suggested that further research should be carried out to explain clearly the role and strategy of communication on this issue in order to boost the success of TQM implementation particularly in manufacturing company.

1 Introduction

This paper explores the pattern of internal communication in Total Quality Management (TQM) implementation at manufacturing sector in Malaysia. It aims to provide a significant insight in preparing for a major empirical study in the associated areas.
Communication is fundamentals to any kinds of organized forms of management in organizational context. It concerns the channel, the forms and conventions and the methods or strategies for achieving communication objectives at work. Communication scholars believed that organizational communication strategies are constructed through a permanent relationship that exist between organization and people and also between people themselves [1]. These processes can come from examining communication practices as revealed within

* Corresponding author: sharina@uum.edu.my

organizational environments, as stated by [2], people create and modify everyday communication strategies based on their interactions, meanings and thoughts while inserting new actors, resources and environments into the process. In fact, effective internal communication gradually will increase employees' trust and commitment, which will lead to employees' engagement [3], and this is very important in successfully implement TQM [4, 5].

In the literature of organizational strategy, communication is well known as a functional mechanism of coordination [6]. This means that people communicate with each other in a functional setting, where they practicing the communication strategies either it is planned or unplanned. If it is planned communications, usually, there are some guidelines in organizational situations that tell the employees how they are supposed to act and communicate. Organizations also provide some resources or means of communication, interpersonal and mediated. All these will create a network or chain, a pattern of internal and external communication. Besides, employees also form informal communication networks usually to compensate the weaknesses of formal networks. [7] added that the so-called 'informal' ones, are very substance of organizing. In other situations, choosing the appropriate communication strategies is difficult because the organizational situations sometimes paralyze employees with guidelines or vague information and insufficient resources.

On the other hand, the need for quality improvement continues unabated in both manufacturing and service sectors. In Malaysia, the manufacturing sector learnt some very painful lessons for the past 20 years because of the competition with other developing industrial countries such as Thailand and China, to improve product quality and remain competitive. Malaysia has also taken itself the challenge of attaining the status of a developed nation by the year 2020. Therefore, Malaysia provides an interesting and important area for empirical study of communication in TQM implementation especially in manufacturing sector. This study is limited to large manufacturing companies operating in Peninsula Malaysia that claimed themselves as TQM adopter with at least one ISO certification because empirical studies done on these segment focused on communication within TQM organizations are still very limited.

2 Literature review

2.1 Identity in intercultural communications

In discussing the role of internal communication, [8] pointed out that, internal communication is crucial for both organizational success and for their day-to-day existence. According to them, this is because internal communication can engage employees' intellectual and creativity to produce value. As [9] mentioned, the role of internal communication is to illuminate the connection between different pieces of information and its job to provide employees with the information they need for their job.

Another issue of internal communication was studied by [10] use a triangulation approach encompassed structured interviews. They adopt the international communication audit questionnaires and a critical incident approach to study the issues of internal communications at Royal Ulster Constabulary (RUC). The result of the interview provided seven categories in terms of how people felt about communication in their sub-division. Among others are, perceived as open to new ideas, willing to share information, able to clarify expectations and able to co-ordinate work. Formal channels as meetings and briefing were perceived as administrative tool and majority indicated a problem in communicating. A little bit difference found by [11] on internal organizational communication. They found moderate support for communication within organization towards lean manufacturing implementation making a positive impact on communication primarily involved communication lines with management. They investigated two specific variables; management support and communication within the organization. Additionally, their findings also showed the majority practice poor communication (64.6 percent) involved communications between departments or communication necessary to facilitate material flow through the factory. They also found strong evidence to prove that the dissemination of information to all employees about lean manufacturing did not occur.

In Thailand, [12] studied about the organizational communication pattern at Thaksin University. The study mentioned that the majority of the respondents agreed that the executives should provide clear information, improve public relations through media and declare any problems occurring in any changes to all members of staff. With regard to increase motivation and confidence of workers, supervisors in each department should hold discussions.

On the other hand, TQM has been so popular across industries until [13] mentioned that TQM rank high on the management agenda in the attempt for business to make more profit, improve product quality, increase customers as well as employees' satisfaction and cost reduction. [14] also asserts that TQM not only promotes the performance of all departments but also joins and integrate them to achieve the objectives of the organization. Long before that, [15], have stated that the proper implementation of TQM could be a powerful vehicle where companies are able to achieve excellence in business performance. Therefore, many organizations in manufacturing and services have devoted considerable attention to the implementation of TQM. However, they also claim that companies that have not achieved TQM potential benefits have begun abandoning the practices. The key principles for its failure is the lack of understanding of what TQM means for each unique organization and how to implement it effectively created skepticism on the effectiveness of TQM.

Previous studies showed that many western firms have adopted integrated quality management strategies had not been fully successful [16, 17, 18, 19]. If the efforts done without really understand on how it can add to quality improvement, it will just incur the overall production cost. According to [20] one of the causes of failure in implementing TQM is the management of communication strategies are not fully in place. His findings showed that in implementing TQM, organizations should take into consideration the communication elements as well as the management.

Therefore the purpose of this study is to identify and describe the pattern of internal communications at manufacturing organizations in Malaysia in order to help improve the TQM implementation.

3 Methodology

This study adapted a survey instrument from European Center for Total Quality Management studied on the Human Resource Management and Total Quality Management relationship. Fifty items related to internal communication were extracted for the purpose of measuring the practice of internal organizational communication in implementing total quality management. These questionnaires were then distributed to identify population using simple random sampling. In this study, 24.3 percent which is 104 companies of the population has responded to the questionnaire distributed, which according to [21] this response rate is similar to other surveys in Malaysia, which usually tend to obtain a response below 20 percent.

Validity test

The 50 items selected from the original instrument were then sent to two experts in organizational communications study to be verified for content validity.

Then, factor analysis was applied to reconfirm the internal communication items related to TQM for construct validity test. Factor analysis provides an empirical solution of the variables structure and summarizes the potential dimensions [22]. All 50 items were subjected to the principal component analysis using SPSS version 19. A review of the correlation matrix showed the presence of many coefficients of more than 0.3 and above, which indicated an appropriate data. The Kaiser-Meyer-Olkin value was 0.72, exceeding the minimum recommended value of 0.6 [23]. The Bartlett's Test of Sphericity reached statistical significance, $\rho < 0.05$, supporting the factorability of the correlation matrix.

Secondly, the principal component analysis was performed and shows the presence of 11 components with eigenvalues exceeding 1. These account for 78.25 percent of the total variance. From these results, a number of components could be retained for the extraction process. Numerous techniques can be applied for determining the number of factors for extraction – eigenvalue exceeding 1, scree plots [22,24] parallel analysis [24], and minimum average partial correlation [25]. In this study, three techniques were applied for the factor extraction, which were the eigenvalue rule, followed by scree plot, and finally parallel analysis. According to [24], this combination for obtaining the result is considered sufficient.

Table 1. The Parallel Analysis Summary Comparing with Principal Component Analysis Eigenvalue

Factor number	Eigenvalue from principal component analysis	Random eigenvalue from parallel analysis	Decision
1	21.03	2.32	Accepted
2	4.28	2.21	Accepted
3	4.12	2.13	Accepted
4	3.02	2.06	Accepted
5	2.26	1.93	Accepted
6	2.02	1.82	Accepted
7	1.58	1.78	Rejected
8	1.46	1.73	Rejected
9	1.23	1.61	Rejected
10	1.12	1.59	Rejected
11	1.08	1.53	Rejected

In factor analysis, the factor loading shows the weight of each item with the related factors where a higher value of factor loading represent greater underlying meaning for describing the factor. According to [23], variables with a minimum loading of 0.32 are acceptable and adequate for interpretation. By referring to the highest loading factor score for each factor, the items were grouped into six factors named, Communication for leadership, Communication for commitment, Communication for customers, Communication for teamwork, Communication for involvement and Communication for training and education.

Therefore, based on the analysis, internal communication items are correlated to these six critical success factors of TQM implementation.

Reliability test

The internal consistencies of the instrument were analyzed using SPSS version 19 by determining the Cronbach's alpha coefficient. [26] recommended a value above 0.6 for good internal consistencies. Table 1 shows the results of the reliability analysis which shows all Cronbach's alpha coefficient values were above 0.7. It was concluded that the measurement tools were consistent, reliable and usable to be used for data collection survey.

Table 2. Reliability Analysis

No	Critical Success Factors	Number of Items	Cronbach alpha coefficient
1.	Leadership	7	0.97
2.	Commitment	8	0.97
3	Customer Focus	8	0.98
4	Teamwork	9	0.94
5	Involvement	9	0.96
6	Training	9	0.92
	All items	50	0.97

4 Pattern of Internal Communication

There are 56 local companies representing 53.85 percent, 30 eastern companies or 28.85 percent of the total sample while 17.31 percent representing 18 western companies. And, majority of the companies took part employ between 150-499 employees. Most of the representative of the organizations were the middle management level and a member of quality circle of the companies. Thus, these group of employees are qualify and knowledgeable about the subject matter of the study.

Table 3. Background Profile

Items	Profile	Frequency	Percentage
1	Continent of Origin		
	a. Local	56	53.85
	b. Eastern	30	28.85
	c. Western	18	17.31
2	Number of employees		
	a. 150 to 499	70	67.31
	b. 500 to 999	16	15.38
	c. 1000 to 1500	16	15.38
	d. More than 2000	2	1.92
3	Representatives' position		
	a. General Manager/ Senior Manager	17	16.35
	b. HR Manager/ HR Personnel	36	34.61
	c. Quality Manager/Quality Executive	28	26.92
	d. Assistant Manager/Executive	14	13.46
	e. Training Manager	9	8.65

Table 4 shows that most of the organizations have specific department or section to manage the quality related issues, which was 94.23 percent compared to only 37.5 percent have section in charge of communication related issues.

Table 4. Section In-Charge of Quality and Communication Related Issues

Items	Section in-charge of	Frequency	Percentage
1	Communication		
	a. Yes	39	37.5
	b. No	65	62.5
2	Quality		
	a. Yes	98	94.23
	b. No	6	5.77

In terms of interpersonal face-to-face communication practiced in the organizations, the most were formal meetings and briefings and the least were informal discussions. These findings revealed that most of the face-to-face communications practices in the organizations were planned and in the form of top-down flow of communication.

For mediated communication, all the organizations agreed on telephone conversations as the main medium of conversations in the organization, followed very closely by e-mails, letters and memos in the third place, notice boards, newsletters/bulletins and suggestion box. Generally, there were differences in terms of frequencies for interpersonal face-to-face communication and mediated communication (document or electronic) used in disseminating information regarding TQM implementation.

Table 5. Flow of Communication and Communication Medium Used

Flow	Medium	Frequency	Percentage
Downward	Formal face-to-face meeting	16	15.38
	Informal	3	2.88
	Written	1	0.96
	Telephone	63	60.57
	Emails	8	7.69
	Other: Notice board	10	9.62

	Morning prayer	3	2.88
Upward	Formal face-to-face meeting	8	7.69
	Informal	12	11.53
	Written	23	22.11
	Telephone	42	40.38
	Emails	17	14.42
	Other: Suggestions box	1	0.96
	Morning prayer	1	0.96
Horizontal	Formal face-to-face meeting	5	4.81
	Informal	12	11.54
	Written	30	28.85
	Telephone	44	42.30
	Emails	11	10.58
	Other: Online forum	2	1.92

Table 6 shows the types of communication that need improvement. The result shows that all direction of internal communication flow still need improvement.

Table 6: Types of Communication Requires Improvement

	Frequency	Percentage
Meetings / Discussions with superior	10	9.62
Meetings / Discussions with subordinate	12	11.54
Meetings / Discussions among peers	14	13.46
Meetings / Discussions inter-dependent	8	7.69
E-Mails	12	11.54
Letters / Memos	9	8.65
Notice boards	8	7.69
Newsletters / Bulletins	13	12.50
Suggestions Boxes	18	17.31

The mean result for the internal organizational communication related to critical success factors (CSFs) of TQM implementations are within the range of four to five (scale is from 1 the least to 6 the most). These findings can be considered as moderate to high except for factor customer focus the mean score is very high. The details of overall mean are presented below:

Table 7. Mean for Degree of Roles of Internal Communication in CSFs (N=104)

No.	CSFs	Mean
1.	Leadership	4.23
2.	Commitment	3.82
3.	Customer focus	4.84
4.	Teamwork	3.71
5.	Involvement	3.66
6.	Training and education	3.66

This could indicate that the organizations could still struggling in practicing effective internal organizational communication in the process of implementing TQM except for the customer focus and leadership factor where the practice of internal organizational communication in majority of the organizations are already established.

5 Conclusion

Although there are consistent answers that the respondents believe internal communication as crucial elements to promote successful TQM implementation but they do not have any guidelines. Companies have provided many types of internal communication channel either for downward, upward or horizontal communication. Anyhow, many still need improvement in terms of the usage and upward flow of communication. Companies should inform and encourage their employees to use and participates in their communication activities.

The respondents agreed that internal communication is one of the core components of TQM implementation. The effectiveness of this core will generate successful leadership, commitment, customer relations, teamwork, training and education and involvement to quality. As mentioned by [27] variables that affected organizational effectiveness most was commitment and satisfaction. Therefore, communication should be included as one of the critical success factors of TQM implementation. And, these then supported by the key management functions of people, processes and procedures, and systems in the organization. All these elements combine and practice together will help increase the success rates of TQM implementation.

More details are needed in order to investigate these findings. Therefore, further study on how communication relates to TQM implementation should be done. In-depth interviews and observations must be carried out to explain clearly on this phenomenon.

References

1. G. Cheney, L.T. Christensen, T.E. Zorn, S. Ganesh. *Organizational Communication in an Age of Globalization: Issues, Reflections, Practices*, 2nd.ed, Waveland Press Inc. (2011)
2. M. Marchiori, S. Bulgacov. International Journal of Strategic Communication, **6** (2012)
3. K. E. Mishra, L.Boynton, A.Mishra. International Journal of Business Communication. 51 ,2 (2014)
4. J. A. N. Shaari. International Review of Business Research, **6** (2010)
5. F. Talib, Z. Rahman, *Identification and Prioritization of Barriers to Total Quality Management Implementation in Service Industry: An Analytic Heirarchy Process Approach. The TQM J.*, **27** (2015)
6. K. Srikanth & P. Puranam. Strategic Manage. J (2011)
7. D. K. Mumby, *Reframing Difference in Organizational Communication Studies: Research, pedagogy, practice*. Thousand Oaks, CA: SAGE (2011)
8. W. R. Kennan, Hazleton, V. *Internal Public Relations, Social Capital, and The Role of Effective Organizational Communication* (2006)
9. B. Quirke. *"Making the Connections: Using Internal Communications to Turn Strategy into Action*. Burlington, VT: Gower Publishing Company (2008)
10. D. Quinn & O. Hargie. An Int. J., **9**. 2 (2004)
11. J. M. Worley & T. L. Doolen. *The Role of Communication and Management Support in a Lean Manufacturing Implementation*. Management Decision, **44**, 2 (2006)
12. C. Noknoi. Eur. J. Sci. Res. **80**, 4 (2012)
13. M. Kolaikuntla, *A Research Study on Estimation of TQM 'Factors Ratings' Through Analytical Hierarchy Process*. Procedia Economics and Finance, **3**, (2012)
14. A. Brun. Int. J. Prod. Econ., **131**, 1 (2011)
15. D. K. Shin & G. A. El-Enein. *Critical Implementation Issues in Total Quality Management. S.A.M. Advance Management Journal,* **63**, 1 (1998)

16. Z. Huq & T.N. Martin. *Workforce cultural factors in TQM/CQI Implementation in Hospitals.* Health Care Manage. R., **25**(3), 80-93 (2000)
17. G. Polat, A. Damci, & Y. Tatar. *Barriers and Benefits of Total Quality Management in the Construction Industry: Evidence from Turkish Contractors.* 7th Research/Expert Conference with International Participations (2011)
18. Tey Lian Seng and Ooi Teck Loon. Research Journal of Applied Sciences, Engineering and Technology. **7**, 16 (2014)
19. A. M. Mosadeghrad. Int J Qual Health Care, **26**, *2* (2013)
20. Gathalian. *People Empowerment: The Key to TQM Success.* In The TQM Magazine, **9**(6), 429-433 (1997)
21. M. Othman. *Penulisan Tesis dalam Bidang Sains Sosial Terapan.* Serdang:Penerbit Universiti Putra Malaysia (2001)
22. J.F.J. Hair, R.E. Anderson, R.L. Tatham, W.C. Black. *Multivariate analysis*, **5** (1998)
23. B.G. Tabachnick, L.S. Fidell, *Using Multivariate Statistics*, Pearson, **6** (2013)
24. J. Pallant, *SPSS Survival Manual*, **12** (2005)
25. E. Ferguson & T. Cox. Int. J. Select. Assess.,**1**, 2 (1993)
26. U. Sekaran. *Research Methods For Business: A Skill Building approach.* 3rd.ed. USA: Hermitage Publishing (2000)
27. R. Mohammed. *Organization-Public Relationships Practices in University Setting.* The International Conference on Communication & Media (i-COME'14) 2014

SHS Web of Conferences **33**, 00079 (2017) DOI: 10.1051/shsconf/20173300079
i-COME'16

Investigative Journalism in Malaysia: The Battle Between Outside and Inside Newsroom Challenges

Ismail Adibah[1,*], *Ahmad Mohd.* Khairie[2], and *Mustaffa* Che Su[3]

[1,2,3]Universiti Utara Malaysia, 06000 Sintok, Kedah, Malaysia.

Abstract. Investigative journalism originated in Western countries but the practice has spread throughout the world, including countries with different political and media environment. Current literature is dominated by Western perspectives, focusing mainly on the developed countries viewpoint. This study is exploring a controlled media environment, focusing on Eastern developing country. This research explores media practitioners' perception on challenges of practicing investigative journalism in Malaysia. In-depth interviews were used in this research. A total of 16 informants were interviewed. Data was examined using thematic analysis. It was found that Malaysian media practitioners acknowledged six outside challenges and eight inside the newsroom challenges. The study concludes that challenges in controlled media environment revolve around press freedom issue and inside the newsroom struggles due to different editorial orientation on the value of investigative journalism.

1 Introduction

Investigative journalism plays an important role as watchdog of the government and corporate world, for the benefit of the public. This highly respected kind of journalism has been exposing secrets, scandals and wrongdoings for more than two centuries. Although investigative journalism is being practiced around the world, much research has been done from the Western developed countries viewpoint. This study attempts to enhance the literature by exploring investigative journalism from a different perspective, of an Eastern developing country. Investigative journalism has often been more celebrated than practiced because there are many obstacles that hinder the journalists from practicing it. The challenges face by investigative reporters "came as much from the outside as within" [1]. Therefore, current research specifically focuses on understanding the outside and within newsroom challenges faced by investigative reporters in Malaysia.

* Corresponding author: adibah@uum.edu.my

 DOI: 10.1051/shsconf/20173300079

2 The outside and inside battle

Investigative journalism practitioners have to face challenges both outside and within their newsroom. The outside challenge involves the unambiguous logic that wrongdoing is difficult to uncover [2] because someone is trying to conceal it. This concealment involves not only the wrongdoers, but sometimes involves the authorities that cover up information on the basis of national security or harmony. For example, Lavrov [3] alleged that in Ukraine, if journalists' investigation involves people in power, the government system would work impeccably to conceal it. He revealed that law enforcement officers even go to the extent of destroying the evidence and denying its existence. Without evidence and supporting documents, proving a story and getting it published is difficult. In the United States, Feldstein [4] remarked that exposing secrets is getting tougher because the government erected a wall of secrecy after the 9/11 terrorist attacks.

Adding to the outside challenges is the laws controlling the media [2,3,5,6,7]. These laws have been seen as responsible for stifling investigative journalism. Britain, Armenia, Poland, Ukraine and Ireland were among those countries where problems with libel laws are evident [8,3,9]. This outside challenge has triggered inside challenges including forcing reporters to exercise self-censorship, sub-editors rewrite to tone down controversial issues, and news editors try to turn a deaf ear to the people's grievances [5]. The risk involved in practicing investigative journalism is another challenge discouraging journalists to pursue investigative reporting [4,10,11,12]. Anderson and Benjaminson [12] have written that investigative journalism is the most risky pursuit in the media. Reporters actually have to face risks before they start an investigation as well as during and after they have finished because people are trying from every angle to stop investigative reporters from bringing the truth to the public.

On the other hand, media ownership concentrations were seen as a challenge within newsroom [4,13,14,11,15,16,17,18,19]. It involves closing of small newspapers, corporate buyouts and media mergers. These developments have increased the concern that too few control too much [16]. Tanner [15] posits that "increasingly media outlets are falling into the hands of people with no tradition of journalism, who see journalism as an extension of marketing and do not have any respect for independent journalism". Moreover, concentrated ownership leads to reduced opportunities for journalists to get their investigative article published because the company's shares are mostly own by people with power and authority like the government. Therefore, media organisations will avoid publishing something that will tarnish their owner's reputation. Media owners also give advertisers and sponsors unwarranted influence over news agendas and even the composition of individual news items [14,11]. De Burgh [17] supported this claim by saying that the concentration of media ownership has had a profound impact on the content of the print and broadcast media and on the very nature of investigative journalism itself. For a corporation that is profit oriented, commercials and advertising space are their first priority, not the news. There is not enough space left for investigative stories to fit in. Another syndrome brought about by ownership is the limited opportunity to publish a story when the organisation's investigative reporters are working for the subsidiary of a large media corporation. This is because, when the top management decides not to publish the story, investigative journalists have no other places to turn to.

Scholars also argued that investigative reports have become expensive to produce [4, 2, 11, 10, 7] since most investigative reporters are senior reporters with high pay. Staffing pressure is another challenge for editors because investigative journalism takes time and very risky. Editors allow reporters to do investigative reporting only in their free time thus, journalists do not have enough time to do research. They are pressured to write about local affairs, telling people what happens throughout the day. When no resources are offered to support these journalists, it is like telling reporters not to do investigative reporting at all. Hargreaves [14] posits that journalists and even editors have become overly dependent on press release. This has become an acceptable culture in news organisations [20]. Feldstein [13] added that journalists are not only touched by partisanship but also by stenography. They just report the news and do not analyse the information and write a more meaningful report. Last but not least, the media is increasingly putting more emphasis on infotainment [4]. It has brought with it the syndrome of tabloidisation or what Franklin [21] called as "broadloid journalism". This is the combination of broadsheet and tabloid. According to Franklin, there has been a general retreat from investigative journalism and the reporting of hard news stories. Broadsheet front pages now mimic those of the tabloids with similar banner headlines, alternative and funny headlines, large print, less text, shorter words, coloured and bigger pictures.

Although significant, these outside and inside newsroom challenges have been analysed from developed Western countries viewpoint. This research will examine these challenges from a different viewpoint involving a different media landscape. Using qualitative in-depth interview technique, this research interviews 16 media practitioners who have experience in investigative journalism to gather a holistic and in-depth understanding on challenges they are facing in practicing investigative journalism in Malaysia. All interviews were transcribed and analysed using thematic analysis, with the help of NVivo software. Informants were identified as Informant 01 to Informant 16 to conceal their identity.

3 Findings and discussions

Interviews yielded two themes of challenges separating the outside and inside challenges faced by investigative journalism practitioners in Malaysia. For outside challenges, six attributes were mentioned by informants which are lawsuits, information availability, issues availability, seeking the truth, weak enforcement and low impact. For inside the newsroom challenges, informants mentioned eight attributes which are journalism skills, culture differences, time, risk, cost, human resources, attitude and media sustainability. Four out of six outside the newsroom challenges mentioned by informants involve the issue of press freedom which has a very close link to government control of the media. The attributed are lawsuits, information availability, seeking the truth and issues availability. Malaysian media practitioners regarded press freedom parameters as values guiding their practice of investigative journalism. This makes it the biggest challenge to conduct investigative journalism as they are already accustomed to the parameters as their boundary. It makes issues availability, information availability and seeking for the truth very challenging tasks.

In a close media system like Malaysia's, it is hard to find whistle blowers since there is no law protecting them. They might need to risk their job by blowing the whistle, where not many are willing to do so. Without whistle blowers, it is a challenging task to find an issue to investigate or a lead to follow up on. Informants also admitted that only a strong reputation will attract whistle blowers. Hence, not many reporters are eligible, especially young reporters who do not have many connections. Other than whistle blowers, reporters have to count on their instinct and inquisitive nature to snoop for issues to investigate.

The second challenge that comes after finding an issue to investigate is getting the information to support the story. According to informants, having a good reputation with sources sometimes helps but it is still hard to get detailed information especially from government agencies. They have to go through much red tape and bureaucracy, even if they are working for the mainstream media which usually publishes positive information about the government in any case. These are the consequences of press freedom parameters in Malaysia which has become accustomed in daily practice. Government officials are so

DOI: 10.1051/shsconf/20173300079

afraid to give information to reporters without clearance from their managers or superiors. This is a big challenge for reporters because they have deadlines to meet.

The third consequence of press freedom parameters in Malaysia is reporters' reliance on online information. Since they have limited access to official sources, reporters turned to the Internet which provides access to the information highway. However, this is not the answer to their information hunger because it is hard to determine the credibility of online information, making finding the truth more challenging than ever. This leads to less confidence in publishing investigative reports because reporters are afraid that they will be sued if they cannot verify their facts. It is also a setback for media organisations to publish investigative reports since they have to bear all the costs if they lose the case, which will also tarnish their credibility and reputation.

Apart from press freedom threats, media organisations also struggle with low impact of investigative reports and weak enforcement from authorities on the consequences of an exposé. These challenges are tight corners for media organisations to either investigate or drop. With the public now turning to the online media, so do advertisers. Therefore, the media cannot allocate large spaces for investigative reports which might not attract public attention. They have to find material that will attract an audience and at the same time pull advertisers. This is the reason why media are publishing more sensational, light, human interest stories compared to serious, heavy and long investigative reports. Audience attitude towards investigative reports has also changed the media organisation, and editors' and reporters' perceptions on practicing investigative journalism. It's worsening when reporters' hard work does not draw authorities' attention. Some informants expressed their disappointment when their reports do bring about a policy change but because of weak enforcement, the wrongdoing continues to be done months after its exposure.

The above mentioned outside challenges will trigger inside the newsroom challenge when it demotivates journalists to pursue investigative journalism. They lost their passion to look for issues, investigate and exposé important public matters. Added with human resource shortage, investigative journalism has been given the back seat where reporters were given daily assignments. Investigative reporting can only be done during their free time. Since investigations can take a very long time to complete, most media organisations do not have the budget to pay journalists who cannot produce stories on a daily basis. Much worse, most good investigative reporters are senior reporters with high salaries. Hence, it hinders the production of investigative reports. In relation to press freedom, reporters require a lot of time to investigate and acquire information from appropriate sources. However, editors are not patient. This time-consuming factor is another big challenge in practicing investigative journalism in Malaysia. Lack of journalism skills adds to the pressure. According to informants who are editors, reporters do not have the requisite skills to sniff for issues. In the current media environment in Malaysia, this is a much needed skill. In addition, reporters are lacking information searching skills. They are more of reporter who report, not journalist who can interpret and elaborate information into something more meaningful to their readers. Writing an investigative report with a closed mind is almost impossible. Editors interviewed in this research also expressed their disappointment in young reporters who are technologically 'savvy' but do not use their skills to search for information. CAR and CAIR are not being utilised properly by Malaysian reporters.

All in all, challenges facing the practice of investigative journalism in Malaysia come from outside and inside media organisation, where some are under their control like human resources, cost, and journalism skills while others like press freedom and media sustainability are beyond their control. However, there are always hope because these situation are not static, there are instead evolving and offering a few bright prospects for investigative journalism to prosper.

References

1. Johnston, D. C. *The best of times, the worst of times for investigative journalism in the US.* In J. Mair, J. & R.L. Keeble., *Investigative journalism: Dead or alive?* United Kingdom: Abramis Academic Publishing. 137-150 (2011).
2. Greenwald, Marilyn, & Joseph Bernt. *The big chill: investigative reporting in the current media environment.* Ames: Iowa State University Press. (2000).
3. Lavrov, V. *Libel laws pose obstacles to Ukraine's investigative journalists.* Nieman Reports, **65**(1), 19 (2011).
4. Feldstein, M. *The challenges and opportunities of 21st century muckraking.* Nieman Reports, Summer. 50-53 (2009).
5. Lent, J. A. *Restructuring of mass media in Malaysia and Singapore – Pounding in the coffin nails?* Bulletin of Concerned Asian Scholars, **16**, 4 (1984).
6. Stapenhurst, R. *The media's role in curbing corruption.* Washington: World Bank Institute. (2000).
7. Willis, J. *The human journalist: Reporters, perspectives and emotions.* United States of America: Greenwood Publishing Group. (2003).
8. Page, B. British Journalism Review, **21**, 1, (2010).
9. Seitz, J. *British libel law: Its ripple effect on journalists worldwide.* Nieman Reports. Spring, **65** (1), 21 (2011).
10. Houston, B. IRE Journal, **27**(6), 4 (2004).
11. Haxton, N. *The Death of Investigative Journalism?* In Tanner (Ed.) Journalism: Investigation and Research. Malaysia: Pearson Education Australia. 20-36 (2002).
12. Anderson, D., & Benjaminson, P. *Investigative reporting. Bloomington*: Indiana University Press. (1976).
13. Feldstein, M. *A muckraking model: Investigative reporting cycles in American history.* Press/Politics, **11**(2), 1- 16 (2006).
14. Hargreaves, I. *Journalism truth or dare?* Oxford: Oxford University Press. (2003).
15. Tanner, S. *Journalism: Investigation and Research.* Malaysia: Pearson Education Australia. (2002).
16. Daniel D.K. *Best of times and worst of times: Investigative reporting in post-Watergate America.* In Greenwald, M.S & Bernt, J. (Eds.), The Big Chill, United States of America: Iowa State University Press. 11-33 (2000).
17. De Burgh, H. *New media and investigative journalists in China.* In J. Mair & R.L. Keeble, (Eds.), Investigative journalism: Dead or alive? United Kingdom: Abramis Academic Publishing. 171-181 (2011).
18. Fleming, C. *Journalism and new technology.* In Hugo de Burgh (Ed.), Investigative journalism: Context and practice. London: Routledge. 169-185 (2000).
19. Waisbord, S. *Watchdog journalism in South America: News, accountability, and democracy.* New York: Columbia University Press. (2000).
20. Waterford J. *The Editor's Position.* In S. Tanner, Journalism: Investigation and research. Malaysia: Pearson Education Australia. 37-47 (2002).
21. Franklin, B. *Newszak and news media.* London: Arnold. (1997).

DOI: 10.1051/shsconf/20173300080

Youth Online Political Participation: The Role of Facebook Use, Interactivity, Quality Information and Political Interest

Shamsu Dauda Abdu[1,*], *Bahtiar* Mohamad[2], and *Suhaini* Muda[3]

[1,2,3]School of Multimedia Technology and Communication, College of Art and Science, Universiti Utara Malaysia.

Abstract. This paper focuses on the role of Facebook use, interactivity with politicians, quality of information and political interest as the predictors of online political participation among youth. The paper argues that Facebook use, interactivity, quality of information and political interests have the potential to engender online political participation among Nigerian youth. Also, by offering them an avenue for acquiring political information necessary for making an informed political decision through deliberations on Facebook; in order to bring new political changes and group mobilization. Therefore, the study was implemented with survey questionnaires collected from youth in Bauchi metropolis (N= 372). Findings reveal that Facebook use, quality of information and political interest positively correlates with online political participation while interactivity with the politicians did not significantly influence online political participation among youth.

1. Introduction

Historically, citizen's participation in the political processes has been considered a fundamental element of an ideal democracy. Thus, participation represents a significant component in political communication and democratic institution. Recent reports from western countries indicate decreasing youth participation in political activities such as low turnout at elections [1]. Some scholars raised an observation on whether all forms of political participation are declining or are youth now shifting paradigm of participation from an institutionalized form to social media sites such as Facebook to affect political and social change in the society. Empirical evidence has shown that the trends which describe an apathetic youth in their contemporary politics can be interpreted by the shift from traditional to expressive online participation [2]. Thus, a study of youth and their online political participation is timely and imperative because political behaviors are often articulated and nurtured at a younger age [3].

* Corresponding author: shamsud.abdul@gmail.com

Giving the above scenario, two decades ago, there has been serious academic concern about the low political participation among youth in many parts of developed and developing nations, thus, some scholars have identified this as a threat to good governance particularly in developing nations like Nigeria and also an impediment to their development agenda [4]. For instance, in Europe, America, Asia and Africa, youth are not stimulated by the traditional media such as radio, television and newspaper which invariably have limited influence on increasing their political interest [5]. Similarly, in Nigeria youth participation over the years has decreased because of the fact that youth in political space has been marginalized and often used as thugs by the selfish politicians who seek personal aggrandizement through violence. Consequently, this trend decreases youth morale to participate in politics [6].

Before the coming of social media, space in newspaper and airtime on radio and television were limited and expensive; thus youth, specifically cannot express their opinion nor participate in politics through the old media. Nevertheless, recent observations in the United States, Europe, Asia, Middle East and Africa shows that youth participation through political activities appears to be shifting paradigm to the online public deliberations [4]. Online social networking sites such as Facebook have brought new hopes and opportunities by connecting youth with politicians and like minds to share information and opinions [7].

Additionally, Facebook is now playing important roles refers to as "migration process" wherein youth around the world are using the platform to address their problem of political apathy, and also engaging in online political discussion that may eventually promote good governance and development in the society [8]. However, what remains unclear is whether Facebook use, interactivity with politicians, quality of information and political interest, may positively influence online political participation among youth and increase their participation.

2. The outside and inside battle

2.1 Online political participation

Obviously, with the increasing role of Facebook in political activities, participation has now taken a new form. Like posting and commenting online, joining and creating online groups, organizing protest events, participation in online discussions with friends and 'liking' on a Facebook profile of a political figure or party indicating political preferences; certainly have become widespread [9]. Thus, Facebook use for online political participation yield significant influence on youth. Conroy, Feezell and Guerrero [10] demonstrate that information exchange on Facebook fosters civic participation, trust, political and life contentment among youth.

Online political participation is online political participatory activities carry out on Facebook such as sending political messages on Facebook to the politicians, commenting on a candidate's Facebook page, seeking donation for a political party or sharing political information on Facebook [11]. Thus, this provides youth more ways to be active in political affairs than what was not possible in the past generations [12].

Furthermore, online activities, especially on Facebook can easily be seen, read and viewed by a large group of youth because of its high connectivity, therefore it has a strong chance of reaching and encouraging them to participate. Gibson and Cantijock [13] observed that attention to political news and also sharing the news have more instrumental quality especially when they are carryout online. In Nigeria, youth have started embracing Facebook to engage in political activities; because they consider it as a medium that make their voice heard in the political activities. In summary, political participation as a concept is dynamic and changing over the time with new form of communication system and it increases youth participatory political activities at the expense of traditional forms of participation [14].

2.2 Facebook usage

Evidently, the use of Facebook recently for online political participation, worldwide has greatly increased, especially among youth and they are particularly using it to influence social and political change. Studies have suggested that youth are now using Facebook to seek for political information, mobilize like minds, create user-generated content and share political views [15-16]. Increasingly, Facebook has been used to mobilize individuals to participate in politics and protest around the world. This, however, suggest that Facebook can help convert or mobilize the people that have been disengaged from offline politics to now participate in an online political activity.

This statement has synergy with the work of Theocharis and Quintelier [2], the authors believe that Facebook provides suitability and accessibility for a sufficient number of people, improved access to information, online opportunities for political expression and political action. Nevertheless, studies have indicated that Facebook is more powerful than traditional media [17]. Facebook provides a similar and advanced features in terms of exposure to information but has the additional benefits and advantages of global reach, better quality and greater speed and also an interactive medium of online political discussion. With these features, Facebook shows a significant role in the formation of political knowledge [18]. Additionally, youth today get their political information from Facebook rather than the legacy traditional media. Facebook may likely motivating online political participation through the access to information particularly among youth in Nigeria.

2.3 Interactivity with politicians

Interactivity is one of the important features of online participation which encourage users to interact with both content and sender of the message [19]. Studies that are mobilization-oriented concentrates on the idea of e- democracy which translate to an interactive online environment where youth for instance can inform themselves about topical political issues and at the same time correspond with political representatives. Facebook provides a great opportunity to politicians to reach out to their constituents and voters. The technology also link and facilitates interaction between community and elected political representatives by providing a public online 'Wall' a space where community members can easily write comments in favour or against their political leaders [20].

While social media in general and Facebook in particular provides important opportunities for youth to interact with politicians and various stockholders, youth can become friends and initiate other linkages with politicians in representatives bodies, government officials and prominent community leaders through Facebook [21]. Consequently, youth 'friendship' and interaction with those important political figures are likely to be vital source of political information and viewpoint. Therefore, these politicians may have the tendencies of mobilizing and influencing youth to participate in online political activities [22].

2.4 Facebook information quality

Studies have reveals that people use a particular channel of communication because of both external and internal factors for instance, the strength and quality of the information, the

DOI: 10.1051/shsconf/20173300080

credibility of source of the information, easy accessibility and understanding of the information, its timeliness, the organization and arrangement of the information and also the completeness and reliability of the information [23]. Similarly, individual personal purpose or intention of having good and reliable information can motivate people to select and use a specific media, therefore this can help in ascertain and justifying Facebook information quality as a motivating factor for youth unparticular for using Facebook for online political participation.

Clearly, what motivate youth to use Facebook is their ability to interact with friends and share quality and accurate information with people. Correa, Hinsley, and de Zúñiga [24], emphasized that the higher the speed and quality of information is and the higher possibility to be politically engaged with other issues and people. This translates that the content of information speed and its authenticity, will determine its acceptability and applicability to people who are already active and could eventually influence people opinions and views. Similarly, Young and Quan-Haase [25] argued that the nature and quality of information is positively related to the large amount of information an individual receive which may satisfy his communication need and propel the person to act or react.

One of the reasons why youth engage in Facebook is motivated by their political information needs and information quality is a significant predictor for online interaction [26], which may also increase information sharing since online communities lack face-to-face interaction Nevertheless, quality of information makes communication easier, clear and understandable. Youth quest for political knowledge and information make them so inquisitive to find out how and where to obtain the right and accurate information.

2.5 Political interest

Superficially, there is a large consensus among scholars that among all other predictors, political interests have a strong influence in terms of motivation to participate in politics. This statement has a synergy with the work of [27] they supported this assertion by emphasizing that political interest is an attitude that provide a stable and sustainable outcome in politics because it often tells whether an individual may be politically active or passive. Therefore, it may be correct to say that Facebook users promote political interest and influence friends in their network are likely to participate in online politics. Additionally, Vissers et.al [28] corroborated that opportunities for political participation are more likely to manifest if the level of political interest among youth specifically is high.

It will also be correct to advance the discussion that political interest is a necessary condition of participation as it measures youth's online political participation. Political interest is significant in political participation because it provide the motivation to get political information, skill or knowledge which is relevant in assessing political alternatives. Consequently, Facebook is now the alternative platform of political expression and political interest appears to be a significant antecedent and predictor of online political participation among youth [29-30]. Facebook political activities such as reading and posting political opinions, interacting with politicians on Facebook, and discussing politics with friends on Facebook groups may likely influence interest and knowledge about positive social and political issues, later facilitating participation in other online political activities [31]. This leads to the following hypotheses:

- ***Hypothesis 1:*** *Level of Facebook use positively influence the level of online political participation among youth.*
- ***Hypothesis 2:*** *Level of interaction with politicians positively influence the level of online political participation among youth.*

- ***Hypothesis 3:*** *Information quality positively influence the level of online political participation among youth.*
- ***Hypothesis 4:*** *Level of political interest positively influence the level of online political participation among youth*

3. Method

In this section, data gathering procedures, respondents and measurement of variables are presented. However, this study involves youth of Bauchi metropolis in Nigeria.

3.1 Respondents

Respondents in this research are the youth comprising both males and females; in Bauchi metropolitan area aged between 18 to 35 years old as recommended by the UNESCO definition of youth. The choice of youth as the focus population is based on the fact that these age groups often are using Facebook for political activities [28]. In Nigeria, youth have been in the frontline of struggle for independence and democracy; they also contribute and ensure good governance as well as demand for credible leadership in the country [32]. Similarly, Bauchi Metropolis has the characteristic of urban area with a high population, high level of development and awareness of politics. Approximately, 70.25% are males between the ages of 18- 22. The majority of the respondents holds Diploma certificate (51.1%).

3.2 Measurement Instrument

The measurement of online political participation in this paper relied primarily on items adapted from the work of Jung, Kim and Gil de Zuniga [33]. This instrument consist of political activities that youth carry out on Facebook such as sending or receiving political information on Facebook. The instrument used to measure Facebook Use includes items adapted from Ellison et al., [34]. This instrument consists of items that describe youth feelings and connection to Facebook. Interactivity with politicians was measured with the item scale adapted from the work Lee and Cho [35]. The items described how Facebook help youth with the opportunity to interact with pelicans and generate feedback from them. Information Quality was adapted from the work of Lim, [36] to measure the quality, speed and accuracy of information on Facebook. While Political Interest was adapted from the work of Whiteley, [37] to measure what motivate youth want to participate in politics online. Thus, each item was measured using 5-point Likert Scale.

Prior to the actual study, we conducted pre-test study among the youth in Nigeria. The pre-test conducted sought to determine the degree of reliability, trustworthiness, dependability of the measurement used in this study. Result of the pre-test shows Cronbach's alpha for online political participation was .824, Facebook use was .859, Interactivity was .863, Information quality was .891, and Political interest Cronbach's alpha was .883 respectively. Before testing the hypotheses, data underwent tests for normality which was conducted for each of the survey items. Tests for normality include kurtosis measures, skewness measures, and visual inspection of histograms. The majority of items appear to be within normality, and analysis indicates normal-shaped histograms.

4. Results

Data was analyzed by using multiple regressions to test the relationship between Facebook use, interactivity with politicians, information quality, political interest and online political participation among youth in Nigeria. Multiple regression technique was chosen because it allows researchers in social science to deal with complexity in human behaviours as adopted in previous studies [38-39]. The central idea of the multiple regression method is that it is used to test the relationship between a dependent variable and two or more independent variables. Thus, by applying a multiple regression technique we can also determine specific independent variables that have a major influence on a dependent variable. The hypotheses deal with the relationship between Facebook use and online political participation among youth respondents. Thus, a positive relationship was predicted in the hypotheses. Results of the multiple regression analysis suggest that there is a significant positive relationship between Facebook use, quality information, political interest and online political participation among youth ($r = .293$, $p < .005$) while there was no significant relationship between interaction with politicians and youth online political participation. Therefore, the results suggest that there is a direct relationship between Facebook use, quality information, political interest and online political participation. Specifically, it indicates that political interest has the highest correlation with online political participation.

Table 1. Descriptive Statistics Correlations and Reliability of Variables

	Variables	α	M	SD	1	2	3	4	5
1.	Online Political Participation.	.701	3.7806	.6730	1.000				
2.	Facebook Use	.827	4.2367	.5783	.293*	.394**	1.000		
3.	Interactivity with Politicians	.789	4.0582	.6594	.427**	.495**	.404**	1.000	
4.	Information Quality.	.725	3.6364	.6110	.378**	.396**	.346**	.496**	1.000
5.	Political Interest.	.776	4.1042	.6119	.492**	.412**	.336**	.505**	.455**

Significance $*p < .05$; $**p < .01$

Table 2. Model Summary

Model	R	R Square	Adjusted R Square	Std. Error of the Estimate	Durbin-Watson
1	.548[a]	.301	.293	.56587	1.930

a. Predictors: (Constant), PI, FU, PQ, IF
b. Dependent Variable: ON

Table 3. ANOVA

Model	Sum of Squares	df	Mean Square	F	Sig.
Regression	50.503	4	12.626	39.429	.000[b]

Residual	117.518	367	.320		
Total	168.021	371			

a. Dependent Variable: ON
b. Predictors: (Constant), PI, FU, PQ, IF

Table 4. Summary of Multiple Regression for Online Political Participation

	Expected sign	***β***	***t-statistics***	***P-value***
Intercept		.759	2.876	.004
Independent variables				
Facebook use	+	.083	1.462	.004**
Interactivity with politicians	+	.179	3.172	.145
Information quality	+	.131	2.245	.002**
Political interest	+	.357	2.182	.025**
Adjusted R^2			.293	
F-statistics *(df 4, 367)*			39.429	.000**
Durbin-Watson			1.93	
N			372	

Notes: ** significant at 0.05 (1-tailed), respectively; *Dependent variables* = ON, Online political participation; *Independent variables* = FU, Facebook use; IF, Interactivity with Politicians; IQ, information quality; PI, Political interest.

The overall pattern of results indicates a significant relationship between online political participation and Facebook use. Results shows that Facebook use, quality information and political interest are significant predictors for online political participation $\beta = .759$, *t-statistics*= 2.876 $p < .05$ and $R^2 = .293$. These results provide support for past research Vissers and Strolle, (2014), Ksiazek, Peer and Lessard, (2014), Carlisle and Patton, (2013), as well as provide support for the current hypothesis. Table 2 indicates that for Facebook use *t-statistics* = 1.462, $\beta = .083$, p-value = .004, $p < .05$, Information quality *t-statistics* = 2.245, $\beta = .131$, p-value = .002, $p < .05$, political interest *t=statistics* =2.182, $\beta = .357$, p-value = .025, $p > .0.5$. Thus, political interest has significant impact on online youth political participation.

5. Discussion

This study suggests that Facebook use, quality information and political interest play an important role in influencing youth online political participation. As can be seen in Table 4, Facebook use, quality information and political interest, show a significant relationship with online political participation. This finding demonstrates an important aspect of political interest on youth online political participation in Nigeria. Thus, for government, politicians or political party to take a meaningful decision or formulate policy in Nigeria, all components of online political participation (Facebook use, information quality, and political interest) must be given a priority in terms of budget allocation and policy implementation. This is because they play a significant role in influencing and perhaps in determining youth online participation. Thus, government, political stakeholders' especially political parties and candidates in Nigeria should focus more on what motivates youth to have an online political interest so as to address youth political apathy in order to increase their participation. Based on the analysis and outcome of the results, the following model of online political participation was developed.

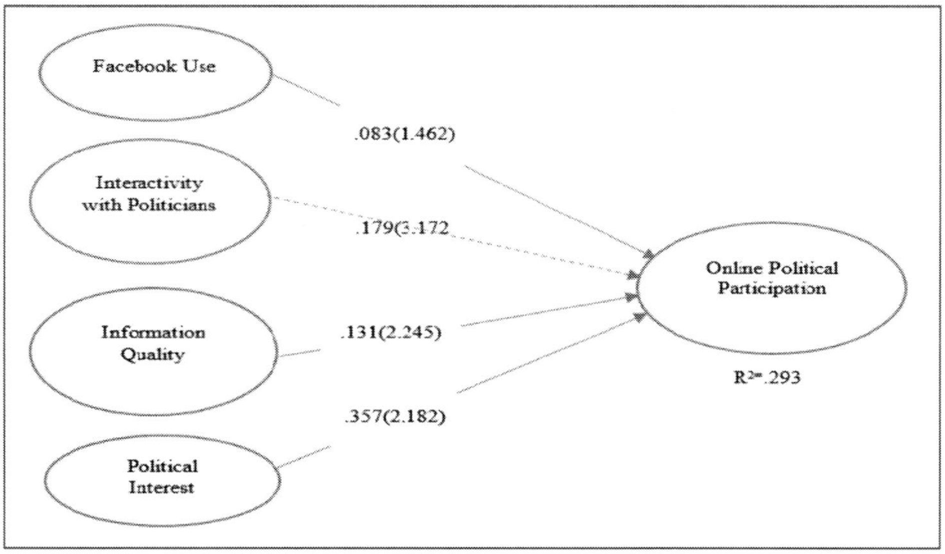

Fig. 1. Model for Online Political Participation

6. Conclusion

To this end, it is imperative to state that online political participation amongst youth is largely dependent on the use of Facebook because more youth are often showcasing stronger reliance on it as their online platform for securing political information they need to make an informed political decision. Hence, targeting and tailoring political messages online to the youth through Facebook and what motivate youth to develop an interest in politics should be an important factor in future campaign strategies. As youth continues to become increasingly dependent on Facebook to keep up with the demands of their socio-political lifestyle and civic engagement, it can be assumed that youth will continue to rely on online forms of social interaction.

References

1. Bakker, T. P., & Vreese, C. H. *Good news for the future? Young people, Internet use, and political participation.* Communication Research, **38**, 451 (2011)
2. Theocharis, Y., & Quintelier, E. *Stimulating citizenship or expanding entertainment? The effect of Facebook on adolescent participation.* New Media & Society (2014)
3. Howard, P. N., & Parks, M. R. *Social media and political change: Capacity, constraint, and consequence.* J. Commun., **62**(2), 359-362 (2012)
4. Waller, L. G. *SAGE Open*, **3**(2), (2013)
5. Dong, Q., Toney, J., & Giblin, P. *Social network dependency and intended political participation.* Human Communication, **13**(1), 13–27,(2010)
6. Olabamiji, O. M. *Use and Misuse of the New Media for Political Communication in Nigeria's 4th Republic.* Developing Country Studies, **4**(2), 44–53 (2014)

7. Skoric, M. M., & Poor, N. *Youth engagement in Singapore: The interplay of social and traditional media*. J. Broadcast. Electron., **57**(2), 187–204 (2013)
8. Chan, M., & Guo, J. *The role of political efficacy on the relationship between Facebook use and participatory behaviors: A comparative study of young American and Chinese adults.* Cyberpsychology, Behavior, and Social Networking, **16**(6), 460–463 (2013)
9. Hooghe, M., Hosch-Dayican, B., & van Deth, J. W. *Conceptualizing political participation.* Acta Politica, **49**(3), 337-348 (2014)
10. Conroy, M., Feezell, J. T., & Guerrero, M. *Facebook and political engagement: A study of online political group membership and offline political engagement.* Computers in Human Behavior, **28**(5), 1535–1546 (2012)
11. Skoric, M. M., & Kwan, G. *eJournal of eDemocracy and Open Government*, **3**(1), 70–79, (2011)
12. Zúñiga, H. G. de, Jung, N., & Valenzuela, S. J Comput Mediat Commun, **17**(3), 319–336 (2012)
13. Cantijoch, M., Gibson, R. K., & Cutts, D. C. *Engagement in the online campaign in the United States and France. A comparative analysis.* EPSA 2013 Annual General Conference Paper, **323** (2013)
14. Vitak, J., Zube, P., Smock, A., Carr, C. T., Ellison, N., & Lampe, C. *It's complicated: Facebook users' political participation in the 2008 election.* Cyberpsychology, Behavior and Social Networking, **14** (3), 107–114 (2011)
15. Towner, T. L. *All political participation is socially networked? New media and the 2012 election.* Social Science Computer Review, **31**(5), 527–541 (2013)
16. Thun, V. Northern Illinois University (2014 to be published)
17. Gromping, M. *'Echo Chambers' Partisan Facebook Groups during the 2014 Thai Election.* Asia Pacific Media Educator, *24*(1), 39–59, (2014)
18. Muntean, A. Aarhus University, (2015 to be published)
19. Ksiazek, T. B., Peer, L., & Lessard, K. *User engagement with online news: Conceptualizing interactivity and exploring the relationship between online news videos and user comments.* New Media & Society, **18**(3), 502-520 (2016)
20. Lahabou and Wok. (2011). *Students' acceptance of an internet campus radio: a test of the technology acceptance model.* Kulliyyah of Islamic Revealed Knowledge and Human Sciences, 1–10 (2011)
21. Tang, G., & Lee, F. L. F. *Facebook use and political participation: The impact of exposure to shared political information, connections with public political actors, and network structural heterogeneity.* Soc. Sci. Comput. Rev., **31**(6), 763–773 (2013)
22. Feezell, J. T., Conroy, M., & Guerrero, (2009)
23. Sundar, S. S., Bellur, S., Oh, J., Jia, H., & Kim, H.-S. *Theoretical importance of contingency in human-computer interaction effects of message interactivity on user engagement.* Commun. Res., 1–**31**. (2014)
24. Correa, T., Hinsley, A. W., & de Zúñiga, H. G. *Who interacts on the Web?: The intersection of users' personality and social media use.* Computers in Human Behavior, **26**(2), 247–253 (2010)
25. Young, A. L., & Quan-Haase, A. *Information revelation and internet privacy concerns on social network sites: a case study of facebook.* C&T '09, 265–274 (2009)
26. Lin, L. C.-S. *Convergence: The International Journal of Research into New Media Technologies*, 1–16. (2014)
27. Moeller, J., de Vreese, C., Esser, F., & Kunz, R. *Pathway to political participation: The influence of online and offline news media on internal efficacy and turnout of first-time voters.* Am. Behav. Sci., ***58***(5), 689–700 (2014)

28. Vissers, S., Hooghe, M., Stolle, D., & Maheo, V. –a. *The impact of mobilization media on off-line and online participation: Are mobilization effects medium-specific?.* Soc. Sci. Comput. Rev., **30**(2), 152–169 (2012)
29. Yamamoto, M., & Kushin, M. J. J Comput Mediat Commun, **19**, 3 (2014)
30. Yang, H. C., & DeHart, J. L. *Social Media + Society*, **2**(1), 1–18. (2016)
31. Himelboim, I., Mccreery, S., & Smith, M. J Comput Mediat Commun, **18,** (2013)
32. National Bureau of Statistics Nigeria, 2012. Retrieved from www.nigerianstat.gov.ng/pages/download/121
33. Jung, N., Kim, Y., & de Zúñiga, H. G. *The mediating role of knowledge and efficacy in the effects of communication on political participation.* Mass Communication and Society, **14**(4), 407–430 (2011)
34. Ellison, N. B., Steinfeld, C., & Lampe, C. *The benefits of Facebook "friends:" Social capital and college students' use of online social network sites.* Journal of Computer-Mediated Communication, **12**(4), 1143–1168 (2007)
35. Lee, S., & Cho, M. *Social Media Use in a Mobile Broadband Environment. Examination of Determinants of Twitter and Facebook Use.* International Journal of Mobile Marketing, **6**(2), 71-78 (2011)
36. Lim, J. (2012 Doctoral dissertation, Brunel University Brunel Business School PhD Theses).
37. Whiteley, P. *Citizenship Education : The Political Science Perspective.* (2005)
38. Cohen, J., Cohen, P., West, S. G., & Aiken, L. S., *Applied multiple regression/correlation Analysis for the behavioural sciences, Mahwah,* (Lawrence Erlbaum Associates, New Jersey 2003)
39. Mohamad, B., Bakar, H. A., & Rahman, N. A. A. *Relationship between corporate identity and corporate reputation: a case of a Malaysian higher education sector.* Jurnal Manajemen Pemasaran, **2**(2), 81-89 (2007)

Agenda Setting Theory and International News: A Comparative Analysis of News Articles Coverage on Iraq War in Malaysian English Newspapers

Hanaa Kadum Kassed[2], and *Che Su* Mustaffa[1,*]

[1]School of Multimedia Technology and Communication, College of Art and Science, Universiti Utara Malaysia.

[2]Ministry of Science and Technology, Baghdad, Iraq

Abstract. The main purpose of the paper is to discuss the application of Agenda Setting Theory in the publication of news about Iraq war in - the New Straits Times press (NSTP) and the Star. This study examines a longitudinal content analysis of 535 news through three periods (before, during and after the occupation of Iraq). The result of this study shows that NSTP gave more emphasis to the issue of the policy of unipolar for the United States while the Star gave more emphasis to the issue of Iraqi suffering. Thus, this study can shed some light on how Eastern country views Iraq and this perhaps could help Iraq reform its image by expanding the scope of diplomatic, business and cultural actions. This study provides support to the theoretical proposition that the media are not successful in telling the audience what to think, but it is stunningly successful in telling the audience what to think about (first level of Agenda Setting). This study also provides support to the theoretical proposition that differences in media organizations have profound impact on how the important issues are framed.

1. Introduction

The mass media played a crucial role in constructing the images of various societies and their people [1]. Mass media news on certain country itself can affect their image on society. The image that will appear could be positive, negative or neutral depending on the content in it. McCombs and Shaw [2] claimed that the mass media pays attention to certain issues. It builds up public images of political figures and it constantly presents objects suggesting what individuals should think about, know about and how they should feel. Perhaps, this hypothesized Agenda Setting Function of the mass media is most succinctly

* Corresponding author: chesu402@uum.edu.my

stated by Cohen [3], that a successful press tells the readers what to think, so the media sets the Agenda Setting for issues.

2. Problem statement

The media greatly affects the image in the minds of people from the outside world. The Agenda Setting Theory effects also have significant implications beyond the images that have been created in people's heads. Originally, the Agenda Setting Theory focused on the salience of issues [2]. Thus, this paper discusses the differences in the portrayal of Iraq in Malaysia by studying two English newspapers, NSTP and the Star, which belong to two different political parties. The United Malays National Organization (UMNO) consists of a majority of Malays, while the Malaysian Chinese Association (MCA) consists of a majority of Chinese. UMNO dominates the political system [4], while MCA is the second largest political party in Malaysia [5]. UMNO owns 80% of NSTP shares while the MCA owns 58% of the Star's shares [6]. In addition, the New Straits Times Press and the Star have the highest circulations in the country.

On the whole, in Malaysia the media reflect the policy of the government and the society in which they operate [7]. Firdaus [8] explained that the Malay political party, UMNO, and the Chinese political party, MCA, are components of the ruling coalition, Barisan Nasional. However it cannot be denied that each party cares for its ethnic interests. A cursory look at the different newspapers should indicate different agenda. These agenda are then relayed back to their readers, "for example, the Malay Mail and the New Straits Times are both owned by the News Straits Times Press but cover different types of stories with the Malay Mail running more human interests and sensationalistic stories. Utusan Malaysia is closely linked to the Malay political party (UMNO), while the Star is closely affiliated to the Chinese political party (MCA)" [8]. Khiang, Ahmad and Kee [9] see the differences of interest and the upholding the rights of the community served by the ethnic newspapers can be seen from its focus and suggest that future researches should consider English newspapers. In conclusion, the ownership of newspapers probably influenced to a greater degree the Agenda Setting Theory.

3. Method

The theory of agenda setting has served as one of the most powerful and fertile concepts in mass media and communication effects research [10]. Agenda-setting theory entails the link between the mass media's emphasis on a particular topic and the corresponding salience of that topic on the public agenda [11]. A more recent definition by Brosius and Kepplinger [12] defines agenda setting as a dynamic process 'in which changes in media coverage lead to or cause subsequent changes in problem awareness of issues'. In other words, when mass media prioritise and frequently cover a topic, the public tends to consider this topic more important.

The origins of agenda-setting theory can be traced back to Walter Lippmann and his 1922 book Public Opinion. Lippmann [13] was the first to ascertain the mass media's critical role in creating and shaping the images and perceptions about the world in the minds of the public, who have little unmediated experience of distant real-world events. Related to Lippmann, another influential idea comes from Cohen [3], who suggests that 'the press may not be successful much of the time in telling people what to think, but it is stunningly successful in telling its readers what to think about'. Hence, rather than injecting ideas and ways of thinking into the public's mind like a hypodermic needle, agenda-setting theory contends that the public learns about the importance of topics in direct proportion to the amount of media coverage of those topics [2]. Since the formal introduction of agenda-setting theory by McCombs and Shaw [11] in the Chapel Hill study on the 1968 US presidential election, a myriad of studies around the world have found correlations between media agendas and public agendas (e.g. [14,15]). On a socio-political level, these US-influenced studies largely ignored non-Western countries with, for example, communist political systems.

This study employed a quantitative content analysis as quantitative content analysis is proven to be a more powerful method than survey and interview because of its unobtrusive nature and its lack of reliance on subjective perception. This study analyses the publication of news stories and articles in the NSTP and the Star newspaper during a specific period. The sample was collected from the published news stories and articles during the three periods (before the occupation 15 January 2002 until 8 April 2003, during the occupation 9 April 2003 until 28 June 2004 and after the occupation 29 June 2004 until 18 September 2005) of both newspapers. The sample for this study was identified through a guided search using library database from Universiti Utara Malaysia (UUM) called Blis BERNAMA.

This study investigates four issues, namely, The US policy and the occupation of Iraq, Iraq's needs and suffering, The Malaysia's policy towards Iraq, and The impact of war on Malaysia. These issues were selected because these were the most covered issues during the periods studied. The four categories are as follows:

- The US policy and the occupation of Iraq: This category refers to invasion of Iraq without a UN mandate and under the excuse of weapons of mass destruction which later led to resist occupation. It also refers to US double-standard policy which was evidenced by its support for Israel's occupation of Palestine, and to the invasion of Iraq as an occupation, not liberation. Furthermore, this category also includes the opposition from the countries toward the US policy, and their concern that the real reason for the invasion was oil and not weapons of mass destruction. Finally, this category also refers to the emergence of the looters under the state of lawlessness in the country.
- Iraq's needs and suffering: This refers to the suffering the Iraqi people faced before, during and after the US occupation. As a result of the military disaster which have taken the lives of a number of innocent civilians (women and children), Iraqis needed food, shelter, medicine, sanitation, electricity, safety and security in addition to the need for medical and humanitarian assistance Moreover, they suffered through what they had seen from the horrors of Abu Ghraib prison and also from the keenness of the occupation forces not to revealed the correct number of victims. Also, there was a need to form an Iraqi government as well as the need to rebuild the country.
- The Malaysia's policy towards Iraq: refers to Malaysia rejecting the US policy towards the Middle East, and to show support for Iraq and Palestine, Malaysia is against the war and occupation of Iraq. This has led to a tensed relationship between Malaysia and the US.
- The impact of war on Malaysia: refers to the impact of the Iraq war on the Malaysian economy and Malaysians living in Iraq and the neighboring countries of Iraq (refers to the Malaysian embassy staff and their families in addition to Malaysian students). For example, the mission to evacuate Malaysians from Iraq and the neighboring countries of Iraq and drop in tourist arrivals from West Asia.

DOI: 10.1051/shsconf/20173300081

4. Findings and discussion

The analysis focused on a comparison of issues relating to Agenda Setting issues in NSTP and the Star descriptively. Figure 1 show an analysis conducted between the two newspapers on Agenda Setting issues for the period before the occupation. In general, the analysis shows, the Agenda Setting issues are displayed in the NSTP more than the Star. The issues related to "The Malaysia's policy towards Iraq" published by NSTP were 44 (53.0%) news, as compared to the Star 39 (47.0%). For instance, Malaysia as the chair of the 116-nation Non- Aligned Movement will continue to speak up against the invasion of Iraq. Foreign Minister Datuk Seri Syed Hamid Albar said it was the nation's responsibility to push for NAM's stance to be heard by the world, NAM has no military might to do anything to stop the war, but it's better than of silence. Although there are various forms of pressures against our stance, but if we are united, we can influence the decisions on the situation in Iraq and also the future of world order (New Straits Times, 28 March, 2003).

Similarly, the issue relating to "The US policy and the occupation of Iraq" covered 63 news (71.6%) in the NSTP compared to 25 (28.4%) in the Star. It shows that NSTP (on 18 July 2002; 24 November 2002 and 13 September 2002) focused about the US desire to control the world as the dominant power, through resolutions of military action against Iraq.

Also, the issue related to "The impact of war on Malaysia" was featured in NSTP in 37(67.3%) articles and news stories, more than the Star which only published 18(32%) articles and news stories for the periods concerned. For example, according to a report in NSTP on 11 February 2003, Malaysia's timber and palm oil exports will be affected if a war breaks out in Iraq, due to the disruption in shipments as well as rising freight and insurance charges.

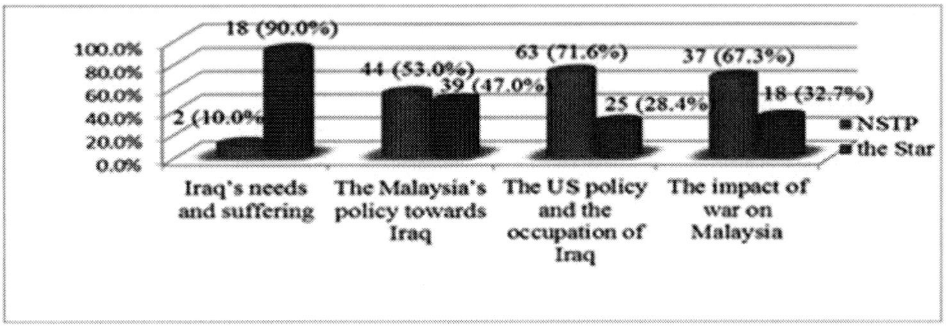

Fig. 1. Comparison of Agenda Setting Issues before the Occupation between NSTP and The Star

The next research question is related to the display of news related to the Agenda Setting issue during the occupation. Figure 2 shows that there is little difference in the Agenda Setting issues published by both newspapers. The Star was found to be more focused on articles and news stories about two issues which are "Iraq's need and Suffering" and "The impact of war on Malaysia" with 30 (61.2%) and 25 (59.5%) news stories published respectively. Meanwhile, NSTP published 19 (38.8%) and 17 (40.5%) articles and news stories respectively. This issue which covers "Iraq's need and Suffering" includes the discussion on how current situation in Iraq was worsening because of many factors, like their existence a political vacuum in Iraq, and many Iraqis feel that the existing political process lacks legitimacy. Thus the Iraqi people need to end the occupation and transfer the power back to Iraqis (New Straits Times, 19 April, 2004). Meanwhile, the issue which covers "The impact of war on Malaysia" includes the discussion on how the Iraq war which led to the decline in gross domestic product (GDP) due to the lower export growth coupled with declining tourism and weaker private consumption, would contribute to a bleaker economic picture for the country (New Straits Times, 2 April, 2003).

The NSTP published more articles and news stories related to issues of "The Malaysia's policy towards Iraq" with 31(66.0 %) news compared to the Star which only published 16 news (34.0%). For example, on 13-21 April 2003, NSTP published news about Malaysia call for an end of the US occupation in Iraq, especially after unseating of Saddam's administration, on the basis that the coalition was not occupation force and that it only wanted to topple Saddam Hussein's administration. The news also reported a peaceful march by Malaysians to protest against the United States-led invasion of Iraq, where many of them held anti-war placards.

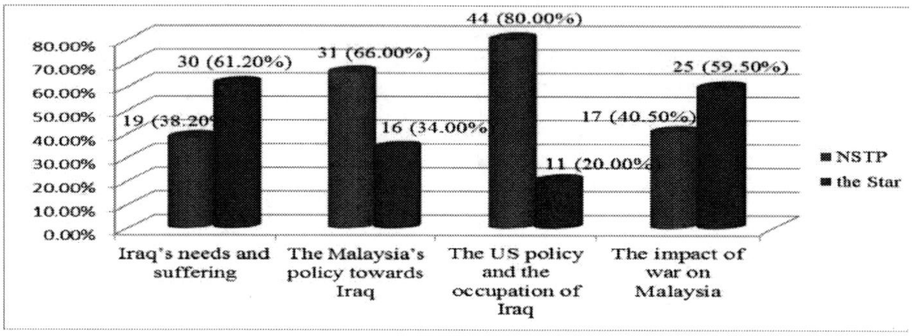

Fig. 2. Comparison of Agenda Setting Issues during the occupation between NSTP and the Star

Looking at the prominent issues constructed by NSTP and the Star for the period after the occupation (Figure 3), the analysis shows NSTP published more news related to the issue of "Iraq's need and Suffering" with 18 (54.5%) news compared to the Star which published 15 (45.5%), where the news were devoted to the percentage of Iraqis who have successfully undergone their first free election with a very encouraging turnout of voters. It was reported that Iraqis wanted to determine their own future with the new government without any direct intervention from the US (New Straits Times, 3 February, 2005). Another news was about the abuse of Iraqi detainees at Abu Ghraib prison in Iraq, suggesting that that the Iraqi people need respect for human rights (New Straits Times, 5 August, 2004).

The issue of "The Malaysia's policy toward Iraq" was published more in NSTP with 21(84.0%) articles and news stories compared to the Star with only 4 (16.0%) articles. Malaysia's policy was reflected in news that reported Malaysia will decide on sending troops for peacekeeping to Iraq only under three conditions which are: first, Ordinary Iraqis want troops from an organization of the Islamic Conference (OIC) member country; second, the troops are not deployed alongside the multinational forces led by the United States; third, a joint decision is reached by the OIC on sending troops to Iraq (New Straits Times, 6 August, 2004).

 DOI: 10.1051/shsconf/20173300081

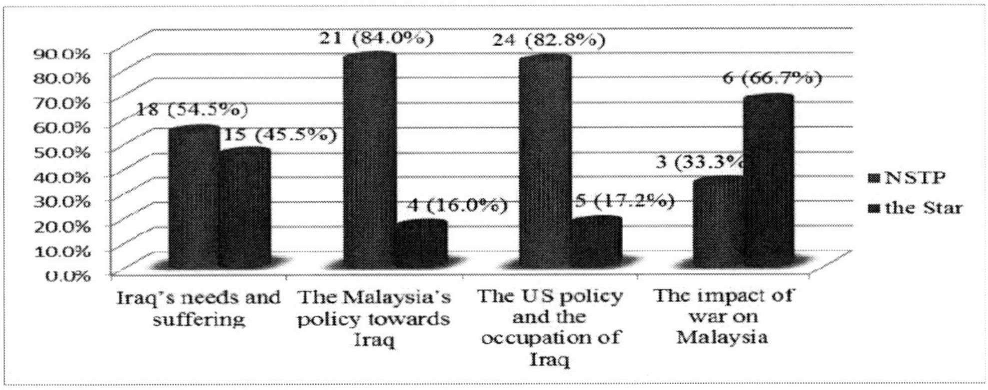

Fig. 3. Comparison of Agenda Setting Issues after the occupation between NSTP and The Star

5. Conclusion

The results of this study contribute toward the understanding of the Agenda Setting process in mass communication. Agenda Setting is a function that the media can play. By deciding what is important, media set the agenda for content within their own pages. For example, it can change a position on an issue such as war. The result of the study proved that the NSTP and the Star focused on different Agenda Setting issues about Iraq through the three periods. NSTP published more articles and news stories about "The Malaysia's policy towards Iraq" and "The US policy and the occupation of Iraq" than the Star. These findings are consistent with the foreign policy which turned Malaysia from having a pro-Western stance in the fifties and sixties to the policy of Non-Aligned Movement in the seventies and after the advent of Tun Dr. Mahathir Mohamad in 1981. In addition, Malaysia continued to support the Iraqi people in the face of UN sanctions and US military strikes. The former Prime Minister of Malaysia, Tun Dr. Mahathir condemned the plans of the United States and Britain to launch an attack against Iraq without a good reason. This finding also support the argument that Malaysia newspapers, in particular the English language newspapers have provide better and healthier reporting's in line with the Malaysia Political Transformation Plan agenda [16].

The findings also indicated that Agenda Setting is the phenomenon of the mass media selecting certain issues and portraying them frequently and prominently, which leads people to perceive those issues as more important than others. The first level Agenda Setting focuses on the amount of coverage of an issue, suggesting that the media decide what issues the public will be aware of. Current study provides evidence that "what to think about" is the first level of the Agenda Setting Theory.

Finally, these findings support Agenda Setting Theory that the amount of coverage devoted to an issue by the media agenda translates into salience about that issue on the public agenda. Further, a contribution is made to the discussion concerning the time frame it takes for an issue to emerge, be covered by the mass media, and translate into public salience. It can be concluded that NSTP reflects the expressive policy of UMNO which was known to be very critical of the US approach, while the Star focused more on sensationalized issues because it is one of the most profitable companies listed on the Kuala Lumpur Stock Exchange.

References

1. N. Saleem, U.S. *Media Framing of Foreign Countries Image: An Analytical Perspective.* Canadian Journal of Media Studies, **2**, 1 (2002).
2. M. McCombs, M. *Setting the Agenda: The Mass Media and Public Opinion.* Cambridge, UK: Polity Press. (2004).
3. B. Cohen. *The press and foreign policy.* New York: Harcourt. (1963).
4. A. Humphreys. *Malaysia's post 9/11 security Strategy: winning "hearts and minds" or legitimising the political status quo?* Kajian Malaysia, **28**, 1 (2010).
5. M. K. Lim, *The State of Media Control in Malaysia.* Paper presented at the annual meeting of the International Communication Association 23 May 2007. US: San Francisco (2007).
6. Z. Nain, & Kim, W. I. *Ownership, control and the Malaysian media.* Who Owns the Media. Penang: Southbound, 249-270 (2004).
7. F. Ibrahim., Mustaffa, N., Kee, C. P., & Ahmad, F. Innov J. **16**, 3 (2011).
8. A. Firdaus. *Ethnic Identity and News Media Preferences in Malaysia.* Paper presented at ARC APFRN Signature Event. Australia, Perth: Curtin University. 27-29 (2006).
9. C. C. Khiang, Ahmad, F., & Kee, C. P. *The Framing of media on contemporary education issue in Malaysia.* Paper presented at the International conference on media and communication 11 – 12 October. 2011. Kuala lumpur: Malaysia. (2011).
10. J., Bryant, J., & Miron, D. J. Commun., **54**, 4 (2004)
11. McCombs, M., & Shaw, D. L. *The evolution of agenda setting research: Twenty-five years in the marketplace of ideas.* J. Commun., **43**, 2 (1972).
12. H. B. Brosius, & Kepplinger, H. M. *The Agenda-Setting Function of Television News: Static and Dynamic Views.* Communication Research, **17**, 2 (1990).
13. W. Lippmann. *Public opinion.* New York: Harcourt. (1922).
14. J. W. Dearing, & Rogers, E. *Agenda-setting.* London: SAGE. (1996).
15. D. Weaver, Graber, D., McCombs, M., & Eyal, C. *Media agenda setting in a presidential election: Issues, images and interest.* Westport, CT: Greenwood Publishing Group. (1981).
16. A. Mohd Khairie & Mohd Baharudin O. *General Election and the Enigma of Free Press in Malaysia.* Procedia - Social and Behavioral Sciences, **155** (2014).

DOI: 10.1051/shsconf/20173300082

Culturally Responsive: Exploring the Attributes of Islamic Health Communication

Mohd Khairie Ahmad[1,*],*Mohd Baharudin* Othman[2] ,*Nor Hayati Mohd* Jalil[3], and *Solahuddin* Ismail[4]

[1,2,3]School of Multimedia Technology and Communication, College of Art and Science, Universiti Utara Malaysia.
[4] School of Government, Universiti Utara Malaysia

Abstract. The purpose of this paper is to explore the attributes (basis and values) of faith-based communication strategy on health communication. Eight series of focus group studies on Muslim community were conducted to gather the data. The finding makes abundantly clear that the tawhidic (the belief in Oneness of God) conception significantly determine the effectiveness of Islamic communication message. In addition, there were another six themes that contributes to Islamic health communication attributes which may influence the receiver's attitudes and behaviours. The insights of this paper may contribute to the further development of health promotion strategies for Muslims community.

1. INTRODUCTION

In acceptance of the reality that many Muslim nations are fronting substantial health-related issues, World Health Organisation (WHO) in collaborations with several Muslim organisations have recognised vision and approaches that culturally sensitive to the Muslim community. Based on the Ottawa Charter, the Amman Declaration in 1989 was marked as an international agenda in promoting health for Muslim community [1]. The purposes and attributes of Islamic communication (IC) are rooted in Islam's social fabric, and are believed to influence on health practices. Islamic communication approaches have put forth as important impacts on the targeted group health behaviour [2]. The purpose of this paper is to explore basis and values of health communication from a community that religiously closes. This article contributes to health communication literature by illustrating an applied case of intersecting local culture (Muslim community) for health issues in general.

2. Literature review

Awareness that culture should be taken into consideration in the ways in which health communication is theorised and practised is growing. There are two distinct yet interrelated approaches to health promotion efforts: the cultural sensitivity approach [3, 4] and the culture-centred approach [5, 6]. The cultural sensitivity approach is conceptualised as

* Corresponding author: khairie@uum.edu.my

developing culturally appropriate health communication practices that would adjust attitude, beliefs, and behaviours of the target audience. More precisely, it is a process of adapting messages to the cultural attributes of the dedicated group. The culture-centred approach pinpoints culture as the core element in theorising about the communication process. By way of explanation, communication theories develop from within the community values and characteristics instead of originating from outside. Explanation of scenario and enunciation of pragmatic solutions based on the landscape of the phenomena emerge from within the culture or subculture that being studied [3]. Yet, both the cultural sensitivity and cultural-centred approaches seek to answer to the call for consolidating culture in health communication practices.

Religiosity relates to the effect of social referents and thus can be understand as comparable to construct from the Theory of Planned Behaviour [7]. Relating health messages to religious matter, or using religious features on messages, may be applicable motivational strategies. This can be done through influence of social elements such as connecting health behaviours to particular sacred commandments or using the norms of the belief as a basis of encouraging or destructive sanctions [8]. As confirmed by Campbell, Demark-Wahnefried, Symons, Kalsbeek, & et al. [9, 10], affirming individual feelings of religious superiority or disgrace can appeal attitudes towards wellbeing behaviours.

Grounded on discussion and dialogue on the Amman Declaration, it was established that there are at least sixty topics of health associated issues, rules, standard or guidance found in the two Muslim main sources (Quran and prophet teachings) which can be employed in producing health promotion messages. The Quran have at least 92 verses in that offer health messages which covered range of health issues [1]. There were also around 158 sunnah (prophet teachings) references found related to the health messages.

Generally, there are three approaches in promoting health from Islamic viewpoint: [11, 12]. Firstly, is the legal approach; it advocates particular rules of the religion that encourage health and prohibits activities that are destructive and risky to individual. Secondly is guiding method that claimed Islam to be a way of life. There is provision of particular protocols and guidelines which monitor individuals on their daily routines in Islam. The healing effect of the Quran is identified as the third approach. Recitation of certain Quran's verses is believed to provide a healing effect of human body system. Conversely, as emphasised by Al-Awwa [13], the Quran is not a medical book. It provides advices to safeguarding happiness and health to mankind through submitting the orders and steering away from the activities it's prohibits. These references offer potential coaching that can stimulus their actions and improve individual health conditions. Islam is equally a religion and a lifestyle that significantly affects Muslim's behaviour. Islam also provides direction not only in divine matters, but also in day-to-day life as well as well-being. The WHO has acknowledged the function of spiritual aspect in encouraging good wellbeing in accordance to Ottawa Charter [9].

3. Study design

This study employed the focus group technique, which is a procedure of group discussion and interaction among participants. Focus group discussions, usually supported with in-depth interviews, are the most prominent tool likely to gain insight into the experiences, beliefs, attitudes and feelings that underlie people's behaviour [15, 16]. Forty eight Muslim participated in this study which formed into eight groups. Every group is set to be six participants. They will be established into primary school children (male), primary school children (female), secondary school children (male), secondary school children (female), adult age 18 to 25 (male), adult age 18 to 25 (female), adult age 26 to 40 (male) and adult age 26 to 40 (female). All focus group sessions were transcribed and analysed using thematic analysis technique. Data were transcribed, coded and analysed using qualitative analysis software, NVivo version 8. The process contributes to identification of themes and trends that developed health communication attributes. Dovetailing with the study's aim, conceptual and relational content analysis techniques were performed on the data [15, 17]. Mentioned analyses offer description and probe the relationship between the concepts identified [18]. Unit analysis for this study could be a word, sentence, sequence of sentences or a complete dialogue related to feelings, attitudes and reactions towards Islamic health messages. A category and coding scheme were developed to classify the content.

4. Findings

Analysis of focus groups found there were 108 references have emerged to explain the characteristics. Thematic analysis of the conversations discovered seven key themes: (i) Quran, (iii) sunnah, (iv) dakwah, (v) ammar makruh nahy munkar, (vi) tawhid, (vii) Muslim's opinion leader, and (vii) media. Analysis also found that education background does not have a significant impact on perceptions of Islamic communication attributes. From the thematic analysis, as illustrated in Figure 1, the key themes of Islamic health promotion were noted by all groups irrespective of their background. This indicates that in general, Muslims regardless of their religiosity had similar perceptions of Islamic communication characteristics.

4.1 Tawhid (attributing oneness of God)

This theme was among the Islamic communication characteristics that were mentioned by all groups. Tawhid which is the core philosophy of Islamic theology plays a very crucial role in persuading Muslims. Participants for the focus groups claimed that this attribute signified an Islamic form of communication. It is because in Islam anything and everything is about their religion and submission to their God. Communication with tawhidic doctrine not only symbolised Islam, but also indicated the first pillar of Islam which is belief in Allah. Psychologically, it creates some persuasion elements that are able to motivate Muslims to act upon the received message. Tawhidic content in the message consciously or unconsciously influences the receiver as it manifests God [19].

4.2 Quran (Muslim holy book)

The analysis showed that the Quran and Muslim opinion leaders appeared to be the most mentioned Islamic communication characteristics. Each of the characteristics represents about twenty per cent of total nodes related to the Islamic communication. All groups felt that the Quran is something that cannot be separated from a message that claims to be Islamic. Participants felt that communication in Islam should contain verses from the Quran or its interpretation or its teachings. Indeed, some participants felt that the narrative in the Quran can motivate Muslims to change their behaviour. What is most important is that the Quran is considered to be Allah's words, and guides Muslims in how to live their lives. Although some might think using the Quran is a conservative approach, it is a Muslim tradition that has not changed for the past 1400 years [20].

DOI: 10.1051/shsconf/20173300082

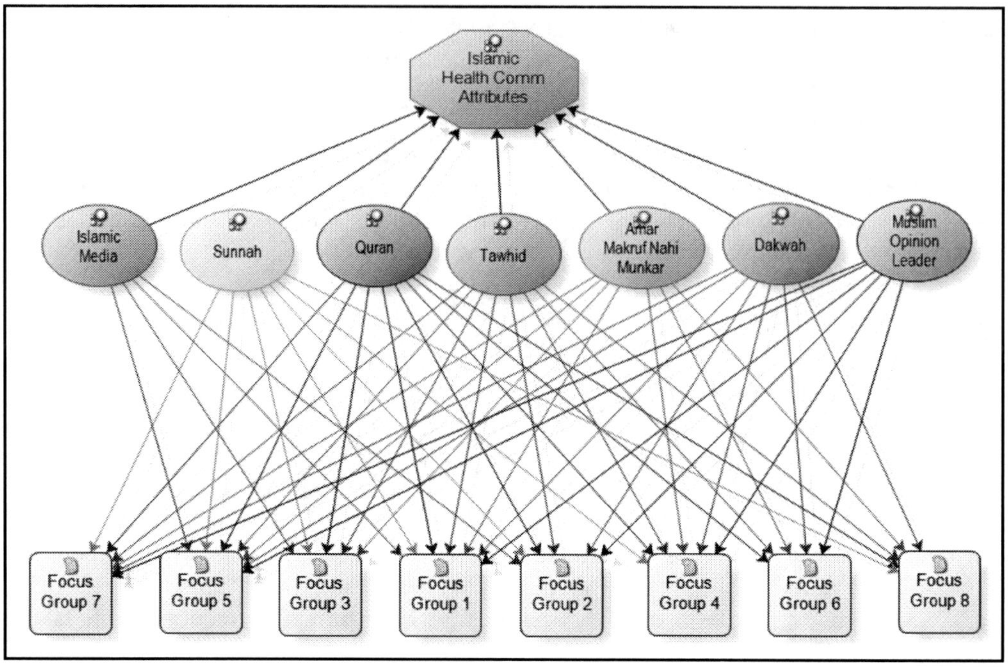

Fig. 1. Islamic Health Communication Attribute Themes

4.3 Sunnah (Prophet Muhammad's sayings and deeds)

As predicted according to previous findings, *sunnah* has been identified by this study as representing an important characteristic of Islamic communication. This third highest ranking attribute was discussed in all groups except group 5. In brief, *sunnah* was the second source of Islamic revelation and it refers to Prophet Muhammad's speeches or actions. Analysis of this attribute reveals that *sunnah* is a source of principles and teachings in Islam related to health issues. This is linked to participants' claims of using the Prophet Muhammad's quotations or actions about health in the campaign's messages. In fact some participants recommended that *sunnah* be used together with clinical or scientific evidence in order to have more impact. Participants of groups 3 and 4 even discussed hadith about the eating practices and food preferences of the Prophet Muhammad when explaining this characteristic.

4.4 Amar Makruf Nahy Munkar (commanding good and prohibiting evil)

This characteristic revealed in this focus group study supported previous study about what shapes Islamic communication [21]. *Amar makruf nahy munkar* (commanding to do good and prohibiting from doing wrong) refers to messages that guide Muslims. It also has been interpreted as the way of communicating ethical doctrine [22]. Based on the analysis, participants have suggested Islamic elements such as *hukm* (Islamic commandments), Islamic principles, treatment method, and preventive messages, to be used when communicating about health.

4.5 Dakwah (Islamic preaching)

DOI: 10.1051/shsconf/20173300082

According to respondents, *dakwah* is linked to the principle of Islam as a way of life. Thus, elements of communication in Islam transmit its religious mission through sacred information and teachings. *Dakwah* can be explained as Islamic teaching, enhanced faith, teaching to be a real Muslim, and the improvement process of a Muslim. The values of *dakwah* are not only to convert others to Islam but also an effort to strengthen Muslims' belief in Islam in all aspects of life.

4.6 Muslim Opinion Leaders

Another most described characteristic is Muslim opinion leader. Participants have classified *ustaz/ustazah* (religious teacher), *imam* (religious leader), and *ulama* (religious scholar) as people who influence the opinions of Muslims. Participants recognised these leaders as being credible and capable of influencing the community's opinion on issues related to Islam. Analysis of participant's conversation explained that these opinion leaders can influence Muslims through their popularity, being role models, their status in the community, their knowledge, and expertise. Furthermore, the Quran tells Muslims to follow these leaders as their authority is only exceeded by the prophet and his companions. The attributes of Muslim opinion leaders in health communication programs include their opinion, advice, decisions, remarks, and personal appearance or popularity as well.

4.7 Media

Media is another characteristic that emerged in the findings. This characteristic was perceived mainly by groups from the Islamic education background. Analysis of the coded conversation related to this character showed participants felt the Islamic health communication should be channelled through what they called 'Islamic media'. From the data gathered, it can be interpreted there are many media channels for health promotion. Among the media mentioned were Friday sermons, religious talk, masjid (*mosque*), and Islamic broadcasting channels. Their argument for the need to communicate through this media was based on credibility and effectiveness. While quite numbers of participants favoured traditional media, the new media has been identified as very potential to reach Muslim community.

5. Conclusions

Generally, this study has recognised the attributes of culturally based health communication (Islamic health communication) of the Muslim community. These attributes emerged across the focus group that representing young and adult Muslims community. This finding is supporting the call for innovation in social change such as in health promotion [23]. Therefore, it is recommended that further study be undertaken of this scenario by investigating if and how Islamic understanding, knowledge and skills affect the implementer's faith-based communication performance. This scope can be expanded to doctors, nurses, and other health professionals, as well as health activists. The study of Islam as culturally sensitive in health promotion is intriguing but as yet has received little research coverage. As this study explains, Islamic communication has potential and is found to positively impact Muslim practices such as health behaviour. This has particular relevance as many Muslim communities are now facing challenges in their health related issues. In this respect, there are still many research topics that can be approached through the Islamic communication framework developed here to mediate social change such as health issues in a Muslim community. The insights of this paper may contribute to the

DOI: 10.1051/shsconf/20173300082

further development of health promotion strategies for Muslims in Islamic nations as well as in non-Islamic nations.

References

1. WHO, *Health Promotion Through Islamic Lifestyles: The Amman Declaration* (Alexandria: World Health Orgaisation, 1996)
2. M.K. Ahmad, & J. Harrison, *Cultural sensitivity in health promotion programs: Islamic persuasive communication.* In Y. Pasadeos (Ed.), Advances in communication and mass media research, 167-180 (Athens: ATINER, 2010)
3. T.L. Thompson, R. Parrott, & J.F. Nussbaum (eds.), *The Routledge handbook of health communication* (Oxon: Taylor & Francis, 2011)
4. K.B. Wright, L.S. Sparks, & H.D. O'Hair. *Health communication in the 21st century* (West Sussex: John Wiley & Sons, 2013)
5. C. Airhihenbuwa, *Health and culture: Beyond the western paradigm* (Thousand Oak: Sage Publications, 1995)
6. M.T. Dutta-Bergman, Communication Theory, **14(3)**, 237-263 (2004)
7. I. Ajzen. *Persuasive communication theory in social psychology: A historical perspective.* Influencing human behaviour , 1-27 (1992)
8. K. Glanz, B.K. Rimer, & F.M. Lewis, (eds.), *Health Behavior and Health Education: Theory, Research and Practice* (San Francisco, CA: Jossey-Bass, 2002)
9. M.K. Campbell, W. Demark-Wahnefried, M. Symons, W.D. Kalsbeek, & et al., *Fruit and vegetable consumption and prevention of cancer: the Black Churches United for Better Health project.* Am. J. Public Health. **89**(9), 1390 (1999)
10. M.H. Al-Khayat, *Health: An Islamic Perspective* (Egypt: World Health Organization, 1997)
11. S. Sipon, M.F. Sakdan, C.S. Mustaffa, N.A. Marzuki, M.S. Khalid, M.T. Ariffin, N.N.N. Nazli, & S. Abdullah, *Spirituality among Flood Victims: A Comparison between Two States.* Procedia-Social and Behavioral Sciences **185,** 357-360 (2015)
12. A. El-Kadi, *What Is Islamic Medicine.* (1996)
13. M. Al-Awwa, *Health in Islamic Literature.* (2002)
14. E. De Leeuw, & A.A. Hussein, *Islamic health promotion and interculturalization.* Health Promot. Int., **14(4)**, 347-353 (1999)
15. F. Rabiee, *Focus-group interview and data analysis.* Proceedings of the Nutrition Society, **63,** 655-660 (2004)
16. D.W. Stewart. & P.N. Shamdasani, *Focus groups: Theory and practice* (Newbury Park: Sage Publications, 1990)
17. Krippendorff, K, *Content analysis: An introduction to its methodology (2nd. ed.)* (Thousand Oaks: Sage, 2004)
18. M.E. Palmquist, K.M. Carley, & T.A. Dale, *Two applications of automated text analysis: Analyzing literary and non-literary texts.* In C. Roberts (Ed.), *Text Analysis for the Social Sciences: Methods for Drawing Statistical Inferences from Texts and Tanscripts* (New Jersey: Lawrence Erlbaum Associates, 1997)
19. Y. Mohamed, The American Journal of Islamic Sciences, **12(1)**, 1-18 (1995)
20. J. Esposito, *What everyone needs to know about Islam* (Oxford: Oxford University Press, 2002)
21. M.K. Ahmad, J. Harrison, & C.L. Davies, *Sacred communication: Exploring the attributes of health promotion programs in the faith-based media'* in M. Buyong, et. al. (eds.). *Potrayal of Islam in the media* (Nilai: University Sains Islam Malaysia, 2010)

22. H. Mowlana, *Theoretical perspectives on Islam and communication.* China Media Research, **3**(4), 23-33 (2007)
23. C. Mason, J. Barraket, S. Friel, K. O'Rourke & C. Stenta, *Social innovation for the promotion of health equity.* Health Promot. Int., **30**(2), 116–125 (2015)

Revisiting Organizational Credibility and Organizational Reputation – A Situational Crisis Communication Approach

Jamilah Jamal[1,*], *and Hassan* Abu Bakar[2]

[1,2]School of Multimedia Technology and Communication, College of Art and Science, Universiti Utara Malaysia.

Abstract. Organizational credibility, the extent of which an organization as the source of messages is perceived as trustworthy and reliable, is one important aspect to determine organization's survival. The perceived credibility of the messages will either strengthen or worsen an organization reputation. The primary objective of this paper is to revisit the concept of organizational credibility and its interaction with organizational outcomes such as organizational reputation. Based on the situational crisis communication theory (SCCT), this paper focuses on the impact of organizational credibility on organizational reputation following a crisis. Even though the SCCT has been widely used in crisis communication research, the theory still has its own limitations in explaining factors that could potentially affect the reputation of an organization. This study proposes a model by integrating organizational credibility in the SCCT theoretical framework. Derived from the theoretical framework, three propositions are advanced to determine the relationships between organizational credibility with crisis responsibility and perceived organizational reputation. This paper contributes to further establishing the SCCT and posits key attributes in the organizational reputation processes..

1. Introduction

How credible is a government agency in communicating crisis messages? Do stakeholders always belief the crisis information coming from public organizations? Recent research on organizational credibility showed that public trust toward public organizations has been consistently decreasing [1]. As such, the role of organizational credibility in re-building public organization reputation following a crisis is noticeable yet empirical evidence on its effect is still absent. Even though researchers have continuously attempt to fill the gap in the study of credibility, there is still a lack of adequate research on the impact it has on organizational reputation [1-2]. A huge body of research suggests that it's highly relevant to address this issue due to the fact that the impact of organization credibility has not yet fully understood [2]. Therefore, this study extends the research on the effect of credibility on organization reputation and its interaction with crisis responsibility in a public organization setting. The primary aim of this study is to assess the impact of crisis on Malaysia's public organizational reputation within the framework of situational crisis communication theory (SCCT) [3-5].

* Corresponding author: jamilah@uum.edu.my

The importance and significance of this study lie in the fact that Malaysian public organizational reputation has been questioned increasingly in the last few years. Previous research indicated that incidences involving regulatory bodies have resulted in declining public trust in government, a sign of unfavorable reputation [1]. More importantly, the decline in trust has implications for the public organizations in regard to the legitimacy of democratic governance [1,6]. As a politically appointed organization, public support and trust are utmost important to ensure the survival of the government in Malaysia. Thus, an empirical study on public organization reputation and indentifying its antecedents is highly relevant and a timely initiative to restore public trust and gain their support by applying a specific mechanism to reduce reputational threats in public organization and create positive outcomes.

This study provides a starting point in advancing the literature on organizational credibility and fills an important void in the crisis communication and public organization reputation literature. This study is also aimed at enriching the situational crisis communication theory and clarifies the roles of organizational credibility while in crisis. By incorporating corporate credibility into the model, it is hoped that this study strengthens the existing model to cater for public sector and offers wider perspective in looking into reputation management during crisis.

2. Review of literature

2.1 Organizational crisis and crisis responsibility

A review on scholars' definition of organizational crisis shows some commonalities, including the fact that a crisis: 1. Crisis is an unplanned event that has potentiality to dismantling the entire structure of an organization; 2. Can affect its internal and external stakeholders; 3. May occur in any organization across all industry nationwide. 4. May affect the survival of an organization [7]. For the purpose of understanding organizational crisis, this study adopts [3,4,8] who define crisis as "the perception of an unpredictable event that threatens important expectancies of stakeholders and can seriously impact an organization performance and generate negative outcomes" (p. 238). In the present study, organizational crisis is operationalized as a situational context in which crisis responsibility is evaluated by participants. Specifically, the relationship between crisis responsibility, organizational reputation, and corporate credibility are perceived in an organizational setting with organization having an ongoing crisis.

Crisis responsibility refers to the degree to which stakeholders attribute responsibility for a crisis to an organization [3-5]. The attribution of crisis responsibility is made by the stakeholders regardless of whether the organization admits or denies the responsibility of the crisis. Thus, crisis responsibility in this study is referred as the attribution of crisis responsibility perceived by stakeholders to organization having a crisis. When the employees perceived the organization actions had triggered the crisis, they will likely attribute the responsibility of the cause to the organization and vice-versa. In this study, participants will make attribution of crisis responsibility based on the crisis type that was chosen for the study.

2.2 Organizational Reputation

In viewing the uniqueness of public organizational reputation, scholars in the political science approach agree with a definition on organizational reputation which is "a set of symbolic beliefs about the unique or separable capacities, roles, and obligations of an organization, where these beliefs are embedded in audience networks" [9]. This definition has been refined slightly to cater for wider scope of public organization and its growing functions as follows: 'Organizational reputation is defined as a set of beliefs about an organization's capacities, intentions, history, and mission that are embedded in a network of multiple audiences' [10]. Reputation uniqueness in this context refers to the demonstration by public organizations that they can create solutions and provide services.

Public organizations rely on strong reputation to achieve delegated autonomy and discretion from politicians. As such, continuous study in this area is crucial to assist public organization to build a strong, favorable reputation [9]. In addition, reputation in public sector has been recognized as "valuable political assets" [6] used to generate public confidence to continuously supporting the government.

Like most public organizations around the globe, public organizations in Malaysia suffer from negative reputation associated with inefficiency, low level of professionalism, financial uncertainties and political unrest [11]. While there is an increasing awareness on the importance of having good reputation among the public organizations in Malaysia literature in this area remain scarce [11]. Due to the nature of public organization is non-profit and service-oriented, thus the role of reputation is often overlooked or given less consideration, both from the public organization key drivers and researchers in this area. Likewise, crisis is often seen as less threaten to public organization than private organization, leaving question with regard to how government organizations in Malaysia manage their reputation following a crisis, remain unanswered. Considering Malaysia, as one of the Asian countries which embraces a different culture from the West, thus the findings from previous study that are predominantly western may not be generalized to the present context [12]. Thus, more theoretical approach is needed on crisis and reputation management research to provide insights for professionals in public sectors in Malaysia.

2.3 Organizational Credibility

The concept of organizational credibility and organizational reputation are closely inter-related to the extent that some scholars argued that credibility derived from reputation [13,14,15]. However, [5] sees credibility as a different construct from reputation, which distinguishes the way the two variables are measured. Organizational reputation is how the public perceives the overall performance of the organization, which will be affected if its credibility is not as strong as its reputation. This study embraces the definition of corporate credibility as 'the organization or corporation that manufactures the product or provides services is seen as a credible source of the communication' [16] (p. 236). In the public sector, organizational credibility is associated with a concept of trust which has usually been viewed from an external environmental or an interorganizational social perspective [17]. In essence, corporate credibility in this study is the perceived expertise and trustworthiness of an organization in times of crisis. Expertise is the organization's knowledge about the subject [18]. An expert organization will appear to be competent, capable and effective and knows the ins and outs of the issue. Trustworthiness demonstrates the honesty and believability of the source and is the organization's goodwill toward or concern for its stakeholders [18]. In this study organization credibility is assumed to have influences on organizational reputation in a crisis situation. In a situation where an organization is seen as not delivering its promises, or conveying messages lacking in believability, it will be perceived as not credible, resulting in an unfavorable reputation.

3. Situational crisis communication theory (SCCT)

The SCCT focuses on the use of communication to preserve and protect an organization's valuable reputation [3-5, 8]. It posits that, as the reputational threat increases, crisis managers should use response strategies that demonstrate a higher level of acceptance of responsibility for the crisis and address the organization's concerns for the victims involved. The SCCT categorized types of crisis into three crisis clusters namely the victim cluster, the accidental cluster and the preventable cluster. Each cluster explains different level of crisis responsibility attributed to the organization. The victim cluster is linked with weak attribution, while the accidental cluster is associated with a reasonable degree of responsibility and the preventable cluster is regarded as high level of attribution to the organization. Thus, crisis responsibility is directly related to the organization reputation; the higher the level of crisis responsibility held by the organization the more severe the impact to its reputation.

Fig. 1. Conceptual model of the study derived from the SCCT [3]

4. Proposition Development

4.1 Crisis responsibility and perceived organization reputation

Crisis scholars have suggested that a direct link exists between an organization's acceptance of crisis responsibility and its reputation [19]. An organization's actions either to assume or to reject crisis responsibility are crucial in rebuilding a reputation damaged by a crisis threat [3-5]. In the event the organization is perceived as responsible for a crisis, the acceptance of crisis responsibility will likely result in a positive reputation. In line with this, it is suggested that more evidence-based research is needed to validate an organization's decision in either accepting or denying responsibility assigned by stakeholders [5, 8]. However, organization needs to identify the cause of the crisis by framing the crisis type and make attribution of crisis responsibility before any decision to employ crisis response strategies can be made. Therefore, the following proposition is advanced:

Proposition 1: Attribution of crisis responsibility is related to public organization reputation

4.2 Crisis responsibility and corporate credibility

Organization's credibility is threatened during a crisis. However, strong organization credibility can protect organization from a severe impact [2-5, 18]. Research on credibility also indicates that communications (crisis messages) affects how stakeholders perceive the organization in crisis. Stakeholders believe organization's side of story when they trust them as credible. Stealing thunder, a quick response by organization increases it credibility. Organizations who are held responsibility are viewed as more credible when they reported the crisis before other sources; that is by stealing thunder [20]. As such the attribution of organization's crisis responsibility is weak when the corporate credibility is strong. In addition, previous findings suggest that the perception of risk (crisis) is related to the level of trust individuals have in organization [1,18,21]. Thus we propose the following proposition:

Proposition 2: Attribution of crisis responsibility is related to public organization's credibility.

4.3 Organizational credibility and perceived organization reputation

Previous research has shown that perceptions of high source credibility influence positive message evaluations, favorable attitude changes and behavioral attentions [21-23]. Likewise, credible sources will be perceived more persuasive in their communication strategies than those with low credibility. In addition, the moderating roles of corporate image and credibility in influencing the effectiveness of advertising and public relations message sequencing in new product introductions found that corporate credibility is shown to attenuate the effectiveness of the PR strategy [24]. Their findings suggested that when a corporation has a more positive image and high credibility, the message sequencing strategy used may be less important. However, the message sequencing strategy will play an important role for a new corporation or one with less than astrophysical image and credibility. The findings spur further research to examine the role of corporate credibility in the organization setting, specifically by looking at the potential it has in influencing public organization reputation. Therefore, the following proposition is advanced:

Proposition 3: Perceived oganizational credibility is related to perceived organization reputation.

5. Conclusions

Considering the magnitude of reputation for public organization, addressing the reputational threats in public organization is highly necessary (25). By focusing on organization's corporate credibility, this study contributes to a growing literature on crisis management in Malaysian public sector. While there are many predictors have been associated with organizational crisis, understanding its relationship with organization credibility is crucial because credibility does not reside in the organization, rather in the mind of its stakeholders and public. Their perception toward organizational curability reflects the organizational expertise, trustworthiness and essentially its reputation.

This paper presents a review of the literature of the key concepts up to their latest development to depict how each concept is relevant to each variable. Then it develops a research framework which is derived from the major themes in the literature review. Three propositions are advanced from the research framework. Due to its limitation, further study may address organizational credibility with other organizational elements such as message sequencing, public relations strategies, organizational performance or even organizational culture.

References

1. D.J. Houston, L.H. Harding, *Public trust in government administrators: Explaining citizen perceptions of trustworthiness and competence.* Public Integrity, **16**(1), 53-75 (2013)
2. B. u. Wissmath, D. Weibel, T. P Reber, *The Impact of Communication Modality on Perceived Credibility of Panel Members.* Electronic News, **4**(1), 23-38. (2010)
3. W.T. Coombs, *Impact of past crises on current crisis communication insights from Situational Crisis Communication Theory.* Journal of Business Communication, **41**(3), 265-289 (2004)
4. W. T. Coombs, *Protecting organization reputations during a crisis: The development and application of situational crisis communication theory.* Corporate Reputation Review, **10**(3), 1-14 (2007)
5. W. T. Coombs, *Ongoing crisis communication* (2014)
6. A. Wæraas, H. Byrkjeflot, *Public sector organizations and reputation management: Five problems.* International Public Management Journal, **15**(2), 186-206 (2012)
7. G. King, J. Bus. Ethics, **41**, 235-249 (2002)
8. W.T. Coombs, S. J. Holladay, *PR strategy and application: Managing influence.* (Wiley-Blackwell, 2010)
9. D. P. Carpenter, *Reputation and Power: Organizational Image and Pharmaceutical Regulation at the FDA.* (Princeton University Press, 2010)
10. D. P. Carpenter, G. A. Krause, *Reputation and public administration.* Public Administration Review, **72**(1), 26-32 (2012)
11. Nik Hazimah, Zaharul Nizal, *International Journal of Business and Management,* **5**, 138-145 (2010)
12. H. A. Bakar, Guowei Jian, G. Fairhurst, *Where do I stand? The interaction of leader–member exchange and performance ratings.* Asian Business & Management, **13**(2), 143–170 (2014)
13. M. E. Goldberg, J. Hartwick, *The effects of advertiser reputation and extremity of advertising claim on advertising effectiveness.* J. Consum. Res., **17**, 172-179 (1990)
14. C. J. Fombrun, C. B. M. Van Riel, *Fame & fortune: How successful companies build winning reputation.(* FT Press, 2004)
15. F. Hansen & L.B. Christensen, *Dimensions in consumer evaluation of corporate brands and the role of emotional response strength (NERS).* Innovative Marketing, **3**(3), 19 (2007)
16. S. J. Newell, R. E. Goldsmith, *The development of a scale to measure perceived corporate credibility.* J. Bus. Res., **52**(3), 235-247 (2001)
17. S. M. Park, *Toward the trusted public organization untangling the leadership, motivation, and trust relationship in US federal agencies.* The American Review of Public Administration, **42**(5), 562–590 (2012)
18. J. M. Kouzes, B. Z. Posner, *Credibility: How leaders gain and lose it, why people demand it.* (John Wiley & Sons, 2012)
19. K.M Pace, T. A. Fediuk, & I.C Botero, *The acceptance of responsibility and expressions of regret in organizational apologies after a transgression.* Corporate Communications: An International Journal, **15**(4), 410-427 (2010)
20. I. M. Arpan, D. R. Roskon-Ewoldsen, *Stealing thunder: Analysis of the effects of proactive disclosure of crisis information.* Public Relations Review, **31**(3), 425-433 (2005)
21. D. Kazoleas, J.J. Teven, *Human Comm. A Publication of the PACA,* **12,** (2009)
22. B. Z. Erdogan, *Celebrity endorsement: A literature review.* Journal of Marketing Management, **15**(4), 291-314 (1999)
23. H. Kim, S. M. Choi, Paper presented at the ICA (2004).
24. B. Ivanov, K. A. Parker, J. D.Sims, Paper presented at the NCA (2009)

DOI: 10.1051/shsconf/20173300083

25. J. Jamal, H. A. Bakar, *The mediating role of charismatic leadership communication in a crisis: A Malaysia Example.* Journal of Business Communication (2015)

User-Generated Contents in Facebook, Functional and Hedonic Brand Image and Purchase Intention

Raji Ridwan Adetunji[1,*], *Mohd Rashid* Sabrina[1], and *Ishak Mohd* Sobhi[1]

[1]Department of Media Management, School of Multimedia Technology and Communication, Universiti Utara Malaysia, Malaysia.

Abstract. User-Generated Contents is a type of brand-related communications on social media platforms. User-Generated Contents (UGC) offers consumers the opportunity of sharing their experiences, contribute their opinion and communicate with other. This implies that brand managers no longer have solitary control in managing the brand images of their brands. Therefore, this study set out to examine the effect of user-generated contents on hedonic brand image and functional brand. Also, the study determines how hedonic and functional brand image influence brand purchase intention. 114 followers of fan pages of Malaysian automotive brands were surveyed. The hypotheses formulated were tested using PLS-SEM. The findings revealed that, UGC have significant and positive impact on both functional and hedonic brand image. Furthermore, functional and hedonic brand image have significant impact on purchase intention. This study provides insight on the importance of UGC, functional and hedonic brand image to brand managers.

1. INTRODUCTION

Social media including Facebook has become an important outlet for building and maintaining successful brand image [1]. Through fan pages or brand profiles that are created on social media, brand managers are offered limitless opportunities to share and post information in form of photos, videos, messages and comments about their brands and companies [2]. Meanwhile, the advent of social media as a channel for disseminating marketing communication messages has given birth to a new type of brand-related communication namely; UGCs which are posted on social media by consumers. Thus, User-Generated Contents on social media platforms allow consumers to share their experiences, contribute an opinion and communicate with other teeming numbers of consumers. This implies that brand managers no longer have solitary control of brand images of their brands [2]. However, this development has generated series of academic concerns which include measuring and understanding how the UGC opportunity alters firm's control of brand equity management [4]. Extant literatures have attempted to

* Corresponding author: rajiridwanadetunji@gmail.com

examine the importance of social media on brand equity. Meanwhile few studies have examined the impact of UGC on functional and hedonic brand image. Furthermore, there is a scarce of empirical justifications on the impact of brand image on brand purchase intention [1].

2. Literature review

2.1 User-generated contents

UGCs are the types of social media contents that are created about a brand by consumers for other consumers' consumption [2]. This is possible because social media offers consumers the opportunity to publish and share self-generated contents between multitude of friends and followers. The possibility of generating and sharing contents about brands, affirms the active and influential role of consumers in managing brands on social media [5]. [6], asserted that before any content can be considered as UGC, it must bear one or all the following three features. Firstly, it must be accessible by the public through any of the social networking sites (such as Facebook, Twitter, Blog and YouTube). Secondly, the content must be created and published with certain amount of creativity, using animations, images, graphics, fictional or fictitious words. Finally, it must be created solely by individuals that are not working for any professional or corporate bodies.

2.2 Functional and Hedonic Brand Image

Bruhn and Schoenmueller [1] reviewed that; brand image signifies consumers' associations with a brand. These associations can be divided into two types namely; functional and hedonic brand image. Hedonic brand image can be explained as the perceptions, feelings, or meaning that consumers created from abstract and imagery-related considerations that are not necessarily related to the brand attributes, performance or utilitarian [7,1]. On the other hand, functional brand image illustrate that consumers' associations with brands serves as informational nodes that are connected to brand nodes in the consumers' mindset and these formulate meaningful information about the quality and value of the brand [8].

2.3 Brand purchase intention

Brand purchase intention is a major concern and the reason why brand managers exert aggressive measures to develop their brand assets such brand equity [9]. Building an attractive brand equity and brand image is absolutely necessary as a strategy for maintaining consumers' satisfaction, acceptance and above all, purchase intentions in a highly competitive market [5]. In the study conducted by [10], purchase intention is explained and measured by consumers' behavioral intentions and willing-to-buy which are consequences of consumers' overall evaluation of a brands and information derived on social media platforms.

2.4 Hypotheses development

2.4.1 The impact of UGC on functional and hedonic brand image

 DOI: 10.1051/shsconf/20173300084

Sequel to the increasing trend of adopting social media for sorting brand-related information, social media platform is increasingly becomes a platform for building brand image. This is because what consumers read on social media affect their judgments and perceptions about their association and nonfunctional attributes of brands. For instance, [11, 12,13] have found significant impact of UGC on either brand image or brand association. Subsequently, the following hypotheses are proposed;

H1: User-generated contents have positive impact on functional brand image
H2: User-generated contents have positive impact on hedonic brand image

2.4.2 The relationship between functional, hedonic brand image and brand purchase intention

Previous studies have implied that, consumers' perceptions of brand image leads to intention to purchase, purchase and permanent purchase of the same brand. Hence, consumers' perceptions that are generated through their judgments of both hedonic and functional attributes of a brand influence their attitudes and behaviors. This notion has been subjected to empirical tests in different context by previous researchers [11,14,15,16,17]. The implication of the findings from these previous studies is that, consumers; mindset and perception of brands enhance the process of making purchase decisions. In view of this, the following hypotheses are formulated;

H3: Functional brand image positively influence purchase intention
H4: Hedonic brand image positively influence purchase intention

2.5 Conceptual Framework

Fig 1 presents the conceptual framework that is proposed in this study. The framework depicts the relationships between UGC, functional and hedonic brand image and purchase intention.

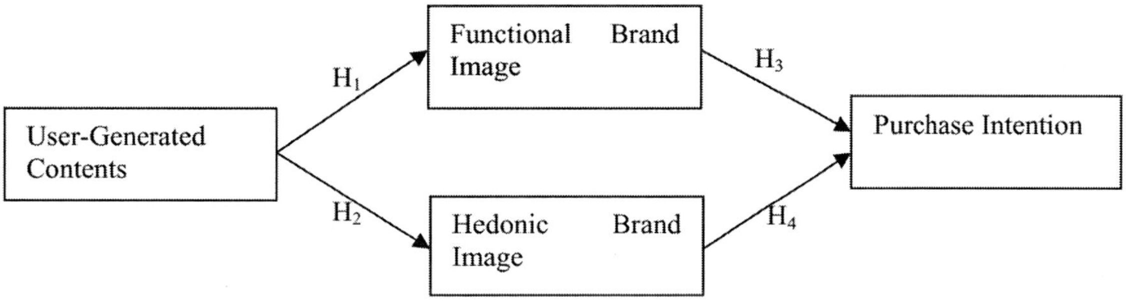

Fig. 1. Conceptual Framework

3. Method and findings

This study adopts a quantitative research method, using a survey research design. Data was gathered from an online survey conducted among followers of automotive brands (both PROTON and PERODUA) on Facebook. The link of online questionnaire was sent in form of personal messages to 300 followers of fan pages of the Malaysian automotive brands. The respondents are those that are actively involve on these fan pages in form of posting contents, commenting and sharing of contents. Out of the 300 that received the questionnaires, 117 answered to the questionnaire. 3 of the received questionnaires were invalid and the analysis is based on 114 respondents. The measurements for measuring the variables understudied in this research were adapted from previous studies. Items for measuring UGC were adapted from [1]. Items for measuring both functional brand image and hedonic brand image were adapted from [1] and [11]. Finally, items were adapted from [12] and [8] for measuring brand purchase intention. A five Likert type scales ranging from 1 "Strongly Disagree" to 5 "Strongly Agree". Certain preliminary statistical assumptions (such as; normality, outliers and linearity) were met before developing the measurement and structural models using PLS-SEM.

3.1 Measurement Model

Table 1 presents the result of the reliability and validity of the data. The findings demonstrate that composite reliability (CR) values of the construct have internal consistency [13]. Also, the values of average variance extracted (AVE) of the entire constructs are above 0.5 which is the minimum threshold value. This explains that, the constructs achieved an acceptable convergent validity [14]. Discriminant validity is achieved as presented in Table 2. The result shows the squared correlations for each construct as lesser than the average variance extracted by the indicators measuring that construct

Table 1. Internal Consistency and Convergent Validity

Construct	Items	Loading	AVE	CR
User-Generated Contents	UGC1	0.9056	0.691	0.8986
	UGC2	0.8934		
	UGC3	0.7996		
	UGC4	0.7115		
Functional Brand Image	FBI1	0.8291	0.7079	0.9061
	FBI2	0.8775		
	FBI3	0.9017		
	FBI4	0.7491		
Hedonic Brand Image	HBI1	0.7705	0.6268	0.8342
	HBI2	0.7734		
	HBI3	0.8297		
Brand Purchase Intention	BPI1	0.9337	0.8975	0.9633
	BPI2	0.9627		
	BPI3	0.9454		

Table 2. Discriminant Validity

	BPI	**FBI**	**HBI**	**UGC**
BPI	***0.947***			
FBI	0.8603	***0.841***		
HBI	0.7695	0.6741	***0.791***	
UGC	0.3203	0.2958	0.392	***0.831***

Notes: The square root of the AVE are marked in italics.

DOI: 10.1051/shsconf/20173300084

3.2 Structural Model

The proposed hypotheses were tested through the assessment of the structural model. Table 3 presents the result of the path coefficient and the decision for every hypothesis. The findings revealed that, all the hypotheses formulated in this study were significant (UGC → FBI, ß = 0.2958, $p < 0.05$; UGC → HBI, ß = 0.392, $p < 0.05$; FBI → BPI, ß = 0.6261, $p < 0.05$; HBI → BPI, ß = 0.3475, $p < 0.05$). In addition, the findings presented in Table 4 shows the values co-efficient of determination (R2), cross-validated redundancy (Q2) and effect size (f2) of exogenous constructs on endogenous constructs. The values of co-efficient of determination shows that both functional brand image and hedonic brand image explain 80.6% variances in brand purchase intention. User-generated contents explain 08.7% of variances in functional brand image and 15.3% of variances in hedonic brand image. The Q2 values reported in Table 4 for brand purchase intention, 0.7157; functional brand image, 0.0616; and hedonic brand image, 0.0932 are all greater than 0, which shows that the exogenous variables have predictive relevance over the endogenous variables [20]. Finally, f2 values explain the importance of each exogenous variables beyond the significance analyzed in Table 3. Table 4 shows that the effect size of functional brand image is large while the effect of hedonic brand image is medium on brand purchase intention.

Table 3. Path Co-efficient Assessment

Hypotheses	Direct Effect (β)	Std. Error	T-Statistics	Decision
UGC -> FBI	0.2958	0.1243	2.3787	Supported
UGC -> HBI	0.392	0.1176	3.3346	Supported
FBI -> BPI	0.6261	0.0544	11.4984	Supported
HBI -> BPI	0.3475	0.0492	7.0649	Supported

Table 4. Determination of co-efficient (R^2), predictive relevance (Q^2) and effect size (f^2)

	R^2	Q^2	f^2	Effect Size
BPI	0.806	0.7157		
FBI	0.0875	0.0616	1.0928	Large
HBI	0.1536	0.0932	0.3351	Medium

4. Discussions and conclusions

The findings of this study demonstrate the significance of user-generated contents on functional brand image and hedonic brand image. Also, the findings also reveal that, both functional and hedonic brand image have significant influence on brand purchase intention. The findings of this study provide an instructive insight for brand managers on the implications of social media communications and especially user-generated contents. This study provide an empirical justification to the notion that, the pervasiveness of social media as a strategic channel for marketing communication channel has enhanced consumers' connection and engagement with brands. Furthermore, consumers' anecdotal comments, experiences and thoughts of a brand have significant influence on other consumers attitude and perception of both functional and hedonic brand image [7] as well as purchase decisions. The finding of this study is consistent with the findings of previous researchers on the impact of social media users' brand-related contents on brand image and brand purchase intention [12]. The study offers a number theoretical contribution which include; the impact of UGC is higher on hedonic brand image than functional brand image. Meanwhile, the findings of this study indicate that, functional brand image influence purchase intention. More so, understanding the impact of UGC on brand equity has a practical importance, considering the level in which consumers are adopting social media platforms to search for brand related information [1].

This study however, has certain limitations that can guide future studies. In view of the nature and features of social media, being a platform that anchor all sorts of communication with numerous contents characteristics (promotion, advertisement and product information) and media type (text, video and photo), researchers should consider determining the differential impacts of these characteristics and media types on brand image. Furthermore, researchers are recommended to determining the impact of UGC and other types of social media contents on other dimensions of brand equity such as brand awareness, brand loyalty and brand preference.

References

1. M. Bruhn, V. Schoenmueller, and D. B. Schafer, *Are social media replacing traditional media in terms of brand equity creation?*. Manag. Res. Rev., **35**, 9 (2012).
2. G. Christodoulides, C. Jevons, and J. Bonhomme. J. Advert. Res., **52**, 1 (2012).
3. OECD, *The Evolution of News and the Internet*, (2010).
4. P. M. Homer, *Perceived quality and image: When all is not 'rosy'*, J. Bus. Res., **61**, 7 (2008).
5. J. Hanaysha and H. Hilman. J. Asian Bus. Strateg., **5**, 7 (2015).
6. K. Goodrich, *Anarchy of effects? Exploring attention to online advertising and multiple outcomes*. Psychol. Mark., **28**, 4 (2011).
7. S. Gensler, F. Völckner, Y. Liu-Thompkins, and C. Wiertz. J. Interact. Mark., **27**, 4 (2013).
8. Y. Zhang. Open J. of Buss. and Manag. **3**, 1 (2015).
9. A. J., Kim, and E. Ko. J. of Busi. Rese., **65**, 10 (2012).
10. S. M., Broniarczyk, and A. D. Gershoff. J. of Mark. Rese., **40**, 2 (2003).
11. B.,Yoo, and N. Donthu. J. of Busi Rese., **52**, 1 (2001).
12. D. Gefen, D. Straub, M. Boudreau, *Structural equation modeling techniques and regression: guidelines for research practice*. Commun. Assoc Inform. Syst., **7**, (2000).
13. J. F., Hair G. T.M., Hult, C. M., Ringle, M. Sarstedt, *A primer on partial least squares structural equation modeling (PLS–SEM)*. Sage Publications,Thousand Oaks. (2014).
14. C. Fornell, J. Cha, *Partial Least Squares. In: Bagozzi RP (ed) Advanced methods of marketing research*. Blackwell, Cambridge, (1994).

Cognitive Engagement and Online Political Participation on Social Media among Youths in Malaysia: The Moderating role of Political Knowledge

Aishat Adebisi Abdulrauf[1,*] , *Norsiah* binti Abdul Hamid[2] *and Mohd Sobhi* bin Ishak[2]

[1]School of Multimedia Technology and Communication, Universiti Utara Malaysia; University of Ilorin, Nigeria.

[2]School of Multimedia Technology and Communication, Universiti Utara Malaysia, Malaysia.

Abstract. The role of political knowledge in engendering political participation among youths in a society cannot be overstated. Yet, there is scarcity of research in this area. Thus, this study assessed the moderating role of political knowledge in the relationship between Access to Political Information on Facebook and Twitter (APIFT), Political Interest (PI) and Policy Satisfaction (PS) with Online Political Participation on Facebook and Twitter (OPPFT). A cross-sectional survey was conducted on 238 undergraduate students of Universiti Utara Malaysia. Questionnaire was used to collect data which was analysed using Partial Least Squares-MultiGroup Analysis (PLS-MGA). Results revealed that political knowledge moderated the relationship between access to political information on Facebook and Twitter and political interest with youth online political participation on Facebook and Twitter. However, it did not moderate the relationship between policy satisfaction with youth online political participation on Facebook and Twitter. Generally, these findings suggests that political knowledge plays a significant role in youth online political participation in Malaysia. Consequently, this study has added empirical evidence to predictors of political participation while at the same time contributing to the body of knowledge on online political participation.

1 Introduction

The use of social media for political participation have grown of recent [1, 2]. Thus, they have provided new avenues for political organization such that citizens can participate in politics [3]. As such political use of social media is one of the fastest growing examples of online political participation [4]. In response to this growth is a steady rise in researches in this area [5] with particular reference to the youths.

Consequently, studies have shown that social media have provided youths with new ways of political participation and democratic citizenship [6]. Specifically, studies conducted in the US [7] and Sweden [8] have shown that social media has significantly influenced political participation among youths. However, some scholars have noted that there is decrease or non-participation of youths in politics whether online or offline [9]. Accordingly, researches in this area have presented mixed results. Yet theoretical attention in previous studies have been inadequate [10]. For instance, social capital [11], civic voluntarism model [12], and uses and gratification theory [13], have been used in previous studies. However, limited studies have examined whether youths may participate more in politics through learning about politics on social media. To address this gap, the present study aims to investigate youth online political participation from a cognitive engagement perspective as recommended by [14]. Thus, the Cognitive Engagement Theory (CET) [15] will be extended to social media with an aim to assess the role of political knowledge in moderating the relationship between access to political information on Facebook and Twitter, political interest, policy satisfaction and online political participation on Facebook and Twitter.

To assess this plausibility, we analyzed data from a cross sectional survey of undergraduate students of Universiti Utara Malaysia between March and May of 2015. We study undergraduates because they constitute a key segment of youths in any society [16]. Also, they are early adopters and heavy users of technology [2] like new

* Corresponding author: nor1911@uum.edu.my

DOI: 10.1051/shsconf/20173300085

media [17] especially social media, like Facebook and Twitter [18]. Moreover, [19] stated that students of tertiary institutions use social media over nine times more than other demographics in the population. Additionally, [20] establishes that CET emphasizes that those with better education, especially higher education are more likely to be able to understand political information and make informed choices on political participation. Therefore, undergraduates have the skills to better access political information on Facebook and Twitter, which in turn increases their knowledge and interest in politics which could subsequently lead to political participation.

Consequently, greater understanding of the effect on online political participation by youths may have important democratic implications, as concerns over non-participation have long being focused on this demographic [21, 2]. The reason for this concern is because non-participation of youths in politics overtime impedes their representation in the democratic process, which may in the long-run undermine the efficacy of the Malaysian democratic system.

Thus, the present study examines the relationship between features of cognitive engagement and online political participation on social media. Therefore, the overall goal of this research is to advance understanding on how political knowledge enabled by social media contributes to the democratic process through youth online political participation in Malaysia.

2 The cognitive engagement theory

The cognitive engagement theory proposes that an individual's political participation is as a result of his education, access to information, political interest, political knowledge and policy satisfaction [15, 22]. Thus, the theory is based on the assumption that the more educated citizens a society have, the better informed they will be and in the long run the more they will participate in politics to show their satisfaction with government policies. Also the lesser the cost of access to information, the more citizens consume information from the media and the higher the level of political knowledge and interest among citizens which further leads to increase in political participation [12]. Originally, CET was proposed in the context of traditional media and traditional political participation. However, the emergence of new media like social media has reduced cost of accessing political information thereby bringing the theory to relevance in the digital age. Thus, in examining CET from social media perspective, it is evident that social media use promotes political participation [23, 24]. It is noteworthy, that of all the social media sites available, the most popularly used for online political participation is Facebook and Twitter [8].

3 Knowledge and online political participation

Political knowledge is information about politics acquired through media particularly social media. As such, [25] noted that one of the most consistent findings in communication literature is that media use is positively linked to higher political knowledge. Thus, media plays an important role in how citizens acquire political knowledge which eventually leads to participation [26].

Consequently, citizens' political knowledge and participation in politics are very important to the democratic process [12, 23] such that a decline in political knowledge is partly to blame for a decline in participation [27]. Hence, social media increases citizen political knowledge [28, 29]. Equally, [23] believes youths who are apathetic to politics would be more willing to access political content through social media. Thus, the more knowledgeable youths are, the more likely they will be interested in politics. This is because it will enables them to acquire greater political knowledge and interest, as a result increase their participation [7, 11]. Hence, informational use of social media for political participation such as reading political news have a significant effect on political knowledge and participation [30]. As a result, citizens with more access to political information in form of news on new media are more likely to participate in politics because it increases their combined knowledge since they use the internet as a means of educating themselves about politics [31].

From the discussion above, it is evident that political knowledge is intricately linked to access to information, political interest and policy satisfaction. Hence, the following hypotheses are formulated:

H_1 There is significant difference between Malaysian youths who do not have political knowledge and those who have political knowledge in terms of the relationship between access to political information on Facebook and Twitter and online political participation on Facebook and Twitter.

H_2 There is significant difference between Malaysian youths who do not have political knowledge and those who have political knowledge in terms of the relationship between political interest and online political participation on Facebook and Twitter.

H_3 There is significant difference between Malaysian youths who do not have political knowledge and those who have political knowledge in terms of the relationship between policy satisfaction and online political participation on Facebook and Twitter.

4 Methods

Data was collected through a cross sectional survey of undergraduates students of Universiti Utara Malaysia during a three month period (March to May 2015). A multistage sampling method was used to select the 238 students who were surveyed. Of those who responded only 168 cases were eventually usable for final analysis. This presented an acceptable response rate of 70.5%. Due to the categorical nature of the moderator political knowledge, Partial Least Squares- MultiGroup Analysis (PLS-MGA) was used to analyze data for this study [32]. This is to determine whether different parameter estimates occur for respondents without political knowledge and those with political knowledge by examining the PLS path model to see if there is significant difference in the study relationships. As a result, the subsamples to be compared were exposed to separate bootstrap analysis and the bootstrap outcomes were used for the hypothesis test of group differences.

5 Measures

5.1 Access to Political Information on Facebook and Twitter

Six items divided into three parts (Attention, Exposure and Reliance) were used to measure access to political information on Facebook and Twitter [33]. Attention to Facebook and Twitter for political information was adapted from [2] and measured on a 7-point semantic differential scale, with 1 for "No Attention" and 7 for "Lots of Attention" as anchors. Respondents were asked two questions on how much attention they pay to political information on Facebook and Twitter. For exposure to political information on Facebook and Twitter, the measurement of [34] was adapted. Respondents were asked two questions on how often they use Facebook and Twitter to access political information on Facebook and Twitter using 7 response categories ranging from "1" for "Never" to "7" for "All the time". The scale for level of reliance on social media for political information was adapted from [28]. Hence, respondents were asked two questions on how much they rely on Facebook and Twitter for political information. The 7-point scale comprised of "1" for "Never Rely" and "7" for "Heavily Rely" as anchors.

5.2 Political Interest

The measures for political interest were adapted from [35]. Eight items were asked on a 7-point scale with "1" for "Strongly Disagree" to "7" for "Strongly Agree". The eight items were; "None of my friends on my social network are interested in politics", "I am too busy to worry about politics", "I often discuss politics with my social network", "Politics makes no difference to people in my social network", "Politics has an impact on anything I do", "I am very interested in politics", "I know less about politics than most people in my social network", and "Sometimes politics seems so complicated I cannot understand".

5.3 Policy Satisfaction

Policy satisfaction was measured as a multidimensional construct with a combination of both outcome and process considerations [36]. For policy process, the measurement of [35] was adapted. As a result, six items were asked on a 7-point semantic differential scale of "1" for "Not at all" and "7" for "A lot" as anchors. The items bordered on if respondents were involved in any of the following; "Have a say in how policies are made in Malaysia", "Have an opportunity to be involved in the policy making process of Malaysia", "Consulted about policy making in Malaysia", "Involved in the policy making process of Malaysia", "Discuss how policies are made among their Facebook and Twitter contacts", and "Have influence on policy making when they and their friends communicate their opinion to government". All these questions were asked in the context of Facebook and Twitter. For policy outcome, the measurements of [37] and [36] were adapted. Hence, on a 7-point scale of "1" for "Very Dissatisfied"

 DOI: 10.1051/shsconf/20173300085

and "7" for "Very Satisfied", respondents were asked one question on how satisfied they are with government policies in Malaysia.

5.4 Online Political Participation on Facebook and Twitter

Online political participation on Facebook and Twitter was measured as a multidimensional construct with two set of questions for Facebook and two set of questions for Twitter as used by [14] for Facebook. Hence, the first two set of questions was on political activities performed and observed on Facebook, while the second two set of questions was on political activities performed and observed on Twitter. These two set of questions formed an index for online political participation on Facebook and Twitter.

5.4.1 Online Political Participation on Facebook.

The scale for online political participation on Facebook was adopted from [14]. Thus, 10 items were used to measure political activities performed on Facebook and 9 items were used to measure political activities observed by respondents on Facebook. One item (discussing political information about a Facebook message) was excluded from political activities observed because it is a political activity that can only be performed and not observed by friends in a network due to its private nature. Accordingly, the 10 items for political activities performed were: "Discussed political information in a Facebook message", "Posted a status update about politics", "Posted a photo about politics" "Posted a video about politics", "Posted a link about politics" "Posted a wall comment in a wall post about politics", "Posted a note about politics" "Joined or left a group about politics", "Clicked 'going' for a political event", and "Liked a political party or candidate page/became a fan of a political candidate or group". These 10 items were measured on a 7-point semantic differential scale ranging from "1" as "Never"" to "7" as "Very Often". The 9 items for political activities observed on Facebook were: "Posting a status update about politics", "Posting a photo about politics" "Posting a video about politics", "Posting a link about politics" "Posting a wall comment in a wall post about politics", "Posting a note about politics" "Joining or leaving a group about politics", "Clicking 'going' for a political event", and " Liking a political party or politician's fan page/ becoming a fan of a political candidate or group". These 9 items were measured on a 7-point scale ranging from "1" as "Never" to "7" as "Very Often".

5.4.2. Online Political Participation on Twitter.

For online political participation on Twitter, 7 items were adopted from [28]. The 7 items were used for both political activities observed on Twitter and political activities performed on Twitter. For political activities performed on Twitter, the items included: "Tweet about politics", "Retweet or quote a tweet about politics", "Mention a politician or a political party", "Reply a tweet about politics", "Join a political discussion", "Join a political debate", and "Follow a politician or political party". These items were measured on a 7-point scale ranging from "1" as "Never" to "7" as "Very Often". For political activities observed on Twitter, the items included: "Tweeting a message about politics", "Retweeting or quoting a tweet about politics", "Mentioning a politician or a political party", "Replying a tweet about politics", "Joining a political discussion", "Joining a political debate" and "Following a politician or political party". These items were measured on a 7-point scale of "1" for "Never" to "7" for "Very Often".

5.5 Political Knowledge

To test for respondent's political knowledge, the scale of [30] was adapted. Consequently, respondent for this study were asked 10 political knowledge questions. These items bordered on political issues in news reports or burning issues that are or were frequently discussed within a year period in Malaysia. The 10 questions were split into information on elections, political parties (ruling parties, opposition leaders), general knowledge, administration, domestic real world events and foreign policy events. Therefore, five response categories were provided including "Don't Know". The "Don't Know" option was included because [30] noted that it minimizes the chance of randomly guessing the correct answer. Consequently, the ten political knowledge items were summed up to form an index ranging from "0" for incorrect answers to "10" for all questions answered correctly. Correct answers were coded as "1" and incorrect answers and "Don't Know" were coded as "0".

5.6 Control variables.

Two variables were used as control. Age was measured in a close-ended format and was limited to the age bracket (16-40). Therefore the options for age comprised: 16-20 years, 21-24 years, 25-29 years, 30-35 years and 36-40 years. Gender was dummy coded as "0" for male and "1" for female.

6 Results and Discussion

The results of the PLS-MGA as shown in Table 1 indicates the group differences for respondents without political knowledge and those with political knowledge in terms of their moderating effect on the relationship between the exogenous variables and the endogenous variable.

Table 1. PLS-MGA for Respondents Without or With Political Knowledge

Relationship	Path Coefficients-(W/O PK Group 1-W PK Group 2)	p-Value (W/O PK Group 1-W PK Group 2)	Findings
APIFT -> OPPFT	.373	.991	Supported
PI -> OPPFT	.344	.011	Supported
PS -> OPPFT	.039	.628	Not Supported

*W/O PK- Without Political Knowledge
W PK - With Political Knowledge

Hypothesis 1 stated that there is significant difference between Malaysian youths who do not have political knowledge and those who have political knowledge in terms of the relationship between access to political information on Facebook and Twitter and online political participation on Facebook and Twitter. This hypothesis was supported as shown by the significant result (β=.991. p>.95) (See Table 1). Hypothesis 2 stated that there is significant difference between Malaysian youths who do not have political knowledge and those who have political knowledge in terms of the relationship between political interest and online political participation on Facebook and Twitter. This hypothesis was supported as evidenced by the significant result (β=.344, p<.10) (See Table 1). Hypothesis 3 stated that there is significant difference between Malaysian youths who do not have political knowledge and those who have political knowledge in terms of the relationship between policy satisfaction and online political participation on Facebook and Twitter. This hypothesis was not supported because the path coefficient between the two subsamples are not significant due the following parameters (β=.039, p>.05).

Therefore in line with the findings of [26], the results of this study indicates that political knowledge moderated the relationship between access to political information on Facebook and Twitter and online political participation on Facebook and Twitter. Similarly, the result of this study which indicates that political knowledge moderated the relationship between political interest and online political participation on Facebook and Twitter is in congruence with the findings of [12]. Additionally, the fact that political knowledge did not moderate the relationship between policy satisfaction and online political participation on Facebook and Twitter is supported by the statement of [22].

7 Conclusion

This study examined the role of social media in youth's online political participation by assessing the potential of political knowledge to enhance this form of participation among Malaysian youths from a cognitive engagement perspective. Thus, this study contributes to the modelling of cognitive engagement theory by introducing political knowledge as a moderator and examining the theory from social media perspective. Furthermore, the study lends empirical support to the relationship between variables of the theory and online political participation thus adding to the body of knowledge on political participation. Also, this study has proved that political knowledge play an important role in engendering political participation among Malaysian youths.

Nevertheless, this study is limited in the nature of its design as the cross sectional survey approach which was used did not allow for causal inferences to be made from the population over a long period of time. Thus, it is recommended that a longitudinal design which could measure online political participation among youths over a long period of time needs to be carried out. Also, online political participation of youths on Facebook and Twitter were assessed using self- report measures. Thus, other measures which could provide a more objective perspective on their activities is recommended. Furthermore, generalizing results from this study may be limited in that participants were from a public university which may not provide adequate representation of youths in Malaysia. Thus, future studies could be expanded to cover youths in other forms of tertiary institutions or those not in school to enhance the generalizability of results of the study. Lastly, since political knowledge did not moderate the relationships on all levels, perhaps it could be introduced as a mediator. Investigating such mediating effect in future research could provide significant results in all cases.

Acknowledgement

This research is funded by Univesiti Utara Malaysia under the postgraduate incentive research grant scheme (S/O Code: 15697).

References

1. S. Boulianne. Info. Comm, & Society., **18** , 5, 524-538 (2015)
2. M. Yamamoto & M. Kushin. Journal of Computer-Mediated Comm. **19**, 3, 430-445 (2013)
3. J. Groshek. International Comm. Gazette, **74,** 8, 750-768 (2012)
4. S. Vissers & D. Stolle, (2014). Journal of Info. and Tech. Politics, **11**, 3, 259-275 (2015)
5. K. Varnali, V. Gorgulu. Info. Comm. and Society. **18**, 1, 1-16 (2014)
6. L. Ke, H. Starkey, London Review of Educ. **12**, 1, 50–62. http://www.ingentaconnect.com/content/ioep. (2014)
7. M.J. Kushin, M. Yamamoto, Mass Comm. and Society, **13**, 5, 608-630 (2010)
8. S. Stieglitz, L. Dang-Xuan. Social Network Analysis and Mining, **3**, 4, 1277-1291 (2013)
9. M.V. Mongkuo, M.L. Lyon, M.B. Hogan, G. Delore, Journal of Political Science and Public Affairs, **2**, 3. (2014)
10. M. Wojcieszak, International Journal of Comm. **6**, 11, 255-256. (2012)
11. H. Gil de Zúñiga, N. Jung, S. Valenzuela, Journal of Computer-Mediated Comm., **17**, 3, 319-336 (2012)
12. M.C.W. Charles. Impact of digital inequality on civic and political participation. Dissertation, ProQuest (2010)
13. N. Okoro, K.A. Nwafor. Global Journal of Arts and Hum and Social Sci. **1**, 3, 29-46. (2013)
14. J. Vitak, P. Zube, A. Smock, C.T. Carr, N. Ellison. C. Lampe. Cyber Psyc., Beh. and Social Networking, **14**, 3, 107-114. (2011)
15. R. Inglehart. The silent revolution: Changing values and political styles among western publics. Princeton, New Jersey: Princeton University (1977)
16. D. Odunlami. New Media and Mass Communication. Retrieved October 28, 2014 from www.iiste.org. (2014)
17. A. Salman, S. Saad. Mediterranean Journal of Social Sci, **6**, 4, 88. (2015)
18. C.E. Uzochukwu. U.E. Patricia. C.A. Ukueze. International Journal of Social Relevance Comm. **2** , 11 (2014)
19. A.T. Titus-Fannie. C.S. Akpan. M.O. Tarnongo. IOSR Journal of Hum and Social Sci. 6(5), 32-44. www.iosrjournals.orgs. (2013)
20. R. J. Dalton. The Journal of Politics. **46**, 264 -284 (1984)
21. Y. Theocharis, E. Quintelier. New Media and Society. 1-20. (2014)
22. C. Pattie, P. Seyd. P. Whiteley. Citizenship in Britain: Values, participation and democracy. Cambridge: Cambridge Uni Press. (2004)
23. I. Rahmawati. ProQuest Dissertation (2014)
24. M.R.Vincente. A. Novo. Govt. Info. Quarterly, **31**, 3, 379-387. (2014)
25. K. Kaufhold, S. Valenzuela., H. Gil de Zuniga. Journalism & Mass Comm. Quarterly, **87**, 3-4, 515–529. (2010)
26. S. Valenzuela. American Behavioral Scientist, **57**, 7, 920-942. (2013)
27. B. O'Neill. Canadian Policy Research Networks. (2006)

28. T.J. Johnson. B.K. Kaye. Journal of Computer Mediated Comm. **19**, 4, 957-974. (2014)
29. J.K. Raoof. H.B. Zaman, A. Ahmad. A. Al-Qaraghuli. International Journal of Phy. Sci., **8**, 12, 1143-1148. (2013)
30. D.V. Dimitrova. A. Shehata, J. Stromback. L.W. Nord. Comm.Research. **41**, 1, 95-118. (2014)
31. C.J. Tolbert. R.S.McNeal. Political research quarterly, **56**, 2, 175-185. (2003)
32. J. Henseler. Challenges at the interface of data analysis, computer science, and optimization, 495-501. (2012)
33. W.P. Eveland. D.A. Scheufele. Pol. Comm. **17**, 215-237. http://www.tanfonline.com/loi/upcp20. (2000)
34. M.A.Bekafigo. A. McBride. Social Science Computer Review, **31**, 5, 625–643. (2013)
35. P.F. Whiteley. The political science perspective research report. RR 631 (2005)
36. P.F. Whiteley. H.D. Clarke. D. Sanders. M. Steward. Presented at the Conference on Citizens and Politics in Britain Today: Still a Civic Culture? London School of Economics (2013)
37. J. Shore. Paper Prepared for the PartiRep Workshop "Conceptualizing Political Participation" Mannheim(2014)
38. A. Tumasjan,T.O Sprenger, P.G. Sandner, I.M. Welpe. Proceeding of the fourth International AAA1 conference on Weblogs and Social Media Predicting, 174-184, (2010)

Preface - Communication: Beyond Nation Cultural Adaption, Images and Identity

Bahtiar Mohamad[1] *& Hassan Abu Bakar*[1]

[1]*Communication Department, School of Multimedia Technology and Communication, Universiti Utara Malaysia, 06010 UUM Sintok, Kedah, Malaysia*

Recent demands in the work environment – changes in lifestyles, challenges in socio-economic and political agenda, issues in cultural integration, planned revamps in organizational systems, corporate restructuring, and business and management reengineering, to name a few – are exerting socio-cultural pressures on the roles communication and medias play in handling these types of scenario. In addition, communication and medias in the 21st century are transforming rapidly through the globalization of organizations and digital technologies. They not only shape organizations and their social structure but they also influence our thinking.

Given this particular context, we believe that communication and media scholars have to examine how their theories and models can be translated into practices. Furthermore, they must ask themselves how their theories, philosophies, and principles can offer comprehensive explanations and strategies in these ever-changing environments. For instance, at the end of 2000, East Asia (South Korea, Malaysia, Thailand, and Indonesia) faced a financial crisis that also affected business worldwide. Prior to the crisis, these countries were the center for international investments. Now, recovering from adversity, these countries have begun to prosper again, and up until now this region has received 25% of the world foreign direct investment.

From a communicational viewpoint, this is an interesting time. Although the crisis has changed the political climate in East Asia, some interesting phenomena have indeed evolved. China and India have eventually become prominent in South East Asian countries' economic policy, while Australia and New Zealand (perceived as Western countries) start to look to South East Asian region as their strategic business partners. These countries, together with ASEAN countries (Association of South East Asian Nation), namely China, Japan, South Korea, India, Australia, New Zealand, and the United States, jointly formed the APEC (Asia-Pacific Economic Cooperation) to facilitate businesses in the Asia-Pacific region. The interesting aspects of these phenomena go beyond questions of (mis-) communication between various participants. Prejudices and the stereotyping of other cultures are already important subjects within the field of communication, and they have led scholars to deliberate about the implications of these phenomena in real settings. These situations not only affect nation's identity, but also create social unrest.

The International Conference on Communication and Media 2016 (i-COME'16), an ICA Regional Conference provides a broad platform to explore how communication and media in contemporary organizations could play pivotal roles in dealing with issues and challenges that have both local and global impacts. The 2016 theme "Communication: Beyond Nation Cultural Adaption, Images and Identity" is aimed toward advancing scholarships in the field of communication which would benefit communities that are experiencing change.

This volume comprises papers that were presented at the i-COME'16: An ICA Regional Conference which was held in Kuala Lumpur, from the 18 to 20 November 2016. Organised by the Communication Department, School of Multimedia Technology and Communication, Universiti Utara Malaysia in collaboration with International Communication Association (ICA), this conference was attended by 200 participants from 26 countries, including those from Asia, America, Europe, Africa and Australia.

As guest editors of the conference proceedings, we would like to express our sincere gratitude to the members of the scientific committee, reviewers and contributors to this volume for their goodwill, support and patience in revising their papers. We are also grateful to the committee members of the i-COME'16: An ICA Regional Conference who have assisted us in bringing this volume to fruition.

GUEST EDITORS:

THE EDITORIAL BOARD MEMBERS:

SCIENTIFIC COMMITTEE

Chair:

Scientific Committee Members:

DOI: 10.1051/shsconf/20173300001

Irwansyah, *Universitas Indonesia, Indonesia*
Isma Rosila Ismail, *Universiti Malaysia Terengganu, Malaysia*
Ismail Sualman, *Universiti Teknologi Mara, Malaysia*
Jamilah Jamal, *Universiti Utara Malaysia, Malaysia*
Joyce Lynn-Sze Cheah, *Universiti Utara Malaysia, Malaysia*
Kalyani Suresh, *Amrita Vishwa Vidyapeetham, India*
Khaizuran Abdul Jalil, *Athromed Sdn Bhd, Malaysia*
Khatijah Omar, *Universiti Malaysia Terengganu, Malaysia*
Kiranjit Kaur Pritam Singh, *Universiti Teknologi Mara , Malaysia*
Lateef Adelakun, *University of Malaya, Malaysia*
Latifah Abd Latib, *Universiti Industri Selangor, Malaysia*
Lucy Supratman, *Telkom University, Indonesia*
Marc Pacoma, *Adamson University, Philippines*
Martha Lestari, *Telkom University, Indonesia*
Masduki, *Indonesian Islamic University, Indonesia*
Melicent Jalova, *Mindanao State University-Iligan Institute of Technology Philippines*
Minho Ju, *Sungkyunkwan University, Republic Of Korea*
Mohamad-Noor Sharipudin, *Universiti Utara Malaysia, Malaysia*
Mohd Baharudin Othman, *Universiti Utara Malaysia, Malaysia*
Mohd Khairie Ahmad, *Universiti Utara Malaysia, Malaysia*
Mohd. Nor Shahizan Ali, *Universiti Kebangsaan Malaysia, Malaysia*
Mudassar Hussain, *University of The Punjab, Pakistan*
Nik Rozaidi, *Rutgers University, United States*
Nor Azura Adzharuddin, *Universiti Putra Malaysia, Malaysia*
Norealyna Misman, *University of Malaya, Malaysia*
Norhafezah Yusof, *Universiti Utara Malaysia*
Normazla Ahmad Mahir, *Universiti Sains Islam Malaysia, Malaysia*
Norzita Yunus, *Infrastructure University Kuala Lumpur*
Olawale Ismail, *Ahmadu Bello University, Zaria, Nigeria*
Patrick Chin Hooi Soh, *Multimedia University, Malaysia*
Pijar Suciati, *Universitas Indonesia, Indonesia*
Pin Lick Soo Hoo, *Universiti Utara Malaysia, Malaysia*
Rizwanah Souket, *UCSI University, Malaysia*
Roro Wulan, *Telkom University, Indonesia*
Saidathul Nizah Mat Tazin, *Universiti Teknologi Mara, Malaysia*
Samuel Hatane, *Petra Christian University, Indonesia*
Shamsu Dauda Abdu, *Federal Polytechnic, Bauchi, Nigeria*
Siti Nor Amalina Ahmad Tajuddin, *The University of Queensland, Australia*
Stephen Croucher, *University of Jyväskylä, Finland*
Sumekar Tanjung, *Islamic University of Indonesia, Indonesia*
Syed Hassan Raza, *Bahauddin Zakariya University, Pakistan*
Thanaseelen Rajasakran, *Universiti Tunku Abdul Rahman, Malaysia*
Thinavan Periyayya, *Universiti Tunku Abdul Rahman, Malaysia*
Wan Norbani Wan Noordin, *Universiti Teknologi Mara, Malaysia*
Wang Yan, *National Chengchi University, Taiwan*
Yeen Lai Khong, *Universtiy Tunku Abdul Rahman Malaysia*
Zalinah Ahmad, *Universiti Utara Malaysia, Malaysia*

The Ideological Struggle of Multicultural Nationalism: Cultural Identity in the 2014 Malaysian Top-Grossing Movie *The Journey*

Wang Changsong[1], and *Chen* Yiming[2,*]

[1]Xiamen University Malaysia, 43900, Kuala Lumpur, Malaysia
[2]School of Journalism and Communication, Jinan University, 510632, Guangzhou, China

Abstract. This study attempts to discover how multiculturalism interacts with sub-state nationalism as portrayed in the Malaysian top-grossing film of 2014 – *The Journey* directed by Chiu Keng Guan. The nature of Malaysian society and its colonial history might suggest that cultural interaction among its three major races (Malays, Chinese, and Indians) is a norm, but this study argues that the culture of these three major races as portrayed in the fictional film is made up of "systems of representations"; such systems of representations do not reflect the intentions of the subjects, but rather, actively construct meanings as conditions and instruments, signifying practice in reality. The application of in-depth interviews (e.g., the film director and the screenplay writer) discovered cultural identity depicted in the film as always having a specific 'positionality' within their representation, and the ideological struggle of multicultural nationalism embedded in the film is considered 'non-subjective'. The authors believe that every culture is local, but no culture is autochthonous, and cultural identity illustrated in the film can be regarded as 'uniqueness of the soil'.

1 Introduction and literature review

The Journey represents something of a box office miracle, but the factual circumstances of Malaysian social, political, and historical nature is more meaningful for the authors of this study. According to the data given by the central government agency for the film industry of Malaysia – FINAS (The National Film Development Corporation Malaysia), the theatrical box-office earnings of the Malaysian Chinese-New-Year-celebration film *The Journey* (directed by Chiu Keng Guan) was 17,170,341.75 Malaysian Ringgit. It statistically assesses the success of this local film, as *The Journey* has become "the all-time top-grossing local film" [1]. The film director of *The Journey* believes that local Chinese filmmaking is promising, but still in its infancy. However, there is another factor: "... the Chinese overwhelmingly watch Hong Kong, Taiwanese and Mainland Chinese films; if not, they watch English-language films, but almost never films made in Bahasa Melayu; the same scenario occurs with Indian audiences; the situation with Malay audiences is more

* Corresponding author: jidayiming@163.com

complex … their first preference is for English-language films, followed by Indian films and only then films in Bahasa Melayu" [2]. The separation along ethnic lines is starkly drawn, and it also suggests that local cinema is a weak cultural force. Chiu, during the interview [1], interprets the audience composition for *The Journey*: there are only six million ethnic Chinese in Malaysia; deducting children and older adults, there are only three million potential Chinese cinemagoers; it would be impossible for this three million to give us seventeen million in box-office earnings. In other words, he meant that this film also successfully attracted a great number of non-ethnic Chinese. In this multi-ethnic country, "the filmmaker provides some illuminating insights into the relationship between film industry and Malaysia's multi-ethnic society" [3]. *The Journey* can be considered as an outstanding example.

The emergence of the multi-ethnic society invented by British colonial policy has resulted in today's Malaysian population being comprised of three major ethnic races: Malay, Chinese, and Indian. The great influx of Chinese into Malaya came after the mid-century had passed and was a result of the expansion of tin mining [4]. Then, another "large wave of newcomers again provide labour, this time for establishment of a new industry, rubber planting", and the "majority was Indians" [4]. In the 1990s, the term transnationalism was adopted to describe the links between these 'migrants' in places of settlement and origin and native communities. In their *Ungrounded Empires : The Cultural Politics of Modern Chinese Transnationalism*, Ong and Nonini stated that Chinese migrants developed a "third culture" in Malaysia, defined as a "modern Chinese transnationalism" that "provides alternative vision in late capitalism to Western modernity and generates new and distinctive social arrangements, cultural discourses, practices, and subjectivities" [5].

1.1 Multiculturalism in Malaysia

Multiculturalism is grounded in multi-sited dialogue among different peoples that result in trivial and nontrivial debates within Malaysia. Lian [6] argued that over the years consociational governments have evolved – more explicitly in Malaysia than in Singapore – that include power-sharing by the charter ethnic communities and the acknowledgement that resources should be distributed proportionate to the ethnic composition of the population [6]. He pointed out that, while consociation underlies the practice of multiculturalism in Malaysia, this is less apparent in Singapore [6]. "Notions of ethnicity and ethnic identity in Malaysia multicultural society are vital for galvanizing ethnic individuals to preserve, protect and promote their cultures, languages and religions, including their economic objectives" [7].

Multiculturalism has been historically embedded in Malaysian society and each of its states has had to manage this problem from the start. In a collection of literature, highlighting issues of migration and multiculturalism in Asia, Malaysia presents a dilemma. "By the 1990s, Malaysian … multiculturalism had to respond to a changing world order" [8]. However, there is also a debate over millennial multiculturalism in Malaysia and its association with Asian forms of neoliberal multiculturalism. Malaysian society is characterized by what Nonini describes as 'imaginaries of desire in tension with cascades of symbolic violence' [9] across different cultural and institutional scales, where localizing and disciplining one group into a bounded place achieves the transnational mobility and power of another group.

Attention to the postcolonial nation-state of Malaysia in the international media frequently focuses on two apparently mutually contradictory areas. According to Goh and Holden, [8] "in economic terms, [Malaysia is] frequently praised as success stories of modernization", and "in cultural terms, [Malaysia is] is often seen as falling away from the principles of liberal democracy that are taken as representatives of modernity" [8]. The authors mean a way of thinking about race that examines differences between groups, aiming to understand and celebrate these differences, to also celebrate diversity. Chiu Keng Guan's *The Journey* coincided with a more general turn of celebrating diversity in Malaysian society, which saw his film enmeshed in a broader politics of cultural representation. *The Journey* was awarded "as '1-Malaysia' film" by Ahmad Shabery Cheek, the former minister of Communication and Multimedia of Malaysia. He regarded multi-racial actors for this film as an industrial phenomenon. Chiu Keng Guan measured Malaysian society with his eyes: randomly take a photo on the street showing different races naturally framed, and this is Malaysia. Considered together, this study is not only comparative across the three major races, but also raises important implications about the multiple ways of thinking, talking and performing race and multiculturalism within the society.

1.2 Cultural identity and Malaysia's multicultural society

Precisely because people construct their identities through discourse, they manifest their understanding of identity through detailed practices and formations at specific historical and institutional sites. The growing awareness of the concept of cultural citizenship embedded in constructions of national identities has emerged as an important analytic distinction in this study between native culture and accommodation to the dominant culture. According to Hall [9], identification is in the end conditional, lodged in contingency. The prevailing belief is that each race has its own culture, one's father-culture is the culture of one's race, and the culture does not obliterate differences [11]. However, another factor is dominance of Malay (Muslim) culture in this multicultural society. Not surprisingly, "it has … served to accentuate the otherness of the non-Malays and to deepen their insecurity as minorities located within the hegemonic Malay scheme of things" [10]. Arguably, the dominant group – Malays – holds "their dominant position in the country". While complex ethnicities are equally applicable to Malaysia, there is no ambiguity or conflict about what the first local film was – the 1933 Malay-language film *Laila Majnun*. Ironically, the local Chinese film industry only began in 2010 [12]. Chiu believed that 2009 signified the opportune year for local Chinese film, following struggle, to create its own identity.

This study uses identity to explore the association between filmic society and 'deconstruction' and 'reconstruction' in terms of its 'historical, genealogical and discursive formations' [9] Hall stated that

> "Identities are, as it were, the positions which the subject is obliged to take up while always 'knowing' (the language of consciousness here betrays us) that they are representations, that representation is always constructed across a 'lack', across a division, from the place of the Other, and thus can never be adequate—identical—to the subject process to a subject-position requires, not only that the subject is 'hailed', but that the subject invests in the position, means that suturing has to be thought of as an *articulation*, rather than a one-sided process, and that in turn places *identification*, if not identities, firmly on the theoretical agenda" [13].

Chiu [12] stated that, since 2010 in Malaysia, the kicks of Chinese-language film started at the hip with a slight knee bend; however the path was not smooth, while the roads were wet and slippery, uneven and rough. The success of *The Journey* is mostly credited for its "message for all Malaysian" [14]. It speaks to or hails Malaysian audiences into a place, which focalizes social subjects of particular discourses. Family is the basic unit of society. Therefore it stands to reason that anything connected to a family would connect with audiences. Chiu [12] figuratively added that the living circles of a Chinese family comprise the Malay and the Indian, and so does a Malay family's. His film provides the meeting point, the point of suture between the discourse of practices and the processes that produce subjectivities, "which construct us as subject which can be 'spoken'" [9-15].

2 Objective of the study

This study aims to investigate how multiculturalism interacts with sub-state nationalism portrayed in the 2014 Malaysian top-grossing film *The Journey* directed by Chiu Keng Guan. This study also inevitably gives attention to particular Malaysian cultural norms and nuances which nurture the local Chinese film industry.

3 Methodology

The in-depth interviews help the authors to understand the filmmakers' thoughts and explore relevant issue in depth. The purpose of in-depth interviewing is not to get answer to questions, but to understand the lived experience of filmmakers and the meaning they make of that experience. The author interviewed the film director – Chiu Keng Guan and the screenplay writer – Ryon Lee respectively at different times. Moreover, content analysis in this study involves drawing representative samples of content to measure or reflect differences in the content of multiculturalism and its association to filmic representation. On the basis of short description of content, interpretations and subjective value-judgements are in ways, which make inter-subjectively valid assessment virtually impossible.

4 Discussion

In a country where the list of top-grossing films was dominated by tales of blood and guts (e.g., *KL Gangster*, 2011), the success of a film as subtle and personal as *The Journey* may signal a growing maturity among cinema-goers. The film, in Chinese dialects with smatterings of Malay and English, demonstrated the nature of a multi-language country. Hassan Abd Muthalib, the local film critic, argued that the popularity and record-breaking box-office earning "was because the film was conceived and made with honesty and sincerity – the vital ingredients for a story full of humanism" [11]. So-called 'humanism' is an important trigger, as cultural identity depicted in the film always represents a specific 'positionality', and the ideological struggle of multicultural nationalism embedded in the film is considered 'non-subjective'. In the interview, Chiu [11] expressed his own experience: "it was unexpected to have cinemagoers of other races though this is based upon a typical Chinese perspective; and ultimately, I was aware that no matter through which race's perspective, the story itself belongs to this country". Ryon Lee, the screenplay writer of *The Journey*, in the interview[2], expressed a different opinion: Astro, the mainstream Malaysian media and entertainment holding company and producer of the film, invested in endless promotional slots, resulting in *The Journey* succeeding at the box-office. The authors believe that both the film's story and its promotional teaser appeal to a sense of humanism during the Chinese New Year. In discussing the films illustrated multiculturalism, we can understand civil groups, citizens, and foreign daughters-in-law portrayed as postcolonial actors. These social actors find themselves in an inextricable position in which they have no choice but to negotiate the colonial legacies of racialization and transform them into postcolonial multiculturalism. This film also directs the authors'

DOI: 10.1051/shsconf/20173300002

attentions to the vernacular of cultural citizenship that articulates rights and differences. According to Rosaldo (2014), "cultural citizenship refers to the right to be different (in terms of race, ethnicity, or native language) with respect to the norms of the dominant national community, without compromising one's right to belong, in the sense of participating in the nation-state's democratic processes" [14].

Although the minister regards it as a good example of '1-Malaysia film', this does not mean the film illustrates events and characters in a manner of true harmony. Daniels [15] argued, "most Malaysians… evoking a conventional sense of togetherness, were convinced that their unity would withstand the scorching effect of economic and political crises". In the film, Uncle Chuan has a childhood friend who is a Malay woman. She had been one of his classmates and the story alludes to a possible (or imaginary) love relationship between her and Uncle Chuan. It is she who tells Benji about Uncle Chuan's birthday according to the Chinese Lunar calendar. It also subtly hints that when one race understands the culture of another race, there will be true understanding and tolerance. However, although the flashback scene of sending his daughter to study abroad seems touching, it implies educational issues. "In Malaysia, … the Chinese population tends to have a greater financial ability to send their children to study overseas", and also, ethnic Chinese are "in a less favourable position to get a public university place due to their ethnicity" [16].

The nature of Malaysian society and its colonial history might suggest that cultural interaction among its three major races is a norm, but this study argues that the cultures of these three major races as portrayed in the fictional film is made up of "systems of representations", and such systems of representations do not reflect the intentions of the subjects, but rather actively construct meanings as conditions and instruments of real, significant practice. According to Chiu [12], amateur actors portray all of the friends Uncle Chuan meets while delivering his daughter's wedding invitation, and they give precise, perfect illustrations of real lifestyles in Malaysia. These may be conceived as the body of Malaysian multicultural nation: the agreement, in this case among the major races, about what the real relationship among social groups is about.

In his *Malaysian Cinema in A Bottle*, Hassan raised an important issue: can we hope for this in Malaysian cinema and celebrate it as something that unites us rather than divides us? Hassan criticised the industry phenomenon, as he believed difference races and different people have their own stories to tell and "we should be tolerant of stories that sometimes are difficult to comprehend or of truth that sometimes hurt" [17]. Also, what he highlights here is actually the way of identifying works of art. He is ideologically opposed to the view that works of arts can be identified "by race, language, culture or ethnicity even though there may be such subjectivities in the works" [17].

5 Conclusion

Above all, this study does not aim to discover the reasons why *The Journey* became the highest-theatrical-box-office-earning Malaysian-Chinese-language film. The cultural identities of the three major races portrayed in *The Journey* are within a system of representations, which express the nature of the ideological struggles of multicultural nationalism in the Malaysian context. Also, this film successfully demonstrates the negotiation of the colonial legacies of racialization. The family depicted in *The Journey* allows viewers to perceive the system of representations of ideology and cultural identity. Obviously, family is always the basic unit of society. Every culture is local, but no culture is autochthonous, and cultural identity illustrated in the film can be regarded as 'uniqueness of the soil'.

References

1. S. K. Yan, The Star, (2014)
2. W. Van der Heide, *Malaysian Cinema, Asian Film: Border Crossings and National Cultures*. (Amsterdam University Press, Amsterdam, 2002).
3. R.W. Hefner, *The Politics of Multiculturalism: Pluralism and Citizenship in Malaysia, Singapore, and Indonesia* (University of Hawai'i Press, Honolulu, 2001)
4. J. A. Bakar, *A Design Guide of Public Parks in Malaysia* (Penerbit Universiti Teknologi Malaysia, Kuala Lumpur, 2002)
5. A. Ong, D. Nonini, *Ungrounded Empires : The Cultural Politics of Modern Chinese Transnationalism* (Routledge, New York, 1997)
6. K. F. Lian, *Multiculturalism, Migration, and the Politics of Identity in Singapore*. (Springer, Singapore, 2015)
7. H. Jamil, H. N. Ismail, Ethnicity and the State of Multiculturalism in Malaysia Education System, *Intercultural and Multicultural Education : Enhancing Global Interconnectedness* (Routledge, Oxon and New York, 2011)
8. D. P. S. Goh, M. Gabrielpillai, P. Holden, G. C. Khoo, *Race and Multiculturalism in Malaysia and Singapore* (Routledge, Oxon and New York, 2009)
9. S. Hall, Who Needs 'Identity'? *Identity: A Reader*, (Sage Publication Int, London, 2000)
10. D. C. L. Lim, *Overcoming Passion for Race in Malaysia Cultural Studies* (Brill, Danvers, 2008)
11. A. M. Hassan, the Journey by Chiu Keng Guan: A Message for All Malaysians, *Astro Awani* (2014)
12. This interview was conducted at one of the authors' – Wang Changsong's – office at the School of Communication, Universiti Sains Malaysia, in Penang, Malaysia, on 18 November 2015.
13. This interview was conducted by one of the authors – Wang Changsong – at the Han Chiang College campus field in Penang, Malaysia, on 5 February 2015
14. J. Clarke, K. Coll, E. Dagnino, Neveu, *Disputing Citizenship* (Policy Press, Clifton, 2014)
15. T. P. Daniels, *Building Cultural Nationalism in Malaysia : Identity, Representation and Citizenship* (Routledge, New York & London, 2005)
16. T. W. W. Yuen, A. C. K. Cheung, *A SWOT Analysis of Exporting Hong Kong's Higher Education to Asian Markets,* Internationalization of Higher Education : The Case of Hong, (Springer, Singapore, 2016)
17. A. M. Hassan, *Malaysian Cinema in A Bottle* (Orange Dove, Selangor, 2013)

Communicating in Collectivistic Culture: Relationship between Interdependent Self-Construal and Cooperative Communication in Cross-Cultural Communication Model

Pin Lick Soo Hoo[1,*]

[1]Communication Department, School of Multimedia Technology and Communication, Universiti Utara Malaysia, Malaysia

Abstract. This concept paper proposes that when companies have expanded their business and operation throughout the Asian countries, the success or failure of a company abroad depends on how effectively its business negotiators can apply their cross-cultural communication skills in a new cultural environment. At the forefront of this change is interdependent self-construal, which stands as communication medium on interaction goals (international business negotiation outcome) in collectivistic culture. The international trade may bring about a lot in terms of cross-cultural communication and international business negotiation, but adopting cooperative communication in the international business negotiations will create more integrative agreements between the international business negotiators. Many scholars believe that if both parties have similarities in communication styles, it will lead to positive interaction (cooperative communication) that contributes to cooperation and influences the interaction goals (negotiation outcome). This paper offers critical insight into the theoretical link between interdependent self-construal and interaction goals. The proposed cross-cultural communication model uses interdependent self-construal and cooperative communication to understand when collectivistic business negotiators develop relationships that promotes positive interaction goals (negotiation outcome).

1 Introduction

Every culture is concerned with the presentation of their face and in negotiation; the impact of face concerns is likely to happen because there are many opportunities for a person's face to be threatened [13]. Despite the extensive studies on negotiation, the literature only focused within the tradition of research on international organizational behaviour focusing on single-culture descriptive studies and multicultural comparative studies rather than examining cross-cultural interaction [10]. In addition, past literature suggested applying

* Corresponding author: eileensoohoo@gmail.com

Hall's low-high context cultural dimension to study East Asian countries in intercultural collaboration as it was claimed that people from different cultural backgrounds have different communication styles [12]. However, Hall's low-high context cultural dimension only focuses on the impact of culture on communication but it fails to explain the face concerns during the cross-cultural interactions. Therefore, at least there are three situations that explain why the need to have inclusive cross-cultural communication model that identified as important step to understand cross-cultural communication especially in international business negotiation. First, in previous interdependent self-construal studies had linked with building a relationship [11]. However, the interdependent self-construal studies yet to recognize how people deal, negotiate and communicate across cultures with their business counterparts with different communication styles in collectivistic culture. Each individual has both independent and interdependent self-construal [11]. Therefore, a model that emphasizes interdependent self-construal that predicts the communicative behaviour of collectivistic culture is needed. Second, the cooperative communication is important to be examined in cross-cultural communication because it explains positive interactions by linking cooperative behaviour to cooperative relationship. Previous studies indicated that interdependence self-construal emphasizes individual goals [6], however it is suggested that priority should be given on how to connect self-construal with relationship, especially in understanding the differences for intergroup relations, business and peaceful coexistence in diverse and interconnected world. Finally, within cross-cultural communication interaction goals that emerge from interaction during cross-cultural negotiation and how conflicting goals shape communicative behavior. A study by [9] indicated that conflicting interaction goals shape interaction tactics and outcome. The link between goals and negotiation outcomes is more complicated when identity of the counterpart is concerned [9]. Therefore, this present paper will address the negotiators' interdependent self-construal and how they predict communicative behavior, communication styles and interaction goals to reach cooperative approach in negotiation.

2 Objectives

This paper aims to provide a proposition model that identifies the impact of differences in communication styles in reaching cooperative outcome in international business negotiations. Specifically, the aim of the article is to answer the following question: What are the underlining mechanisms of the relationship between interdependent self-construal and interaction goals? By answering this question, it will enable us to understand towards advancing the current knowledge of cross-cultural communication especially in negotiation within collectivistic cultural context.

3 Review of Literature

3.1 Relationship of Interdependent Self-construal and Cooperative Communication

Self-construal is one's self-image and consists of independent and interdependent self-construal [7]. Self-construal is used in previous studies to explain differences in communication styles [7]. The study of interdependent self and cooperative communication, however, has two other highly significant consequences for the field of cross-cultural communication in collectivistic context, and they are the focus here. First, the study of interdependent self and cooperative communication has renewed and extended cross-cultural communication understands of interdependent self or culture and casts it as central to analyse communication behaviours of business negotiators in international business negotiations. Second, the study of interdependent self and cooperative communication has led to the realization that people are not separated but instead they are linked together. H.R. Markus [6] suggested that we require each other to complete each other even though culture shapes our thinking, feeling and action. Although [6] claimed that there was no direct link between face or face work and self-construal, it was found that when communicating with others, there was indirect influence of individualism-collectivism with personal goals, communication styles and face concerns. Thus, independents tend to be more self-face oriented than other-face oriented while high interdependent people value other-face and mutual-face concerns. For that reason, high interdependent people prefer to appeal to other-face concerns in conflict situations in order to maintain the harmony of the relationship. This study investigates the relationship with others which defines the interdependent self-construal.

There are limited evidence that provide explanation about relationship between interdependent self-construal and perceived cooperative communication in collectivistic culture. Past literature defined cooperative communication as communicative behaviours lead to sharing of achievement of workgroup goals [8]. Scholars (e.g., [4]) asserted that cooperative communication is an interaction in a way that encourages mutual goals and solve issues for mutual benefit. In our model, cooperative communication, directly influenced by interdependent self-construal exerts an immediate impact on group's interaction goals.

In conclusion, interdependent self-construal serves as an important factor influencing the cooperative communication but remain minimal attention in past literature, it was proposed that individuals with an interdependent self-construal begin building a relationship [11] which is important in managing, negotiating and communicating across cultures to achieve the corporations' goals in global business [3]. Within the framework of face-negotiation theory studies, it was suggested that face concerns are very important as it determines the communicative behaviour of international business negotiation. However, there are many significant queries remain unaddressed in the literature of face-negotiation theory and one of the questions is how and whether similarities and differences in communication styles will influence the cooperative communication which produce positive interaction goals and vice versa. Therefore, this current study addresses this gap contributes to an enhanced understanding of the mediating relationship of the cooperative communication with the interdependent self-construal and interaction goals (negotiation outcome) in collectivistic culture. A consideration of the relationship between interdependent self-construal and cooperative communication may open doors to new understanding of the role of the self in cross-cultural communication and business negotiation in collectivistic culture.

On the basis of the above literature review, it may be postulated that interdependent self-construal and cooperative communication have direct dynamic link to define the relationship of cultural norms and communication styles as it remains an uncertain issue in the research. Hence, if the relationship of interdependent self-construal and cooperative negotiation is positive, it positively affects the communicative behaviours. Likewise, if it is negative, it will negatively affect the communicative behaviours. Thus, this paper proposes the following preposition.

Preposition 1: Interdependent self-construal has positive association with cooperative communication.

DOI: 10.1051/shsconf/20173300003

3.2 Relationship of Dimensions of Culture and Interaction Goals

There is limited literature that provided the link between cooperative communication and interaction goals in collectivistic culture. In previous studies, cooperative communication generally focused on workgroup research as a communication technique which considers cultural sensitivities, values and practices [5]. Cooperative communication is also considered as the key objective of international business negotiations as it leads to positive negotiation outcome to the negotiators. The cooperative communication is defined as individuals become more cooperative in reaching work-related goals with other group members by exchanging more information, sharing ideas and resources, showing concern and interest in what others want to achieve and give assistance, tend to be more responsive, supportive, and open to each other's needs, and consult and discuss issues to reach mutually fulfilling agreements [1]. Tjosvold explained cooperative as collaborators who are open to being influenced by one another, use their power to facilitate each other's goal accomplishment, accurately, perceive each other's needs, view the relationship as being characterized by trust and mutual helpfulness and like each other [1]. Therefore, this is also the aim of the international business negotiators to enhance cooperative communication as it can probably increase the frequency of communication and collaboration.

On the other hand, interaction goals are considered as negotiation outcome of international business negotiations in this study. Interaction goals are defined as capability to engage in acting, thinking or behaving in certain ways with others that will also influence the other party's attitudes or behaviours [9]. Generally, when people enter negotiation, the negotiators believe they hold incompatible goals but to produce positive outcome, both parties must cooperate for mutual agreement [9]. These studies demonstrated that goals were the key to motivation in negotiation [2] and interaction goals were the negotiation outcome [9]. However, past literature remains less explanatory in term of international business negotiations link between both concepts of cooperative communication and interaction goals. Research has shown that a step onward is needed to provide explanation and identification of emotions on the interaction goals which indicates that how negotiators' trust and distrust affect an individual's positive and negative perceptions on willingness to cooperate [9]. Despite the fact that cooperative communication impacts on interaction goals is less examined in many previous studies, this study improves the initiative to examine the role of cooperative communication on interaction goals (negotiation outcome).

Based on the extensive reviews, previous studies had not measured the relationship between cooperative communication and international business negotiation outcomes (i.e. interaction goals) in a collectivistic business culture and environment. Therefore, it is necessary to examine interdependent self-construal, cooperative communication and interaction goals from collectivistic cultures among the international business negotiators. Hence, if the relationship of cooperative communication and interaction goals is positive, it positively affects the communication behaviours. Likewise, if it is negative, it will negatively affect the communication behaviours. Based on this, this paper proposes that:-

Preposition 2: Cooperative communication associates with interaction goals in international business negotiations.

3.3 Cooperative Communication as Mediation

Cooperative communication is the key factor of the interaction goals as it serves as the significant predictor of it as suggested in communication styles across culture. In addition, it was found that in previous studies by many researchers that cooperative communication

DOI: 10.1051/shsconf/20173300003

is a significant fact that cooperative communication also have role to explain the relation of the cooperative behavior [5]. However, there is no empirical compilation and enough studies in the past have been conducted to explain the mediating role of cooperative communication in the connection of the exposure of interdependent self-construal to interaction goals which is proposed in this present study. Based on this, this paper proposes that:-

Preposition 3: Cooperative communication mediates the relationships between interdependent self-construal and interaction goals.

On the basis of the above arguments, we put forth the idea of the following preposition.

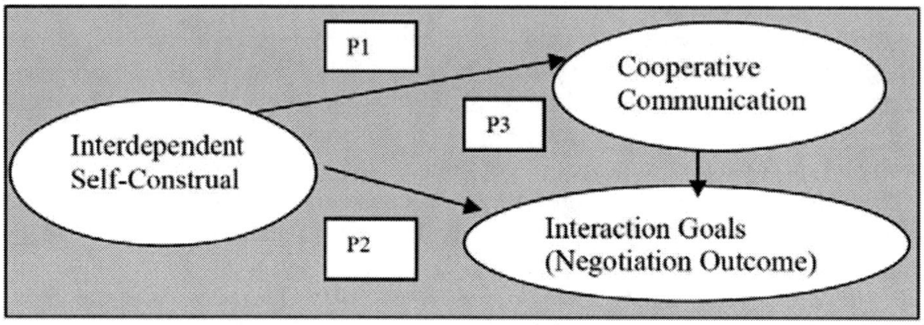

Fig.1. Relationship between Interdependent Self-construal, Cooperative Communication and Interaction Goals (Negotiation Outcome)

4 Limitations and Suggestions

In spite of the fact that this paper is a theoretical contribution, but this is still not without the some prospective limitations. The key limitation of this paper is conceivably it is focused on cross-cultural communication in international business negotiations, in explaining the link of the interdependent self-construal as the main factor that impacts the international business negotiation outcome but it is not the only predictors for the negotiation outcome.

Secondly, literature recommended some other imperative factors which also need explanation of their relationship with antecedents like cooperative communication and goal-oriented group behaviour [5]. It is proposed that cooperative communication should be treated as the mediator and interaction goals as the outcome behaviour. Thus, it would be enviable for future research to study the influence of cooperative communication other than interdependent self-construal in defining and clarifying certain other components which have link with the communication behaviour.

Thirdly, this paper is only temp to explain the relationship of exposure of interdependent self-construal and interaction goals in literature it remains in consistent to explain some elements like interdependency between the two parties for the purpose of effective communication processes which are operationally defined as equal participation, cooperation and respect [7]. Even though study had proven that interdependence was positively associated with cooperation, there were other important factors which were being acknowledged that will influence the communication process which are ethnic, gender and age [7]. Thus, it would be suggested that future research should focus on relationship among situation, self-construal, and communication process. The reason is it may be that cooperation and respecting a group lead to a feeling of interdependence with the other members or the feelings of interdependence lead to cooperation and respect [7].

Lastly, this paper is only a theoretical exploration as it further deepens our understanding of the impact of interdependent self-construal on international business negotiations. It is expected that this work leads towards explanation of the role of interdependent self-construal, cooperative communication and interaction goals in collectivistic culture to address the relation of exposure of the cross-cultural communication which may be studied further to determine the empirical findings about the other dynamics of cross-cultural communication in international business negotiations. According to [12] international negotiators require additional skills and cultural competence compared to those who are involved in domestic business negotiations. When there is frequent communication between two parties, both parties already adjust to each other communication styles [7].

References

1. D. Tjosvold, W. Johnson, W., R.T. Johnson, JRL, **116** (1984)
2. D.W. Hendon, R.A. Hendon, P. Herbig, *Cross-cultural Business Negotiation*. America: Greenwood Publishing Group, Inc. (1998)
3. E. Okoro, IJBMS, **7**, 130 (2012)
4. G. Chen, C. Liu, D. Tjosvold, JMS. **42** (2005)
5. H.A. Bakar, V. Sheer, Manag. Commun. Q, **27** (2013)
6. H.R. Markus, S. Kitayama, Perspect Psychol Sci, **5** (2010)
7. J.G. Oetzel, Hum Commun Res, **25** (2001)
8. J. Lee, Manag. Commun. Q, **11** (1997)
9. M. Liu, S.R. Wilson, Comm. Res, **38** (2010)
10. N. J. Adler, J. L Graham, JIBS, **20** (1989)
11. N. Agrawal, D. Maheswaran, J. Consum. Res., **31** (2005)
12. N. Reynolds, A. Simintiras, E. Vlachou, Int Market, **20** (2003)
13. S. Ting-Toomey, *Constructive intercultural conflict management*. In Ting-Toomey, S. (Ed.). Communicating across cultures. New York: The Guilford Press (1999)

AUTHOR INDEX

AUTHOR INDEX